S0-ASR-590

Handbook of Experimental Pharmacology

Volume 94/I

Chemical Carcinogenesis and Mutagenesis I

Contributors

F. A. Beland · L. G. Cain · J. S. Felton · J. D. Groopman · P. L. Grover
M. Hall · S. S. Hecht · K. Hemminki · D. Hoffmann · F. F. Kadlubar
M. G. Knize · M. A. Knowles · P. D. Lawley · W. Lijinsky
D. B. Ludlum · G. P. Margison · P. J. O'Connor · D. H. Phillips
H. C. Pitot · C. E. Searle · O. J. Teale

Editors

C. S. Cooper and P. L. Grover

Springer-Verlag Berlin Heidelberg New York
London Paris Tokyo Hong Kong

Colin S. Cooper, Ph. D.
Philip L. Grover, D. Sc.

The Institute of Cancer Research:
Royal Cancer Hospital
Chester Beatty Laboratories
Fulham Road
London SW3 6JB, Great Britain

With 86 Figures

ISBN 3-540-51182-2 Springer-Verlag Berlin Heidelberg New York
ISBN 0-387-51182-2 Springer-Verlag New York Berlin Heidelberg

Library of Congress Cataloging-in-Publication Data. Chemical carcinogenesis and mutagenesis/contributors, F.A. Beland ... [et al.]; editors, C.S. Cooper and P.L. Grover. p. cm. – (Handbook of experimental pharmacology; v. 94) Includes bibliographies and index.
ISBN 0-387-51182-2 (v. 1: U.S. alk. paper)
1. Carcinogenesis. 2. Carcinogens. 3. Chemical mutagenesis. 4. Mutagens. I. Beland, F.A. (Frederick A.) II. Cooper, C.S. (Colin S.), 1954– . III. Grover, Philip L. IV. Series.
[DNLM: 1. Carcinogens. 2. Mutagens. W1 HA51L v. 94/QZ 202 C5174] QP905.H3 vol. 94 [RC268.5] 615'.1 s–dc19 [616.99'4071] DNLM/DLC

Printed in Germany

Typesetting, printing and bookbinding: Brühlsche Universitätsdruckerei, Giessen
2127/3130-543210 – Printed on acid-free paper

List of Contributors

F. A. Beland, Division of Biochemical Toxicology, National Center for Toxicological Research, Jefferson, AR 72079, USA

L. G. Cain, Boston University School of Public Health, Environmental Health Section, 80 E. Concord Street, Boston, MA 02118, USA

J. S. Felton, Biomedical Sciences Division, Lawrence Livermore National Laboratory, University of California, L452, POB 5507, Livermore, CA 94550, USA

J. D. Groopman, Johns Hopkins University, School of Hygiene and Public Health, Department of Environmental Health Sciences, 615 North Wolfe Street, Baltimore, MO 21205 USA

P. L. Grover, Institute of Cancer Research, Chester Beatty Laboratories, 237 Fulham Road, London SW2 6JB, Great Britain

M. Hall, Department of Metabolism and Pharmacokinetics, Huntingdon Research Center, Huntingdon, Cambs. PE18 6ES, Great Britain

S. S. Hecht, American Health Foundation, Naylor Dana Institute, Valhalla, NY 10595, USA

K. Hemminki, Institute of Occupational Health, Topeliuksenkatu 41 a A, SF-00250 Helsinki, Sweden

D. Hoffmann, American Health Foundation, Naylor Dana Institute, Valhalla, NY 10595, USA

F. F. Kadlubar, Office of Research, National Center for Toxicological Research, Jefferson, AR 72079, USA

M. G. Knize, Biomedical Sciences Division, Lawrence Livermore National Laboratory, University of California, L452, POB 5507, Livermore, CA 94550, USA

M. A. Knowles, Marie Curie Research Institute, The Chart, Oxted, Surrey RH8 0TL, Great Britain

P. D. Lawley, Institute of Cancer Research, Royal Cancer Hospital, Chester Beatty Laboratories, 237 Fulham Road, London SW3 6JB, Great Britain

W. Lijinsky, Frederick Cancer Research Facility, BRI-Basic Research Program, Frederick, MD 21701, USA

D. B. LUDLUM, Department of Pharmacology, University of Massachusetts, Medical Center, 55 Lake Avenue North, Worcester, MA 01655-2397, USA

G. P. MARGISON, Cancer Research Campaign Laboratories, Department of Carcinogenesis, Paterson Institute for Cancer Research, Christie Hospital and Holt Radium Institute, Withington, Manchester M20 9BX, Great Britain

P. J. O'CONNOR, Cancer Research Campaign Laboratories, Department of Carcinogenesis, Paterson Institute for Cancer Research, Christie Hospital and Holt Radium Institute, Withington, Manchester M20 9BX, Great Britain

D. H. PHILLIPS, Institute of Cancer Research, Royal Cancer Hospital, Chester Beatty Laboratories, 237 Fulham Road, London SW3 6JB, Great Britain

H. C. PITOT, McArdle Laboratory for Cancer Research, Departments of Oncology and Pathology, University of Wisconsin Medical School, 450 N. Randall Avenue, Madison, WI 53706, USA

C. E. SEARLE, Cancer Research Campaign Laboratories, Department of Cancer Studies, The Medical School, University of Birmingham, Birmingham B15 2TJ, Great Britain

O. J. TEALE, Cancer Research Campaign Laboratories, Department of Cancer Studies, The Medical School, University of Birmingham, Birmingham B15 2TJ, Great Britain

Dr. E. C. Miller

This volume is respectfully dedicated by its contributors to the memory of Dr. Elizabeth Cavert Miller, Professor of Oncology at the McArdle Laboratory for Cancer Research of the University of Wisconsin, who died on October 14, 1987, from the very disease to which she had devoted a lifetime of outstanding scientific research.

Foreword

I have been privileged to witness and participate in the great growth of knowledge on chemical carcinogenesis and mutagenesis since 1939 when I entered graduate school in biochemistry at the University of Wisconsin-Madison. I immediately started to work with the carcinogenic aminoazo dyes under the direction of Professor CARL BAUMANN. In 1942 I joined a fellow graduate student, ELIZABETH CAVERT, in marriage and we soon commenced a joyous partnership in research on chemical carcinogenesis at the McArdle Laboratory for Cancer Research in the University of Wisconsin Medical School in Madison. This collaboration lasted 45 years. I am very grateful that this volume is dedicated to the memory of Elizabeth. The important and varied topics that are reviewed here attest to the continued growth of the fields of chemical carcinogenesis and mutagenesis, including their recent and fruitful union with viral oncology. I feel very optimistic about the application of knowledge in these fields to the eventual solution of numerous problems, including the detection and estimation of the risks to humans of environmental chemical carcinogens and related factors.

JAMES A. MILLER
Van Rensselaer Potter
Professor Emeritus of Oncology
McArdle Laboratory for Cancer Research
University of Wisconsin Medical School
Madison, Wisconsin

Preface

In order to understand and hopefully to prevent the processes by which chemicals induce cancer in man, it will be necessary to achieve three goals. Firstly, the classes of chemicals that are responsible for human chemical carcinogenesis must be identified. Secondly, the detailed metabolism of these chemicals should be examined and their targets within susceptible cells defined. Thirdly, it is important to identify the cancer-specific changes that are induced in cells by chemical exposure and ultimately to discover how these changes lead to tumour induction. In this book we have attempted to bring together these three areas of cancer research which span several disciplines, ranging from epidemiology, through studies on metabolic activation, to cellular and molecular biology.

The International Agency for Cancer Research has listed many chemicals or mixtures of chemicals for which there is considered to be sufficient evidence of carcinogenicity in man. Despite this substantial tally, it remains a disturbing fact that the environmental, dietary or endogenous agents that are responsible for most of the major types of human cancer have still to be identified. In addition, even when carcinogenic substances, such as tobacco smoke, have been clearly implicated, there is still debate regarding the contributions of individual components of these complex mixtures to their overall biological effects. It is therefore not at all surprising that continued efforts will have to be made to identify potential carcinogens using a combination of approaches, including bacterial and mammalian cell mutation assays, epidemiological studies and the range of methods that can now be used to detect chemicals that have already become covalently attached to cellular macromolecules. From amongst this armoury the new ^{32}P-postlabelling procedure (Part I, Chapter 13) that enables extremely low levels of carcinogen-DNA adducts to be detected deserves particular mention as a technique that has the potential to provide completely fresh insights into the identity of chemicals that contribute to cancer development in man.

It is generally accepted that DNA is an important cellular target for chemical carcinogens. This belief arose, in part, because of the attractiveness of the somatic mutation model of cancer development that is reviewed in the introductory chapter and, in part, because of the correlations observed between the extents of covalent binding of chemicals to DNA and their carcinogenic potencies. As a consequence of the extensive interest in this area we now possess, for many important classes of chemical carcinogens, a detailed knowledge of their pathways of metabolic activation and of the mechanisms by which activated metabolites modify DNA. Both the specificity of interaction of the activated carcinogens with DNA and the cellular machinery responsible for repairing the lesions thus introduced have come under close scrutiny. Such studies have, in particular, demonstrated the central importance of DNA repair enzymes in determining the biological consequences of carcinogen exposure, as dramatically illustrated by the high incidence of some types of cancer in individuals with deficiencies in specific repair enzymes.

It is well established that the consequences of exposure to chemical carcinogens can be influenced by a variety of factors that may be collectively referred to as "modifiers of chemical carcinogenesis". These include the tumour promoters, which interact with specific cellular proteins, as well as a whole range of substances that can act as inhibitors of carcinogenesis. There are also occasions when chemicals may act synergistically with other classes of cancer-causing agents. Two examples are provided by (a) the 'cooperation' between fungal toxins and hepatitis B virus in the induction of hepatocellular carcinoma and (b) the proposed interaction between cigarette smoking and papilloma viruses in the development of cancer of the cervix. Despite the potential importance of such interactions in the induction of human cancer, surprisingly little is known about the molecular mechanisms involved in cooperation between viral and chemical agents. This is perhaps an area that should be targeted for particular attention in the future. Furthermore, an individual's genetic make-up, in addition to determining the status of the DNA repair enzymes as mentioned above, may have a key role in controlling that person's susceptibility to chemical exposure.

The new technologies of DNA transfection and molecular biology have resulted in significant advances (a) in the identification of cellular genes that are potential targets for chemical carcinogens and (b) in the characterization of the specific types of genetic alterations that may be involved in tumour induction. The potential targets now include several well-characterized protooncogenes, which may be activated by mutation to form oncogenes, as well as the less well characterized tumour-suppressor genes or anti-oncogenes that may be inactivated during tumour development. These molecular studies have lead to the unification of several areas of cancer research and cellular biology. Of particular note is the observation that the same genes are activated in a range of tumour types, indicating that there may exist common mechanisms of cancer development amongst histologically-diverse groups of tumours. Moreover the recognition that many of the protooncogenes encode proteins, such as growth factors and growth factor receptors, that are the normal components of cellular control pathways has caused a revolution in our perception of the ways in which alterations in cellular genes lead to transformation.

The newer molecular genetics studies also have implications for the more traditional areas of chemical carcinogenesis and mutagenesis. The confirmation that specific genetic alterations may be directly involved in tumour development justifies the extensive analysis of the interactions of carcinogens with DNA that has occurred over the past 20 years and provides a firm foundation for many of the bacterial and mammalian mutagenicity tests, which hitherto has rested precariously on the observed correlations between mutagenicity and carcinogenicity. It is equally important to appreciate the limitations of the molecular approaches used for the analysis of cancer induction. For example, these techniques, although extremely important, have so far provided few insights into the nature of the environmental and dietary components that contribute to the incidence of cancer in man, an observation that simply underscores the continued importance of the more traditional epidemiological and mutational studies. Indeed, it is our hope that the bringing together of chapters reviewing widely-separated aspects of chemical carcinogens in one book will serve to highlight both the virtues and the limitations of each area and will help to identify those lines of research that may, in the future, prove most fruitful.

Finally we wish to extend our gratitude to HELEN ANTON and AUDREY INGLEFIELD for their assistance with the organization and preparation of this book and to DORIS M. WALKER of Springer-Verlag for all her help and advice.

London

COLIN S. COOPER
PHILIP L. GROVER

Contents

Part I. Theories of Carcinogenesis

Part II. Exposure to Chemical Carcinogens

CHAPTER 3

Advances in Tobacco Carcinogenesis

CHAPTER 11

N-Nitroso Compounds
P. D. LAWLEY. With 16 Figures 409

CHAPTER 12

Heterocyclic-Amine Mutagens/Carcinogens in Foods
J. S. FELTON and M. G. KNIZE. With 6 Figures 471

Contents of Companion Volume 94, Part II

Part I. Theories of Carcinogenesis

CHAPTER 1

Mechanisms of Chemical Carcinogenesis: Theoretical and Experimental Bases

H. C. PITOT

A. Introduction

Although ionizing radiation, infectious biological agents, and genetic determinants play significant roles in the causation of human cancer, environmental chemical agents and mixtures of such agents, both exogenous and endogenous to the living organism, are causally involved, too (DOLL and PETO 1981; WYNDER and GORI 1977). While lifestyle factors such as diet, sexual mores, and reproductive history may not be considered under the heading of chemical carcinogenesis in the usual sense, in all of these (just as in cigarette smoke) chemicals and mixtures of chemicals are the predominant components that cause human cancer. A potential exception in this group are those neoplasms developing as a result of sexual promiscuity, where the human papillomavirus may play a major role in genital cancers, both male and female (ZUR HAUSEN 1987). Until the past decade, the chemical induction of cancer was the principal experimental tool of carcinogenesis. Within the past 10 years experimental investigations of viral carcinogenesis have equalled or exceeded those in the chemical induction of cancer. However, our understanding of the mechanisms of the chemical induction of neoplasia preceded our understanding of such mechanisms in the viral induction of cancer (MILLER 1978).

It has become increasingly evident that our knowledge of the correlation of environmental factors with the development of neoplasia in humans and the carefully controlled chemical induction of cancer in experimental animals does not give a complete picture of this disease. Even with a knowledge of the mechanisms potentially critical in such causative relationships, our understanding is still incomplete. A critical element that has been missing in most etiologic concepts of the neoplastic process is the natural history or pathogenesis of the disease itself. Today, an understanding of the process of carcinogenesis must involve not only identification of the mechanisms of action of chemicals on cells that convert them to neoplasia but also the process of the development of cancer that results from the initial interaction of a chemical with a cell. Therefore, the etiology, mechanisms, and pathogenesis of neoplasia are inseparable in our understanding of the theoretical bases of chemically induced cancer.

B. Historical Aspects

It was the chemical causation of cancer that first became apparent to physicians. Perhaps RAMAZZINI (cf. WRIGHT 1940) was the first to delineate in 1700 a chemical causation of cancer by pointing out the high incidence of breast cancer in nuns. Although unknown to him at that time, hormonal interactions are involved in the high incidence of breast cancer in celibate nuns. Later, astute clinical observations related the use of tobacco in the form of snuff to the cause of nasal polyps, as reported by HILL in London in 1761 (cf. REDMOND 1970). Shortly thereafter, Percivall POTT (1775) observed the occurrence of cancer of the scrotum in a number of his patients, all of whom had been employed as chimney sweeps as young boys. From these common histories, POTT with remarkable insight concluded (a) that the occupation of these men as young boys was directly and causally related to their malignant disease and (b) that the soot to which they were excessively exposed in their work was the causative agent of the cancer. Further support for POTT's proposals came more than a century later when BUTLIN (1892) reported the relative rarity of scrotal cancer in chimney sweeps on the European continent compared with those in England. The lower incidence of the disease appeared to be the result of frequent bathing and protective clothing worn by continental workers. That excessive exposure to other chemicals could be causally related to human cancer was also reported during the last century (cf. MILLER 1978). In particular, REHN's (1895) observations on the association of cancer of the urinary bladder following chronic exposure to aromatic amines in the manufacture of "aniline" dyes eventually led to the identification of 2-naphthylamine, benzidine, and 4-aminobiphenyl as relatively refined chemicals that could induce specific cancers in humans. Studies on experimental chemical carcinogenesis began in 1915 when YAMAGAWA and ICHIKAWA demonstrated the chemical induction of skin carcinomas on the ears of rabbits after repeated topical applications of coal tar for extended periods. These experimental studies, coupled with epidemiologic observations in the nineteenth century, more than justified the proposals Dr. POTT had made almost 150 years earlier.

The experimental chemical induction of cancer by crude coal tar opened the avenue to the potential for isolation of specific chemical compounds from such a mixture. The earlier observations on the "aniline" dye cancers had already indicated that relatively pure chemical compounds could be carcinogenic. Several studies (e.g. KENNAWAY 1925; HIEGER 1930) led to the demonstration by KENNAWAY and HIEGER (1930) of the carcinogenicity of a synthetic polycyclic hydrocarbon, dibenz[*a,h*]anthracene (1,2,5,6-dibenzanthracene), which they synthesized. A few years later a related chemical, 3,4-benzpyrene (benzo[*a*]pyrene) was isolated from coal tar by COOK et al. (1933) and shown to be highly carcinogenic.

Since aromatic amines were identified as putative human carcinogens by the epidemiologic data mentioned above (REHN 1895), it was also natural to investigate such "aniline" dyes and their derivatives for carcinogenicity in animals. Such an action of 2-naphthylamine was not demonstrated in experimental animals until HUEPER and his associates (1938) succeeded in inducing cancer of the urinary bladder in dogs by feeding them this chemical. However, the experimental observation that chemicals other than polycyclic hydrocarbons could induce cancer in controlled experiments in animals was first made by YOSHIDA (1933) after the administration of *o*-aminoazotoluene or 2′,3-dimethyl-4-aminoazobenzene by mouth to rats and mice. Liver cancer was induced in these species when these chemicals were fed for extended periods of time. Another experimen-

Chemical compounds:

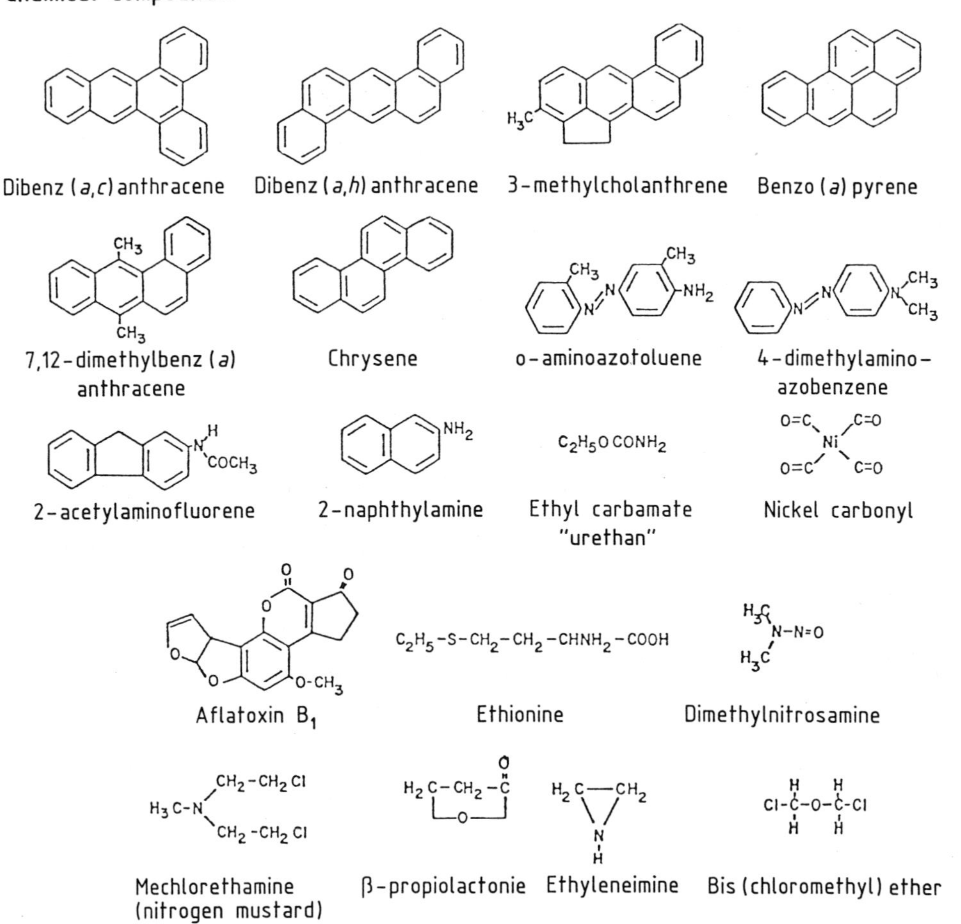

Chemical mixtures:

Alcoholic beverages
Tobacco smoke
Chemical from isopropyl alcohol manufacture
Betel "quid" and "chewing" tobacco

"Life-Style" chemical environments:

Dietary factors (caloric level, fat, protein)
Reproductive factors (late age at 1st pregnancy, low parity)

Fig. 1. Chemical structures and listings of chemicals, mixtures of chemicals, and life-style situations involving exposure to multiple chemicals inducing the neoplastic state

tal observation of chemical carcinogenesis that justified the observations of RAMAZZINI was the induction of mammary cancer in male mice treated with estrone (LACASSAGNE 1932).

Since 1940 a large number of synthetic and naturally occurring chemicals and chemical mixtures have been found to be carcinogenic in experimental animals or, by epidemiologic investigations, in humans. Structures and descriptions of representative chemicals and chemical mixtures in this category are seen in Fig. 1. It is not within the scope of this discussion to give a detailed description of the carcinogenic action of each of these agents. Rather, the figure is presented to the reader to indicate the extremely disparate nature of the structures of these chemical agents. This disparity presented a serious dilemma to the experimental oncologist of the 1950s: did these agents induce cancer by one or a few mechanisms or by numerous alternative mechanisms, perhaps unique for each chemical or class of chemicals? The major breakthrough in our understanding of the mechanisms involved in chemical carcinogenesis was the demonstration that reactive metabolites of carcinogenic chemicals were critical intermediates in the induction of neoplasia (MILLER and MILLER 1969). An equally important experimental advance was the demonstration by a number of investigators nearly 2 decades prior to the findings of the MILLERS (ROUS and KIDD 1941; MOTTRAM 1944; BERENBLUM and SHUBIK 1947) of the multistage nature of epidermal carcinogenesis in the mouse. Unfortunately, the impact of these experiments at that time was not appreciated by the majority of investigators in the field of oncology, at least in relation to our understanding of the molecular pathogenesis of the neoplastic process.

C. Mechanistic Theories of Chemical Carcinogenesis

Mechanistic theories of chemical carcinogenesis probably had their beginning with the physicochemical considerations by the PULLMANS on the electronic structure and carcinogenic activity of carcinogenic polycyclic hydrocarbons (cf. PULLMAN and PULLMAN 1955). Their studies related the electron density of specific regions of these molecules, termed the K and L regions, with the carcinogenicities of the polycyclic hydrocarbons. Since that time a number of other methods for the theoretical prediction of carcinogenicity based on the chemical structure of carcinogens have been proposed (cf. SMITH et al. 1978; LOWE and SILVERMAN 1984; NORDEN et al. 1978; JERINA and LEHR 1977). More recent theoretical considerations in this area have been based on information unavailable to the PULLMANS at the time of their initial proposals. However, their pioneering efforts were among the first to look for a common molecular basis for the carcinogenicity of diverse chemicals.

In a more restricted sense, but later shown to be widely applicable, were the studies of the MILLERS (MILLER and MILLER 1947), from which they proposed the deletion hypothesis in 1947. This hypothesis, put forth at a time when the genetic role of DNA was still in question, argued that neoplasia of the liver was the result of the "deletion" of one or more proteins caused by their interaction with the chemical carcinogen itself. Although the theoretical basis for the original hypothesis may not have been entirely accurate by present day knowledge, the MILLERS later demonstrated the scientific veracity of the binding aspect of their initial proposal (see below). It is from their demonstration of the critical importance of specific metabolic pathways in the activation of chemical carcinogens for their effectiveness that the theories of chemical carcinogenesis were placed on a firm scientific foundation.

I. Electrophilicity of Chemicals as a Determinant of Their Carcinogenicity

An understanding of the mechanism by which a diversity of chemical structures were each capable of inducing the malignant process was first elucidated by the MILLERS in their demonstration that the metabolism of the carcinogen *N*-acetylaminofluorene to its *N*-hydroxyl derivative, with subsequent esterification of the *N*-hydroxyl group, yields a highly reactive compound capable of nonenzymatic reaction with nucleophilic sites on proteins and nucleic acids (MILLER and MILLER 1969). They then proposed that a majority of chemical carcinogens are or are converted by metabolism into electron-pair-deficient electrophilic reactants that exert their biological effects by covalent interaction with cellular macromolecules, the critical target most probably being DNA (cf. MILLER 1970; MILLER 1978). These model studies with acetylaminofluorene led to the proposal that most chemical carcinogens could be considered as "procarcinogens," which require metabolism to "proximate" and finally to "ultimate" carcinogens. The ultimate form of the carcinogen is that form which actually interacts with cellular constituents to induce the neoplastic transformation. A number of structures of procarcinogens, proximate and ultimate carcinogens may be seen in Fig. 2. In some instances, the structure of the ultimate form of a carcinogenic chemical is still not clear. In other cases there may be more than one ultimate carcinogenic metabolite of the same procarcinogen, while some chemicals actually are in the ultimate form (e.g., alkylating agents) or are converted to the ultimate form by nonenzymatic reactions in the organism.

After the demonstration by the MILLERS of the critical significance of electrophilic metabolites in chemical carcinogenesis, the ultimate forms of a number of carcinogenic chemicals were described (Fig. 2). However, the ultimate form of carcinogenic polycyclic hydrocarbons was still unknown. As early as 1950, BOYLAND proposed that the formation of epoxides as intermediates in the metabolism of these chemicals might be important in the carcinogenic process. However, it was not until 1970 that JERINA and his associates detected the formation of such intermediates in biological systems. Further investigations demonstrated that epoxides of polycyclic hydrocarbons could react with nucleic acids and proteins in the absence of xenobiotic enzymes. Surprisingly, K-region epoxides of a number of carcinogenic polycyclic hydrocarbons were weaker carcinogens than the parent hydrocarbons. In 1974 SIMS and his associates proposed that a diol-epoxide of benzo[*a*]pyrene is the ultimate form of this carcinogen. Subsequent studies by a number of investigators (YANG et al. 1976; cf. CONNEY 1982; HARVEY 1981) demonstrated that the structure of this ultimate form is (+)*anti*-benzo[a]pyrene-7,8-dihydrodiol-9,10-epoxide.

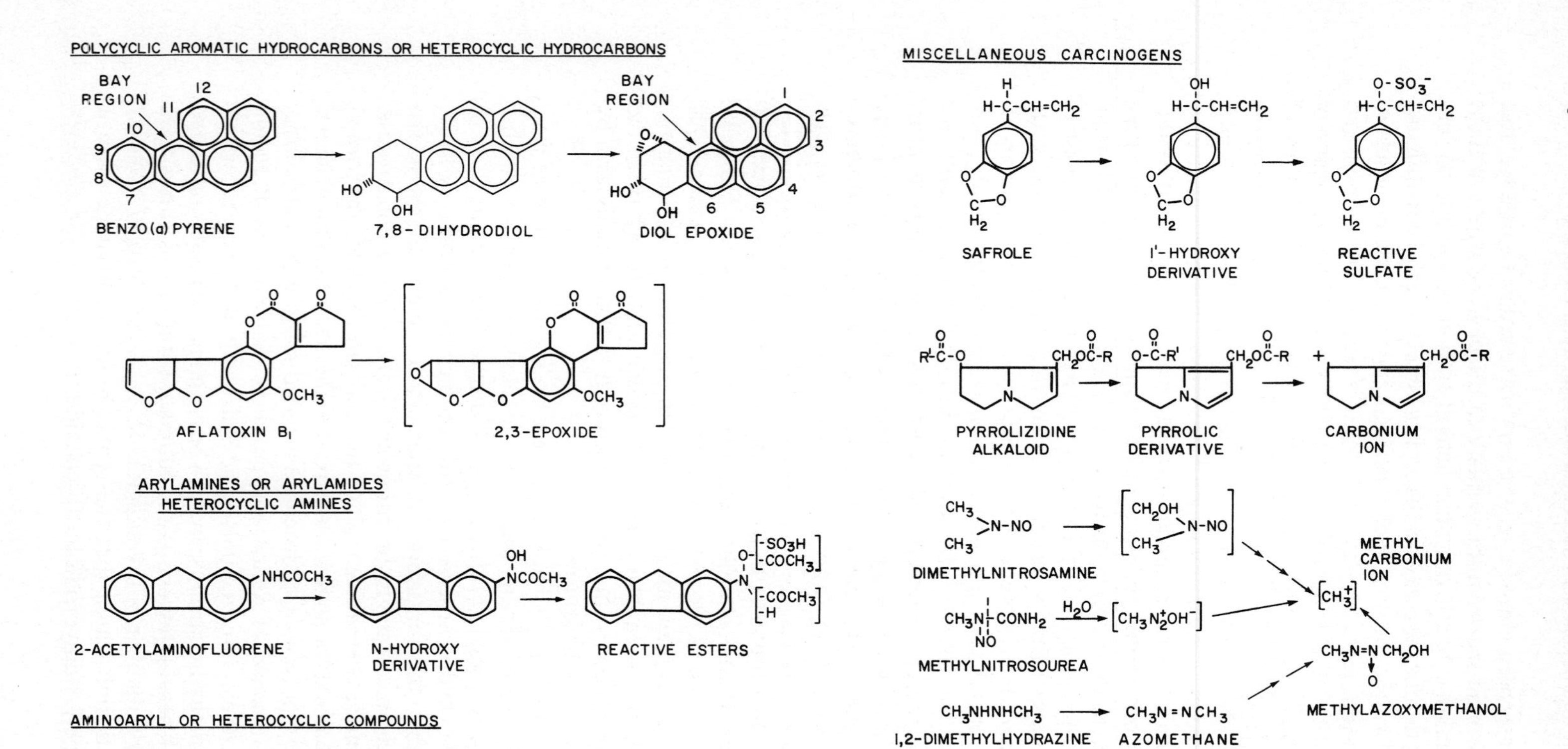

Fig. 2. Structures of representative chemical carcinogens and their metabolic derivatives, the proximate and ultimate carcinogenic forms. (From PITOT 1986, with permission of the author and publisher)

One of the more interesting theories of chemical carcinogenesis that resulted from these findings is the importance of the oxidation state of the carbons of the "bay region" of potentially carcinogenic hydrocarbons. The bay region of benzo[*a*]pyrene can be seen in Fig. 2, and analogous bay regions may be identified in a number of other polycyclic hydrocarbons. JERINA, CONNEY and their associates as well as others (JERINA and LEHR 1977; CONNEY 1982) have proposed that epoxidation of a dihydro, angular benzo ring that forms part of the bay region of a polycyclic hydrocarbon results in the most likely ultimate carcinogenic form of the molecule. As seen from Fig. 2, the bay region is the sterically hindered region formed by the angular benzo ring. While an occasional exception has been found, most carcinogenic polycyclic hydrocarbons conform to this concept.

Since the cancer phenotype is heritable from one cell to another, one of the essential foundations of any theory of chemical carcinogenesis has always been considered to be a genetic or structural alteration of DNA as a result of the reaction of the chemical. The concept of an ultimate form of a carcinogen capable of spontaneous reaction with DNA to produce adducted, structural alterations in the genome, i.e., the MILLERS' concept, appears quite suitable in this respect. To establish this point even further, they and their associates (MAHER et al. 1968) demonstrated through the use of native transforming DNA that only the ultimate forms of 2-acetylaminofluorene and *N*-methyl-4-aminoazobenzene are capable of inducing mutations in the DNA, as evidenced by its loss of ability to transform appropriate bacterial strains. Furthermore, the induced mutations are spontaneously reversible; this suggests that their structure represents single base-pair changes. Therefore, with the demonstration of the mutagenicity of ultimate reactive forms of chemical carcinogens as well as their increased carcinogenicity in many cases (cf. MILLER 1978), the closely related nature of carcinogenesis and mutagenesis became clarified to an extent not possible prior to the studies of the MILLERS (KAWACHI et al. 1979). However, it has become increasingly clear over the past decade that not all chemicals induce neoplastic transformation by mechanisms involving ultimate electron-pair-deficient electrophilic forms and metabolites.

II. Potential Role of Free Radicals as Ultimate Carcinogens

Another highly reactive class of chemical intermediates are the single electron-deficient or free radicals derived from chemical carcinogens. During the past decade, both direct and indirect evidence has accumulated that such free radical derivatives of chemical carcinogens may be produced both metabolically and nonenzymatically during the metabolism of such agents (cf. NAGATA et al. 1982). Free radicals carry no charge but possess a single, unpaired electron, such a structure being highly reactive at temperatures compatible with mammalian life. Free radicals, especially those of oxygen, have long been considered as important critical intermediates in the biological effects of ionizing radiation, including its carcinogenic effects (cf. BIAGLOW 1981). The idea that free radical forms of chemical carcinogens may be important in the induction of the neoplastic transformation by chemicals has developed from two general lines of evidence: (a) The inhibition of chemical carcinogenesis by the simultaneous and/or subsequent administration of antioxidants (ITO and HIROSE 1987; KAHL 1986) and (b) the direct demonstration of the formation of free radicals in vivo as monitored by their ultimate reaction with macromolecules (WISE et al. 1984; O'BRIEN 1985).

The indirect evidence developed from the demonstration of the inhibition of chemical carcinogenesis by antioxidants has recently become much more complicated by the variety of effects of such compounds on the carcinogenic process and on metabolic pathways important in chemical carcinogenesis. It is now clear that most classic antioxidants such as butylated hydroxytoluene, butylated hydroxyanisole, tertiary-butylhydroquinone, propylgallate, DL-α-tocopherol, sodium L-ascorbate, and ethoxyquin as well as other such compounds exert inhibitory, enhancing, or no effect on the first two stages of carcinogenesis, depending on the tissue and species under investigation (ITO and HIROSE 1987; KAHL 1986). Furthermore, some antioxidants [e.g. butylated hydroxyanisole (ITO et al. 1983) and butylated hydroxytoluene (DANIEL 1986)] are actually carcinogens themselves when administered at extremely high levels for prolonged periods of time. Such compounds are excellent inducers of the synthesis of a variety of xenobiotic enzymes, a number of which are involved in the metabolism of chemicals to their ultimate carcinogenic form (CHA et al. 1982). In fact, butylated hydroxyanisole and butylated hydroxytoluene, as well as several other antioxidants, specifically induce the enzyme quinone reductase in several cell types (DELONG et al. 1986; WEFERS et al. 1984). Since this enzyme tends to inhibit the formation of free radicals during normal metabolic reactions, it is quite likely that these antioxidants exert their effect through inducing the activity of this enzyme. Therefore, the exact mechanism whereby antioxidants alter the process of carcinogenesis is not clear, but it is likely that various mechanisms are involved, some of which probably affect the formation of free radicals by chemical carcinogens, other xenobiotics, and normal metabolic pathways.

Although there is no question that free radical intermediates may be formed during the metabolism of chemical carcinogens and other xenobiotics, it was largely during the past decade that specific metabolic reactions of these agents were shown to proceed through free radical intermediates. WONG and his associates (1982) presented evidence for the metabolic activation of *N*-hydroxy-2-acetylaminofluorene by rat mammary gland cells concomitant with the endoperoxidation of arachidonic acid to prostaglandin (PGH_2). This metabolic pathway is shown in Fig. 3, in which the peroxidation of the prostaglandin peroxide (PGG_2) occurs concomitantly with the formation of a nitroso free radical that spontaneously dismutates to nitrosofluorene and *N*-acetoxy-acetylaminofluorene. Similar activation of *p*-dimethylaminoazobenzene has also been reported (VASDEV et al. 1982). β-Naphthylamine (FISCHER and MASON 1986), benzidine (JOSEPHY et al. 1983), and 2-aminofluorene (KRAUSS and ELING 1985) appear to undergo similar metabolic reactions or are capable of such reactions. Furthermore, the endoperoxidation of arachidonic acid is not the only pathway leading to the formation of free radical intermediates of chemical carcinogens, since peroxides of unsaturated fatty acids within membranes appear to be capable of carrying out similar chemical reactions (FLOYD 1981; MCNEILL and WILLS 1985).

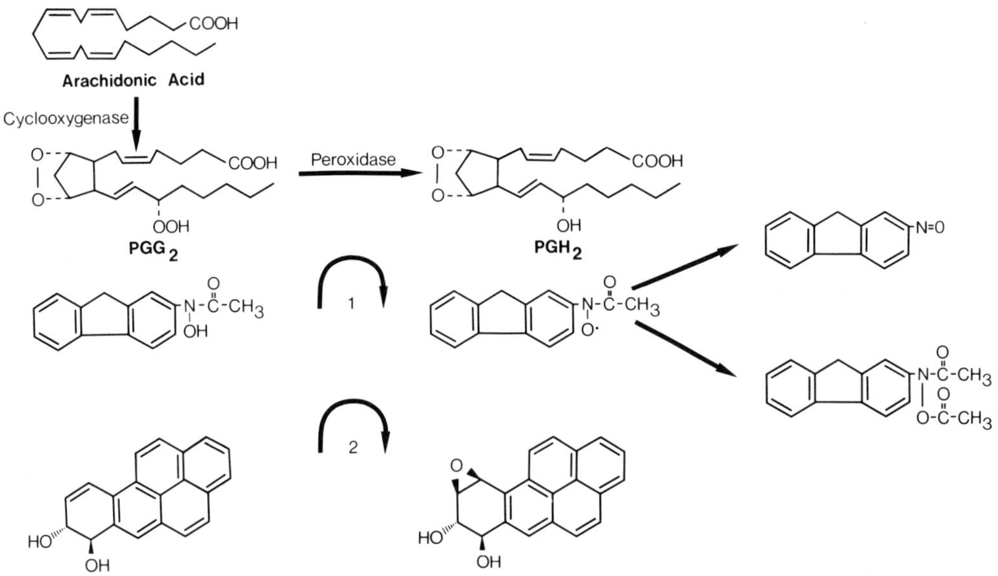

Fig. 3. Metabolic scheme for the metabolism of arachidonic acid to prostaglandin (PGH_2) with associated reactions of carcinogens (2-acetylaminofluorene, reaction *1*; benzo[*a*]pyrene-1,8-diol, reaction *2*) leading to the ultimate forms of these chemical carcinogens. PGG_2, prostaglandin peroxide

MARNETT and his associates (REED and MARNETT 1982; DIX and MARNETT 1983) demonstrated similar metabolic activation of dihydrodiols of carcinogenic polycyclic hydrocarbons. Such a representative pathway is noted in reaction (2) of Fig. 3. In this instance, however, the pathway noted in (2) leads predominantly to a different stereoisomer of the diol-epoxide than that resulting from the action of mixed function oxidases on the diol (Fig. 2). MARNETT has proposed that the amount of the *anti*-diol-epoxide formed, predominantly by the lipid peroxidation pathway, would be a measure of the importance of this pathway in the metabolic activation of the polycyclic hydrocarbon, since mixed function oxidases convert the diol predominantly to the *syn*-diol-epoxide (DIX and MARNETT 1983).

That such pathways are important in the activation of chemical carcinogens to their ultimate forms has been demonstrated in several instances. WISE et al. (1984) reported that in dog bladder and kidney the prostaglandin pathway for the metabolic activation of carcinogenic aromatic amines is predominant, whereas in liver such reactions appear to occur mostly by a mixed function oxidase mechanism. In vitro studies by KRAUSS and ELING (1985) showed the formation of unique arylamine: DNA adducts from 2-aminofluorene via the prostaglandin pathway. A number of these adducts, as yet unidentified, appear to be different from those formed by the electron-pair-deficient electrophilic attack of aminofluorene derivatives on DNA. Polycyclic aromatic hydrocarbon-induced epidermal carcinogenesis in a two-stage mechanism may also involve free radical intermediates, as demonstrated by TRUSH et al. (1985). These workers showed that the formation of carcinogen-DNA adducts results, at least in part, from the oxidant-dependent metabolic activation of benzo[*a*]pyrene by phorbol ester-stimulated polymorphonuclear leukocytes, thus suggesting a possible link between the inflammatory process that occurs during carcinogenesis and the development of neoplasia. Thus, there is substantial in vitro and in vivo evidence for the role of free radical intermediates in significant mechanisms of carcinogenesis by chemicals.

III. Altered DNA Methylation as a Theoretical Mechanism of Chemical Carcinogenesis

A number of chemical carcinogens have now been shown to be capable of inhibiting the normal methylation of some deoxycytidine residues in DNA by *S*-adenosylmethionine, as catalyzed by nuclear transmethylases in liver, brain, and other tissues in vivo (Riggs and Jones 1983). Methylation of DNA cytosines is probably involved in the heritable expression or repression of specific genes in eukaryotic cells (Doerfler 1983; Holliday 1987). The inhibition of DNA methylation may occur by several mechanisms, including the formation of covalent adducts or single-strand breaks in the DNA and the direct inactivation or inhibition of the enzyme DNA:*S*-adenosylmethionine methyltransferase, which is responsible for normal methylation. In support of this theoretical concept is the demonstration of the hypomethylation of DNA from some neoplasms (Goelz et al. 1985; Diala et al. 1983) as well as abnormal methylation during the process of carcinogenesis (Boehm et al. 1983; Shivapurkar et al. 1984; Shank 1984). Alterations in the methylation of specific genes within neoplasms have also been described (Vedel et al. 1983; Baylin et al. 1986). Furthermore, the inhibitor of DNA methylation, 5-azacytidine, has been reported to be carcinogenic in mice (Cavaliere et al. 1987) and acts as a cocarcinogen in chemical hepatocarcinogenesis (Denda et al. 1985). Thus, the alteration of DNA methylation by chemical carcinogens represents another theoretical pathway by which these agents may induce neoplastic transformation.

IV. Aberrations of DNA Repair as Theoretical Mechanisms of Chemical Carcinogenesis

In the three theoretical mechanisms of chemical carcinogenesis indicated thus far, changes in the structure of DNA are the end result of each mechanism. Although altered methylation of the 5-position of cytidine in DNA itself is a potential mechanism that alters gene expression, methylation induced by electrophilic (carbonium ion) or free radical methyl groups represents an extension of any or all three of the above mechanisms. Thus, inherent in these discussions is the formation of DNA damage, which potentially may be required by one or more of a number of mechanisms (cf. Teoule 1987). It is not within the purview of this discussion to present a detailed analysis of the mechanisms of DNA repair. Rather, it is my intent to demonstrate the importance and possible relationship of DNA repair mechanisms, both normal and altered, to the process of carcinogenesis. Substantial experimental evidence has demonstrated the existence of chemical damage to DNA following the administration of chemical carcinogens even at low doses (BRAMBILLA et al. 1983; STEWART et al. 1985). Such damage may include single-strand breaks or other structural features, as well as the formation of specific adducts between the ultimate form of the chemical carcinogen and the DNA itself (KRIEK et al. 1984). Furthermore, such damage can persist within the DNA molecule for extended periods of time, presumably without significant repair (STOUT and BECKER 1980; BELAND and KADLUBAR 1985). Thus far, it has not been possible to equate unequivocally the presence or persistence of a single type of damage or single DNA adduct with the site of and formation of ultimate neoplasms (SINGER 1985), although certain adducts by correlative studies appear to be closely associated with the neoplastic transformation. This is evidenced by adduct persistence in the damaged DNA (SWENBERG et al. 1984) and the formation and removal of specific adducts under controlled experimental situations (SCHERER et al. 1977; SWENBERG et al. 1985). The classic studies of GOTH and RAJEWSKI (1974) demonstrating the persistence of the O^6-ethylguanine in rat brain DNA and its loss from rat liver DNA following administration of the carcinogen ethylnitrosourea correlate closely with the production of neoplasms of the nervous system, but not of the liver, by this agent. Similar studies have now been reported in different tissues (BEDELL et al. 1982) and in different species (BECKER and SHANK 1985).

Although there is some evidence that carcinogenic chemicals and radiation may activate some form of "error-prone" mechanism of DNA repair or replication (SARASIN et al. 1982), most of the available evidence suggests that the persistence of unrepaired DNA damage including DNA-chemical adducts may well be the cause of mutations that can potentially result in initiation of the altered cell (MAHER and MCCORMICK 1984; DONIGER et al. 1985; WINTERSBERGER 1982). Furthermore, a variety of factors, both exogenous and endogenous, may interfere with normal DNA repair, potentially resulting in abnormal repair and/or mutational events (BERNHEIM and FALK 1983). Perhaps the best evidence that faulty DNA repair may lead to carcinogenesis is seen in the high incidence of skin cancer in patients suffering from xeroderma pigmentosum, an autosomal recessive disease involving primarily the skin (KRAEMER 1980). Individuals affected by this condition have a marked inability to repair damage to DNA caused by ultraviolet light. Although the exact mechanism of the increased incidence of skin cancer seen in these patients is not known, the extensive DNA damage from ultraviolet light, which is largely unrepaired, appears to correlate with the high incidence of neoplasia in these individuals. Thus there appears to be strong associative evidence that faulty DNA repair and/or persistence of chemical adducts of DNA is a theoretical mechanism in the development of neoplasia.

V. Other Theoretical Mechanisms of Chemical Carcinogenesis

Many other mechanistic theories of chemical carcinogenesis have been proposed. It will not be possible here to discuss all such theories, but several will be mentioned, since they deal with a chemical agent usually not explicitly covered by the mechanisms described above. This group of chemicals are the inorganic chemical carcinogens. Although it is quite possible that metals and organometallic compounds may alter DNA structure through metabolic reactions identical with or analogous to those discussed above, alternate explanations have also been sought. Paramount among these is the proposal by SIROVER and LOEB (1976) that carcinogenic metals induce infidelity of DNA polymerase during DNA synthesis. Such a mechanism would result in mutations and/or changes in the expression of genes within the affected cell. The evidence for this comes primarily from in vitro studies and thus is as yet more theoretical than those indicated above.

Other inorganic, complex carcinogens in both humans and animals are certain forms of asbestos and related fibres. Although it is possible that asbestos contains adsorbed contaminated materials including carcinogenic hydrocarbons and metals, it stimulates the generation of free radicals, especially those of oxygen (GOODGLICK and KANE 1986). By a similar or different mechanism, asbestos's interaction with cells in culture can induce cytogenetic abnormalities (OSHIMURA et al. 1984; LECHNER et al. 1985). Since the crystal structure and size of carcinogenic asbestos moieties are critical for their carcinogenic action, their mechanisms have been likened to the induction of sarcomas by inert plastic and metal films, which have been studied extensively by BRAND and his associates (1975). In this latter instance it is clear that the chemical nature of the implant is not the critical factor inducing the neoplastic change, but rather its physical characteristics are most important. BRAND has theorized that a few cells attached to the film, which are themselves "preneoplastic," are present in the tissue prior to the implantation of the physical carcinogen, and he proposed that the implant appears to "create the conditions" required for carcinogenesis of these cells.

Another class of chemicals that appear to induce neoplasms by mechanisms different from those described above are hormones, both steroid and polypeptide. Until recently there was little or no evidence that such hormones interacted in a covalent manner directly with DNA or altered DNA repair, although DNA methylation during development can be modulated by certain hormones (cf. DOERFLER 1983; HOLLIDAY 1987). Recently, however, using highly sensitive techniques to determine the qualitative presence of DNA adducts, RANDERATH and his associates (LIEHR et al. 1986, 1987) reported evidence for the formation of some type of DNA adduct, the structure of which is not known, during estrogen-induced renal cancer in Syrian hamsters. On the other hand, a specific synthetic hormone, diethylstilbesterol, has been shown to be carcinogenic both in vivo and in vitro in hamsters. Furthermore, this chemical induces chromosome abnormalities in vitro (TSUTSUI et al. 1983). There is now evidence that both diethylstilbesterol and certain steroids may be metabolised to one or more active forms including epoxides (METZLER 1982; KADIS 1978). Still, the level of interaction of natural steroid hormones with DNA is extremely low (several orders of magnitude) as compared with typical chemical carcinogens (LUTZ et al. 1982). Therefore, in order to place hormones into one of the mechanisms described above, one must either argue that they are extremely potent in their carcinogenic effect compared with numerous other chemical carcinogens or look for other mechanisms whereby these chemicals produce their effects.

Since all of the theoretical mechanisms discussed thus far have varying degrees of experimental support, one is inclined to suggest that each plays some role in the mechanism of chemical carcinogenesis. The problem facing the experimental oncologist is to partition these roles appropriately in order to get a clearer overall picture of the induction of cancer by chemicals. In order to do this, it will be necessary to view these theories of chemical carcinogenesis in the light of the natural history of the development of the neoplastic process discussed below.

D. Pathogenesis of Malignancy: Natural History of Neoplastic Development

Only relatively recently have concepts of pathogenesis been related to the critical actions of chemical carcinogens. The pioneering studies referred to earlier (Rous and Kidd 1941; Mottram 1944; Berenblum and Shubik 1947), which demonstrated in one model system – epidermal carcinogenesis in the mouse – that cancer develops in stages, were a critical breakthrough in our understanding of the pathogenesis of neoplasia. Since then, the original two-stage concept has been modified (Boutwell 1964; Kinzel et al. 1986). A third stage, termed progression by many investigators, has been demonstrated through the use of protocols that have been altered from the original experiments of 4 decades ago (Potter 1983; Hennings and Yuspa 1985; O'Connell et al. 1986). Of even greater significance is the fact that the multistage concept of the development of neoplasia has been found to be a relatively ubiquitous characteristic in the development of neoplasms in most, if not all, tissues under appropriate circumstances (cf. Pitot 1986).

Some of the more important characteristics of the three major stages in the development of neoplasia are given in Table 1. For a more detailed description of the evidence supporting the various characteristics noted in the table, the reader is referred to previous publications from this and other laboratories (Pitot et al. 1987a, 1988; Slaga 1983; Weinstein et al. 1984).

Table 1. Biological characteristics of the stages of initiation, promotion, and progression in carcinogenesis[a]

Initiation	Promotion	Progression
Irreversible, with constant "stem cell" potential	Reversible	Irreversible. Measurable and/or morphologically discernible alteration in cell genome's structure
Efficacy sensitive to xenobiotic and other chemical factors	Promoted cell population's existence dependent on continued administration of the promoting agent	Growth of altered cells sensitive to environmental factors during early phase
Spontaneous (fortuitous) occurrence of initiated cells can be demonstrated	Efficacy sensitive to dietary and hormonal factors	Benign and/or malignant neoplasms characteristically seen
Requires cell division for "fixation"	Dose response exhibits measurable threshold and maximal effect dependent on dose of initiating agent	"Progressor" agents act to advance promoted cells into this stage but may not be initiating agents
Dose response does not exhibit a readily measurable threshold	Relative effectiveness of promoters depends on their ability with constant exposure to cause an expansion of the progeny of the initiated cell population	Spontaneous (fortuitous) progression can be demonstrated
Relative effect of initiators depends on quantitation of focal lesions following defined period of promotion		

[a] Adapted from Pitot et al. (1987).

I. Initiation

The first stage in the natural history of neoplastic development, initiation, is irreversible, as judged by an extended separation between the time of initiation or first application of the chemical carcinogen and the beginning of the second stage, promotion (cf. BOUTWELL 1964; PITOT 1978). However, the effectiveness of initiation depends on its relationship in time to cellular replicative DNA synthesis and cell division (YING et al. 1982; ISHIKAWA et al. 1980; HENNINGS et al. 1978). That the period of DNA synthesis itself is critical for the "fixation" and thus irreversibility of initiation has been demonstrated in other systems both in vivo and in vitro (WARWICK 1971; FREI and RITCHIE 1963; MCCORMICK and BERTRAM 1982; BOREK and SACHS 1968). As yet, single initiated cells cannot be unequivocally identified by known methodologies, although a recent report by MOORE and his associates (1987) suggested that single hepatocytes exhibiting the presence of the marker placental glutathione-*S*-transferase in liver represent initiated hepatocytes. However, it is not clear that all such single, altered hepatocytes produce clones during the stage of promotion (see below). Much more significant is the question of "spontaneous" or "fortuitous" initiated cells whose presence has been identified in the tissues of experimental animals (EMMELOT and SCHERER 1980; POPP et al. 1985; ETHIER and ULLRICH 1982); comparable changes have been noted in humans, at least in mammary tissue (VAN BOGAERT 1984). In the absence of exogenous promotion, the number of identifiable clones derived from spontaneously initiated cells and the total number of cells occupying such clones (volume %) increase with the age of the animal, at least in the liver (SCHULTE-HERMANN et al. 1983; OGAWA et al. 1981). However, when exogenous promoting agents are administered, the number of such focal changes in this organ increases during the first 2 months of life but remains unchanged thereafter (PITOT et al. 1985). The interpretation of this latter finding is that spontaneous initiation in the liver, dependent on fixation by cell division like any initiation process, occurs early in life when the DNA synthetic rate of the organ is relatively high, but when this decreases to low levels, further initiation is extremely rare.

The stage of initiation can be altered by exogenous and probably endogenous factors. The metabolism of chemicals to their ultimate forms capable of initiation can be inhibited by a variety of chemicals in several different tissues (cf. WATTENBERG 1978; WATTENBERG et al. 1983; THOMPSON et al. 1984). 5-Azacytidine (DENDA et al. 1985) and inhibitors of poly(ADP)ribosylation (TAKAHASHI et al. 1984) enhance the process of initiation during multistage hepatocarcinogenesis in the rat. The presence or absence of a threshold or no-effect level for initiating agents has been determined only by extrapolation in most studies. These investigations rely on the appearance of neoplasms as a measure of the process of initiation (EHLING et al. 1983). However, the direct measurement of initiation can be made most accurately in multistage hepatocarcinogenesis by quantitatively evaluating the number of focal lesions induced by the initiating agent. With this system, the process of initiation appears to be a linear, dose-related phenomenon that does not exhibit a readily measurable threshold (SCHERER and EMMELOT 1975; PITOT et al. 1987a, b). On this basis, it has recently been possible to evaluate quantitatively the relative potency of agents as initiators in multistage hepatocarcinogenesis (PITOT et al. 1987a). However, in such calculations and, for that matter, in evaluating complete carcinogens for their initiating capabilities, it is critical that the dose employed be nontoxic and nontumorigenic (PITOT et al. 1987b).

II. Promotion

The principal characteristic of the stage of tumor promotion that distinguishes it clearly from the stages of initiation and progression is that of reversibility. In the several model systems of multistage carcinogenesis that have been studied most extensively, the reversibility of promotion is characteristic (TATEMATSU et al. 1983; GLAUERT et al. 1986; STENBÄCK 1978; DIPAOLO et al. 1981). Furthermore, both in multistage hepatocarcinogenesis (HENDRICH et al. 1986) and multistage epidermal carcinogenesis (REDDY et al. 1987), focal lesions that disappear upon removal of the promoting agent can be shown to reappear on readministration of the promoting agent. Interestingly, REDDY and his associates (1987) have presented evidence that many of the papillomas reappearing on readministration of the promoting agent 12-O-tetradecanoyl-phorbol-13-acetate (TPA) do not appear to develop from those lesions which regressed upon removal of TPA. The regression or reversion of focal lesions during the stage of promotion, at least in the liver, may be due to the "remodeling" of hepatocytes from their altered appearance within enzyme-altered foci to normal-appearing hepatocytes (TATEMATSU et al. 1983); alternatively, the loss of focal lesions during the stage of promotion may be due to individual cell death (BURSCH et al. 1984). HANIGAN and PITOT (1985) published evidence that cells derived from enzyme-altered foci during the stage of promotion in multistage hepatocarcinogenesis in the rat can be transplanted into syngeneic hosts, with the subsequent development of foci from the transplanted cells occurring only if the host is continuously treated with the promoting agent, in this case phenobarbital. This study indicates that the existence of altered hepatocytes during the stage of promotion is dependent on the continued presence of the promoting agent. This phenomenon is quite analogous to the "dependent" neoplasms of endocrine tissues described by FURTH many years ago (1968).

Unlike initiation, which when once fixed by cell division (see above) is irreversible and heritable, the stage of promotion can be modulated by a variety of environmental alterations. BOUTWELL (1964) demonstrated that altering the frequency with which the promoting agent is administered drastically affects the appearance of papillomas during promotion in epidermal carcinogenesis. Similarly, aging also may inhibit the process of promotion (VAN DUUREN et al. 1975) as does the diet of the animal, both in amount (BOUTWELL 1964) and composition (HENDRICH et al. 1988). Unlike initiating agents, the dose response to promoting agents exhibits a threshold or no-effect level and a maximal response; the latter response is to be expected if only a finite number of cells have been initiated at the beginning of the experiment and if the promoting agent possesses no initiating activity (VERMA and BOUTWELL 1980; GOLDSWORTHY et al. 1984). A threshold or no-effect level of response to promoting agents could be predicted from their characteristic of reversibility, by analogy to the reversible effects of many xenobiotics that have pharmacological activities and that produce their effects through receptor mediators (ALDRIDGE 1986). In fact, many promoting agents exert their effects on gene expression through the mediation of receptor mechanisms (cf. PITOT 1986). The relative potency of promoting agents can be related to their effectiveness in expanding the progeny of initiated cell populations, as has been done in a model of multistage hepatocarcinogenesis (PITOT et al. 1987 a). Again, such measurement of promoting efficacy depends on the use of an initiating agent that has essentially no promoting action or toxic effects at the dose employed, and the measurement of relative effectiveness is carried out in that part of the dose-response curve between the threshold and maximal effect levels (PITOT et al. 1987 b).

III. Progression

The stage of progression as defined in modern concepts is somewhat different from that originally proposed by FOULDS (1965). While FOULDS considered the phenomenon of progression as encompassing the entire natural history of neoplastic development, in a more modern sense the stage of progression follows promotion in multistage carcinogenesis, and it is in this stage that irreversible, benign and/or malignant neoplasms are characteristically seen (SCHULTE-HERMANN 1985). Progression has been defined (PITOT 1986) as that stage of carcinogenesis exhibiting measurable (by recombinant DNA technology or related methods) and/or morphologically identifiable (karyotypic) changes in the structure of the cell genome. Such changes are directly related to the increased growth rate, invasiveness, metastatic capability, and biochemical changes in the neoplastic cell during this stage. It is the latter characteristic of progression that had been emphasized by FOULDS, but with a greater understanding of the molecular nature of changes in the cell genome and their effects on the expression of the genome, a relationship as defined above can now be made. Evidence for the presence and continued evolution of karyotypic changes during the stage of progression has been described during the development of a number of different types of neoplasms, including experimental epidermal carcinogenesis (ALDAZ et al. 1987), leukemia/lymphoma (cf. NOWELL 1986), and most recently in our own laboratory during multistage hepatocarcinogenesis (Y. XU, L. SARGENT, and H. C. PITOT, unpublished observations).

The characteristics of progression as defined in multistage carcinogenesis can be seen in Table 1. The irreversibility of this stage is assumed because of the obvious accompanying alterations in the cell genome distinguishing it from the reversible, preceding stage of promotion. However, it is clear that under certain circumstances cells in the stage of progression may be induced, by treatment with specific chemicals, into terminal differentiation, thereby removing them from continued progression to an even more malignant state (REISS et al. 1986). Other environmental alterations can produce changes in gene expression, growth rate, and functional processes within cells during progression (HORSFALL et al. 1986; NOBLE 1977). Agents that act only during the stage of progression, or at least to advance a cell from promotion to progression, have not been definitely characterized as yet in most systems, although the free radical generator, benzoylperoxide, appears to act as a "progressor" agent in experimental epidermal carcinogenesis (O'CONNELL et al. 1986). Theoretically, such progressor agents should be capable of inducing the genetic changes characteristic of the stage of progression; examples of such agents would be clastogenic agents and complete carcinogens, i.e., chemicals capable of inducing the malignant transformation from initiation through progression. That such agents can act in this manner has now been demonstrated with an "initiation-promotion-initiation" format earlier proposed by POTTER (1981) and experimentally demonstrated in the mouse epidermis by HENNINGS et al. (1985), who found that when the usual initiation-promotion format was subsequently followed by the application of a second complete carcinogen, such as an alkylating agent, a high incidence of carcinoma rapidly resulted. This was in contrast to the standard initiation-promotion format in this tissue, which gave primarily benign neoplasms during the time span of the experiment. In experimental multistage hepatocarcinogenesis in the rat, a similar regimen has been used to induce "foci-within-foci," which have been proposed as representing the earliest, morphologically discernible beginning of the stage of progression in this tissue (SCHERER et al. 1984; FARBER 1973; HIROTA and YOKOYAMA 1985). In this system, SCHERER et al. (1984) proposed that the foci-within-foci are the direct precursors of malignant neoplasms. The occurrence of foci-within-foci in a number of the initiation-promotion protocols for rat liver, such as that described by PITOT et al. (1978), is quite infrequent. However, this would be expected if the alteration actually represents the earliest demonstrable lesion at the interface of promotion and progression, since this model system results in very few malignant neoplasms compared with the large number of altered hepatic foci seen during the stage of promotion. Thus, the occasional focus-within-a-focus in this system probably represents spontaneous or fortuitous progression by analogy with spontaneous initiation, both of which are likely to be genetic events.

From the characteristics and previous definitions of the three stages of neoplastic development, it is the stages of initiation and progression that appear to involve changes in the structure of the genome of the cell. Such changes are clearly demonstrable during the stage of progression, and there is overwhelming evidence that initiation results in alterations in the genetic material of the cell (MILLER 1978). Thus, the natural history of the development of neoplasia in the multistage concept is quite analogous to the two-hit theory of KNUDSON (1986) developed from studies of neoplasms in humans that exhibit a clear Mendelian pattern of inheritance. Furthermore, with our present knowledge of the multistage development of neoplasia, we are now in a position to place the theoretical basis of chemical carcinogenesis in a framework of the natural history of the development of neoplasia such that our understanding of the process of carcinogenesis, its causes, mechanisms, and ultimately its prevention may be on a much firmer basis than in the past.

E. Reconciliation of the Theoretical Bases of Chemical Carcinogenesis with the Natural History of Neoplastic Development

Given the marked increase in our knowledge of molecular biology and molecular genetics during the past several decades, it is not surprising that today a basic precept of our understanding of the nature of the neoplastic process is that genetic change is at the heart of any theoretical basis of the cause and nature of malignancy. Therefore, the mechanistic theories of chemical carcinogenesis have evolved largely from studies of the effects or potential effects of chemicals on the genome. The pervasiveness of this concept has led investigators to seek for genetic alterations resulting from the actions of chemical carcinogens when there is little or no evidence of their altering the structure of the genome in a permanent, heritable manner. Examples of this include naturally occurring estrogens (LIEHR et al. 1986), polypeptide hormones, asbestos (OSHIMURA et al. 1984), dioxin (POLAND and GLOVER 1979), and TPA (DZARLIEVA-PETRUSEVSKA and FUSENIG 1985). Despite such attempts, a large number of chemicals are clearly capable of inducing malignant transformation in vivo upon chronic administration, although there is little or no evidence that either the substance itself or a metabolite is capable of altering the structure of DNA (cf. WEISBURGER and WILLIAMS 1981). On the other hand, some clastogenic compounds such as trimethylphosphate (LEGATOR et al. 1973), urea, and sodium benzoate (ISHIDATE and ODASHIMA 1977) have not been shown to be carcinogenic. Furthermore, some mutagenic chemicals are noncarcinogenic in rodents (SHELBY and STASIEWICZ 1984). Thus, it is clear that the mechanistic theories of chemical carcinogenesis on their own cannot account for the varied effects of chemicals shown to be carcinogenic in mammals. When, however, one considers the various mechanistic theories of chemical carcinogenesis and the bases upon which they have been proposed in the light of the natural history of the development of neoplasia, then a more integrated picture of chemical carcinogenesis evolves.

Although the classification of chemical carcinogens has been proposed in a number of ways (Fig. 1; WEISBURGER and WILLIAMS 1981; HECKER 1976), we are now in a better position to classify chemical carcinogens by their role in effecting one or more of the stages in the development of neoplasia:

Initiating agent (incomplete carcinogen) – a chemical capable of initiating cells only

Promoting agent – a chemical capable of causing the expansion of initiated cell clones

Progressor agent – a chemical capable of converting an initiated cell or a cell in the stage of promotion to a potentially malignant cell

Complete carcinogen – a chemical possessing the capability of inducing cancer from normal cells, usually possessing properties of initiating, promoting, and progressor agents

As pointed out previously (PITOT 1986), chemical agents capable only of inducing the stage of initiation appear to be relatively rare, although some, as yet untested, mutagenic noncarcinogens may fit into this category. In practical terms, all complete carcinogens can be given at doses at which only their initiating capacity is effective, and thus they act, de facto, as incomplete carcinogens. Promoting agents, on the other hand, are by definition incapable of initiating cells but may promote spontaneously initiated cells already present in the target tissues (PITOT et al. 1988), thus giving the appearance of complete carcinogenicity. While progressor agents include all complete carcinogens, again by definition, as pointed out earlier, those with their primary effect in inducing one or more cells in the stage of promotion to enter progression are likely to exist. Agents bringing about chromosomal alterations are likely candidates to be classified as progressor agents. These will also obviously include some agents with promoting activity as well, such as TPA and peroxisomal proliferating agents (REDDY et al. 1987), that are capable of indirect induction of DNA damage and clastogenesis through free radical mechanisms. It is now possible to test for progressor activity in multistage hepatocarcinogenesis through the initiation-promotion-initiation protocol, in which the final initiation is carried out by a progressor agent. Furthermore, there is substantial evidence that metal carcinogenesis may be associated with a high level of clastogenesis (VAINIO and SORSA 1981), as are the effects of synthetic estrogens (TSUTSUI et al. 1983) and asbestos (OSHIMURA et al. 1984; LECHNER et al. 1985). Although we do not yet know the nature of the genetic change involved in the induction of the stage of progression, it would appear that such changes are major rather than simple point mutations. It is interesting to speculate that the process of initiation may involve primarily the latter, whereas progression may be concerned with structural chromosomal abnormalities, major deletions, translocations, and/or insertions of genetic material. Clearly, if progression is the second allelic mutation required for the development of cancer (KNUDSON 1986), then more extensive DNA alterations would favor such a second mutation in a gene discretely altered during initiation.

The roles of proto-oncogenes and oncogenes in the stage of tumor formation by chemical carcinogens is beyond the scope of this review. These aspects of chemical carcinogenesis are addressed elsewhere in this volume.

F. Conclusions

Although it is possible to reconcile many of the mechanistic concepts of chemical carcinogenesis with the natural history of neoplastic development during chemical carcinogenesis, many questions are still unanswered in our knowledge of oncology. At this point, however, we can make use of our knowledge and this reconciliation of mechanistic theories and biological observations. If a number of chemical carcinogens do not act by structurally altering DNA and the genetic apparatus, then it is possible that the risk of effects of reversibly acting promoting agents to the human being will be significantly different from that of the irreversible effects of complete carcinogens. One can not only expect thresholds of exposure, but since the effects are reversible and can be modulated by environmental factors, active prevention of the process of promotion in human neoplasia is a clear reality. Such chemoprevention is already being tested (BERTRAM et al. 1987), and further research on modification of the stage of promotion is clearly warranted. The major environmental factors causing human cancer appear to act predominantly during the stage of promotion, as has been pointed out previously from this laboratory (Pitot 1986). Therefore, the actual prevention of more than 50% of human cancer by voluntary actions by the individual is a present-day reality.

On the other hand, with the identification of the stage of progression and the possibility that agents can act selectively, or predominantly, at this stage, further methods for the identification of, and considerations for the regulation of, progressor agents in the environment are now required and certainly will be necessary in the future. Prevention is still the most efficient, most cost effective, and, in many ways, the most convenient method by which to eliminate human disease. Through the reconciliation of mechanistic theories with the evident disease process, a dramatic decrease in the incidence of and mortality from human cancer is now within our grasp.

References

Aldaz CM, Conti CJ, Klein-Szanto AJP, Slaga TJ (1987) Progressive dysplasia and aneuploidy are hallmarks of mouse skin papillomas: relevance to malignancy. Proc Natl Acad Sci USA 84:2029–2032

Aldridge WN (1986) The biological basis and measurement of thresholds. Annu Rev Pharmacol Toxicol 26:39–58

Baylin SB, Höppener JWM, de Bustros A, Steenbergh PH, Lips CJM, Nelkin BD (1986) DNA methylation patterns of the calcitonin gene in human lung cancers and lymphomas. Cancer Res 46:2917–2922

Becker RA, Shank RC (1985) Kinetics of formation and persistence of ethylguanines in DNA of rats and hamsters treated with diethylnitrosamine. Cancer Res 45:2076–2084

Bedell MA, Lewis JG, Billings KC, Swenberg JA (1982) Cell specificity in hepatocarcinogenesis: preferential accumulation of O^6-methylguanine in target cell DNA during continuous exposure of rats to 1,2-dimethylhydrazine. Cancer Res 42:3079–3083

Beland FA, Kadlubar FF (1985) Formation and persistence of arylamine DNA adducts in vivo. Environ Health Perspect 62:19–30

Berenblum I, Shubik P (1947) A new, quantitative approach to the study of the stages of chemical carcinogenesis in the mouse's skin. Br J Cancer 1:383–386

Bernheim NJ, Falk H (1983) Cancer and the environment. J Am Coll Toxicol 2:23–54

Bertram JS, Kolonel LN, Meyskens FL Jr (1987) Rationale and strategies for chemoprevention of cancer in humans. Cancer Res 47:3012–3031

Biaglow JE (1981) The effect of ionizing radiation on mammalian cells. J Chem Educ 58:144–156

Boehm TLJ, Grunberger D, Drahovsky D (1983) Aberrant de novo methylation of DNA after treatment of murine cells with *N*-acetoxy-*N*-2-acetylaminofluorene. Cancer Res 43:6066–6071

Borek C, Sachs L (1968) The number of cell generations required to fix the transformed state in x-ray-induced transformation. Proc Natl Acad Sci USA 59:83–85

Boutwell RK (1964) Some biological aspects of skin carcinogenesis. Prog Exp Tumor Res 4:207–250

Boyland E (1950) The biological significance of metabolism of polycyclic compounds. Biochem Soc Symp 5:40–54

Brambilla G, Carlo P, Finollo R, Bignone FA, Ledda A, Cagelli E (1983) Viscometric detection of liver DNA fragmentation in rats treated with minimal doses of chemical carcinogens. Cancer Res 43:202–209

Brand KG, Buoen LC, Johnson KH, Brand I (1975) Etiological factors, stages, and the role of the foreign body in foreign body tumorigenesis: a review. Cancer Res 35:279–286
Bursch W, Lauer B, Timmermann-Trosiener I, Barthel G, Schuppler JH, Schulte-Hermann R (1984) Controlled death (apoptosis) of normal and putative preneoplastic cells in rat liver following withdrawal of tumor promoters. Carcinogenesis 5:453–458
Butlin HT (1892) Cancer of the scrotum in chimney-sweeps and others. II. Why foreign sweeps do not suffer from scrotal cancer. Br Med J 2:1–6
Cavaliere A, Bufalari A, Vitali R (1987) 5-Azacytidine carcinogenesis in BALB/c mice. Cancer Lett 37:51–58
Cha Y-M, Heine HS, Moldeus P (1982) Differential effects of dietary and intraperitoneal administration of antioxidants on the activities of several hepatic enzymes of mice. Drug Metab Dispos 10:434–435
Conney AH (1982) Induction of microsomal enzymes by foreign chemicals and carcinogenesis by polycyclic aromatic hydrocarbons. G.H.A. Clowes Memorial Lecture. Cancer Res 42:4875–4917
Cook JW, Hewett CL, Hieger I (1933) The isolation of a cancer-producing hydrocarbon from coal tar. Parts I, II, and III. J Chem Soc, pp 395–405
Daniel JW (1986) Metabolic aspects of antioxidants and preservatives. Xenobiotica 16:1073–1078
De Long MJ, Prochaska HJ, Talalay P (1986) Induction of NAD(P)H:quinone reductase in murine hepatoma cells by phenolic antioxidants, azo dyes, and other chemoprotectors: a model system for the study of anticarcinogens. Proc Natl Acad Sci USA 83:787–791
Denda A, Rao PM, Rajalakshmi S, Sarma DSR (1985) 5-Azacytidine potentiates initiation induced by carcinogens in rat liver. Carcinogenesis 6:145–146
Diala ES, Cheah MSC, Rowitch D, Hoffman RM (1983) Extent of DNA methylation in human tumor cells. JNCI 71:755–764
DiPaolo JA, DeMarinis AJ, Evans CH, Doniger J (1981) Expression of initiated and promoted stages of irradiation carcinogenesis in vitro. Cancer Lett 14:243–249
Dix TA, Marnett LJ (1983) Metabolism of polycyclic aromatic hydrocarbon derivatives to ultimate carcinogens during lipid peroxidation. Science 221:77–79
Doerfler W (1983) DNA methylation and gene activity. Annu Rev Biochem 52:93–124
Doll R, Peto R (1981) The causes of cancer. Oxford University Press, New York
Doniger J, Day RS, DiPaolo JA (1985) Quantitative assessment of the role of O^6-methylguanine in the initiation of carcinogenesis by methylating agents. Proc Natl Acad Sci USA 82:421–425
Dzarlieva-Petrusevska RT, Fusenig NE (1985) Tumor promoter 12-*O*-tetradecanoylphorbol-13-acetate (TPA)-induced chromosome aberrations in mouse keratinocyte cell lines: a possible genetic mechanism of tumor promotion. Carcinogenesis 6:1447–1456
Ehling UH, Averbeck D, Cerutti PA, Friedman J, Greim H, Kolbye AC Jr, Mendelsohn ML (1983) Review of the evidence for the presence or absence of thresholds in the induction of genetic effects by genotoxic chemicals. Mutat Res 123:281–341
Emmelot P, Scherer E (1980) The first relevant cell stage in rat liver carcinogenesis. A quantitative approach. Biochim Biophys Acta 605:247–304
Ethier SP, Ullrich RL (1982) Detection of ductal dysplasia in mammary outgrowths derived from carcinogen-treated virgin female BALB/c mice. Cancer Res 42:1753–1760
Farber E (1973) Hyperplastic liver nodules. Methods Cancer Res 7:345–375
Fischer V, Mason RP (1986) Formation of iminoxyl and nitroxide free radicals from nitrosonaphthols: an electron spin resonance study. Chem Biol Interact 57:129–142
Floyd RA (1981) Free-radical events in chemical and biochemical reactions involving carcinogenic arylamines. Radiat Res 86:243–263
Foulds L (1965) Multiple etiologic factors in neoplastic development. Cancer Res 25:1339–1347
Frei JV, Ritchie AC (1963) Diurnal variation in the susceptibility of mouse epidermis to carcinogen and its relationship to DNA synthesis. JNCI 32:1213–1220
Furth J (1968) Hormones and neoplasia. Thule International Symposia on Cancer and Aging. Nordiska Bokhandelns Förlag, Stockholm, pp 131–151

Glauert HP, Schwarz M, Pitot HC (1986) The phenotypic stability of altered hepatic foci: effect of the short-term withdrawal of phenobarbital and of the long-term feeding of purified diets after the withdrawal of phenobarbital. Carcinogenesis 7:117–121
Goelz SE, Vogelstein B, Hamilton SR, Feinberg AP (1985) Hypomethylation of DNA from benign and malignant human colon neoplasms. Science 228:187–190
Goldsworthy T, Campbell HA, Pitot HC (1984) The natural history and dose-response characteristics of enzyme-altered foci in rat liver following phenobarbital and diethylnitrosamine administration. Carcinogenesis 5:67–71
Goodglick LA, Kane AB (1986) Role of reactive oxygen metabolites in crocidolite asbestos toxicity to mouse macrophages. Cancer Res 46:5558–5566
Goth R, Rajewsky MF (1974) Persistence of O^6-ethylguanine in rat-brain DNA: correlation with nervous system-specific carcinogenesis by ethylnitrosourea. Proc Natl Acad Sci USA 71:639–643
Hanigan M, Pitot HC (1985) Growth of carcinogen-altered rat hepatocytes in the liver of syngeneic recipients promoted with phenobarbital. Cancer Res 45:6063–6070
Harvey RG (1981) Activated metabolites of carcinogenic hydrocarbons. Accounts Chem Res 14:218–226
Hecker E (1976) Definitions and terminology in cancer (tumor) etiology. An analysis aiming at proposals for a current internationally standardized terminology. Z Krebsforsch 86:219–230
Hendrich S, Glauert HP, Pitot HC (1986) The phenotypic stability of altered hepatic foci: effects of withdrawal and subsequent readministration of phenobarbital. Carcinogenesis 7:2041–2045
Hendrich S, Glauert HP, Pitot HC (1988) Dietary effects on initiation and promotion of hepatocarcinogenesis in the rat. J Cancer Res Clin Oncol 114:149–157
Hennings H, Yuspa SH (1985) Two-stage tumor promotion in mouse skin: an alternative interpretation. JNCI 74:735–740
Hennings H, Michael D, Patterson E (1978) Croton oil enhancement of skin tumor initiation by *N*-methyl-*N′*-nitro-*N*-nitrosoguanidine: possible role of DNA replication. Proc Soc Exp Biol Med 158:1–4
Hennings H, Shores R, Mitchell P, Spangler EF, Yuspa SH (1985) Induction of papillomas with a high probability of conversion to malignancy. Carcinogenesis 6:1607–1610
Hieger I (1930) LVIII. The spectra of cancer-producing tars and oils of related substances. Biochem J 24:505–511
Hirota N, Yokoyama T (1985) Comparative study of abnormality in glycogen storing capacity and other histochemical phenotypic changes in carcinogen-induced hepatocellular preneoplastic lesions in rats. Acta Pathol Jpn 35(5):1163–1179
Holliday R (1987) The inheritance of epigenetic defects. Science 238:163–170
Horsfall DJ, Tilley WD, Orell SR, Marshall VR, Cant ELMcK (1986) Relationship between ploidy and steroid hormone receptors in primary invasive breast cancer. Br J Cancer 53:23–28
Hueper WC, Wiley FH, Wolfe HD (1938) Experimental production of bladder tumors in dogs by administration of beta-naphthylamine. J Ind Hyg Toxicol 20:46–84
Ishidate M Jr, Odashima S (1977) Chromosome tests with 134 compounds on Chinese hamster cells in vitro – a screening for chemical carcinogens. Mutat Res 48:337–354
Ishikawa T, Takayama S, Kitagawa T (1980) Correlation between time of partial hepatectomy after a single treatment with diethylnitrosamine and induction of adenosine triphosphatase-deficient islands in rat liver. Cancer Res 40:4261–4264
Ito N, Hirose M (1987) The role of antioxidants in chemical carcinogenesis. Jpn J Cancer Res 78:1011–1026
Ito N, Fukushima S, Hagiwara A, Shibata M, Ogiso T (1983) Carcinogenicity of butylated hydroxyanisole in F344 rats. JNCI 70:343–352
Jerina DM, Lehr RE (1977) The bay-region theory: a quantum mechanical approach to aromatic hydrocarbon-induced carcinogenicity. In: Ullrich V, Roots I, Hildbrant AJ, Estrabrook RW, Conney AH (eds) Microsomes in drug oxidation. Pergamon, Oxford

Jerina DM, Daly JW, Witkop B, Zaltzman-Niremberg P, Udenfriend S (1970) 1,2-Naphthaleneoxide as an intermediate in the microsomal hydroxylation of naphthalene. Biochemistry 9:147–156
Josephy PD, Eling TE, Mason RP (1983) Co-oxidation of benzidine by prostaglandin synthase and comparison with the action of horseradish peroxidase. J Biol Chem 258:5561–5569
Kadis B (1978) Steroid epoxides in biologic systems: a review. J Steroid Biochem 9:75–81
Kahl R (1986) The dual role of antioxidants in the modification of chemical carcinogenesis. J Environ Sci Health 1:47–92
Kawachi T, Nagao M, Yahagi T, Takahashi Y, Sugimura T, Matsushima T, Kawakami T, Ishidate M (1979) Our view on the relation between mutagens and carcinogens. In: Miller EC et al. (eds) Naturally occurring carcinogens-mutagens and modulators of carcinogenesis. Proc 9th International Symposium of The Princess Takamatsu Cancer Research Fund. University Park Press, Baltimore, pp 337–344
Kennaway EL (1925) Experiments on cancer-producing substances. Br Med J 2:1–4
Kennaway EL, Hieger I (1930) Carcinogenic substances and their fluorescence spectra. Br Med J 1:1044–1046
Kinzel V, Fürstenberger G, Loehrke H, Marks F (1986) Three-stage tumorigenesis in mouse skin: DNA synthesis as a prerequisite for the conversion stage induced by TPA prior to initiation. Carcinogenesis 7:779–782
Knudson AG Jr (1986) Genetics of human cancer. Annu Rev Genet 20:231–251
Kraemer KH (1980) Xeroderma pigmentosum. In: Demis DJ, Dobson RL, McGuire J (eds) Clinical dermatology. Harper & Row, Philadelphia
Krauss RS, Eling TE (1985) Formation of unique arylamine: DNA adducts from 2-aminofluorene activated by prostaglandin H synthase. Cancer Res 45:1680–1686
Kriek E, Engelse L, Scherer E, Westra JG (1984) Formation of DNA modifications by chemical carcinogens. Identification, localization and quantification. Biochim Biophys Acta 738:181–201
Lacassagne A (1932) Apparition de cancers de la mamelle chez la souris mâle, soumise à des injections de folliculine. C R Acad d Sc 195:630–632
Lechner JF, Tokiwa T, LaVeck M, Benedict WF, Banks-Schlegel S, Yeager H Jr, Banerjee A, Harris CC (1985) Asbestos-associated chromosomal changes in human mesothelial cells. Proc Natl Acad Sci USA 82:3884–3888
Legator MS, Palmer KA, Adler I-D (1973) A collaborative study of in vivo cytogenetic analysis. I. Interpretation of slide preparations. Toxicol Appl Pharmacol 24:337–350
Liehr JG, Avitts TA, Randerath E, Randerath L (1986) Estrogen-induced endogenous DNA adduction: possible mechanism of hormonal cancer. Proc Natl Acad Sci USA 83:5301–5305
Liehr JG, Hall ER, Avitts TA, Randerath E, Randerath K (1987) Localization of estrogen-induced DNA adducts and cytochrome P-450 activity at the site of renal carcinogenesis in the hamster kidney. Cancer Res 47:2156–2159
Lowe JP, Silverman BD (1984) Predicting carcinogenicity of polycyclic aromatic hydrocarbons. Accounts Chem Res 17:332–338
Lutz WK, Jaggi W, Schlatter C (1982) Covalent binding of diethylstilbesterol to DNA in rat and hamster liver and kidney. Chem Biol Interact 42:251–257
Maher VM, McCormick JJ (1984) Role of DNA lesions and repair in the transformation of human cells. Pharmacol Ther 25:395–408
Maher VM, Miller EC, Miller JA, Szybalski W (1968) Mutations and decreases in density of transforming DNA produced by derivatives of the carcinogens 2-acetylaminofluorene and *N*-methyl-4-aminoazobenzene. Mol Pharmacol 4:411–426
McCormick PJ, Bertram JS (1982) Differential cell cycle phase specificity for neoplastic transformation and mutation to ouabain resistance induced by *N*-methyl-*N*′-nitro-*N*-nitrosoguanidine in synchronized C3H10T1/2 C18 cells. Proc Natl Acad Sci USA 79:4342–4346
McNeill JM, Wills ED (1985) The formation of mutagenic derivatives in benzo[*a*]pyrene by peroxidising fatty acids. Chem Biol Interact 53:197–207
Metzler M (1982) Diethylstilbestrol: hormonal or chemical carcinogen? Trends Pharmacol Sci 3:174–175

Miller EC (1978) Some current perspectives on chemical carcinogenesis in humans and experimental animals: presidential address. Cancer Res 38:1479–1496
Miller EC, Miller JA (1947) The presence and significance of bound aminoazo dyes in the livers of rats fed *p*-dimethylaminoazobenzene. Cancer Res 7:468–480
Miller JA (1970) Carcinogenesis by chemicals: an overview. G.H.A. Clowes memorial lecture. Cancer Res 30:559–576
Miller JA, Miller EC (1969) The metabolic activation of carcinogenic aromatic amines and amides. Prog Exp Tumor Res 11:273–301
Moore MA, Nakagawa K, Satoh K, Ishikawa T, Sato K (1987) Single GST-P positive liver cells – putative initiated hepatocytes. Carcinogenesis 8:483–486
Mottram JC (1944) A developing factor in experimental blastogenesis. J Pathol Bacteriol 56:181–187
Nagata C, Kodama M, Ioki Y, Kimura T (1982) Free radicals produced from chemical carcinogens and their significance in carcinogenesis. In: Floyd RA (ed) Free radicals and cancer. Dekker, New York
Noble RL (1977) Hormonal control of growth and progression in tumors of Nb rats and a theory of action. Cancer Res 37:82–94
Norden B, Edlund U, Wold S (1978) Carcinogenicity of polycyclic aromatic hydrocarbons studied by SIMCA pattern recognition. Acta Chem Scand 32:602–608
Nowell PC (1986) Mechanisms of tumor progression. Cancer Res 46:2203–2207
O'Brien PJ (1985) Free-radical-mediated DNA binding. Environ Health Perspect 64:219–232
O'Connell JF, Klein-Szanto AJP, DiGiovanni DM, Fries JW, Slaga TJ (1986) Enhanced malignant progression of mouse skin tumors by the free-radical generator benzoyl peroxide. Cancer Res 46:2863–2865
Ogawa K, Onoe T, Takeuchi M (1981) Spontaneous occurrence of gamma-glutamyl transpeptidase-positive hepatocytic foci in 105-week-old Wistar and 72-week-old Fischer 344 male rats. JNCI 67:407–412
Oshimura M, Hesterberg TW, Tsutsui T, Barrett JC (1984) Correlation of asbestos-induced cytogenetic effects with cell transformation of Syrian hamster embryo cells in culture. Cancer Res 44:5017–5022
Pitot HC (1979) Drugs as promoters of carcinogenesis. In: Estabrook RW, Lindenlaub E (eds) The induction of drug metabolism symposium, Ashford Castle, Ireland, May 24–27, 1978. Schattauer, Stuttgart, p 471–483
Pitot HC (1986) Fundamentals of oncology, 3rd edn. Dekker, New York
Pitot HC, Barsness L, Goldsworthy T, Kitagawa T (1978) Biochemical characterisation of stages of hepatocarcinogenesis after a single dose of diethylnitrosamine. Nature 271:456–457
Pitot HC, Grosso LE, Goldsworthy T (1985) Genetics and epigenetics of neoplasia: facts and theories. Carcinogenesis 10:65–79
Pitot HC, Beer DG, Hendrich S (1987a) Multistage carcinogenesis of the rat hepatocyte. Banbury Report 25: Nongenotoxic Mechanisms in Carcinogenesis. Cold Spring Harbor Laboratory, Cold Spring Harbor, New York, pp 41–53
Pitot HC, Goldsworthy TL, Moran S, Kennan W, Glauert HP, Maronpot RR, Campbell HA (1987b) A method to quantitate the relative initiating and promoting potencies of hepatocarcinogenic agents in their dose-response relationships to altered hepatic foci. Carcinogenesis 8:1491–1499
Pitot HC, Beer D, Hendrich S (1988) Multistage carcinogenesis: the phenomenon underlying the theories. In: Iversen O (ed) Theories of carcinogenesis. Hemisphere Press, Washington, pp 159–177
Poland A, Glover E (1979) An estimate of the maximum in vivo covalent binding of 2,3,7,8-tetrachlorodibenzo-*p*-dioxin to rat liver protein, ribosomal RNA, and DNA. Cancer Res 39:3341–3344
Popp JA, Scortichini BH, Garvey LK (1985) Quantitative evaluation of hepatic foci of cellular alteration occurring spontaneously in Fischer-344 rats. Fundam Appl Toxicol 5:314–319

Pott P (1775) Chirurgical observations relative to the cataract, the polypus of the nose, the cancer of the scrotum, the different kinds of ruptures, and the mortification of the toes and feet. Hawes, Clarke, and Collins, London
Potter VR (1981) A new protocol and its rationale for the study of initiation and promotion of carcinogens in rat liver. Carcinogenesis 2:1375–1379
Potter VR (1983) Alternative hypotheses for the role of promotion of chemical carcinogenesis. Environ Health Perspect 50:139–148
Pullman A, Pullman B (1955) Electronic structure and carcinogenic activity of aromatic molecules. New developments. Adv Cancer Res 3:117–169
Reddy AL, Caldwell M, Fialkow PJ (1987) Studies of skin tumorigenesis in PGK mosaic mice: many promoter-independent papillomas and carcinomas do not develop from pre-existing promoter-dependent papillomas. Int J Cancer 39:261–265
Redmond DE Jr (1970) Tobacco and cancer: the first clinical report, 1761. N Engl J Med 282:18–23
Reed GA, Marnett LJ (1982) Metabolism and activation of 7,8-dihydrobenzo[*a*]pyrene during prostaglandin biosynthesis. J Biol Chem 257:11368–11376
Rehn L (1895) Blasengeschwulste bei Fuchsin-Arbeitern. Arch Klin Chir 50:588–600
Reiss M, Gamba-Vitalo C, Sartorelli AC (1986) Induction of tumor cell differentiation as a therapeutic approach: preclinical models for hematopoietic and solid neoplasms. Cancer Treat Rep 70:201–218
Riggs AD, Jones PA (1983) 5-Methylcytosine, gene regulation, and cancer. Adv Cancer Res 40:1–30
Rous P, Kidd JG (1941) Conditional neoplasms and subthreshold neoplastic states. A study of tar tumors in rabbits. J Exp Med 73:365–376
Sarasin A, Bourre F, Benoit A (1982) Error-prone replication of ultraviolet-irradiated simian virus 40 in carcinogen-treated monkey kidney cells. Biochemie 64:815–821
Scherer E, Emmelot P (1975) Kinetics of induction and growth of precancerous liver-cell foci, and liver tumour formation by diethylnitrosamine in the rat. Eur J Cancer 11:689–696
Scherer E, Steward AP, Emmelot P (1977) Kinetics of formation of O^6-ethylguanine in, and its removal from liver DNA of rats receiving diethylnitrosamine. Chem Biol Interact 19:1–11
Scherer E, Feringa AW, Emmelot P (1984) Initiation-promotion-initiation. Induction of neoplastic foci within islands of precancerous liver cells in the rat. Models Mech Etiol Tumor Prom 56:57
Schulte-Hermann R (1985) Tumor promotion in the liver. Arch Toxicol 57:147–158
Schulte-Hermann R, Timmermann-Trosiener I, Schuppler J (1983) Promotion of spontaneous preneoplastic cells in rat liver as a possible explanation of tumor production by nonmutagenic compounds. Cancer Res 43:839–844
Shank RC (1984) Toxicity-induced aberrant methylation of DNA and its repair. Pharmacol Rev 36:19S-24S
Shelby MD, Stasiewicz S (1984) Chemicals showing no evidence of carcinogenicity in long-term, two-species rodent studies: the need for short-term test data. Environ Mutagen 6:871
Shivapurkar N, Wilson MJ, Poirier LA (1984) Hypomethylation of DNA in ethionine-fed rats. Carcinogenesis 5:989–992
Sims P, Grover PL, Swaisland A, Pal K, Hewer A (1974) Metabolic activation of benzo(a)pyrene proceeds by a diol-epoxide. Nature 252:326–328
Singer B (1985) In vivo formation and persistence of modified nucleosides resulting from alkylating agents. Environ Health Perspect 62:41–48
Sirover MA, Loeb LA (1976) Metal-induced infidelity during DNA synthesis. Proc Natl Acad Sci USA 73:2331–2335
Slaga TJ (1983) Overview of tumor promotion in animals. Environ Health Perspect 50:3–14
Smith IA, Berger GD, Seybold PG, Serve MP (1978) Relationships between carcinogenicity and theoretical reactivity indices in polycyclic aromatic hydrocarbons. Cancer Res 38:2968–2977

Stenbäck F (1978) Tumor persistence and regression in skin carcinogenesis. Z Krebsforsch 91:249–259
Stewart BW, Hristoforidis C, Haber M (1985) Dose-dependent persistence of alkylation-induced single-stranded regions in rat liver DNA in vivo. Cancer Lett 28:27–33
Stout DL, Becker F (1980) Progressive DNA damage in hepatic nodules during 2-acetylaminofluorene carcinogenesis. Cancer Res 40:1269–1273
Swenberg JA, Dyroff MC, Bedell MA, Popp JA, Huh N, Kirstein U, Rajewsky MF (1984) O^4-Ethyldeoxythymidine, but not O^6-ethyldeoxyguanosine, accumulates in hepatocyte DNA of rats exposed continuously to diethylnitrosamine. Proc Natl Acad Sci USA 81:1692–1695
Swenberg JA, Richardson FC, Boucheron JA, Dyroff MC (1985) Relationships between DNA adduct formation and carcinogenesis. Environ Health Perspect 62:177–183
Takahashi S, Nakae D, Yokose Y, Emi Y, Denda A, Mikami S, Ohnishi T, Konishi Y (1984) Enhancement of DEN initiation of liver carcinogenesis by inhibitors of NAD^+ ADP ribosyl transferase in rats. Carcinogenesis 5:901–906
Tatematsu M, Nagamine Y, Farber E (1983) Redifferentiation as a basis for remodeling of carcinogen-induced hepatocyte nodules to normal appearing liver. Cancer Res 43:5049–5058
Teoule R (1987) Radiation-induced DNA damage and its repair. Int J Radiat Biol 51:573–589
Thompson HJ, Chasteen ND, Meeker LD (1984) Dietary vanadyl(IV) sulfate inhibits chemically-induced mammary carcinogenesis. Carcinogenesis 5:849–851
Trush MA, Seed JL, Kensler TW (1985) Oxidant-dependent metabolic activation of polycyclic aromatic hydrocarbons by phorbol ester-stimulated human polymorphonuclear leukocytes: possible link between inflammation and cancer. Proc Natl Acad Sci USA 82:5194–5198
Tsutsui T, Maizumi H, McLachlan JA, Barrett JC (1983) Aneuploidy induction and cell transformation by diethylstilbestrol: a possible chromosomal mechanism in carcinogenesis. Cancer Res 43:3814–3821
Vainio H, Sorsa M (1981) Chromosome aberrations and their relevance to metal carcinogenesis. Environ Health Perspect 40:173–180
van Bogaert L-J (1984) Mammary hyperplastic and preneoplastic changes: taxonomy and grading. Breast Cancer Res Treat 4:315–322
Van Duuren BL, Sivak A, Katz C, Seidman I, Melchionne S (1975) The effect of aging and interval between primary and secondary treatment in two-stage carcinogenesis on mouse skin. Cancer Res 35:502–505
Vasdev S, Tsuruta Y, O'Brien PJ (1982) Prostaglandin synthetase mediated activation of *p*-dimethylaminoazobenzene (butter yellow). Prostaglandins and Cancer: First International Conference. Liss, New York, pp 155–158
Vedel M, Gomez-Garcia M, Sala M, Sala-Trepat JM (1983) Changes in methylation pattern of albumin and α-fetoprotein genes in developing rat liver and neoplasia. Nucleic Acids Res 11:4335–4354
Verma AK, Boutwell RK (1980) Effects of dose and duration of treatment with the tumor-promoting agent, 12-*O*-tetradecanoylphorbol-13-acetate, on mouse skin carcinogenesis. Carcinogenesis 1:271–276
Warwick GP (1971) Effect of the cell cycle on carcinogenesis. Fed Proc 30:1760–1765
Wattenberg LW (1978) Inhibition of chemical carcinogenesis. JNCI 60:11–18
Wattenberg LW, Borchert P, Destafney CM, Coccia JB (1983) Effects of *p*-methoxyphenol and diet on carcinogen-induced neoplasia of the mouse forestomach. Cancer Res 43:4747–4751
Wefers H, Komai T, Talalay P, Sies H (1984) Protection against reactive oxygen species by NAD(P)H:quinone reductase induced by the dietary antioxidant butylated hydroxyanisole (BHA). Decreased hepatic low-level chemiluminescence during quinone redox cycling. FEBS Lett 169:63–66
Weinstein IB, Gattoni-Celli S, Kirschmeier P, Lambert M, Hsiao W, Backer J, Jeffrey A (1984) Multistage carcinogenesis involves multiple genes and multiple mechanisms. J Cell Physiol Suppl 3:127–137

Weisburger JH, Williams GM (1981) Carcinogen testing: current problems and new approaches. Science 214:401–407
Wintersberger U (1982) Chemical carcinogenesis – the price of DNA repair? Naturwissenschaften 69:107–113
Wise RW, Zenser TV, Kadlubar FF, Davis BB (1984) Metabolic activation of carcinogenic aromatic amines by dog bladder and kidney prostaglandin H synthase. Cancer Res 44:1893–1897
Wong PK, Hampton MJ, Floyd RA (1982) Evidence for lipoxygenase-peroxidase activation of *N*-hydroxy-2-acetylaminofluorene by rat mammary gland parenchymal cells. Prostaglandins and Cancer: First International Conference. Liss, New York, pp 167–179
Wright WC (1940) De morbis artificum (author: Bernardino Ramazzini). The Latin text of 1713. University of Chicago Press, Chicago
Wynder EL, Gori GB (1977) Contribution of the environment to cancer incidence: an epidemiologic exercise. JNCI 58:825–832
Yamagawa K, Ichikawa K (1915) Experimentelle Studie über die Pathogenese der Epithelialgeschwulste. Mitteilungen Med Facultat Kaiserl Univ Tokyo 15(2):295–344
Yang SK, McCourt DW, Roller PP, Gelboin HV (1976) Enzymatic conversion of benzo[*a*]pyrene leading predominantly to the diol-epoxide *r*-7,*t*-8-dihydroxy-*t*-9,10-oxy-7,8,9,10-tetrahydrobenzo[*a*]pyrene through a single enantiomer of *r*-7,*t*-8-dihydroxy-7,8-dihydrobenzo[*a*]pyrene. Proc Natl Acad Sci USA 73:2594–2598
Ying TS, Enomoto K, Sarma DSR, Farber E (1982) Effects of delays in the cell cycle on the induction of preneoplastic and neoplastic lesions in rat liver by 1,2-dimethylhydrazine. Cancer Res 42:876–880
Yoshida T (1933) Über die serienweise Verfolgung der Veränderungen der Leber der experimentellen Hepatomerzeugung durch o-Aminoazotoluol. Trans Jpn Pathol Soc 23:636–638
zur Hausen H (1987) Papillomaviruses in human cancer. Cancer 59:1692–1696

Part II. Exposure to Chemical Carcinogens

CHAPTER 2

Environmental Carcinogens

K. Hemminki

A. Introduction

There are many ways by which to define "environmental" carcinogens. Here the epithet "environmental" is taken to cover any unwanted, environmentally derived chemicals that enter the human body via food, drink or air and that have been shown to cause, or are suspected of causing, cancer in humans and/or experimental animals. As a further qualification, only man-made materials will be considered, with the exception of radon. The definition includes environmental contaminants in food but excludes food additives and natural ingredients. Mytotoxins, passive smoking and occupational carcinogens will not be dealt with because they are discussed elsewhere in this book. Furthermore, the effects of chlorofluorocarbons on stratospheric ozone levels, and thus indirectly on ultraviolet radiation and skin cancer, are beyond the scope of this chapter.

Environmental carcinogens are discussed here by chemical group rather than by source of exposure. However, in the context of each chemical the different sources of exposure are presented, and quantitative data will be given when feasible. Before individual chemicals are considered, the fate of chemicals in the environment, classification of carcinogens, variation in the populations exposed and principles of risk estimation will be discussed.

Environmental carcinogenesis is a very wide field covering environmental hygiene, human exposure assessment, carcinogenic properties of chemicals and some degree of risk assessment. For this reason several sources have been consulted in preparing this chapter. The World Health Organization has published valuable reference material in this area such as *Guidelines for Drinking-Water Quality* (WHO 1984a), *Air Quality Guidelines* (WHO 1987a) and *Environmental Health Criteria* on a number of compounds; the Environmental Protection Agency of the United States of America (U.S.EPA) has assessed human exposure to and risks from a variety of environmental carcinogens, and many of these risk estimates are quoted here. The Monographs series of the International Agency for Research on Cancer offers authoritative, qualitative assessments on the carcinogenic properties of chemicals and data on the occurrence of these chemicals in the environment; furthermore, the periodicals *Environmental Carcinogenesis Reviews* and *Environmental Health Perspectives* are invaluable sources of reference data in the areas to be covered in this chapter.

Sampling and analytical techniques in environmental hygiene have developed considerably during the past 2 decades. The accuracy of the methods has improved, and sensitivities have increased by orders of magnitude for some measurements. The older results thus need to be treated with due caution. The greatest problem is, however, the paucity of measurements. Rarely are representative measurements available, even for one subpopulation, that cover all the possible sources of exposure. This being the case, extrapolations even to a national level may be tenuous and extrapolations to a global level are sheer guesswork.

B. Fate of Chemicals in the Environment

Chemicals are released into the environment from natural and man-made (anthropogenic) sources, of which only the latter are considered in this chapter. Emissions into the environment started with the use of fire and have greatly increased as a consequence of industrialization, involving particularly the chemical industry, energy production and transportation.

The global fate of chemicals is not well understood. Release into the air or dumping into a waste site does not remove the chemical from the ecosystem. In the air it may undergo chemical reactions, some of which are catalysed by light (WHO 1987a). Typical reactions in the air include oxidation by ozone, hydroxyl radicals or other oxidising agents to yield many types of products, such as CO, CO_2, H_2O, H_2O_2, aldehydes, peroxyacylnitrates, organic acids and phosgene. The unreacted chemicals may be deposited in water or into the ground, which can form a more stable compartment for some chemicals. The chemicals deposited in dump sites have a tendency to leak out into ground or surface water, or to volatilise and penetrate into the atmosphere. Release into the sewage can cause the eventual passage of the chemical into water reservoirs.

The global fate of chemicals is schematically shown in Fig. 1. Extraction of raw materials, e.g. mining of asbestos, may cause some environmental releases. Major releases take place during manufacturing, e.g. refining of oil or synthesis of chemicals through intermediary stages. The uses of chemicals may also entail major releases into the environment, e.g. production of energy. Finally, disposal by incineration, dumping or release to the sewage system determines the form in which the chemical enters the environment. At any intermediary stage, storage and transportation may entail releases into the environment. Furthermore, by-products and reaction vehicles are removed at various stages of the refinement process and will be released into the environment. It is important to realise that there are no isolated pockets in the ecosystem. Once a chemical has been extracted or synthesised, it will remain in the ecosystem unless it is converted into something else through chemical reactions (spontaneously, enzymatically or by burning). This aspect has been repeatedly overlooked due to ignorance and/or neglect and has been the cause of most known environmental health problems.

Many environmental health problems relate to the persistence of certain types of chemicals in the environment. DDT and chemicals such as PCBs were used in large quantities after the Second World War, resulting in high environmental concentrations all over the globe in the course of 3 decades. These compounds bioaccumulate, for example fish can concentrate DDT 10000 times from the surrounding water (WHO 1984a). Human milk contains so much of these com-

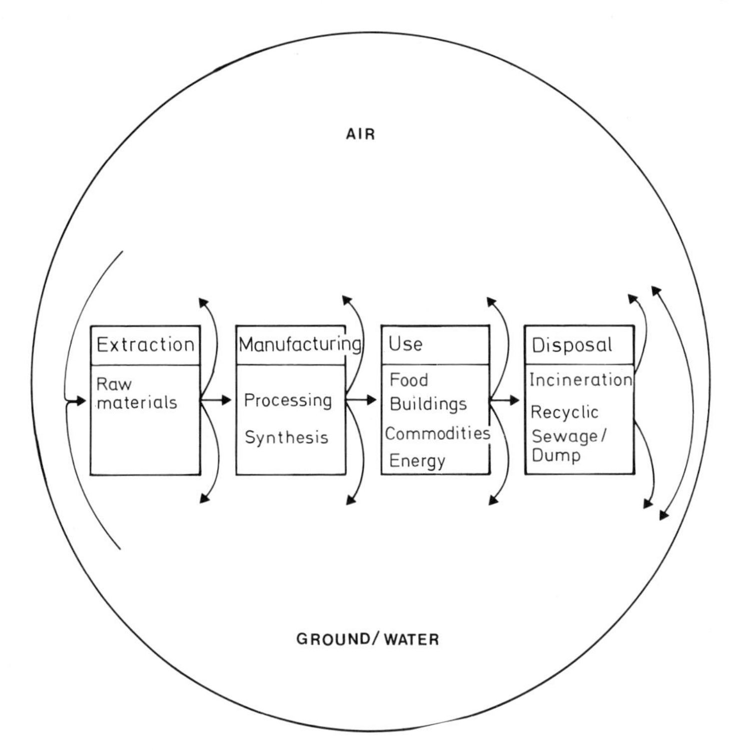

Fig. 1. Global fate of chemicals. *Arrows* indicate transfer to the next stage of use or release to the environment

pounds in Western countries that infants ingest an average of about 0.5 mg of DDT and PCBs during their first half year of life (see Sect. XII). Cadmium compounds provide another example of the problems that arise due to persistence. Even though the metal has a limited use, municipal incinerators put out cadmium in such quantities into the environment that the internal organs of some long-lived animals are considered to be inedible by humans. Asbestos is a further example of what was originally thought to be principally an occupational problem. Even though the use of asbestos has been discontinued in some countries, the occupational problems persist in the demolition of structures in which asbestos was used. Asbestos has now become an environmental problem as well. The general public is exposed to asbestos fibres used, for example, in construction materials. Asbestos-coated surfaces release the fibres to an extent that has prompted expensive reconstruction work in many countries. Concrete water pipelines contain asbestos that is released into the tap water, causing widespread concern. The last but not least, as an example of the lack of precautions taken in the handling of environmental chemicals, is chemical dump sites. These have been used in most industrialised countries for decades and are still in use even now, without any regard to the fate of the chemicals in the dumps. Recently the USA has started massive and immensely expensive clear-up operations at its dump sites, whilst in Europe concern is building up, but action is delayed.

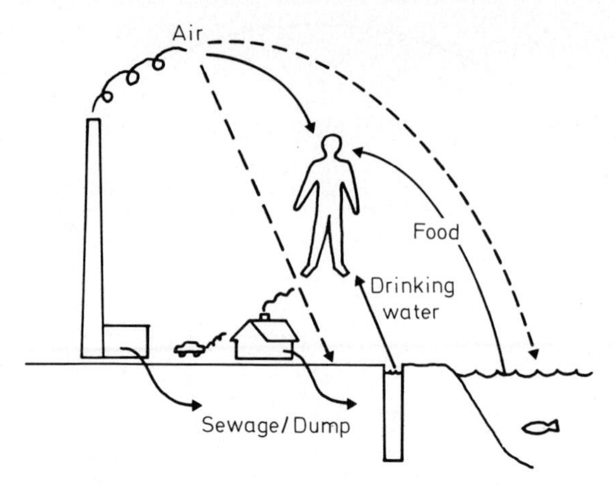

Fig. 2. Sources of human exposure to environmental contaminants. *Broken lines* indicate deposition to water and soil

Figure 2 illustrates the sources of exposure to environmental carcinogens. The direct routes of exposure are through air and drinking water, while enrichment in the food chain and exposure through food is an indirect route. As discussed earlier, contaminants in the atmosphere are deposited on the ground and in the water and may thus contribute to exposure through drinking water and food.

Table 1 indicates some main sources of human exposure to selected environmental carcinogens. Food is the main source for metals, polycyclic aromatic hydrocarbons, polyhalogenated hydrocarbons and pesticides, even though all of them are found in smaller quantities in drinking water and air. For halogenated hydrocarbons, drinking water and air are about equally important as sources. Air is the main source of formaldehyde and also of benzene. Further quantitation of the various sources of exposure is given in the context of the individual chemicals discussed.

Table 1. Sources of human exposure to environmental carcinogens

Source	Type of contaminant
Food	Metals
	Polycyclic aromatic hydrocarbons
	Pesticide residues
Drinking water	Halogenated hydrocarbons (trihalomethanes)
Air	Formaldehyde
	Benzene
	Halogenated hydrocarbons

C. Classification of Environmental Carcinogens

The International Agency for Research on Cancer (IARC) has evaluated in its Monograph programme the level of evidence for the carcinogenicity of chemicals. Recently, the evidence was updated and summarised, and this source was used as the qualitative basis for carcinogenicity in the present chapter (IARC 1987). The following lists the compounds of environmental interest that IARC has reviewed:

Group 1: Agents carcinogenic to humans
Arsenic and arsenic compounds
Asbestos
Benzene
Chromium compounds, hexavalent
Erionite
Nickel and nickel compounds
Soots
Tobacco smoke
Vinyl chloride

Group 2 A: Agents probably carcinogenic to humans
Acrylonitrile
Benz[*a*]anthracene
Benzo[*a*]pyrene
Cadmium and cadmium compounds
Dibenz[*a,h*]anthracene
Ethylene dibromide
Ethylene oxide (ethylene)
Formaldehyde
Polychlorinated biphenyls
Propylene oxide (propylene)
Styrene oxide

Group 2 B: Agents possibly carcinogenic to humans
Acetaldehyde
Benzo[*b*]fluoranthene
Benzo[*j*]fluoranthene
Benzo[*k*]fluoranthene
Carbon tetrachloride
Chloroform
Chlorophenols
Dichlorodiphenyltrichloroethane (DDT)
Dibenz[*a,h*]acridine
Dibenz[*a,j*]acridine
7H-Dibenzo[*c,g*]carbazole
Dibenzo[*a,e*]pyrene

Dibenzo[*g,h*]pyrene
Dibenzo[*a,i*]pyrene
Dibenzo[*a,l*]pyrene
para-Dichlorobenzene
1,2-Dichloroethane
Dichloromethane (methylene chloride)
Di(2-ethylhexyl)phthalate
Hexachlorobenzene
Hexachlorocyclohexanes
Lead and lead compounds
5-Methylchrysene
5-Nitroacenaphthene
Polybrominated biphenyls
Styrene
2,3,7,8-Tetrachlorodibenzo-para-dioxin(TCDD)
Tetrachloroethylene
Toxaphene

Group 1 contains agents that are causally linked to cancer in humans. Asbestos, benzene and tobacco smoke (not discussed in this chapter) are particularly important environmental carcinogens among these 9 agents. Group 2A lists 11 environmental chemicals, which IARC considers probably carcinogenic to humans. Group 2B contains 29 agents considered possibly carcinogenic to humans.

The IARC lists contain several structural analogues, such as many polycyclic aromatic hydrocarbons. These do not appear individually in any environmental exposure situation, and, hence, they are discussed later under one heading. As is the case with polycyclic aromatic hydrocarbons, many other environmental contaminants contain a number of structural analogues, e.g. air samples contain benzene, toluene, xylenes, ethylbenzenes, trimethylbenzenes, etc. in relatively equal amounts, as they contain many aliphatic hydrocarbons, halogenated hydrocarbons, etc. In such cases, reference may be made to the concentrations of derivatives, even if their carcinogenicity has not been settled.

The IARC classification is qualitative, and the grouping indicates the strength of evidence rather than carcinogenic potency. For risk estimation purposes it is useful to have quantitative data on the potency of carcinogens. Such data have been systematically compiled by GOLD and coworkers (1984), who calculate tumorigenic dose 50 (TD_{50}) values for carcinogens from animal bioassays. This figure is the dose of the carcinogen (mg/kg per day) given for a lifetime which causes tumours in 50% of the animals (if the controls are tumourfree). Thus, potent carcinogens have low TD_{50} values (e.g. TCDD 100 ng/kg per day) and weak carcinogens, a high TD_{50} value (e.g. DDT 100 mg/kg per day). Figure 3 shows the vast range of carcinogenic potencies as calculated for experimental animals (GOLD et al. 1984). Application of such potency figures is useful in risk estimation, but it should be done with the understanding that potency data have limitations even in extrapolations between two strains of a single species, let alone between species.

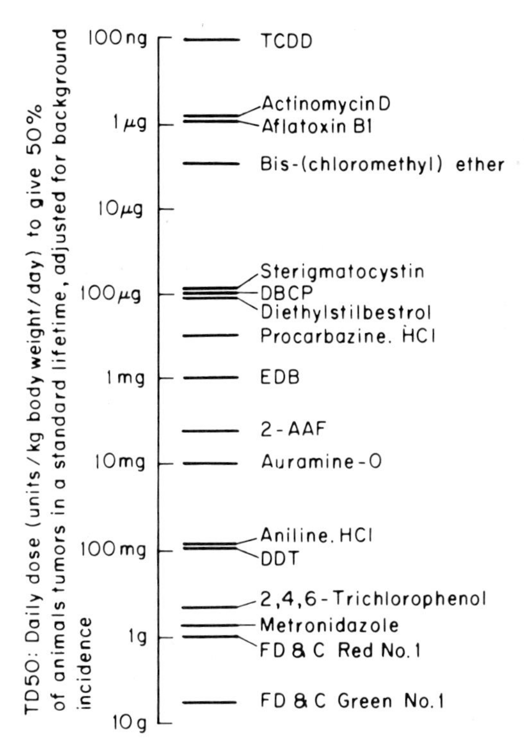

Fig. 3. Examples of carcinogenic potency values from male rats, given as TD_{50} (from GOLD et al. 1984). *DBCP*, dibromochloropropane; *EDB*, ethylene bromide; *FD&C red no. 1* and *FD&C green no. 1* are food dyes; *TCDD*, 2,3,7,8-tetrachlorodibenzo-para-dioxin; *2-AAF*, 2-acetylaminofluorene

D. Exposed Populations

Exposure to environmental carcinogens varies extensively by subpopulation. (Exposure is considered here as the amount of chemical inhaled or ingested, with no regard to the fraction absorbed. Absorbed dose is the exposure times fraction absorbed.) In previous comparisons of exposure levels between workers exposed occupationally and the total population, it was noted that the exposure levels ranged between 1 and 4 orders of magnitude (HEMMINKI and VAINIO 1984). The comparisons become more complicated when different types of exposure are considered for the subpopulations. As an example, four types of exposure are considered in Table 2. Because of the inherent variability of such exposure figures, they should be taken illustratively rather than factually.

Asbestos is released into tap water from asbestos cement pipelines and may cause a daily exposure of the order of 10^4–10^8 fibres, mainly by the oral route (Table 2). This is high compared with the average inhalation exposure of some 200 fibres/day. People living in busy urban centres are exposed to 10 times higher levels of 2000 fibres/day. These are, in turn, a few orders of magnitude less than moderate (10^4–10^6 fibres/day) or heavy (10^6–10^7 fibres/day) occupational exposure levels.

Table 2. Occupational and environmental exposure levels (μg/day) to selected contaminants

	Source of exposure				
	Ingestion	Inhalation			
		Occupational		Environmental	
		Heavy	Moderate	Polluted	Average
Asbestos (F/m^3)[a]	–[b]	10^5–10^6	10^3–10^5	100	10
Lead	100	1 500	200	20	1
Styrene	1–10	3 000 000	2 000 000	400	5
Tetrachloro-ethylene	10–100	500 000	70 000	8 000	50

[a] F/m^3 = fibres of > 5 μm/m^3, daily inhalation volume is taken as 20 m^3.
[b] Drinking water may contain between 10^4–10^8 asbestos fibres per liter.
Source: HEMMINKI and VAINIO (1984).

Exposure to lead through food and drink approximates to 100 μg/day (Table 2). Inhaled lead levels are lower: about 1 μg/day on average and about 20 μg/day by populations living in busy urban areas. Moderate occupational exposure is one order and heavy occupational exposure two orders of magnitude higher for inhaled lead as compared with that of people living in polluted environments.

The exposure patterns for styrene and tetrachloroethylene are quite similar (Table 2). The quantities ingested approximate to those inhaled by the average population. Polluted environments entail exposures of 100 times the average levels. Heavy occupational exposures exceed the average environmental exposures by 10^5–10^6 times for styrene and 10^4 times for tetrachloroethylene.

The above examples convey the futility of reference to an "average" exposure, because exposures are individual and range over orders of magnitude. However, for most compounds the paucity of measurements invalidates exact approaches to the assessment of true exposures. Thus, I will be later referring to "average" exposure for a simplistic description of the assumed order of magnitude of exposure from environmental sources.

It is an interesting question as to what extent the high-exposure populations actually weight the means. This depends on the size of the high-exposure population and the level of its exposure. If exposures are due to a few stationary sources, e.g. industrial facilities releasing some intermediates into their immediate vicinity, the high-exposure populations may account for most of the total exposure. If, on the other hand, the exposures are due to numerous sources such as automobiles or individual households, the high-exposure populations first of all do not differ dramatically from the average exposure and second, do not account for a large fraction of the total exposure.

E. Risk Assessment

In risk assessment, information available from very specific study citations (e.g. occupational studies) is applied to the general population in order to calculate its risk. Therefore, quantitative risk assessment or dose-exposure assessment generally includes the extrapolation of risks from relatively high-dose levels, at which cancer responses can be measured, to relatively low-dose levels, which are of concern in environmental protection as such risks are too small to be measured directly either through animal studies or epidemiological studies (WHO 1981).

Many models of risk assessment have been developed. The choice of the extrapolation model depends on the current understanding of the mechanisms of carcinogenesis (ANDERSON 1983, 1985; U.S.EPA 1980; WHO 1984a, 1987a). No single mathematical procedure can be regarded as the most appropriate for low-dose extrapolation. However, methods based on a linear, non-threshold assumption have been used at the international (WHO 1981, 1983a; IARC 1982a, b) and national (various health assessment documents produced by U.S.EPA) levels, mainly because they are tangible and scientifically reasonable.

In this chapter the risk associated with lifetime exposure to a certain concentration of a carcinogen in the air has been estimated by linear extrapolation and the carcinogenic potency expressed as the incremental unit risk estimate. The incremental unit risk estimate for an air pollutant "is defined as the additional lifetime cancer risk occurring in a hypothetical population in which all individuals are exposed continuously from birth throughout their lifetimes to a concentration of 1 $\mu g/m^3$ of the agent in the air they breathe" (U.S.EPA 1985). Calculations expressed in unit risk estimates provide the opportunity to compare the carcinogenic potency of different agents and can help to set priorities in pollution control according to the existing exposure situation.

Risk estimation models as discussed above can be used to calculate the numbers of expected cases of cancer. If, for example, there is an occupational subpopulation accounting for 1% of the total, and it is exposed to 10^5 asbestos fibres/m^3 as compared with an average exposure in the general population of 10 fibres/m^3, we would expect roughly 10^3 times more cases of asbestos-related respiratory cancer in the occupational population (10^5 times higher exposure, 1/100 of the population = 10^3).

Such considerations can be extended to examine the likelihood of finding an excess risk of cancer due to environmental exposure. If, in the above example, the occupational population experiences a tenfold higher risk of lung cancer (as compared with the total population exposed to 10 asbestos fibres/m^3), how large a risk can we then expect in the total population as compared with a hypothetical control population (not exposed to asbestos). The answer is that the total population has an infinitesimally increased relative risk of respiratory cancer due to asbestos. Such a small difference could never be detected by epidemiological means even if a non-exposed control population were available. This does not mean that the risk would be irrelevant.

In fact only two or three environmental contaminants have been shown by epidemiological means to cause cancer. Asbestos has been shown to cause mesothelioma through environmental exposure, detected in the wives of asbestos workers. Populations living in the vicinity of smelters emitting arsenic experience according to several studies an excess risk of lung cancer (WHO 1981). The third example, the association of lung cancer to the soot of urban centres is a debated issue because other causes cannot be excluded (IARC 1984a, b). Nevertheless, the paucity of hard epidemiological data on environmental carcinogens illustrates the relative insensitivity of epidemiology as a means of assessing environmental cancer risk.

F. Examples of Important Environmental Carcinogens

In this section some important environmental carcinogens are reviewed, including their presence in the environment, the extent and sources of human exposure, their carcinogenic properties and the magnitude of human risk. Because of space limitations, it is not possible to review all the compounds listed previously. For example, alkenes such as ethene and propene have attracted recent interest as air pollutants but they will not be discussed. The air and drinking-water guidelines of WHO (1984a, 1987a) have been important sources of reference in this section.

I. Arsenic

Background levels of As in air are 1–10 ng/m^3 in rural areas. Concentrations can reach several hundred ng/m^3 in some cities and exceed 1000 ng/m^3 near non-ferrous metal smelters and near some power plants, depending on the As content of the burnt coal. As in air is present mainly in particulate form as inorganic As. Methylated As is a minor component in the air of suburban, urban and industrial areas (WHO 1981).

Particulate As compounds may be inhaled, deposited in the respiratory tract and absorbed into the blood. For an assumed breathing rate of 20 m^3/day and retention and absorption of the order of 30% of the intake, the absorbed doses may be estimated in rural areas (1–10 ng/m^3) to be 0.0006–0.06 μg/day and in urban areas (10–200 ng/m^3) to be 0.06–1 μg/day. Tobacco smoke may contain As especially when the tobacco plants have been treated with lead arsenate insecticide. Although the use of As-containing pesticides is now prohibited in most countries, the natural content of As in tobacco may still result in some exposure. At present, it is estimated that about 6 μg of As may be inhaled per pack of cigarettes smoked, of which about 2 μg would be retained in the lungs (WHO 1987a).

Drinking water may contribute significantly to oral intake in certain regions with high As concentrations in well water or in mine drainage areas, e.g. in Taiwan drinking water may contain 0.5 mg As/l (VELEMA 1987). More commonly drinking water sources contain less than 10 μg As/l, yet drinking water is usually the main form of inorganic As to which humans are exposed (WHO 1984a).

The As levels in foods are usually well below 1 mg/kg wet weight, and it is mainly present in organic forms. The use of organic As compounds as feed additives for poultry and swine may lead to increased levels in meat. Wine made from grapes sprayed with As-containing insecticides or fungicides may contain appreciable amounts (up to 0.5 mg/l) in trivalent inorganic form. Much higher levels of As, but in organic forms (e.g. arsenobetaine) are present in seafood, particularly marine fish (1–10 mg/kg), and values over 100 mg/kg have been measured in certain bottom-feeding fish and crustaceans (WHO 1987a).

The representative intake of As is of the order of 40 μg/day in foods of terrestrial origin and 80 μg/day in seafoods. Both inorganic and organic arsenic compounds are readily absorbed from the gastrointestinal tract.

Lung cancer is the critical outcome of the inhalation of As. An increased incidence of lung cancer has been seen in several occupational groups exposed to inorganic As compounds. Some studies have also shown that populations near emission sources of inorganic As, such as smelters, have a moderately elevated risk of lung cancer (WHO 1987a). At an air concentration of 1 $\mu g/m^3$, the upper bound lifetime risk estimate is 2×10^{-3} (U.S.EPA 1984a). There are at least two studies in which As in drinking water correlates either with skin cancer or with lung and bladder cancer (VELEMA 1987).

II. Cadmium

The concentrations of Cd in air range in rural areas from <1 to 5 ng/m^3, whereas in urban areas 5–15 ng/m^3 and in industrialized areas 15–50 ng/m^3 have been reported. Much higher concentrations, of up to 300 ng/m^3, have been measured near metal-processing industries (WHO 1987a).

Assuming an air concentration of Cd 50 ng/m^3, a daily inhalation of 20 m^3 of air, and that indoor concentrations are similar to outdoor concentrations (even in industrialized areas), the average daily intake of Cd via inhalation would not be more then 1 μg. Less than 50% of the inhaled amount of Cd is expected to be absorbed from the lungs (WHO 1987a).

Tobacco contains Cd, and smoking may contribute significantly to its uptake. Cigarettes may contain from 0.5 to 3 μg Cd/g tobacco, depending on the country of origin. The cigarettes smoked in Europe generally contain Cd 1–2 μg/g (IARC 1988). Smoking a pack of 20 cigarettes a day may result in the inhalation of 1–6 μg/day, since about 10% of the Cd is in the mainstream smoke. If 50% is absorbed via the lungs, the absorbed dose is up to 3 μg/day.

Drinking water normally contains very low concentrations of Cd. Concentrations range normally between 0.1 and 2.0 μg/l; occasionally, levels up to 5 μg/l have been reported, and on rare occasions levels as high as 10 μg/l have been detected. Estimated daily exposure to Cd via water, based on a water consumption of 2 l per day, ranges from substantially less than 1 μg to over 10 μg per day (WHO 1984a).

The daily intake via food has been well documented. In European countries and in North America the average intake is 10–30 μg/day, but there may be large individual variation depending on age and dietary habits. In Japan the average intake is generally 40–50 μg/day but may be much higher in severely polluted areas. About 5% of the ingested Cd is normally absorbed by adults (WHO 1987a).

Carcinogenicity of Cd compounds is well established in experimental animals, and the evidence has been found to be sufficient by IARC. As regards carcinogenic risk to humans, IARC considers the evidence to the limited, and Cd has been classified in group 2A, i.e. probably carcinogenic to humans (IARC

1987a). The U.S.EPA made a quantitative assessment and estimated the incremental unit cancer risk from continuous lifetime exposure to 1 $\mu g/m^3$ of cadmium to be 1.8×10^{-3} (U.S.EPA 1981).

III. Chromium

Cr is a hard, grey metal most commonly found in the trivalent state in nature, but hexavalent compounds are found in small quantities.

An important route of exposure to Cr compounds is through inhalation of Cr-containing particles. The airborne Cr is predominantly in a trivalent state. Cr intake from inhalation, based on a 24-h respiratory volume of 20 m^3 and an absorption rate of about 5%, is 50 ng in most urban areas where an average chromium concentration in air is 0.05 $\mu g/m^3$. However, near emission sources, where concentrations of airborne Cr can exceed 10 $\mu g/m^3$, daily intake may reach 10 µg (WHO 1987a).

Cr has been determined to be a component of American cigarette tobacco with a concentration varying from 0.24 to 6.3 mg/kg.

The Cr concentration in waterways and ground water varies with the type of surrounding industrial sources and the type of underlying soils. An analysis of tap waters in American cities showed a Cr concentration ranging from 0.4 to 8 µg/l. In chlorinated and aerated water Cr is predominantly in the hexavalent form, and it may be an important source of this form, even though in terms of total Cr, drinking water is a minor source (WHO 1984a).

Cr levels in soils vary with soil origin and the degree of contamination from anthropogenic Cr sources. Tests on soils have shown Cr concentrations ranging from 1 to 100 mg/kg, with the average concentration ranging from 14 to about 70 mg/kg (WHO 1987a).

The Cr intake from a typical North American diet was estimated to be about 60–90 µg/day and may generally be in the range of 50–200 µg/day, which agrees with European estimates (HEMMINKI et al. 1983).

Hexavalent Cr compounds are carcinogenic, although the various compounds have a wide range of potencies. As the bronchial tree is the major target organ for the carcinogenic effects of hexavalent Cr compounds, and cancer primarily occurs after inhalation exposure, uptake in the respiratory organs is of the greatest importance to the cancer risk in humans. At an air concentration of 1 $\mu g/m^3$ of hexavalent Cr, the upper bound lifetime risk is estimated to be 6×10^{-2} (U.S.EPA 1984b).

IV. Nickel

The major routes of Ni intake for humans are inhalation, ingestion and percutaneous absorption. Assuming a daily ventilation rate of 20 m^3, the amount of airborne nickel entering the respiratory tract is in the range of 0.2 to 0.4 µg/day when concentrations in ambient air are 10–20 ng/m^3. Cigarette smoking, however, increases these figures: smoking two packs of cigarettes will result in the inhalation of 3–15 µg/day. Assuming a concentration of 5 µg/l, a daily consumption of 2 l drinking water would result in a daily Ni intake of 10 µg. A mean intake of Ni from food ranges between 200 and 300 μg/day. Vegetable products generally have higher Ni contents than animal products; thus a vegetarian will ingest more Ni than the above figures indicate (WHO 1987a).

Several epidemiological studies of workers exposed to Ni or Ni compounds have clearly demonstrated an excess incidence of cancer of the nasal cavity, the lung and possibly the larynx. However, it is still not possible to state with certainty which specific Ni compounds are carcinogenic to humans, although the Ni ion itself and probably also metallic Ni are the most likely candidates. At an air concentration of 1 $\mu g/m^3$ of nickel dust, the upper bound lifetime risk estimate is 4.4×10^{-4} (U.S.EPA 1985).

V. Asbestos

The commercial term asbestos refers to a group of fibrous serpentine and amphibole minerals that have high tensile strength, are poor heat conductors and are relatively resistant to chemical attack. The principal varieties of asbestos are chrysotile, crocidolite and amosite. Anthophyllite, tremolite and actinolite asbestos are rare, and the commercial use of anthophyllite asbestos has been discontinued.

Once emitted into the atmosphere, the asbestos fibres can travel considerable distances. Because no chemical breakdown of the fibres takes place, washout by rain or snow is the only cleansing mechanism.

Various subgroups of the population are exposed to different fibre concentrations for varying lengths of time. The fibre burden is the number of accumulated critical fibres ($F > 5$ μm) = fibre concentration (F/m^3) × years of exposure × air volume inhaled each year at place of exposure (m^3/year). The risks are roughly comparable, if the accumulated fibre burdens are the same (WHO 1987a).

Typical rural air samples from industrial countries contain less than 10 F ($> 5\ \mu m/m^3$). Urban air contains more, 100–1000 F/m^3. In communities where asbestos mining is carried out an even higher concentration of asbestos may be detected, e.g. 10000 F/m^3. Indoor air exposures are generally not known in sufficient detail, but they may range from 400 to 500 F/m^3. If such estimates can be generalised, indoor sources of asbestos would be at least as important to the population as outdoor exposures. However, the indoor concentrations depend critically on to what extent asbestos has been used in the indoor construction materials.

Drinking water (and food) may contain asbestos fibres from natural sources (e.g. rock) and man-made sources (e.g. asbestos-cement pipes); in the latter case, the total fibre content in drinking water can vary from 10^4 F/l to more then 10^8 F/l. In Finland drinking water samples contained $0.5–10 \times 10^6$ F/l. It has been estimated that 5% of the population in North America consumes drinking water containing in excess of 10^7 F/l (VELEMA 1987). Wine sieved through asbestos filters may contain up to 10^8 F/l.

Asbestos is a proven human carcinogen. Assuming a lifelong exposure to inhaled asbestos at 100 F/m^3 (asbestos fibres longer than 5 μm, optically measured), the lifetime risk for mesothelioma would be $1/10^5$ and probably somewhat higher for lung cancer among smokers (WHO 1987a).

Evidence concerning the health risk from the ingestive uptake of asbestos fibres is not as strong as that for inhalation. Although several studies show increased gastrointestinal cancer rates, a quantitative assessment of the health risks is not feasible. VELEMA (1987) reviewed nine studies comparing cancer rates in populations exposed to asbestos in drinking water with non-exposed populations. Most studies found an increased risk of stomach cancer, the highest relative risk being 1.70.

VI. Acrylonitrile

Acrylonitrile uptake in non-occupationally exposed persons living in the vicinity of plants may be close to 20 μg/day, and 20–40 μg/day in people smoking acrylonitrile-fumigated cigarettes. An average consumption of butter and soft margarine in acrylonitrile-copolymer containers amounts to an ingestion of 1–10 μg/day. Acrylonitrile is carcinogenic in animals, and as there is limited evidence of carcinogenicity in humans, IARC has classified acrylonitrile as a probable human carcinogen (group 2A). At an air concentration of 1 μg/m^3, the lifetime risk is estimated to be 1.7×10^{-5} (U.S.EPA 1983).

VII. Benzene

Air is one of the primary sources of benzene. About 50% of inhaled benzene in air is absorbed. Benzene intake, based on a 24-h respiratory volume of 20 m^3 at rest will be 10 μg per day for each 1 μg/m^3 in the air. The daily adult intake at a typical ambient benzene level of 16 μg/m^3 will therefore be about 160 μg.

Cigarette smoke contains relatively high benzene concentrations (150–204 mg/m^3) and represents an important source of exposure for smokers. Estimates for the uptake of benzene from smoking range from 10–30 μg per cigarette, which would constitute an additional intake of up to 600 μg per day for smokers consuming 20 cigarettes per day.

Benzene has been identified as a contaminant in drinking water at levels of 0.1–1.0 μg/l with the highest reported concentration at 300 μg/l (WHO 1984a).

Benzene has been detected in several foods, i.e. in eggs at high levels although the origin is unclear; in irradiated beef, 19 μg/kg; and in canned beef, 2 μg/kg. Benzene has also been detected but not quantitated in fish, cooked chicken, roasted nuts, various fruits, vegetables and dairy produce. It appears that normal cooking methods may cause an increase in the benzene content of food. Dietary intake of benzene may be as high as 250 μg/day, but the data available do not allow for precise estimates (WHO 1984a, 1987a).

In summary, non-smokers living in rural areas have a total intake of about 0.3 mg of benzene per day, whereas heavy smokers living in urban areas may receive as much as 5 times this amount. Benzene causes leukaemia in humans. At an air concentration of 1 μg/m^3 of benzene, the estimated lifetime leukaemia risk is 4×10^{-6} (WHO 1987a).

VIII. 1,2-Dichloroethane

1,2-Dichloroethane (DCE) is one of the largest-volume, synthetic organic chemicals manufactured in the USA (GOLD 1980). Its main uses are in the chemical industry for the production of vinyl chloride, 1,1,1-trichloroethane, tetrachloroethylene and trichloroethylene. It is also used as a lead scavenger and fumigant.

The intake from urban air in Western Europe and the USA was estimated to be between 8 and 80 μg per day, with an average of about 20 μg per day. The intake from rural air would be about one-tenth of this (WHO 1987b). Average levels found in drinking water are usually below the detection limit of 1 μg/l (WHO 1987b).

Reports on residues in food are scarce and show levels below 1 μg/kg. Significant residues of DCE in food (spice, grains) are possible after use of the compound as an extractant or fumigant. In fumigated grain, levels up to 300 mg/kg have been detected (WHO 1987b).

DCE is carcinogenic in rats and mice. By linear extrapolation of the results of an oral study in rats, one additional case of cancer per million persons exposed over a lifetime would result from a dose of 6.9 μg/day taken orally. As absorption of DCE is rapid and complete whether exposure is by ingestion or inhalation, the same amount might come from inhalation of 20 m^3 air/day containing about 0.4 μg/m^3 (WHO 1987a). U.S.EPA (1979) has estimated that the daily intake through drinking water of 14 μg of DCE would cause a lifetime cancer risk of 1/100000. This kind of estimate was used in the WHO *Drinking-Water Guidelines* (1984a), but it is about 5 times higher than the estimate used for *Air Quality Guidelines* (WHO 1987a). As the estimates are based on animal data, they are well within the acceptable variation.

IX. Formaldehyde

The possible routes of exposure to formaldehyde (HCHO) are ingestion, inhalation and dermal absorption. If it is assumed that normal work exposures are similar to home exposures, the daily exposure resulting from breathing is about 1 mg/day, with rare exposures reaching a maximum of 5 mg/day.

Concentrations of 60–130 mg/m^3 were measured in mainstream smoke. For someone who smokes 20 cigarettes per day this would lead to an exposure of 1 mg/day. Except for accidental ingestion of formaldehyde-contaminated water, concentrations in drinking water can be expected to be less than 0.1 mg/l water, thus being a minor source (below 0.2 mg/day) (WHO 1987a).

There is some natural formaldehyde in raw food (such as fruits, vegetables and meat) and some accidental contamination from fumigation (e.g. in grain). However, formaldehyde is usually present in a bound and unavailable form. Yet it should be noted that the primary binding products of formaldehyde, i.e. methylols and Schiff bases, are readily reversible and may thus liberate free formaldehyde.

There is sufficient evidence for the carcinogenicity of formaldehyde in experimental animals (IARC 1982a, b). Formaldehyde was considered a probable human carcinogen by IARC (group 2A). The calculation of risk estimates is problematic because the available data on animals are non-linear and do not allow a reasonable use of the linearized risk estimation models.

X. Methylene Chloride (Dichloromethane)

The global average concentration of dichloromethane can be estimated at about 0.10 μg/m^3 (WHO 1987a). In surveys involving urban areas in the United States of America, average levels varied from about 1 to 13 μg/m^3. The same concentration range has been measured at three locations with differing levels of air pollution in the Netherlands (WHO 1984b). In instances involving exposure of people living near waste disposal sites, the levels have been much higher. In non-occupational indoor environments during the use of dichloromethane-containing paint removers, the time-weighted averages in a room without ventilation varied between 460 and 2980 mg/m^3. The concentrations of dichloromethane in drinking water are usually at or below 1 μg/l (WHO 1984b). It was considered as a possible human carcinogen by an IARC working group (IARC 1987).

XI. Polycyclic Aromatic Hydrocarbons and Nitrated Forms

Polynuclear (or polycyclic) aromatic hydrocarbons (PAH) are a large group of organic compounds with two or more benzene rings. They have a relatively low solubility in water but are highly lipophilic. Since PAH have a low vapour pressure, almost all the PAH of 4 rings and larger present in the air are adsorbed onto particles. PAH dissolved in water and adsorbed on particulate matter in air can undergo photodecomposition when exposed to UV light from solar radiation. Some microorganisms in soil can degrade PAH.

PAH are mainly formed as a result of pyrolytic processes, especially incomplete combustion of organic materials, as well as in natural processes such as carbonisation. There are several hundred PAH. In addition, a number of heterocyclic aromatic compounds (e.g. carbazole, acridine) as well as PAH with one or more NO_2 groups (nitro-PAH) can be generated by incomplete combustion. Nitro-PAH are readily formed from PAH through a nitration reaction, and they are usually present in all environmental samples in which PAH are found but in lesser amounts (ROSENKRANZ and MERMELSTEIN 1985). Thus, nitro-PAH may contribute to any biological effects of environmentally derived PAH.

About 500 PAH have been detected in the air (but most measurements have been made on benzo[*a*]pyrene (BaP) or on a few other "indicator" compounds. Data prior to the mid-1970s may only be comparable to a certain extent with later figures because of different sampling and analytical procedures. Moreover, methods of energy consumption and transportation have changed extensively, affecting the magnitude and pattern of release into the environment.

The natural background level of BaP (not including forest fires and volcanic eruptions) might be nearly zero. In the USA in the 1970s the annual average value of BaP in urban areas without coke ovens was less than 1 ng/m^3 and in other cities between 1–5 ng/m^3. In several European cities in the 1960s, the annual average BaP values were above 100 ng/m^3. PAH are one of the few air pollutants, whose concentrations have decreased in most western countries during the past decades.

Exposure to a relatively high concentration of BaP in ambient air, e.g. 50 ng/m^3 and a deposition rate of 50% from 20 m^3 air inhaled per day, can be estimated to result in an intake of 500 ng/day. However, the BaP intake in clean rural areas may be no more than 1% of this amount, and even in the big cities of Europe and North America, where adequate smoke control has been achieved, it may be only a few percent of this figure.

The average total BaP content in the mainstream smoke of one cigarette was 35 ng before 1960 and 18 ng in 1978/1979; modern "low-tar" cigarettes deliver 10 ng BaP (WHO 1988). The concentration of BaP in restaurants and public places ranges between 0.25–760 ng/m^3; in rooms extremely polluted with cigarette smoke, the levels are in excess of 20 ng/m^3 (IARC 1986).

Examination of a number of drinking water supplies for six PAH (fluoranthene, benzo[*b*]fluoranthene, benzo[*k*]fluoranthene, BaP, benzo[*ghi*]-perylene, indenol[1,2,3-*cd*]pyrene) indicated that the collective concentrations generally did not exceed 0.1 μg/l. The concentrations of these six PAH were between 0.001 and 0.01 μg in 90% of the samples. PAH are mainly particle bound even in water, and a proper removal of particles from drinking water is an efficient way of removing the PAH (WHO 1984a).

PAH are found in substantial quantities in some foods, depending on the method of cooking, preservation and storage, and are detected in a wide range of meat, fish, vegetables and fruits. American sources indicate an intake of total PAH from food on the order of 1.6–16 μg/day. The content of benzo[*a*]pyrene in various processed foods (refined, broiled, smoked) was reported to be as high as 50 μg/kg. However, nearly all the cited data were published between 1965 and 1975 when methods of PAH measurement were less sophisticated.

As a rule of thumb it is assumed that 99% of exposure to PAH is contributed by food, 0.9% by inhalation and 0.1%–0.3% by drinking water (WHO 1984a). However, there are serious concerns about the accuracy to these figures due to the analytical techniques that were applied earlier and changing exposure patterns. Also, individual circumstances may lead to entirely different exposure estimates.

For several PAH, heterocyclic aromatic compounds and nitro-PAH there is sufficient evidence of carcinogenicity in animals (Rosenkranz and Mermelstein 1985). Because human exposure to PAH is always in combination with many other agents, it has not been possible to single out PAH as human carcinogens (IARC 1984). Yet the increased cancer risks of chimney sweeps, coke oven workers, aluminum workers, foundry workers and people exposed to diesel exhaust are very likely to be caused, in part, by PAH (IARC 1984).

A number of different risk estimates for PAH have been made, based primarily on using BaP as the index compound. The US Environmental Protection Agency has presented an upper bound lifetime cancer risk estimate of 62 per 100000 exposed people per 1 μg benzene-soluble coke oven emission per m^3 ambient air. Based on 0.7% content of BaP in these emissions, a lifetime risk estimate for 1 ng BaP/m^3 would predict 9 instances of cancer in 100000 people (WHO 1987a).

XII. Polyhalogenated Aromatic Compounds

1. Polychlorinated Biphenyls

PCBs are a group of 209 isomers differing in the number and site of chlorination. The biological properties of the isomers differ extensively. The most toxic PCB congeners 3,3′,4,4′-tetra-, 3,3′,4,4′,5-penta-, and 3,3′,4,4′,5,5′-hexachlorobiphenyls are approximate isostereomers of the supertoxic 2,3,7,8-tetrachlorodibenzo-*p*-dioxin (TCDD), which suggests a common mechanism of toxicity (SAFE et al. 1985). Upon heating, oxygen may attack PCBs, forming polychlorinated dibenzofurans (PCDFs) (HUTZINGER et al. 1985a).

PCBs have been used extensively since the 1940s. However, once their stability in nature was understood, attempts were made to restrict their use. Today PCBs are found mainly as transformer and capacitator fluids. In fires of electrical equipment PCBs and PCDFs are spread and sometimes cause extensive evacuation and clean-up operations. After restriction of the use of PCBs, their concentrations have levelled off in the environment, for instance in the Baltic Sea.

The persistence of PCBs in the food chain has caused exposures measurable in humans all over the world. Studies carried out in Finland and in many other countries show that human tissues contain about 2 mg PCB/kg fat, and human milk about 0.5–2 mg PCB/kg fat (STORACH and VAZ 1985). The main dietary source of PCBs is fish. Freshwater fish have been found to contain 56 µg PCB/kg wet weight, and the concentration is somewhat higher in saltwater fish. The average daily human exposure can be estimated at 5–10 µg. Commercial derivatives of PCBs have been found to be carcinogenic in experimental animals.

2. Chlorinated Phenols

Pentachlorophenol, tetrachlorophenol and trichlorophenol are environmental contaminants enriched in the food chain. They originate from the bleaching of cellulose, the disinfection of water and their use as pesticide and anti-fungal agents in wood and board. Chlorinated phenols are relatively stable, and concentrations of 2–11 µg/kg have been measured in freshwater fish in Finland. The daily exposure can be estimated at 5 µg. 2,4,6-Trichlorophenol has been found to be carcinogenic in experimental animals.

Commercial preparations of chlorinated phenols may be contaminated by TCDD, a potent carcinogen, or by dimeric and trimeric chlorophenols, which upon heating may be converted to polychlorinated dibenzo-*p*-dioxins (PCDDs) such as TCDD. PCDDs are produced in small quantities by combustion processes including residential wood-burning furnaces (CZUCZWA and HITES 1985). However, the main contributions come from contamination in commercial preparations of 2,4,5-trichlorophenoxyacetic acid and polychlorinated phenols; furthermore, bleaching of cellulose causes the formation of PCDDs (HUTZINGER et al. 1985b). CZUCZWA and HITES (1985) have shown that the amounts of PCDD and PCDF in the sediment of the Great Lakes in North America have increased concomitantly with the production of chloro-aromatic compounds in the chemical industry, while there is no correlation with the utilisa-

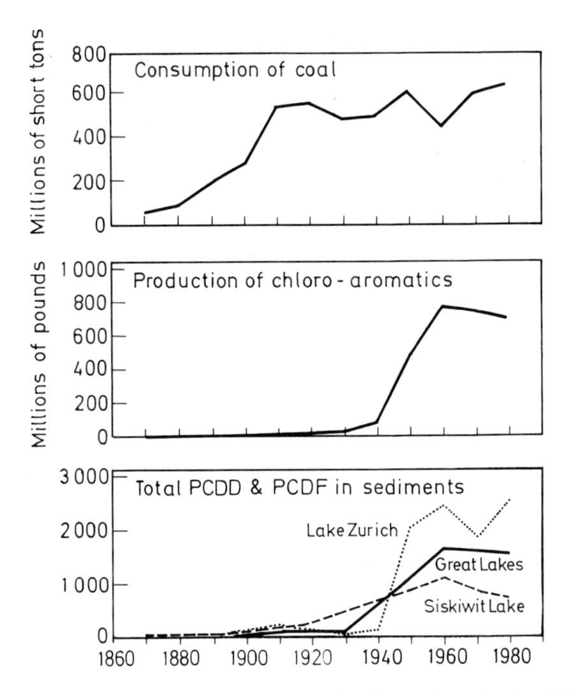

Fig. 4. Time trends in consumption of coal, production of synthetic chlorinated organic chemicals, and total polychlorinated dibenzo-p-dioxins (*PCDD*) and polychlorinated dibenzofurans (*PCDF*) in the sediment of three lakes. (From CZUCZWA and HITES 1985)

tion of coal (Fig. 4). Data from Lake Zurich in Europe agree with those from North America. Such data suggest that environmental contamination by PCDDs and PCDFs is related, at least indirectly, to the chemical industry (CZUCZWA and HITES 1985). The level of exposure to PCDDs and PCDFs can be estimated at 0.1–1 ng/day, the main source being fish.

3. Hexachlorobenzene

Hexachlorobenzene is an ubiquitous environmental contaminant earlier thought to pass into the environment via its use as a pesticide. Later it was found that the quantities in the environment exceed its use as a pesticide to such a large extent that there must be other sources, probably the combustion of chlorinated compounds. The average daily exposure in Finland, based on the consumption of fish, can be estimated at 0.5 μg (HEMMINKI et al. 1983). Hexachlorobenzene has been found to cause tumours in experimental animals.

4. Dichlorodiphenyltrichloroethane, Toxaphene and Chlordane

In addition to hexachlorobenzene, environmental contamination by other chlorinated pesticides and their derivatives may lead to human exposure, primarily through fish. In Finland the estimated daily exposure to DDT deriva-

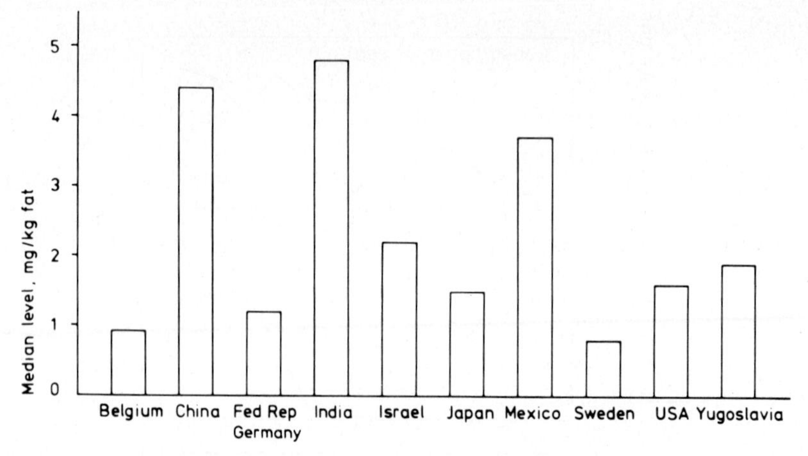

Fig. 5. Median levels of p,p′-DDE, a metabolite of dichlorodiphenyltrichloroethane (DDT) in human milk from different countries. (From the UNEP/WHO Pilot Project as shown by Slorach and Vaz 1986)

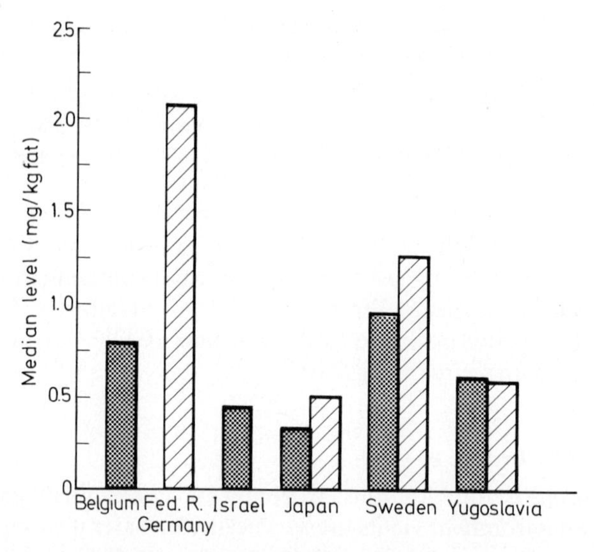

Fig. 6. Median levels of polychlorinated biphenyls (PCBs) in human milk fat. The samples from China, India and Mexico had no detectable levels of PCBs. The different *bar thicknesses* indicate different methods of determination. (From the UNEP/WHO Pilot Project as shown by Slorach and Vaz 1986)

tives is about 2 μg and about 4 μg each of toxaphene and chlordane. All of these compounds have been found to cause tumours in experimental animals.

Polyhalogenated aromatic hydrocarbons are relatively stable in the environment, and they bioaccumulate. They are therefore causing environmental hazards never detected before. The most dramatic example of a new type of environmental health problem is the contamination of human milk all over the world by DDT, PCBs and many other polyhalogenated aromatic hydrocarbons to such an extent that it would not meet the standards set by WHO for commercial milk preparations. Figures 5 and 6 show the concentrations of DDT derivatives and PCBs in human milk from various countries (SLORACH and VAZ 1985). It is doubtful whether such compounds would ever have been used, if subsequent human exposure could have been predicted. The moral of the story is that the bioaccumulation of new chemicals must be studied before they enter the market. Bioaccumulation must not be allowed because it is a potential time bomb. Even though we do not have scientific evidence on the magnitude of human health hazards due to environmental exposure to polyhalogenated aromatic compounds (or due to destruction of stratospheric ozone levels by halogenated hydrocarbons, such as chlorofluorocarbons), circumstantial evidence is certainly strong enough for it to be used to prevent the marketing of bioaccumulating compounds.

XIII. Radon

Radon-222 is a member of the radioactive decay chain of uranium-238, and radon-220 (often referred to a thoron) is a member of the decay chain of thorium-232. The contribution made by thoron to human exposure in indoor environments is usually small; the decay chain of uranium-238 is shown in Fig. 7.

The half-life of radon-222 is 3.8 days, and it decays into short-lived isotopes of polonium, lead, bismuth and thallium, which are together referred to as radon daughters. In a closed space radon and its daughters are in an equilibrium: if there is 1 unit of radon, there are 0.3–0.5 units of its daughters (UNSCEAR 1982).

Uranium and radium occur widely in the earth's crust. The average level of radon gas concentration in the atmosphere at ground level is given as 3 Bq/m^3 with a range from 0.1 (over oceans) to 10 Bq/m^3 (1 Bq = 27 pCi). There are wide variations in radon concentration in different parts of the world. High radon concentrations are usually noted in areas where terrestrial gamma radiation is high. In Europe and North America the mean radon daughter concentrations range between 10 and 30 Bq/m^3 in residential dwellings; in Scandinavia the levels are higher, some 50 Bq/m^3 (CASTREN et al. 1985). However, in each country a small fraction of dwellings has concentrations that exceed 10 times the national average for that country (UNSCEAR 1982; WHO 1987a).

The potential for radon entry into dwellings from the ground depends mainly on the concentration of radium in the soil and on the permeability of the soil to radon. Examples of ground with high radon potential are alum shales and some granites, due to high radon-226 concentrations, and the presence of eskers (deposits of gravel, sand and rounded stone from subglacial streams during the ice ages) characterised by high permeability. The ground could also be contaminated with tailings with enhanced concentrations. For those who live close to the ground, e.g. in detached houses or on the ground floor of multi-family buildings without cellars, the most important radon source is the ground. The inflow of radon from the soil takes place mainly by pressure-driven flow, with diffusion playing a minor role. The magnitude of the inflow varies with several

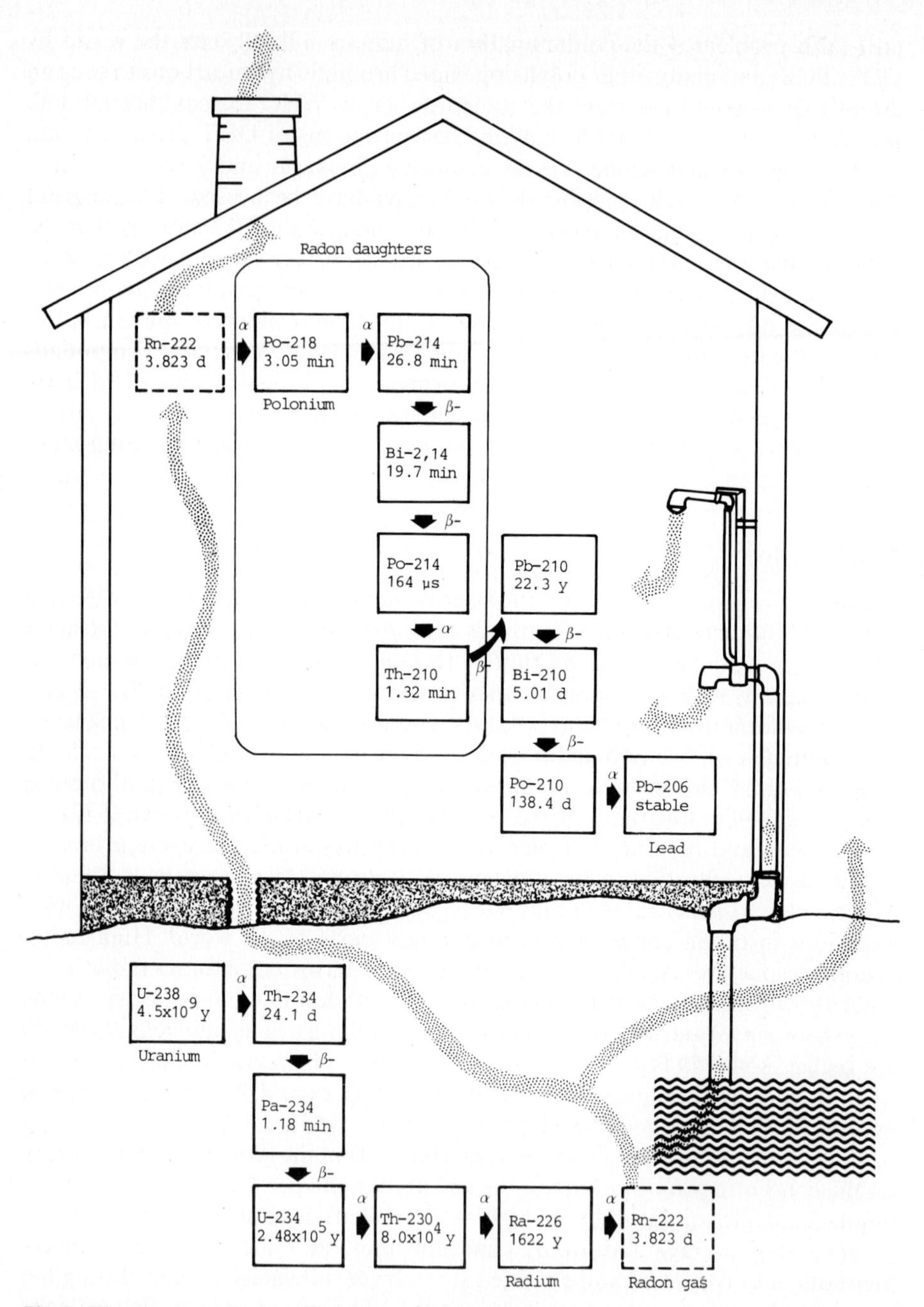

Fig. 7. The decay pathway of uranium showing the formation of radon daughters and sources of human exposure. The *times under the isotope name* indicate half-lives of decay

parameters, the most important being the air pressure difference between soil air and indoor air, the tightness of the surfaces in contact with the soil on the site and the radon exhalation rate of the underlying soil (MOSSMAN et al. 1986; WHO 1987a).

Another potential source of radon is from building materials. Radon exhalation from building materials depends not only on the radium concentration but also on factors such as the fraction of radon produced which is released from the material, the porosity of the material and the surface preparation and finish of the walls. Building materials containing the by-product gypsum, and concrete containing alum shale may have elevated radium concentrations. The concentrations of radon in brick and concrete may also be high if the raw materials have been taken from locations with high levels of natural radioactivity (UNSCEAR 1982; WHO 1987a).

Radon also diffuses into underground water reservoirs and causes exposure when such water is used. In wells drilled in rock the radium and radon concentration of the water may be high. When such water is used in the household, radon will be released into the indoor air and cause an increase in the average radon concentration. Usually radon daughters released into the air cause higher exposures than radon daughters remaining in the drinking water. However, the parent radium-226 remains in water and is present at 0.1 Bq/l levels in areas where water is used from wells drilled in rock of high radium content, e.g. in Maine and Iowa, USA, and in Scandinavia and Finland (VELEMA 1987). Natural gas may also contain some radon which is released into room air when burned.

Radon gas or its daughters, usually bound to particles, enter the body through inhalation, or to a lesser extent, through ingestion of radon-bearing water. Bronchial deposition of radon daughters is by far the most important source of cancer risk: a very small amount of tissue receives a concentrated exposure due to the deposition pattern of the particles to which the radon daughters are bound. Soluble radon daughters may be carried away by the pulmonary circulation and be distributed to other organs. In animal experiments exposure to radon has induced lung cancer (WHO 1987a).

Also in humans, the increased risk of lung cancer among miners, especially uranium miners, has been well documented in a number of large epidemiological studies (UNSCEAR 1982; MOSSMAN et al. 1986). These results are consistent with the theory that there is a linear relationship between cumulative radon daughter exposure and excess lung cancer frequency. There are a few studies that have shown a relationship between lung and some other forms of cancer and radioactivity (radium and radon) in drinking water in the Iowa and Maine municipalities (VELEMA 1987).

As radon is a well-documented carcinogen in humans, it is likely that risk estimations for the total population are reasonably accurate. In the UNSCEAR report (1982) a mean indoor radon daughter concentration of 15 Bq/m^3 has been estimated, averaged over the whole population in the temperate region. Assuming an occupancy factor of 0.8 for indoor environments (amounting to a daily dose of some 240 Bq), we could reasonably expect an annual incidence of 10 to 40 cancer cases per million persons attributable to radon daughter exposure. It can also be estimated that under these conditions about 5%–15% of the observed lung cancer frequency or of the lifetime risk may be attributable to indoor radon daughters (WHO 1987a). This relative risk is nearly equal for males and females, and for smokers and non-smokers. There is no doubt that radon is the most important environmental carcinogen in large parts of the world among the compounds discussed in this chapter. It should be pointed out that the inhaled amount of radioactivity as radon daughters (some 240 Bq) is much higher than that ingested in drinking water even in high activity areas (some 0.2 Bq).

XIV. Styrene

Styrene is present in unpolluted rural air in low concentrations. In urban atmospheres the concentration is around 0.3 μg/m^3, leading to a daily intake of about 6 μg/person. In polluted urban air and within 1 km of styrene polymerization units, the concentration can be 20–30 μg/m^3. Persons living in such areas inhale 400–600 μg of styrene daily. Indoor sources of styrene may also contribute to the level of exposure and may be high if styrene is released from the materials used in the interior (WHO 1987a).

Styrene has been identified in cigarette smoke condensates. The reported levels range from 20 to 48 μg per cigarette. Styrene has also been detected in drinking water in the USA at concentrations of less than 1 μg/l.

Polystyrene and its co-polymers have been widely used as packaging materials for foodstuffs. The ability of the styrene monomer to migrate from polystyrene food packaging into the food has been reported in a number of publications and probably accounts for the greatest contamination of foods by styrene monomer. In a study that was designed to detect styrene monomer at 0.05 mg/kg, no migration of the monomer was detected in milk samples stored in polystyrene containers for up to 8 days (WHO 1983b).

Styrene has been found in yogurt and other milk products packaged in polystyrene containers, at concentrations of 2.5–80 μg/kg. The content of styrene in the products increases during storage.

Studies on experimental animals have provided limited evidence that styrene may be carcinogenic. In one study oral administration of styrene to mice induced a significant increase in pulmonary tumours at a dose of 1350 mg/kg and a doubtful increase in another strain of mice at a dose of 300 mg/kg. Styrene-7,8-oxide, the primary metabolite of styrene, was carcinogenic in two studies on rats following its oral administration. An IARC working group classified styrene as a possible, and styrene oxide as a probable, human carcinogen (WHO 1983b; IARC 1987).

XV. Tetrachloroethylene

Air pollution usually represents the major source of exposure. Global background levels are in the range of 0.2 μg/m^3. Levels in urban air may rise to 70 μg/m^3 and in the vicinity of waste disposal sites to 400 μg/m^3. Mean urban air levels have been found to be between 1 and 6 μg/m^3 in Western Europe and the USA (WHO 1984c). Indoor levels up to 250 μg/m^3 have been measured. Indoor pollution may be equal to, or more significant than, exposure to ambient air.

Intake via food can be an important source of exposure and has previously been calculated to be as much as 160 μg per day in the FRG and Switzerland (WHO 1984c). However, current food processing operations are believed to have resulted in a marked decrease in tetrachloroethylene in food (WHO 1987a). Drinking water is a minor source as the levels measured are at or below 1 μg/l (WHO 1984c).

Tetrachloroethylene was considered a possible human carcinogen by an IARC working group.

XVI. Trihalomethanes

Trihalomethanes are common by-products of the chlorination of drinking water, and they include chloroform, bromodichloromethane, dibromochloromethane and tribromomethane. They are present in normal surface waters as pollutants. In Finnish lakes the combined concentrations of trihalomethanes range from 0–20 μg/l. However, the main source of trihalomethanes is chlorination of drinking water. The concentrations correlate with high levels of the precursor substances, e.g. chlorine, naturally occurring bromine and humic acid derivatives. In Finland most of the population has been exposed to trihalomethanes at levels up to 100 μg/l in drinking water, causing a mean daily exposure of 50 μg/d (Hemminki et al. 1983); of this, 25 μg was thought to be chloroform. Recently, chlorination practices have been changed, and trihalomethane levels have been reduced.

Chloroform has been shown to be carcinogenic in experimental animals (IARC 1979a). The other trihalomethanes have been subjected to proper carcinogenicity testing only recently, and complete data are not yet available. Their activity in short-term tests may serve to predict their carcinogenicity.

There have been many epidemiological studies that have analysed the effects of the contaminants in drinking water on cancer incidence and mortality (Velema 1987). Although many of the studies suggest that the presence of trihalomethanes and/or other chlorinated contaminants may be associated with an increased risk of cancer of the colon, rectum and urinary bladder, the evidence is not strong for a number of reasons. The data cannot, however, be entirely overlooked, and every effort should be made to secure drinking water containing minimal amounts of suspected carcinogens.

XVII. Vinyl Chloride

Currently, general population exposure to vinyl chloride results overwhelmingly from industrial production sources, inhalation being the primary route of entry. Assuming a daily inhalation of 20 m^3 of air, the vast majority of the population would inhale from 2 to 10 μg daily.

Individuals living within 5 km of well-controlled production sources could be exposed to 10 to 100 times as much (WHO 1987a).

Vinylchloride has been found in the smoke of cigarettes at a level of 1.3–16 ng/cigarette (IARC 1986). There is very little information on current concentrations of vinyl chloride in water systems. Because of its volatility it is not likely to remain in significant concentrations in drinking water.

It has been identified as a contaminant of foods and liquids packaged in polyvinyl chloride (PVC) materials. However, with the implementation of stringent manufacturing specifications for PVC, such contamination has decreased substantially, and it is estimated that the maximum intake per person in foods and liquids is now less than 0.1 μg/day (WHO 1988).

Vinyl chloride is a well-established human carcinogen (IARC 1979b). The critical concern for environmental exposures is the risk of malignancy. Estimates based on human studies indicate a lifetime risk for exposure to 1 $\mu g/m^3$ in the range of $0.5–1 \times 10^{-6}$. Estimates based on animal studies are in relatively good agreement with this (WHO 1987a).

G. Conclusions

Exposure data on the individual compounds discussed are presented in Table 3. The exposure levels of "average" non-smokers differ extensively by compound – from 50 pg of TCDD to 1 mg of formaldehyde. For metals, asbestos, PAH, DDT, TCDD and trihalomethanes the oral route of exposure is quantitatively the largest; for benzene the oral route and inhalation are equally important; for acrylonitrile, 1,2-dichloroethane, formaldehyde, methylene chloride, radon, styrene, tetrachloroethylene and vinyl chloride inhalation usually accounts for the largest intake.

Table 3 also includes calculations on the number of cancer cases that the inhaled doses may cause, based on the risk estimates discussed in the text. It should be emphasised that such estimates, if based on animal data, are generally thought to carry an error margin of 100 or greater. Thus, the figures presented must not be taken literally. One figure is clearly higher than the others: radon results in 30 cases of lung cancer/10^6 persons per year. The risk estimate for radon is based on human data and can be considered well founded. The calculations for Cr are given as a range of 0.5–5 cases/10^6 persons per year as the estimate assumes airborne Cr to be 1%–10% in the hexavalent form (it is assumed that only hexavalent Cr is carcinogenic). Inhalation of asbestos is assumed to cause 1 lung cancer and 0.7 mesotheliomas/10^6 persons per year. As, Cd, asbestos, benzene and PAH are thought to cause about 1 case of cancer/10^6 persons per year when inhaled as environmental contaminants. For formaldehyde, methylene chloride, DDT, TCDD, styrene, tetrachloroethylene and trihalomethanes the exposure routes or the experimental data on carcinogenic potency are difficult to interpret, and no estimates are given for these compounds. The levels of inhaled formaldehyde are such that it may rank as an important environmental carcinogen.

Inhalation exposures have been considered above. However, for many of the compounds, "environmental" exposure is mainly via the oral route, and for some substances there is human evidence, weak as it may be, of carcinogenic effects. There are at least suggestive data on ingested As, asbestos and trihalomethanes

Table 3. Exposure to environmental carcinogens and estimates of the resulting number of cancers caused through inhalation

	Total exposure [a] (μg/day)	Inhaled dose [b] (μg/day)	Type of cancer	Number of patients [c] (patients/10^6/year)
Arsenic	60 [e]	0.5	Lung	0.7
Cadmium	20 [e]	0.2	Lung	0.6
Chromium	100 [e]	0.05	Lung	0.5–5 [d]
Nickel	200 [e]	0.1	Lung	0.06
Asbestos	10^5F [e]	5×10^3F	Mesothelioma	0.7
			Lung	1.0
Acrylonitrile	2	1	Lung	0.02
Benzene	300	160	Leukaemia	0.8
1,2-Dichloroethane	20	20	All cancers	0.04–0.2
Formaldehyde	1000	1000	Lung	?
Methylene chloride	100	100	?	?
PAH	10 [e]	1	Lung	0.9
DDT	2 [e]	?	Liver	?
TCDD	0.00005 [e]	?	Liver	?
Radon	15 Bq/m^3	7.5 Bq/m^3	Lung	30
Styrene	10	5	?	?
Tetrachloroethylene	200	100	?	?
Trihalomethanes	10 [e]	?	?	?
Vinyl chloride	5	2.5	Angiosarcoma	0.3

[a] Total "average" exposure is estimated for non-smokers based on the data presented in the test.
[b] Dose calculated for non-smokers assuming 5% absorption for chromium, 100% absorption for dichloroethane, formaldehyde and methylene chloride, and 50% absorption for the remaining compounds.
[c] Number of cases is calculated from the risk estimation data presented in the test and from the inhaled dose.
[d] The risk estimates assumed that 1% or 10%, respectively, of chromium is hexavalent.
[e] Most of the exposure is by the oral route.
PAH, polycyclic aromatic hydrocarbons; DDT, dichlorodiphenyltrichloroethane; TCDD, 2,3,7,8-tetrachlorobenzo-*para*-dioxin.

implicating a carcinogenic effect. Furthermore, there is no a priori reason to exclude carcinogenic effects of absorbed organic compounds, such as benzene and PAH, when they are ingested.

Assuming that there would be, as in a typical Western population, about 2000 cases of cancer/10^6 persons per year, we therefore see that environmental chemicals are only responsible for a small fraction of all cancers; radon would account for 1.5% and all the remaining inhalation exposures would account for 0.2%–0.5%. If we additionally assume that oral ingestion of As, Cd, asbestos, benzene, PAH and trihalomethanes accounts for another 0.5% of all cancers, the total effect of the environmental carcinogens considered here would be about 2%. These figures are based upon the limited number of carcinogens considered. Furthermore, many of the risk estimates used here are based on animal experiments from single-agent exposures. Environmental carcinogens may have promotional or other interactive effects on other carcinogens which have not so far been studied in experimental animals and thus cannot be taken into account.

Doll and Peto (1981) estimated, based on epidemiological reasoning for the USA, that air pollutants cause 2% of all cancers (range of acceptable estimates 1%–5%). They additionally included "industrial products" with a <1% proportion as a separate category. Considering the crudeness of my and their estimates, the results are amazingly close to each other. Irrespective of the magnitude of the known cancer risk associated with environmental chemicals, there are very many reasons why environmental pollution can and should be minimised.

References

Anderson EL (1985) Quantitative approaches in use in the United States to assess cancer risk. In: Vouk VB, Butler GC, Hoel DG, Peakall DP (eds) Methods for estimating risk of chemical injury: human and non-human biota and ecosystems. SCOPE 26. Wiley, New York, pp 405–436

Anderson EL and the Carcinogen Assessment Group of the U.S. Environmental Protection Agency (1983) Quantitative approaches in use to assess cancer risk. Risk Analysis 3:277–295

Castren O, Voutilainen A, Winquist K, Mäkelainen I (1985) Studies of high indoor radon areas in Finland. Sci Total Environ 45:311–318

Czuczwa JM, Hites RA (1985) Dioxins and dibenzofurans in air, soil and water. In: Kamrin MA, Rodgers PW (eds) Dioxins in the environment. Hemisphere, Washington D.C., pp 85–99

Doll R, Peto R (1981) The causes of cancer: quantitative estimates of avoidable risks of cancer in the United States today. JNCI 66:1191–1308

Gold LS (1980) Human exposures to ethylene dichloride. Banbury Rep 5:209–225

Gold LS, Sawyer CB, Magaw R, Backman GM, De Veciana M, Levison R, Hooper NK, Havender WR, Bernstein L, Peto R, Pike MC, Ames BN (1984) A carcinogenic potency data base of the standardized results of animal bioassays. Environ Health Perspect 58:9–319

Hemminki K, Vainio H (1984) Human exposure to potentially carcinogenic compounds. In: Berlin A, Draper M, Hemminki K, Vainio H (eds) Monitoring human exposure to carcinogenic and mutagenic agents. IARC scientific publications No. 59. International Agency for Research on Cancer. Lyon, pp 37–45

Hemminki K, Vainio H, Sorsa M, Salminen S (1983) An estimation of the exposure of the population in Finland to suspected chemical carcinogens. Environ Carcinog Rev C1:55–95

Hutzinger O, Choudry GG, Brock GC, Johnston LE (1985a) Formation of polychlorinated dibenzofurans and dioxins during combustion, electrical equipment fires and PCB incineration. Environ Health Perspect 60:3–9

Hutzinger O, Berg MVD, Olie K, Opperhuizen A, Safe S (1985b) Dioxins and furans in the environment: evaluating toxicological risk from different sources by multi-criteria analysis. In: Kamrin MA, Rodgers PW (eds) Dioxins in the environment. Hemisphere, Washington D.C., pp 9–32

IARC (1979a) IARC monographs in the evaluation of the carcinogenic risk of chemicals to humans, vol 20. Some halogenated hydrocarbons. International Agency for Research on Cancer, Lyon

IARC (1979b) IARC monographs on the evaluation of the carcinogenic risk of chemicals to humans, vol 19. Some monomers, plastics and synthetic elastomers, and acrolein. International Agency for Research on Cancer, Lyon

IARC (1982a) IARC monographs on the evaluation of the carcinogenic risk of chemicals to humans, vol 29. Some industrial chemicals and dyestuffs. International Agency for Research on Cancer, Lyon

IARC (1982b) IARC monographs on the evaluation of the carcinogenic risk of chemicals to humans, vol 29, supplement 4. Chemicals, industrial processes and industries associated with cancer in humans. International Agency for Research on Cancer, Lyon

IARC (1984) IARC monographs on the evaluation of the carcinogenic risk of chemicals to humans, vol 34. Polynuclear aromatic compounds, part 3. Industrial exposures in aluminum production, coal gasification, coke production, and iron and steel founding. International Agency for Research on Cancer, Lyon
IARC (1986) IARC monographs on the evaluation of the carcinogenic risk of chemicals to humans, vol 38. Tobacco smoking. International Agency for Research on Cancer, Lyon
IARC (1987) IARC monographs on the evaluation of the carcinogenic risk of chemicals to humans, suppl 7. International Agency for Research on Cancer, Lyon
Mossman KL, Thomas DS, Dritschilo A (1986) Environmental radiation and cancer. Environ Carcinog Rev C4(2):119–161
Rosenkranz HS, Mermelstein R (1985) The genotoxicity, metabolism and carcinogenicity of nitrated polycyclic aromatic hydrocarbons. J Environ Sci Health 3C:221–272
Safe S, Bandiera S, Sawyer T, Robertson L, Safe L, Parkinson A, Thomas PE, Ryan DE, Reik LM, Levin W, Denomme MA, Fujita T (1985) PCBs: structure-function relationships and mechanism of action. Environ Health Perspect 60:47–56
Slorach SA, Vaz R (1985) PCB levels in breast milk: data from the UNCP/WHO pilot project on biological monitoring and some other recent studies. Environ Health Perspect 60:121–126
U.S. EPA (1979) Chlorinated ethanes. Ambient water quality criteria (draft). Criteria and Standards Division, Office of Water Planning and Standards, EPA, Washington D.C.
U.S. EPA (1980) Guidelines and methodology used in the preparation of health effects assessment chapters of the consent decree water quality criteria. Fed Reg 45:79347–79357
U.S. EPA (1981) Updated mutagenicity and carcinogenicity assessment of cadmium. Appendum to the health assessment document for cadmium. EPA/600/8-81/23. EPa/6/8-83/25F. June 1985. Final report. EPA, Washington D.C.
U.S. EPA (1983) Health assessment document for acrylonitrile. Final report. EPA-600/8-82-007F. Office of Health and Environmental Assessment, Washington D.C.
U.S. EPA (1984a) Health assessment document for inorganic arsenic. Final report. EPA-600/8-83-021F. Office of Health and Environmental Assessment, Washington D.C.
U.S. EPA (1984b) Health assessment document for chromium. Final report. EPA-600/8-83-014F. Environmental Criteria and Assessment Office, Research Triangle Park, North Carolina
U.S. EPA (1985) Health assessment document for nickel. Final report. EPA-600/8-83/021F. Office of Health and Environmental Assessment, Washington D.C.
UNSCEAR report to the general assembly with annexes (1982) Ionizing radiation: sources and biological effects. United Nations, New York
Velema JP (1987) Contaminated drinking water as a potential cause of cancer in humans. Environ Carcinog Rev C5:1–28
WHO (1981) Environmental health criteria 18. Arsenic. World Health Organization, IPCS, Geneva
WHO (1983a) Environmental health criteria 27. Guidelines on studies in environmental epidemiology. World Health Organization, Geneva
WHO (1983b) Environmental health criteria 26. Styrene. World Health Organization, Geneva, p 123
WHO (1984a) Guidelines for drinking-water quality, vol 1. Recommendations. World Health Organization, Geneva, p 130
WHO (1984b) Environmental health criteria 32. Methylene chloride. World Health Organization, Geneva, p 55
WHO (1984c) Environmental health criteria 31. Tetrachloroethylene. World Health Organization, Geneva, p 48
WHO (1987a) Air quality guidelines, vol 1–2. World Health Organization, Copenhagen
WHO (1987b) Environmental health criteria 62. 1,2-Dichloroethane. World Health Organization, Geneva, p 90

CHAPTER 3

Advances in Tobacco Carcinogenesis

D. HOFFMANN and S. S. HECHT

A. Introduction

In their first reports on smoking and disease both the ROYAL COLLEGE OF PHYSICIANS OF LONDON (1962) and the US SURGEON GENERAL OF THE PUBLIC HEALTH SERVICE (1964) concluded that cigarette smoking is causally related to lung cancer in humans and is associated with cancer of the oral cavity, larynx and urinary bladder. These conclusions were based on epidemiological data and were supported by laboratory studies. Today, 25 years later and after extensive research, epidemiological reports from more than 20 countries have led the US SURGEON GENERAL (1986a) and the INTERNATIONAL AGENCY FOR RESEARCH ON CANCER (1986) to the conclusion that smoking of cigarettes is causally related to cancer of the respiratory tract, the upper digestive tract, pancreas, renal pelvis and bladder and that cigarette smokers also face an increased risk for cancer of the cervix. Cigar and pipe smoking are also causally related to cancer of the respiratory tract, oral cavity and esophagus, although, in the case of lung cancer, not to the same extent as cigarette smoking (US SURGEON GENERAL 1986a; IARC 1986). In addition to active smoking, involuntary smoking, i.e., the exposure to environmental tobacco smoke, has been incriminated as a risk factor for cancer of the lung in nonsmokers (IARC 1986; US NATIONAL RESEARCH COUNCIL 1986; US SURGEON GENERAL 1986a). Furthermore, chewing of tobacco and especially the oral use of snuff were found to be associated with cancer of the oral cavity (IARC 1985a; US SURGEON GENERAL 1986b) and possibly with cancer of the nasal cavity, kidney and bladder (BRINTON et al. 1984; US SURGEON GENERAL 1986a; KABAT et al. 1986; GOODMAN et al. 1986).

The rise in lung cancer in industrialized countries has been directly correlated with the increase in the manufacture and consumption of cigarettes (DOLL and PETO 1981). In fact, it has been estimated that 85%–90% of all lung cancer deaths in American males in 1978 were caused by tobacco smoking and that about 30% of all cancers in the USA and in the UK can be attributed to tobacco use (WYNDER and GORI 1977; HIGGINSON and MUIR 1979; DOLL and PETO 1981).

In view of the convincing epidemiologic evidence and widespread awareness of the role of tobacco products as causes of cancer, one is sometimes asked why there is a need for further studies in tobacco carcinogenesis. After all, if tobacco use were to cease, the problem would disappear in a few decades. The hazards of tobacco usage are well-known by the public and are taught in many educational institutions. In fact, the Surgeon General of the US Public Health Services deserves the full support of the medical and scientific community in his quest for a "smoke-free society" by the year 2000 (KOOP 1986). Unfortunately, the recent statistics on tobacco use are not supportive of this goal. In 1985, for example, approximately 600 billion cigarettes were sold in the USA alone and the annual per capita consumption of cigarettes for individuals aged 18 years and older was approximately 3400 (TOBACCO JOURNAL INTERNATIONAL 1987). The consumption of cigarettes in many Asian, African and South American countries has also sharply risen in recent years. For example, between 1976 and 1986 the cigarette production in the People's Republic of China increased by 84.4% to 1.296 billion, in Egypt by 112.9% to 49.5 billion cigarettes, and in Brazil by 44.4% to 168.9 billion cigarettes (IARC 1986; TOBACCO JOURNAL INTERNATIONAL 1987). In the USA and Sweden there has also been a constant rise in the consumption of snuff tobacco, at least until 1985 (MAXWELL 1986; TOBACCO JOURNAL INTERNATIONAL 1988).

This review is intended to document the progress achieved in tobacco toxicology during the past 2 decades (WYNDER and HOFFMANN 1967). New knowledge in this field has contributed much to our understanding of the epidemiologic findings and to tobacco carcinogenesis and environmental carcinogenesis in general. New methods and concepts have been developed in tobacco carcinogenesis and in chemical carcinogenesis and have provided new insights into both fields of research (HOFFMANN and HARRIS 1986).

B. Tobacco and Tobacco Smoke

In most parts of the world, more than 60 species of *Nicotiana* can be found, but only *N. tabacum* is commercially cultivated on a large scale. *N. rustica* is grown in some areas of China, India and the USSR. Tobacco leaves are usually dried, cured, aged, and, in some instances, fermented. The leaves of the bright (Virginia) varieties of *N. tabacum* are flue cured in steam-heated barns, which results in tobacco with high sugar content and relatively low levels of nitrate ($<0.1\%$). Burley leaves, on the other hand, are simply air-cured and are low in sugars and relatively high in nitrate content ($<5\%$), and oriental leaves which are sun-cured, have medium sugar content and are low in nitrate ($<0.6\%$). For cigars, for some types of pipe tobaccos, and for use as smokeless tobaccos, leaves are not only cured but also fermented (TSO 1972).

Processed, unadulterated tobacco contains at least 2550 known compounds (DUBE and GREEN 1982). The bulk of the tobacco consists of carbohydrates ($\simeq 50\%$) and proteins. Other significant constituents are alkaloids (0.5%–5%) with nicotine as the predominant compound (90%–95% of total alkaloids), terpenes (0.1%–3.0%), polyphenols (0.5%–4.5%), phytosterols (0.1%–2.5%), carboxylic acids (0.1%–0.7%), alkanes (0.1%–0.4%), aromatic hydrocarbons, aldehydes, ketones, amines, nitriles, *N*- and *O*-heterocyclic compounds, pesticides, alkali nitrates (0.01%–5%), and at least 30 metallic compounds (WYNDER and HOFFMANN 1967; IARC 1986).

The burning of tobacco generates mainstream smoke (MS) during puff-drawing, and sidestream smoke (SS) during smouldering between puffs. The physicochemical nature of these smoke types is a function of various factors.These include the type of tobacco, the temperatures prevailing during puff-drawing (860°–900° C) or smouldering (500°–650° C), the reducing atmosphere characteristic of the burning zone and the physical design of the tobacco product (e.g. length, diameter, paper, wrapper or pipe bowl, variety of the cigarette paper and filter tip).

The composition of the processed tobacco in cigarettes has a profound influence on the chemistry and toxicity of the smoke. Cigarette manufacture in the USA, Japan, and most European countries utilizes blends of bright, burley, and oriental tobaccos, whereas cigarettes sold in the UK and Finland contain exclusively bright tobaccos. Both types of cigarettes deliver a weakly acidic mainstream smoke (MS) (pH 5.5–6.2) in which nicotine occurs in protonated form in the particulate matter. In France and some parts of Italy, North Africa, and South America, a high percentage of the cigarette brands contain only burley tobaccos. In the smoke of these cigarettes which is neutral to weakly alkaline (pH 6.8–7.5), a significant proportion of the nicotine is found in the vapor phase in unprotonated form. The smoke of cigars is neutral to alkaline (pH 6.5–8.0), and, like the smoke of burley cigarettes, it contains unprotonated nicotine in the vapor phase. The pH of sidestream smoke (SS) of cigarettes and cigars ranges between 6.8–8.5; thus, it contains free nicotine (BRUNNEMANN and HOFFMANN 1974). Unprotonated nicotine is more quickly absorbed through the buccal mucosa than protonated nicotine (ARMITAGE and TURNER 1970).

The 400–500 mg of MS freshly emerging from the mouthpiece of a cigarette is an aerosol which contains about 1×10^{10} particles per ml; these range in diameter from 0.1–1.0 µm (mean diameter 0.2 µm) and are dispersed in a vapor phase (INGEBRETHSEN 1986). About 95% of the MS effluent of a non-filter cigarette is comprised of 400–500 individual gaseous components with nitrogen, oxygen, and carbon dioxide as major constituents. As of our state of knowledge, the particulate matter contains at least 3500 individual compounds (Fig. 1; DUBE and GREEN 1982).

All combustion products contain free radicals; in the case of tobacco smoke these are highly reactive oxygen- and carbon-centered types in the vapor phase, and relatively stable radicals in the particulate phase. The principle of the latter appears to be a quinone/hydroquinone complex which is capable of reducing molecular oxygen to superoxide and, eventually, to hydrogen peroxide and hydroxyl radicals (NAKAYAMA et al. 1984; CHURCH and PRYOR 1985).

The generation of MS and SS components follows different pathways. The compounds are either transferred structurally intact from the tobacco into the smoke (e.g. nicotine, phytosterols, long-chain paraffins), or they are completely pyrosynthesized in the hot zones without specific precursors (e.g. carbon monoxide, phenols, benzene, benzo[*a*]pyrene), or they are partially transferred and partially pyrosynthesized (e.g. *N'*-nitrosonornicotine, certain volatile aldehydes). The majority of the smoke components are pyrosynthesized either by partial degradation or by oxidation of specific tobacco precursors (e.g. furans, indoles, flavor components derived from tobacco terpenoids), or they are totally synthesized from specific constituents (e.g. hydrogen cyanide, nitrogen oxides, ammonia, catechols; GREEN 1977; SCHMELTZ and HOFFMANN 1977; JOHNSON 1977; CARMELLA et al. 1984).

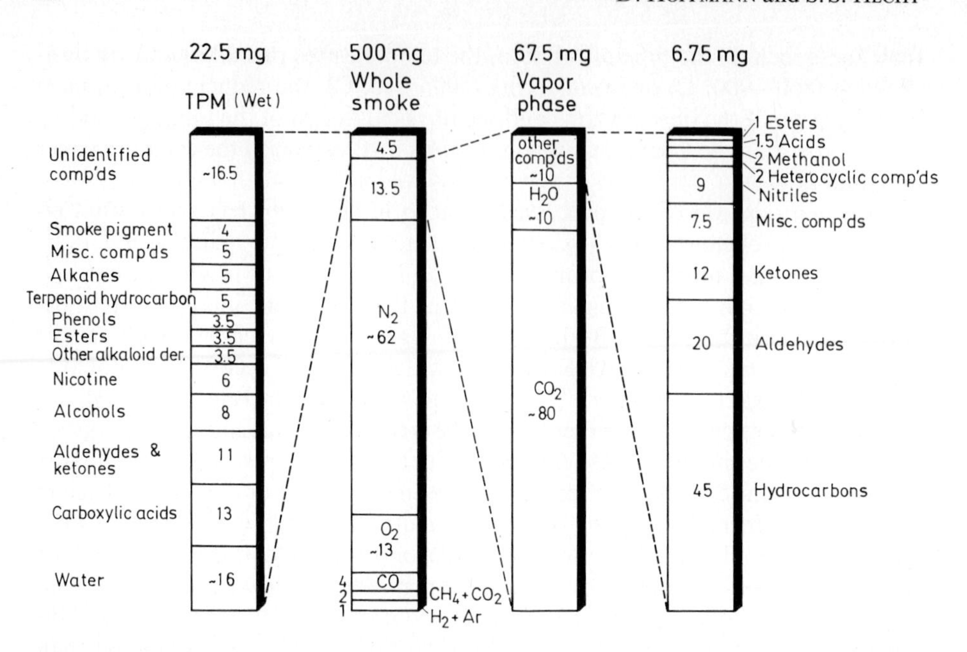

Fig. 1. Total cigarette smoke composition (% w/w) (DUBE and GREEN 1982) *TPM*, total particulate matter

For chemical analysis, the smoke is arbitrarily separated into a vapor phase and a particulate phase. Those individual smoke components of which more than 50% appear in the vapor phase of fresh MS are considered volatile smoke components; all others are particulate phase components (Fig. 1). Tables 1 and 2 list the major types of components identified and their estimated concentration in the smoke of one cigarette (WYNDER and HOFFMANN 1967; TSO 1972; GREEN 1977; ENZELL et al. 1977; US SURGEON GENERAL 1982; IARC 1986; WAHLBERG and ENZELL 1987). These tables present data which are important with regard to bioactivity of smoke constituents but are by no means to be regarded as a complete analysis of cigarette smoke. The quantitative data in this review are derived from cigarettes that were machine-smoked under standardized laboratory conditions (BRUNNEMANN et al. 1976). Therefore, the data do not fully reflect the human setting. This applies especially to smokers of low yield cigarettes who tend to compensate for the low nicotine and low tar delivery by drawing smoke more intensely and inhaling it more deeply (HERNING et al. 1981; HALEY et al. 1985).

Tobacco is known to contain at least 30 metals (NORMAN 1977). For example, the tobacco of one cigarette was found to contain 38 mg of potassium, 22 mg of calcium, and 5.5 mg of magnesium as the major metals. Since less than 1% of the metals is transferred from the tobacco into the smoke (JENKINS et al. 1985), these elements form too minute a proportion to be listed in Table 2.

Tables 1 and 2 also omit information about the chemical nature and concentrations in cigarette smoke of agricultural chemicals and pesticides, which originate from the residues of such compounds on the tobacco (WYNDER and HOFFMANN 1967; IARC 1985a, 1986). We have not included this information be-

Table 1. Major constituents of the vapor phase of the mainstream smoke of nonfilter cigarettes

Compound[a]	Concentration/cigarette	(% of total effluent)
Nitrogen	280 – 320 mg	(56 –64 %)
Oxygen	50 – 70 mg	(11 –14 %)
Carbon dioxide	45 – 65 mg	(9 –13 %)
Carbon monoxide	14 – 23 mg	(2.8– 4.6%)
Water	7 – 12 mg	(1.4– 2.4%)
Argon	5 mg	(1.0)
Hydrogen	0.5– 1.0 mg	
Ammonia	10 – 130 μg	
Nitrogen oxides [NO_x]	100 – 600 μg	
Hydrogen cyanide	400 – 500 μg	
Hydrogen sulfide	20 – 90 μg	
Methane	1.0– 2.0 mg	
Other volatile alkanes (20)	1.0– 1.6 mg[b]	
Volatile alkenes (16)	0.4– 0.5 mg	
Isoprene	0.2– 0.4 mg	
Butadiene	25 – 40 μg	
Acetylene	20 – 35 μg	
Benzene	12 – 50 μg	
Toluene	20 – 60 μg	
Styrene	10 μg	
Other volatile aromatic hydrocarbons (29)	15 – 30 μg	
Formic acid	200 – 600 μg	
Acetic acid	300 –1 700 μg	
Propionic acid	100 – 300 μg	
Methyl formate	20 – 30 μg	
Other volatile acids (6)	5 – 10 μg[b]	
Formaldehyde	20 – 100 μg	
Acetaldehyde	400 –1 400 μg	
Acrolein	60 – 140 μg	
Other volatile aldehydes (6)	80 – 140 μg	
Acetone	100 – 650 μg	
Other volatile ketones (3)	50 – 100 μg	
Methanol	80 – 180 μg	
Other volatile alcohols (7)	10 – 30 μg[b]	
Acetonitrile	100 – 150 μg	
Other volatile nitriles (10)	50 – 80 μg[b]	
Furan	20 – 40 μg	
Other volatile furans (4)	45 – 125 μg[b]	
Pyridine	20 – 200 μg	
Picolines (3)	15 – 80 μg	
3-Vinylpyridine	10 – 30 μg	
Other volatile pyridines (25)	20 – 50 μg[b]	
Pyrrole	0.1– 10 μg	
Pyrrolidine	10 – 18 μg	
N-Methylpyrrolidine	2.0– 3.0 μg	
Volatile pyrazines (18)	3.0– 8.0 μg	
Methylamine	4 – 10 μg	
Other aliphatic amines (32)	3 – 10 μg	

[a] Numbers in parentheses represent the individual compounds identified in a given group.
[b] Estimate.

Table 2. Major constituents of the particulate matter of the mainstream smoke of nonfilter cigarettes

Compound[a]	µg/cigarette
Nicotine	1000 –3000
Nornicotine	50 – 150
Anatabine	5 – 15
Anabasine	5 – 12
Other tobacco alkaloids (17)	n.a.
Bipyridyls (4)	10 – 30
n-Hentriacontane [n-$C_{31}H_{64}$]	100
Total nonvolatile hydrocarbons (45)[c]	300 – 400[c]
Naphthalene	2 – 4
Naphthalenes (23)	3 – 6[c]
Phenanthrenes (7)	0.2 – 0.4[c]
Anthracenes (5)	0.05– 0.1[c]
Fluorenes (7)	0.6 – 1.0[c]
Pyrenes (6)	0.3 – 0.5[c]
Fluoranthenes (5)	0.3 – 0.45[c]
Carcinogenic polynuclear aromatic hydrocarbons (11)[b]	0.1 – 0.25
Phenol	80 – 160
Other phenols (45)[c]	60 – 180[c]
Catechol	200 – 400
Other catechols (4)	100 – 200[c]
Other dihydroxybenzenes (10)	200 – 400[c]
Scopoletin	15 – 30
Other polyphenols (8)[c]	n.a.
Cyclotenes (10)[c]	40 – 70[c]
Quinones (7)	0.5
Solanesol	600 –1000
Neophytadienes (4)	200 – 350
Limonene	30 – 60
Other terpenes (200–250)[c]	n.a.
Palmitic acid	100 – 150
Stearic acid	50 – 75
Oleic acid	40 – 110
Linoleic acid	60 – 150
Linolenic acid	150 – 250
Lactic acid	60 – 80
Indole	10 – 15
Skatole	12 – 16
Other indoles (13)	n.a.
Quinolines (7)	2 – 4
Other aza-arenes (55)	n.a.
Benzofurans (4)	200 – 300
Other *O*-heterocyclic compounds (42)	n.a.
Stigmasterol	40 – 70
Sitosterol	30 – 40
Campesterol	20 – 30
Cholesterol	10 – 20
Aniline	0.36
Toluidines	0.23
Other aromatic amines (12)	0.25
Tobacco-specific *N*-nitrosamines (4)[b]	0.34– 2.7
Glycerol	120

[a] Numbers in parentheses represent individual compounds identified. [b] For details, see Table 3. [c] Estimate. n.a., Not available.

cause of the many variations in the nature and the amounts of these agents in tobaccos from country to country and from year to year (WITTEKINDT 1985). Nevertheless, it is fairly certain that commercial tobacco contains up to a few parts per million of DDT, DDD, and maleic hydrazide; less than 20% of these amounts are transferred into the MS.

The increasing market share of cigarettes with low smoke yields has only been attained because flavor additives made these products "consumer-acceptable". Flavor compounds are usually derived from extracts of tobacco or other plant products but may also be synthetic in nature (LEFFINGWELL et al. 1972). Except for menthol (0–500 µg in the smoke of a cigarette; PERFETTI and GORDIN 1985) the flavor additives are trade secrets; thus, there is little information in the literature about their presence and levels in commercial tobacco products. However, it is known that manufacturers in many countries have discontinued the use of coumarin (a carcinogen in rats; IARC 1976).

C. The Changing Cigarette

Epidemiological studies have documented a dose-response relationship between the number of cigarettes smoked and the development of cancer of the lung, oral cavity, larynx, esophagus, bladder, and kidney (US SURGEON GENERAL 1982; IARC 1986). Bioassays with whole smoke and with tar have also demonstrated a dose-response relationship (WYNDER and HOFFMANN 1967; DONTENWILL 1974; BERNFELD et al. 1974). Thus, a reduction of tar and nicotine was considered as one step towards the reduction of cancer risk for those smokers who were not willing to give up smoking (US SURGEON GENERAL 1981). In addition to tar and nicotine, several toxic and tumorigenic agents such as carbon monoxide, volatile *N*-nitrosamines, and carcinogenic PAH were also significantly reduced (HOFFMANN et al. 1980, 1984; US SURGEON GENERAL 1981). Although smokers of low yield cigarettes tend to compensate for reduced intake of nicotine (HERNING et al. 1981; HALEY et al. 1985), they do not, in general, compensate fully for low smoke yields. Studies on smokers indicate that prolonged use of low-yield cigarettes reduces the risk for cancer to some extent. However, the reduction in risk is only minor compared with giving up cigarette smoking altogether (US SURGEON GENERAL 1982; IARC 1986).

Figure 2 shows the reduction in sales-weighted tar and nicotine delivery of the average American cigarette. Arrows pinpoint the introduction of technical changes during various years which had a profound influence on the sales-weighted average nicotine and tar deliveries (NORMAN 1982). Since 1981 the tar delivery has varied between 14.0 and 12.7 mg, and the nicotine values have remained stable at 0.9 mg per cigarette. These data indicate that the reduction in nicotine has not occurred to the same extent as the reduction in tar. This trend is even more pronounced for cigarettes in the United Kingdom (Fig. 3; JARVIS and RUSSELL 1985). Since nicotine is the habituating agent in tobacco products, it is of major concern that further reduction of its smoke yield has not been implemented.

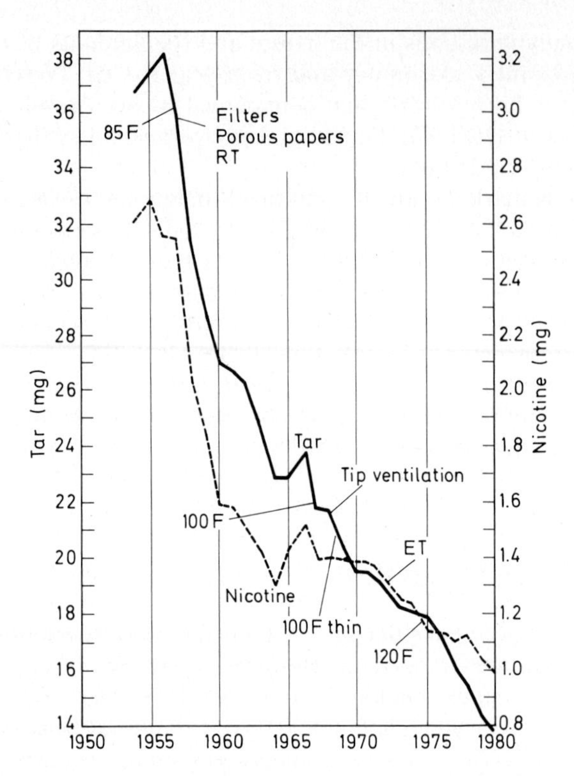

Fig. 2. Sales-weighted average tar and nicotine yields of American cigarettes

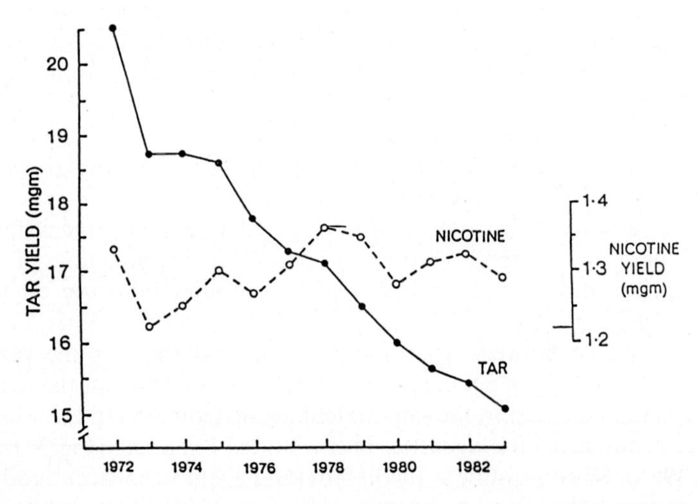

Fig. 3. Sales-weighted tar and nicotine yield of British cigarettes, 1972–1983 (JARVIS and RUSSELL 1985)

Some modifications in the make-up of commercial cigarettes have also led to a selective reduction of certain toxic and tumorigenic agents. Cellulose acetate filters, the most common cigarette filter tips, can selectively reduce phenols and volatile *N*-nitrosamines; perforated filter tips effect a reduction of smoke yields by air dilution and, in addition, a selective reduction of carbon monoxide and hydrogen cyanide. Charcoal filter tips are capable of selectively reducing volatile aldehydes and hydrogen cyanide. The utilization of reconstituted tobacco, expanded tobacco, and tobacco ribs in the manufacture of cigarettes has led to a selective reduction of carcinogenic PAH in the smoke. When tar from such cigarettes is evaluated for tumorigenicity on mouse skin, one observes a reduction of its biological activity by comparison with the same dose of other cigarette tars (WYNDER and HOFFMANN 1967; BERNFELD et al. 1974; US SURGEON GENERAL 1981; HALEY et al. 1985). However, the incorporation of ribs and stems into the cigarette blend and the utilization of more burley varieties as cigarette fillers have caused an increase in the nitrate content of the American blended cigarette from $\simeq$0.5% to 1.2%–1.5%. While this development has led to a reduction of the smoke yields of tar, phenols, and carcinogenic PAH, it has, on the other hand, increased nitrogen oxides (NO_x) and carcinogenic nitrosamines in the smoke (WYNDER and HOFFMANN 1967; US SURGEON GENERAL 1981, 1982; HOFFMANN et al. 1980, 1984). The biological activity of *N*-nitrosamines is not reflected in bioassays on mouse skin and requires evaluation by other assays in mice, rats, or hamsters.

It needs to be stressed that the modified cigarettes have somewhat reduced toxicity and tumorigenic activity in bioassays; however, these reductions are in no way equal to the reduction of cancer risk which can be achieved by cessation of smoking.

D. Carcinogenic Compounds in Tobacco and Tobacco Smoke

The recent IARC monographs on *Tobacco Smoking* and *Tobacco Habits Other Than Smoking* have presented comprehensive reviews of the carcinogenic and toxic components of tobacco and tobacco smoke (IARC 1985a, 1986). Table 3, condensed from these monographs, summarizes the known carcinogens in tobacco and tobacco smoke and gives the ranges of their concentrations along with the evaluations of their carcinogenic activities where available. Structures of representative carcinogens are shown in Fig. 4. The diversity of carcinogenic compounds in tobacco and tobacco smoke may cause ambiguity as to which among them are most important. In the following section we will discuss the likely role of the various types of carcinogens in cancer induction by tobacco and tobacco smoke (see Table 4).

I. Polynuclear Aromatic Hydrocarbons (PAH)

Inhalation studies with laboratory animals have demonstrated that the particulate matter of tobacco smoke induces malignant tumors of the respiratory tract, most notably in the larynx of the Syrian golden hamster (DONTENWILL 1974;

Table 3. Tumorigenic agents in tobacco and tobacco smoke

Compounds	In processed tobacco (per g)	In mainstream smoke (per cigarette)	IARC evaluation of evidence of carcinogenicity [a]	
			In laboratory animals	In humans
PAH				
Benz[*a*]anthracene		20 – 70 ng	Sufficient	
Benzo[*b*]fluoranthene		4 – 22 ng	Sufficient	
Benzo[*j*]fluoranthene		6 – 21 ng	Sufficient	
Benzo[*k*]fluoranthene		6 – 12 ng	Sufficient	
Benzo[*a*]pyrene	0.1 – 90 ng	20 – 40 ng	Sufficient	Probable
Chrysene		40 – 60 ng	Sufficient	
Dibenz[*a,h*]anthracene		4 ng	Sufficient	
Dibenzo[*a,i*]pyrene		1.7 – 3.2 ng	Sufficient	
Dibenzo[*a,l*]pyrene		present	Sufficient	
Indeno[1,2,3-*cd*]pyrene		4 – 20 ng	Sufficient	
5-Methylchrysene		0.6 ng	Sufficient	
Aza-arenes				
Quinoline		1 – 2 μg		
Dibenz[*a,h*]acridine		0.1 ng	Sufficient	
Dibenz[*a,j*]acridine		3 – 10 ng	Sufficient	
7*H*-Dibenzo[*c,g*]-carbazole		0.7 ng	Sufficient	
N-Nitrosamines				
N-Nitrosodimethylamine	ND– 215 ng	0.1 – 180 ng	Sufficient	
N-Nitrosoethylmethyl-amine		3 – 13 ng	Sufficient	
N-Nitrosodiethylamine		ND – 25 ng	Sufficient	
N-Nitrosopyrrolidine	ND– 360 ng	1.5 – 110 ng	Sufficient	
N-Nitrosodiethanolamine	ND–6900 ng	ND – 36 ng	Sufficient	
N'-Nitrosonornicotine	0.3 – 89 μg	0.12– 3.7 μg	Sufficient	
4-(Methylnitrosoamino)-1-(3-pyridyl)-1-butanone	0.2 – 7 μg	0.08– 0.77 μg	Sufficient	
N'-Nitrosoanabasine	0.01– 1.9 μg	0.14– 4.6 μg	Limited	
N-Nitrosomorpholine	ND– 690 ng		Sufficient	
Aromatic amines				
2-Toluidine		30 – 200 ng	Sufficient	Inadequate
2-Naphthylamine		1 – 22 ng	Sufficient	Sufficient
4-Aminobiphenyl		2 – 5 ng	Sufficient	Sufficient
Aldehydes				
Formaldehyde	1.6 – 7.4 μg	70 – 100 μg [b]	Sufficient	
Acetaldehyde	1.4 – 7.4 μg	18 –1400 μg [b]	Sufficient	
Crotonaldehyde	0.2 – 2.4 μg	10 – 20 μg		
Miscellanous organic compounds				
Benzene		12 – 48 μg	Sufficient	Sufficient
Acrylonitrile		3.2 – 15 μg	Sufficient	Limited
1,1-Dimethylhydrazine	60 – 147 μg		Sufficient	
2-Nitropropane		0.73– 1.21 μg	Sufficient	
Ethylcarbamate	310 – 375 ng	20 – 38 ng	Sufficient	
Vinyl chloride		1 – 16 ng	Sufficient	Sufficient

Table 3 (continued)

Compounds	In processed tobacco (per g)	In mainstream smoke (per cigarette)	IARC evaluation of evidence of carcinogenicity[a]	
			In laboratory animals	In humans
Inorganic compounds				
Hydrazine	14 – 51 ng	24 – 43 ng	Sufficient	Inadequate
Arsenic	500 – 900 ng	40 – 120 ng	Inadequate	Sufficient
Nickel	2000 –6000 ng	0 – 600 ng	Sufficient	Limited
Chromium	1000 –2000 ng	4 – 70 ng	Sufficient	Sufficient
Cadmium	1300 –1600 ng	41 – 62 ng	Sufficient	Limited
Lead	8 – 10 μg	35 – 85 ng	Sufficient	Inadequate
Polonium-210	0.2 – 1.2 pCi	0.03– 1.0 pCi		

[a] No designation indicates that an evaluation by IARC has not been carried out.
[b] The 4th report of the independent scientific committee on smoking and health (1988) published values for the 14 leading British cigarettes in 1986 (51.4% of the market) of 20–105 μg/cigarette (mean 59 μg) for formaldehyde and 550–1150 μg/cigarette (mean 910 μg) for acetaldehyde.
PAH, polynuclear aromatic hydrocarbons, ND, not detected.

Table 4. Likely causative agents for tobacco-related cancers

Organ(s)	Initiator or carcinogen	Enhancing agents
	PAH	Catechol (cocarcinogen), weakly acidic tumor promoters
Lung, larynx	NNK Polonium-210 (minor factor), acetaldehyde, formaldehyde	Acrolein, crotonaldehyde(?)
Esophagus	NNN	
Pancreas	NNK	
Bladder	4-Aminobiphenyl 2-Naphthylamine	
Oral cavity (smoking)	PAH NNK, NNN	Ethanol
Oral cavity (snuff dipping)	NNK, NNN Polonium-210	Irritation(?) *Herpes simplex*(?)

PAH, polynuclear aromatic hydrocarbons; NNK, 4-(methylnitrosamino)-1-(3-pyridyl)-1-butanone; NNN, N′-nitrosonornicotine.

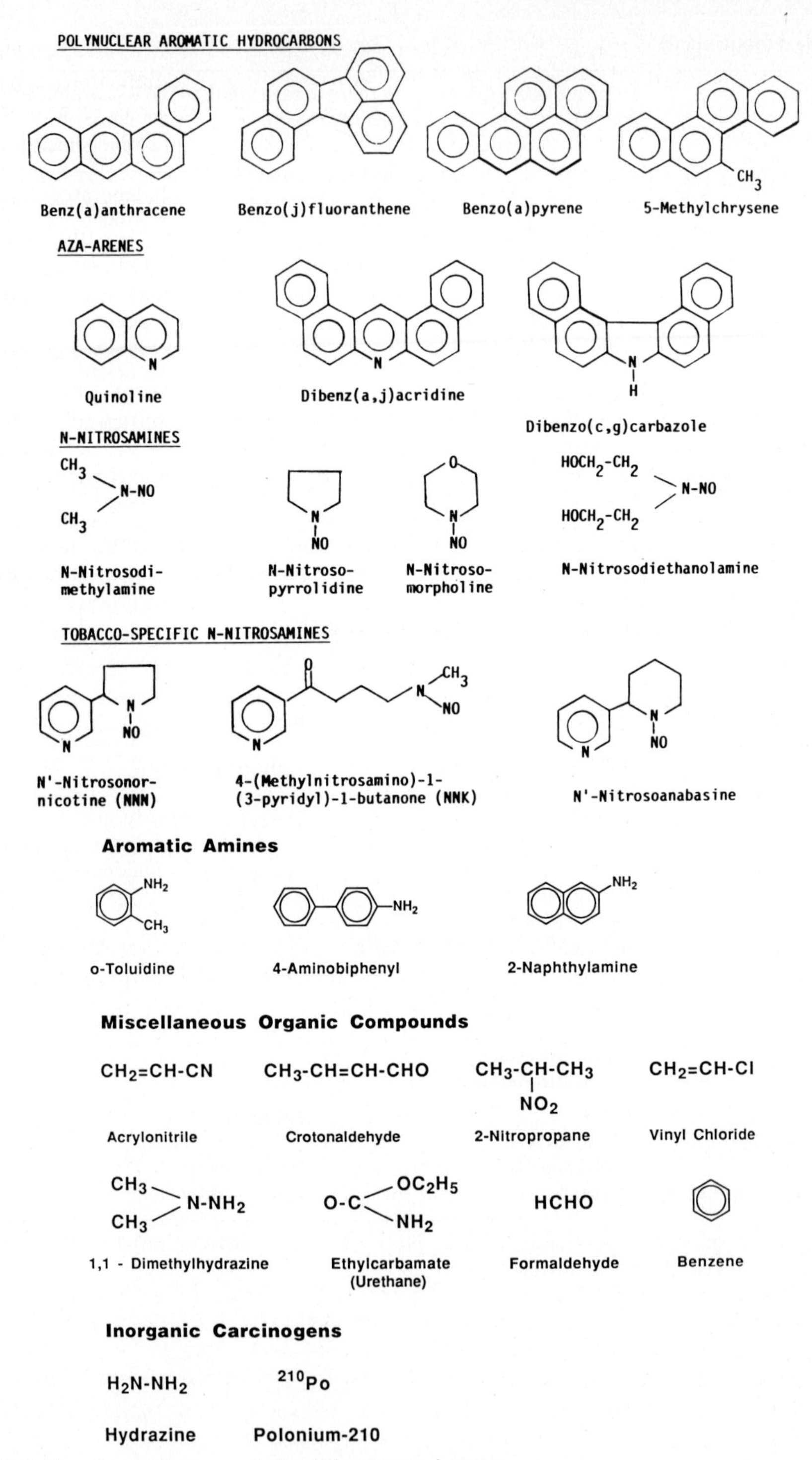

Fig. 4. Structures of representative tobacco carcinogens

BERNFELD et al. 1974; IARC 1986). The particulate matter is more carcinogenic than the gas phase. Fractions and subfractions of the particulate matter have been extensively assayed for tumorigenic activity on mouse skin. It has been clearly demonstrated that the most tumorigenic fractions in these assays are those with highly concentrated PAH (HOFFMANN and WYNDER 1971; HOFFMANN et al. 1978). However, PAH by themselves do not account for the tumorigenic activity on mouse skin induced by the total particulate matter. Most of the skin tumor activity can again be demonstrated when the PAH-enriched fractions are tested together with the weakly acidic fraction in either an initiation-promotion protocol or a cocarcinogenesis protocol (HOFFMANN and WYNDER 1971; HOFFMANN et al. 1978). This approach has indicated that PAH, together with weakly acidic tumor promoters and carcinogens such as catechol, are crucial factors in mouse skin tumorigenesis induced by the particulate matter of tobacco smoke. However, it must be emphasized that mouse skin is particularly responsive to PAH tumorigenesis. It is not equally responsive to other important classes of carcinogens such as *N*-nitrosamines or aromatic amines.

PAH are recognized as contact carcinogens, and in this respect it is likely that the neoplasms obtained in mouse skin would also be observed in other tissues which are in direct contact with tobacco smoke. It is well established that PAH induce tumors in the respiratory tract of hamsters upon intratracheal instillation (SAFFIOTTI et al. 1985). PAH also induce lung tumors upon implantation in the lung with beeswax as a carrier. This protocol has also been successfully employed with tobacco smoke condensates (STANTON et al. 1972; DEUTSCH-WENZEL et al. 1983). The neutral subfraction of cigarette smoke condensate in which the PAH are concentrated is the only fraction which upon repeated intratracheal instillation induces squamous tumors in the lung of rats (DAVIS et al. 1975).

The tumorigenicity of inhaled BaP has also been established in Syrian golden hamsters (THYSSEN et al. 1981). Intratracheal instillation in Syrian golden hamsters of 7,12-dimethylbenz[*a*]anthracene – a highly tumorigenic, synthetic PAH which does not occur in tobacco smoke – followed by tobacco smoke exposure leads to a high incidence of respiratory tract tumors consistent with an initiation : promotion model (KOBAYASHI et al. 1974). These findings, taken together with the results of the bioassays on mouse skin, provide strong evidence for the role of PAH as tumor initiators in tobacco-related respiratory carcinogenesis.

The levels of exposure to PAH as experienced by smokers are not inconsistent with their potential role as causative agents for respiratory tract cancer. BaP, as a representative PAH, typically induces a high incidence of tumors on mouse skin after topical application of 5 μg three times weekly for 60 weeks (HECHT et al. 1976). This corresponds to a total dose of approximately 36 mg/kg body weight. Tumors of the respiratory tract are induced upon intratracheal instillation of a single dose of 5 mg BaP on Fe_2O_3, corresponding to approximately 50 mg/kg body weight (0.2 mmol/kg), or upon chronic administration of a total dose of 7.5 mg BaP on Fe_2O_3 (SAFFIOTTI et al. 1972). A smoker who smokes 40 cigarettes per day for 40 years would be exposed to approximately 12 mg of BaP, or about 0.16 mg/kg (0.61 μmol/kg). Applying these calculations on the basis of mg/kg body weight may be too conservative for a locally acting carcinogen such as BaP.

The calculated doses delivered locally to the target tissue areas may be comparable in the animal models and human systems. These exposure estimates and the determinations of the tumorigenic potential of PAH in bioassays strongly suggest that PAH play a significant role in the induction of respiratory tract cancer in smokers.

II. *N*-Nitrosamines

Among the various carcinogenic *N*-nitrosamines that have been detected in tobacco and tobacco smoke, *N'*-nitrosonornicotine (NNN) and 4-(methylnitrosamino)-1-(3-pyridyl)-1-butanone (NNK) are consistently the most prevalent (IARC 1985a, 1986). NNK induces tumors of the lung, nasal cavity, pancreas, and liver in F344 rats, as well as tumors of the lung, trachea, and nasal cavity in Syrian golden hamsters, and lung tumors in mice. NNN gives tumors of the esophagus and nasal cavity in rats, tumors of the trachea and nasal cavity in Syrian golden hamsters, and lung tumors in mice (HOFFMANN and HECHT 1985; HECHT and HOFFMANN 1988; RIVENSON et al. 1988). The carcinogenic potency of NNK is particularly notable. In hamsters, a single dose of 1 mg (5 µmol) induced respiratory tract tumors in 6 of 20 animals (HECHT et al. 1983a). A comparative study of NNK and *N*-nitrosodimethylamine demonstrated that NNK is more tumorigenic, especially in the lung and nasal cavity (HECHT 1986a).

The organospecificity of NNK for the lung is consistent with its role in tobacco smoke-induced respiratory carcinogenesis. The lung is the main target organ for NNK administered either p.o. or s.c. to rats and hamsters (HECHT and HOFFMANN 1988; RIVENSON et al. 1988). Lung tumors have also been induced in mice after topical applications of high doses of NNK (LAVOIE et al. 1987a). It has not been tested by inhalation. In contrast to NNK, NNN seldom induces lung tumors. However, it is the most prevalent nitrosamine in tobacco smoke known to induce tumors of the esophagus (HECHT and HOFFMANN 1988).

Human exposure to NNK in tobacco smoke is consistent with its potential role as a causative agent for lung cancer. In the MS of an American nonfilter cigarette, purchased in 1986, NNK amounted to 425 ng (ADAMS et al. 1987). On the basis of 40 cigarettes per day, cumulative exposure to NNK in 40 years of smoking would be about 250 mg, or approximately 3 mg/kg (0.015 mmol/kg). A single dose of 0.05 mmol/kg of NNK induces a significant incidence of respiratory tract tumors in Syrian golden hamsters. These calculations, which ignore the probable endogenous formation of NNK (HOFFMANN et al. 1984), point to a significant risk for the smoker and strongly support the role of NNK as an important etiologic factor in lung cancer.

NNK and NNN are likely causative agents for oral cancer induced by snuff. Their levels in snuff tobaccos are typically in the range of 1–100 µg/g and are thus generally 1000 times above those of BaP. The only other carcinogens known to be present in snuff are ^{210}Po, formaldehyde, acetaldehyde, and crotonaldehyde (HOFFMANN et al. 1987). A mixture of NNK and NNN applied to the oral mucosa of rats (total dose, 1.6 mmol/kg) induced tumors at or near the site of application in 8 of 30 animals (HECHT et al. 1986b). Snuff dippers who use the most popular products presently marketed in the USA would be exposed to 67 µg/g tobacco of NNK and NNN (HOFFMANN et al. 1987). A user of 10 g of snuff per day would be exposed to 670 μg or 3.5 μmol. In 40 years of snuff dipping, total estimated exposure would be approximately 9.8 g or 130 mg/kg (about 0.7 mmol/kg). Disregarding the possible endogenous formation of tobacco-specific *N*-nitrosamines (TSNA), this dose fairly approximates the dose used in the animal bioassay, indicating that NNK and NNN are important etiologic factors for oral cancer induction by snuff.

III. Aromatic Amines

Among the aromatic amines identified in cigarette smoke, 4-aminobiphenyl and 2-naphthylamine are the most carcinogenic compounds and are recognized as human bladder carcinogens (IARC 1972, 1974a). Because their concentration in cigarette MS is relatively low, there is uncertainty about their role in human bladder cancer induced by smoking, although DOLL has postulated that they may be involved in the etiology of bladder cancer among cigarette smokers (DOLL 1971). Recent data on the levels of 4-aminobiphenyl-hemoglobin adducts in smokers support DOLL's concept. Levels of 4-aminobiphenyl-hemoglobin adducts correlated with relative risk for bladder cancer among groups of Italians who were either nonsmokers or smoked cigarettes made from bright tobacco or black burley tobacco. Use of black tobacco cigarettes was associated with the highest risk and the highest levels of adducts (BRYANT et al. 1988).

Of the known carcinogenic pyrolysis products of the amino acids, so far only 2-amino-3-methylimidazo(4,5-*f*)quinoline has been detected in trace amounts of 0.26 ng in the smoke of a Japanese filter cigarette (YAMASHITA et al. 1986).

Among the compounds identified in tobacco smoke, only the aromatic amines are associated with bladder cancer in experimental animals and humans. None of the *N*-nitrosamines which are consistently found in tobacco smoke have been shown to be bladder carcinogens in laboratory animals.

IV. Aldehydes

Studies of the chronic inhalation of formaldehyde (14 ppm) and acetaldehyde (1000–3000 ppm) have conclusively demonstrated that these compounds cause significant incidences of nasal cavity tumors in rats (IARC 1982a). Since rats are obligatory nose breathers, these results indicate that the aldehydes are contact carcinogens, which might be expected to affect the lung in humans. In these bioassays the actual exposure of the rat nasal tissues to formaldehyde and acetaldehyde is not known, and it is therefore difficult to compare the dose received to that of a smoker. However, due to the high levels of formaldehyde and acetaldehyde in cigarette smoke, a role in respiratory tract carcinogenesis may be surmised. In 40 years of smoking at a rate of 40 cigarettes/day, exposure to formaldehyde (100 μg/cigarette) and acetaldehyde (1000 μg/cigarette) would amount to about 58 g (26 mmol/kg) and 580 g (177 mmol/kg), respectively. These doses are 1000–10000-fold higher than those of the PAH and *N*-nitrosamines.

A total dose of approximately 17 mmol/kg of crotonaldehyde administered to F344 rats in the drinking water induces liver tumors (Chung et al. 1986). Although it is a relatively weak carcinogen, it occurs in cigarette MS in amounts up to 10 μg/cigarette and, consequently, could play a role in tobacco carcinogenesis. High levels of acrolein are also found in cigarette MS. While it has not been shown to be carcinogenic, its ciliatoxic effects are likely to play an indirect role in tobacco smoke-related respiratory carcinogenesis (IARC 1974b).

V. Miscellaneous Organic Compounds

Significant amounts of benzene are found in cigarette MS (up to 50 μg/cigarette). Sufficient evidence exists that this aromatic hydrocarbon causes leukemia in humans (IARC 1974b). On the basis of analytical data obtained for exhaled breath, it has been calculated that a smoker inhales about 2 mg of benzene per day while a nonsmoker inhales only 0.2 mg per day (Wallace et al. 1987). Former epidemiological studies have not demonstrated a strong association of smoking and leukemia (IARC 1982b). However, a recent prospective study among 248,000 U.S. veterans indicates that cigarette smokers have a significant increase in mortality from leukemia (Kinlen and Rogot 1988).

On the basis of levels of ethylene oxide-hemoglobin adducts found in smokers, it has recently been estimated that up to 15% of lung cancer caused by smoking could be due to endogenous formation of ethylene oxide from ethylene (Törnqvist et al. 1986). However, there is no evidence that ethylene causes tumors in animals (IARC 1979a). Cigarette smoke contains also traces of ethylene oxide (0.02 μg/cigarette; Binder and Lindner 1972), a known animal carcinogen and a probable human carcinogen (IARC 1985b).

Aza-arenes such as dibenz[*a,j*]acridine, dibenz[*a,h*]acridine, and 7*H*-dibenzo[*c,g*]carbazole are recognized as strong carcinogens (IARC 1986), but their levels in cigarette smoke are low. Quinoline, a liver carcinogen in rats (Shinohara et al. 1977) and in newborn mice (La Voie et al. 1987b), is present in cigarette smoke at a concentration of 1–2 μg/cigarette (Dong et al. 1978). Sufficient evidence exists for the carcinogenicity of acrylonitrile in animals (IARC 1979b). It induces primarily tumors of the CNS in rats. Although it is present in cigarette MS, its potential role in tobacco carcinogenesis is difficult to evaluate due to the lack of data.

Hydrazine causes tumors of the lung and liver upon oral administration to mice and rats and gives nasal tumors in rats upon inhalation (IARC 1974c). Hydrazine in cigarette smoke may originate partly from maleic hydrazide (Liu et al. 1974). Data on hydrazine levels in cigarettes marketed in 1987 are not available.

2-Nitropropane ($\simeq$1.0 μg/cigarette; Hoffmann and Rathkamp 1968) induces hepatocellular tumors in rats upon inhalation or oral exposure (IARC 1982c). Its organospecificity for liver suggests that it does not play a major role in tobacco carcinogenesis.

Ethyl carbamate (urethane) is carcinogenic to a variety of tissues including the respiratory tract of mice, rats, and hamsters (IARC 1974d). Its levels in cigarette smoke are similar to those of hydrazine (0.03 µg/cigarette; SCHMELTZ et al. 1978). Its potential role in tobacco carcinogenesis is difficult to evaluate.

Vinyl chloride, a human carcinogen, causes angiosarcoma of the liver and produces a variety of tumors in rats, mice, and hamsters (IARC 1979c). Its low levels in cigarette MS do not support a major role in tobacco carcinogenesis.

VI. Inorganic Carcinogens

Carcinogenic metals occur in both unburned tobacco and in tobacco MS (WYNDER and HOFFMANN 1967; NORMAN 1977; JENKINS et al. 1985). Evidence for carcinogenicity in humans or experimental animals exists for arsenic, nickel, chromium, cadmium, and lead.

Levels of arsenic in tobacco have decreased since 1952, when its use as a pesticide was discontinued. Arsenic levels in tobacco are between 0.5–0.9 ppm. Some 7%–18% of this amount is found in MS (US SURGEON GENERAL 1982). Arsenic is known to cause skin and lung cancer in humans, but data in laboratory animals are limited (IARC 1980a). However, a simple intratracheal instillation into rats of an arsenical mixture induces bronchiogenic carcinoma (IVANKOVIC et al. 1979).

Levels of nickel in cigarette tobacco range from 2.0–6.2 µg/cigarette. From 10% to 20% of the nickel is transferred into MS (NATIONAL RESEARCH COUNCIL 1975). It has been suggested that part of the nickel in cigarette smoke may exist as nickel carbonyl, but this was not proven experimentally (ALEXANDER et al. 1983). A variety of nickel compounds are carcinogenic in experimental animals, giving local as well as systemic tumors. Nickel subsulfide produces lung cancer in rats upon inhalation. Epidemiologic data have demonstrated that workers in nickel refineries have an excess incidence of cancer of the lung and nasal cavity (IARC 1973). Taken together, these data suggest a possible role for nickel in tobacco carcinogenesis.

Chromium is present in ppm quantities in tobacco and from 4–70 ng/cigarette in MS (IARC 1986). Increased incidences of lung cancer have been observed in workers in the chromate-producing industry. Calcium chromate is carcinogenic to rats after administration by several routes, including intrabronchial instillation. Several other chromium compounds produce local tumors (IARC 1980b).

Levels of cadmium have been determined to be 1.3–1.6 µg/g in tobacco and 41–62 ng/cigarette in tobacco smoke (PERINELLI and CARUGNO 1978). Cadmium chloride, oxide, sulfate, and sulfide cause local tumors in rats upon s.c. injection. Long-term exposure of rats to aerosols of cadmium chloride (12.5, 25, and 50 $\mu g/m^3$) produces a dose-dependent incidence of primary lung carcinomas (adenocarcinoma and squamous cell carcinoma; TAKENAKA et al. 1983). Evidence for carcinogenicity in humans is limited (IARC 1980c).

Levels of lead in tobacco range from 8 to 10 µg/g tobacco and 34 to 85 ng/cigarette in cigarette MS. Lead acetate and subacetate produce a variety of tumors in rats and mice, but evidence for human carcinogenicity of lead is considered inadequate (IARC 1980c). The possible roles of chromium, cadmium, and lead in tobacco carcinogenesis are difficult to evaluate given the present data base.

Taken together, the evidence for a major role of these materials as etiologic factors in tobacco carcinogenesis is not compelling.

Polonium-210 exists in unburned tobacco (0.2–1.2 pCi/g) and cigarette MS (0.03–1.0 pCi/cigarette; IARC 1985a, 1986). This α-particle-emitting element is strongly carcinogenic, producing tumors of the lung upon inhalation in rats and upon intratracheal instillation in Syrian golden hamsters (US SURGEON GENERAL 1982). The quantities of polonium-210 found in the lungs of smokers are generally about three times higher than those in nonsmokers. However, the significance of polonium-210 in tobacco-induced lung cancer has been questioned upon comparison of these data with those obtained in miners (HARLEY et al. 1980). In 1987, the US NATIONAL COUNCIL ON RADIATION PROTECTION AND MEASUREMENT ascribed about 1% of the risk of lung cancer after 50 years of cigarette smoking to the role of polonium-210 inhaled from the smoke (NATIONAL COUNCIL ON RADIATION PROTECTION AND MEASUREMENT 1987). The role of polonium-210 (0.2–1.2 pCi/g snuff) as a potential etiologic factor in oral cancer induction by snuff dipping requires further evaluation (HOFFMANN et al. 1987).

E. Smokeless Tobacco

Smokeless tobacco is being used in various forms throughout the world. Its composition varies depending on the regional availability of tobaccos and on local customs. In North America and in Western Europe smokeless tobaccos can be purchased as plug tobacco, loose leaf, twist tobacco, and snuff. Between 1978 and 1985, sales in the USA of plug and twist tobacco, pipe tobacco, cigars, and cigarettes decreased significantly while loose leaf tobacco sales increased by 12.6% to 32500 tons and snuff sales by 35% to 22000 tons per year (MAXWELL 1981, 1986). Of all tobacco products sold in Sweden since 1983, only snuff sales have increased (1987, 4695 tons; TOBACCO JOURNAL INTERNATIONAL 1988).

Tobacco chewing entails placing a “chaw” of loose leaf or a “quid” of plug tobacco in the gingival buccal area, where the material is held. Moderate chewers hold the tobacco up to 200 min per day; heavy chewers are known to hold each “chaw” or “quid” for longer times and to use up to eight “portions” per day.

The steep rise in snuff consumption in the USA is directly correlated with the increasing popularity of snuff dipping, especially among male adolescents and young men. In the USA, snuff dipping is practised by at least 10 million people and in Sweden by 17% of all men (IARC 1985a; US SURGEON GENERAL 1986b). Snuff dipping is the practice of placing a pinch of moist or dry snuff or a “tea bag” (sachet) containing snuff into the gingival fold. Young snuff dippers hold the tobacco 100–250 min per day and consume on the average 10 g of snuff per day (US SURGEON GENERAL 1986b; PALLADINO et al. 1986).

The natives of Iran and of the Soviet Central Asian Republics practise oral use of nass which is usually composed of local tobaccos, ash, cotton oil or sesame oil, and lime. Most nass users consume about 10–15 portions of the tobacco mixture each day (IARC 1985a; ZARIDZE et al. 1985).

Chewing of betel quid with or without tobacco is practised throughout Asia, especially in India, Pakistan, Sri Lanka, Indonesia, and Singapore. Worldwide, at least 200 million people are estimated to be betel quid chewers. Although betel quid is chewed in several different ways in various countries, its composition is relatively consistent. The mixture usually contains pieces of areca nut, catechu, and lime as major ingredients and often also tobacco. The components are wrapped in the betel leaf, often together with spices and/or flavoring agents. The betel leaf is folded over its contents, and it is placed into the mouth and chewed. Generally, the quid is chewed after meals; however, habitual users chew 15–20 quids per day (ARJUNGI 1976; IARC 1985a).

I. Epidemiology

A number of case control studies have shown the proportion of tobacco chewers among patients with cancer of the oral cavity, pharynx, and larynx to be 2–3 times higher than among controls. Although some of these studies did not separate tobacco chewers from snuff dippers, and confounding by smoking and/or alcohol consumption was not always excluded, the epidemiological data as a whole incriminate chewing of tobacco as a significant risk factor for cancer of the upper digestive tract including cancer of the esophagus (IARC 1985a; US SURGEON GENERAL 1986b; CONNOLLY et al. 1986).

Four case-control studies have implicated snuff dipping in the etiology of cancer of the oral cavity and, to a lesser extent, cancer of the pharynx. In a study conducted by WINN et al. in North Carolina and published in 1981, the relative risk of oral and pharyngeal cancer for white women who were snuff dippers was four times that of women who did not use tobacco in any form. In addition, they observed a strong dose-response relationship (IARC 1985a). The US SURGEON GENERAL and the IARC concluded their reviews on the association of snuff dipping and oral cancer by stating that oral use of snuff of the types commonly used in North America and Western Europe is carcinogenic to humans (IARC 1985a; US SURGEON GENERAL 1986b).

There is only limited epidemiologic evidence that chewing of nass alone is associated with an increased risk of cancer of the oral cavity or esophagus (IARC 1985a; ZARIDZE et al. 1985).

A large number of case-control studies, primarily from India, have clearly demonstrated that chewing of tobacco-containing betel quid is causally associated with cancer of the oral cavity, pharynx, and esophagus. The evidence is supported by two dose-response studies and a prospective study from India (HIRAYAMA 1966; GUPTA et al. 1982; IARC 1985a; US SURGEON GENERAL 1986b). Evidence for a role of tobacco-free betel quids in the etiology of human cancer is less clear since most studies have not separated the habits of chewing these quids from smoking and from chewing quids with tobacco (GUPTA et al. 1982; US SURGEON GENERAL 1986b).

II. Bioassays

Gavage feeding of an alcohol extract of tobacco induces tumors of the lung and of the liver in mice (IARC 1985a). Swabbing of the oral cavity of mice, rats, and hamsters with extracts of chewing tobacco did not lead to a significant number of tumors of the oral cavity in any of these animals (IARC 1985a; US SURGEON GENERAL 1986b).

Most bioassays with snuff and with snuff extracts have failed to elicit significant tumor response in laboratory animals (IARC 1985a; US SURGEON GENERAL 1986b; HECHT et al. 1986b). Insertion of snuff into a surgically created lip canal of rats has induced small, but statistically insignificant, numbers of tumors in the lip canal and oral cavity (HIRSCH and JOHANNSON 1983; HECHT et al. 1986b). However, application of snuff extract to the cheek pouches of hamsters, repeatedly infected with *herpes simplex*, results in a high percentage of invasive squamous cell carcinoma in the animals (PARK et al. 1986).

Treatment of the oral cavity of hamsters with tobacco-containing betel quid, tobacco-free betel quid, areca nut with tobacco, as well as extracts thereof has led to significant numbers of benign and malignant tumors of the cheek pouch and/or the forestomach (RANADIVE et al. 1976).

III. Carcinogens

The composition of processed, unadulterated tobacco has been discussed in Sect. B. In addition, various flavor additives are found in chewing tobacco, snuff, and betel quid preparations (NAIR et al. 1986a; LAVOIE et al. 1989).

It is of special significance that the preparation of smokeless tobacco products, which entails curing, fermenting, and aging, leads to the formation of TSNA from nicotine and other tobacco alkaloids such as nornicotine, anatabine, and anabasine (Fig. 5). So far, six TSNA have been identified in smokeless tobacco. These are the carcinogens NNN, NNK, *N'*-nitrosoanabasine (NAB)

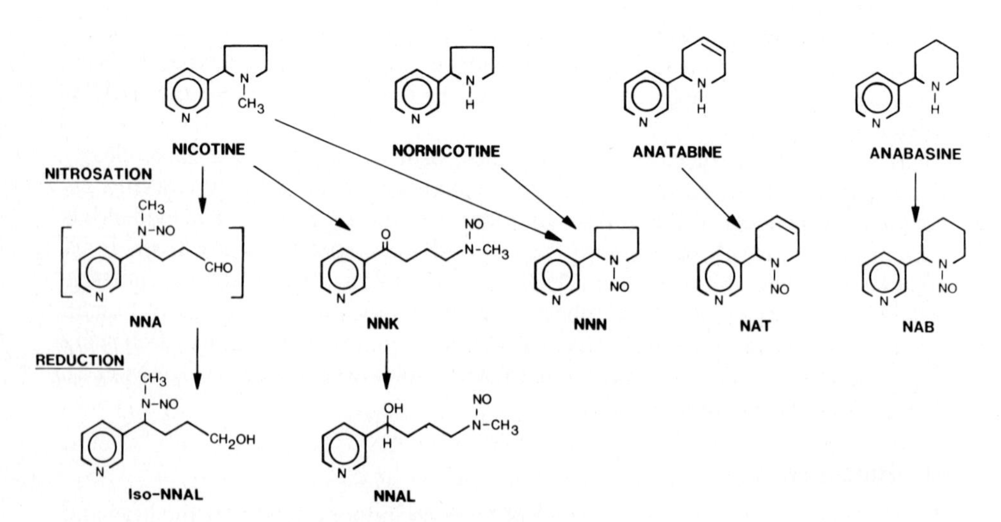

Fig. 5. Formation of tobacco-specific *N*-nitrosamines

Table 5. Tobacco-specific *N*-nitrosamines in smokeless tobacco (ppb)[a]

Product	NNN	NNK	NAT	NAB
USA				
Loose leaf	670– 8 200(6)	380(1)	2 300(1)	140(1)
Plug tobacco	3 400– 4 300(3)			
Snuff – moist	3 120–135 000(26)	100–13 600(25)	1 340–339 000(20)	10– 6 700(16)
Snuff – dry	9 000– 52 000(3)	1 800–13 000(3)	18 000– 38 000(3)	600–60 000(3)
Sweden				
Snuff	2 260–154 000(18)	870– 2 950(18)	1 840–21 400(18)	130–150(3)
Plug tobacco	2 090(1)	240(1)	1 580(1)	100(1)
Canada				
Snuff	50 400– 79 000(2)	3 200– 5 800(2)	152 000–170 000(2)	4 000– 4 800(2)
Denmark				
Snuff	4 460– 8 000(3)	1 350– 7 030(3)	2 680– 6 170(3)	
Germany				
Plug tobacco	1 420– 2 130(2)	30– 40(2)	330– 500(2)	30– 50(2)
Snuff	6 080– 6 700(2)	1 500– 1 540(2)	3 920– 4 370(2)	
United Kingdom				
Snuff – moist	11 800(1)	1 820(1)	3 020(1)	86(1)
USSR				
Nass	120– 520(4)	20– 130(4)	32– 300(4)	8– 30(4)
India				
Chewing tobacco	470– 2 400(5)	130– 230(4)	300– 450(4)	30– 70(4)
Belgium				
Chewing tobacco	7 380(1)	130(1)	970(1)	

[a] Number in parentheses, number of samples analyzed.
NNN, *N*′-nitrosonornicotine; NNK, 4-(methylnitrosamino)-1-(3-pyridyl)-1-butanone; NAT, *N*′-nitrosoanatabine; NAB, *N*′-nitrosoanabasine.

and 4-(methylnitrosamino)-1-(3-pyridyl)-1-butanol (NNAL), the noncarcinogenic *N*′-nitrosoanatabine (NAT), and 4-(methylnitrosamino)-4-(3-pyridyl)-1-butanol, (iso-NNAL), which is now being bioassayed. Swabbing of the oral cavity of rats with a solution containing a mixture of low doses of NNN and NNK induces a significant number of tumors of the mouth and lung, indicating that these TSNA are active contact carcinogens as well as organ-specific systemic carcinogens (Hecht et al. 1986 b).

Table 5 lists the currently available quantitative analytical data on TSNA in various smokeless tobacco products (Brunnemann et al. 1986; Hoffmann et al. 1987, 1988). The concentrations of TSNA in smokeless tobacco exceed by far the permissible limits of 5 ppb set by US governmental agencies for individual *N*-nitrosamines in beer and bacon (US FDA 1980; USDA 1983). Several studies have shown that the tobacco chewer and snuff dipper do indeed extract TSNA from the tobacco (Hoffmann and Adams 1981; Palladino et al. 1986; Nair et al. 1986 b). It is also very likely that during chewing and snuff dipping additional amounts of TSNA are formed endogenously from the tobacco alkaloids and nitrosating agents (Hoffmann and Hecht 1985; Nair et al. 1986 b).

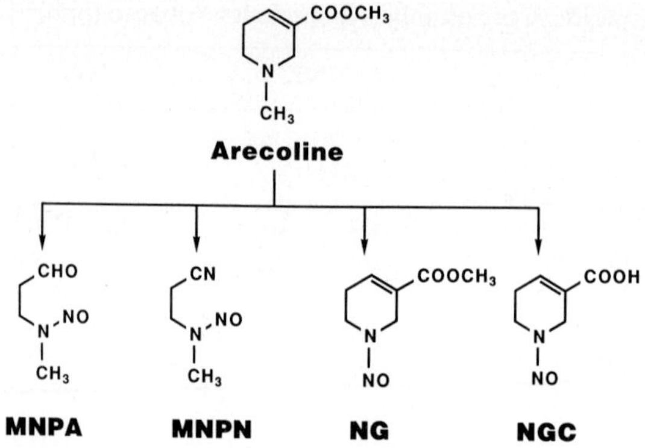

Fig. 6. Formation of *N*-nitrosamines from arecoline

Other carcinogens also identified in smokeless tobacco are volatile *N*-nitrosamines (*N*-nitrosodimethylamine up to 4 ppb in chewing tobacco and up to 215 ppb in snuff), *N*-nitrosomorpholine (up to 40 ppb in snuff), *N*-nitrosodiethanolamine (up to 680 ppb in chewing tobacco and up to 6800 ppb in snuff), formaldehyde (up to 7.4 ppm in snuff), acetaldehyde (up to 7.4 ppm in snuff), crotonaldehyde (up to 2.4 ppm in snuff), benzo[*a*]pyrene (up to 80 ppb in chewing tobacco and 90 ppb in snuff), as well as traces of the radioelement polonium-210 (up to 0.6 pCi/g in snuff) (IARC 1985a; Nair et al. 1986b; Hoffmann et al. 1987; LaVoie et al. 1989).

Chewing of betel quid with tobacco leads not only to the formation of TSNA but also to the formation of arecoline-derived nitrosamines, namely *N*-nitrosoguvacine (NGC) and *N*-nitrosoguvacoline (NG), and the highly carcinogenic 3-(methylnitrosamino)propionitrile (MNPN) (Fig. 6; Brunnemann et al. 1986; Nair et al. 1986b; Prokopczyk et al. 1987).

The chemical and biochemical assays described in the foregoing strongly support the human data which incriminate smokeless tobacco as a human carcinogen. This applies especially to oral snuff and to betel quid with tobacco.

A major obstacle to an appropriate bioassay of chewing tobacco and snuff is the difficulty of repeating the administration of fresh snuff several times each day, assuring its retention in the oral cavity over several hours, and conducting such an assay over a period of up to 2 years. At this time, the surgically created lip canal, or a modification thereof, in the rat which allows for longer retention of the snuff product, appears to be the most approximate simulation of the human habit. A lifetime assay with this model is expected to lead to a significant tumor incidence in the oral cavity of rats.

F. Environmental Tobacco Smoke

Epidemiological studies have incriminated environmental smoke exposure as a risk factor for lung cancer in nonsmokers (IARC 1986; US NATIONAL RESEARCH COUNCIL 1986; US SURGEON GENERAL 1986a). In fact, it has been estimated that nonsmokers living with smoking spouses have an about 30% higher risk for lung cancer than nonsmokers with to nonsmoking spouses (US NATIONAL RESEARCH COUNCIL 1986). However, this conclusion has been challenged by several investigators (IARC 1986; US NATIONAL RESEARCH COUNCIL 1986; LEE 1987). In 1985 IARC considered the available epidemiologic data as inconclusive (IARC 1986). Nevertheless, a biological basis for an association between environmental smoke exposure and lung cancer clearly exists. Compounds resulting from the combustion of tobacco, which are known carcinogens, are inhaled as pollutants of ambient air and are retained by the nonsmoker. IARC concluded that because of its physicochemical nature and in view of known concepts of tobacco carcinogenesis, "passive smoking gives rise to some risk of cancer" (IARC 1986). This judgement emphasizes that epidemiologic methods may not be sufficiently sensitive for establishing a risk factor for cancer from environmental smoke exposure. Therefore, highly sensitive chemical and biochemical methods are needed to assay this exposure and the uptake of tumorigenic agents by nonsmokers.

The smoke generated during smouldering of tobacco products between puff drawing is SS. When it is obtained under standardized laboratory conditions, undiluted SS contains far higher concentrations of toxic and tumorigenic agents than MS, which is drawn puff by puff. Table 6 presents data for those agents in SS that are known to be carcinogens, tumor promoters, or cocarcinogens. The release of volatile *N*-nitrosamines and aromatic amines into SS is remarkably high (IARC 1985a; US NATIONAL RESEARCH COUNCIL 1986; US SURGEON GENERAL 1986a; HOFFMANN and WYNDER 1986; GUERIN 1987). Whereas filter tips, especially perforated filter tips, can significantly reduce the concentration of toxic and tumorigenic agents in MS (see Sect. C), they have no reducing effect on the agents released in SS (ADAMS et al. 1987). Thus, it does not matter whether the source for environmental tobacco smoke is the SS of a nonfilter cigarette, or the SS of a cigarette with a highly active filter tip. SS appears to be slightly more genotoxic in the Ames test than MS (LEWTAS et al. 1987), and on a gram to gram basis the particulate matter of SS is more carcinogenic on mouse skin than the particulate matter of MS (WYNDER and HOFFMANN 1967).

SS represents the major source for environmental tobacco smoke; the smoke diffusing through the cigarette paper, that escaping from the burning cone during active smoking, and that portion of MS which is exhaled are other contributors. Table 7 presents some data for toxic agents in indoor environments (IARC 1986; US NATIONAL RESEARCH COUNCIL 1986; US SURGEON GENERAL 1986a). The concentration of toxic agents in environmental tobacco smoke appears low by comparison with their levels in undiluted MS, but one needs to take into consideration that the active inhalation of tobacco smoke is limited to the time it takes to smoke each cigarette, while the involuntary inhalation of environmental tobacco smoke can occur over several hours each day. This is reflected in

Table 6. Some toxic and tumorigenic agents in undiluted cigarette sidestream smoke

Compound	Type of toxicity	Amount in sidestream smoke per cigarette	Ratio of sidestream: mainstream smoke
Vapor phase			
Carbon monoxide	T	26.8 – 61 mg	2-5 – 14.9
Carbonyl sulfide	T	2 – 3 μg	0.03– 0.13
Benzene	C	240 – 490 μg	8 – 10
Formaldehyde	C	1500 μg	50
3-Vinylpyridine	SC	330 – 450 μg	24 – 34
Hydrogen cyanide	T	14 – 110 μg	0.06– 0.4
Hydrazine	C	90 ng	3
Nitrogen oxides (NO_x)	T	500 –2000 μg	3.7 – 12.8
N-Nitrosodimethylamine	C	200 –1040 ng	20 –130
N-Nitrosopyrrolidine	C	30 – 390 ng	6 –120
Particulate phase			
Tar	C	14 – 30 mg	1.1 – 15.7
Nicotine	T	2.1 – 46 mg	1.3 – 21
Phenol	TP	70 – 250 μg	1.3 – 3.0
Catechol	CoC	58 – 290 μg	0.67– 12.8
o-Toluidine	C	3 μg	18.7
2-Naphthylamine	C	70 ng	39
4-Aminobiphenyl	C	140 ng	31
Benz[*a*]anthracene	C	40 – 200 ng	2 – 4
Benzo[*a*]pyrene	C	40 – 70 ng	2.5 – 20
Quinoline	C	15 – 20 μg	8 – 11
NNN	C	0.15– 1.7 μg	0.5 – 5.0
NNK	C	0.2 – 1.4 μg	1.0 – 22
N-Nitrosodiethanolamine	C	43 ng	1.2
Cadmium	C	0.72 μg	7.2
Nickel	C	0.2 – 2.5 μg	13 – 30
Polonium-210	C	0.5 – 1.6 pCi	1.06– 3.7

C, carcinogenic; CoC, cocarcinogenic; SC, suspected carcinogen; T, toxic; TP, tumor promoter.

comparative measurements of the uptake of nicotine by active and passive smokers. In blood serum and in the urine of passive smokers, the concentration of cotinine, a major metabolite of nicotine, amounts to about 1% of its concentration in the physiologic fluids of an active cigarette smoker (GREENBERG et al. 1984; US NATIONAL RESEARCH COUNCIL 1986; US SURGEON GENERAL 1986a; RUSSELL 1987; SEPKOVIC et al. 1988). Other indicators for the uptake of environmental tobacco smoke constituents such as carboxyhemoglobin, thiocyanate, and 4-aminobiphenyl adducts are not significantly elevated in physiologic fluids of exposed individuals, primarily because levels of these pollutants are low and can be derived from sources other than tobacco combustion (US NATIONAL RESEARCH COUNCIL 1986; US SURGEON GENERAL 1986a). There is a great need for other biological markers as indicators of the uptake of carcinogenic agents from environmental tobacco smoke by nonsmokers. The determination of ad-

Table 7. Some toxic and tumorigenic agents in indoor environments polluted by tobacco smoke[a]

Pollutant	Location	Concentration/m³
Nitric oxide	Workrooms	50 – 440 μg
	Restaurants	17 – 270 μg
	Bar	80 – 520 μg
	Cafeteria	2.5– 48 μg
Nitrogen dioxide	Workrooms	68 – 410 μg
	Restaurants	40 – 190 μg
	Bar	2 – 116 μg
	Cafeteria	67 – 200 μg
Hydrogen cyanide	Living room	8 – 122 μg
Benzene	Public places	20 – 317 μg
Formaldehyde	Living room	23 – 50 μg
Acrolein	Public places	30 – 120 μg
Acetone	Public places	360 –5 800 μg
Phenols (volatile)	Coffee houses	7.4– 11.5 ng
N-Nitrosodimethylamine	Restaurants, public places	0 – 240 ng
N-Nitrosodiethylamine	Restaurants, public places	0 – 200 ng
Nicotine	Public places	1 – 6 μg
	Restaurants	3 – 10 μg
	Workrooms	1 – 13.8 μg
Benzo[*a*]pyrene	Restaurants, public places	3.3– 23.4 ng

[a] References: KLUS and KUHN (1982); IARC (1986); US NATIONAL RESEARCH COUNCIL (1986); KLUS et al. (1987).

ducts of NNN and NNK with globin appears to be one promising approach to this problem of dosimetry (CARMELLA and HECHT 1987).

G. Recent Studies on Mechanisms of Tobacco Carcinogenesis and Their Application to Dosimetry

By understanding the mechanisms involved in cancer causation by tobacco, tobacco smoke, its subfractions, and constituents, one can develop a rational hypothesis applicable to cancer prevention not only in tobacco users but also in the general population as a whole. The epidemiology of tobacco use provides leads for mechanistic studies relevant to carcinogenesis in several tissues. The mechanistic studies, in turn, provide insights for preventive approaches.

One important area that emerges from an understanding of metabolic activation and detoxification of tobacco smoke constituents is human dosimetry. Although measurements of carboxyhemoglobin, thiocyanate, nicotine and cotinine are objective indicators of an individual's uptake of tobacco smoke, and measurements of parameters such as urinary mutagenicity and sister chromatid exchanges in peripheral blood lymphocytes provide an indication of biological response to tobacco smoke, these measures do not, in themselves, delineate the individual's response to these specific environmental carcinogens (IARC 1986). That information is best obtained by assessing levels of macromolecular adducts with carcinogens or metabolites of carcinogens. Development of such assays is based on examining the mechanisms of metabolic activation and detoxification of tobacco smoke carcinogens. We will briefly summarize some recent studies on the mechanisms of tobacco carcinogenesis, with emphasis on those components that, according to our present knowledge, appear to play major roles in cancer induction – PAH, tobacco-specific *N*-nitrosamines, and aromatic amines – and we will note the applications of these studies to human dosimetry where appropriate.

I. Polynuclear Aromatic Hydrocarbons

The mechanisms by which PAH interact with DNA, activate oncogenes, and initiate the carcinogenic process are described in detail in other chapters in this book. These studies have shown that diol epoxides with one carbon terminus of the epoxide ring in the bay region (bay region diol epoxides) such as (+)7a,8β-dihydroxy-9β,10β-epoxy-7,8,9,10-tetrahydrobenzo[*a*]pyrene [(+)-*anti*-BPDE] are major ultimate carcinogens of several of the carcinogenic PAH which occur in tobacco smoke. Similar mechanisms of activation are seen with carcinogenic methylated PAH in tobacco smoke, such as 5-methylchrysene, with the additional requirement that highly tumorigenic bay region diol epoxides have a methyl group and epoxide ring in the same bay region (HECHT et al. 1986c, 1987). These studies are important with respect to tobacco carcinogenesis because they provide a rationale for the high tumor-initiating activity on mouse skin of the PAH-enriched subfractions of tobacco smoke condensate (HOFFMANN and WYNDER 1971). However, an apparent contradiction exists in ascribing the mouse skin tumorigenicity of PAH such as benzo[*a*]pyrene (BaP) to the formation of *anti*-BPDE as an ultimate carcinogen. Assays of *anti*-BPDE on mouse skin consistently show that it is less tumorigenic than BaP. To investigate factors responsible for this apparent contradiction, the disposition, metabolism, and DNA binding in mouse epidermis of *anti*-BPDE and BaP were compared (MELIKIAN et al. 1987). The results indicate that there are remarkable differences in the penetration of *anti*-BPDE and BaP through the epidermis. Whereas BaP removal from the epidermis is slow, 60%–65% of *anti*-BPDE disappears within 3 min of application, and a second, slower phase of removal is observed between 8 min and 2 h after application. During this second phase, *anti*-BPDE is apparently protected from hydrolysis and from reaction with DNA, such that formation of adducts is more efficient from BaP than from *anti*-BPDE. Thus, the disposition and reactivity of *anti*-BPDE that is topically applied to mouse skin differs from that observed with the diol epoxide that is generated intracellularly from topically applied BaP. This difference may account for the relatively low tumorigenicity of *anti*-BPDE on mouse skin.

An important aspect of the role of PAH in tobacco carcinogenesis is their tumor-initiating or cocarcinogenic effect in co-application with the weakly acidic fraction of tobacco smoke condensate (HOFFMANN et al. 1978). Extensive fractionation studies have clearly shown that the acidic fraction has both promoting and cocarcinogenic activities and that the majority of the tumorigenic activity of tobacco smoke condensate, as measured on mouse skin, can be ascribed to the combined effects of PAH, weakly acidic cocarcinogens, and tumor promoters (HOFFMANN et al. 1978). The cocarcinogenic activity of tobacco smoke condensate on mouse skin can largely be attributed to catechol (VAN DUUREN and GOLDSCHMIDT 1976; HECHT et al. 1981). Recent investigations have focused on the mechanism underlying this observation (MELIKIAN et al. 1986). According to these studies catechol has several major effects on BaP metabolism in mouse epidermis. Important among these is that the ratio of *anti*- to *syn*-BPDE-DNA adducts in mouse epidermis increases in the presence of catechol. Thus, the major effect of catechol as a cocarcinogen is exerted during the terminal activation of (−)BaP-7,8-diol to (+)*anti*-BPDE.

Although subfractions of the weakly acidic fraction of tobacco smoke condensate are known to contain tumor promoters, and a number of components of these subfractions have been identified, none has shown significant activity (HECHT et al. 1975). The characterization of the tumor promoters in tobacco smoke is a continuing challenge. Studies of the effects of cigarette smoke condensate subfractions on normal human bronchial epithelial cells have led to the conclusion that the methanol-extracted neutral fraction is likely to contain compounds with promoting activities similar to those of the phorbol esters, indole alkaloids, and certain polyacetates, since this fraction affects growth, morphology, epidermal growth factor binding, and plasminogen activator activity, and causes single-strand breaks similar to those induced by such promoters (WILLEY et al. 1987).

Antibodies developed against the major BPDE-DNA adduct have been used to assess its presence in surgical specimens of lung tissue, in human placenta, and in peripheral blood lymphocytes (PERERA et al. 1982; HARRIS et al. 1985; EVERSON et al. 1986). Although evidence has been obtained for the presence of such adducts in samples from smokers, significant differences between smokers and nonsmokers have not been observed. These analyses are complicated by the fact that the antibodies crossreact with DNA adducts formed from PAH other than BaP, and by uncertainties in quantitation introduced by the method of raising the antibodies (see Chap. 13).

II. Tobacco-Specific *N*-Nitrosamines

Pathways of NNK and NNN metabolism have been investigated in vivo in laboratory animals and in vitro in subcellular fractions, cultured cells, and cultured tissues from animals and humans (HECHT et al. 1983b; IARC 1985a). These studies have identified a variety of metabolic transformations, among which α-hydroxylation appears to be the most important, leading to the formation of DNA adducts and protein adducts. Figure 7 summarizes these pathways for NNK and NNN. The fact that formation of 7-methylguanine, O^6-methylguanine, and O^4-methylthymidine occurs in the lung, liver, and nasal mucosa of F344 rats treated with NNK but not in the esophagus, spleen, kidney, and brain has been noted in several studies (CASTONGUAY et al. 1985; HECHT et al. 1986a; BELINSKY et al. 1986). It is remarkable and perhaps significant that DNA methylation has been consistently detected only in target tissues of NNK-treated rats. The organospecificity of NNK for the lung of rats, mice, and hamsters has been noted. Studies of O^6-methylguanine levels in lung DNA during chronic dosing with NNK have shown that this promutagenic base accumulates and persists, in part due to inhibition by high doses of NNK of the repair enzyme, O^6-methylguanine-DNA methyl-transferase (BELINSKY et al. 1986, 1987). It is also significant that the repair enzyme is inhibited by acrolein and other aldehydes which occur in high concentrations in cigarette smoke (KROKAN et al. 1985). The efficiency of O^6-methylguanine formation was particularly high in Clara cells. These studies with NNK, which create a mechanistic link between nicotine exposure and the formation of promutagenic DNA adducts, suggest that smokers should have methylated DNA. This has recently been examined using a monoclonal antibody against O^6-methyldeoxyguanosine. The adduct was detected in 6 of 20 samples of human placental DNA, but no relationship to smoking was observed (FOILES et al. 1988).

A second pathway of NNK α-hydroxylation has been observed, yielding the intermediate 4-(3-pyridyl)-4-oxobutyl diazohydroxide (compound 7 of Fig. 7). This intermediate is also formed by 2'-hydroxylation of NNN. The formation of the keto alcohol (compound 9) upon neutral thermal or acid hydrolysis of DNA from rats treated with NNK or NNN has been demonstrated and is consistent with 4-(3-pyridyl)-4-oxobutylation of DNA by compound 7 (HECHT et al. 1988). Although the biological significance of such adducts is presently unknown, they would appear to have potential for dosimetry studies because of their unique structural relationship to nicotine. Thus, it is significant that 4-(3-pyridyl)-4-oxobutylation of hemoglobin is observed in rats treated with NNK or NNN, as indicated by the release of compound 9 upon base or acid hydrolysis (CARMELLA and HECHT 1987). The development of sensitive analytical methods for compound 9, as released from globin or DNA, is likely to provide an approach to human dosimetry of TSNA.

III. Aromatic Amines

4-Aminobiphenyl and 2-naphthylamine are the most likely cigarette smoke components to be involved in bladder cancer induction in smokers, according to presently available data. The mechanisms by which these compounds are metabolically activated and produce DNA adducts in the bladder epithelium have been extensively studied and are discussed elsewhere (BELAND and KADLUBAR, this volume). These studies have shown that the corresponding hydroxylamines are key intermediates in DNA and protein modification. The hydroxylamines also react with hemoglobin to form, in the case of 4-aminobiphenyl, a sulfonic acid amide of β-cysteine (GREEN et al. 1984; NEUMANN 1984; BRYANT et al. 1987). This adduct readily releases 4-aminobiphenyl upon treatment with dilute acid. A method was developed to analyze the released 4-aminobiphenyl by gas chromatography with detection by negative ion chemical ionization mass spectrometry (BRYANT et al. 1987). Application of this method to smokers shows that adduct levels are higher than in nonsmokers and decrease after quitting. This method should be useful in further assessing the role of aromatic amines in bladder cancer induction by tobacco smoke.

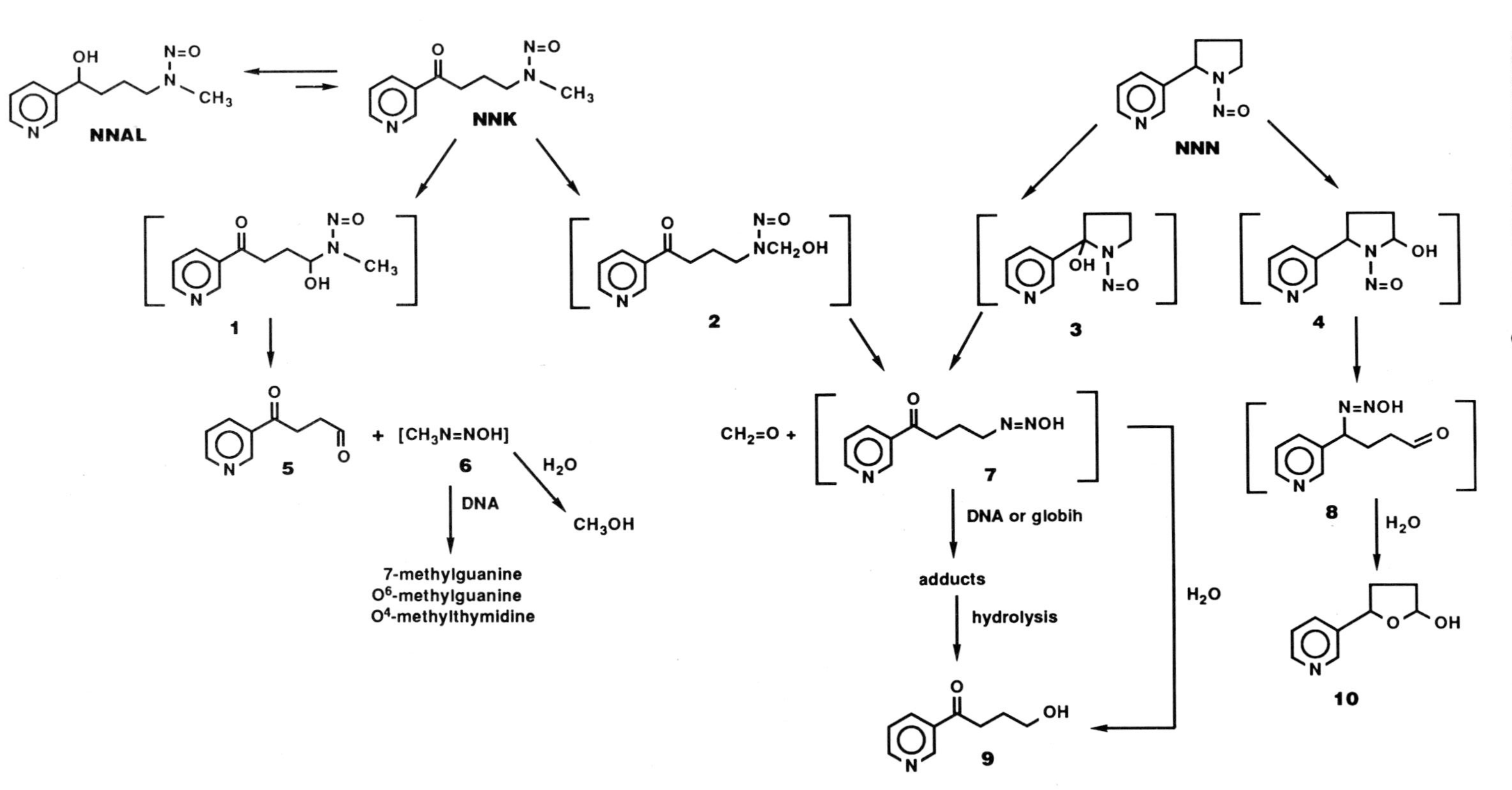

Fig. 7. Metabolic activation of 4-(methylnitrosamino)-1-(3-pyridyl)-1-butanone (*NNK*) and *N'*-nitrosonornicotine (*NNN*) to hypothetical intermediates (shown in *brackets*) which bind to DNA and protein. Structures of aldehydes and alcohols resulting from decomposition or reaction with H_2O of the intermediates are also indicated. Further metabolic transformation of these compounds as well as other metabolic pathways of NNK and NNN are summarized in references: IARC (1985, 1986); Carmella and Hecht (1987). *NNAL*, 4-(methylnitrosamino)-1-(3-pyridyl)-1-butanol

Recent studies have also shown that single ring aromatic amines, including the weak bladder carcinogen *o*-toluidine, are present in human urine (EL-BAYOUMY et al. 1986). The available data do not indicate that there are significant differences between smokers and nonsmokers.

IV. DNA Damage Induced by Unknown Constituents of Tobacco Smoke

The ^{32}P-postlabeling assay for DNA adducts, developed by RANDERATH and coworkers, is described in detail in the chapter in this volume by D. PHILLIPS. This exceedingly sensitive assay has been applied to DNA isolated from the placentas of smoking and nonsmoking mothers and has shown the presence of tobacco smoke-related DNA adducts (EVERSON et al. 1986). The occurrence of such adducts has also been observed in DNA from the bronchus and larynx of smokers (RANDERATH et al. 1986). DNA from mice treated with cigarette smoke condensate shows the presence of a number of adducts, one of which appears to be chromatographically indistinguishable from the major smoking-related adduct seen in human DNA samples (RANDERATH et al. 1986). Although the source of this adduct is not known, it does not appear to be derived from BaP or several other PAH, nor from tobacco-specific *N*-nitrosamines or aromatic amines. This suggests that there are other agents in tobacco smoke leading to DNA damage.

Cigarette smoke generates active oxygen species such as hydrogen peroxide, superoxide radical, and hydroxyl radical (NAKAYAMA et al. 1984; CHURCH and PRYOR 1985; COSGROVE et al. 1985). These agents can induce single-strand breaks in human cells or cell-free systems (BORISH et al. 1985; NAKAYAMA et al. 1986). Cigarette smoke trapped in phosphate buffered saline induces single-strand breaks ten times more efficiently than the hydrogen peroxide-generated ones, indicating that nonvolatile constituents of smoke can also cause single-strand breaks (KAO-SHAN et al. 1987). The identities of these agents are not known nor is there strong evidence that such intermediates are involved in carcinogenesis.

Most investigations on the prevalence of chromosomal aberrations in peripheral blood lymphocytes of cigarette smokers and nonsmokers have found increased levels in smokers (IARC 1986). Similarly, the overall evidence indicates that frequencies of sister chromatid exchanges are higher in smokers than in nonsmokers (IARC 1986). Compared with nonsmokers, smokers were also found to have a higher level of fragile sites, an increased number of metaphases with extensive breakage, and elevated expression of fragile sites at cancer break-points and oncogene sites in chromosomal DNA of peripheral blood lymphocytes (KAO-SHAN et al. 1987). The components of tobacco smoke which are responsible for such changes are not known.

H. Perspectives

The Centers for Disease Control have calculated that 84400 cancer deaths in men and 35900 cancer deaths in women in the USA in 1984 were directly attributable to tobacco smoking (CENTERS FOR DISEASE CONTROL 1987). In other industrialized countries, especially in Europe, the deaths from smoking-related cancers have reached comparable proportions (IARC 1986). This fact alone places continued emphasis not only on smoking withdrawal clinics and health education but also calls for further research into tobacco carcinogenesis. As we see it, high priorities should be assigned to research programs in the areas discussed in the following paragraphs.

I. Inhalation Bioassays

The high toxicity of whole smoke, primarily due to CO, presents a major obstacle for the successful induction of lung carcinoma and lung adenocarcinoma in rats and hamsters in inhalation assays with tobacco smoke. However, the use of markers of deposition of smoke particulates, such as decachlorobiphenyl (HOFFMANN et al. 1979), should lead to the design of inhalation devices and smoking cycles for long-term studies which will enable a higher degree of deposition of tar in the respiratory tracts of the animals. This in turn should elicit a significant tumor response. Reports from recent inhalation bioassays do, in fact, show progress in methodology (HAYES et al. 1988). Improved methodologies and specially designed experimental cigarettes will assist in evaluating the contribution to the carcinogenicity of whole smoke by such agents as PAH, catechols, tobacco-specific *N*-nitrosamines, and volatile aldehydes.

II. Flavor Additives

The increased market share of low yield cigarettes in many countries has led to the design of cigarette filters which deliver a rich "flavor bouquet". This has been achieved partially by selecting tobacco varieties that are rich in flavor and/or by adding natural or synthetic flavoring agents to the cigarette tobaccos. However, the possible contribution of such flavoring materials to the overall toxicity and tumorigenicity of the smoke of low yield cigarettes is not known at this time. Knowledge of the structure of the flavor compounds and their breakdown products in the smoke and the utilization of new inhalation methodologies should allow estimation of their possible contribution to the overall toxicity and carcinogenicity of low yield cigarettes. Presently available in vitro assays and in vivo bioassays other than inhalation studies can be used as screening methods and could provide some meaningful information.

III. Bioassays with Smokeless Tobacco

As was discussed earlier in this review, there is an urgent need for new bioassay methods which will lead to the induction of oral tumors in laboratory animals with smokeless tobacco products. The evaluation of snuff and betel quid tobacco mixtures is particularly important. Such bioassays should define the contribution of tobacco-specific *N*-nitrosamines to the carcinogenicity of smokeless tobaccos and of *Areca*-derived *N*-nitrosamines to betel quid with tobacco mixtures as well as the extent of endogenous formation of both types of *N*-nitrosamines during chewing and snuff dipping. The chemical nature of the tumor initiators in smokeless tobacco and the tumorigenic effects of the high amounts of flavoring agents added to these tobacco products also need to be determined.

IV. Nutrition and Tobacco Carcinogenesis

A number of laboratory studies support the concept that certain types of cancer are greatly influenced by macro- and micronutrients (REDDY et al. 1980; HAYASHI et al. 1986; IP et al. 1986). Epidemiological data, supported by laboratory studies, have indicated that nutrition can also play a role in tobacco carcinogenesis (ARMSTRONG and DOLL 1975; WYNDER et al. 1987; BIRT and POUR 1983; BEEMS and VAN BEEK 1984). A systematic follow-up of these leads by further laboratory studies is needed.

V. Tobacco Smoke and Indoor Radon Levels

Several studies are envisioned to answer the question whether there is any synergism between the carcinogenic effects of active smoking and/or involuntary smoking and elevated indoor levels of radon (US NATIONAL RESEARCH COUNCIL 1986). Case control studies on lung cancer should be combined with determination of the background α-radiation in the homes of lung cancer patients and with measurements of tobacco smoke exposure by specific markers of uptake. Inhalation assays with concurrent exposure to radon and tobacco smoke lead to a better understanding of the mechanisms involved in the suspected synergistic effects of radon exposure and active and/or passive smoking.

VI. Biochemistry of Tobacco Carcinogenesis

In Sect. G we discussed some recent developments in the metabolic activation and macromolecular binding of tobacco carcinogens. It is expected that at least some of the chemical lesions formed between metabolites of TSNA or *Areca*-derived carcinogens and DNA will be identified. We also need to determine how DNA alkylation by tobacco carcinogens is affected by smoke constituents such as catechols and other tobacco cocarcinogens and tumor promoters. Do such agents enhance DNA binding and/or inhibit the repair of chemical lesions in DNA?

Currently, highly sensitive methods are being developed for assaying the uptake of carcinogenic agents from tobacco smoke and from environmental tobacco smoke. These types of studies should also lead to an estimate of the endogenous formation of carcinogens in smokers or in environmentally exposed persons. It is known that endogenous nitrosation reactions in cigarette smokers lead to higher urinary excretion of nitrosoproline and nitrosothioproline compared with nonsmokers (HOFFMANN and HARRIS 1986). However, little is known about the possible endogenous formation of TSNA, ethylene oxide, and other carcinogens from specific precursors upon smoke inhalation.

It has been hypothesized that smokers detoxify tobacco carcinogens in a more efficient manner than do nonsmokers (REMMER 1987). In this regard, we need methods which will enable us to determine rates of metabolic activation and DNA binding of tobacco carcinogens in humans as a function of nutritional factors, age, and other characteristics which may modify the susceptibility to tobacco carcinogens. It is hoped that our increased awareness of the nature and biochemical fate of major tobacco carcinogens will provide answers to at least some of the questions concerning the genotoxic effects of tobacco smoke exposure in smokers as well as in nonsmokers with prolonged heavy exposure to environmental tobacco smoke.

Acknowledgements. The inspiration and support given to us by Ernst L. Wynder and the extensive contributions of our colleagues John D. Adams, Shantu Amin, Klaus D. Brunnemann, Steven G. Carmella, Fung-Lung Chung, Karam El-Bayoumy, Peter Foiles, Edmond J. LaVoie, Assieh A. Melikian, Bogdan Prokopczyk, Abraham Rivenson, and Neil Trushin are gratefully acknowledged. We thank Ilse Hoffmann, and Bertha Stadler for editorial assistance. Our studies are supported by grants no. CA-29580, CA-32391 and CA-44377 from the National Cancer Institute.

References

Adams JD, O'Mara-Adams KJ, Hoffmann D (1987) Toxic and carcinogenic agents in undiluted mainstream smoke and sidestream smoke of different types of cigarettes. Carcinogenesis 8:729–731

Alexander AJ, Goggin PL, Cooke MA (1983) A Fourier-transform infrared spectrometric study of the pyrosynthesis of nickel tetracarbonyl and iron pentacarbonyl by combustion of tobacco. Anal Chim Acta 151:1–12

Arjungi KN (1976) Areca nut: a review. Arzneimittel Forsch 26:951–956

Armitage AK, Turner DM (1970) Absorption of nicotine in cigarette and cigar smoke through the oral mucosa. Nature 226:1231–1232

Armstrong B, Doll R (1975) Environmental factors and cancer incidence and mortality in different countries, with special reference to dietary practices. Int J Cancer 15:617–631

Beems RB, Van Beek L (1984) Modifying effect of dietary fat on benzo(a)pyrene-induced respiratory tract tumors in hamsters. Carcinogenesis 5:413–417

Belinsky SA, White CM, Boucheron GA, Richardson FC, Swenberg JA, Anderson M (1986) Accumulation and persistence of DNA adducts in respiratory tissue of rats following multiple administrations of the tobacco specific carcinogen 4-(*N*-methyl-*N*-nitrosamino)-1-(3-pyridyl)-1-butanone. Cancer Res 46:1280–1284

Belinsky SA, White CM, Devereux TR, Swenberg JA, Anderson MW (1987) Cell selective alkylation of DNA in rat lung following low dose exposure to the tobacco specific carcinogen 4-(*N*-methyl-*N*-nitrosamino)-1-(3-pyridyl)-1-butanone. Cancer Res 47:1143–1148

Bernfeld P, Homburger F, Russfield AB (1974) Strain differences in the response of inbred Syrian golden hamsters to cigarette smoke inhalation. JNCI 53:1141–1151

Binder H, Lindner W (1972) Bestimmung von Aethylenoxyd im Rauch garantiert unbegaster Zigaretten. Fachliche Mitt Osterr Tabakregie 13:215–220

Birt DF, Pour PM (1983) Increased tumorigenesis induced by *N*-nitroso-*bis*-(2-oxopropyl)amine in Syrian golden hamsters fed high-fat diets. JNCI 70:1135–1139

Borish ET, Cosgrove JP, Church DF, Deutsch WA, Pryor WA (1985) Cigarette tar causes single-strand breaks in DNA. Biochem Biophys Res Commun 133:780–786

Brinton LA, Blot WJ, Becker JA, Winn DM, Browder JP, Farmer JC, Fraumeni JF (1984) A case control study of cancers of the nasal cavity and paranasal sinuses. Am J Epidemiol 119:896–906

Brunnemann KD, Hoffmann D (1974) The pH of tobacco smoke. Food Cosmet Toxicol 12:115–124

Brunnemann KD, Hoffmann D, Wynder EL, Gori GB (1976) Determination of tar, nicotine, and carbon monoxide in cigarette smoke. A comparison of international smoking conditions. Department of Health, Education, and Welfare (DHEW) publ no (NIH) 76-1221:441–449
Brunnemann KD, Prokopczyk B, Hoffmann D, Nair JH, Ohshima H, Bartsch H (1986) Laboratory studies on oral cancer and smokeless tobacco. Banbury Rep 23:197–213
Brunnemann KD, Hornby AP, Stich HF (1987) Tobacco-specific nitrosamines in the saliva of Inuit snuff dippers in the Northwest Territories of Canada. Cancer Lett 37:7–16
Bryant MS, Skipper PL, Tannenbaum SR, Maclure M (1987) Hemoglobin adducts of 4-aminobiphenyl in smokers and nonsmokers. Cancer Res 47:602–608
Bryant MS, Vineis P, Skipper PL, Tannenbaum SR (1988) Haemoglobin adducts of aromatic amines in people exposed to cigarette smoke. IARC Sci Publ 89:133–136
Carmella SG, Hecht SS (1987) Formation of hemoglobin adducts upon treatment of F344 rats with the tobacco-specific nitrosamines 4-(methylnitrosamino)-1-(3-pyridyl)-1-butanone and *N'*-nitrosonornicotine. Cancer Res 47:2626–2630
Carmella SG, Hecht SS, Tso TC, Hoffmann D (1984) Roles of tobacco cellulose, sugars, and chlorogenic acid as precursors to catechol in cigarette smoke. J Agric Food Chem 32:267–273
Castonguay A, Foiles PG, Trushin N, Hecht SS (1985) A study of DNA methylation by tobacco-specific *N*-nitrosamines. Environ Health Perspect 62:197–202
Centers for Disease Control, Atlanta, Georgia (1987) Smoking attributable mortality and years of potential life lost – United States, 1984. MMWR 36:693–697
Chung FL, Tanaka T, Hecht SS (1986) Induction of liver tumors in F344 rats by crotonaldehyde. Cancer Res 46:1285–1289
Church DF, Pryor WA (1985) Free-radical chemistry of cigarette smoke and its toxicological implications. Environ Health Perspect 64:111–126
Cohen LA (1987) Diet and cancer. Sci Am 257:42–48
Connolly GN, Winn DM, Hecht SS, Henningfield JE, Walker B Jr, Hoffmann D (1986) The reemergence of smokeless tobacco. N Engl J Med 314:1020–1027
Cosgrove JP, Borish ET, Church DF, Deutsch WA, Pryor WA (1985) The metal-mediated formation of hydroxy-radical by aqueous extracts of cigarette tar. Biochem Biophys Res Commun 132:390–396
Davis BR, Whitehead JK, Gill ME, Lee PN, Butterworth AD, Roe FJC (1975) Response of rat lung to tobacco smoke condensate or fractions derived from it administered repeatedly by intratracheal instillation. Br J Cancer 31:453–461
Deutsch-Wenzel RP, Brune H, Grimmer G, Dettbarn G, Misfeld J (1983) Experimental studies in rat lungs on the carcinogenicity and dose-response relationships of eight frequently occurring environmental polycyclic aromatic hydrocarbons. JNCI 71:539–544
Doll R (1971) Cancers related to smoking. Proc 2nd world conf, Smoking and health. Pitman Med, London, pp 10–23
Doll R, Peto R (1981) The causes of cancer. Quantitative estimates of available risks of cancer in the United States today. JNCI 66:1191–1308
Dong M, Schmeltz I, Jacobs E, Hoffmann D (1978) Aza-arenes in tobacco smoke. J Anal Chem 2:21–25
Dontenwill WP (1974) Tumorigenic effects of chronic cigarette smoke inhalation on Syrian golden hamsters. In: Karbe E, Park JF (eds) Experimental lung cancer – carcinogenesis and bioassays. Springer, Berlin Heidelberg New York, pp 331–359
Dube MF, Green CR (1982) Methods of collection of smoke for analytical purposes. Recent Adv Tobacco Sci 8:42–102
El-Bayoumy K, Donahue J, Hecht SS, Hoffmann D (1986) Identification and quantitation of aniline and toluidines in urine of smokers and nonsmokers. Cancer Res 46:6064–6067
Enzell CR, Wahlberg I, Aasen AJ (1977) Isoprenoids and alkaloids of tobacco. Progr Chem Org Nat Products 34:1–79
Everson RB, Randerath E, Santella SA, Cefalo RC, Avitts TA, Randerath K (1986) Detection of smoking-related covalent DNA adducts in human placenta. Science 231:54–57
Foiles PG, Miglietta LA, Akerkar SA, Everson RB, Hecht SS (1988) Detection of O^6-methyldeoxyguanosine in human placental DNA. Cancer Res 48:4184–4188

Goodman MT, Morgenstern H, Wynder EL (1986) A case-control study of factors affecting the development of renal cell cancer. Am J Epidemiol 124:926–941
Green CR (1977) Some relationship between tobacco leaf and smoke composition. Proc Am Chem Soc Symp, Recent advances in the chemical composition of tobacco and tobacco smoke, New Orleans, pp 426–470
Green LC, Skipper PL, Juresky RJ, Bryant MS, Tannenbaum SR (1984) In vivo dosimetry of 4-aminobiphenyl in rats via a cysteine adduct in hemoglobin. Cancer Res 44:4254–4259
Greenberg RA, Haley NJ, Etzel RA, Loda FA (1984) Measuring the exposure of infants to tobacco smoke: nicotine and cotinine in saliva and urine. N Engl J Med 310:1075–1078
Guerin MR (1987) Formation and physicochemical nature of sidestream smoke. IARC Sci Publ 81:11–23
Gupta PC, Pindborg JJ, Mehta FS (1982) Comparison of carcinogenicity of betel-quid with and without tobacco. An epidemiological review. Ecol Dis 1:213–219
Haley NJ, Sepkovic DW, Hoffmann D, Wynder EL (1985) Cigarette smoking as a risk factor for cardiovascular disease Part VI. Compensation with nicotine available as a single variable. Clin Pharmacol Ther 38:164–170
Harley NH, Cohen BS, Tso TC (1980) Polonium-210. A questionable risk factor in smoking-related carcinogenesis. Banbury Rep 3:93–104
Harris CC, Vahakangas K, Newman MJ, Trivers GE, Shamsuddin A, Sinopoli N, Mann D, Wright WE (1985) Detection of benzo(a)pyrene diol epoxide-DNA adducts in peripheral blood lymphocytes and antibodies to the adducts in serum from coke oven workers. Proc Natl Acad Sci 82:6672–6676
Hayashi Y, Nagao M, Sugimura T, Takayama S, Tomatis L, Wattenberg LW, Wogan GN (eds) (1986) Diet, nutrition and cancer. Japan Sci Soc Press, Tokyo, p 345
Hayes AW, Coggins CRE, Ayres PH, Burger GT, Mosberg AT (1988) Ninety-day inhalation study in rats comparing smoke from cigarettes with burned or only heated tobacco. Toxicol Abstr 8:252–254
Hecht SS, Hoffmann D (1988) Tobacco-specific nitrosamines, an important group of carcinogens in tobacco and tobacco smoke. Carcinogenesis 9:875–884
Hecht SS, Thorne RL, Maronpot RR, Hoffmann D (1975) Tumor promoting subfractions of the weakly acidic fraction. JNCI 55:1329–1336
Hecht SS, Loy M, Maronpot RR, Hoffmann D (1976) A study of chemical carcinogenesis: comparative carcinogenicity of 5-methylchrysene, benzo(a)pyrene, and modified chrysenes. Cancer Lett 1:147–154
Hecht SS, Carmella S, Mori H, Hoffmann D (1981) Role of catechol as a major cocarcinogen in the weakly acidic fraction of smoke condensate. JNCI 66:163–169
Hecht SS, Adams JD, Numoto S, Hoffmann D (1983a) Induction of respiratory tract tumors in Syrian golden hamsters by a single dose of 4-(methylnitrosamino)-1-(3-pyridyl)-1-butanone (NNK) and the effect of smoke inhalation. Carcinogenesis 4:1287–1290
Hecht SS, Castonguay A, Rivenson A, Mu B, Hoffmann D (1983b) Tobacco-specific nitrosamines: carcinogenicity, metabolism, and possible role in human cancer. J Environ Health Sci CI 1:1–54
Hecht SS, Trushin N, Castonguay A, Rivenson A (1986a) Comparative carcinogenicity and DNA methylation in F344 rats by 4-(methylnitrosamino)-1-(3-pyridyl)-1-butanone and *N*-nitrosodimethylamine. Cancer Res 46:498–504
Hecht SS, Rivenson A, Braley J, DiBello J, Adams JD, Hoffmann D (1986b) Induction of oral cavity tumors in F344 rats by tobacco-specific nitrosamines and snuff. Cancer Res 46:4162–4166
Hecht SS, Melikian AA, Amin S (1986c) Methylchrysenes as probes for the mechanisms of metabolic activation of carcinogenic methylated polynuclear aromatic hydrocarbons. Acc Chem Res 19:4162–4166
Hecht SS, Amin S, Huie K, Melikian AA, Harvey RG (1987) Enhancing effects of a bay region methyl group on tumorigenicity of enantiomeric bay region diol epoxides formed stereoselectively from methylchrysenes in mouse epidermis. Cancer Res 47:5310–5315

Hecht SS, Spratt TE, Trushin N (1988) Evidence for 4-(3-pyridyl)-4-oxobutylation of DNA in F344 rats treated with the tobacco-specific nitrosamines 4-(methylnitrosamino)-1-(3-pyridyl)-1-butanone and *N'*-nitrosonornicotine. Carcinogenesis 9:161–165

Herning RI, Jones RT, Bachman J, Mines AH (1981) Puff volume increases when low-nicotine cigarettes are smoked. Br J Med 283:187–189

Higginson J, Muir CS (1979) Environmental carcinogenesis: misconceptions and limitations to cancer control. JNCI 63:1291–1298

Hirayama T (1966) An epidemiological study of oral and pharyngeal cancer in central and South-East Asia. Bull WHO 34:41–69

Hirsch JM, Johannson SL (1983) Effect of long-term application of snuff on oral mucosa. An experimental study in the rat. J Oral Pathol 12:187–189

Hoffmann D, Adams JD (1981) Carcinogenic tobacco-specific *N*-nitrosamines in snuff and in the saliva of snuff dippers. Cancer Res 41:4305–4308

Hoffmann D, Harris CC (eds) (1986) Mechanisms in tobacco carcinogenesis. Banbury Rep 23. Cold Spring Harbor Laboratories, New York

Hoffmann D, Hecht SS (1985) Nicotine-derived *N*-nitrosamines and tobacco-related cancer: current status and future directions. Cancer Res 45:935–944

Hoffmann D, Rathkamp G (1968) Primary and secondary nitroalkanes in cigarette smoke. Beitr Tabakforsch 4:124–134

Hoffmann D, Wynder EL (1971) A study of tobacco carcinogenesis. XI. Tumor initiators, tumor accelerators, and tumor promoting activity of condensate fractions. Cancer 27:848–864

Hoffmann D, Wynder EL (1986) Chemical constituents and bioactivity of tobacco smoke. IARC Sci Publ 74:145–165

Hoffmann D, Schmeltz I, Hecht SS, Wynder EL (1978) Polynuclear aromatic hydrocarbons in tobacco carcinogenesis. In: Gelboin G, Ts'o PO (eds) Polynuclear hydrocarbons and cancer, vol 1. Chemistry, molecular biology and environment. Academic, New York, pp 85–117

Hoffmann D, Rivenson A, Hecht SS, Hilfrich J, Kobayashi N, Wynder EL (1979) Model studies in tobacco carcinogenesis with the Syrian golden hamster. Progr Exp Tumor Res 24:370–390

Hoffmann D, Tso TC, Gori GB (1980) The less harmful cigarette. Prev Med 9:287–296

Hoffmann D, Brunnemann KD, Adams JD, Hecht SS (1984) Formation and analysis of *N*-nitrosamines in tobacco products and their endogenous formation in consumers. IARC Sci Publ 57:743–762

Hoffmann D, Adams JD, Lisk D, Fisenne I, Brunnemann KD (1987) Toxic and carcinogenic agents in dry and moist snuff. JNCI 79:1281–1286

Hoffmann D, Brunnemann KD, Venitt S (1988) Carcinogenic nitrosamines in oral snuff. Lancet I:1232

Ingebrethsen BJ (1986) Aerosol studies of cigarette smoke. Recent Adv Tobacco Sci 12:54–142

International Agency for Research on Cancer (1972) 4-Aminobiphenyl. IARC Monogr 1:74–79

International Agency for Research on Cancer (1973) Nickel and inorganic nickel compounds. IARC Monogr 3:126–149

International Agency for Research on Cancer (1974a) 2-Naphthylamine. IARC Monogr 4:97–111

International Agency for Research on Cancer (1974b) Benzene. IARC Monogr 7:203–221

International Agency for Research on Cancer (1974c) Hydrazine. IARC Monogr 4:127–136

International Agency for Research on Cancer (1974d) Urethane. IARC Monogr 7:111–140

International Agency for Research on Cancer (1976) Coumarin. IARC Monogr 10:113–119

International Agency for Research on Cancer (1979a) Ethylene and polyethylene. IARC Monogr 19:157–186

International Agency for Research on Cancer (1979b) Acrylonitrile. IARC Monogr 19:73–113
International Agency for Research on Cancer (1979c) Vinyl chloride. IARC Monogr 19:377–438
International Agency for Research on Cancer (1980a) Arsenic and arsenic compounds. IARC Monogr 23:39–141
International Agency for Research on Cancer (1980b) Chromium and chromium compounds. IARC Monogr 23:205–323
International Agency for Research on Cancer (1980c) Lead and lead compounds. IARC Monogr 23:325–415
International Agency for Research on Cancer (1982a) Formaldehyde (gas). IARC Monogr [Suppl] 4:131–132
International Agency for Research on Cancer (1982b) Benzene. IARC Monogr 29:93–148
International Agency for Research on Cancer (1982c) 2-Nitropropane. IARC Monogr 29:331–343
International Agency for Research on Cancer (1985a) Tobacco habits other than smoking; betel-quid and areca-nut chewing; and some related nitrosamines. IARC Monogr 37
International Agency for Research on Cancer (1985b) Ethylene oxide. IARC Monogr 36:189–226
International Agency for Research on Cancer (1986) Tobacco smoking. IARC Monogr 38
Ip C, Birt DF, Rogers AE, Mettlin C (eds) (1986) Dietary fat and cancer. Liss, New York
Ivankovic S, Eisenbrand G, Preussmann R (1979) Lung carcinoma induction in BD rats after single intratracheal instillation of an arsenic-containing pesticide mixture formerly used in vineyards. Int J Cancer 24:766–768
Jarvis MJ, Russell MAH (1985) Tar and nicotine yield of U.K. cigarettes, 1972–1983: sales-weighted estimates from non-industry sources. Br J Addict 80:429–434
Jenkins RW Jr, Goldey C, Williamson TG (1985) Neutron activation analysis in tobacco and tobacco smoke studies: 2R1 cigarette composition smoke transference and butt filtration. Beitr Tabakforsch 13:59–65
Johnson WR (1977) The pyrogenesis and physicochemical nature of tobacco smoke. Recent Adv Tobacco Sci 3:1–27
Kabat GC, Dieck GS, Wynder EL (1986) Bladder cancer in nonsmokers. Cancer 57:362–367
Kao-Shan C-S, Fine RL, Whang-Peng J, Lee EC, Chabner BA (1987) Increased fragile sites and sister chromatid exchanges in bone marrow and peripheral blood of young cigarette smokers. Cancer Res 47:6278–6282
Kinlen LJ, Rogot E (1988) Leukemia and smoking habits among United States Veterans. Br Med J 297:657–659
Klus H, Kuhn H (1982) Verteilung verschiedener Tabakrauchbestandteile auf Haupt- und Nebenstromrauch. Eine Übersicht. Beitr Tabakforsch 11:229–265
Klus H, Begutter H, Ball M, Introp I (1987) Environmental tobacco smoke in real life situations. Proc 4th internat conf indoor air quality and climate, vol 2:137–141 Berlin (West)
Kobayashi N, Hoffmann D, Wynder EL (1974) A study of tobacco carcinogenesis. XII. Epithelial changes in the upper respiratory tracts of Syrian golden hamsters by cigarette smoke. JNCI 53:1085–1089
Koop CE (1986) The quest for a smoke-free young America by the year 2000. J School Health 56:8–9
Krokan H, Grafstrom RC, Sundqvist K, Esterbauer H, Harris CC (1985) Cytotoxicity, thiol depletion and inhibition of O^6-methylguanine-DNA methyltransferase by various aldehydes in cultured human bronchial fibroblasts. Carcinogenesis 6:1755–1759
LaVoie EJ, Prokopczyk G, Rigotty J, Czech A, Rivenson A, Adams JD (1987a) Tumorigenic activity of the tobacco-specific nitrosamines 4-(methylnitrosamino)-1-(3-pyridyl)-1-butanone (NNK), 4-(methylnitrosamino)-4-(3-pyridyl)-1-butanol (iso-NNAL) and *N'*-nitrosonornicotine (NNN) on topical application to Sencar mice. Cancer Lett 37:277–283
LaVoie EJ, Shigematsu A, Rivenson A (1987b) The carcinogenicity of quinoline and benzoquinolines in newborn CD-1 mice. Gann 78:139–143

LaVoie EJ, Tucciarone P, Kagan M, Adams JD, Hoffmann D (1988) Quantitative analyses of steam distillates and aqueous extracts of smokeless tobacco. J Agr Food Chem 31:154–157

Lee PN (1987) Lung cancer and passive smoking: association an artifact due to misclassification of smoking habits. Tox Lett 35:157–162

Leffingwell JC, Young HJ, Bernasek E (1972) Tobacco flavoring for smoking products. RJ Reynolds Tobacco Company, Winston Salem, North Carolina

Lewtas J, Williams K, Loefroth G, Hammond K, Laederer B (1987) Environmental tobacco smoke: mutagenic emission rates and their relationship to other emission factors. Proc 4th internat conf on indoor quality and climate, vol 2:8–12 Berlin (West)

Liu YY, Schmeltz I, Hoffmann D (1974) Quantitative analyses of hydrazine in tobacco and cigarette smoke. Anal Chem 46:885–889

Maxwell JC Jr (1981) Smokeless tobacco: the economy stunted growth. Tobacco Reporter 108 (7):34–35

Maxwell JC Jr (1986) Smokeless sales drop in US. Tobacco Reporter 113 (8):36–37

Melikian AA, Leszczynska JM, Hecht SS, Hoffmann D (1986) Effects of the cocarcinogen catechol on metabolism and DNA formation by benzo(a)pyrene in mouse skin in vivo. Carcinogenesis 7:9–15

Melikian AA, Bagheri K, Hecht SS (1987) Contrasting disposition and metabolism of topically applied benzo(a)pyrene, *trans*-7,8-dihydroxy-7,8-dihydrobenzo(a)pyrene, and 7α,8β-dihydroxy-9β,10β-epoxy-7,8,9,10-tetrahydrobenzo(a)pyrene in mouse epidermis in vivo. Cancer Res 47:5354–5360

Nair JH, Ohshima H, Malaveille C, Friesen IK, O'Neill AA (1986a) Identification, occurrence and mutagenicity in *S. typhimurium* of two synthetic nitroarenes, musk ambrette and musk xylene, in Indian chewing tobacco and betel quid. Food Chem Toxicol 24:27–31

Nair J, Ohshima H, Pignatelli B, Friesen M, Malaveille C, Calmels S, Bartsch H (1986b) Modifiers of endogenous carcinogen formation: studies on in vivo nitrosation in tobacco users. Banbury Rep 23:45–61

Nakayama T, Kodama M, Nagata C (1984) Generation of hydrogen peroxide and superoxide anion radical from cigarette smoke. Gann 75:95–98

Nakayama T, Kaneko M, Kodama M (1986) Volatile gas components contribute to cigarette-smoke-induced DNA single strand breaks in cultured human cells. Agr Biol Chem 50:1319–1320

National Council on Radiation Protection and Measurement (1987) Ionizing radiation exposure of the population of the United States. NCRP rep no 93, Bethesda, MD

National Research Council. Committee on Medical and Biological Effects of Environmental Pollutants (1975) Nickel. Natl Acad Sci, Washington DC, publ no 75-1475

Neumann HG (1984) Analysis of hemoglobin as a dose monitor for alkylating and arylating agents. Arch Toxicol 56:1–6

Norman V (1977) An overview of the vapor phase semi-volatile and nonvolatile components of cigarette smoke. Recent Adv Tobacco Sci 3:28–58

Norman V (1982) Changes in smoke chemistry of modern day cigarettes. Recent Adv Tobacco Sci 8:144–177

Palladino G, Brunnemann KD, Adams JD, Haley NJ, Hoffmann D (1986) Snuff-dipping in college-students: a clinical profile. Military Med 151:342–346

Park NH, Sapp JP, Herbosa EG (1986) Oral cancer induced in hamsters with herpes simplex infection and simulated snuff dipping. Oral Surg Oral Med Oral Pathol 62:164–168

Perera FP, Poirier MC, Yuspa SH, Nakayama J, Jaretzki A, Curnen MM, Knowles DM, Weinstein IB (1982) A pilot project in molecular cancer epidemiology: determination of benzo(*a*)pyrene-DNA adducts in animals and human tissues by immunoassays. Carcinogenesis 3:1405–1410

Perfetti TA, Gordin HH (1985) Just noticeable difference studies of mentholated cigarette products. Tobacco Sci 29:57–66

Perinelli MA, Carugno N (1978) Determination of trace metals in cigarette smoke by flameless atomic absorption spectrometry. Beitr Tabakforsch 9:214–217

Prokopczyk B, Rivenson A, Bertinato P, Brunnemann KD, Hoffmann D (1987) 3-(Methylnitrosamino)propionitrile: occurrence in saliva of betel quid chewers, carcinogenicity, and DNA methylation in F344 rats. Cancer Res 47:467–471
Ranadive KJ, Gothoskar SV, Rao AR, Tezabwella BU, Ambaye RY (1976) Experimental studies on betel-nut and tobacco carcinogenicity. Int J Cancer 17:469–476
Randerath E, Avitts TA, Reddy MV, Miller RH, Everson RB, Randerath K (1986) Comparative ^{32}P-analysis of cigarette smoke induces DNA damage in human tissues and mouse skin. Cancer Res 46:5869–5877
Reddy BS, Cohen LA, McCoy GD, Hill P, Weisburger JH, Wynder EL (1980) Nutrition and its relationship to cancer. Adv Cancer Res 32:237–345
Remmer H (1987) Passively inhaled tobacco smoke: a challenge to toxicology and preventive medicine. Arch Toxicol 61:85–104
Rivenson A, Hoffmann D, Prokopczyk B, Amin S, Hecht SS (1988) Induction of lung and exocrine pancreas tumors in F344 rats by tobacco-specific and areca-derived *N*-nitrosamines. Cancer Res 48:6912–6917
Royal College of Physicians (1962) Smoking and Health. Pitman, London
Russell MAH (1987) Estimation of smoke dosage and mortality of nonsmokers from environmental tobacco smoke. Tox Lett 35:9–18
Saffiotti U, Montesano R, Sellakumar AR, Cefis F, Kaufman DG (1972) Respiratory tract carcinogenesis induced in hamsters by different dose levels of benzo(*a*)pyrene and ferric oxide. JNCI 49:1199–1204
Saffiotti U, Stinson SF, Keenan KP, McDowell EM (1985) Tumor enhancement factors and mechanisms in the hamster respiratory tract carcinogenesis model. In: Mass MJ et al. (eds) Carcinogenesis, a comprehensive survey, vol 8. Raven, New York, pp 63–92
Schmeltz I, Hoffmann D (1977) Nitrogen-containing compounds in tobacco and tobacco smoke. Chem Rev 77:295–311
Schmeltz I, Chiong KG, Hoffmann D (1978) Formation and determination of ethyl carbamate in tobacco and tobacco smoke. J Anal Toxicol 2:265–268
Sepkovic DW, Axelrad CM, Colosimo SG, Haley NJ (1987) Measuring tobacco smoke exposure: clinical applications and passive smoking. 80th annual meeting. An exhibition of the Air Pollution Control Assoc, New York
Shinohara Y, Ogiso T, Hananouchi M, Nakanishi K, Yoshimura T, Ito N (1977) Effect of various factors on the induction of liver tumors in animals by quinoline. Gann 68:785–796
Stanton MF, Miller E, Wrench C, Blackwell R (1972) Experimental induction of epidermoid carcinoma in the lungs of rats by cigarette smoke condensate. JNCI 49:867–877
Takenaka S, Oldiges H, Koenig H, Hochrainer D, Oberdoerster G (1983) Carcinogenicity of cadmium chloride aerosols in W rats. JNCI 70:367–373
Thyssen J, Althoff J, Kimmerle G, Mohr U (1981) Inhalation studies with benzo(a)pyrene in Syrian golden hamsters. JNCI 66:575–577
Tobacco Journal International (1987) Production of cigarettes for selected countries. Tobacco J Int (1980):377–379 and (1987):358–360
Tobacco Journal International (1988) Only consumption of snuff is rising in Sweden. Tobacco J Int (1988):180–182
Törnqvist M, Osterman-Golkar S, Kautiainen A, Jensen S, Farmer PB, Ehrenberg L (1986) Tissue doses of ethylene oxide in cigarette smokers determined from adduct levels in hemoglobin. Carcinogenesis 7:1519–1521
Tso TC (1972) Physiology and biochemistry of tobacco plants. Dowden, Hutchinson and Ross, Stroudsburg, Pennsylvania
US Dept of Agriculture, Food Safety, and Quality Service (1983) Nitrates, nitrites and ascorbates (as isoascorbates) in bacon. Fed Regist 256:347–351
US Food and Drug Administration (1980) Dimethylnitrosamines in malt beverages; availability of guide. Fed Reg 45:39341–39342
US National Research Council (1986) Environmental tobacco smoke-measuring exposures and assessing health effects. National Academy Press, Washington DC
US Surgeon General (1964) Smoking and health. PHS publ no 1103 Washington DC

US Surgeon General (1981) The health consequences of smoking – the changing cigarette. DHHS (PHS) 81-50156 Washington DC
US Surgeon General (1982) The health consequences of smoking – cancer. DHHS publ no (OHS) 82-50179 Washington DC
US Surgeon General (1986a) The health consequences of involuntary smoking. DHHS (CDC) 87-8398 Washington DC
US Surgeon General (1986b) The health consequences of using smokeless tobacco. NIH publ no 86-2874 Washington DC
Van Duuren BL, Goldschmidt BM (1976) Cocarcinogenic and tumor promoting agents in tobacco carcinogenesis. JNCI 56:1237–1242
Wahlberg I, Enzell CR (1987) Tobacco isoprenoids. Natural Product Rep 1987:237–276
Wallace L, Pellizari E, Hartwell TD, Perritt R, Ziegenfus R (1987) Exposures to benzene and other volatile compounds from active and passive smoking. Arch Environ Health 42:272–279
Willey JC, Grafstrom RC, Moser CE Jr, Oyanne C, Sundqvist K, Harris CC (1987) Biochemical and morphological effects of cigarette smoke condensate and its fractions on normal human bronchial epithelial cells in vitro. Cancer Res 47:2045–2049
Winn DM, Blot WJ, Shy CM, Pickle LW, Toledo A, Fraumeni JF Jr (1981) Snuff-dipping and oral cases among women in the Southern United States. N Engl J Med 304:745–749
Wittekindt W (1985) Changes in recommended plant protection agents for tobacco. Tobacco J Int 5:390–394
Wynder EL, Gori GB (1977) Contribution of the environment to cancer incidence. An epidemiologic exercise. JNCI 58:825–832
Wynder EL, Hoffmann D (1967) Tobacco and tobacco smoke – studies in experimental carcinogenesis. Academic, New York
Wynder EL, Hebert JR, Kabat GC (1987) Association of dietary fat and lung cancer. JNCI 79:631–637
Yamashita M, Wakabashi K, Nagao M, Sato S, Yamaizumi Z, Takahashi M, Kinea N, Tomita I, Sugimura T (1986) Detection of 2-amino-3-methylimidazo-[4,5-*f*]quinoline in cigarette smoke condensate. Gann 77:419–422
Zaridze DG, Blettner M, Trapeznikov NN, Kuvshinov JP, Matiakin EG, Poljakov DP, Poddubni BK, Parshikova SM, Rottenberg VI, Chamrakulov FS, Chodjaeva MC, Stich HF, Rosin MP, Thurnham DI, Hoffmann D, Brunnemann KD (1985) Survey of a population with a high incidence of oral and oesophageal cancer. Int J Cancer 36:153–158

CHAPTER 4

Occupational Carcinogens

C. E. SEARLE and O. J. TEALE

A. Historical Introduction

I. Early Occupational Cancer

Bernardino RAMAZZINI (1633–1714) has, with good reason, been called the Father of Occupational Medicine. He was deeply moved by the appalling conditions under which many in his native Italy had to earn their living and made detailed observations of working conditions in a great range of occupations from cleaning out cesspits and other dirty trades to singers, nuns and notaries (HUNTER 1978). He described the ailments suffered by many as a result of their callings, recommended measures for their prevention, and advised doctors when visiting workers in their homes to add another question to those recommended by Hippocrates over 2000 years earlier – What is your occupation? A translation of his book *De Mortis Artificum Diatriba* (RAMAZZINI 1713) has been published.

RAMAZZINI was, however, not the first to describe occupational disease and accidents. Agricola's *De Re Metallica* of 1556 dealt in 12 volumes with the current practices of mining and refining metals and the accidents and lung diseases commonly encountered by the workers. The eccentric physician Paracelsus described in 1531 the lung disease "mala metallorum" which killed many miners at early ages. Some of this lung disease was almost certainly lung cancer (HAERTING and HESSE 1879), together with tuberculosis and silicosis. Though not recognised as cancer at the time, these were probably the first accounts to describe occupational cancer.

The mining area chiefly involved was the Erzgebirge (Ore Mountains) at Schneeberg in present-day German Democratic Republic and Jáchymov (Joachimsthal) in Czechoslovakia. Mining for various metals has been carried out there since precious metals were discovered about the twelfth century, so doubtless cancer and other occupational diseases had occurred long before Agricola and Paracelsus. The actual carcinogenic factors will have been variable and complex and are still uncertain, but radiation from radon gas and its decay products will probably have been of particular importance.

The first clear recognition of an occupational cancer was by the London surgeon POTT (1775), who described the common occurrence of cancer of the scrotum in chimney sweeps. He recognised the malignant nature of the disease and correctly attributed it to prolonged contact of the skin with soot. The first identification of carcinogenic chemicals, the polycyclic aromatic hydrocarbons, around 1930 can be clearly traced back to POTT's observations.

Over a century after POTT's publication, occupational cancer of the urinary bladder was recognised in a small group of German dyestuff workers by REHN (1895), and this in turn led to the pioneering epidemiological studies of CASE and coworkers (1954), which identified the aromatic amines responsible for the disease. The range of known occupational carcinogens now covers chemicals of many organic and inorganic types as outlined in this review, while there are various occupations with known or suspected carcinogenic risks where the causative factors have not been identified.

II. Prevention of Occupational Cancer

Even the early demonstrations that some types of cancer could have occupational causes imply that these cancers at least should be preventable. This was well appreciated by BUTLIN (1892), following his visits to sweeps and doctors in Belgium, France and some other countries. Though its incidence was then declining, scrotal cancer was still occurring in sweeps and some other workers in Britain, whereas elsewhere it was apparently unknown. BUTLIN believed this to be due to the widespread use of protective clothing, together with frequent bathing or thorough washing, and he recommended the adoption of similar protective measures in Britain (though expressing doubts as to whether such foreign practices would be acceptable!).

Today the view that much cancer originates from the operation of environmental factors and should therefore be capable of prevention is widely accepted by the scientific and medical fraternities, if not yet by much of the public. With the intention of reducing present-day high levels of both cancer and circulatory disease in the Western world, the public has been given much generally well-founded advice on the avoidance of smoking and healthier eating and drinking habits. Very great progress has already been made in the prevention of occupational cancers, sometimes by the replacement of identified carcinogens by safer materials, often by the imposition of strict controls on how potential carcinogens are handled in the workplace. Nevertheless, many difficulties remain in the way of ensuring that people do not contract cancer as a result of their work, arising both from the inherent limitations of epidemiological studies and from our still inadequate understanding of the possible carcinogenic risks of many substances of commercial importance.

III. Occupational Carcinogens

Of the many hundreds of known chemical carcinogens, only a very few have clearly been shown to be causes of human cancer, some through their use as drugs and some as occupational carcinogens. There are also, however, various processes that have had associated cancer risks for workers without our knowing the agents responsible, as well as chemicals that have come under suspicion of having caused some cases of occupational cancer, but so far without firm evidence. It is not, therefore, surprising that there are wide differences between the many published lists of occupational chemicals or processes with carcinogenic hazards, according to the strength of the evidence that is accepted.

Table 1. Known and suspected occupational carcinogens

Carcinogen	Occupation	Cancer
Polynuclear aromatic compounds in soots, tars, some mineral oils	Various including sweeps, tar workers, cotton spinners, roofers, boat-builders and repairers, fishermen, tool setters	Scrotum, other skin
2-Naphthylamine, 4-aminobiphenyl	Chemicals, rubber, cable-making	Bladder
Benzidine	Chemicals, dyestuffs, laboratory reagent	Bladder
? Michler's ketone	Auramine manufacture	Bladder
Mustard gas	Chemicals, warfare	Lung, larynx
bis(Chloromethyl) ether, technical chloromethyl methyl ether	Chemicals	Lung
? Diisopropyl sulphate	Isopropanol manufacture by strong acid process	Nasal sinuses
? Benzotrichloride	Manufacture of benzoyl chloride, etc.	Lung
Vinyl chloride (monomer)	Polyvinyl chloride manufacture	Liver angiosarcoma, ? other sites
Benzene	Chemicals, solvent	Leukaemia
Unknown ? via immuno-suppression	Professional chemists	Lymphoma, ? brain
Arsenic compounds	Manufacture, use of arsenical pesticides, mining, smelting of various metals	Lung, skin
Nickel subsulphide, oxide, etc.	Nickel refining	Nasal sinuses, lung
Zinc chromate, other Cr(VI) compounds	Production and use of chromates	Lung
Beryllium compounds	Mining and various uses	Lung
Cadmium oxide, etc.	Battery manufacture, alloying, plating, etc.	? Prostate, ? lung
Radon + ?	Underground mining of haematite and other ores	Lung
Asbestos dust	Mining, multiple uses	Lung (syngergism with smoking), mesothelioma of pleura, peritoneum
Wood dusts	Furniture manufacture	Nasal sinuses
Leather dusts	Shoe-making	Nasal sinuses

Thus Davies (1982) gives a short list of occupations with widely agreed cancer risks, and Saracci (1984) a more extensive list including many with reported risks and their suspected causes. Merletti et al. (1984) tabulate over 100 chemicals (including drugs) and processes with symbols indicating the firmness of the evidence for human carcinogenicity at each body site, as judged by working groups at IARC and published in the extensive Monographs series.

Table 1 lists well-established and suggested chemical causes of occupational cancers, without attempting to cover the great range of occupations which have given more tentative indications of carcinogenic risks. Some are now fortunately largely of historical interest. The outstanding carcinogen listed, both in terms of its scale of use and numbers of workers involved, is asbestos, which will remain a serious problem for many years to come.

For many commercial chemicals there is evidence of carcinogenicity from tests in animals but little or none of carcinogenicity in humans. With some, for example certain halogenated solvents, significant numbers of tumours have only developed following administration of extremely high levels during most of the animals' lifespan. One suspects that in such cases human risks of cancer are virtually non-existent, though this can be controversial. It is always good practice to minimise human exposures to chemicals, and this is often essential for quite different reasons such as short-term toxic actions. There is, of course, no doubt over the need for strict precautions in dealing with potent carcinogens such as 1,2-dibromoethane, even in the absence of any evidence for their being human occupational carcinogens.

This chapter deals mainly with well-established or strongly suspected carcinogens according to their chemical class (aromatic amines, alkylating agents, etc.). Apart from reference to the likely role of radon in mining, it will not cover carcinogenesis by radiation. The reader is referred to HUNTER's classic work (1978) *The Diseases of Occupations* for the early history of carcinogenesis by X-rays and radio-elements, and for more recent reviews to BOICE and LAND (1982), SMITH (1985) and several chapters in BANNASCH (1987).

B. Polycyclic Aromatic Hydrocarbons

I. Occupational Skin and Scrotal Cancer

Apart from their use as research chemicals, carcinogens of this type are encountered only in the complex mixtures of compounds generated during incomplete combustion of organic matter, but this results in their very widespread distribution in the environment. It was their presence in soot, a product of the highly inefficient combustion of bituminous coal in open grates, that was responsible for the scrotal cancer of chimney sweeps described by POTT (1775).

Just a century later, the same disease was being reported in a number of other occupations, similarly involving body contact with these carcinogens in coal tar and paraffin (VOLKMANN 1875), oil and pitch (MANOUVRIEZ 1876) and shale oil (BELL 1876). In these occupations, tumours were commonly multiple, affecting also the skin of the trunk and legs, but cancer of the scrotal skin is a much more serious condition owing to the ease of its spread to the testes and internal organs.

Oils of this type became widely used industrially, displacing the unstable natural oils used previously as lubricants. This resulted in scrotal cancer becoming a serious problem also in the British cotton mule-spinning industry, in which the operatives' clothing and groin become soaked in oil thrown from the machinery, and HENRY (1946) reported on the occurrence of 615 fatalities between 1911 and 1938. It was found experimentally that the lubricating oils used in cotton spinning were carcinogenic in mice, but that "white" oils, treated to remove the polycyclic contaminants, were not (WOODHOUSE 1950). In the UK the

Mule-Spinning (Health) Special Regulations (1953) specified the use of these more expensive white oils in the cotton industry, since when mule-spinning and its associated disease have become obsolete.

More recently, scrotal cancer was found in some workers, particularly tool-setters, in the Midlands engineering industry centred on Birmingham (CRUICKSHANK and SQUIRE 1950), and WALDRON et al. (1984) reported a marked increase in scrotal cancer in this area from 1955 to the 1970s. Of the patients with known occupations, 61% were attributable to oil and 8% to tar or pitch. Of 792 workers in Göteborg, Sweden, exposed to cutting fluids, 7 turners but no grinders developed carcinoma of the scrotum (JÄRVHOLM and LAVENIUS 1987). In engineering, the disease results from contamination of the groin with cutting oils, generally "neat" (undiluted) mineral oils. Heat generated in metal working may also cause carcinogenic contaminants to appear in previously non-carcinogenic oils. Scrotal cancer in the Midlands engineering declined again from the 1970s, but occasional cases still arise.

Reports of mineral oil cancers have come also from outside the UK, particularly the French Savoy Alps. The fact that many more cases of occupational cancers due to pitch, tar and oil have been reported from the UK than elsewhere may be partly attributable to the UK's long record of notification and compensation and not simply to poorer hygienic standards (KIPLING and COOKE 1984).

The literature contains many reports of skin cancer arising in occupations involving contact with tar and tar-treated products such as ropes. Occupational skin cancer has also arisen from contact with tarry byproducts from manufacture of 4,4′-bipyridyl, an intermediate for the herbicide paraquat (BOWRA et al. 1982). On the other hand, risks that might perhaps have been expected, for example from contact with various oils in vehicle maintenance, appear not to be significant.

II. Cancer at Other Sites

Skin cancer is the characteristic but not the only cancer attributable to materials containing polycyclic aromatic compounds. Increased rates of lung cancer were reported in gas and coke workers by DOLL (1952) and DOLL et al. (1965) in the UK and by LLOYD (1971) in the USA, though the increases were only about twofold, very much less than those attributable to smoking. A possible increase in bladder cancer in these workers was probably caused by aromatic amines (see Sect. C). Increased incidences of upper respiratory and digestive tract tumours have been attributed to inhalation of oil mists in cotton spinning (HENRY 1947) and in engineering by HOLMES et al. (1970) and WATERHOUSE (1972).

III. Recognition and Occurrence of Polycyclic Hydrocarbons

Much experimental work on carcinogenic soots and oils culminated in the synthesis of the first pure chemical carcinogen known, dibenz[*a,h*]anthracene (KENNAWAY and HIEGER 1930) and isolation of the closely related benzo[*a*]pyrene from coal-tar pitch (COOK et al. 1933). Benzo[*a*]pyrene is the main contributor to the carcinogenic action of the complex mixtures of similar compounds in combustion products, and on account of the ease of its fluorimetric determination it is commonly used as indicator of the levels of these compounds in, for example, air pollution. Very many tons are released annually into the atmosphere from incineration of rubbish, power generation and many other sources. However, apart from their contribution to occupational skin cancer, little is known regarding the nature of the human risks from their widespread distribution at low levels in many foodstuffs, air pollution, etc.

The range of known compounds of this type is now very large, and some have for long played a very important part in developing our understanding of carcinogenesis, but for occupational health the problems are rather of ensuring hygienic working conditions that will prevent their again being causes of human cancer. In view of the past occurrence of skin cancer in the British shale-oil industry, it is reassuring that a recent study of shale-oil workers in the USA found fewer instances of skin cancer than in a control population of coal miners (ROM et al. 1985).

Reviews of polycyclic hydrocarbon carcinogens and their associated occupational hazards have been given by PHILLIPS (1983), IARC (1983, 1984b, 1984c), DIPPLE et al. (1984), KIPLING and COOKE (1984) and HARVEY (1985).

C. Aromatic Amines and Related Compounds

I. Occupational Bladder Cancer

The Frankfurt surgeon REHN (1895) realised that the occurrence of bladder cancer in 3 of some 45 workers engaged for 15–29 years in the manufacture of fuchsin (magenta) indicated the likelihood of these cancers being a consequence of their chemical exposure. As aniline was known to cause haemorrhagic cystitis, REHN believed aniline to be the likely cause of the cancers, and although LEICHTENSTERN (1898) suggested that naphthylamines might cause bladder cancer, the disease became known for many years as "aniline cancer". However, when REHN (1906) later reported a further 38 patients he did suggest that naphthylamines and benzidine might have been involved. Aniline dye workers in Basel were later found by LEUENBERGER (1912) to have a risk of bladder cancer 33 times greater than comparable industrial workers elsewhere.

With the growth of dyestuffs manufacture in other countries, many more reports linking the industry with bladder cancer began to appear, for example by WIGNALL (1929) in the UK and by GEHRMANN (1934) in the USA, who stated, "Combined experiences indicate that probably aniline, alpha and beta naphthylamine and benzidine are the etiological compounds". By 1947 GOLDBLATT of ICI regarded 2-naphthylamine and benzidine as certain causes of occupational bladder cancer, and aniline and 1-naphthylamine as possible or probable causes. By this time 2-naphthylamine had been shown by HUEPER et al. (1938) to induce bladder cancer in dogs.

Various industrial organisations took steps to reduce exposure to aromatic amines and to introduce medical checks for workers, but it was not until the pioneering epidemiological studies begun by CASE in 1948 that the principal amine carcinogens responsible for causing bladder cancer were established beyond doubt.

II. Recognition of Human Bladder Carcinogens

CASE's study, under the auspices of the Association of British Chemical Manufacturers, traced 4622 workers in participating companies with at least 6 months' exposure to the suspect amines. Of these, 341 had developed bladder cancer, 30 times the expected number, and it was found that the most hazardous amine was indeed 2-naphthylamine, followed by benzidine (CASE et al. 1954). The manufacture of magenta and auramine also carried a carcinogenic risk, without incriminating the dyes themselves (CASE and PEARSON 1954). WILLIAMS (1958) later found that, of 18 men heavily exposed through distilling 2-naphthylamine and benzidine for over 5 years, 17 developed bladder cancer, some arising after only 1–2 years. Meanwhile, manufacture of 2-naphthylamine in Britain, but not elsewhere, had been discontinued in 1949.

Various studies have found 1-naphthylamine to be also associated with bladder cancer, but this amine has commonly been contaminated with up to 10% of the 2-isomer. Pure 1-naphthylamine has been found to be virtually free of carcinogenic activity when administered to dogs for 9 years (PURCHASE et al. 1981). A derivative, 1-naphthylthiourea has, however, caused some cases of bladder cancer in its former use as a rodenticide (ANTU). While ANTU also contains 2-naphthylamine as an impurity, DAVIES et al. (1982) suggested that it may be carcinogenic in its own right.

An unexpected finding in the UK epidemiological study was of a bladder cancer risk in the rubber industry also, attributed to the presence of 2-naphthylamine in an antioxidant formulation (CASE and HOSKER 1954). When in 1971 two workmen were awarded substantial damages against both the rubber company which employed them and the supplier of the antioxidant held responsible for their developing bladder cancer, it set an important precedent regarding a company's responsibility for the safety of its products (PARKES and EVANS 1984).

A third aromatic amine is well-established as a human bladder carcinogen. 4-Biphenylamine (4-aminobiphenyl; 4-aminodiphenyl; xenylamine) was used as a rubber antioxidant in the USA, but experimental demonstration of its carcinogenicity by WALPOLE et al. (1952) prevented its use in the UK in more than very small amounts. In the USA, MELICK et al. (1955) found 19 patients with bladder cancer in 171 workers within 20 years of its introduction. The proportion of affected workers later rose from 11% to 16.8% (MELICK et al. 1971), and this amine was evidently even more hazardous than 2-naphthylamine.

Outside the chemical and rubber industries there has also been reason to suspect risks of bladder cancer, but on much less firm evidence. In a study of bladder cancer cases in the Leeds area, ANTHONY and THOMAS (1970) suggested raised risks in a variety of occupations such as medical work, especially in nurses, and some other occupations, though on the basis of rather small numbers of patients. A likely cause of bladder cancer in nurses would have been benzidine, widely used as a reagent in hospital and many other laboratories. Chemists might similarly have been expected to suffer increased risks of bladder cancer, but several epidemiological studies, summarised by SEARLE (1984), unexpectedly indicated increased numbers of lymphomas but no increase in bladder cancer.

From experimental work a very large number of carcinogenic aromatic amines, nitro compounds and azo dyes are now known, and many have played an important part in developing our understanding of chemical carcinogenesis. They have been reviewed in considerable detail by GARNER et al. (1984), while many individual compounds have been covered in the IARC Monograph series and submitted to testing by the National Cancer Institute and National Toxicology Program (NCI/NTP) in the USA. The epidemiology of occupational bladder cancer, reviewed by PARKES and EVANS (1984), is still a topic of importance, and the need to monitor workers with past exposure to the carcinogens will continue for the foreseeable future. The possibility of exposure to carcinogenic amines during reclamation of old rubber also needs to be borne in mind (WHITTY 1987).

III. Control of Amine Carcinogen Hazards

In the UK the CARCINOGENIC SUBSTANCES REGULATIONS (1967) prohibited the use or possession of 2-naphthylamine, benzidine, 4-biphenylamine and its precursor 4-nitrobiphenyl. Controls were also placed on a number of related industrial amines, discussed below, and in the manufacture, but not the use, of the dyes auramine and magenta. The regulations applied only to industrial premises, and thus not to situations such as hospital laboratories in which benzidine was widely used in testing for blood in urine and faeces. Safer procedures are now available for such tests. Action by education authorities has also been necessary to prevent use of these and other carcinogens in school laboratories (SEARLE 1984).

It is reassuring that controls appear to have been effective in eliminating the excess risk of bladder cancer from the British rubber industry, but a recent study involving over 36000 workers suggests that there are still raised risks of stomach, liver and lung cancer (SORAHAN et al. 1986), possibly associated with workers' exposure to dust and fumes.

In the USA, new Occupational Health and Safety Standards, published in 1974, applied not only to the carcinogenic aromatic amines but also to carcinogens of different types, including some alkylating agents and nitrosamines. There is now a continually developing scene of regulations on different types of hazardous chemicals, which is outside the scope of this chapter. It is important that the emotive nature of cancer does not deflect attention from the many other, generally much shorter term ways in which chemical misuse can and does harm humans, both at work and environmentally.

IV. Other 2- and 3-Ring Aromatic Amines

1. Substituted Benzidines and Other Industrial Compounds

Under the UK Carcinogenic Substances Regulations 1967, controls were imposed on the uses of o-tolidine (3,3′-dimethylbenzidine), o-dianisidine (3,3′-dimethoxybenzidine), 3,3′-dichlorobenzidine and on the employment of personnel in contact with them. For each amine there is some evidence of carcinogenicity in animals, though so far this is only slight for o-tolidine. It has been suggested that 3,3′-dichlorobenzidine, which was commonly handled in the same works as benzidine, might have contributed to bladder carcinogenesis, but, according to MacIntyre (1975), combined European experience from several countries has shown no instances attributable to the chloro compound.

The important industrial chemical 4,4′-methylenebis(2-chloroaniline) (MBOCA; MOCA; DACPM) has had long use in polyurethane manufacture. There is strong experimental evidence for its carcinogenicity (HSE 1983), but it appears still to be widely employed. 4,4′-Methylenedianiline appears weakly carcinogenic (Lamb et al. 1986). It has not been incriminated as a cause of occupational cancer, but an accidental poisoning caused severe liver toxicity in the "Epping jaundice" incident of 1966.

There is little evidence for the carcinogenicity of *N*-phenyl-2-naphthylamine, though results of full tests are awaited. Very little metabolism to 2-naphthylamine has been detected, and epidemiological evidence for carcinogenicity has been negative, despite earlier contamination with up to 100 ppm of 2-naphthylamine (Walpole et al. 1952). No carcinogenic hazards from diphenylamine would be expected when pure, but in the past it has contained small amounts of the hazardous 4-biphenylamine.

2. 2-Fluorenylacetamide

2-Fluorenamine was patented in 1940 as a potentially valuable insecticide, but when its biologically equivalent acetyl derivative 2-fluorenylacetamide (2-acetylaminofluorene) was subjected to testing, it was found to be an extremely potent carcinogen for many organs of the rat. This amide has since had extensive experimental use in cancer research for elucidating mechanisms of chemical carcinogenesis, but it would have been disastrous had it, or the amine, gone into large-scale manufacture and use for the purpose originally intended (Weisburger and Weisburger 1958).

V. Single-Ring Aromatic Amines

1. Aniline and Derivatives

Since identification of the aromatic amines primarily responsible for industrial bladder cancer, interest has largely centred on the activity of amines with 2 or 3 rings in the molecule. However, carcinogenicity has now been demonstrated in a number of amines with a single aromatic ring, including aniline, once thought responsible for "aniline cancer". Aniline is largely inactive in the bacterial mutagenicity and other short-term tests used as indicators of possible carcinogenicity, and it was inactive on feeding as the hydrochloride to mice. Rats fed the hydrochloride at up to 0.6% in the diet, however, developed significant numbers of sarcomas of the spleen and other abdominal sites (NCI 1978). Bus and Popp (1987) have suggested that the splenic carcinogenicity of aniline and some related compounds (o-toluidine, dapsone) may have a threshold above which erythrocytes are unable to cope with the toxic insult. These tests do not suggest that aniline has after all a direct role in bladder carcinogenesis, but it seems possible that, under conditions resulting in haemorrhagic cystitis, it might have a co-carcinogenic action, potentiating the action of the later recognised carcinogens. WALPOLE et al. (1952) have suggested that early industrial aniline was contaminated with the carcinogenic 4-biphenylamine.

There is also experimental evidence for the carcinogenicity of o-toluidine (NCI 1979 a), but only circumstantial evidence for its having been an occupational carcinogen. The m- and p-isomers are substantially free of carcinogenic activity. The methoxyl-containing p-cresidine is, however, a carcinogen of considerable potency for several rat and mouse organs (NCI 1979 b), and a potential human hazard in its use as a dye intermediate.

One aniline derivative associated with human cancer is the analgesic phenacetin (4-ethoxyacetanilide), formerly widely used in combination with aspirin and caffeine. Many reports link abuse of such mixtures with kidney toxicity and renal pelvis cancer (IARC 1980 b), but the amounts ingested by such patients were enormously greater than any likely occupational exposure.

2. Phenylenediamines

A variety of phenylenediamines have large-scale use in synthetic hair-dye preparations. Many have been found to be active bacterial mutagens, but only 2,4-diaminotoluene and 2,4-diaminoanisole have shown significant carcinogenicity in animals, leading to restrictions on their use. Concern over possible carcinogenic risks to hairdressers and home users has led to a number of epidemiological studies, but the results have been far from clear. Given their very extensive use, carcinogenic risks do not appear very significant, though considerable care is needed to avoid sensitisation reactions.

VI. Dyes

1. Magenta and Auramine

Although there were carcinogenic hazards associated with dye manufacture from the nineteenth century, there do not appear to have been significant risks from the dyes themselves that cannot be attributed to contaminating amine intermediates. Though its manufacture is hazardous, magenta itself has never been shown to be carcinogenic, and the human and animal evidence that auramine is carcinogenic is not very strong. However, Michler's ketone, an intermediate in the manufacture of auramine and some other dyes, is much more definitely carcinogenic (NCI 1979 c) and may have been responsible for some positive reports of auramine activity. It is worth noting that very many dyes are marketed at a much lower standard of purity than are most other organic chemicals.

2. Azo Dyes

Many azo dyes, such as 4-dimethylaminoazobenzene (butter yellow), are well-recognised experimental carcinogens, hence the strict limitations in various countries on the use of synthetic food dyes to a short positive list of dyes believed to be free of significant risk.

There has been particular concern over possible occupational hazards of a range of dyes made from benzidine and substituted benzidines, such as Direct Blue 6, and warnings have been issued from US safety bodies (OSHA 1980). Possible risks derive both from the animal carcinogenicity demonstrated with some dyes and from contamination by the parent amines. One would expect risks to be greatest during actual manufacture of the dyes, some of which cannot legally be made in the UK.

Some azo dyes such as chrysoidine are also aromatic amines. A strange but very widespread use of chrysoidine (2,4-diaminoazobenzene), particularly in the UK, has been for surface colouring of maggots used as bait by coarse fishermen, supposedly to make them more attractive to fish. Reasons for suspecting some risk of bladder cancer from the extensive and prolonged body contamination caused by this practice were put forward by SEARLE and TEALE (1982), and there is a little epidemiological support for this (SOLE and SORAHAN 1985). Evidence for animal carcinogenicity has been confused, possibly because the term chrysoidine is also applied to methyl-containing dyes, which are more mutagenic and probably carcinogenic than the methyl-free dye (SOLE and CHIPMAN 1986). Misuse of chemicals in sport in ways unacceptable in industry is clearly undesirable, particularly with regard to the large numbers of people who are involved, often from only 10 years of age.

VII. Nitro Compounds

Though so far of relatively little concern as occupational carcinogens, many nitro compounds are carcinogenic and often extremely potent. Being metabolised to the same proximate carcinogens as the corresponding carcinogenic aromatic amines, it is not surprising that compounds such as 2-nitro naphthalene and some nitrobiphenyls are carcinogenic. 4-Nitrobiphenyl, for example, induced bladder tumours in two of four dogs on repeated oral administration (DEICHMANN et al. 1958). This compound was one of the four prohibited in British industry by the Carcinogenic Substances Regulations 1967 and conceivably could have contributed to the human carcinogenicity of 4-biphenylamine during its use in the USA (MELICK et al. 1971).

Technical dinitrotoluenes are rat-liver carcinogens, but a recent study concluded that while 2,6-dinitrotoluene is a complete carcinogen the 2,4-isomer has only promoting activity (LEONARD et al. 1986). Pentachloronitrobenzene, known as quintozene, is a widely used agricultural fungicide. From reviewing carcinogenicity and other test data, WHO (1984) concluded it to be substantially free of hazard in normal use but recommended control on impurities, particularly the carcinogenic hexachlorobenzene.

Many nitroarenes, such as nitro derivatives of the polycyclic aromatic hydrocarbon pyrene, have been found to be mutagenic and carcinogenic. There is growing interest in their possible importance as environmental contaminants in, for example, carbon black and vehicle exhausts (ROSENKRANZ and MERMELSTEIN 1985).

Among heterocyclic nitro compounds, many nitrofurans are potent carcinogens, which severely limits their potential usefulness as antibacterial drugs and food preservatives. The potent experimental carcinogen 4-nitroquinoline 1-oxide was originally intended for use as an anti-fungal agent.

The aliphatic 2-nitropropane is a relatively volatile solvent and chemical intermediate, with extensive industrial use and possibilities for human exposure; it also occurs in tobacco smoke. However, warnings about its hazards have been issued, following demonstration that high-level inhalation exposure induces liver tumours in rats (LEWIS et al. 1979). In industrial use 2-nitropropane has caused acute toxic effects and a number of fatalities, but there is at present no evidence for its carcinogenicity in humans.

D. Alkylating Agents

This group of chemicals includes many of industrial importance as well as some of the earliest drugs found useful for cancer chemotherapy. They are commonly reactive and toxic chemicals, containing structures such as a labile carbon-halogen bond, for example in mustard gas, or a strained ring system, as in β-propiolactone and the aziridines (ethyleneimines). They have also been widely used experimentally as mutagens and carcinogens, which introduce alkyl groups into biological macromolecules without the prior metabolism necessary to activate other classes of genotoxic carcinogens (LAWLEY 1984).

Several alkylating agents are known or suspected to be causes of occupational cancer. Others, outside the scope of this review, have caused second cancers as an unfortunate effect of their use in treating cancer patients. The risks of occupational exposure to anticancer drugs have been reviewed by FISHBEIN (1987).

I. Mustard Gas; 2,2′-Bis(chloroethyl) Sulphide

Mustard gas has been suspected of causing respiratory cancer in survivors of gas attacks in the First World War, but the evidence has been inconclusive. There is little doubt, however, that it has caused cancer in some workers with heavy and prolonged exposure in war gas factories. WADA et al. (1968) found that 33 of 322 workers in a Japanese factory making mustard gas and some other toxic agents died of respiratory cancer, over 30 times the expected number. Many sites in the respiratory tract were affected. This finding has been borne out by more recent studies in Japan by NISHIMOTO et al. (1986). Increased incidences of respiratory cancer in British mustard gas workers have been reported by MANNING et al. (1981) and EASTON et al. (1988).

In Europe, dismantling war gas factories after the Second World War with inadequate safety measures has been found to result in multiple skin tumours of several types and refractory ulcers (KLEHR 1984). The possible late effects on the respiratory and other systems are still unknown. Exposure had been primarily to mustard gas and nitrogen mustard.

II. Chloromethyl Ethers

A series of papers (see, for example, LEMEN et al. 1976) have reported raised incidences of lung cancers in workers exposed to high levels of chloromethyl methyl ether (CMME) after 1962 in a chemical plant in Philadelphia. Technical CMME is known to contain high levels of the highly carcinogenic bis(chloromethyl) ether (BCME), but the relative contributions of the two agents are unknown. Lung cancer in men exposed to chloromethyl ethers has been reported from Germany and Japan. A study in two British factories was negative, apart from a South Wales works prior to plant modification in 1972 (McCallum et al. 1983). A recent study of US chloromethyl ether workers (Maher and Defonso 1987) found a respiratory cancer rate three times that expected, with a clear dose-response pattern, but it also showed a continuing decline compared with previous studies of the same plant. The occupational carcinogenicity of chloromethyl ethers has been briefly summarised by Alderson (1986).

Roe (1985a, b) has drawn attention to the risks of inadvertent exposure to BCME in the course of chloromethylation reactions using mixtures of methanol, hydrogen chloride and formaldehyde and reported the occurrence of three patients with lung cancer under the age of 40 among a small group of chloromethylation workers. Inadvertent exposure to BCME may also occur as a consequence of its reversible formation from hydrogen chloride and formaldehyde. The conditions under which this may occur have been reviewed in depth by Travenius (1982).

III. Benzoyl Chloride Manufacture

Chlorination of toluene gives sequentially benzyl chloride, benzal (benzylidene) chloride and benzotrichloride, from which benzoyl chloride is obtained. In two small Japanese plants making benzoyl chloride under poor conditions, 5 of 41 workers developed lung cancer and another developed lymphoma; 2 instances of lung cancer occurred in relatively young non-smokers (Sakabe and Fukuda 1977). In a British study (Sorahan et al. 1983), raised incidences of both respiratory and digestive cancers were found but only in workers exposed before 1951.

There is little reason to suspect benzoyl chloride itself as the cause of the excess cancers. The most likely candidate on present evidence is benzotrichloride, found by Fukuda et al. (1981) to be a very potent mouse skin carcinogen compared with benzyl chloride, benzal chloride and benzoyl chloride and by Yoshimura et al. (1986) to be carcinogenic by inhalation.

IV. Isopropanol Manufacture and Alkyl Sulphates

Manufacture of isopropanol from propylene by the now superseded "strong acid" process was associated in two factories in the USA with a raised incidence of nasal and pharyngeal tumours, though the numbers were again small (Alderson 1986). "Isopropyl oil", largely polymeric propylenes, is formed as a by-product in isopropanol manufacture, but the intermediate di-isopropyl sulphate appears the most likely carcinogenic agent (Wright 1979).

Diethyl sulphate may similarly have been responsible for an excess of upper respiratory cancer in workers in a plant making ethanol by the strong acid process (Lynch et al. 1979). Druckrey et al. (1966) suggested that several cases of bronchial cancer among a small group of German workers may have been caused by heavy exposure to dimethyl sulphate, which WHO (1985 b) considers to be a potential human carcinogen on the basis of its inhalation carcinogenicity in rats and other evidence.

V. Epoxides

Many epoxides (oxiranes) are of considerable industrial importance. Their mutagenicity and carcinogenicity has naturally led to anxieties over possible carcinogenic risks for epoxide-exposed workers, but to date there is little firm evidence for occupational carcinogenicity except, perhaps, for ethylene oxide. These agents are analogous to the primary metabolites formed during biological activation of carcinogens such as PAH, aflatoxins and vinyl chloride.

1. Ethylene Oxide

This important alkylating agent, a gas at ambient temperatures, is used on a large scale as a chemical intermediate, with smaller amounts being employed as a sterilant for some food and medical products. There is an extensive recent literature on its carcinogenicity and possible human hazards, summarised briefly by ALDERSON (1986) and the LANCET (1986) and in greater depth by LANDRIGAN et al. (1984), IARC (1985 b), GOLBERG (1986), and KOLMAN et al. (1986).

Ethylene oxide is highly toxic and undoubtedly carcinogenic in a number of test systems. Brain tumours were among the tumours obtained in long-term rat inhalation tests (GARMAN et al. 1985). Several publications from Sweden have indicated increases in leukaemia and stomach cancer in workers exposed during sterilisation of hospital equipment (HOGSTEDT et al. 1986), but a study of workers in an American manufacturing plant showed no excess cancer incidence (MORGAN et al. 1981).

With other anxieties over possible increases in abortion rates among women exposed to ethylene oxide during pregnancy and over possible increases in chromosome abnormalities, it is clear that workplace levels of this agent must be minimised.

2. Other Epoxides

A large number of epoxides have been found to be carcinogenic in animal tests, though among some 80 epoxides listed in the review by MANSON (1980) many others had shown no activity. Active ones include, however, a number of important industrial chemicals for which there is nevertheless very little evidence of occupational carcinogenesis.

Propylene oxide, another chemical intermediate and fumigant, is considerably less toxic than ethylene oxide and, though carcinogenic, does not appear very potent (LYNCH et al. 1984; WHO 1985 a; RENNE et al. 1986). An epidemiological study of German propylene oxide workers was negative (THIESS et al. 1982), but this study was criticised by IARC (1985 c) on several counts, including workers' exposure also to ethylene oxide and other chemicals.

Styrene oxide, a metabolite of styrene as well as an industrial intermediate, induces forestomach carcinomas in rats after repeated oral administration of 250 mg/kg doses (MALTONI et al. 1979). There are no epidemiological data (IARC 1985d), but the relatively weak carcinogenicity and low volatility suggest that carcinogenic risks would be very small.

Epichlorohydrin is a toxic chemical intermediate and solvent, with genotoxic activity in a number of test systems (SRÁM et al. 1983). In rats it is a forestomach carcinogen at a repeated oral dose of 10 mg/kg (WESTER et al. 1985), and some nasal cavity tumours are induced in inhalation tests at 30 and 100 ppm (LASKIN et al. 1980). The only suggestion of human carcinogenicity comes from a study of American workers exposed to relatively high levels prior to 1965 (ENTERLINE 1982). There was a small but significant increase in respiratory cancer, but some workers had also had exposure to the putative respiratory carcinogen in isopropanol manufacture.

A range of glycidyl ethers, such as phenyl glycidyl ether, are of considerable importance in resin production. They show mutagenic activity, as would be expected. Some, not always chemically defined, have given evidence of relatively weak carcinogenicity in animals. HOPKINS (1984) suggests that the industrial hygiene standards necessary to deal with their toxic effects should avoid any likelihood of occupational carcinogenesis.

VI. β-Propiolactone

β-Propiolactone, with a strained 4-membered ring, is a potent alkylating mutagen and carcinogen (BRUSICK 1977) used in the chemical industry and also for thc inactivation of bacteria and viruses in blood and hospital specimens (SPIRE et al. 1984).

β-Propiolactone has considerable volatility, and there could be significant carcinogenic risks with repeated use unless precautions against inhalation and skin contact are observed. It readily polymerises and in contact with aqueous media is also rapidly hydrolysed to the innocuous hydracrylic acid.

VII. Aziridines

A range of simple aziridines (ethyleneimines) were found carcinogenic by WALPOLE et al. (1954), who pointed out possible hazards of their industrial use. Some aziridines also have uses in cancer chemotherapy.

The simplest compound of the series, aziridine, is a chemical intermediate used particularly in the manufacture of flocculating agents. It is a mouse carcinogen (IARC 1975a), but there is no evidence that this or other aziridines have actually caused occupational cancer.

Triethylenephosphoramide (TEPA) had former industrial use, and for chemotherapy it was superseded by the less toxic thioTEPA. It is one of the cytotoxic drugs suspected of giving rise to leukaemia in some treated patients (IARC 1975b).

E. Halocarbons

A very wide range of halocarbons have large-scale commercial use, including chemicals such as the many halogenated solvents and intermediates, fluorocarbons, pesticides and polyhalogenated biphenyls (PCBs and PBBs). In many the carbon-halogen bond is extremely stable, as in vinyl chloride and the PCBs. In others, such as 1,2-dibromoethane and some other simple halocarbons, it is relatively labile, and there is no clear dividing line between these and compounds discussed in Sect. D.

Vinyl chloride is the most clearly recognised occupational carcinogen in this group, but a high proportion of the compounds have shown some degree of carcinogenicity in animal tests. 1,2-Dibromomethane is a potent carcinogen, but in many others activity has only been demonstrated after long-term administration at very high dose levels. In the absence of genotoxic activity there then appears to be little relevance to any likely conditions of human exposure. The review by GREIM and WOLFF (1984) points out the particular difficulties of assessing carcinogenicity data for many halocarbons.

I. Vinyl Chloride

The gas vinyl chloride, more correctly chloroethene, is the raw material for the massive PVC industry. Workmen exposed to extremely high levels, some estimates being much over 1000 ppm, during cleaning out of polymerisation reactors, have suffered various toxic effects, particularly acro-osteolysis affecting the fingers. However, in 1974 CREECH and JOHNSON reported three patients with the normally extremely rare cancer, angiosarcoma of the liver, in men from a single US PVC plant. The disease was clearly occupational, and cases from many other countries have since been reported, though the total numbers are not large. The angiosarcoma register recently contained 99 confirmed cases, and pre-1974 exposure may give rise to a further 150–300 cases (PURCHASE et al. 1987). When vinyl chloride was only regarded as having a narcotic action on humans, exposure limits were 500 ppm, but these are now generally controlled to under 10 ppm.

Vinyl chloride is carcinogenic for many sites in animals (IARC 1979d) and would be expected to affect other sites in humans. Epidemiological studies suggest it has also caused tumours of the brain, lung, lymphatic and haematopoietic systems (INFANTE 1981; BEAUMONT and BRESLOW 1981; WAGONER 1983). However, the risk of brain carcinogenesis appears lower than that for the liver, and for other sites the evidence has been inconsistent (PURCHASE et al. 1987).

II. Vinylidene Chloride and Chloroprene

Related industrially important monomers are vinylidene chloride (1,1-dichloroethene) and chloroprene (2-chloro-1,3-butadiene). Both compounds were reviewed by IARC (1979e, f), and vinylidene chloride more recently by HSE (1985b). Various carcinogenicity tests have, however, been regarded as inadequate, and there is still a dearth of sound experimental or human evidence on possible carcinogenic risks.

Incomplete tests of vinylidene chloride reported by MALTONI et al. (1977) resulted in high mortality. However, many male mice developed kidney tumours, though only at a level (25 ppm) which also caused severe nephrotoxicity. Rat inhalation tests at levels up to 75 ppm revealed both increases and decreases in tumours, but these were not judged attributable to the test chemical (QUAST et al. 1986). Oral administration tests by PONOMARKOV and TOMATIS (1980) and NTP (1982) were also negative. It is interesting, however, that in a 1-year inhalation test at 75 ppm two rats developed angiosarcoma, the characteristic tumour associated with the related vinyl chloride (IARC 1979e).

Tests of chloroprene, including those by PONOMARKOV and TOMATIS (1980) with oral administration, have found no evidence of carcinogenicity. Results of more relevant NTP tests using inhalation cannot be expected for several years. No sound evidence for occupational carcinogenicity has come from several studies (IARC 1979f), but one chloroprene worker with no known exposure to vinyl chloride has developed liver angiosarcoma (INFANTE 1977).

III. 1,2-Dibromoethane and Related Compounds

1,2-Dibromoethane (ethylene dibromide) has had large-scale uses as a fumigant, chemical intermediate and lead scavenger in petrol, but it has been found to induce tumours at many sites in rats when administered orally or by inhalation. There is no evidence that it causes human tumours, but concern over its undoubted hazards has led to bans or restrictions on its use.

The closely related 1,2-dibromo-3-chloropropane is also carcinogenic, but the particular concern with this fumigant has been over the incidence of sterility in exposed men. A recent study has indicated that some men suffer permanent destruction of germinal epithelium in the testes (EATON et al. 1986). NTP tests of both compounds have been summarised by HUFF (1983a, b).

1,2-Dichloroethane (ethylene dichloride) is important industrially, particularly for the manufacture of vinyl chloride, and is also carcinogenic but is much less potent than the bromo analogue, being largely inactive in inhalation tests (STORER et al. 1984). Although there is no evidence of carcinogenicity in humans, there has been considerable concern over its health risks (AMES et al. 1980), which have included many cases of acute poisoning.

An accident causing heavy exposure and poisoning by the soil fumigant 1,3-dichloropropene may have caused the death of two men from lymphoma and one from leukaemia within a few years (MARKOVITZ and CROSBY 1984). An increased risk of lymphoma has also been attributed to frequent use of herbicides, especially 2,4-D (2,4-dichlorophenyoxyacetic acid) and 2,4,5-T (2,4,5-trichlorophenoxyacetic acid) (HOAR et al. 1986). The absence of increased risk of soft-tissue sarcoma was in line with some other American and Swedish studies but at variance with work on herbicide manufacturing workers and Vietnam war veterans.

IV. Tetrachloromethane

Tetrachloromethane (carbon tetrachloride) is carcinogenic for the liver in rats and mice, by mechanisms possibly involving free radical formation and lipid peroxidation. It is important industrially, particularly for fluorocarbon manufacture. While there are no epidemiological studies of exposed workers, IARC (1979a) records three case reports of deaths from liver cancer within 7 years of episodes of acute poisoning. One patient had, however, previously suffered from jaundice. The chemical's high toxicity has resulted in many poisonings and fatalities, necessitating the setting of low exposure limits which should also avoid any significant risk of carcinogenesis.

V. Other Chlorinated Solvents

Chloroform similarly has important industrial uses and formerly was very extensively used as an anaesthetic. It is a liver toxin and experimental carcinogen, but there is no human evidence of carcinogenesis (IARC 1979b). Its carcinogenicity in rodents has led to its removal from various medicinal and cosmetic products, and its toxicity also necessitates low exposure limits in the workplace. There has been much concern in the USA over the presence of chloroform and other halocarbons in chlorinated drinking water (RAM et al. 1986).

Trichloroethylene is considerably less toxic, though readily decomposed to highly toxic products. High level administration to mice induces liver tumours, but these are attributed to metabolism to the weak peroxisome proliferator, trichloroacetic acid. There appears to be no significant carcinogenic risk to humans (KIMBROUGH et al. 1985). Some epidemiological studies, involving particularly dry-cleaning workers, have provided no evidence of carcinogenicity (ALDERSON 1986). Tetrachloroethylene has now largely replaced tetrachloromethane and trichloroethylene as the dry-cleaning solvent. Once again there is some limited evidence of rodent carcinogenicity by non-genotoxic mechanisms, not suggestive of any human risk.

VI. Polychlorinated Pesticides

DDT was the forerunner of a range of stable and persistent polychlorinated compounds, introduced to replace arsenicals as pesticides after the Second World War. Some have been found carcinogenic, predominantly for the liver, in rats and mice. Others, including DDT, are carcinogenic in mice only, and others have shown no activity (GREIM and WOLFF 1984). As they are generally inactive in short-term tests, any carcinogenicity is by non-genotoxic mechanisms. Restrictions have been placed on a number of such pesticides on account of their generally weak carcinogenicity in rodents and, in some cases, their toxicity. Misuse has led to episodes of occupational and non-occupational poisoning. In general there is no evidence of human carcinogenicity, though isolated reports have linked cases of leukaemia or aplastic anaemia with exposure to γ-hexachlorocyclohexane (gammexane, lindane) or the mixed isomers (IARC 1979c). A considerably raised incidence of lung cancer was reported among pesticide workers by BARTHEL (1976), but there had been exposure to a wide range of chemicals including arsenicals.

VII. Polychlorinated and Polybrominated Biphenyls

Liquid mixtures of polychlorinated biphenyls (PCBs) have had large-scale uses in, for example, electrical and hydraulic equipment. Being extremely stable, they have become widely distributed in the environment, including food chains and human tissues and fluids.

Liver carcinogenicity has been found in rodent tests of the more highly chlorinated biphenyls and possibly premalignant changes in the intestine of treated monkeys. Several studies have shown that PCBs increase the incidence of liver tumours induced by various carcinogens, indicating that their promoting potential is greater than their initiating potential (GREIM and WOLFF 1984).

Human information on the effects of PCBs comes chiefly from a serious poisoning episode in Japan in 1968, caused by contamination of rice oil with PCBs which also contained, however, polychlorinated terphenyls (PCTs) and polychlorinated dibenzofurans. Some 2000 "Yusho" patients had estimated intakes of 600 mg of PCBs and PCTs and 3 mg of polychlorinated dibenzofurans, which are analogues of the highly toxic chlorinated dibenzodioxins (dioxins). In addition to various skin, neural and hormonal disorders in the patients, cancer accounted for 38% of 31 deaths occurring within 10 years, compared with 21% in an unexposed population (MASUDA 1985). An increased incidence of malignant melanoma in 72 petrochemical workers exposed to the PCB mixture Aroclor 1254 was reported by BAHN et al. (1976), but other studies have not confirmed this observation (ALDERSON 1986).

The polybrominated biphenyls (PBBs), made primarily for use as flame retardants, are also liver carcinogens for rats and mice (KIMBROUGH et al. 1981; ANONYMOUS 1983). Accidental incorporation into animal feed in 1973 caused widespread poisoning of livestock in Michigan, and ingestion of PBBs by farming and other families in the state. Extensive monitoring of the exposed population and PBB workers has shown a variety of symptoms, without clarifying the part played by PBBs in causing them (DURHAM 1983; IARC 1986c). There is no evidence of human carcinogenicity to date.

Proceedings of two symposia on health aspects of this range of compounds have been published (ENVIRONMENTAL HEALTH PERSPECTIVES 1985a, b), and IARC has also reviewed PCBs (1978) and PBBs (1986c).

F. Nitrosamines

N-Nitroso carcinogens are probably now the chemical carcinogens of predominant research interest. Despite the fact that they have not so far been unequivocally linked with any human cancers, there are good reasons for suspecting the involvement of exogenously or endogenously formed nitrosamines or nitrosamides in causing some cancers, such as those of the stomach, oesophagus and nasopharynx. Various tobacco products contain exceptionally high levels of tobacco-specific nitrosamines, which are probably the main agents responsible for oral "snuff-dipper's cancer" in the USA (HOFFMANN et al. 1986).

Most nitrosamines and nitrosamides submitted to carcinogenicity testing, now over 300, have proved carcinogenic. Potency is generally high, and tumours are induced in many body organs in a remarkably specific manner (PREUSSMANN and STEWART 1984; PREUSSMANN and WIESSLER 1987). All of the many species tested have proved susceptible to nitrosamine carcinogenesis, and humans are undoubtedly also susceptible. Environmental and occupational aspects of these carcinogens are reviewed by PREUSSMANN and EISENBRAND (1984) and in several IARC publications (O'NEILL et al. 1984; BARTSCH et al. 1987).

I. Nitrosamines in Industry

1. Intentional Use

Simpler volatile nitrosamines such as *N*-nitrosodimethylamine (dimethylnitrosamine) have properties that could have been useful industrially, for example as solvents and synthetic intermediates. WEISBURGER and WEISBURGER (1966) drew attention to a number of American and European patents granted between 1958 and 1963 for uses of dialkyl and other nitrosamines in a variety of commercial processes, which could well have carried high risks of causing occupational cancer. Fortunately, deliberate use of these carcinogens appears to have been avoided. An exception was a synthetic process using *N*-nitrosodimethylamine, which resulted in levels approaching 1 ppm in the ambient air of Baltimore and subsequently the enforced closure of the process (FINE et al. 1976).

2. Inadvertent Formation

Even without intentional use, nitrosamines can be formed inadvertently in the workplace, as in many other situations. The occurrence of *N*-nitrosodimethylamine and *N*-nitrosomorpholine in the rubber industry was reported by FAJEN et al. (1979), and nitrosamines can apparently be released from products such as tyres upon storage. Atmospheric levels can be greatly reduced by improved ventilation, but a better solution is to identify the sources and prevent nitrosamine formation wherever possible.

Some epidemiological studies have shown the existence of raised levels of several cancers associated with particular operations and chemical exposures in the American rubber industry (MCMICHAEL et al. 1976; MONSON and FINE 1978), but there is no apparent reason for attributing these to nitrosamines rather than other substances. Several studies have also shown the occurrence of *N*-nitrosodimethylamine and *N*-nitrosomorpholine in leather tanneries, leading to calculated daily intakes of 440 and 20 µg, respectively (FINE 1980).

The existence of carcinogenic contaminants in mineral oils used in engineering was referred to earlier. For some purposes, however, these oils have been replaced as metal-working fluids by oils based on di- and triethanolamine. The use of nitrite as a corrosion inhibitor in these fluids has led to the formation of *N*-nitrosodiethanolamine (FAN et al. 1977). Though considerably less potent than *N*-nitrosodiethylamine, this nitrosamine is undoubtedly carcinogenic (LIJINSKY and KOVATCH 1985). One small-scale study of workers exposed to synthetic cutting oils has found no evidence of carcinogenicity (JÄRVHOLM et al. 1986), but the possibility of nitrosamine exposure clearly needs to be guarded against in engineering as well as in other industrial and environmental situations.

G. Inorganic Carcinogens

Exposure to compounds of arsenic, nickel and chromium has long been associated with occupational cancer, and more recently beryllium has been added to the list. There is only weak evidence for cadmium and iron compounds having caused human cancer, although these and compounds of various other metals are experimental carcinogens and mutagens (IARC 1980a; GILMAN and SWIERENGA 1984; SUNDERMAN 1984a; FURST and RADDING 1984). Cancers have also resulted from the use of radioactive metallic compounds, as in the former luminous dial painters who ingested radium-series metals and in patients receiving the X-ray contrast medium Thorotrast, which contains thorium dioxide. Being a consequence of long-term body irradiation, these examples are regarded as being outside the scope of this chapter.

I. Arsenic

Arsenic compounds have clearly caused human cancer in occupational and non-occupational settings. Tumours induced have been chiefly but not exclusively of the skin and lungs.

As arsenic is present in many ores, smelting processes for various metals can release arsenic trioxide, leading to inhalation by workers. Several studies of copper smelters exposed before 1956 showed relatively small increases in lung cancer mortality (LUBIN et al. 1981; ENTERLINE and MARSH 1982a, b), but six lung cancer deaths in a heavily exposed group represented a 14-fold increase over expectation (HIGGINS et al. 1981). A Swedish study indicated a possible synergism with smoking (PERSHAGEN et al. 1981). ENTERLINE et al. (1987) recently reported that the lung cancer risk to copper smelter workers appears greater than earlier thought.

Other occupational exposure to arsenic formerly occurred through the agricultural use of arsenical sprays, particularly in vineyards. Bordeaux mixture contains copper sulphate, lime and calcium arsenite and was associated with skin and lung carcinogenesis in exposed workers (NEUBAUER 1947; GALY et al. 1963). Non-occupational exposures to arsenic associated with carcinogenesis have been through pollution of drinking water (YEH 1973; CHEN et al. 1986) and from medicinal use of Fowler's solution and other preparations (SCHMÄHL et al. 1977; REYMANN et al. 1978).

Many past tests of arsenic compounds for carcinogenicity in animals have been negative, but IVANKOVIC et al. (1979) have induced lung carcinoma in some rats by intratracheal instillation of a discontinued arsenical pesticide mixture. Since then a few other experiments have also shown some evidence of activity with arsenic trioxide or arsenical dusts (SUNDERMAN 1984a). Although SUNDERMAN suggested that arsenic may act as a tumour promoter rather than as an initiator, it has given positive results in a number of short-term tests for genotoxic activity; the mechanisms of its undoubted carcinogenic action in humans remain obscure.

II. Nickel

The carcinogenicity of nickel has had very extensive epidemiological and experimental study (SUNDERMAN 1984a, b). In 1932 an alert physician's observations of nasal sinus cancer in two workers at a South Wales nickel refinery led to the recognition of an excess risk of nasal and lung cancer in the workers (DOLL 1984). Nickel carcinogenesis has been subsequently established among nickel refinery workers in several other countries, some studies also suggesting increased risks of cancer at other sites such as the stomach in the USA (ENTERLINE and MARSH 1982a, b) and USSR (SAKNYN and SHABYNINA 1973).

A study covering 54000 workers at two Canadian nickel plants also found increased risks of cancer in workers in sintering operations, with an 80-fold increase in nasal cancer among those at one plant (ROBERTS et al. 1980). A more recent prospective study of Canadian nickel workers, however, found small increases in lung and laryngeal cancer, but no cases of nasal cancer (SHANNON et al. 1984).

Nickel refining involves the formation and decomposition of the toxic gas nickel carbonyl, $Ni(CO)_4$, but though this is carcinogenic in animals it is not thought to have been responsible for carcinogenesis in nickel workers. More likely causes are relatively insoluble nickel compounds present in inhaled dusts, including the potent experimental carcinogen nickel subsulphide, Ni_3S_2, and nickel oxides (IARC 1982b).

III. Chromium

Epidemiological studies from several countries have now demonstrated raised risks of lung cancer in chromium workers, largely in those engaged in chromate production (LANGÅRD 1983; GILMAN and SWIERENGA 1984; SUNDERMAN 1984a). In some studies the lung cancer risk was raised two- to threefold over expectation (HAYES et al. 1979; ALDERSON et al. 1981), while SATOH et al. (1981) reported a standardised mortality ratio of 9.2 among Japanese workers engaged in chromium work between 1918 and 1935.

Workers in chromium industries have been exposed to a complex range of tervalent and hexavalent compounds which differ widely in their solubilities. The risk of carcinogenesis is greatest with sparingly soluble Cr(VI) compounds such as calcium and zinc chromates, rather than with the still less soluble lead chromate or the highly soluble sodium and potassium chromates. Thus, in the study of the mortality of chromium pigment workers in Britain reported by DAVIES (1984), a significantly increased risk of lung cancer occurred with moderate to heavy exposure to zinc chromate; some cases occurred after relatively brief exposure and with unusually short latent periods. There was no evidence for carcinogenesis by lead chromate even under conditions associated with signs of lead poisoning.

Chromium plating is often carried out in small organisations and involves exposure to aerosols of chromium trioxide, also containing hexavalent chromium. Some epidemiological studies and case reports have suggested raised cancer risks among chromium plating workers (LANGÅRD 1983), while SORAHAN et al. (1987) have reported significant increases in several cancers among 2689 nickel and chromium platers employed in Britain between 1946 and 1983, particularly lung cancer in relation to chromium bath work.

There is no evidence of occupational cancer caused by trivalent chromium materials. Experimentally, significant carcinogenicity has only been found with sparingly soluble Cr(VI) compounds, though it is difficult to devise animal experiments which mimic conditions of human exposure. However, using a technique of implanting pellets containing test materials within the bronchi of rats, LEVY and VENITT (1986) and LEVY et al. (1986) obtained tumours only with a few Cr(VI) materials. Small yields of tumours were obtained with the industrial materials $K_2CrO_4 \cdot 3\,ZnCrO_4 \cdot Zn(OH)_2$ and $CaCrO_4$. Two commercial samples of $SrCrO_4$ induced the most tumours, but none were obtained with soluble chromate or dichromate or with Cr(III) materials such as chromite ore. Animal experiments are in accord with the many short-term tests carried out which show activity confined almost entirely to Cr(VI) compounds.

Hexavalent chromium in the body is readily reduced to the trivalent form, and this also occurs within cells. Only Cr(VI) compounds pass easily through the cell wall, so that carcinogenicity is apparently confined to hexavalent chromium compounds with solubility characteristics favouring long retention and slow release in the body. Within the cell, however, Cr(III) may be the form that interacts with DNA and leads eventually to malignant change (LEVY and VENITT 1986). NORSETH (1986) considers that all forms of chromium are potentially carcinogenic, with their risks being governed by their bioavailability, a view that was criticised in later correspondence.

IV. Beryllium

Beryllium became important industrially in the 1940s, but one important use in fluorescent lighting was discontinued after about 1950 when beryllium compounds became recognised as the cause of serious lung disease in workers. Beryllium is an experimental carcinogen, inducing lung or bone tumours in rodents depending on the route of administration, and positive results have been reported from a range of short-term tests with bacteria and mammalian cells (GILMAN and SWIERENGA 1984; LÉONARD and LAUWERYS 1987).

Human evidence from several individual studies has been inconclusive, with small numbers of excess cancer cases being reported from some studies but not others (ALDERSON 1986). KUSCHNER (1981) and a working group chaired by DOLL (1981) concluded that beryllium had caused some cases of occupational cancer, while IARC (1982b) called the evidence "limited". Nevertheless, it is clear that beryllium exposure needs to be carefully controlled on account of its various other adverse effects on health.

V. Cadmium

Cadmium is a toxic metal with large-scale uses in the manufacture of nickel-cadmium batteries and alloys and in electroplating. It has given rise to local sarcomas and testicular tumours following injection into rodents, but other animal tests and many short-term tests have been negative (IARC 1982b; GILMAN and SWIERENGA 1984).

A raised incidence of prostate cancer was reported in nickel-cadmium battery workers by KIPLING and WATERHOUSE (1967), but numbers were small (4 observed, 0.58 expected), and exposure of these workers to cadmium was then very heavy. Small excesses of prostate or lung cancer have been found in some, but not all, other studies (ALDERSON 1986). With one exception the working group chaired by DOLL (1981) considered that there were indications of a link between cadmium exposure and prostate cancer. According to PISCATOR (1981) any such effect would only derive from excessive exposure interfering with the zinc-hormone relationship in the prostate gland and not from any direct action on the cells. Since then SORAHAN and WATERHOUSE (1985) have reported that new cases of prostate cancer among battery workers were almost exactly the number expected. ARMSTRONG and KAZANTZIS (1985) have suggested a marginally increased risk from high to medium cadmium exposure, but high exposure is also associated with bronchitis, emphysema, possibly lung cancer and nephritis.

VI. Iron

Underground mining of haematite iron ore in a number of countries carries an increased risk of lung cancer, which is not found in surface haematite miners. Both processes involve exposure to dusts of iron(III) oxide, silica and various other materials (IARC 1982b). Iron(III) oxide (Fe_2O_3) has not been found carcinogenic or active in very limited short-term tests, but it may possibly potentiate a carcinogenic action of the radioactive gas radon, to which only the underground miners are exposed.

Some of the many studies of iron and steel foundry workers have also shown an increased risk of lung cancer (IARC 1984a). The causes are unknown, but foundry atmospheres contain a very wide range of constituents, including various known carcinogens and promoting agents.

VII. Other Metals

A number of other metals of industrial importance have given some evidence of carcinogenicity in animals, without so far providing any indication of their being occupational carcinogens for humans (GILMAN and SWIERENGA 1984; SUNDERMAN 1984a). Perhaps of particular interest is lead, which has been found carcinogenic for the kidney in rats and mice, though not in hamsters or dogs. Lead compounds are, of course, highly toxic, and there has been particular concern and controversy over possible harmful effects of environmental lead on children. Despite some heavy occupational exposures, however, limited epidemiological studies have not demonstrated that such exposures have caused cancer in humans.

H. Asbestos and Other Mineral Fibres

I. Introduction

Asbestos has been in use since at least 450 BC, and its harmful effects on slaves were noted by Pliny the Younger about 100 AD. It now poses the most outstanding problem in occupational cancer, on account of the huge scale of its use (1976 world production approx. 5 million tonnes) and the clear association between inhalation of asbestos dust and occupational lung cancer and mesothelioma of the pleura and peritoneum. Recognition of its hazards has led to great improvements in conditions surrounding its many uses and to bans on use of the more hazardous forms. Nevertheless, huge amounts are still present as insulation and fireproofing in buildings, ships and railway locomotives and will pose problems for many years to come.

The term "asbestos" covers a range of fibrous minerals. Of the 1976 production 97% was chrysotile (white asbestos), the remainder being amosite (brown asbestos) and crocidolite (blue asbestos). Some human exposure may also occur to non-commercial forms present in some minerals and soils. All are silicates, the metal being largely magnesium in chrysotile while much iron is present in amosite and crocidolite. Differences in the chemistry of the different forms are reflected in marked variations in the shape and dimensions of the fibres as revealed by electron microscopy, and these appear particularly important in carcinogenesis. With respect to induction of experimental mesothelioma, carcinogenesis is associated with fibres of diameter <0.25 µm and length >5 µm (WAGNER 1984). However, there is increasing reason to believe that chemical surface characteristics are also of great importance (DUNNIGAN 1984).

The literature in this field is enormous. Useful brief reviews are given by HOWARD (1984) and WAGNER (1984), while recently ALDERSON (1986) has provided a valuable outline of epidemiological aspects. The last IARC monograph to review asbestos carcinogenesis is now over 10 years old, but the proceedings of a 1979 IARC/INSERM symposium on asbestos and other mineral fibres have been published (WAGNER 1980). Ingestion of asbestos occurs through swallowing of inhaled dust as well as from its presence in some drinking water supplies (ENVIRONMENTAL HEALTH PERSPECTIVES 1987).

II. Asbestos and Lung Cancer

Even if asbestos dust were not carcinogenic, stringent control of asbestos dust exposure would be required to prevent the serious condition of asbestosis, which progressively destroys lung function. However, from the 1930s reports began to appear linking asbestos exposure with lung cancer. Lung cancer in British asbestos workers was reported by DOLL (1955), and ALDERSON (1986) tabulates many recent studies of lung cancer in asbestos mining, milling, textile and cement manufacture, insulation, etc., with ratios of observed to expected cases ranging up to 8–9.

1. Synergism with Smoking

Exposure of cigarette smokers to asbestos dust provides the clearest example of synergism in human carcinogenesis, a number of studies reporting risks of lung cancer in smoking asbestos workers much greater than expected from simple summation of the individual risks of smoking and asbestos dust. Compared with non-smoking, non-asbestos workers in the USA, SELIKOFF and HAMMOND (1979) reported lung cancer death rates in one study increased 5 times for non-smoking asbestos workers, 11 times for smokers without asbestos exposure and 53 times for smoking asbestos workers. Not all studies find such a striking multiplicative effect (ALDERSON 1986), but undoubtedly smoking is exceptionally hazardous for those also exposed to asbestos dust.

III. Asbestos and Mesothelioma

Apart from some early case reports, mesothelioma consequent on asbestos exposure was first reported from the crocidolite mining area of South Africa by WAGNER et al. (1960), and this occupational cancer has now been found also in various other countries. Families and others living near asbestos works and near crocidolite but not other mines are also affected as well as workers. The average latent period from first exposure to diagnosis is 40 years, and in this case the smoking history is not important.

Mesothelioma of the pleura or peritoneum is a very serious but fortunately rare form of cancer, encountered almost entirely in association with asbestos workers. Thus, an atlas of cancer mortality in England and Wales shows a remarkable concentration of high-incidence areas around ports, where there has been high exposure to crocidolite in ship insulation, as well as certain asbestos works (GARDNER et al. 1983). Cases of mesothelioma have more unexpectedly arisen also from occupations such as the manufacture of gas masks in the 1940s, often using crocidolite asbestos padding (ACHESON et al. 1982). A case of mesothelioma in a 28-year-old man was attributed to short but heavy domestic exposure to crocidolite (BOOTH and WEAVER 1986), but this interpretation was questioned on account of the very short latent period and the occasional cases of mesothelioma that do arise in young people.

1. Erionite and Mesothelioma

A tragic exception to the usually accepted association of mesothelioma with asbestos and industry has recently been recognised in some small villages in central Turkey. Here, very high incidences of mesothelioma and lung cancer evidently derive from the presence of fine fibres of erionite, a form of zeolite, in the local volcanic tuff used for building and other purposes (ROHL et al. 1982; ARTVINLI and BARIS 1985). Samples of erionite dust have been found more efficient than any form of asbestos dust in inducing mesothelioma in rats (WAGNER et al. 1985). Erionite dust exposure from an early age thus appears the most probable cause of mesothelioma in these areas and poses particularly difficult problems in preventive medicine.

IV. Asbestos and Other Cancers

Many of the studies listed by ALDERSON (1986) have also reported on the incidence of other cancers in asbestos workers. Very wide variations have been found. For gastrointestinal cancer only a few studies have shown significantly raised risks. Some high relative risks have been found for laryngeal cancer, but with very small numbers of patients. A few studies have suggested raised levels of ovarian cancer among female asbestos workers in England.

V. Asbestos Controls

Asbestos has been of outstanding importance for modern industrial societies, but recognition of its serious health hazards has necessitated increasingly stringent controls on its use and the development of safer substitutes. Thus in the UK the Asbestos Regulations 1969 replaced regulations enacted 36 years earlier to prevent asbestosis but have themselves been superseded by tighter controls. By 1984 HSE control limits in the UK were 0.2 fibres per ml of workplace air for crocidolite and amosite, and 0.5 fibres per ml for chrysotile. Former uses of crocidolite and amosite are now not permitted, but large amounts will still be encountered during ship-breaking, demolition and maintenance work for many years to come.

VI. Man-Made Mineral Fibres

These are made by various processes from molten glass, slag or rock. Unlike asbestos they have an amorphous structure and tend to fracture transversely rather than longitudinally. Concern over their possible health hazards has led to extensive testing in animals and epidemiological studies of exposed workers (WHO/IARC 1983; Saracci 1986). Some studies have suggested somewhat increased risks of lung cancer, but combining data from several countries shows only a very small increase of borderline significance (ALDERSON 1986). There is concern over the possibility of newer "superfine" fibres (diameter <3 μm) proving more hazardous, and in the UK there is currently a provisional recommended limit of 1 fibre per ml for such fibres (HSE 1987). Studies on possible hazards of newer ceramic fibres are being initiated in the UK.

I. Wood and Leather Dust

I. Nasal Cancer in Woodworkers

An association between exposure to wood dust and adenocarcinoma of the nasal passages was first reported from the High Wycombe area, a centre of the furniture-making industry in England, by ACHESON et al. (1967). Similar findings were subsequently made in other areas of the UK (ACHESON et al. 1981) and in many other countries with also, in some cases, raised risks of laryngeal cancer and Hodgkin's disease (see concise review by ALDERSON 1986, and an in-depth survey by IARC 1981).

Conditions leading to nasal cancer appear to have been worst in the 1920s and 1930s, with latent periods of some 30 years. Some cases have appeared after relatively short exposures to wood dust. The carcinogenic factors are unknown but may be naturally occurring carcinogens, more particularly in hardwoods. As with asbestos, long retention in the body of readily inhaled insoluble dust is probably an important factor leading to carcinogenesis.

Other important wood industries have also received extensive epidemiological study. Though IARC did not consider the evidence for carcinogenesis as sufficient, some increased incidences of cancer have been reported, including nasal cancer and Hodgkin's disease in lumber, sawmill, pulp and paper workers. Some increases in cases of soft-tissue sarcoma and lymphoma have been attributed to chlorophenols encountered in preservatives, but very many other chemicals are also used in the various wood industries studied (IARC 1981).

II. Nasal Cancer in Leather Workers

The presence of an important boot- and shoe-making centre not far from High Wycombe facilitated recognition of a high incidence of nasal carcinoma presumably resulting from inhalation of leather dust (ACHESON et al. 1970). This association has since been confirmed in Italy and elsewhere, but leather tanning does not carry this risk. Very large numbers of chemicals are used here also; the association between leukaemia and benzene used in shoe-making is discussed in Sect. J. The various leather industries were similarly reviewed by ALDERSON (1986) and IARC (1981).

J. Benzene

Benzene is an extremely important commercial chemical, with an estimated annual world production approaching 7 million tonnes in 1980. Most is used for conversion to styrene, phenol, cyclohexane and other chemicals, with solvent use now relatively small. Not possessing alkyl substituents to permit ready oxidation and elimination from the body, benzene has considerably greater toxicity than toluene or xylenes, and a variety of metabolites are thought to derive from its initial oxidation to benzene epoxide.

In humans chronic benzene toxicity affects particularly the haematopoietic system and may lead to irreversible anaemias and leukaemia. Much of the early evidence for human carcinogenicity came from case reports of leukaemia following high-level exposure to benzene being used as a solvent under conditions of bad ventilation. One important series of such reports recorded cases of leukaemia among shoe-makers in Turkey who were exposed to high levels of benzene in rubber cements. New cases ceased to appear soon after the use of benzene was phased out (AKSOY 1985). Shoe-makers in Italy and France have been similarly affected. Exposures in shoe-makers may have been in the range 100–600 ppm (HSE 1982).

Less clear evidence has generally come from studies of workers in large industrial organisations. Here, atmospheric levels have dropped considerably from the 10–100 ppm common in the 1940s (FISHBEIN 1984), and workers have been exposed to many other chemicals also. WONG (1987) recently reported on cancer incidence among 7676 workers in seven American plants. Mortality rates from lymphoid and haematopoietic cancers were only slightly above national rates but were significantly above those in unexposed controls in the same plants (relative risk 3.20). For those with benzene exposures of at least 720 ppm-months, the estimated relative risk was 3.93. RINSKY et al. (1987) calculated relative risks of 3.37 for leukaemia and 4.09 for multiple myeloma among workers in three plants manufacturing natural rubber film from rubber solution in benzene.

A recent study from China, however, showed a sevenfold increase in mortality from leukaemia among 28460 workers exposed to benzene in 233 relatively small factories, with an average latent period of 11.4 years (YIN et al. 1987). Estimated atmospheric levels were generally 20–170 ppm but occasionally over 300 ppm.

The leukaemogenic action of benzene has generated much controversy as well as a large and growing literature. In the 1970s the US Occupational Health and Safety Administration recommended that the permissible exposure limit for benzene should be reduced from 10 to 1 ppm, but the proposed new standard was overturned by the US Supreme Court in 1980, and the higher limit still stands. In correspondence following a 1983 international conference on benzene (MEHLMAN 1985), INFANTE (1987) strongly criticised the approach which he termed, "Take no protective action until definitive evidence becomes available". To us it seems that relatively little attention is paid to the serious toxic effects of benzene other than leukaemia in determining permissible levels.

Many early animal tests of benzene for carcinogenicity were inconclusive, but the activity of benzene administered orally in olive oil or by inhalation has been demonstrated in a large series of experiments summarised by MALTONI et al. (1985). Benzene was also found to be carcinogenic in NTP tests using rats and mice given benzene orally in corn oil (1986). Tumours were induced at a number of sites, particularly in mice. Benzene is inactive in bacterial and other mutagenicity tests but induces chromosome abnormalities in some animal species and in exposed humans (DEAN 1985).

Benzene carcinogenicity is reviewed briefly by ALDERSON (1986) and in depth by IARC (1982a) and MARCUS (1987) The toxicity and carcinogenicity of benzene are not only industrial problems; care is also needed to avoid its use as a solvent in situations such as educational establishments and hospitals.

K. Formaldehyde and Other Aldehydes

Formaldehyde (methanal) is a reactive gas, generally encountered as the stabilised 37% aqueous solution formalin, which has very many large-scale uses as a chemical intermediate, especially in resin manufacture, and as a sterilant and preservative. In recent years it has been the source of much laboratory and epidemiological research and continuing controversy over its possible carcinogenic hazards.

I. Animal Bioassays

The evidence that formaldehyde is carcinogenic comes mainly from two series of long-term tests in which rats inhaled levels up to about 14 ppm in air. At this level a high proportion of rats developed carcinomas of the nasal cavities. In one series of tests (SWENBERG et al. 1980; KERNS et al. 1983), inhalation of formaldehyde at 14.3 ppm for 24 months led to nasal cavity carcinomas in 44% of animals. At 5.6 ppm, however, less than 1% of rats developed these tumours, and very few mice developed tumours even at the higher level.

In the other tests, 14 ppm formaldehyde was inhaled by rats together with 10 ppm of hydrogen chloride (ALBERT et al. 1982; SELLAKUMAR et al. 1985), which treatment also led to many animals developing nasal carcinomas. These were attributed largely or entirely to the action of formaldehyde, with no significant contribution from the hydrogen chloride or any bis(chloromethyl) ether that might have been formed from the chloride-formaldehyde reaction.

In view of the non-linear dose-response of rats to formaldehyde inhalation and the association of nasal tumours with considerable tissue damage, the significance of the rat tumours for humans is far from clear; they may be a consequence of overloading protective mechanisms of detoxication and DNA repair (SWENBERG et al. 1983). Hamsters, which have inhaled 10 and 30 ppm in lifespan tests without developing nasal tumours, also show minimal changes in their nasal tissues (DALBEY 1982).

II. Epidemiological Studies

Many cohort and case-control studies of cancer in relation to formaldehyde exposure in a wide range of occupations and in the home environment have now been published. Evidence prior to a 1984 symposium has been summarised by BLAIR et al. (1985) and by O'BERG (1985). Both reviews comment on the raised incidence of brain cancer and leukaemia among certain exposed professional groups of anatomists, embalmers and pathologists, but not all studies have shown this. The main British study, for example, which covered 7680 men employed before 1965 in six chemical or plastics factories, found only a small excess of lung cancer in one factory when compared with national, but not local, lung cancer rates (ACHESON et al. 1984). A relatively small proportion of workers, however, had had high level exposure for more than 5 years and had been followed for 20 years. A later and larger study of 26561 workers in 10 American factories also provided "little evidence to suggest that formaldehyde exposures affected the mortality experience of these industrial workers" (BLAIR et al. 1986).

There has been considerable concern over possible hazards from formaldehyde liberated from home construction materials and urea-formaldehyde foam insulation. VAUGHAN et al. (1986a) report a strong association between living in mobile homes in west Washington state and cancer of the nasopharynx but not cancer of the oropharynx or nasal sinuses. Numbers were small, and the significance of any formaldehyde exposure is unknown, but apparent small increases in nasopharyngeal tumours were also seen in the parallel occupational study by VAUGHAN et al. (1986b) and in that of BLAIR et al. (1986).

In areas in which nasopharyngeal carcinoma is prevalent, particularly south China, the disease is well-known to be associated with the widespread Epstein-Barr virus. Contrasting with these observations, a Danish study has reported evidence for increased cancer of the nasal cavity and paranasal sinuses associated with formaldehyde exposure (OLSEN and ASNAES 1986).

There are thus wide discrepancies even between those studies that have reported some increases in cancer among subjects exposed to formaldehyde, and the question of formaldehyde carcinogenicity for humans remains confused and controversial. In the USA the Environmental Protection Agency has attempted to estimate possible numbers of cancer cases attributable to formaldehyde and may introduce regulations to extend control over formaldehyde exposures (MARSHALL 1987).

The literature in this area is very large and still growing. Useful reviews have been given by FLAMM and FRANKOS (1985), ACHESON (1985), ALDERSON (1986), and NELSON et al. (1986).

III. Other Aldehydes

FERON et al. (1982) reported that inhalation by hamsters of extremely high levels of acetaldehyde (2500 ppm reducing to 1650 ppm) induces benign or malignant tumours of the larynx or nasal cavities in 17%–28% of animals. The tests were extended to rats, most of which developed nasal carcinomas after inhalation of levels up to (initially) 3000 ppm for 28 weeks (WOUTERSEN et al. 1986). As with formaldehyde, treatment at carcinogenic levels also results in considerable damage and regeneration in nasal tissues. These perhaps unexpected results do not seem to indicate significant human risks from occupationally acceptable levels of under 100 ppm, but there is no reliable evidence available at this time. Acetaldehyde carcinogenicity has been reviewed by IARC (1985a).

Glutaraldehyde has been receiving increasingly extensive use as a sterilising and fixing agent in, for example, hospitals, dairy and other food industries, and biological laboratories. It is highly reactive and irritant, and the question of its possible carcinogenic risks has been raised. While the results of unpublished NTP tests cannot be anticipated, glutaraldehyde has given no evidence of activity in various mutagenicity tests in which formaldehyde and acetaldehyde are positive (SLESINSKI et al. 1983).

L. Some Other Suspected Occupational Hazards

Various other important industrial chemicals have shown some degree of experimental carcinogenicity, leading to suspicions of possible carcinogenic risks in the workplace. In some cases epidemiological studies have been carried out, but so far without giving clear evidence on which to judge possible risks to personnel.

I. Acrylonitrile

This important chemical for the polymer industry is a neurotoxin which has given positive results in some tests for mutagenicity, probably dependent on metabolism to the epoxide. Some tests suggest weak carcinogenicity, particularly for the central nervous system, though results were only significant when microscopic tumours were included in the assessment (BIGNER et al. 1986). While early human studies suggested possible increased risks of respiratory cancer from high exposures in the 1950s, a recent American study found only an increase in prostate cancer (CHEN et al. 1987). However, the numbers were small, and the question of carcinogenic risk remains open.

II. Acrylamide

This related chemical is extensively used for polymerisation to polyacrylamides, in biological laboratories as well as in the manufacturing industry. It has toxic effects on the nervous system, testes and skin, and though showing little evidence for genotoxicity in short-term tests, it has given evidence of weak carcinogenicity in rodents (JOHNSON et al. 1986). A recent study of acrylamide workers found only 4 deaths due to cancer against 6.5 expected (SOBEL et al. 1986), but its use requires careful control on account of its undoubted toxic effects and suspicions of germ-cell mutagenicity.

III. Amitrole

Amitrole (3-amino-1,2,4-triazole) is a non-mutagenic herbicide constituent of low toxicity, but it shows some anti-thyroid activity. While feeding at >2000 ppm induces thyroid and liver tumours in mice (INNES et al. 1969), levels up to 100 ppm induce thyroid tumours in rats but are not carcinogenic for mice or hamsters (STEINHOFF et al. 1983).

The carcinogenicity of amitrole appears to be largely a poorly understood consequence of hormonal imbalance, as seen with some other chemicals with anti-thyroid activity and with some hormones. Restricting administration of such compounds to 5 days a week, already common practice with test compounds given by gavage, would allow 2 days for thyroid recovery and might give results more relevant to any likely conditions of human exposure.

There is virtually no epidemiological evidence to judge human risk. A Swedish study found a small excess of cancer deaths in herbicide-exposed railway workers, but numbers for those exposed only to amitrole were very small (5 cancers, expected 3.3) (AXELSON et al. 1980). Amitrole has been reviewed by IARC (1986a).

IV. 1,3-Butadiene

Butadiene, used on a huge scale in synthetic rubber manufacture, is a gas with very low acute toxicity that is assumed to be metabolised via its epoxides. It has now been found to be carcinogenic for rats and mice on inhalation at high levels comparable to those which were, until recently, acceptable for industrial atmospheres (though in practice levels were generally very much lower). Long-term inhalation by rats of butadiene at 1000 and 8000 ppm induced tumours of the pancreas, uterus, Zymbal gland, mammary gland, thyroid and testis (OWEN et al.

1987). NTP tests with mice at 625 and 1250 ppm were terminated at 61 weeks because of high mortality, but a variety of tumours were also seen by this time in mice (HUFF et al. 1984).

There is no evidence that butadiene has been carcinogenic for humans (HSE 1985 a), but industrial limits have now been drastically reduced to 10 ppm in the UK and the USA in view of the animal carcinogenicity data and anxieties that butadiene might be a germ-cell mutagen.

V. Hydrazine

The highly reactive and toxic hydrazine is widely used in the production of herbicides, medicinals, plastics and also in rocket propellants and water treatment. In experimental animals it has induced liver and lung tumours on oral administration and nasal tumours by inhalation (CABRAL 1985).

Though suspected of carcinogenic potential for humans, no evidence of carcinogenicity was found in a small-scale study of workers exposed to hydrazine under very poor conditions in a British factory between 1945 and 1971 (WALD et al. 1984). It was estimated that general work areas then had hydrazine levels of 1–10 ppm, with perhaps 100 ppm near open storage vessels. Considering its relatively weak mutagenic and carcinogenic action and the current maximum levels of 0.1 ppm in air, any significant carcinogenic risk now appears unlikely.

VI. Di(2-ethylhexyl) Phthalate and Related Compounds

Di(2-ethylhexyl) phthalate (DEHP), di(2-ethylhexyl) adipate (DEHA) and some related esters have large scale use as plasticisers and for many lesser purposes, leading to widespread distribution in the environment. A cause of particular concern has been the migration that can occur from plastic film wrapping into foodstuffs.

NTP tests of DEHP fed to rats at 6000 and 12000 ppm and mice at 3000 and 6000 ppm led to liver tumours in a significant proportion of the animals, particularly in females (KLUWE 1986). Tumours were not induced at other sites, and DEHP is not mutagenic or active as an initiator or promoter in two-stage carcinogenicity tests.

This very weak carcinogenicity of DEHP appears to be a consequence of its inducing proliferation of peroxisomes within cells, a phenomenon seen particularly with hypolipidaemic drugs such as clofibrate, and characteristic of rodents but not primates. From the extensive and growing literature in this field, it does not now appear that agents such as DEHP or DEHA (which is less active in rodents) represent any significant carcinogenic hazard to humans, even though their ingestion does seem inherently undesirable (TURNBULL and RODRICKS 1985; HSE 1986; GANGOLLI 1986; BRIDGES 1987).

VII. Tobacco Smoke

Interactions between occupational and environmental carcinogens and smoking have been reviewed by SARACCI (1987). Apart from the well-known asbestos-smoking synergism, there is evidence for interactions of smoking with other occupational exposures. Limited evidence for nickel is consistent with an additive relationship, but several studies indicate interactions approaching multiplicative for exposure of copper smelters to arsenic and for carcinogenic aromatic amine exposure.

Tobacco smoke is among the many potential carcinogenic factors encountered in the workplace, those exposed including office workers and many others with no exposure to industrial chemicals. For many, of course, smoke inhaled at work is in addition to that in public places and at home. Tobacco smoke is an extremely complex mixture which contains several thousand identified chemicals, including many polycyclic hydrocarbons and other well-known carcinogens (IARC 1986b). Sidestream smoke, emitted directly into the atmosphere when tobacco smoulders between puffs, differs markedly from the mainstream smoke inhaled by the smoker, and in particular contains much higher levels of several highly carcinogenic nitrosamines (BRUNNEMANN et al. 1977; see HOFFMANN and HECHT, this volume).

Epidemiological studies of the effects of environmental tobacco smoke on "passive smokers" have so far concentrated on the risks of lung cancer in non-smoking women in relation to smoking by their husbands. A number of studies, though not all, have reported small but significant increases in lung cancer risk to the non-smoking spouse. The findings have generated much controversy, but various recent reviews have concluded that the relationship is real (BLOT and FRAUMENI 1986; SURGEON GENERAL 1986; VAINIO 1987).

In addition to strong pressure against smoke pollution of public buildings and transport, a smoke-free work environment is increasingly recognised as desirable, not only for protecting workers' health but also economically in terms of lower absenteeism, fire risks, cleaning costs and increased productivity. If a health hazard is now accepted, companies may under existing legislation be obliged to impose controls on smoking in the workplace. With ever stricter controls being applied to the presence of harmful chemicals at work, even those likely to pose, at the most, extremely small risks, it is now quite illogical for tobacco smoke to enjoy continued exemption from control.

M. Continuing Problems and Prospects

I. Proportion of Cancer Attributable to Occupation

The most clearly identified causes of cancer in the Western world are cigarette smoking and the range of occupational factors outlined above, but there are great differences in the numbers of victims with cancer attributable to them. Tobacco was estimated to cause 30% of all cancer deaths in the USA (DOLL and PETO 1981) and comparable, though generally smaller proportions in various countries. At the other extreme are some accepted occupational carcinogens such as mustard gas, which has caused a small number of lung cancer cases among a relatively small exposed population. Many other occupational carcinogens, particularly asbestos, have of course caused very many more than this, though still in relatively small numbers compared with cancer incidence in the general population, particularly in the period since the identification of important carcinogenic factors and the introduction of effective steps to prohibit or strictly control their uses.

As various epidemiologists had suggested the proportion of total cancer incidence attributable to occupational factors to be less than 5% or 10%, considerable surprise and controversy was generated when an unpublished discussion paper suggested that (for the USA) "estimates of at least 20% appear more reasonable, and may be conservative" (BRIDBORD et al. 1978).

The methodology on which their very high estimates were based attracted severe criticism, particularly from DOLL and PETO (1981), who considered that the document "should not be regarded as a serious contribution to scientific thought". In a later comprehensive report for the US Office of Technology Assessment, they made quantitative estimates of current avoidable risks of cancer in the USA. As noted above, they concluded that "by far the largest reliably known percentage is the 30% of current US cancer deaths that are due to tobacco", a figure which was expected to increase by a further 2%–3% by the mid-1980s. A figure of the order of 35% was suggested for dietary factors, with a very much greater degree of uncertainty.

The proportion of cancer deaths attributable to occupational factors in the USA was provisionally estimated as 4%, lung cancer accounting for the major proportion of this. On the basis of estimates of this order, complete success in preventing occupational cancer would have a much smaller impact on national cancer rates than would significant reductions in tobacco use or major improvements in dietary habits. However, 4%, or even 1%, of all cancer deaths nationally still represents a large number of people, and a considerably larger percentage in that section of the community actually exposed to potential risks from carcinogens at work. Prevention of such cancers must clearly remain an important goal for many years to come.

II. Identification of Carcinogenic Hazards

1. Identification Through Epidemiology

Past successes in identifying causes of occupational cancer have been achieved despite the great difficulties occasioned by the generally long latent period between exposure to a carcinogenic agent and the diagnosis of cancer, as well as by the number of confounding factors such as differences in smoking habits and socioeconomic class. Carcinogenesis is a prolonged multi-stage process, and an occupational cancer may well have derived from exposure 20 or more years prior to diagnosis of the disease, perhaps in a process long discontinued or in an earlier employment.

The choice of adequate control populations to compare with the workers under study presents many difficulties, and a "healthy worker effect" may be found when, as often happens, members of the general population used as controls are in poorer health than those fit to be employed. Inevitably, numbers of workers are often too small for any but a major carcinogenic effect to be observed, and it is not surprising that there are wide discrepancies between different studies of workers in comparable jobs.

Occupational risks have shown up most clearly when the exposed populations are at very greatly increased risk, as from early contact with mineral oils or aromatic amines under highly unhygienic conditions, or if the cancer induced has been one rarely seen in the general population, such as mesothelioma of the pleura and peritoneum or angiosarcoma of the liver. Other high relative risks may remain to be discovered, but in the more advanced technological societies, occupational cancer is now more likely to involve relatively small increases in cancers already common in the general population, particularly lung cancer, and these will be much more difficult to recognise.

A large epidemiological study, such as that in the US bladder cancer study covering nearly 3000 workers in the chemical industry (ZAHM et al. 1987), may still have insufficient statistical power to detect increased cancer risks because of the small proportion of the study population engaged in specific operations. In Denmark the existence of a long-established national cancer registry together with a personal identification system with details of employments permits studies not possible in other countries. The analysis by OLSEN and JENSEN (1987) of over 93000 cancer cases in Denmark between 1970 and 1979 in relation to occupation showed many generally small increases in cancer risks in a variety of occupations. However, even here the increases involved generally small numbers of actual patients, and some apparent cancer increases will have been due to chance. (These data are to be further refined and supplemented by another 100000 cases up to 1984.)

Many occupations involve exposure to a variety of chemicals, and studies in relation to their risks of cancer can then only give pointers to the factors responsible for any observed cancer increase, though the field may be greatly reduced where there is knowledge of the carcinogenic properties of the materials involved. Recent reviews of the very active field of occupational epidemiology, with its economic and social aspects, have been given by NICHOLSON (1984), SCHOTTENFELD (1984), SARACCI (1984), ROE (1985 b) and ALDERSON (1986).

A large volume on the significance of industrial carcinogens has been published (MALTONI and SELIKOFF 1988). HUNTER's classic "Diseases of Occupations" has now appeared as a multi-author work edited by RAFFLE et al. (1987). In the UK the comprehensive Control of Substances Hazardous to Health Regulations 1988 come into force from 1 October 1989.

2. Experimental Identification of Carcinogens

Occupational epidemiology shows the past existence of hazards that have already caused cancer, but the ideal solution is, of course, to prevent disease arising in the first place through recognition of potentially hazardous agents and processes, with appropriate control and steady improvement in working conditions.

In the past, many animal carcinogenicity tests have been quite inadequate for detecting carcinogens that were not of high potency, but in recent years many industrial and environmental chemicals have been subjected to much more thorough tests, particularly in the very large-scale programme initiated in the USA by the National Cancer Institute (NCI) and continued in the National Toxicology Program (NTP). Large groups of animals, mostly rats and mice of both sexes, are treated at two dose levels, one approaching the highest that does not cause severe chronic toxic effects, for much of their lifespan. This results in extremely high level dosing for compounds of very low toxicity. Some 300 detailed reports on the carcinogenicity of tested substances had been published up to NTP's management status report for October 1987, with many other tests projected, in progress or under evaluation.

In terms of dose required to induce tumours in 50% of animals in long-term tests, the potency of carcinogens ranges over some seven orders of magnitude, from aflatoxin B_1, carcinogenic for rats at 1 μg/kg per day, to agents which give barely significant increases in tumours in animals receiving high doses in NTP and comparable tests. Even though animal test results cannot be directly extrapolated to the conditions of human exposure, carcinogenic potency must be an important factor in risk assessment. There is, for example, much more cause for concern over a potent carcinogen such as 1,2-dibromoethane than, say, 1,4-dichlorobenzene, despite the small excess of tumours after long-term oral administration of extremely high levels which led NTP to regard the latter as clearly carcinogenic. Quantitative estimates of carcinogen potency, based on acceptable published tests, have been developed recently by AMES and his colleagues for many industrial and environmental materials (GOLD et al. 1986; AMES et al. 1987).

There are humane and practical objections to the NTP and similar lifespan carcinogenicity bioassays, which are extremely lengthy, costly and unable to cope with the wide range of substances for which carcinogenicity data would be desirable. Newer methodologies, as in the "decision point" approach (WEISBURGER and WILLIAMS 1984; WILLIAMS and WEISBURGER 1986), make extensive use of information from bacterial mutagenicity tests, assays for mammalian cell transformation and other short-term in vitro tests. These can be followed if needed by limited, shorter animal tests such as the induction of lung tumours in mice and breast tumours in rats. Full-scale animal tests are then only a last resort when uncertainties remain over an important chemical with extensive human exposure.

The preliminary short-term tests in this type of approach show whether a compound found carcinogenic is active by a genotoxic or epigenetic (non-genotoxic) mechanism. For genotoxic carcinogens no threshold dose level can be demonstrated experimentally, and the dose-response relationship at very low levels is unknown. Somewhere, however, there must be an ill-defined level at which, for all practical purposes, carcinogenic risks become negligible (as one hopes is the case in respect of the ppb levels of potent polycyclic hydrocarbon and *N*-nitroso carcinogens present in many common foodstuffs).

Many carcinogens are now regarded as acting through epigenetic mechanisms, not involving a direct attack on DNA. They include tumour promoters, immunosuppressive agents and agents which cause disturbance of hormone balance, and these probably do have a threshold below which there is no carcinogenic risk. It is also somewhat reassuring that some compounds reported carcinogenic, such as trichloroethylene and various peroxisome proliferators, are active in animals by mechanisms which appear not to be relevant for humans.

Better understanding of these mechanistic aspects of carcinogenesis will be invaluable for pinpointing "the important causes of human cancer among the vast number of minimal risks" (Ames et al. 1987) and utilising available resources more effectively in eliminating carcinogenic hazards from the workplace and environment.

Acknowledgements. We thank Dr. J. A. H. Waterhouse for valuable comments, Miss D. Williams for efficient secretarial assistance, and the Cancer Research Campaign for financial support.

References

Acheson ED (1985) Formaldehyde: epidemiological evidence. IARC Sci Publ 65:91–95

Acheson ED, Hadfield EH, Macbeth RG (1967) Carcinoma of the nasal cavity and accessory sinuses in woodworkers. Lancet i:311–312

Acheson ED, Cowdell RH, Jolles B (1970) Nasal cancer in the Northamptonshire boot and shoe industry. Br Med J 1:385–393

Acheson ED, Cowdell RH, Rang E (1981) Nasal cancer in England and Wales: an occupational survey. Br J Ind Med 38:218–224

Acheson ED, Gardner MJ, Pippard EC, Grime LP (1982) Mortality of two groups of women who manufactured gas masks from chrysotile and crocidolite asbestos: a 40 year follow-up. Br J Ind Med 39:344–348

Acheson ED, Barnes HR, Gardner MJ, Osmond C, Pannett B, Taylor CP (1984) Formaldehyde in the British chemical industry. Lancet i:611–616

Aksoy M (1985) Malignancies due to occupational exposure to benzene. Am J Ind Med 7:395–402

Albert RE, Sellakumar AR, Laskin S, Kuschner M, Nelson N, Snyder CA (1982) Gaseous formaldehyde and hydrogen chloride induction of nasal cancer in the rat. JNCI 68:597–603

Alderson M (1986) Occupational cancer. Butterworths, London

Alderson MR, Rattan NJ, Bidstrup L (1981) Health of workmen in the chromate-producing industry in Britain. Br J Ind Med 38:117–124

Ames BN, Infante P, Reitz R (eds) (1980) Ethylene dichloride: a potential health risk? Banbury Report 5. Cold Spring Harbor, New York

Ames BN, Magaw R, Gold LS (1987) Ranking possible carcinogenic hazards. Science 236:271–280

Anonymous (1983) PBB carcinogenicity in rats and mice. Food Chem Toxicol 21:688–689

Anthony HM, Thomas GM (1970) Tumors of the urinary bladder: an analysis of the occupations of 1030 patients in Leeds, England. JNCI 45:879–895

Armstrong BG, Kazantzis G (1985) Prostatic cancer and chronic respiratory and renal disease in British cadmium workers: a case control study. Br J Ind Med 42:540–545

Artvinli M, Baris YI (1985) Erionite-related diseases in Turkey. Nato Asi Ser G 3:515–519

Axelson O, Sundell L, Andersson K, Edling C, Hogstedt C, Kling H (1980) Herbicide exposure and tumor mortality: an updated epidemiologic investigation on Swedish railroad workers. Scand J Work Environ Health 6:73–79

Bahn AK, Rosenwaite I, Herrmann N, Grover P, Stellman J, O'Leary K (1976) Melanoma after exposure to PCB's. N Engl J Med 295:450

Bannasch P (ed) (1987) Cancer risks: strategy for elimination. Springer, Berlin Heidelberg New York

Barthel E (1976) Gehäuftes Vorkommen von Bronchialkrebs bei beruflicher Pestizidexposition in der Landwirtschaft. Z Erkrank Atm-Organe 146:266–274

Bartsch H, O'Neill IK, Schulte-Hermann R (eds) (1987) The relevance of *N*-nitroso compounds to human cancer: exposures and mechanisms. IARC Sci Publ 84

Beaumont JJ, Breslow NE (1981) Power considerations in epidemiological studies of vinyl chloride workers. Am J Epidemiol 114:725–734

Bell J (1876) Paraffin epithelioma of the scrotum. Edin Med J 22:135–137
Bigner DD, Bigner SH, Burger PC, Shelburne JD, Friedman HS (1986) Primary brain tumors in Fischer 344 rats chronically exposed to acrylonitrile in their drinking water. Food Chem Toxicol 24:129–137
Blair A, Walrath J, Malker H (1985) Review of epidemiologic evidence regarding cancer and exposure to formaldehyde. In: Turoski V (ed) Formaldehyde: analytical chemistry and toxicology. Adv Chem Ser 210. American Chemical Society, Washington DC, pp 261–273
Blair A, Stewart P, O'Berg M, Gaffey W, Walrath J, Ward J, Bales R, Kaplan S, Cubit D (1986) Mortality among industrial workers exposed to formaldehyde. JNCI 76:1071–1084
Blot WJ, Fraumeni JF Jr (1986) Passive smoking and lung cancer. JNCI 77:993–1000
Boice JD Jr, Land CE (1982) Ionising radiation. In: Schottenfeld D, Fraumeni JF Jr (eds) Cancer epidemiology and prevention. Saunders, Philadelphia, pp 231–253
Booth SJ, Weaver EJM (1986) Malignant pleural mesothelioma five years after domestic exposure to blue asbestos. Lancet i:435
Bowra GT, Duffield DP, Osborn AJ, Purchase IFH (1982) Premalignant and neoplastic lesions associated with occupational exposure to "tarry" byproducts during manufacture of 4,4'-bipyridyl. Br J Ind Med 39:76–81
Bridbord K, Decouflé P, Fraumeni JF Jr, Hoel DG, Hoover RN, Rall DP, Saffiotti U, Schneiderman MA, Upton AC (1978) Estimates of the fraction of cancer in the United States related to occupational factors. NCI/NIEHS/OSHA (Unpublished discussion paper)
Bridges JW (1987) Use of toxicity data – a case study of di-(2-ethylhexyl) phthalate. In: Richardson ML (ed) Toxic hazard assessment of chemicals. Royal Society of Chemistry, London, pp 233–246
Brunnemann KD, Yu L, Hoffmann D (1977) Assessment of carcinogenic volatile *N*-nitrosamines in tobacco and in mainstream and sidestream smoke from cigarettes. Cancer Res 37:3218–3222
Brusick DJ (1977) The genetic properties of beta-propiolactone. Mutat Res 39:241–256
Bus JS, Popp JA (1987) Perspectives on the mechanism of action of the splenic toxicity of aniline and structurally related compounds. Food Chem Toxicol 25:619–626
Butlin HT (1892) Cancer of the scrotum in chimney-sweeps and others. Br Med J 1:1341–46; 2:1–6, 66–71
Cabral JRP (1985) Hydrazine: laboratory evidence. In: Wald NJ, Doll R (eds) Interpretation of negative epidemiological evidence for carcinogenicity. IARC Sci Publ 65:71–73
Carcinogenic Substances Regulations (1967) Statutory instrument 1967, no 879. Her Majesty's Stationery Office, London
Case RAM, Hosker ME (1954) Tumours of the urinary bladder as an occupational disease in the rubber industry in England and Wales. Br J Prev Soc Med 8:39–50
Case RAM, Pearson JT (1954) Tumours of the urinary bladder in workmen engaged in the manufacture of certain dyestuff intermediates in the British chemical industry. Part II. Further consideration of the role of aniline and of the manufacture of auramine and magenta (fuchsin) as possible causative agents. Br J Ind Med 11:213–216
Case RAM, Hosker ME, McDonald DB, Pearson JT (1954) Tumours of the urinary bladder in workmen engaged in the manufacture and use of certain dyestuff intermediates in the British chemical industry. Part I. The role of aniline, benzidine, alpha-naphthylamine and beta-naphthylamine. Br J Ind Med 11:75–104
Chen C-J, Chuang Y-C, You S-L, Lin T-M, Wu H-Y (1986) A retrospective study on malignant neoplasms of bladder, lung and liver in black-foot disease endemic area in Taiwan. Br J Cancer 53:399–405
Chen JL, Walrath J, O'Berg MT, Burke CA, Pell S (1987) Cancer incidence and mortality among workers exposed to acrylonitrile. Am J Ind Med 11:157–163
Cook JW, Hewett CL, Hieger I (1933) The isolation of a cancer-producing hydrocarbon from coal tar. J Chem Soc 395–405
Control of Substances Hazardous to Health Regulations (1988) Statutory Instrument 1988 No 1657. Her Majesty's Stationery Office, London

Creech JL Jr, Johnson MN (1974) Angiosarcoma of the liver in the manufacture of vinyl chloride. J Occup Med 16:150–151
Cruickshank CND, Squire JR (1950) Skin cancer in the engineering industry from the use of mineral oil. Br J Ind Med 7:1–11
Dalbey WE (1982) Formaldehyde and tumors in hamster respiratory tract. Toxicology 24:9–14
Davies JM (1982) The prevention of occupational cancer. In: Alderson M (ed) The prevention of cancer. Arnold, London, pp 184–209
Davies JM (1984) Lung cancer mortality among workers making lead chromate and zinc chromate pigments at three English factories. Br J Ind Med 41:158–169
Davies JM, Thomas HF, Manson D (1982) Bladder tumours among rodent operatives handling ANTU. Br Med J 285:927–931
Dean BJ (1985) Recent findings on the genetic toxicology of benzene, toluene, xylenes and phenols. Mutat Res 154:153–181
Deichmann WB, MacDonald WM, Coplan MM, Woods RM, Anderson WA (1958) Para nitrobiphenyl, a new bladder carcinogen in the dog. Industr Med Surg 27:634–637
Dipple A, Moschel RC, Bigger CAH (1984) Polynuclear aromatic carcinogens. In: Searle CE (ed) Chemical carcinogens, 2nd edn. ACS Monograph 182, vol 1. American Chemical Society, Washington DC, pp 41–163
Doll R (1952) The causes of death among gasworkers with special reference to cancer of the lung. Br J Ind Med 9:180–185
Doll R (1955) Mortality from lung cancer in asbestos workers. Br J Ind Med 12:81–86
Doll R (1984) Nickel exposure: a human health hazard. In: Sunderman FW Jr (ed) Nickel in the human environment. IARC Sci Publ 53:3–21
Doll R, Peto R (1981) The causes of cancer: quantitative estimates of avoidable risks of cancer in the United States today. JNCI 66:1191–1308; also published separately by Oxford University Press, Oxford
Doll R and workgroup (1981) Problems of epidemiological evidence. Environ Health Perspect 40:11–20
Doll R, Fisher REW, Gammon EJ (1965) Mortality of gasworkers with special reference to cancers of the lung and bladder, chronic bronchitis, and pneumoconiosis. Br J Ind Med 22:1–12
Druckrey H, Preussmann R, Nashed N, Ivankovic S (1966) Carcinogene alkylierende Substanzen. 1. Dimethylsulfat, carcinogene Wirkung an Ratten und wahrscheinliche Ursache von Berufskrebs. Z Krebsforsch 68:103–111
Dunnigan J (1984) Biological effects of fibers: Stanton's hypothesis revisited. Environ Health Perspect 57:333–337
Durham PJN (1983) Human health effects of PBB. Food Chem Toxicol 21:515–518
Easton DF, Peto J, Doll R (1988) Cancers of the respiratory tract in mustard gas workers. Br J Ind Med 45:652–659
Eaton M, Schenker M, Whorton MD, Samuels S, Perkins C, Overstreet J (1986) Seven-year follow-up of workers exposed to 1,2-dibromo-3-chloropropane. J Occup Med 28:1145–1150
Enterline PE (1982) Importance of sequential exposure in the production of epichlorohydrin and isopropanol. Ann NY Acad Sci 381:344–349
Enterline PE, Marsh GM (1982a) Cancer among workers exposed to arsenic and other substances in a copper smelter. Am J Epidemiol 116:895–911
Enterline PE, Marsh GM (1982b) Mortality among workers in a nickel refinery and alloy manufacturing plant in west Virginia. JNCI 68:925–933
Enterline PE, Henderson VL, Marsh GM (1987) Exposure to arsenic and respiratory cancer: a reanalysis. Am J Epidemiol 125:929–938
Environmental Health Perspectives (1985a) Potential health effects of polychlorinated biphenyls and related persistent halogenated hydrocarbons: US symposium. Environ Health Perspect 60:3–221
Environmental Health Perspectives (1985b) Workshop on occupational hazards caused by polychlorinated biphenyls (PCBs) and chlorobenzenes in capacitors and transformers: Finland-US symposium. Environ Health Perspect 60:223–352

Environmental Health Perspectives (1987) Report on cancer risks associated with the ingestion of asbestos. Environ Health Perspect 72:253–265

Fajen JM, Carson GA, Rounbehler DP, Fan TY, Vitar R, Goff VE, Wolf MH, Edwards GS, Fine DH, Reinhold V (1979) *N*-Nitrosamines in the rubber and tire industry. Science 205:1262–1264

Fan TY, Morrison J, Rounbehler DP, Ross R, Fine DH (1977) *N*-Nitrosodiethanolamine in synthetic cutting fluids: a part-per-hundred impurity. Science 196:70–71

Feron VJ, Kruysse A, Woutersen RA (1982) Respiratory tract tumours in hamsters exposed to acetaldehyde vapour alone or simultaneously to benzo[*a*]pyrene or diethylnitrosamine. Eur J Cancer Clin Oncol 18:13–31

Fine DH (1980) Exposure assessment to preformed environmental *N*-nitroso compounds from the point of view of our own studies. Oncology 37:199–202

Fine DH, Rounbehler DP, Belcher NM, Epstein SS (1976) *N*-Nitroso compounds: detection in ambient air. Science 192:1328–1330

Fishbein L (1984) An overview of environmental and toxicological aspects of aromatic hydrocarbons. I. Benzene. Sci Total Environ 40:189–218

Fishbein L (1987) Perspectives on occupational exposure to antineoplastic agents. Arch Geschwulstforsch 57:219–248

Flamm WG, Frankos V (1985) Formaldehyde: laboratory evidence. IARC Sci Publ 65:85–90

Fukuda K, Matsushita H, Sakabe H, Takemoto K (1981) Carcinogenicity of benzyl chloride, benzal chloride, benzotrichloride and benzoyl chloride in mice by skin application. Gann 72:655–664

Furst A, Radding SB (1984) New developments in the study of metal carcinogenesis. J Environ Sci Health C2:103–133

Galy P, Touraine R, Brune J, Gallois P, Roudier R, Lorie R, Lheureux P, Wissendanger T (1963) Les cancers broncho-pulmonaires de l'intoxication arsenicale chronique chez les viticulteurs du Beaujolais. Lyon Med 210:735–744

Gangolli SD (1986) Is any hazard associated with the use of Clingfilm for wrapping food? Br Med J 292:1112

Gardner MJ, Winter PD, Taylor CP, Acheson ED (1983) Atlas of cancer mortality in England and Wales 1968–1978. Wiley, Chichester

Garman RH, Snellings WM, Maronpot RR (1985) Brain tumours in F344 rats associated with chronic inhalation exposure to ethylene oxide. Neurotoxicology 65:117–137

Garner RC, Martin CN, Clayson DB (1984) Carcinogenic aromatic amines and related compounds. In: Searle CE (ed) Chemical carcinogens, 2nd edn. ACS Monograph 182, vol 2. American Chemical Society, Washington DC, pp 175–276

Gehrmann GH (1934) The carcinogenetic agent – chemistry and industrial aspects. J Urol 31:126–137

Gilman JPW, Swierenga SHH (1984) Inorganic carcinogenesis. In: Searle CE (ed) Chemical carcinogens, 2nd edn. ACS Monograph 182, vol 1. American Chemical Society, Washington DC, pp 577–630

Golberg L (1986) Hazard assessment of ethylene oxide. CRC Press, Boca Raton

Gold LS, De Veciana M, Backman GM, Magaw R, Lopipero P, Smith M, Blumenthal M, Levinson R, Bernstein L, Ames BN (1986) Chronological supplement to the carcinogenic potency database: standardised results of animal bioassays published through December 1982. Environ Health Perspect 67:161–200

Goldblatt MW (1947) Occupational cancer of the bladder. Br Med Bull 4:405–416

Greim H, Wolff T (1984) Carcinogenicity of organic halogenated compounds. In: Searle CE (ed) Chemical carcinogens, 2nd edn. ACS Monograph 1982, vol 1. American Chemical Society, Washington DC, pp 525–575

Haerting FH, Hesse W (1879) Der Lungenkrebs, die Bergkrankheit in den Schneeberger Gruben. Wochenschr Gerichtl Med 31:102–132

Harvey RG (ed) (1985) Polycyclic hydrocarbons and carcinogenesis. ACS Symposium 283. American Chemical Society, Washington DC

Hayes RB, Lilienfeld AM, Snell LM (1979) Mortality in chromium chemical production workers: a prospective study. Int J Epidemiol 8:365–374

Henry SA (1946) Cancer of the scrotum in relation to occupation. Oxford University Press, London
Henry SA (1947) Occupational cutaneous cancer attributable to certain chemicals in industry. Br Med Bull 4:389–401
Higgins I, Welch K, Oh M, Bond G, Hurwitz P (1981) Influence of arsenic exposure and smoking on lung cancer among smelter workers: a pilot study. Am J Ind Med 2:33–41
Hoar SK, Blair A, Holmes FF, Boysen CD, Robel RJ, Hoover R, Fraumeni JF Jr (1986) Agricultural herbicide use and risk of lymphoma and soft tissue sarcoma. J Am Med Assoc 256:1141–1147
Hoffmann D, Harley NH, Fisenne I, Adams JD, Brunnemann KD (1986) Carcinogenic agents in snuff. JNCI 76:435–437
Hogstedt C, Aringer L, Gustavsson A (1986) Epidemiologic support for ethylene oxide as a cancer causing agent. J Am Med Soc 255:1575–1578
Holmes JG, Kipling
MD, Waterhouse JAH (1970) Subsequent malignancies in men with scrotal epithelioma. Lancet ii:214–215
Hopkins J (1984) Genotoxicity and carcinogenicity of glycidyl ethers. Food Chem Toxicol 22:780–783
Howard JK (1984) Relative cancer risks from exposure to different asbestos fibre types. New Zealand Med J 97:646–649
HSE (1982) Benzene. Toxicity review 4. Her Majesty's Stationery Office, London
HSE (1983) Trimellitic anhydride (TMA), 4,4′-methylenebis(2-chloroaniline) (MBOCA), *N*-nitrosodiethanolamine (NDELA). Toxicity review 8. Her Majesty's Stationery Office, London
HSE (1985a) 1,3-Butadiene. Toxicity review 11. Her Majesty's Stationery Office, London
HSE (1985b) Vinylidene chloride. Toxicity review 13. Her Majesty's Stationery Office, London
HSE (1986) Review of the toxicity of the esters of *o*-phthalic acid (phthalate esters). Toxicity review 14. Her Majesty's Stationery Office, London
HSE (1987) Occupational exposure limits 1987. Guidance note EH40/87. Her Majesty's Stationery Office, London, p 25
Hueper WC, Wiley FH, Wolfe HD (1938) Experimental induction of bladder tumours in dogs by administration of beta-naphthylamine. J Ind Hyg Toxicol 20:46–84
Huff JE (1983a) 1,2-Dibromoethane (ethylene dibromide). Environ Health Perspect 47:359–363
Huff JE (1983b) 1,2-Dibromo-3-chloropropane. Environ Health Perspect 47:365–369
Huff JE, Melnick RL, Solleveld HA, Haseman JK, Powers M, Miller RA (1984) Multiple organ carcinogenicity of 1,3-butadiene in $B6C3F_1$ mice after 60 weeks of inhalation exposure. Science 227:548–549
Hunter D (1978) The diseases of occupations, 6th edn. Hodder and Stoughton, London, pp 33–37
IARC (1975a) Aziridine. IARC Monogr 9:37–46
IARC (1975b) Tris(1-aziridinyl)phosphine sulphide. IARC Monogr 9:85–94
IARC (1978) Polychlorinated biphenyls. IARC Monogr 18:43–103; (1982) [Suppl 4]: 217–219
IARC (1979a) Carbon tetrachloride. IARC Monogr 20:371–399
IARC (1979b) Chloroform. IARC Monogr 20:401–427
IARC (1979c) Hexachlorocyclohexane (technical HCH and lindane). IARC Monogr 20:195–239
IARC (1979d) Vinyl chloride, polyvinyl chloride and vinyl chloride-vinyl acetate copolymers. IARC Monogr 19:377–438
IARC (1979e) Vinylidene chloride and vinylidene chloride-vinyl chloride copolymers. IARC Monogr 19:439–459
IARC (1979f) Chloroprene and polychloroprene. IARC Monogr 19:131–156
IARC (1980a) Some metals and metallic compounds. IARC Monogr 23:39–415
IARC (1980b) Phenacetin. IARC Monogr 24:135–161
IARC (1981) Wood, leather and some associated industries. IARC Monogr 25
IARC (1982a) Benzene. IARC Monogr 29:93–148

IARC (1982b) Chemicals, industrial processes and industries associated with cancer in humans: IARC Monographs, vol 1 to 29. IARC Monogr [Suppl 4]
IARC (1983) Polynuclear aromatic compounds, part 1. Chemical, environmental and experimental data. IARC Monogr 32
IARC (1984a) Iron and steel founding. IARC Monogr 34:133–190
IARC (1984b) Polynuclear aromatic compounds, part 2. Carbon blacks, mineral oils and some nitroarenes. IARC Monogr 33
IARC (1984c) Polynuclear aromatic compounds, part 3. Industrial exposures in aluminium production, coal gasification, coke production, and iron and steel founding. IARC Monogr 34
IARC (1985a) Acetaldehyde. IARC Monogr 36:101–132
IARC (1985b) Ethylene oxide. IARC Monogr 36:189–226
IARC (1985c) Propylene oxide. IARC Monogr 36:227–243
IARC (1985d) Styrene oxide. IARC Monogr 36:245–263
IARC (1986a) Amitrole. IARC Monogr 41:293–317
IARC (1986b) Chemistry and analysis of tobacco smoke. IARC Monogr 38:83–126
IARC (1986c) Polybrominated biphenyls. IARC Monogr 41:261–292
Infante PF (1977) Carcinogenic and mutagenic risks associated with some halogenated olefins. Environ Health Perspect 21:251–254
Infante PF (1981) Observations on the site-specific carcinogenicity of vinyl chloride to humans. Environ Health Perspect 41:89–94
Infante PF (1987) Benzene toxicity: studying a subject to death. Am J Ind Med 11:599–604
Innes JRM, Ulland BM, Valerio MG, Petrucelli L, Fishbein L, Hart ER, Pallotta AJ, Bates RR, Falk HL, Gart JJ, Klein M, Mitchell I, Peters J (1969) Bioassay of pesticides and industrial chemicals for tumorigenicity in mice: a preliminary note. JNCI 42:1101-1114
Ivankovic S, Eisenbrand G, Preussmann R (1979) Lung carcinoma induction in BD rats after a single intratracheal instillation of an arsenic-containing pesticide mixture formerly used in vineyards. Int J Cancer 24:786–788
Järvholm B, Lavenius B (1987) Mortality and cancer morbidity in workers exposed to cutting fluids. Arch Environ Health 42:361–366
Järvholm B, Lavenius B, Sällsten G (1986) Cancer morbidity in workers exposed to cutting fluids containing nitrites and amines. Br J Ind Med 43:563–565
Johnson KA, Gorzinski SJ, Bodner KM, Campbell RA, Wolf CH, Friedman MA, Mast RW (1986) Chronic toxicity and oncogenicity study on acrylamide incorporated in the drinking water of Fischer 344 rats. Toxicol Appl Pharmacol 85:154–168
Kennaway EL, Hieger I (1930) Carcinogenic substances and their fluorescence spectra. Br Med J II:1044–1046
Kerns WD, Pavkov KL, Donofrio DJ, Gralla EJ, Swenberg JA (1983) Carcinogenicity of formaldehyde in rats and mice after long-term inhalation exposure. Cancer Res 43:4382–4392
Kimbrough RD, Groce DF, Korver MP, Burse VW (1981) Induction of liver tumors in female Sherman strain rats by polybrominated biphenyls. JNCI 66:535–542
Kimbrough RD, Mitchell FL, Houk VN (1985) Trichloroethylene: an update. J Toxicol Environ Health 15:369–383
Kipling MD, Cooke MA (1984) Soots, tars, and oils as causes of occupational cancer. In: Searle CE (ed) Chemical carcinogens, 2nd edn. ACS Monograph 182, vol 1. American Chemical Society, Washington DC, pp 165–174
Kipling MD, Waterhouse JAH (1967) Cadmium and prostatic carcinoma. Lancet I:730–731
Klehr MW (1984) Spätmanifestationen bei ehemaligen Kampfgasarbeitern unter besonderer Berücksichtigung der cutanen Befunde. Z Hautkr 59:1161–1164
Kluwe WM (1986) Carcinogenic potential of phthalic acid esters and related compounds: structure-activity relationships. Environ Health Perspect 65:271–278
Kolman A, Näslund M, Calleman CJ (1986) Genotoxic effects of ethylene oxide and their relevance to human cancer. Carcinogenesis 7:1245–1250
Kuschner M (1981) The carcinogenicity of beryllium. Environ Health Perspect 40:101–105

Lamb JC, Huff JE, Haseman JK, Murthy ASK, Lilja H (1986) Carcinogenesis studies of 4,4′-methylenedianiline hydrochloride given in drinking water to F344/N rats and B6C3F$_1$ mice. J Toxicol Environ Health 18:325–337

Lancet (1986) Ethylene oxide – a human carcinogen? Lancet ii:201–202 (editorial)

Landrigan PJ, Meinhardt TJ, Gordon J, Lipscomb JA, Burg JR, Mazzuckelli LF, Lewis TR, Lemen RA (1984) Ethylene oxide: an overview of toxicologic and epidemiologic research. Am J Ind Med 6:103–115

Langård S (1983) The carcinogenicity of chromium compounds in man and animals. In: Burrows D (ed) Chromium – metabolism and toxicity. CRC Press, Boca Raton, pp 13–29

Laskin S, Sellakumar AR, Kuschner M, Nelson N, La Mendola S, Rusch GM, Katz GV, Dulak NC, Albert RE (1980) Inhalation carcinogenicity of epichlorohydrin in non-inbred Sprague-Dawley rats. JNCI 65:751–757

Lawley PD (1984) Carcinogenesis by alkylating agents. In: Searle CE (ed) Chemical carcinogens, 2nd edn. ACS Monograph 182, vol 1. American Chemical Society, Washington DC, pp 325–484

Leichtenstern O (1898) Über Harnblasenentzündung und Harnblasengeschwülste bei Arbeitern in Farbfabriken. Dtsch Med Wochenschr 24:709–713

Lemen RA, Johnson WM, Wagoner JK, Archer VE, Saccomanno G (1976) Cytologic observations and cancer incidence following exposure to BCME. Ann NY Acad Sci 271:71–80

Léonard A, Lauwerys R (1987) Mutagenicity, carcinogenicity, and teratogenicity of beryllium. Mutat Res 186:35–42

Leonard TB, Adams T, Popp JA (1986) Dinitrotoluene isomer-specific enhancement of the expression of diethylnitrosamine-initiated hepatocyte foci. Carcinogenesis 7: 1797–1803

Leuenberger SC (1912) Die unter dem Einfluß der synthetischen Farbindustrie beobachtete Geschwulstentwicklung. Beitr Klin Chir 80:208–316

Levy LS, Venitt S (1986) Carcinogenicity and mutagenicity of chromium compounds: the association between bronchial metaplasia and neoplasia. Carcinogenesis 7:831–835

Levy LS, Venitt S, Bidstrup L (1986) Investigation of the potential carcinogenicity of a range of chromium containing materials on rat lung. Br J Ind Med 43:243–256

Lewis TR, Ulrich CE, Busey WM (1979) Subchronic inhalation toxicity of nitromethane and 2-nitropropane. J Environ Pathol Toxicol 2:233–249

Lijinsky W, Kovatch RM (1985) Induction of liver tumours in rats by nitrosodiethanolamine at low doses. Carcinogenesis 6:1679–1681

Lloyd JW (1971) Long-term mortality study of steelworkers. V. Respiratory cancer in coke plant workers. J Occup Med 13:53–68

Lubin JH, Pottern LM, Blot WJ, Tokudome S, Stone BJ, Fraumeni JF Jr (1981) Respiratory cancer among copper smelter workers: recent mortality statistics. J Occup Med 23:779–784

Lynch DW, Lewis TS, Moorman WJ, Burg JA, Groth DH, Khan A, Ackerman LJ, Cockrell BY (1984) Carcinogenic and toxicologic effects of inhaled ethylene oxide and propylene oxide in F344 rats. Toxicol Appl Pharmacol 76:69–84

Lynch J, Hanis NM, Bird MG, Murray KJ, Walsh JP (1979) An association of upper respiratory tract cancer with exposure to diethyl sulphate. J Occup Med 21:333–341

MacIntyre I (1975) Experience of tumors in a British plant handling 3,3′-dichlorobenzidine. J Occup Med 17:23–26

Maher KV, deFonso LR (1987) Respiratory cancer among chloromethyl ether workers. JNCI 78:839–843

Maltoni C, Selikoff IJ (1988) Living in a chemical world: occupational and environmental significance of industrial carcinogens. Ann NY Acad Sci 534:1–1045

Maltoni C, Cotti G, Morisi L, Chieco P (1977) Carcinogenicity bioassays of vinylidene chloride. Research plan and early results. Med Lav 68:241–262

Maltoni C, Failla G, Kassapidis G (1979) First experimental demonstration of the carcinogenic properties of styrene oxide. Med Lavoro 5:358–362

Maltoni C, Conti B, Cotti G, Belpoggi F (1985) Experimental studies on benzene at the Bologna Institute of Oncology: current results and ongoing research. Am J Ind Med 7:415–446

Manning KP, Skegg DC, Stell PM, Doll R (1981) Cancer of the larynx and other occupational hazards of mustard gas workers. Clin Otolaryngol 6:165–170

Manouvriez A (1876) Maladies et hygiène des ouvriers, travaillant à la fabrication des agglomérés de houille et de brai. Ann Hyg Publ (Paris) 45:459–482

Manson MM (1980) Epoxides – is there a human health problem? Br J Ind Med 37:317–336

Marcus WL (1987) Chemicals of current interest – benzene. Toxicol Ind Health 3:205–266

Markovitz A, Crosby WH (1984) Chemical carcinogenesis: a soil fumigant, 1,3-dichloropropene, as possible cause of hematologic malignancies. Arch Int Med 144:1409–1411

Marshall E (1987) EPA indicts formaldehyde, 7 years later. Science 236:381

Masuda Y (1985) Health status of Japanese and Taiwanese after exposure to contaminated rice oil. Environ Health Perspect 60:321–325

McCallum RI, Woolley V, Petrie A (1983) Lung cancer associated with chloromethyl methyl ether manufacture: an investigation at two factories in the United Kingdom. Br J Ind Med 40:384–389

McMichael AJ, Spirtas R, Gamble JF, Tousey PM (1976) Mortality among rubber workers: relationship to specific jobs. J Occup Med 18:178–185

Mehlman MA (ed) (1985) Benzene: scientific update. Am J Ind Med 7:361–492

Melick WF, Escue HM, Naryka JJ, Mezera RA, Wheeler EP (1955) The first reported cases of human bladder tumours due to a new carcinogen – xenylamine. J Urol 74:760–766

Melick WF, Naryka JJ, Kelly RE (1971) Bladder cancer due to exposure to paraminobiphenyl: a 17-year followup. J Urol 106:220–226

Merletti F, Heseltine E, Saracci R, Simonato L, Vainio H, Wilbourn J (1984) Target organs for carcinogenicity of chemicals and industrial exposures in humans: a review of the results in the IARC Monographs on the evaluation of the carcinogenic risk of chemicals to humans. Cancer Res 44:2244–2250

Monson RR, Fine LJ (1978) Cancer mortality and morbidity among rubber workers. JNCI 61:1047–1053

Morgan RW, Claxton KW, Divine BJ, Kaplan SD, Harris VB (1981) Mortality among ethylene oxide workers. J Occup Med 23:767–770

Mule Spinning (Health) Special Regulations (1953) Statutory instrument 1953 no 1545. Her Majesty's Stationery Office, London

NCI (1978) Bioassay of aniline hydrochloride for possible carcinogenicity. Tech Rep Ser No 130: DHEW Publ No (NIH) 78-1385

NCI (1979a) Bioassay of *o*-toluidine hydrochloride for possible carcinogenicity. Tech Rep Ser No 153: DHEW Publ No (NIH) 79-1709

NCI (1979b) Bioassay of *p*-cresidine for possible carcinogenicity. Tech Rep Ser No 142: DHEW Publ No (NIH) 79–1397

NCI (1979c) Bioassay of Michler's ketone for possible carcinogenicity. Tech Rep Ser No 181: DHEW Publ No (NIH) 79–1737

Nelson N, Levine RJ, Albert RE, Blair AE, Griesemer RA, Landrigan PJ, Stayner LT, Swenberg JA (1986) Contribution of formaldehyde to respiratory cancer. Environ Health Perspect 70:23–35

Neubauer O (1947) Arsenical cancer: a review. Br J Cancer 1:192–251

Nicholson WJ (1984) Research issues in occupational and environmental cancer. Arch Environ Health 39:190–202

Nishimoto Y, Yamakido M, Shigenobu T, Takutake M, Matsusaka S (1986) Cancer of the respiratory tract observed in workers having retired from a poison gas factory. Gan To Kagaku Ryoko 13:1144–1148 (Abstr. ICRDB Cancergram CK02 1986, 86/12 No 42)

Norseth T (1986) The carcinogenicity of chromium and its salts. Br J Ind Med 43:649–651

NTP (1982) Carcinogenesis bioassay of vinylidene chloride (CAS No 75-35-4) in F344/N rats and $B6C3F_1$/N mice (gavage study). NTP Tech Rep 228: NIH Publ No 82-1784

NTP (1986) Toxicology and carcinogenesis studies of benzene (CAS No. 71-43-2) in F344/N rats and B6C3F$_1$ mice (gavage studies). NTP Tech Rep 289: NIH Publ No 86-2545
O'Berg MT (1985) Formaldehyde and cancer: an epidemiologic perspective. In: Turoski V (ed) Formaldehyde: analytical chemistry and toxicology. Adv Chem Ser 210. American Chemical Society, Washington DC, pp 289–295
Olsen JH, Asnaes S (1986) Formaldehyde and the risk of squamous cell carcinoma of the sinonasal cavities. Br J Ind Med 43:769–774
Olsen JH, Jensen OM (1987) Occupation and risk of cancer in Denmark: an analysis of 93,810 cancer cases, 1970–1979. Scand J Work Environ Health 13 [Suppl 1]:1–91
O'Neill IK, von Borstel RC, Miller CT, Long J, Bartsch H (eds) (1984) N-Nitroso compounds: occurrence, biological effects and relevance to human cancer. IARC Sci Publ 57
OSHA (1980) Benzidine, *o*-tolidine, and *o*-dianisidine-based dyes. Health Hazard Alert, DHHS (NIOSH) Publ No 81-106
Owen PE, Glaister JR, Gaunt IF, Pullinger DH (1987) Inhalation toxicity studies with 1,3-butadiene: 3. Two-year toxicity/carcinogenicity study in rats. Am Ind Hyg Assoc J 48:407–413
Parkes HG, Evans AEJ (1984) Epidemiology of aromatic amine cancers. In: Searle CE (ed) Chemical carcinogens, 2nd edn. ACS Monograph 182, vol 1. American Chemical Society, Washington DC, pp 277–301
Pershagen G, Wall S, Taube A, Linnman L (1981) On the interaction between occupational arsenic exposure and smoking and its relation to lung cancer. Scand J Work Environ Health 7:302–309
Phillips DH (1983) Fifty years of benzo[*a*]pyrene. Nature 303:468–472
Piscator M (1981) Role of cadmium in carcinogenesis, with special reference to cancer of the prostate. Environ Health Perspect 40:107–120
Ponomarkov V, Tomatis L (1980) Long-term testing of vinylidine chloride and chloroprene for carcinogenicity in rats. Oncology 37:136–141
Pott P (1775) Chirurgical observations relative to the cataract, the polypus of the nose, the cancer of the scrotum (etc). Hawes, Clarke & Collins, London, pp 63–68 (Reproduced in Natl Cancer Inst Monogr (1963) 10:7–13)
Preussmann R, Eisenbrand G (1984) *N*-Nitroso carcinogens in the environment. In: Searle CE (ed) Chemical carcinogens, 2nd edn. ACS Monograph 182, vol 2. American Chemical Society, Washington DC, pp 829–868
Preussmann R, Stewart B (1984) *N*-Nitroso carcinogens. In: Searle CE (ed) Chemical carcinogens, 2nd edn. ACS Monograph 182, vol 2. American Chemical Society, Washington DC, pp 643–828
Preussmann R, Wiessler M (1987) The enigma of the organ-specificity of carcinogenic nitrosamines. Trends Pharmacol Sci 8:185–189
Purchase IFH, Kalinowski AE, Ishmael J, Wilson J, Gore CW, Chart IS (1981) Lifetime carcinogenicity study of 1- and 2-naphthylamine in dogs. Br J Cancer 44:892–901
Purchase IFH, Stafford J, Paddle GM (1987) Vinyl chloride: an assessment of the risk of occupational exposure. Food Chem Toxicol 25:187–202
Quast JF, McKenna MJ, Rampy LW, Norris JM (1986) Chronic toxicity and oncogenicity study on inhaled vinylidine chloride in rats. Fundam Appl Toxicol 6:105–144
Raffle PAB, Lee WR, McCallum RI, Murray R (1987) Hunter's Diseases of Occupations. Hodder and Stoughton, London
Ram NM, Calabrese EJ, Christman RF (eds) (1986) Organic carcinogens in drinking water: detection, treatment, and risk assessment. Wiley, New York
Ramazzini B (1713) De morbis artificum diatriba. Translated by Wright WC (1964) History of medicine, Ser 23. New York Academy of Medicine, New York
Rehn L (1895) Blasengeschwülste bei Fuchsin-Arbeitern. Arch Klin Chir 50:588–600
Rehn L (1906) Über Blasenerkrankungen bei Anilinarbeitern. Verh Dtsch Ges Chir 35:313–316
Renne RA, Giddens WE, Boorman GA, Kovatch R, Haseman JE, Clarke WJ (1986) Nasal cavity neoplasia in F344/N rats and (C57BL/6 X C3H)F$_1$ mice inhaling propylene oxide for up to two years. JNCI 77:573–582
Reymann F, Moller R, Nielsen A (1978) Relationship between arsenic intake and internal malignant neoplasms. Arch Dermatol 114:378–381

Rinsky RA, Smith AB, Hornung R, Filloon TG, Young RJ, Okun AH, Landrigan PJ (1987) Benzene and leukaemia. N Engl J Med 316:1044–1050
Roberts RS, Julian JA, Shannon HS, Muir DCF (1980) Mortality studies in Ontario nickel workers: the INCO/JOHC study. In: Brown SS, Sunderman FW Jr (eds) Nickel toxicology. Academic, New York, pp 27–30
Roe FJC (1985a) Chloromethylation: three lung cancer cases in young men. Lancet ii:268
Roe FJC (1985b) Occupational cancer: Where now and where next? Scand J Work Environ Health 11:181–187
Rohl AN, Langer AM, Moncure G, Selikoff IJ, Fischbein A (1982) Endemic pleural disease associated with exposure to mixed fibrous dust in Turkey. Science 216:518–520
Rom WN, Krueger G, Zone J, Attfield MD, Costello J, Burkart J, Turner ER (1985) Morbidity study of US oil shale workers employed during 1948–1969. Arch Environ Health 40:58–62
Rosenkranz HS, Mermelstein R (1985) The genotoxicity, metabolism and carcinogenicity of nitrated polycyclic aromatic hydrocarbons. J Environ Sci Health 3:221–272
Sakabe H, Fukuda K (1977) An updating report on cancer among benzoyl chloride manufacturing workers. Ind Health 15:173–174
Saknyn AV, Shabynina NK (1973) Epidemiology of malignant neoplasms in nickel undertakings (in Russian). Gig Tr Prof Zabol 17:25–29
Saracci R (1984) Occupation. In: Vessey MP, Gray M (eds) Cancer risks and prevention. Oxford University Press, Oxford, pp 99–118
Saracci R (ed) (1986) Contributions to the IARC study on mortality and cancer incidence among man-made mineral fiber production workers. Scand J Work Environ Health 12 [Suppl 1]:1–93
Saracci R (1987) The interaction of tobacco smoking and other agents in cancer etiology. Epidemiol Revs 9:175–193
Satoh K, Fukuda Y, Tarii K, Katsuno N (1981) Epidemiological study of workers engaged in the manufacture of chromium compounds. J Occup Med 23:835–838
Schmähl D, Thomas C, Auer R (1977) Iatrogenic carcinogenesis. Springer, Berlin Heidelberg New York, pp 4–26
Schottenfeld D (1984) Chronic disease in the workplace and cancer. Arch Environ Health 39:150–157
Searle CE (1984) Chemical carcinogens as laboratory hazards. In: Searle CE (ed) Chemical carcinogens, 2nd edn. ACS Monogr 182, vol 1. American Chemical Society, Washington DC, pp 303–323
Searle CE, Teale J (1982) Chrysoidine-dyed bait: a possible carcinogenic hazard to anglers? Lancet i:564
Selikoff IJ, Hammond EC (1979) Asbestos and smoking. J Am Med Assoc 242:458–459 (editorial)
Sellakumar AR, Snyder CA, Solomon JJ, Albert RE (1985) Carcinogenicity of formaldehyde and hydrogen chloride in rats. Toxicol Appl Pharmacol 81:401–406
Shannon HS, Julian JA, Roberts RS (1984) A mortality study of 11,500 nickel workers. JNCI 73:1251–1258
Slesinski RS, Hengler WC, Guzzie PJ, Wagner KJ (1983) Mutagenicity evaluation of glutaraldehyde in a battery of in vitro bacterial and mammalian test systems. Food Chem Toxicol 21:621–629
Smith P (1985) Radiation. In: Vessey MP, Gray M (eds) Cancer risks and prevention. Oxford University Press, Oxford, pp 119–148
Sobel W, Bond GG, Parsons TW, Brenner FE (1986) Acrylamide cohort mortality study. Br J Ind Med 43:785–788
Sole GM, Chipman JK (1986) The mutagenic potency of chrysoidines and Bismarck brown dyes. Carcinogenesis 7:1921–1923
Sole G, Sorahan T (1985) Coarse fishing and risk of urothelial cancer. Lancet i:1477–1479
Sorahan T, Waterhouse JAH (1985) Cancer of the prostate among nickel-cadmium battery workers. Lancet i:459
Sorahan T, Burges DC, Waterhouse JA (1987) A mortality study of nickel/chromium platers. Br J Ind Med 44:250–258

Sorahan T, Waterhouse JAH, Cooke MA, Smith EMB, Jackson JR, Temkin L (1983) A mortality study of workers in a factory manufacturing chlorinated toluenes. Ann Occup Hyg 27:173–182
Sorahan T, Parkes HG, Veys CA, Waterhouse JAH (1986) Cancer mortality in the British rubber industry 1946–80. Br J Ind Med 43:363–373
Spire B, Barré-Sinoussi F, Montagnier L, Chermann JC (1984) Inactivation of lymphadenopathy associated virus by chemical disinfectants. Lancet ii:899–901
Srám RJ, Tomatis L, Clemmesen J, Bridges BA (1983) ICPEMC publ no 7: An evaluation of the genetic toxicology of epichlorohydrin. A report of an expert group of the International Commission for Protection against Environmental Mutagens and Carcinogens. Biol Zbl 102:603–620
Steinhoff D, Weber H, Mohr U, Boehme K (1983) Evaluation of amitrole (aminotriazole) for potential carcinogenicity in orally dosed rats, mice, and golden hamsters. Toxicol Appl Pharmacol 69:161–169
Storer RD, Jackson NM, Conolly RB (1984) In vivo genotoxicity and acute hepatotoxicity of 1,2-dichloroethane in mice: comparison of oral, intraperitoneal, and inhalation routes of exposure. Cancer Res 44:4267–4271
Sunderman FW Jr (1984a) Recent advances in metal carcinogenesis. Ann Clin Lab Sci 14:93–122
Sunderman FW Jr (ed) (1984b) Nickel in the human environment. IARC Sci Publ 53
Surgeon General (1986) The health consequences of involuntary smoking: a report of the Surgeon General. US Department of Health and Human Services, Rockville, Maryland
Swenberg JA, Kerns WD, Mitchell RI, Gralla EJ, Pavkov KL (1980) Induction of squamous cell carcinomas of the rat nasal cavity by inhalation exposure to formaldehyde vapor. Cancer Res 40:3398–3402
Swenberg JA, Barrow CS, Boreiko CJ, Heck Hd'A, Levine RJ, Morgan KT, Starr TB (1983) Non-linear biological responses to formaldehyde and their implications for carcinogenic risk assessment. Carcinogenesis 4:945–952
Thiess AM, Frentzel-Beyme R, Link R, Stocker WG (1982) Mortality study on employees exposed to alkylene oxides (ethylene oxide/propylene oxide) and their derivatives. In: Prevention of occupational cancer – international symposium, Helsinki 1981. International Labour Office, Geneva, pp 249–259
Travenius SZM (1982) Formation and occurrence of bis(chloromethyl) ether and its prevention in the chemical industry. Scand J Work Environ Health 8 [Suppl 3]:1–86
Turnbull D, Rodricks JV (1985) Assessment of possible carcinogenic risks to humans resulting from exposure to di(2-ethylhexyl) phthalate (DEHP). J Am Coll Toxicol 4:111–145
US Department of Labor, Occupational Safety and Health Administration (1974) Carcinogens. Fed Register 39(20):3755–3797
Vainio H (1987) Is passive smoking increasing cancer risk? Scand J Work Environ Health 13:193–196
Vaughan TL, Strader C, Davis S, Daling JR (1986a) Formaldehyde and cancers of the pharynx, sinus and nasal cavity: I. Residential exposures. Int J Cancer 38:685–688
Vaughan TL, Strader C, Davis S, Daling JR (1986b) Formaldehyde and cancers of the pharynx, sinus and nasal cavity. II. Occupational exposures. Int J Cancer 38:677–683
Volkmann R (1875) Über Theer-, Paraffin- und Russkrebs (Schornsteinfegerkrebs). Beitr Chirurg, Leipzig, pp 370–381
Wada S, Miyanishi M, Nishimoto Y, Kambe S, Miller RW (1968) Mustard gas as a cause of respiratory neoplasia in man. Lancet i:1161–1163
Wagner JC (ed) (1980) Biological effects of mineral fibres. IARC Sci Publ 30: INSERM Symp Ser 92: vols I, II. International Agency for Research on Cancer, Lyon
Wagner JC (1984) Mineral fiber carcinogenesis. In: Searle CE (ed) Chemical carcinogens, 2nd edn. ACS Monograph 182, vol 1. American Chemical Society, Washington DC, pp 631–641
Wagner JC, Sleggs CA, Marchand P (1960) Diffuse pleural mesothelioma and asbestos exposure in North Western Cape Province. Br J Ind Med 17:260–271

Wagner JC, Skidmore JW, Hill RJ, Griffiths DM (1985) Erionite exposure and mesotheliomas in rats. Br J Cancer 51:727–730
Wagoner JK (1983) Toxicity of vinyl chloride and poly(vinyl chloride): a critical review. Environ Health Perspect 52:61–66
Wald N, Boreham J, Doll R, Bonsall J (1984) Occupational exposure to hydrazine and subsequent risk of cancer. Br J Ind Med 41:31–34
Waldron HA, Waterhouse JA, Tessema N (1984) Scrotal cancer in the West Midlands 1936–76. Br J Ind Med 41:437–444
Walpole AL, Williams MHC, Roberts DC (1952) The carcinogenic action of 4-aminodiphenyl and 3:2′-dimethyl-4-aminodiphenyl. Br J Ind Med 9:255–263
Walpole AL, Roberts DC, Rose FL, Hendry JA, Homer RF (1954) Cytotoxic agents. IV. The carcinogenic actions of some monofunctional ethyleneimine derivatives. Br J Pharmacol 9:750–761
Waterhouse JAH (1972) Lung cancer and gastro-intestinal cancer in mineral oil workers. Ann Occup Hyg 15:43–44
Weisburger EK, Weisburger JH (1958) 2-Fluorenamine and related compounds. Adv Cancer Res 5:331–431
Weisburger JH, Weisburger EK (1966) Chemicals as causes of cancer. Chem Eng News, Feb 7:124–142
Weisburger JH, Williams GM (1984) Bioassay of carcinogens: in vitro and in vivo tests. In: Searle CE (ed) Chemical carcinogens, 2nd edn. ACS Monograph 182, vol 2. American Chemical Society, Washington DC, pp 1323–1373
Wester PW, Van der Heijden CA, Bisschop A, Van Esch GJ (1985) Carcinogenicity study with epichlorohydrin (CEP) by gavage in rats. Toxicology 36:325–339
Whitty F (1987) Bladder cancer in rubber workers. Br J Ind Med 44:647
WHO (1984) Quintozene. Environ Health Criteria 41:1–38
WHO (1985a) Propylene oxide. Environ Health Criteria 56:1–53
WHO (1985b) Dimethyl sulfate. Environ Health Criteria 48:1–55
WHO/IARC (1983) Biological effects of man-made mineral fibres: report on a WHO/IARC meeting, 1982. EURO Reports and Studies 81. WHO Regional Office for Europe, Copenhagen
Wignall TH (1929) Industrial diseases affecting the bladder. Br Med J 2:258–259
Williams GM, Weisburger JH (1986) Chemical carcinogens. In: Klaasen CD, Amdur MO, Doull J (eds) Casarett and Doull's toxicology: the basic science of poisons, 3rd edn. Macmillan, New York, pp 99–173
Williams MHC (1958) Occupational tumours of the bladder. In: Raven RW (ed) Cancer, vol 3. Butterworths, London, pp 337–380
Wong O (1987) An industry wide mortality study of chemical workers occupationally exposed to benzene. I. General results. II. Dose response analyses. Br J Ind Med 44:365–395
Woodhouse DL (1950) The carcinogenic activity of some petroleum fractions and extracts. Comparative results in tests on mice repeated after an interval of eighteen months. J Hyg (Lond) 48:121–134
Woutersen RA, Appelman LM, Van Garderen-Hoetmer A, Feron VJ (1986) Inhalation toxicity of acetaldehyde in rats. III. Carcinogenicity study. Toxicology 41:213–231
Wright M (1979) The hidden carcinogen in the manufacture of isopropyl alcohol. In: Deichmann WB (ed) Toxicology in occupational medicine. Elsevier, Amsterdam, pp 93–98
Yeh S (1973) Skin cancer in chronic arsenicism. Human Pathol 4:469–485
Yin S-N, Li G-L, Tain F-D, Fu Z-I, Jin C, Chen Y-J, Luo S-J, Ye P-Z, Zhang J-Z, Wang G-C, Zhang X-C, Wu H-N, Zhong Q-C (1987) Leukaemia in benzene workers: a retrospective cohort study. Br J Ind Med 44:124–128
Yoshimura H, Takemoto K, Fukuda K, Matsushita H (1986) Carcinogenicity in mice by inhalation of benzotrichloride and benzoyl chloride. Sangyo Igaku 28:352–359; Chem Abstr 106:80137
Zahm SH, Hartge P, Hoover R (1987) The national bladder cancer study: employment in the chemical industry. JNCI 79:217–222

CHAPTER 5

Therapeutic Agents as Potential Carcinogens

D. B. LUDLUM

A. Introduction

In spite of the antiquity of medicine as a profession, an appreciation of the fact that therapeutic agents can cause cancer is relatively recent. Interestingly enough, much of this understanding evolved concurrently with the development of cancer chemotherapy. The use of nitrogen mustard for the treatment of malignancies began in the 1940s, and various congeners of this compound were synthesised and introduced into practice shortly thereafter (CALABRESI and PARKS 1985). One of these, chlornaphazine, proved to be a bladder carcinogen (THIEDE and CHRISTIENSEN 1969). Subsequently, concerns have arisen that other therapeutic agents, including those used in general practice, might also cause malignancies.

In the same year that the chlornaphazine paper appeared, the International Agency for Research on Cancer (IARC) began an evaluation of the literature on compounds which might be carcinogenic to humans. In this continuing program, literature data are evaluated by working groups which meet in Lyon, and the conclusions are published in a series of monographs. The approach used by the IARC is described in a preamble which is included in each monograph; monographs also contain a cumulative index of compounds that have been considered up to that time. A recent revision of the preamble and cumulative index has appeared in volume 42 of the monograph series (IARC 1987 a).

Working groups consider a wide range of data, from laboratory investigations through epidemiological studies in humans. Data on the synthesis and purity of the compounds, on their absorption, distribution, metabolism, and excretion, on their mutagenicity and generalized toxicity, on their carcinogenicity to animals, on their interactions with DNA, and on human exposure are all considered. The purpose of the program is to evaluate und publish reviews of data on carcinogenicity, but the monographs do not make any recommendations concerning risk-to-benefit ratios.

Since therapeutic agents fall within the range of compounds considered in this program, data concerning many such compounds have been evaluated. Table 1, which has been compiled from a recent cumulative index, lists those therapeutic agents for which monographs are available. Many of these agents are known by several different names, and a list of synonyms has also been published recently (IARC 1985).

Other useful compilations of data are available. An early volume by SCHMÄHL et al. (1977) reviewed therapeutic agents and treatment modalities which seemed to be associated with an increased risk of cancer. The classic paper on cancer

Table 1. Therapeutic agents for which IARC Monographs are available

Therapeutic agents
Acriflavinium chloride
Actinomycin
Adriamycin
Anesthetics, volatile
Cyclopropane
Diethyl ether
Enflurane
Fluroxene
Halothane
Isoflurane
Methoxyflurane
Nitrous oxide
Azaserine
Azathioprine
N,N-Bis(2-chloroethyl)-2-naphthylamine (chlornaphazine)
Bis(2-chloroethyl) nitrosourea (BCNU)
Bleomycins
Chlorambucil
Chloramphenicol
1-(2-Chloroethyl)-3-cyclohexyl-1-nitrosourea (CCNU)
Cisplatin
Coal tar preparations
Conjugated oestrogens
Coumarin
Cyclophosphamide
Dacarbazine
Daunomycin
Diazepam
Diethylstilbestrol
Dithranol
Estradiol-17β
Estriol
Estrone
Ethinylestradiol
5-Fluorouracil
Hycanthone
8-Hydroxyquinoline
Iron-dextran complex
Isonicotinic acid hydrazide (isoniazid)
Megestrol acetate
Melphalan
6-Mercaptopurine
Mestranol
Methotrexate
5-Methoxypsoralen
Metronidazole
Mitomycin C
Niridazole
Nitrogen mustard
Noresthisterone
Norgestrel
Oral contraceptives
Phenacetin
Phenobarbital
Phenoxybenzamine
Phenylbutazone
Phenytoin
Prednisone
Procarbazine
Proflavine
Progesterone
Pronetalol
Propylthiouracil
Reserpine
Rifampicin
Streptozotocin
Testosterone
Thiouracil
Tris(1-aziridinyl)phosphine sulphide (thiotepa)
Uracil mustard
Vinblastine
Vincristine

aetiology by DOLL and PETO (1981) includes a review of therapeutic agents. LAWLEY (1984) reviewed carcinogenesis by alkylating agents, PREJEAN and MONTGOMERY (1984) reviewed carcinogenesis by antineoplastic agents, and CONNORS (1984) reviewed carcinogenesis by a wide range of therapeutic agents. Recently, a volume by SCHMÄHL and KALDOR (1986) has reviewed the alkylating cytostatic drugs. A comprehensive review and analysis of the literature on various chemicals, including some therapeutic agents, was published by NESNOW et al. (1986); this same review describes the establishment of a computerized GENE-TOX carcinogen data base.

As an example of legislative action in this area, the Secretary of the United States Department of Health and Human Services is required to publish an annual report which contains "a list of all substances (i) which either are known to

Table 2. Therapeutic agents included in the Fourth Annual Report on Carcinogens as substances known or anticipated to be human carcinogens [a]

A. Anticancer agents

- Adriamycin
- Bis(2-chloroethyl)nitrosourea (BCNU)
- 1,4-Butanediol dimethylsulfonate (myleran)
- Certain combined chemotherapy for lymphomas
- Chlorambucil
- 1-(2-Chloroethyl)-3-cyclohexyl-1-nitrosourea (CCNU)
- Cyclophosphamide
- Dacarbazine
- Melphalan
- Nitrogen mustard
- Procarbazine
- Tris(1-aziridinyl)phosphine sulphide (thiotepa)

B. Hormonal agents

- Conjugated estrogens
- Diethylstilbestrol
- Estrogens (not conjugated)
  - Estradiol-17β
  - Estrone
  - Ethinylestradiol
  - Mestranol
  - Norethisterone
  - Oxymetholone
  - Progesterone

C. Other compounds and treatment modalities

- Analgesic mixtures containing phenacetin
- Azathioprine
- *N,N*-Bis(2-chloroethyl)-2-naphthylamine (chlornaphazine)
- Iron dextran complex
- Methoxsalen with ultraviolet A therapy (PUVA)
- Metronidazole
- Phenacetin
- Phenazopyridine hydrochloride
- Phenytoin
- Propylthiouracil
- Reserpine
- Selenium sulfide
- Streptozotocin

[a] US DHHS (1985).

be carcinogens or which may reasonably be anticipated to be carcinogens and (ii) to which a significant number of persons residing in the United States are exposed." The list of therapeutic agents included in the Fourth Annual Report on Carcinogens (United States Government DHHS 1985) is given in Table 2.

B. Classification of Potentially Carcinogenic Therapeutic Agents

The monographs and reviews mentioned above provide a wealth of information on the potential carcinogenicity of therapeutic agents. Thus, the challenge here is to review the mechanisms by which certain therapeutic agents can act as carcinogens and to provide an overall approach that can be used to consider other compounds. Studies with antineoplastic agents provide guidance along these lines because information is available both on their activity as carcinogens and on their interactions with DNA.

Viewing the literature as a whole, those therapeutic agents which are carcinogenic fall into two major classes: compounds that modify DNA, and compounds that act through a hormonal mechanism. This is clearly illustrated by the grouping of therapeutic agents in Table 2. Compounds in the first group are antineoplastic agents which are known to modify DNA. Compounds in the second group are hormones. Most of the agents in the third group are known either to modify DNA or to produce a hormonal effect. Some of the DNA-modifying agents are complete carcinogens, but the hormones act primarily as promoters.

Much of the research in this area has focused on the apparent initiating step, the reaction of therapeutic agents with DNA, and a consideration of these reactions forms a major part of this review. As pioneered and described by Miller and Miller (1981), some compounds, including certain therapeutic agents, act directly to modify DNA while others require metabolic activation. Quite likely, it is at this point that species differences and local tissue factors operate to influence the frequency with which different tumors occur in test animals. Consequently, examples of metabolic activation of both antineoplastic and other therapeutic agents are included below.

The general plan herein is to review DNA modifications by antineoplastic agents first since the evidence for carcinogenicity is strong for many of these; furthermore, the information which is available about them provides some indication as to what DNA modifications are associated with tumor initiation. Next, the action of hormones as promoting agents will be considered and, finally, some metabolic pathways by which other therapeutic agents are converted to DNA-reactive intermediates will be reviewed.

Before taking these subjects up, however, it is important to comment on the role of host defenses in providing protection against exposure to potentially carcinogenic therapeutic agents. Protection can occur at any level, from detoxification or lack of activation of the agent to the cytotoxic action of various immunological defenses against transformed cells. In particular, DNA repair could play an important role in defending the cell against initiation; Samson (1986) has recently reviewed the adaptation of mammalian cells to alkylation damage. However, the range of DNA adducts which would be recognized by the various repair enzymes has not been elucidated.

Means of evaluating the defenses against carcinogenesis in an animal system are not well-developed at the present time. Emphasizing the importance of such mechanisms, however, is the recent finding that a common metabolite, *S*-

adenosylmethionine, is able to modify DNA chemically (RYDBERG and LINDAHL 1982). It would appear that a protective mechanism has been evolved against this alkylation in the form of O^6-alkylguanine-DNA alkyltransferase. This protein removes alkyl groups from the O^6-position of guanine, thereby correcting a lesion which might otherwise be carcinogenic (DEMPLE et al. 1985). It is very likely that other such mechanisms exist, but nothing is known as yet about the level of protection that they would provide against DNA-damaging therapeutic agents.

C. Carcinogenicity of Antineoplastic Agents

I. Nitrogen Mustards

After the successful introduction of nitrogen mustard into clinical practice, various congeners of this compound were synthesized and tested for antitumor activity; a few nitrogen mustards were also tested for other therapeutic activities. The chemical structures of some are shown in Fig. 1. Cyclophosphamide, chlorambucil, and melphalan are all bifunctional agents used successfully in the treatment of various malignancies. Phosphoramide mustard, also bifunctional, is a metabolite of cyclophosphamide and is probably responsible to most of the cytotoxic activity of this compound (COLVIN et al. 1973; CONNORS et al. 1974; STRUCK et al. 1975).

The use of chlornaphazine for the treatment of polycythemia vera in combination with ^{32}P was accompanied by an increased incidence of bladder tumors, while no such increase was found with the use of ^{32}P alone (THIEDE and CHRISTIENSEN 1969). The prototype compound, nitrogen mustard, was shown to be carcinogenic in mice shortly after its introduction into clinical use (HESTON

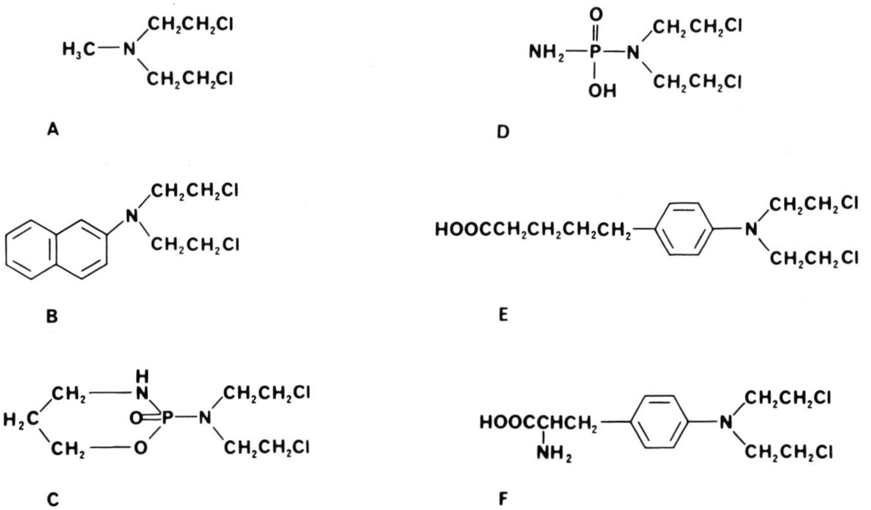

Fig. 1. Structures of nitrogen mustards which are or have been used as therapeutic agents. **A** nitrogen mustard; **B** chlornaphazine; **C** cyclophosphamide; **D** phosphoramide mustard; **E** chlorambucil; **F** melphalan

1949). In that study, 29 out of 29 test mice developed lung tumors after intravenous injection of the compound.

These studies have been confirmed repeatedly in animals, but as with most therapeutic agents, the data available on human exposure provide only limited evidence for the carcinogenicity of nitrogen mustard as a single agent in humans. In large part, this is because nitrogen mustard and similar compounds are usually used in combination with other therapeutic agents in a clinical setting. Combination chemotherapy with regimens which include nitrogen mustard, especially the MOPP regimen (nitrogen mustard, vincristine, prednisone, and procarbazine), has been associated with an increase in the incidence of acute nonlymphocytic leukemia as well as in a variety of solid tumors (ARSENEAU et al. 1977; AUCLERC et al. 1979; COLTMAN and DIXON 1982; GLICKSMAN et al. 1982; KYLE 1984).

There is considerable evidence that chlorambucil is carcinogenic to both animals and humans. Intraperitoneal administration of chlorambucil produces adenomas and adenomacarcinomas of the lung in mice (SHIMKIN et al. 1966; WEISBURGER et al. 1975) and tumors of hematopoietic origin in rats (WEISBURGER et al. 1975). Since chlorambucil has been used therapeutically as a single agent, evidence for its carcinogenicity in humans is also quite strong. It has produced acute leukemia in patients treated for nonmalignant disease (AYMARD et al. 1980; BLANC et al. 1981; MÜLLER and BRANDIS 1981) as well as in patients treated for malignancies (DUMONT et al. 1980; BERK et al. 1981; HAROUSSEAU et al. 1980).

The evidence for the carcinogenicity of cyclophosphamide to animals and humans is also considered sufficient. After oral or intravenous administration of cyclophosphamide to rats, benign and malignant tumors occurred at various sites including the bladder (SCHMÄHL and HABS 1979; SCHMÄHL 1974). Similarly, both benign and malignant tumors were produced in mice after its subcutaneous administration (SCHMÄHL and OSSWALD 1970). Finally, there have been numerous reports of malignancies occurring in patients treated with cyclophosphamide for malignant and nonmalignant diseases (IARC 1981). These have included both acute leukemias and solid tumors, particularly of the bladder.

Presumably, these malignancies were initiated by the active form of cyclophosphamide, phosphoramide mustard. Literature reports on the carcinogenicity of phosphoramide mustard do not seem to be available, but this compound is known to be teratogenic (MIRKES et al. 1981).

The data linking melphalan to carcinogenicity in both animals and humans are also generally considered sufficient. Its intraperitoneal injection produces lymphosarcomas and a dose-related increase in lung tumors in mice (SHIMKIN et al. 1966; WEISBURGER et al. 1975). When melphalan was administered intraperitoneally to rats, it produced peritoneal sarcomas (WEISBURGER et al. 1975). The agent has sometimes been used alone for the treatment of ovarian cancer and multiple myeloma, and the increased rates of leukemia which are seen in such patients have been judged to be significant (LAW and BLOM 1977; REIMER et al. 1977; EINHORN 1978).

These and other studies indicate that the bifunctional nitrogen mustards are, as a class, carcinogenic. Since the reactions of these compounds with DNA apparently initiate carcinogenesis, an analysis of DNA modification by the nitrogen mustards should provide some important leads in understanding the process.

With the exception of cylophosphamide, the nitrogen mustards shown in Fig. 1 have sufficiently strong electrophilic centers to react directly with DNA. Cyclophosphamide, however, must first be metabolized by a mixed-function oxidase in the liver, where it is converted to 4-hydroxycyclophosphamide (BROCK and HOHORST 1963; COLVIN et al. 1973; CONNORS et al. 1974; STRUCK et al. 1975). This compound is in equilibrium with aldophosphamide which decomposes spontaneously to phosphoramide mustard. Other enzyme systems inactivate the compound entirely, and it is apparently the balance between activation and detoxification that is important in producing the useful therapeutic activity of cyclophosphamide.

A variety of data indicates that nitrogen mustards react with DNA, but relatively few of the adducts have been characterized chemically. MEHTA et al. (1980) used ultraviolet spectrometry and field desorption mass spectrometry to characterize an adduct of phosphoramide mustard with guanosine and deoxyguanosine. The guanosine adduct, which corresponds to adduct A in Fig. 2, was found to be very unstable in comparison with 7-methylguanosine. This instability has been confirmed by CHETSANGA et al. (1982) and by KALLAMA and HEMMINKI (1984). Since the more stable 7-substituted guanines appear to base-pair normally with cytosine (LUDLUM 1970), this instability may be an important factor in the biological activity of the nitrogen mustards.

VU et al. (1981) investigated the reactions of phosphoramide mustard with guanylic acid and characterized the three structures shown in Fig. 2. Adduct A is the nucleotide analogue of the nucleoside adduct described above. Adduct B is

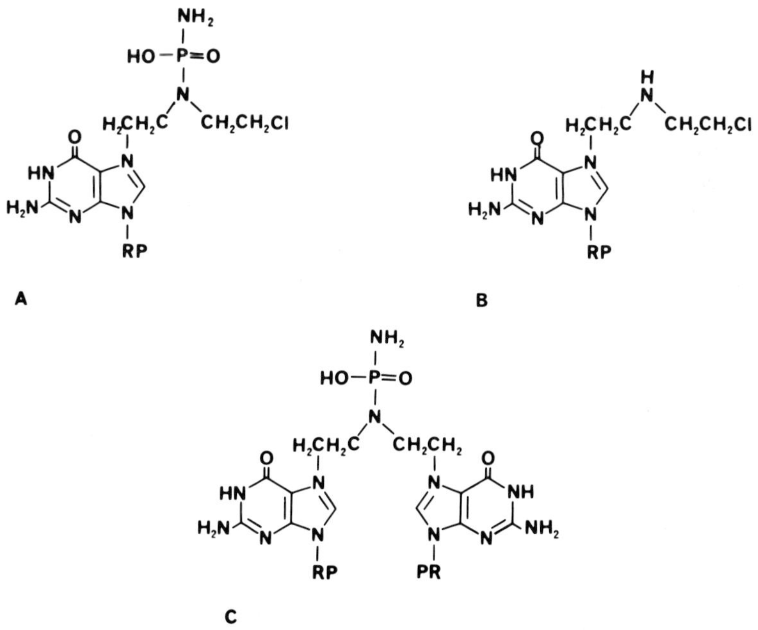

Fig. 2. Structures of guanine adducts formed by reaction with phosphoramide mustard. *RP*, ribose phosphate

very similar except that it is an adduct of nornitrogen mustard, a decomposition product of phosphoramide mustard. Adduct C is the result of bifunctional reaction of phosphoramide mustard with two guanylic acid moieties and probably corresponds to the structure responsible for DNA-DNA interstrand cross-linking.

Other nitrogen mustards evidently react with guanine in a similar manner. LINDEMANN and HARBERS (1980) obtained evidence for the reaction of cyclophosphamide, ifosfamide, and trofosphamide with guanosine in the presence of microsomes, although the primary site of reaction with DNA was with the phosphate groups. KALLAMA and HEMMINKI (1984) showed that the reactions of nitrogen mustard, chlorambucil, and phosphoramide mustard with guanosine which was labelled with ^3H in the C-8 position released the label, indicating that the alkylating agents had reacted with the N-7 position of guanosine (TOMASZ 1970).

II. Myleran

Evidence for the carcinogenicity of myleran to humans has also generally been judged to be sufficient. A large study was undertaken of patients who had had lung tumors removed surgically and were then treated with myleran or other therapy; the myleran patients had an increased incidence of acute non-lymphocytic leukemia (STOTT et al. 1977). The evidence for carcinogenicity in animals is more limited, but intravenous administration of myleran to mice significantly increased the incidence of thymus and ovarian tumors (IARC 1974).

Myleran, whose structure is shown in Fig. 3, is also sufficiently electrophilic to react with DNA without prior metabolic activation. Again, alkylation of the 7-position of guanine has been reported, and the structures of the two adducts shown in Fig. 4 have been elucidated by a combination of ultraviolet and mass spectrometry (TONG and LUDLUM 1980). As with nitrogen mustards, sequential reaction of two guanine moieties can result in a DNA-DNA interstrand cross-link. However, as noted below, some antineoplastic agents which do not cause cross-linking are also carcinogenic, which implies that cross-linking reactions are not required for the initiation of carcinogenesis by therapeutic agents.

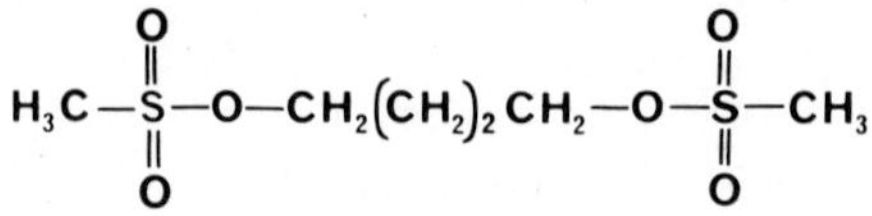

Fig. 3. Structure of myleran

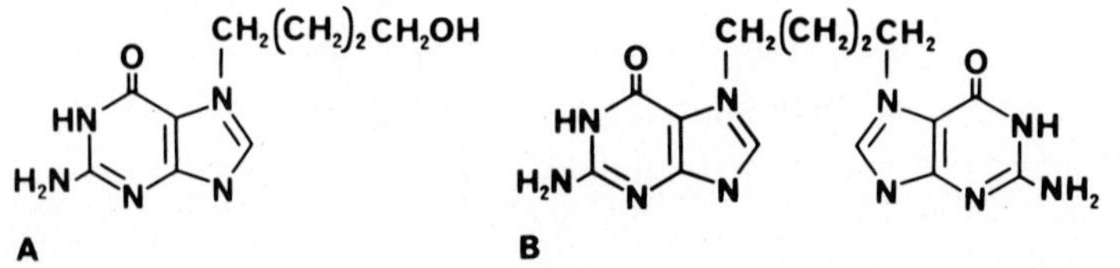

Fig. 4. Adducts of guanine isolated from DNA which had been reacted with myleran. **A** 7-Hydroxybutylguanine; **B** 1,4-di(7-guanyl)butane

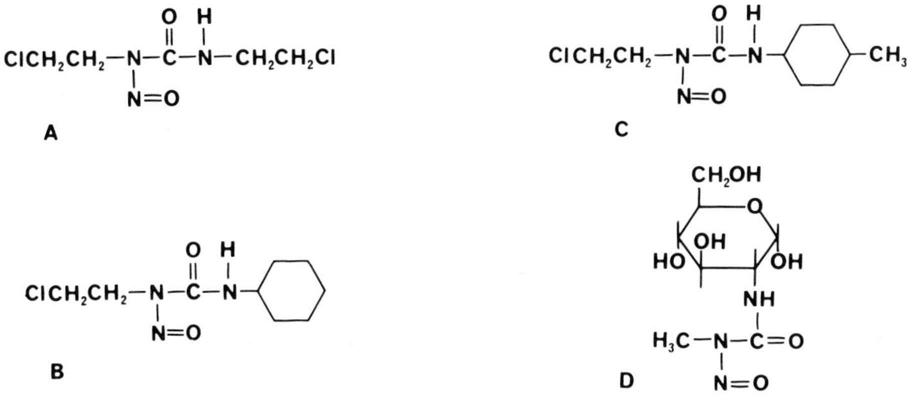

A

C

B

D

Fig. 5. Structures of some nitrosoureas which have been used therapeutically. **A** *N,N'*-bis(2-chloroethyl)-*N*-nitrosourea (BCNU); **B** *N*-(2-chloroethyl)-*N'*-cyclohexyl-*N*-nitrosourea (CCNU); **C** *N*-(2-chloroethyl)-*N'*-(4-methylcyclohexyl)-*N*-nitrosourea (methyl-CCNU); **D** streptozotocin

III. Therapeutic Nitrosoureas

The structures of four therapeutically effective nitrosoureas are shown in Fig. 5. Three of the compounds, *N,N'*-bis(2-chloroethyl)-*N*-nitrosourea (BCNU), *N*-(2-chloroethyl)-*N'*-cyclohexyl-*N*-nitrosourea (CCNU), and *N*-(2-chloroethyl)-*N'*-(4-methylcyclohexyl)-*N*-nitrosourea (methyl-CCNU) are synthetic products (Montgomery 1981), while streptozotocin is a naturally occurring compound. The haloethylnitrosoureas have been used in the treatment of a variety of malignancies including central nervous system tumors, but the use of streptozotocin has been much more limited.

All four of these therapeutic nitrosoureas are carcinogenic in animal systems. Intraperitoneal administration of BCNU to rats produces tumors in numerous organs including the lungs (Weisburger 1977); intraperitoneal administration of CCNU also produces lung tumors in rats; intraperitoneal administration of streptozotocin produces malignancies of the liver, kidney, and pancreas (Weisburger et al. 1975; Weisburger 1977). Intravenous administration of methyl-CCNU produces lung tumors in rats (Habs and Schmähl 1984). An increased incidence of acute nonlymphocytic leukemia has been associated with the administration of BCNU (Greene et al. 1985) and of methyl-CCNU to humans (Boice et al. 1983, 1986).

Comparison of the reactions of the haloethylnitrosoureas with DNA with the reactions of streptozotocin with DNA is instructive. The haloethylnitrosoureas transfer chloroethyl and hydroxyethyl groups to a variety of nucleophilic sites in DNA, as shown in Fig. 6 (Gombar et al. 1980; Ludlum and Tong 1981 a, b; Tong et al. 1982 b). The haloethyl groups can react a second time with another nucleophilic site to form cross-links (Tong and Ludlum 1981; Tong et al. 1982 a). Streptozotocin, on the other hand, transfers methyl groups to the nucleophilic sites in DNA (Masiello et al. 1981; Bennett and Pegg 1981), and

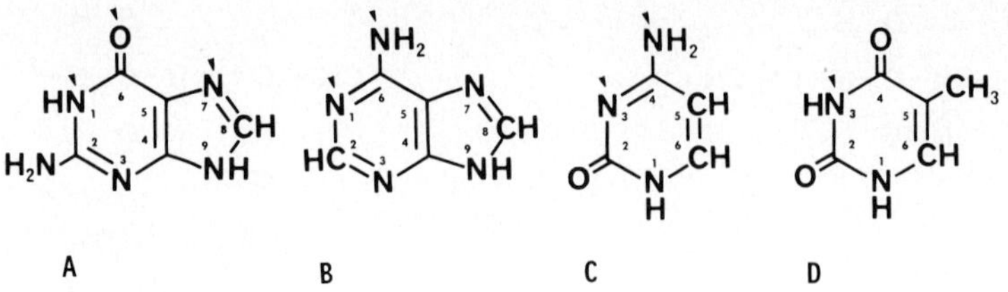

A B C D

Fig. 6. Sites (*arrowheads*) of base substitution by the haloethylnitrosoureas. **A** guanine; **B** adenine; **C** cytosine; **D** thymine

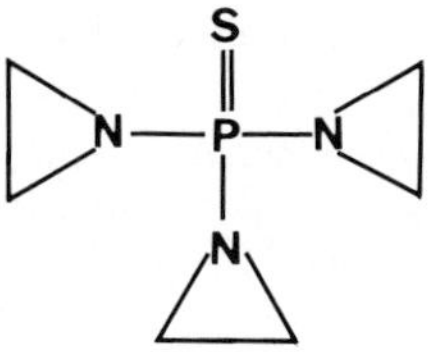

Fig. 7. Structure of thiotepa

these would not cause cross-linking. Consequently, it would appear that cross-linking reactions, although they may contribute to cytotoxicity, are not necessary for carcinogenesis by therapeutic agents.

IV. Thiotepa

The structure of thiotepa is shown in Fig. 7. It is included here as another antineoplastic agent which is carcinogenic and which can presumably react with DNA. Intraperitoneal administration of thiotepa produces a dose-related increase in the incidence of lung adenomas in mice (STONER et al. 1973). Intravenous administration of thiotepa produces a wide variety of tumors including sarcomas in rats (SCHMÄHL and OSSWALD 1970).

There have been numerous case reports of acute nonlymphocytic leukemia following the administration of thiotepa to humans, but interpretation of these data is complicated by the fact that the patients frequently received other therapeutic agents as well. Thus, the evidence for carcinogenicity of thiotepa to humans has been considered inadequate, while the evidence for thiotepa carcinogenicity in animals is generally considered sufficient.

There is little doubt that the ethyleneimine rings of thiotepa would react with DNA in much the same way that the activated immonium forms of nitrogen mustard react. Although adducts of thiotepa have not been characterized, HEMMINKI (1984) showed that unsubstituted ethyleneimine reacts with the N-7 position of guanosine and deoxyguanosine. The related compound, triethylenemelamine, has also been shown to react with the N-7 position of guanosine (TOMASZ 1970).

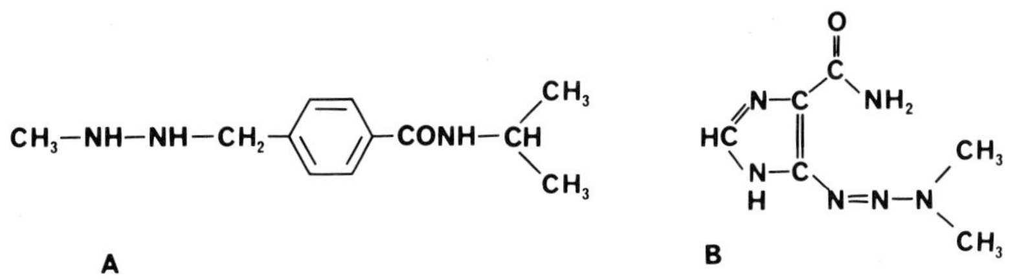

Fig. 8. Structures of **A** procarbazine and **B** dacarbazine

V. Procarbazine and Dacarbazine

These two compounds, whose structures are shown in Fig. 8, were synthesized independently according to entirely different rationales. However, they are both effective antitumor agents, carcinogenic in animal models, and probably able to methylate DNA as described below.

Oral administration of procarbazine produces pulmonary tumors in mice (Bacci et al. 1982) and rats (Bacci et al. 1984). Its intraperitoneal administration to mice increases the incidence of tumors in the lungs, hematopoietic system, and nervous system, as well as of the uterus in female animals (National Cancer Institute 1979). Intraperitoneal administration in rats produces tumors of the hematopoietic system, nervous system, and mammary glands (National Cancer Institute 1979). Because procarbazine is usually used with other therapeutic agents, evidence linking its administration to carcinogenicity in humans is considered inadequate. However, there is an increased incidence of acute leukemia following the use of regimens, particularly the MOPP regimen, which incorporate procarbazine (Arseneau et al. 1977; Auclerc et al. 1979; Coltman and Dixon 1982; Glicksman et al. 1982; Kyle 1984).

Oral administration of dacarbazine to rats produces mammary adenocarcinomas, lymphosarcomas of the thymus gland, and a variety of other tumors (Beal et al. 1975). Its intraperitoneal administration to mice produces lung tumors and lymphomas (Weisburger et al. 1975; Weisburger 1977). Again, however, the data on humans are inadequate partly because the compound is usually used in combination.

Unfortunately, the reactions of procarbazine and dacarbazine with DNA are complex and not completely elucidated. Both compounds undergo considerable metabolism, but radioactivity from [^{14}C]methyl procarbazine is transferred to RNA (Kreis 1970) and could presumably be transferred to DNA by the same mechanism. When [^{14}C]methyl dacarbazine is administered to rats, ^{14}C-labeled 7-methyl guanine is found in both DNA and RNA (Skibba and Bryan 1971). These data, added to what is known about the reaction of streptozotocin with DNA, would indicate that DNA methylation by antitumor agents can be an initiating event in carcinogenesis.

VI. Summary

Taken together, the data reviewed above indicate that DNA-reactive antitumor agents are carcinogenic and that a sufficient initiating event is monofunctional alkylation of DNA. Clearly, it would be helpful in evaluating other agents if the

DNA modifications which produce this effect could be identified. In considering the carcinogenic action of ethylating agents, attention has focused recently on alkylation of DNA oxygens (SINGER et al. 1981). However, several of the DNA-reactive antineoplastic agents mentioned above, including the nitrogen mustards and myleran, are evidently much more selective for the N-7 position of guanine. Considering the fact that 7-methylguanine appears to base-pair normally, the instability of some of the adducts formed by antineoplastic agents may be an important factor in producing their carcinogenic effect. However, it would appear that any therapeutic compound which can modify DNA is a potential initiating carcinogen, although even this realization must be weighed against evidence that most cells possess some ability to repair such damage.

D. Hormones as Carcinogenic Agents

Hormones constitute the other major class of therapeutic agents which have been associated with carcinogenesis in both animal experiments and in epidemiological studies. Estrogenic agents as a group may all be capable of inducing tumors in estrogen-dependent tissue under certain circumstances; Table 2, which lists compounds singled out by the National Toxicology Program, contain both naturally occurring and synthetic estrogenic agents. Two progestins and one anabolic steroid are also listed in this table.

The actions of these hormones, and their interactions with DNA-modifying agents, are clearly very complex. Beyond the obvious fact that they are growth stimulants for hormone-responsive tissues, they have broad effects on the intact organism. Thus, they have effects on intermediary metabolism and on the transformation of exogenous compounds, effects on the immune system, and possible effects on virus production. Furthermore, the hormones themselves can be metabolized, and some of the metabolites may play an important role. However, since hormone-related tumor production is most common in hormone-dependent tissue, attention is focused on their growth-stimulating characteristics.

Estradiol-17β is the prototype naturally occurring estrogenic hormone. Oral administration of this compound to female mice increases the incidence of mammary tumors (WELSCH et al. 1977; HIGHMAN et al. 1977). Subcutaneous administration of estradiol to male mice produces interstitial cell tumors of the testes (HOOKER and PFEIFFER 1942). In another study, subcutaneous administration produces an increased incidence of pituitary tumors and mammary carcinomas in both male and female mice (GARDNER 1941). Subcutaneous implantation of estradiol pellets in rats increases the incidence of mammary carcinomas (MACKENZIE 1955). Subcutaneous implants in normal and castrated male hamsters, as well as in ovariectomized female hamsters, produces renal tumors (KIRKMAN 1959). Subcutaneous administration of estradiol to young female mice produces tumors of the cervix or vagina (PAN and GARDNER 1948). These and other studies have led the IARC working group to conclude that the evidence for carcinogenicity of estradiol-17β in animals is sufficient; however, no data are available on this compound as a single agent in humans (IARC 1982).

Data linking the synthetic estrogen, diethylstilbestrol, to carcinogenesis are considered sufficient in both animals and human (IARC 1982). LACASSAGNE (1938) first demonstrated the ability of diethylstilbestrol to produce mammary tumors by injecting the compound subcutaneously into male mice. When diethylstilbestrol was administered orally, female mice developed adenocarcinomas of the mammary gland, uterus, and cervix (HIGHMAN et al. 1977). Subsequently, parenteral administration of diethylstilbestrol to a range of animal species has been shown to result in the development of mammary, cervical, uterine, and testicular tumors (IARC 1979).

Prenatal administration of diethylstilbestrol is causally related to the development of adenocarcinomas of the vagina and cervix in human females (HERBST and SCULLY 1970; HERBST et al. 1972; NOLLER et al. 1972). These results have also been confirmed by numerous other investigators.

Progesterone can be considered the prototype progestational agent. Subcutaneous implants of this compound in mice produced mammary carcinomas in one study (TRENTIN 1954) and ovarian and uterine tumors in other studies (LIPSCHÜTZ et al. 1967a, b). When progesterone is given in combination with known carcinogens, the tumor incidence increases and the latent period decreases (POEL 1969; GLUCKSMANN and CHERRY 1968). No epidemiological data are available for evaluating the effects of human exposure to progestational agents as single compounds.

Other growth-stimulating steroids have been associated with an increased incidence of tumors, however. Thus, numerous case reports have appeared relating the administration of oxymetholone, a synthetic androgenic steroid which has been used for its anabolic activity, to the development of hepatic tumors (IARC 1982).

HENDERSON et al. (1988), reviewing the mechanisms by which estrogens may cause cancer in humans, suggest that the increased frequency of mitotic activity in the target organ is of primary importance. After an initiating event, this tumor-promoting activity would increase the likelihood that a clone of cells could achieve independent growth. A second mechanism is proposed to explain the development of tumors after fetal exposure. This mechanism postulates an arrest of normal maturation to leave abnormal cells which could develop into tumors later under the stimulus of the increased hormonal levels accompanying puberty.

To summarize this topic, it is apparent that naturally occurring and synthetic hormones, and presumably other growth-regulating factors as well, can act as promoting agents. This action can lead to the development of tumors which are initiated spontaneously or by environmental factors. Consequently, in evaluating other therapeutic agents, it is important to examine their potential for stimulating mitosis, as well as for interacting with DNA.

E. Other Therapeutic Agents as Potential Carcinogens

Except for agents like thorotrast that are carcinogenic by virtue of their physical effects, the potential carcinogenicity of any therapeutic agent can be considered in the light of the two questions raised above. That is, can the agent modify DNA, or does it have a hormonal or growth-regulating effect? Continuing the emphasis on mechanisms in this review, these possibilities will be illustrated with a few therapeutic agents chosen because they illustrate particular mechanisms, not because of the level of risk that they present.

Currently, less information is available about the potential growth-stimulating activities of most therapeutic agents than on their metabolism and ability to modify DNA. However, one agent which might produce a carcinogenic effect through a hormonal action is reserpine. This compound is a naturally occurring substance which has been used for its psychoactive and antihypertensive effects.

Oral administration of reserpine produces mammary tumors in female mice and carcinomas of the seminal vesicles in male mice, while oral administration of the compound to rats produces pheochromocytomas (National Cancer Institute 1980; Muradyan 1986 b). Subcutaneous administration of reserpine also results in an increased tumor incidence in both mice and rats (Muradyan 1986 b). However, when reserpine is administered in combination with the known carcinogens 3-methylcholanthrene and *N*-methyl-*N*-nitrosourea, it decreases the incidence of mammary tumors in rats (Gerard et al. 1980; Verdeal et al. 1983). Epidemiological studies have provided some evidence for the carcinogenicity of reserpine in humans, but variations among these studies led the IARC working group to conclude that the evidence for its carcinogenicity in humans was inadequate (IARC 1987 b).

Reserpine increases the level of prolactin in the serum of patients receiving the drug for the treatment of hypertension (Lee et al. 1976). Since increased levels of prolactin have been associated with increased frequency of mammary tumors (Smithline et al. 1975), this suggests a hormonal mechanism by which reserpine could increase the incidence of tumors.

The number of therapeutic agents which could produce a carcinogenic effect by interacting with DNA is much larger. The compound shown in Fig. 9, phenoxybenzamine, is a single-armed nitrogen mustard which has been used as a blocking agent for receptors in the sympathetic nervous system. Although no studies establishing this point appear to have been published, this agent could presumably react directly with DNA. Phenoxybenzamine produces lung tumors in mice (Stoner et al. 1973) and peritoneal sarcomas in rats after intraperitoneal injection (National Cancer Institute 1978).

Relatively few other therapeutic agents besides those used for the treatment of malignancies are sufficiently reactive to modify DNA directly, but metabolic transformations can convert many of them into DNA-reactive forms. This topic is approached here by considering the metabolism of therapeutic agents with particular functional groups which can be activated. Even so, complete coverage of this subject lies outside the scope of this review because it would appear that

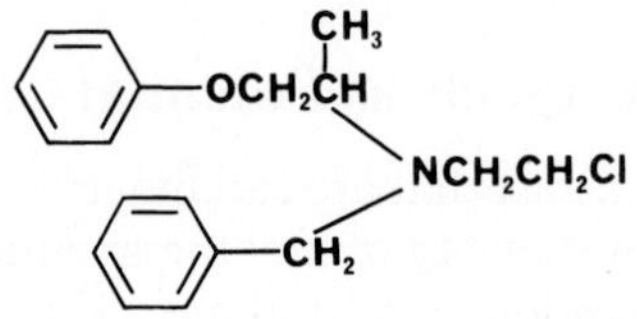

Fig. 9. Structure of phenoxybenzamine

many drug-metabolizing pathways can generate electrophiles in the process of converting substances into more polar, water-soluble compounds. Furthermore, although many metabolites have been characterized which are capable of reacting with DNA, adduct formation as a potentially initiating event has been demonstrated in only a few cases.

Considerable progress has been made in studies of the metabolism of the aromatic amide phenacetin and its reaction with DNA. This compound has been used for many years, either alone or in combinations with other agents, as a mild analgesic. Oral administration of phenacetin produces tumors of the urinary tract in mice and rats (NAKANISHI et al. 1982; MURADYAN 1986a) and in the nasal cavity of rats (ISAKA et al. 1979). Oral administration of the metabolite of phenacetin, *N*-hydroxyphenacetin, produces hepatocellular carcinomas in rats (CALDER et al. 1976). In humans, an association has been found between tumors of the urinary tract and the use of phenacetin-containing analgesics (IARC 1987b).

Phenacetin is extensively metabolized in most animal species and can be de-acetylated, de-ethylated, or hydroxylated on the aromatic ring. However, by analogy with the metabolism of the well-known carcinogen *N*-acetylamino-fluorene, N-hydroxylation to form *N*-hydroxyphenacetin (Fig. 10) may be the most significant route as far as carcinogenesis is concerned (HINSON and MITCHELL 1976). *N*-Hydroxyphenacetin can be activated further by transfer of the acetyl group or by sulphate conjugation and has been shown to react with transfer RNA, apparently through these intermediates (VAUGHT et al. 1981). More recently, MULDER et al. (1984) have shown that *N*-hydroxyphenacetin binds to DNA, although the nature of the adducts has not yet been determined.

Therapeutic agents that contain nitro groups are also apparently activated so that they are able to modify DNA through the formation of *N*-hydroxy intermediates; this possibility has been studied for the chemotherapeutic agent metronidazole. Oral administration of metronidazole increases the incidence of lymphomas in female mice and of lung tumors in both male and female mice (RUSTIA and SHUBIK 1972; CAVALIERE et al. 1983). Oral administration of metronidazole to rats produces a variety of neoplasms including mammary tumors (COHEN et al. 1973; RUSTIA and SHUBIK 1979; CAVALIERE et al. 1984).

Although no direct evidence for the modification of DNA has been obtained for metronidazole in vivo, LARUSSO et al. (1977) found that radiolabeled

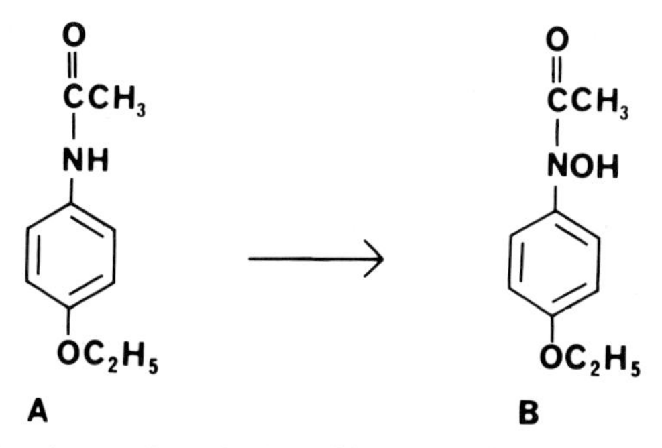

Fig. 10. Structures of **A** phenacetin and **B** its oxidized metabolite, *N*-hydroxyphenacetin

metronidazole became covalently bound to DNA when it was reduced in the presence of DNA in vitro. A variety of other nitroimidazoles can be reduced to the reactive hydroxylamino form (McClelland et al. 1984; Chrystal et al. 1980). Varghese and Whitmore (1981, 1983) have obtained evidence that the related nitroimidazole, misonidazole, reacts with DNA after reduction to the hydroxylamino form. The analogous reduction of metronidazole would yield the reactive intermediate shown in Fig. 11. Recently, Ludlum et al. (1988) obtained evidence that metronidazole forms an adduct with the N-2 position of guanosine when it is reduced, presumably to the hydroxylamino intermediate. No data are available on whether or not these reactions occur in vivo, however.

Other nitro-containing compounds could also be reduced to a reactive hydroxylamino form, but evidence for this reaction is generally lacking. One therapeutic agent that contains a nitro group, azathioprine, has been associated with an increased risk of neoplasms (Penn 1979; Kinlen 1985; Fries et al. 1985; Isomäki et al. 1978). However, this agent may act by decreasing host defences against malignancies through its suppression of the immune system.

The extensive literature on carcinogenic nitroso-containing compounds has stimulated interest in therapeutic agents which might be nitrosated in the gastrointestinal tract. Amines, amides, and amidines are all possible targets for nitrosation, and therapeutic agents contain members of all three classes; a compilation of these potentially nitrosatable agents has been prepared by Rao (1980). The other ingredients necessary for nitrosation reactions to occur, nitrite and an acid environment, are both present in the stomach, and studies performed in vitro have demonstrated that a variety of therapeutic agents can be nitrosated (Lijinsky et al. 1972; Lijinsky 1974). However, the concentrations of acid and nitrite present in the stomach depend on diet, animal species, and other factors, so that the amount of nitrosated drug which would form in vivo in any individual case is hard to predict.

Oral administration of the secondary amines, morpholine or *N*-methylbenzylamine, together with nitrites produces malignant tumors in rats (Sander and Bürkle 1969). Since then, numerous experiments have demonstrated tumor induction after the simultaneous administration of nitrites and nitrosatable compounds. Studies of this sort with therapeutic agents have been summarized by IARC (1980).

Cimetidine, a compound which is potentially nitrosatable in vivo, has been studied extensively because of its importance in clinical medicine. Nitrosocimetidine, synthesized in vitro, has been shown to cause DNA modifications, also in

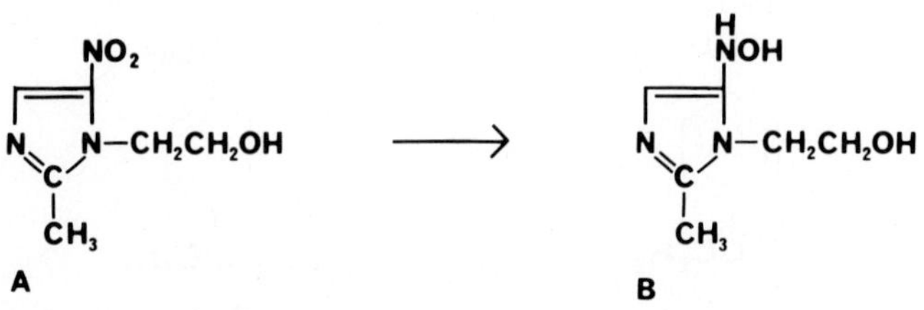

Fig. 11. Structures of **A** metronidazole and **B** its reduced form, 1-(2-hydroxyethyl)-2-methyl-5-hydroxyaminoimidazole

vitro, similar to those produced by the well-known carcinogen *N*-methyl-*N'*-nitro-*N*-nitrosoguanidine (JENSEN and MAGEE 1981). However, nitrosation of cimetidine has not been shown to occur in vivo; furthermore, exogenously prepared nitrosocimetidine is detoxified rapidly in a variety of animal species (JENSEN et al. 1987). These studies, of course, demonstrate the importance of detoxification as another defense mechanism against the potential carcinogenicity of therapeutic agents.

Recent studies of 1,2-dibromoethane by OZAWA and GUENGERICH (1983) suggest a route by which halogenated therapeutic agents could be activated to modify DNA. Dibromoethane reacts with glutathione to form a bromoethyl derivative of glutathione which is capable of alkylating the N-7 position of guanine; similar reaction with therapeutic agents that contain two chemically reacting groups might also lead to DNA modification through a glutathione intermediate.

To summarize this section, many drug-metabolizing reactions are potentially capable of generating DNA-reactive intermediates. More studies are needed, however, to determine the extent to which such reactions occur in vivo and to establish the ability of cellular mechanisms to repair the DNA damage so caused.

F. Conclusions

The studies reviewed above show that certain therapeutic agents are carcinogenic and suggest that they produce this effect either through modification of cellular DNA or through a hormonal or growth-regulating mechanism. Although many of the metabolic routes which can activate drugs to DNA-reactive forms have been investigated, relatively few of the DNA adducts have been characterized; studies of cellular defenses against these modifications are just beginning. In comparison with what is known about initiation, relatively little is known about mechanisms by which therapeutic agents could produce a tumor-promoting effect.

Although the risks of potential carcinogenicity from a therapeutic agent must be weighed against the benefits in each individual case, they are often judged to be acceptable in the treatment of malignant disease. Judgments become much more difficult when relatively benign conditions are involved, especially when chronic treatment is required, and the evidence for potential risk or benefit is less clear. However, it is a safe prediction that therapeutic agents will continue to be monitored carefully for potential carcinogenic effects in the future and that these studies will contribute to an understanding of the importance of environmental exposure in general. This is particularly true since therapeutic agents are administered to humans in pure form in known amounts, conditions which are rarely fulfilled in studies of other kinds of environmental exposure.

Acknowledgement. I gratefully acknowledge support of this work from the American Cancer Society through grant BC-457.

References

Arseneau JC, Canellos GP, Johnson R, DeVita VT (1977) Risk of new cancers in patients with Hodgkin's disease. Cancer 40:1912–1916

Auclerc G, Jacquillat C, Auclerc MF, Weil M, Bernard J (1979) Post-therapeutic acute leukaemia. Cancer 44:2017–2025

Aymard JP, Frustin J, Witz F, Columb JN, Lederlin P, Herbeuval R (1980) Acute leukaemia after prolonged chlorambucil treatment for non-malignant disease: report of a new case and literature survey. Acta Haematol 63:283–285

Bacci M, Cavaliere A, Fratini D (1982) Lung carcinogenesis by procarbazine chlorate in BALB/c mice. Carcinogenesis 3:71–73

Bacci M, Cavaliere A, Amorosi A (1984) Procarbazine hydrochlorate carcinogenesis in Osborne-Mendel rats. Oncology 41:106–108

Beal DD, Skibba JL, Croft WA, Cohen SM, Bryan GT (1975) Carcinogenicity of the antineoplastic agent, 5-(3,3-dimethyl-1-triazeno)-imidazole-4-carboxamide, and its metabolites in rats. JNCI 54:951–957

Bennett RA, Pegg AE (1981) Alkylation of DNA in rat tissues following administration of streptozotocin. Cancer Res 41:2786–2790

Berk PD, Goldberg JD, Silverstein MN, Weinfeld A, Donovan PB, Ellis JT, Landaw SA, Laszlo J, Najean Y, Pisciotta AV, Wasserman LR (1981) Increased incidence of acute leukemia in polycythemia vera associated with chlorambucil therapy. N Engl J Med 304:441–447

Blanc AP, Gastaut JA, Sebahoun G, Dalivoust P, Murisasco A, Carcassone Y (1981) Naissance d'une leucémie aiguë au décours d'un traitement immunosuppresseur par le chlorambucil: une observation. Nouv Presse Med 10:1717–1719

Boice JD Jr, Greene MH, Killen JY Jr, Ellenberg SS, Keehn RJ, McFadden E, Chen TT, Fraumeni JF Jr (1983) Leukemia and preleukemia after adjuvant treatment of gastrointestinal cancer with semustine (methyl-CCNU). N Engl J Med 309:1079–1084

Boice JD Jr, Greene MH, Killen JY Jr, Ellenberg SS, Fraumeni JF Jr (1986) Leukemia after adjuvant chemotherapy with semustine (methyl-CCNU). Evidence of a dose-response effect. N Engl J Med 314:119–120

Brock N, Hohorst H-J (1963) Über die Aktivierung von Cyclophosphamid in vivo und in vitro. Arzneimittelforsch 1021:1–31

Calabresi P, Parks RE Jr (1985) Chemotherapy of neoplastic diseases. In: Gilman AG, Goodman LS, Rall TW, Murad F (eds) Goodman and Gilman's The pharmacological basis of therapeutics, 7th edn. MacMillan, New York, pp 1240–1306

Calder IC, Goss DE, Williams PJ, Funder CC, Green CR, Ham KN, Tange JD (1976) Neoplasia in the rat induced by *N*-hydroxyphenacetin, a metabolite of phenacetin. Pathology 8:1–6

Cavaliere A, Bacci M, Amorosi A, Del Gaudio M, Vitali R (1983) Induction of lung tumors and lymphomas in BALB/c mice by metronidazole. Tumori 69:379–382

Cavaliere A, Bacci M, Vitali R (1984) Induction of mammary tumors with metronidazole in female Sprague-Dawley rats. Tumori 70:307–311

Chetsanga CJ, Polidori G, Mainwaring M (1982) Analysis and excision of ring-opened phosphoramide mustard-deoxyguanosine adducts in DNA. Cancer Res 42:2616–2621

Chrystal JT, Koch RL, Goldman P (1980) Metabolites from the reduction of metronidazole by xanthine oxidase. Mol Pharmacol 18:105–111

Cohen SM, Ertürk E, Van Esch AM, Crovetti AJ, Bryan GT (1973) Carcinogenicity of 5-nitrofurans, 5-nitroimidazoles, 4-nitrobenzenes, and related compounds. JNCI 51: 403–417

Coltman CA, Dixon DO (1982) Second malignancies complicating Hodgkin's disease: a Southwest Oncology Group 10-year followup. Cancer Treat Rep 66:1023–1033

Colvin M, Padgett CA, Fenselau C (1973) A biologically active metabolite of cyclophosphamide. Cancer Res 33:915–918

Colvin M, Brundrett RB, Kan M-NN, Jardine I, Fenselau C (1976) Alkylating properties of phosphoramide mustard. Cancer Res 36:1121–1126

Connors TA (1984) Carcinogenicity of medicines. In: Searle CE (ed) Chemical carcinogens, 2nd edn, vol II. ACS Monograph 182, Washington DC, pp 1241–1278

Connors TA, Cox PJ, Farmer PB, Foster AB, Jarman M (1974) Some studies of the active intermediates formed in the microsomal metabolism of cyclophosphamide and isophosphamide. Biochem Pharmacol 23:115–129

Demple B, Sedgwick B, Robins P, Totty N, Waterfield MD, Lindahl T (1985) Active site and complete sequence of the suicidal methyltransferase that counters alkylation mutagenesis. Proc Natl Acad Sci USA 82:2688–2692

Doll R, Peto R (1981) The causes of cancer: quantitative estimates of avoidable risks of cancer in the United States today. JNCI 66:1191–1308

Dumont J, Thiery JP, Mazabraud A, Natali JC, Trapet P, Vilcoq JR (1980) Acute myeloid leukemia following non-Hodgkin's lymphoma: danger of prolonged use of chlorambucil as maintenance therapy. Nouv Rev Fr Hematol 22:391–404

Einhorn N (1978) Acute leukemia after chemotherapy (melphalan). Cancer 41:444–447

Fries JF, Bloch D, Spitz P, Mitchell DM (1985) Cancer in rheumatoid arthritis: a prospective long-term study of mortality. Am J Med (Suppl 1A) 78:56–59

Gardner WU (1941) The effect of estrogen on the incidence of mammary and pituitary tumors in hybrid mice. Cancer Res 1:345–358

Gerard SS, Gardner B, Patti J, Husain V, Shouten J, Alfonso AE (1980) Effects of triiodothyronine and reserpine on induction and growth of mammary tumors in rats by 3-methylcholanthrene. J Surg Oncol 14:213–218

Glicksman AS, Pajak TF, Gottlieb A, Nissen N, Stutzman L, Cooper MR (1982) Second malignant neoplasms in patients successfully treated for Hodgkin's disease: a Cancer and Leukemia Group B study. Cancer Treat Rep 66:1035–1044

Glucksmann A, Cherry CP (1968) The effect of oestrogens, testosterone and progesterone on the induction of cervico-vaginal tumours in intact and castrate rats. Br J Cancer 22:545–562

Gombar CT, Tong WP, Ludlum DB (1980) Mechanism of action of the nitroso-ureas IV: reactions of BCNU and CCNU with DNA. Biochem Pharmacol 29:2639–2643

Greene MH, Boice JD Jr, Strike TA (1985) Carmustine as a cause of acute nonlymphocytic leukemia. N Engl J Med 313:579

Habs M, Schmähl D (1984) Long-term toxic and carcinogenic effects of cytostatic drugs. Dev Oncol 15:201–209

Harousseau JL, Andrieu JM, Dumont J, Montagnon B, Asselain B, Daniel MT, Flandrin G (1980) Leucémies aiguës myéloblastiques survenant au cours de l'évolution de lymphomes malins nonhodgkiniens. Nouv Presse Med 9:3513–3516

Hemminki K (1984) Reactions of ethyleneimine with guanosine and deoxyguanosine. Chem Biol Interact 48:249–260

Henderson BE, Ross R, Bernstein L (1988) Estrogens as a cause of human cancer: the Richard and Hinda Rosenthal Foundation Award Lecture. Cancer Res 48:246–253

Herbst AL, Scully RE (1970) Adenocarcinoma of the vagina in adolescence. A report of 7 cases including 6 clear-cell carcinomas (so-called mesonephromas). Cancer 25:745–757

Herbst AL, Kurman RJ, Scully RE, Poskanzer DC (1972) Clear-cell adenocarcinoma of the genital tract in young females. Registry report. N Engl J Med 287:1259–1264

Heston WE (1949) Induction of pulmonary tumors in strain A mice with methylbis(β-chloroethyl)amine hydrochloride. JNCI 10:125–140

Highman B, Norvell MJ, Shellenberger TE (1977) Pathological changes in female C3H mice continuously fed diets containing diethylstilbestrol or 17β-estradiol. J Environ Pathol Toxicol 1:1–30

Hinson JA, Mitchell JR (1976) *N*-Hydroxylation of phenacetin by hamster liver microsomes. Drug Metab Dispos 4:430–435

Hooker CW, Pfeiffer CA (1942) The morphology and development of testicular tumors in mice of the A strain receiving estrogens. Cancer Res 2:759–769

International Agency for Research on Cancer (1974) Evaluation of the carcinogenic risk of chemicals to humans: some aromatic amines, hydrazine and related substances, *N*-nitroso compounds and miscellaneous alkylating agents. IARC Monographs 4:247–252

International Agency for Research on Cancer (1976) Evaluation of the carcinogenic risk of chemicals to man: some miscellaneous pharmaceutical substances. IARC Monographs 13
International Agency for Research on Cancer (1979) Evaluation of the carcinogenic risk of chemicals to man: sex hormones II. IARC Monographs 21:173–231
International Agency for Research on Cancer (1980) Evaluation of the carcinogenic risk of chemicals to man: some pharmaceutical drugs. IARC Monographs 24:297–314
International Agency for Research on Cancer (1981) Evaluation of the carcinogenic risk of chemicals to humans: some antineoplastic and immunosuppressive agents. IARC Monographs 26:165–202
International Agency for Research on Cancer (1982) Evaluation of the carcinogenic risk of chemicals to humans: chemicals, industrial processes and industries associated with cancer in humans. IARC Monographs [Suppl] 4:173–205
International Agency for Research on Cancer (1985) Evaluation of the carcinogenic risk of chemicals to humans: cross index of synonyms and trade names in volumes 1 to 36. IARC Monographs [Suppl] 5
International Agency for Research on Cancer (1987a) Evaluation of the carcinogenic risk of chemicals to humans: silica and some silicates. IARC Monographs 42:13–32, 265–289
International Agency for Research on Cancer (1987b) Evaluation of the carcinogenic risk of chemicals to humans: overall evaluations of carcinogenicity: an updating of IARC Monographs vol 1–42. IARC Monographs [Suppl] 7:310–312, 330–332
Isaka H, Yoshii H, Otsuji A, Koike M, Nagai Y, Koura M, Sugiyasu K, Kanabayashi T (1979) Tumors of Sprague-Dawley rats induced by long-term feeding of phenacetin. Gann 70:29–36
Isomäki HA, Hakulinen T, Joutsenlahti U (1978) Excess risk of lymphomas, leukemia and myeloma in patients with rheumatoid arthritis. J Chron Dis 31:691–696
Jensen DE, Magee PN (1981) Methylation of DNA by nitrosocimetidine in vitro. Cancer Res 41:230–236
Jensen DE, Stelman GJ, Spiegel A (1987) Species differences in blood-mediated nitrosocimetidine denitrosation. Cancer Res 47:353–359
Kallama S, Hemminki K (1984) Alkylation of guanosine by phosphoramide mustard, chloromethine hydrochloride and chlorambucil. Acta Pharmacol Toxicol 54:214–220
Kinlen LJ (1985) Incidence of cancer in rheumatoid arthritis and other disorders after immunosuppressive treatment. Am J Med 78 [Suppl 1A]:44–49
Kirkman H (1959) Estrogen-induced tumors of the kidney. IV. Incidence in female Syrian hamsters. Natl Cancer Inst Monogr 1:59–75
Kohn KW, Spears CL, Doty P (1966) Interstrand cross-linking of DNA by nitrogen mustard. J Mol Biol 19:266–288
Kreis W (1970) Metabolism of an antineoplastic methylhydrazine derivative in a P815 mouse neoplasm. Cancer Res 30:82–89
Kyle RA (1984) Second malignancies associated with chemotherapy. In: Perry MC, Yarbro JW (eds) Toxicity of chemotherapy. Grune & Stratton, New York, pp 479–506
Lacassagne A (1938) Apparition d'adénocarcinomes mammaires chez des souris males traitées par une substance oestrogène synthétique. C R Soc Biol (Paris) 129:641–643
LaRusso NF, Tomasz M, Müller M, Lipman R (1977) Interaction of metronidazole with nucleic acids in vitro. Mol Pharmacol 13:872–882
Law IP, Blom J (1977) Second malignancies in patients with multiple myeloma. Oncology 34:20–24
Lawley PD (1984) Carcinogenesis by alkylating agents. In: Searle CE (ed) Chemical carcinogens, 2nd edn, vol I. ACS Monograph 182, Washington DC, pp 325–484
Lee PA, Kelly MR, Wallin JD (1976) Increased prolactin levels during reserpine treatment of hypertensive patients. J Am Med Assoc 235:2316–2317
Lijinsky W (1974) Reaction of drugs with nitrous acid as a source of carcinogenic nitrosamines. Cancer Res 34:255–258
Lijinsky W, Conrad E, Van de Bogart R (1972) Carcinogenic nitrosamines formed by drug/nitrite interactions. Nature 239:165–167

Lindemann H, Harbers E (1980) In-vitro-Reaktion der drei alkylierenden Pharmaka Cyclophosphamid, Ifosfamid und Trofosfamid mit DNS und DNS-Bausteinen. Arzneimittelforsch 30:2075–2080

Lipschütz A, Iglesias R, Panasevich VI, Salinas S (1967a) Granulosa-cell tumours in mice by progesterone. Br J Cancer 21:144–152

Lipschütz A, Iglesias R, Panasevich VI, Salinas S (1967b) Pathological changes induced in the uterus of mice with the prolonged administration of progesterone and 19-nor-contraceptives. Br J Cancer 21:160–165

Ludlum DB (1970) Properties of 7-methylguanine-containing templates for ribonucleic acid polymerase. J Biol Chem 245:477–482

Ludlum DB, Tong WP (1980) Crosslinking of DNA by busulfan. Formation of diguanyl derivatives. Biochim Biophys Acta 608:174–181

Ludlum DB, Tong WP (1981a) Modification of DNA and RNA bases. In: Prestayko AW, Crooke ST, Baker LH, Carter SK, Schein PS (eds) Nitrosoureas: current status and new developments. Academic, New York, pp 85–94

Ludlum DB, Tong WP (1981b) Modification of DNA and RNA bases by the nitrosoureas. In: Serrou B, Schein P, Imbach J-L (eds) Nitrosoureas in cancer treatment. Elsevier, Amsterdam, pp 21–31

Ludlum DB, Colinas RJ, Kirk MC, Mehta JR (1988) Reaction of reduced metronidazole with guanosine to form an unstable adduct. Carcinogenesis 9:593–596

MacKenzie I (1955) The production of mammary cancer in rats using oestrogens. Br J Cancer 9:284–299

Masiello P, Karunanayake EH, Bergamini E, Hearse DJ, Mellows G (1981) [^{14}C] Streptozotocin: its distribution and interaction with nucleic acids and proteins. Biochem Pharmacol 30:1907–1913

McClelland RA, Fuller JR, Seaman NE, Rauth AM, Battistella R (1984) 2-Hydroxylaminoimidazoles: unstable intermediates in the reduction of 2-nitroimidazoles. Biochem Pharmacol 33:303–309

Mehta JR, Przybylski M, Ludlum DB (1980) Alkylation of guanosine and deoxyguanosine by phosphoramide mustard. Cancer Res 40:4183–4186

Miller EC, Miller JA (1981) Searches for ultimate chemical carcinogens and reactions with cellular macromolecules. Cancer 47:2327–2345

Mirkes PE, Fantel AG, Greenaway JC, Shepard TH (1981) Teratogenicity of cyclophosphamide metabolites: phosphoramide mustard, acrolein, and 4-ketocyclophosphamide in rat embryos cultured in vitro. Toxicol Appl Pharmacol 58:322–330

Montgomery JA (1981) The development of the nitrosoureas: a study in congener synthesis. In: Prestayko AW, Crooke ST, Baker LH, Carter SK, Schein PS (eds) Nitrosoureas: current status and new developments. Academic, New York, pp 3–8

Mulder GJ, Kadlubar FF, Mays JB, Hinson JA (1984) Reaction of mutagenic phenacetin metabolites with glutathione and DNA: possible implications of toxicity. Molec Pharmacol 26:342–347

Müller W, Brandis M (1981) Acute leukemia after cytotoxic treatment for nonmalignant disease in childhood: a case report and review of the literature. Eur J Pediatr 136:105–108

Muradyan RY (1986a) Experimental studies of phenacetin carcinogenicity. Vopr Onkol 32:63–70

Muradyan RY (1986b) A study of possible carcinogenicity of reserpine. Vopr Onkol 32:76–81

Nakanishi K, Kurata Y, Oshima M, Fukushima S, Ito N (1982) Carcinogenicity of phenacetin: long-term feeding study in B6C3F$_1$ mice. Int J Cancer 29:439–444

National Cancer Institute (1978) Bioassay of phenoxybenzamine hydrochloride for possible carcinogenicity. NCI Carcinog Tech Rep Ser 72

National Cancer Institute (1979) Bioassay of procarbazine for possible carcinogenicity. NCI Carcinog Tech Rep Ser no 19

National Cancer Institute (1980) Bioassay of reserpine for possible carcinogenicity. Dept of Health, Education, and Welfare pub no (NIH) 80-1749

Nesnow S, Argus M, Bergman H, Chu K, Frith C, Helmes T, McGaughty R, Ray V, Slaga TJ, Tennant R, Weisburger E (1986) Chemical carcinogens. A review and analysis of the literature of selected chemicals and the establishment of the gene-tox carcinogen data base. Mutat Res 185:1–195
Noller KL, Decker DG, Lanier AP, Kurland LT (1972) Clear-cell adenocarcinoma of the cervix after maternal treatment with synthetic estrogens. Mayo Clin Proc 47:629–630
Ozawa N, Guengerich FP (1983) Evidence for formation of an *S*-[2-(N^7-guanyl)ethyl]glutathione adduct in glutathione-mediated binding of the carcinogen 1,2-dibromoethane to DNA. Proc Natl Acad Sci USA 80:5266–5270
Pan SC, Gardner WU (1948) Carcinomas of the uterine cervix and vagina in estrogen- and androgen-treated hybrid mice. Cancer Res 8:337–341
Penn I (1979) Tumor incidence in human allograft recipients. Transplant Proc 11: 1047–1051
Poel WE (1969) Bioassays with inbred mice: their relevance for the random-bred animal. Prog Exp Tumor Res 11:440–460
Prejean JD, Montgomery JA (1984) Structure-activity relationships in the carcinogenicity of anticancer agents. Drug Metab Rev 15:619–646
Rao GS (1980) *N*-Nitrosamines from drugs and nitrite: potential source of chemical carcinogens in humans? Pharm Int 1:187–190
Reimer RR, Hoover R, Fraumeni JF Jr, Young RC (1977) Acute leukemia after alkylating-agent therapy of ovarian cancer. N Engl J Med 297:177–181
Rustia M, Shubik P (1972) Induction of lung tumors and malignant lymphomas in mice by metronidazole. JNCI 48:721–726
Rustia M, Shubik P (1979) Experimental induction of hepatomas, mammary tumors, and other tumors with metronidazole in noninbred Sas:MRC(W1)BR rats. JNCI 63:863–868
Rydberg B, Lindahl T (1982) Nonenzymatic methylation of DNA by the intracellular methyl group donor *S*-adenosyl-L-methionine is a potentially mutagenic reaction. Eur Molec Biol Organization 1:211–216
Samson L (1986) The adaptive response of mammalian cells to alkylating damage and repair: see my implications for carcinogenesis and risk assessment. Plenum, New York, p 327
Sander J, Bürkle G (1969) Induktion maligner Tumoren bei Ratten durch gleichzeitige Verfütterung von Nitrit und sekundären Aminen. Z Krebsforsch 73:54–66
Schmähl D (1974) Investigation on the influence of immunodepressive means on the chemical carcinogenesis in rats. Z Krebsforsch 81:211–215
Schmähl D, Habs M (1979) Carcinogenic action of low-dose cyclophosphamide given orally to Sprague-Dawley rats in a lifetime experiment. Int J Cancer 23:706–712
Schmähl D, Kaldor JM (1986) Carcinogenicity of alkylating cytostatic drugs. IARC publications no 78, Lyon
Schmähl D, Osswald H (1970) Experimentelle Untersuchungen über carcinogene Wirkungen von Krebs-Chemotherapeutica and Immunosuppressiva. Arzneimittelforsch 20: 1461–1467
Schmähl D, Thomas C, Auer R (1977) Iatrogenic carcinogenesis. Springer-Verlag, Berlin Heidelberg New York
Shimkin MB, Weisburger JH, Weisburger EK, Gubareff N, Suntzeff V (1966) Bioassay of 29 alkylating chemicals by the pulmonary-tumor response in strain A mice. JNCI 36:915–935
Singer B, Spengler SJ, Bodell WJ (1981) Tissue-dependent enzyme-mediated repair or removal of *O*-ethyl pyrimidines and ethyl purines in carcinogen-treated rats. Carcinogenesis 2:1069–1073
Skibba JL, Bryan GT (1971) Methylation of nucleic acids and urinary excretion of ^{14}C-labeled 7-methylguanine by rats and man after administration of 4(5)-(3,3-dimethyl-1-triazeno)imidazole 5(4)-carboxamide. Toxicol Appl Pharmacol 18:707–719
Smithline F, Sherman L, Kolodny HD (1975) Prolactin and breast carcinoma. N Engl J Med 292:784–792
Stoner GD, Shimkin MB, Kniazeff AJ, Weisburger JH, Weisburger EK, Gori GB (1973) Test for carcinogenicity of food additives and chemotherapeutic agents by the pulmonary tumor response in strain A mice. Cancer Res 13:3069–3085

Stott H, Fox W, Girling DJ, Stephens RJ, Galton DAG (1977) Acute leukaemia after busulfan. Br Med J 2:1513–1517
Struck RF, Kirk MC, Witt MH, Laster WR Jr (1975) Isolation and mass spectral identification of blood metabolites of cyclophosphamide: evidence for phosphoramide mustard as the biologically active metabolite. Biomed Mass Spectrom 2:46–52
Thiede T, Christiensen BC (1969) Bladder tumors induced by chlornaphazine. Acta med Scand 185:133–137
Tomasz M (1970) Novel assay of 7-alkylation of guanine residues in DNA. Application to nitrogen mustard, triethylenemelamine and mitomycin C. Biochem Biophys Acta 213: 288–295
Tong WP, Ludlum DB (1980) Crosslinking of DNA by busulfan: formation of diguanyl derivatives. Biochim Biophys Acta 608:174–181
Tong WP, Ludlum DB (1981) Formation of the cross-linked base, diguanylethane, in DNA treated with *N,N'*-bis(2-chloroethyl)-*N*-nitrosourea. Cancer Res 41:380–382
Tong WP, Kirk MC, Ludlum DB (1982a) Formation of the crosslink, 1-[N^3-deoxycytidyl],2-[N^1-deoxyguanosinyl]ethane, in DNA treated with *N,N'*-bis(2-chloroethyl)-*N*-nitrosourea (BCNU). Cancer Res 42:3102–3105
Tong WP, Kohn KW, Ludlum DB (1982b) Modifications of DNA by different haloethylnitrosoureas. Cancer Res 42:4460–4464
Trentin JJ (1954) Effect of long-term treatment with high levels of progesterone on the incidence of mammary tumors in mice. Proc Am Assoc Cancer Res 1:50
United States Government Department of Health and Human Services (1985) Fourth annual report in carcinogens. National Toxicology Program, Bethesda, MD, 78-1317
Varghese AJ, Whitmore GF (1981) Cellular and chemical reduction products of misonidazole. Chem Biol Interactions 36:141–151
Varghese AJ, Whitmore GF (1983) Modification of guanine derivatives by reduced 2-nitroimidazoles. Cancer Res 43:78–82
Vaught JB, McGarvey PB, Lee MS, Garner CD, Wang CY, Linsmaier-Bednar EM, King CM (1981) Activation of *N*-hydroxyphenactin to mutagenic and nucleic acid-binding metabolites by acyl transfer, deacylation, and sulfate conjugation. Cancer Res 41:3424–3429
Verdeal K, Ertürk E, Rose DP (1983) Effects of reserpine administration on rat mammary tumors and uterine disease induced by *N*-nitrosomethylurea. Eur J Can Clin Oncol 19:825–834
Vu VT, Fenselau CC, Colvin OM (1981) Identification of three alkylated nucleotide adducts from the reaction of guanosine 5'-monophosphate with phosphoramide mustard. J Am Chem Soc 103:7362–7364
Weisburger EK (1977) Bioassay program for carcinogenic hazards of cancer chemotherapeutic agents. Cancer 40:1935–1951
Weisburger JH, Griswold DP, Prejean JD, Casey AE, Wood HB, Weisburger EK (1975) The carcinogenic properties of some of the principal drugs used in clinical cancer chemotherapy. Recent Results Cancer Res 52:1–17
Welsch CW, Adams C, Lambrecht LK, Hassett CC, Brooks CL (1977) 17β-Oestradiol and enovid mammary tumorigenesis in C3H/HeJ female mice: counteraction by concurrent 2-bromo-α-ergocryptine. Br J Cancer 35:322–328

Part III.
In Vivo and In Vitro Carcinogenesis

CHAPTER 6

In Vivo Testing for Carcinogenicity

W. LIJINSKY

A. Introduction

Until this century, cancer was a rare disease, and there was no great interest in its nature or origin. Physicians were properly most concerned with the infectious diseases which carried off large proportions of the population, particularly the very young. Concern with those infectious diseases which kill a large portion of the population before middle age is still overwhelming in most of the world, especially in Asia and Latin America. With a few exceptions, cancer is a disease of older people and, therefore, a major problem only in the industrialized countries of Europe and North America. As survival rates improve in any country, it is likely that the incidence of cancer will increase and the urge to prevent this disease will become stronger, since cure is usually difficult and frequently impossible.

It is ironic that improvements in health in industrialized countries during the past 100 years or so have led to a huge increase in the incidence of cancer and other degenerative diseases, which often entail protracted misery. For a long time, the process of carcinogenesis has been known to be different in kind from the course of common infectious diseases. Even in the earliest association of cancer of the scrotum with exposure to a chemical agent, noted by POTT more than two centuries ago, the induction of cancer was known to be a long-term process. In that case, cancer arose in fairly young men who had been exposed to coal soot from an early age through their occupation as chimney sweeps. The causative agents of the skin cancer in the soot were unknown until 50 years ago, when carcinogenic polycyclic aromatic hydrocarbons were isolated from coal tar pitch by KENNAWAY, COOK, HIEGER and their associates (COOK et al. 1933). In the meantime, YAMAGIWA and ICHIKAWA (1915) had reported the induction of skin cancer in rabbits' ears following the painting of coal tar, thereby establishing an animal model for the skin cancer found among coal tar workers. Yet earlier work of ROUS (1911) had shown that a virus had the capacity to induce tumors in animals (and presumably in humans also), and cancer due to exposure to radiation, an equally new phenomenon, was observed in workers who painted watch dials with paint containing radium. From these observations arose the idea that cancer is not a spontaneous phenomenon but a disease that has an external cause. From those small beginnings has developed a substantial research community devoted to the prevention of cancer through the identification of causative agents and investigation of their mechanisms of action.

Since the 1950s, several individual compounds, widely used and entailing considerable human exposure, have been defined as carcinogens through chronic tests in animals. Some of these are diethylstilbestrol, vinyl chloride, bis-chloromethyl ether, and ethylene oxide. Other carcinogens identified in this way have been examples of large groups of chemically similar compounds, such as nitrosamines (MAGEE et al. 1976), aflatoxins (BUTLER et al. 1969), and epoxides (VAN DUUREN et al. 1972). These groups of carcinogens have created great interest because of human exposure to them and because, due to the diversity of chemical structures, they present an invaluable opportunity for the investigation of mechanisms of carcinogenesis. The novelty of these carcinogenic structures emphasizes the comment by SHUBIK and SICÉ (1956) that "chemical structure alone cannot be relied upon to predict the absence of carcinogenicity in any substance."

Emphasis on this approach to dealing with human cancer has seemed to diminish somewhat lately, in favor of more fashionable – and esoteric – studies in genetics and molecular biology. Nevertheless, the strong National Toxicology Program in the USA and the efforts of private industry and of individual investigators ensure that the examination of chemicals for cancer-causing properties will continue. There has been considerable co-ordination of these efforts through the International Agency for Research on Cancer in Lyon, France, and there has been a gradual evolution of standards by which the quality of such tests are judged. The results of tests of thousands of substances for carcinogenic activity have been summarized in a series of volumes, United States Public Health Service publication number 149, begun by Jonathan HARTWELL and Philippe SHUBIK in the 1950s (HARTWELL 1955). This valuable resource now occupies almost 2 m of shelf space and includes many well-conducted studies, together with many that are less reliable.

One of the most important criteria now accepted is that a substance cannot be claimed to be free of carcinogenic potential based on experiments in which small groups of animals are used, or in which low doses of the test agent are used, or which are terminated within too short a time for development of tumors, for example, a year or less. It is difficult to produce a false positive result in a chronic carcinogenicity assay in animals, but it is quite easy to produce a false negative result by neglecting to follow some of the guidelines for the proper conduct of such assays.

A corollary of the need to conduct chronic toxicity assays in animals according to rules regarding group size, species, and longevity of the assay is that they are time-consuming and expensive, and require the commitment of scarce talent, such as pathologists, veterinarians, chemists, and others. This disadvantage has given impetus to the development of less expensive and faster assays for carcinogenic potential that could replace, totally or in part, the in vivo carcinogenesis assay. The best of these, the *Salmonella typhimurium* histidine reversion assay developed by B. N. AMES (MCCANN et al. 1975), detects as mutagens a high proportion of those carcinogens that have been examined in it, and most noncarcinogens have been negative in the assay. There is, of course, no quantitative correlation between mutagenic potency and carcinogenic potency. There have been many carcinogens which were negative in the AMES assay and in other short-term assays; human exposure to some of these nonmutagenic carcinogens has been considerable (REDDY and QURESHI 1979; LIJINSKY et al. 1980a, b), which suggests that relying upon mutagenicity assays for assurance of lack of carcinogenic properties in a substance is unwise. Recent surveys of the results of a number of short-term assays conducted with substances that have been tested in chronic bioassays in rodents (ZEIGER 1987; TENNANT et al. 1987) reach essentially the same conclusion that short-term assays, individually or as a group, cannot replace chronic studies in animals for identification of carcinogens.

Only retrospectively, through epidemiology can agents be discovered that cause cancer in humans. The tools of epidemiology are coarse, and the number of epidemiologists is small, so that only a handful of agents are known that can, with reasonable certainty, be claimed to cause – or make major contributions to – certain human cancers. Almost all of them have induced tumors of some type in experimental animals. Humans have been exposed to many others which have been tested in vivo in animals. Those that have been positive in such tests must be considered more likely to contribute to increased risk of cancer in humans than those that are negative.

The alternative is to believe that other animals are so different from humans that experiments in animals are not a reliable guide to toxicological effects in humans. This would bring to a rapid end developments in pharmacology and drug development, including the development of cancer therapeutic agents. The findings at about the same time that carcinogens such as vinyl chloride, diethylstilbestrol, 2-naphthylamine, benzidine, and polycyclic aromatic compounds are carcinogenic in humans and in animals gives encouragement to the use of in vivo experiments. This is true even though the results of such assays cannot indicate with certainty which type of tumor the carcinogen will induce in humans – or, indeed, in any other species. A carcinogen known to induce cancer in humans when tested in experimental animals can sometimes give rise to the same tumor in one species and, to a different tumor in a further test species.

To continue the tradition of attempting to deal with the cancer problem in humans through prevention, that is, by identifying carcinogens and reducing or eliminating human exposure to them, in vivo carcinogenesis testing must be improved and refined. The present discussion is not exhaustive but examines the criteria governing such experimental studies and the interpretation of the results.

B. Development and Use of In Vivo Carcinogenesis Tests

I. History of Carcinogenicity Testing

The earliest observations of carcinogenesis were in humans, as long ago as 1775, when the surgeon Percival POTT ascribed the unusual scrotal skin cancer he saw in chimney sweeps to their contact with soot. Similarly, REHN in 1895 showed a relationship between the bladder cancer in his patients and their exposure to aromatic amines similar to aniline in the dyestuff factories in which they worked. Interest in the experimental induction of cancer in animals developed later, in keeping with the modern concept of studying human disease in animal models. The studies by ROUS (1911) of induction of cancer by transmissible factors or viruses and by YAMAGIWA and ICHIKAWA (1915) of induction of skin cancer by painting coal tar on rabbits' ears date from early in the twentieth century and were followed by a slow development of interest in the process of producing cancer in animals by artificial means, or carcinogenesis. The focus in the 1920s was on the aromatic and fluorescent components of coal tar and associated hydrocarbon materials, including mineral oils, which were related to the high incidence of skin cancer in people who worked with these materials. A group of scientists headed by KENNAWAY in London were pioneers in the effort to identify pure compounds responsible for the production of skin cancer and their mechanism of action, which has remained an abiding interest. In 1933, benzo[*a*]pyrene was isolated from coal tar pitch – an enormous task. At about the same time, a group of chemists (I. HIEGER, J. W. COOK, E. CLAR) associated with the project synthesized a number of polycyclic aromatic hydrocarbons which produced cancer when painted on the skin of mice; mice became the preferred species for such experiments. Several of these hydrocarbons were later identified in coal tar and petroleum products.

Several years later, an interest developed on the part of a few pathologists into the effects in animals of aromatic amines, such as 2-naphthylamine and benzidine, which were suspected of causing occupational bladder cancer in humans (HUEPER et al. 1938), and of azo dyes, such as dimethylaminoazobenzene (butter yellow), which caused liver tumors in rats (MILLER and MILLER 1953), when given in the diet. At about the same time, the demonstration that *N*-acetyl-2-aminofluorene, proposed for use as an insecticide, produced tumors of several types when fed to rats (WILSON et al. 1941) led to its withdrawal as a commercial product. This was an important decision, made without any direct information about the effect of the compound in humans. Similar decisions have been made about many other substances based on their carcinogenic effects in animals, one of the most recent being the antihistamine methapyrilene. The latter decision is interesting because methapyrilene appears to lack the mutagenic and other DNA-damaging properties of most carcinogens (LIJINSKY et al. 1980a). There has been increasing awareness that demonstration of carcinogenic properties of a substance in animals suggests that it might have similar effects in humans, and caution is needed in its use. More than 40 years ago, Druckrey stated that there is no reason for assuming that there is any dose of a carcinogen that is without effect in humans, a belief that has often been echoed by others since.

Until 30 years ago, there was no formal approach to testing a substance for carcinogenic properties. In an important review, SHUBIK and SICÉ (1956) discussed guidelines for such tests, with comparatively few examples to draw upon, and those were the results of experiments which had been carried out by individual investigators in a usually logical, albeit ad hoc manner. SHUBIK and SICÉ discussed the importance of uniformity and known purity of the substance to be tested, the minimum number of animals to be used, the choice of species, strain, sex, route of exposure, and vehicle, and, most important, the need for high doses and lifetime exposure to avoid the failure to detect a weak carcinogenic effect leading to tumors manifesting late in life. Good animal husbandry and thorough histopathological examination of the animals were mandatory; this is sometimes neglected, even today. These authors made two points which are as valid today as then. Firstly, there is no relationship between lifespan of the animal and the time needed to develop tumors, which depends upon dose. Secondly, our knowledge is insufficient to decide, on the basis of chemical structure, that a substance is not a carcinogen, although it is possible to make a guess that it might be carcinogenic from its chemical structure. However, a recently developed system developed to predict carcinogenicity of chemicals (Rosenkranz and Klopman 1987) failed by predicting that nitrosothiazolidine is carcinogenic, and it was noncarcinogenic in an animal test (Lijinsky et al. 1988).

II. The Bioassay Program (National Cancer Institute)

Based on those early thoughts, a reasonably sound system for testing substances for carcinogenic activity has been developed. Drs. Shubik, Saffiotti, and many others have contributed to this development, which culminated in the Bioassay Program of the National Cancer Institute (later the National Toxicology Program). The development of principles on which testing of chemicals for carcinogenicity were to be based is described in several reports, including those of the Food and Drug Administration (FDA 1971) and the National Cancer Institute (Ushew 1971). The thoughts behind this effort were that identification of carcinogenic substances would lead to their removal, or at least to the control of human exposure to them. It was felt necessary only to devise experiments of adequate size in suitable animals of two species, using both sexes and two dose levels of the substance, the higher being the maximum tolerated dose (MTD) having no apparent adverse effect on the animals. The second concentration was half that, to ensure that the assay could be completed, in the event that the MTD was cumulatively toxic and caused death of the animals before there was time for tumors to develop.

Many hundreds of substances, compounds, or mixtures have been tested in this way during the past 20 years. Approximately half were not carcinogenic (Haseman et al. 1987), because they did not cause a statistically significant increase in the incidence of any tumor at either concentration, in either sex, of either of the two species of animal used (usually rats and mice, selected for low spontaneous incidence of tumors of liver and lungs). Conversely, a substance that did produce an increased incidence of tumors in at least one sex of one species must be considered a carcinogen, by definition. Clearly, production of tumors in both sexes or in both species strengthens the evidence. This well-laid plan has run into difficulties partly because the high standards for conducting the assay have not always been maintained and partly because of the Delaney Amendment, passed in 1958 by the US Congress, which requires that any substance found to be carcinogenic be banned as a food additive. Many substances used as food additives were classified as "generally recognized as safe" (or GRAS), a list compiled from the answers to a questionnaire sent to a group of experts about that time. Some of these turned out to be carcinogenic when properly tested. Knowledge of carcinogenesis and its mechanisms was weaker then than it is now – and today it can hardly be called adequate – so it was natural that some mistakes were made. However, most GRAS substances have not turned out to be carcinogenic. Nevertheless, because the outcome in a few cases was unfavorable to some parties, there has been a continuing attack on the concept of the bioassay and a demand for revisions of the relatively simple and straightforward concept.

As more subtances are tested, there is a tendency to discover more carcinogens of unusual structures, that is, not resembling chemically any of the previously recognized groups of carcinogens. For example, a number of hypolipidemic drugs, such as clofibrate, have induced liver tumors in rats (REDDY et al. 1980), as have the drugs methapyrilene (LIJINSKY et al. 1980a) and pyrilamine (LIJINSKY 1984a), and a group of plasticizers, including diethylhexyl phthalate and diethylhexyl adipate. These carcinogens have "unusual" structures and are not mutagens in the usual assays, but that they produce liver tumors cannot be denied. It must be assumed that the ultimate effect these compounds have in the chain of events that leads to tumor formation and progression is the same as that by the more "usual" carcinogens, such as nitrosodiethylamine or aflatoxin B1. There is no basis for classifying them as tumor promotors simply because relatively large doses could be used in the bioassay, as their toxicity is low. In the case of methapyrilene, lower doses than those originally used have also induced tumors (LIJINSKY 1984a), some of which have all the usual characteristics of malignancy, such as invasion and metastasis. The problem seems to be one of nomenclature or classification.

III. Mechanisms of Carcinogenesis

From the earliest days of cancer research much attention has been given to the mechanisms by which carcinogens could exert their biological effects.

At first, there was interest in binding to protein in vivo as a means by which the carcinogen could interfere with regulation of cell replication. Such reactions of polycyclic hydrocarbons, aromatic amines, and azo dyes were studied by prominent investigators such as Charles HEIDELBERGER, James and Elizabeth MILLER, and others. Together with other investigators, including BOYLAND, SIMS, GROVER, and DIPPLE, they were interested in the metabolic conversion of well-known types of carcinogen to reactive electrophiles capable of interacting with macromolecules. When the list of carcinogens was expanded to include simple alkylating agents such as nitrosamines and other *N*-nitroso compounds, alkylhalides, such as vinyl chloride, and epoxides, the idea gained momentum that these carcinogens had their effect through alkylation of DNA. It then became popular to believe in a single mechanism, namely, conversion of the carcinogen into an electrophile that alkylated DNA and thus caused mutations. This attractive hypothesis relegates to a separate category those carcinogens that fail to fit. Some have classified nonmutagenic carcinogens as tumor promotors, although most of them do not behave as do the few, more widely accepted tumor promotors. In a curious perversion, carcinogens that are not mutagenic are felt not to be "real carcinogens", and mutagens that are not carcinogenic are ignored.

Among the *N*-nitroso group can be found large numbers of compounds that fulfill the criteria of having been tested adequately in animals, as well as many carcinogens that fit well with the hypothesis that they could act through formation of an agent that alkylates nucleic acids, especially DNA, in vivo. These compounds are also mutagenic in a variety of short-term assays, including the bacterial mutagenesis systems developed by AMES and his colleagues (McCANN et al. 1975). There are also a number of potent rat liver carcinogens, such as azoxyalkanes, which are not bacterial mutagens even with rat liver microsomal activation but which alkylate liver DNA extensively in vivo; this is also true of nitrosomethyl-2-oxopropylamine, for example (LIJINSKY 1988a, b). On the other hand, there are a number of carcinogenic nitrosamines, such as nitrosomorpholine and nitrosopyrrolidine, which are bacterial mutagens but which have not been shown to alkylate rat liver DNA measurably in vivo (and which are potent liver carcinogens for rats at quite low doses). There are other nitrosamines that are bacterial mutagens that are not detectably carcinogenic. From this mixture of findings it has not been possible to deduce any basic principle to guide us in determining the carcinogenic properties of *N*-nitroso compounds, other than by testing in animals. Even so, mistakes can be made by choosing the wrong species or the wrong route of administration (cf. formaldehyde, carcinogenic only by inhalation, inactive by feeding).

IV. Carcinogenesis as a Toxicity Test

Carcinogenesis is different from most other forms of toxicity in that the effects of the agent are cumulative over a period of time, and the results appear only after a long time interval; morphological changes in cells characterizable as neoplastic are seen weeks or months after treatment begins. This is illustrated by an old experiment (LIJINSKY et al. 1976) in which Wistar rats were treated with nitrosomorpholine in drinking water for 30 weeks, beginning at 8 weeks of age. The first rat died with a liver tumor at week 17 of the experiment, and obviously it took less than 17 weeks for neoplastic changes to be observable, had examination been possible. Other rats died at intervals with the same tumor until final killing at 104 weeks. At this time, two rats had liver tumors which did not kill the animal, and several rats had no liver tumor. Assuming that all of the tumors grew at the same rate, there was an enormous difference in the time it took to establish the initial transformation, although the chemical change produced by the carcinogenic nitrosamine must have been complete shortly after the termination of treatment at 30 weeks. It is quite unusual to be able to detect the nitrosamine, or even a metabolite, in an animal as little as several days after treatment.

It is possible, of course, that the same tumor grows at different rates, even in animals of the same stock and very closely related, and this complicates understanding of carcinogenesis enormously. The question of how the process of carcinogenesis begins, perhaps as "initiation" of one or many cells, followed by proliferation of one or many "transformed" cells until a tumor is established, is unresolved. The transformed cells could proliferate spontaneously or under the influence of a proliferative agent, perhaps one identified as a "tumor promotor." Also unresolved is at what stage and how rapidly these cells lose their normal properties and cellular controls, and whether they pass through a "benign" stage before becoming malignant. The last point is important in the interpretation of carcinogenesis studies in which the test animals develop only benign tumors. These should be considered evidence of carcinogenicity of the test substance, because within the short lifespan of rats, for example, there might have been insufficient time for malignant tumors to develop to detectable size; individual cancer cells cannot be recognized, and a sizeable mass is needed for diagnosis.

As well as inducing tumors which develop at different rates, even after induction by the same treatment, carcinogens differ in potency. In fact, carcinogenic treatments differ in potency, which is why there is usually an increasing response to higher doses of a carcinogen. However, it is rare that a single dose, however large, of any carcinogen gives rise to a tumor; multiple doses are necessary. This might mean that the early stages of carcinogenesis take so long after treatment with "weak" carcinogens or low doses of "strong" carcinogens that too little of the lifespan remains for progression of the tumors to a malignant state, in which invasion and metastasis are manifest. The philosophical question whether potent carcinogens are initiators and promoters whereas weak carcinogens are mainly promotors is difficult to resolve in our present ignorance of the reasons for the long interval between the application of the carcinogen and the appearance of the tumor. Apparently, this interval is required for certain chemical or physical changes to take place and not simply the mathematics of cell duplication, which can be very rapid. For example, some virally induced tumors can be detected 2 or 3 weeks following treatment with the agent, an interval much shorter than that following treatment with the most powerful chemical carcinogen. Transplanted tumor cells (and metastases, which are analogous) proliferate very rapidly. A chemically induced thymic lymphoma, for example, is large enough to kill a rat within 3 weeks after inoculation with a few thousand cells of the primary tumor. It is the changes following carcinogen treatment, when the target cells appear normal, that are probably the most important in propelling cells irreversibly to the cancerous state, but these changes are the least understood (and little investigated).

In the Bioassay Program several hundred compounds were tested according to the strict criteria established, and approximately half of them were found to be carcinogenic. This certainly does not mean that half of all substances are carcinogenic, since most of the compounds selected were already under suspicion for one reason or another or are so widely used that their safety is a major concern. Furthermore, not all of the assays were conducted well (the contracting organizations varied in quality, as do all organizations), and little or no flexibility was allowed. This led to an inability to overcome, for example, unexpected long-term toxicity, and also, because cost containment was necessary, valuable additional information that could have been gained at small cost was forfeited. There were many results that were equivocal, but many that showed clear-cut evidence of carcinogenicity; many results were of considerable consequence, particularly to the manufacturers of the substances found, unexpectedly, to be carcinogenic. In general the findings had desirable results, in that removal of carcinogenic impurities from a suspect material was possible, as in the case of the nitrosodipropylamine responsible for the carcinogenic activity of the herbicide trifluralin. In other cases, substitutes have been found for the identified carcinogen, so that the public at large was not deprived.

Some of the results posed difficulties even to avid proponents of experimental bioassays. For example nitrilotriacetic acid, a detergent builder and seemingly innocuous, gave rise to bladder tumors in rats when fed at high concentrations (only achievable because its toxicity is so low) of 2% in the diet. There is a possibility that physical effects were largely responsible for the outcome. Nevertheless, a few dubious outcomes are a small price to pay for identification of large numbers of substances that can jeopardize human health, particularly when the alternative is to expose the human population to them and then count the bodies of those who develop cancer. It is equally absurd to claim that a substance that induces tumors in experimental animals is devoid of risk to the health of humans as to claim that such a substance certainly will induce cancer in anyone exposed to it. The results must be accepted as an indication of probability and nothing more. Carcinogenic substances that are absolutely needed may, of course, continue to be used but with caution; there are very few substances in this category, and saccharin is definitely not included. In the case of carcinogens to which exposure is limited to workers in a particular industry, it is probably adequate to reduce their exposure to a minimum by using protective devices and warning them, of course. There is a natural reluctance for any manufacturer or purveyor of a material identified as a carcinogen to believe the result and take the appropriate action. Therefore, governments are obliged to make decisions about the sale and use of such carcinogenic materials, in the public interest.

V. Short-Term Assays as Substitutes for In Vivo Carcinogenesis

There has been considerable pressure to circumvent the need to test chemicals in animals for carcinogenic effects, and this has led to the development of a large number of short-term assays, which are faster and cheaper, and do not destroy animals. Most of these assays are based on the premise that the mechanism of action of carcinogens is through induction of mutations in somatic cells, which leads to expression of the mutation as the uncontrolled cell replication we call cancer. This is an old and attractive hypothesis, although evidence that is other than circumstantial is lacking. Apart from mutagenesis assays which have become very popular, there are several tests involving transformation of fibroblasts and other nonepithelial cells, whose mechanism is quite unknown. The mutagenesis assays are complementary to studies of the mechanisms of carcinogenesis, which usually involve a search for adducts in DNA. Of course, a very low frequency of adducts is needed to bring about a mutation, and the failure in many cases to find adducts or to identify them is hardly serious, if the hypothesis is correct. However, that the matter is in doubt is the principal reason for continuing in vivo assays for carcinogenicity.

Two analyses have recently appeared of the correlation between the results of several short-term assays and carcinogenicity in rodents of a large number of substances tested in the standard bioassay. The *Salmonella* mutagenesis assay (McCann et al. 1975) correlates as well, or better, with these carcinogenesis results, positive and negative, than any other test. Even so, the correlation is not good, there being similar numbers of carcinogens that are mutagenic and carcinogens that are not mutagenic (Zeiger 1987). There is also a substantial number of mutagens that are not carcinogenic and a satisfying number of nonmutagens without carcinogenic activity. It would be difficult, as the authors point out (Tennant et al. 1987), to devise a strategy to replace in vivo assays with short-term tests, based on this objective analysis. It is especially troublesome that, of the handful of substances known to be carcinogenic in humans, several are not mutagenic; two notable ones are diethylstilbestrol and benzene. Another is asbestos, which is probably better considered as a special case. It must be concluded that in our present state of knowledge (or ignorance) about the mechanisms of cancer causation there is no possibility of replacing the in vivo carcinogenesis experiment for identifying carcinogens. We must look to development of alternative assays, but more understanding of carcinogenesis is required, and it seems that assays based on the mutational hypothesis of cancer induction hold little promise. Even among the *N*-nitroso compounds and their relatives, which have been my concern for 2 decades, there has been no support for a simple, unifying, mutational explanation of carcinogenesis. This is in spite of the finding that most *N*-nitroso compounds are carcinogens, and most are mutagens – but not all mutagens are carcinogens and vice versa.

VI. In Vivo Assays with *N*-Nitroso Compounds as Examples

A survey of studies with *N*-nitroso compounds illustrates the advantages and disadvantages of in vivo carcinogenesis assays. Most of the experiments have been carried out in rats, mainly of one strain, but including some experiments in Sprague-Dawley and Wistar rats. Some of the compounds examined have also been studied in hamsters and mice, mainly by skin painting in the latter species. The animals used have been bred and raised "in house" and were genetically very uniform. Both sexes were used, but not for all compounds, and the maximum tolerated dose was not usually ascertained; obviously, when it was exceeded, the experiment aborted itself. The aims of these studies, which have involved almost 200 compounds, have been not only to discover whether or not a compound is carcinogenic, but to relate the effectiveness or potency and the types of tumor induced to the chemical structure of the compound. By induction, this might lead to sufficient understanding of the similarities and differences between compounds to permit the planning of meaningful biochemical studies which could shed light on the mechanism of carcinogenesis.

There have been more than a thousand publications dealing with nitrosodimethylamine (the most studied nitrosamine) and its action as a methylating agent of DNA, yet it is still not plausible that such methylation is the sole reason for the induction of tumors. This is of more than academic interest, since there is considerable human exposure to this nitrosamine, and it is most improbable that it poses no increased carcinogenic risk to humans, even though there is as yet no epidemiological evidence that it leads to the induction of cancer in humans. Similar compounds can form the same methylating agent, assumed to be a methyldiazonium ion or similar entity, and these include azoxymethane, nitrosomethylethylamine, nitrosomethyl-*n*-propylamine, nitrosomethyl-2-oxopropylamine, and nitrosomethylurea. Several of these compounds given to rats at similar doses have given rise to a very similar pattern of methylation of DNA, yet they differ in their carcinogenic effects, sometimes enormously. As an example, nitrosodimethylamine (NDMA) induces tumors of the liver, lung, kidney, and nasal mucosa when given to F344 rats by gavage; yet the same dose regimen of the isomeric azoxymethane, which gives rise to a very similar pattern of DNA methylation, produces no tumors of the liver, lung, or nasal mucosa. It did, however, induce kidney tumors and a high incidence of tumors of the colon, which neither NDMA nor any other nitrosamine has induced in more than low incidence. This underscores the need to know more about the chemistry and toxicology of these compounds than just their ability to form a methylating agent in vivo. Nitrosomethylurea is a directly acting carcinogen and mutagen, yet its behavior as a carcinogen is quite different from NDMA, even though its methylating properties are similar. The action of nitrosomethylethylamine is complicated by the possibility of its yielding an ethylating agent as well as a methylating agent. Its methylating properties are almost identical to those of NDMA (as would be predicted), but it induces no kidney tumors in rats, although it induces tumors of the liver, lung, and nasal mucosa (LIJINSKY et al. 1987b). The same is true of a variety of compounds that become ethylating agents or hydroxyethylating agents, and so on. The alkylation of DNA might be

Table 1. Carcinogenesis by methylating agents in rats and hamsters

Compound	Species and route of administration	Total dose (mmole)	Sex	Median week of death	Tumors induced (%)
Nitroso-dimethyl-amine	R-Water	1.3	F	31	Liver 100
	R-Gavage	1.6	M	45	Lung 80, kidney 50, liver 50, nasal 15
	H-Gavage	0.1	M	43	Liver 78, nasal 14
Nitroso-methyl-ethyl-amine	R-Water	5	M	31	Liver 100, esophagus 35
		1	M	63	Liver 45, nasal 20
	R-Gavage	1.6	M	38	Liver 95, nasal 50, lung 44
	H-Gavage	0.15	M	70	Liver 70, nasal 20
Azoxy-methane	R-Water	1.6	M	67	Liver 80, kidney 55, colon 40
	R-Gavage	1.6	M	60	Colon 75, kidney 38
	H-Gavage	0.1	M	41	Liver 100, colon 67
Nitroso-methylurea	R-Gavage	0.8	M	35	Forestomach 100, nervous system 50
		0.4	F	33	Forestomach 100, nervous system 67
	H-Gavage	0.2	M	31	Forestomach 92, spleen 83
		0.2	F	27	Forestomach 92, spleen 92
Nitrosobis-(2-oxo-propyl)-amine	R-Water	1.5	M	60	Lung 100, thyroid 70, kidney 25, liver 20
		1.7	F	61	Liver 90, lung 75
	R-Gavage	1.1	M	44	Lung 100, thyroid 90, bladder 60
		1.1	F	59	Liver 92, lung 83, thyroid 25
	H-Gavage	0.4	F	25	Liver 100, pancreas 50, lung 20
Nitroso-methyl-2-oxo-propyl-amine	R-Water	1.8	M	23	Esophagus 100, trachea 55 nasal 25
		1.8	F	24	Esophagus 95, liver 75, nasal 65, trachea 30
	R-Gavage	1.7	M	14	Esophagus 60, nasal 20, trachea 10
	H-Gavage	0.2	F	26	Liver 80, nasal 95

R, rats; H, hamsters.

Table 2. Common neoplasms in F344 rats

Untreated	Carcinogen treated
Liver, hepatocellular adenoma or neoplastic nodule	Liver, hepatocellular adenoma/carcinoma
Mammary gland fibroadenoma/adenocarcinoma	Liver, hemangiosarcoma
Mononuclear cell leukemia	Liver, cholangioma/cholangiocarcinoma
Pancreas, acinar cell adenoma	Lung, alveolar-bronchiolar adenoma/carcinoma
Pituitary, adenoma/carcinoma	Lung, squamous cell carcinoma
Thyroid, C-cell adenoma/carcinoma	Trachea, papilloma/carcinoma
Pancreas, islet cell adenoma/carcinoma	Kidney, tubular cell adenoma/carcinoma
Adrenal medulla, pheochromocytoma	Kidney, mesenchymal tumor
Uterus, endometrial stromal polyp	Kidney, pelvis transitional cell papilloma/carcinoma
Testis, interstitial cell tumor	Urinary bladder, transitional cell papilloma/carcinoma
Skin, squamous or basal cell tumor, keratoacanthoma	Thyroid, follicular cell adenoma/carcinoma
Subcutaneous fibroma/fibrosarcoma	Esophagus, papilloma/carcinoma
Testis, tunic mesothelioma	Pharynx, papilloma/carcinoma
	Tongue, papilloma/carcinoma
	Forestomach, papilloma/carcinoma
	Glandular stomach, adenoma/carcinoma
	Duodenum, adenoma/carcinoma
	Ileum, adenoma/carcinoma
	Colon, adenoma/carcinoma
	Nasal mucosa, adenoma/carcinoma
	Thymus, lymphoblastic lymphoma
	Mammary gland, adenocarcinoma
	Uterus/cervix, adenoma/adenocarcinoma
	Testis, tunic mesothelioma
	Brain/spinal cord, astrocytoma, mixed glioma, Schwannoma
	Zymbal gland, adenoma/carcinoma
	Pancreas, acinar cell adenoma/carcinoma
	Osteosarcoma

necessary, but, clearly, the particular carcinogen expresses other properties that profoundly affect the additional changes necessary to give rise to tumors. These additional properties are not understood, but they determine which cells of which organs develop into tumors. The differences in carcinogenic effect between one methylating agent and another, for example, are not marginal but are dramatic, as shown in Table 1. It can be readily seen that chemical structure has a strong influence on both potency and on the types of tumor which appear, but the nature of this effect of chemical structure is beyond our understanding at present.

When different routes of administration of the carcinogen are employed, or when the compound is administered at different rates, as for example in drinking water at different concentrations versus gavage at different concentrations, there are often profound differences in result, both in potency and in the type of tumors induced. This is also illustrated in Table 1 and demonstrates the importance of pharmacokinetics in carcinogenesis, an aspect which, unfortunately, has been little explored. A handful of investigators, such as L. K. KEEFER (MICO et al. 1985), have begun to scratch the surface of this facet of carcinogenesis, but even the investigation of the simplest compounds, NDMA and nitrosomethylethylamine, is an enormous task. Much more effort is needed in this area, for, until our understanding of the pharmacology of carcinogens is greatly increased, it will be meaningless to talk of the relative potency of carcinogens, even in a single species. The distribution of the carcinogen in the body determines the concentration in various organs and compartments, which in turn determines which enzymes come into play. There is a balance between several metabolic pathways in all likelihood, and the resolution of this determines the effectiveness of the dose and the tumorigenic outcome. In view of the apparently small difference between one treatment and another, it is surprising that the differences between the patterns of tumors that appear can be so sharp. Studies with *N*-nitroso compounds have led to the induction of almost every possible type of tumor in rats or hamsters (Tables 2 and 3), many of them closely resembling human tumors, so that it cannot be claimed that one organ or another is resistant to the action of carcinogens as a whole. Therefore, the reason for the failure of a particular carcinogen to induce all types of tumor must lie in the particular and involved chemistry of that compound in that species.

The problem, complicated enough in a single species, becomes much more complex when one or more additional species are compared, for example, rats and Syrian hamsters. *N*-Nitroso compounds are certainly the most versatile and broadly acting group of carcinogens we know. Yet, even amongst them, there are sharp differences in activity between one compound and another and between one species and another, although most *N*-nitroso compounds are carcinogenic

Table 3. Common neoplasms in Syrian Hamsters

Untreated	Carcinogen treated
Skin, melanoma	Skin, melanoma, squamous cell carcinoma/papilloma
Colon, polyp	Mammary gland, adenocarcinoma
Liver, hepatocellular adenoma	Forestomach, papilloma/carcinoma
Adrenal cortex, adenoma/carcinoma	Glandular stomach, adenoma/carcinoma
	Colon, polyp/adenocarcinoma
	Oral mucosa, papilloma/carcinoma
	Tongue, papilloma/carcinoma
	Liver, hepatocellular adenoma/carcinoma
	Liver, hemangiosarcoma
	Liver, cholangioma/cholangiocarcinoma
	Pancreas, duct adenoma/carcinoma
	Lung, alveolar-bronchiolar adenoma/carcinoma
	Lung, squamous cell carcinoma
	Trachea, papilloma/carcinoma
	Larynx, papilloma/carcinoma
	Nasal mucosa, papilloma/carcinoma
	Kidney, renal cell adenoma/carcinoma
	Urinary bladder, transitional cell papilloma/carcinoma
	Uterus-cervix, adenoma/carcinoma
	Spleen, haemangiosarcoma

in all. The explanation for these differences must lie in the interaction of a particular chemical structure with the particular biochemical makeup of a species. This would have great bearing on carcinogenic risk assessment in humans, if only we understood the nature of the biological makeup of humans. It is probably here that genetics plays a major role, but we do not know what that role is. Suffice it to say that there are dramatic differences even between rodent species in response to particular types of carcinogen. Insofar as it has been explored, there is comparatively little difference in the pattern of alkylation produced in, say, the liver between rats and hamsters by a given alkylating *N*-nitroso compound, yet there are large differences in tumor response, both in tumor pattern and in potency. For example, tumors of the esophagus are the most common ones induced in rats by nitrosamines, but no nitrosoalkylurea induces those tumors in rats, and no nitrosamine induces those tumors in hamsters (LIJINSKY 1988b, c).

Many fewer studies have been made in species other than the small rodents (rats, hamsters, and mice). In general, rats and hamsters are of similar susceptibility, while mice seem to be less susceptible to carcinogenesis by most *N*-nitroso compounds, although producing tumor patterns more similar to those of rats than of hamsters. In contrast, guinea pigs seem resistant to several nitrosamines (CARDY and LIJINSKY 1980) and respond, if at all, only with tumors of the liver. This is reminiscent of the resistance of guinea pigs to induction of tumors by aromatic amines or by polynuclear compounds, apparently because of lack of the necessary activating enzymes to metabolize them (MILLER et al. 1964), although this might not be the only explanation in the light of recent observations with *N*-nitroso compounds. Few *N*-nitroso compounds have been tested in more than two or three species, but those that have, especially nitrosodiethylamine (SCHMÄHL et al. 1978), have induced tumors of one or more types in all of them, the liver being the most common site in the case of this carcinogen. Therefore, it does not seem that the resistance of a certain species to carcinogenesis by a particular compound is due to any innate overall resistance to carcinogenesis but instead is a particular failure of that carcinogen to effect the necessary biological and biochemical changes in that species.

VII. Methods of In Vivo Carcinogenesis

Once the appropriate species for the test is chosen, a reliable source of uniform animals is needed. They should, of course, be free from disease and preferably should not have been exposed to animals (including humans) other than their mothers and siblings. There does not seem to be any particular advantage in using inbred animals, unless it is desired to transplant any tumors that might appear. If the aim is to represent human populations, then randomly bred animals are the appropriate model. If the animals are of the same strain and the same stock, their response can be expected to be quite uniform, as has been found in many experiments at the Frederick Cancer Research Facility.

Rats seem to be one animal of choice for these studies, since they display a very broad spectrum of organs susceptible to carcinogenesis (Table 2). Several strains have been widely used, the most frequent currently being Fischer 344 rats, which were chosen for the work of the National Toxicology Program. These rats have high spontaneous incidences of mononuclear cell leukemia, tumors of the anterior pituitary, interstitial cell neoplasms of the testis, fibroadenomas of the mammary gland, and pheochromocytomas of the adrenal medulla, together with much lower incidences of islet cell tumors of the pancreas, polyps of the uterus, C-cell tumors of the thyroid, and others (Table 2). It is rare to find that any carcinogen treatment increases the incidence of any of these tumors or advances their appearance. They can, in effect, be ignored. F344 rats have the advantage of a low incidence of spontaneous tumors of the liver, nervous system, gastrointestinal tract, and urinary tract, which are sites at which many carcinogens induce tumors. Sprague-Dawley rats are equally useful but grow very large, and females frequently develop huge benign mammary tumors, which are not lethal but place the animals in distress. Wistar and Donryu rats have no particular advantages but are certainly reliable test animals. One great advantage of F344 rats is the huge amount of information about the background incidence of tumors, because so many experiments have been conducted with large numbers of controls, in the NTP Bioassay Program, for example (SOLLEVELD et al. 1984).

The same argument would encourage the use of the B6C3F1 hybrid mouse, which is very uniform and has been widely used for bioassays, thus providing voluminous information about the background incidence of spontaneous tumors. Some of the strains of mice that have been commonly used, but not in the large Bioassay Program, are less acceptable because they have high spontaneous incidences of tumors of the liver and lung, organs in which carcinogens frequently induce tumors, making evaluation of the small effects of weak carcinogens impossible. Mice, like rats, show a broad range of organs susceptible to the induction of tumors. However, Syrian hamsters are not very suitable for studying the effects of miscellaneous carcinogens, although they have been used in some institutions quite extensively for research with particular types of carcinogen including, for example, those inducing lung tumors by instillation of carcinogens in the trachea (SAFFIOTTI et al. 1968), or those inducing bladder tumors (aromatic amines), or for studies of nitrosamines that induce tumors of the pancreatic ducts. The range of organs in the hamster responding to carcinogens is quite small (Table 3), the gastrointestinal tract other than the stomach being usually unresponsive, for example, to *N*-nitroso compounds (LIJINSKY 1987a). One of the enigmas of carcinogenesis is the induction of tumors in the esophagus by almost half of all nitrosamines that have been tested in rats and by a large proportion of those tested in mice, but not by a single nitrosamine that has been tested in hamsters. This insusceptibility of the hamster esophagus might be put down to the absence of appropriate activating enzymes, although this has not been demonstrated as the reason.

Much more difficult to explain, however, is the lack of susceptibility of most organs of the hamster to the induction of tumors by the directly acting nitrosoalkylureas (LIJINSKY 1988c). Among these compounds, which require no metabolic activation, are found the most broadly acting carcinogens in rats and mice, many of them, depending on their structure, inducing tumors in a dozen or more organs, often in a single animal (LIJINSKY and REUBER 1983). Some of these tumors, for example in the nervous system, small intestine, glandular stomach, follicular cells of the thyroid, or osteosarcomas, represent the best models of these human tumors in experimental animals. In sharp contrast, all of these broadly acting, carcinogenic nitrosoalkylureas uniformly induce only hemangio-endothelial sarcomas of the spleen and, sometimes, squamous cell tumors of the forestomach in Syrian hamsters. This they do regardless of the structure of the alkyl group. This result is not due, as might be expected, to lack of alkylation of DNA by the nitrosoalkylureas in other organs of the hamster, which, on the contrary, is just as prevalent as in the same organs of the rat (LIJINSKY 1988a). Likewise, it is not easily understood why the nitrosoalkylureas transform endothelial cells of the hamster spleen into tumors but not endothelial cells in the liver or aorta or blood vessels elsewhere in the body.

Most other species, including guinea pigs, cats, dogs, and monkeys, although longer lived than rats and mice, are less suitable for experimental studies of carcinogenesis. They are larger and require treatment with greater quantities of perhaps dangerous chemicals, they are more difficult and more expensive to maintain, and there is much less information about their background incidence of tumors. So, unless exceptional reasons arise, there is no need to use other species, since we are reasonably sure that the finding of tumors in rats or mice suggests that the substance is more likely than not to increase the cancer risk of humans exposed to it.

Having decided to test the substance in rats or mice, it is necessary to know the purity of the substance and its homogeneity. It is not unreasonable to test a substance that is not highly pure, if that is the state to which humans are exposed, but homogeneity is important, so that the effects on one animal are the same as the effects on another.

VIII. Routes of Administration

Next, it is important to decide on the appropriate mode of administration. Drinking water is appropriate if the substance is water soluble, but it is best to supply the substance in a controlled way, rather than allowing ad libitum consumption, which involves measurement of the volumes ingested to assess the dose. This is a time-consuming process. Instead, if the volume of solution is controlled to 20 ml per rat per day, or 5 ml per mouse per day, this is all consumed, and plain water can be supplied 2 or 1 days of each week to allow the animals to compensate for any water deficit they have incurred. This procedure has been followed for many years in these laboratories and has led to very reproducible responses to carcinogen administration. An alternative, if the substance is not soluble in water, is to mix it with food and allow animals to ingest it ad libitum. This is a wasteful procedure and allows spreading of a potentially dangerous material. Again, the dose received by the animals cannot be well quantified, because the animals will eat the food reluctantly if the substance is not palatable. Disposal of unconsumed material also becomes a problem. However, with substances that are not stable in water, mixture in food might be a convenient means of treatment, the alternative to which is gavage in solution in vegetable oil or a similar edible medium. This method of pulsed dosing has been criticized because of its artificiality and because suggestions have been made that the vehicle, frequently corn oil, itself gives rise to tumors (CARR et al. 1983). However, there is no good evidence to support this criticism; the tumors of the pancreas that are sometimes seen occur spontaneously, and any increase might be an artefact or due to impurities in some batches of corn oil. Frequently, in experiments using gavage, very large volumes of the oil vehicle are administered, 5 ml per kg body weight three to five times a week (ANDERSON 1987). These volumes are needlessly large, and successful bioassays using gavage have been conducted using much smaller volumes of corn oil, for example, with benzyl chloride (LIJINSKY 1986a) and styrene oxide (LIJINSKY 1986b). The use of water as a vehicle for soluble compounds delivered as pulsed doses seems not to be popular. In my experience with nitrosamines, a larger tumor response is often elicited when the compound is given in drinking water than by gavage to rats (LIJINSKY 1984b), but in the case of nitrosodialkylureas (which are stable in aqueous solution) the opposite seems to be the case, and gavage treatment is always more effective than administration in drinking water.

Inhalation is not a common route of administration, because of the costly facilities needed and because only a small number of compounds are sufficiently volatile. On the other hand, there are several substances to which human exposure is by inhalation, and this might seem the most appropriate mode of treatment of experimental animals. There are very few substances, however, which have shown results using inhalation superior to those that could have been obtained by other means. One of the most notable is formaldehyde, which gives rise to tumors of the nasal mucosa in rats inhaling near-toxic doses (SWENBERG et al. 1980), whereas when given by other routes, i.e. in drinking water, there had been no indication of carcinogenicity. Hexamethylenetetramine, a condensation product of formaldehyde and ammonia, which can be considered a stabilized form of formaldehyde, is without carcinogenic effect even when given to rats at quite high concentrations in drinking water (LIJINSKY and TAYLOR 1977). Tumors of the nasal mucosa in rodents, like tumors of the forestomach and Zymbal's gland, have no true equivalent in humans. Nevertheless, a substance which induces these tumors has demonstrated carcinogenic activity and must be classified as a carcinogen. The mechanism of induction of tumors in the nasal mucosa or forestomach or zymbal gland in rodents is unknown as, for that matter, is the mechanism of induction of tumors in any other organ, by those carcinogens or by others. It must be borne in mind that it would be as difficult to classify NDMA as it is to classify formaldehyde, based only on the results of inhalation studies, since NDMA induces only tumors of the nasal mucosa in rats by inhalation, although it is an extremely powerful inducer of tumors of the liver, kidney, and lung when given by the oral route, particularly when administered in pulsed doses by gavage.

The main deficiency of inhalation exposure in assessing the carcinogenic activity of a substance is that the dose administered to the animals is strictly limited by the vapor pressure of the substance, the volume of air that the animal can breathe, and the tolerance of the animal for the compound. These are often severe limitations which might reduce the amount of substance to which the animals are exposed to below that which can be effective in inducing tumors within their short lifetime. In other words, it might not be possible by inhalation to exaggerate the exposure of the animals sufficiently above that of humans to compensate for small groups of animals and an exposure of 2 years' duration compared with 50 years or more for humans. Therefore, while a test by inhalation that results in induction of tumors is an acceptable result, a negative result must be evaluated carefully and cannot be accepted as evidence of lack of carcinogenicity.

IX. Assessment of Results

It is worth emphasizing that a substance which induces tumors in animals at any dose and by any route of administration is by definition a carcinogen and obviously a greater risk in this regard than a substance that does not induce tumors at any dose and by any route of administration. The nature and magnitude of the risk posed to humans by such carcinogenic substances can be debated, but that is a separate question requiring additional knowledge of their pharmacology, biochemistry, and toxicology. This information is lacking in almost every case, even in those instances in which cancer in humans has been induced by the substance.

Most of the short-term tests are of no help to us, since they cannot model the complexity of even a simple multicellular organism and depend on the invocation of dogma to relate those results to events involved in tumor induction. Mammals, even small ones such as rats and mice, resemble each other biologically much more closely than they resemble bacteria or protozoa or yeast or cells in culture, whatever their source. It would seem that insects would be good facsimiles of mammals, except that they usually do not live long enough to develop tumors. The main difficulty with using short-term tests to assess the carcinogenic potential of substances is that the dynamics of progression of tumors cannot be represented in these systems, even granting that a mutational event might be involved in the initial stages of action of the carcinogen. This is obviously not true of the many carcinogens which have failed to exhibit mutagenic activity or DNA-damaging activity in any of a large number of short-term assays (e.g., methapyrilene, clofibrate).

Assuming that the bioassay has been well conducted, the survival of the animals adequate, and the pathology of high quality, showing a statistically significant increase in tumors of a certain type at only one dose level, what determination should be made of the importance of the outcome? If the tumor is one that occurs at a considerable incidence in untreated, aging animals, such as lung adenomas in mice, liver adenomas in some strains of rat, and thyroid tumors in rats, then the result is less impressive, although not without interest or importance. For example, our finding that high doses of sodium nitrite increase the incidence of benign and malignant liver neoplasms in female, but not in male, rats did not lead us to conclude that sodium nitrite is a carcinogen (LIJINSKY et al. 1983a). On the other hand, a substance that has such an effect in both sexes of rats or mice or in both species would provide strong evidence. If the tumor induced by treatment with the substance is one that rarely or never appears in control animals of that sex and species, then the evidence is even stronger that the substance represents a carcinogenic risk to people exposed to it. As mentioned before, such a result provides no indication that the same tumor would be induced in humans by that substance (e.g., the controversy over saccharin and the futile effort to find an increase in bladder tumors in people who use that carcinogenic chemical). The interspecies differences in response to carcinogens can be very great, as in the case of nitrosoalkylureas, which induce one common tumor in both rats and hamsters, that of the forestomach, but totally different tumors of other types in the two species. Aromatic amines, vinyl chloride, and many nitrosamines also show these marked differences in their effects in rats and hamsters.

An overwhelming advantage of chronic tests in animals over short-term assays is that many normal and abnormal physiological effects can be modelled that might be important in modulating carcinogenesis, for example the importance of additive or synergistic effects, which might manifest themselves only over long periods of time. Also the effects of tumor promotors and inhibitors or modifiers can be observed only in animal systems. However, the importance of tumor promotors in general in human cancer has yet to be demonstrated (SCHMÄHL 1985), and even in experimental animals the evidence for their role is often weak. The experiments are often poorly conducted and the outcomes of dubious significance.

Also important in carcinogenesis in animals might be the effect of the age of the animals on their susceptibility to carcinogens. It appears that very young animals are often more susceptible to a given carcinogen than young adults, and older animals are frequently less susceptible than young adults (LIJINSKY and KOVATCH 1986). Sex and the hormones related to it are often a modifying factor, so that males are more susceptible to some carcinogens and females more susceptible to others. Furthermore, the tumors induced in males by a carcinogen can be quite different from those induced in females, apart from the obvious tumors of the reproductive systems. This emphasizes the importance of testing a substance in both sexes. These effects are very important in understanding mechanisms of carcinogenesis and have no counterpart or substitute in a short-term test.

C. Criteria for In Vivo Assays

I. Group Size

The end point of a chronic toxicity assay for carcinogenesis is the induction of neoplasms. These induced neoplasms must appear in the treated animals in high enough incidence for a statistically significant elevation over that in untreated controls to be calculated. Whether the neoplasms are benign or malignant matters less than the number of animals bearing them in the treated group. According to the estimates of BOYLAND (1957), published by the National Research Council in 1959, 4 animals with a tumor in a group of 100 treated constitute a significant difference from 0 among 100 controls; similarly, 4 animals with a tumor among a group of 50 represents a significant incidence compared with 0 in controls. However, for 1 animal with a tumor in the control, 6 or 7 animals with that tumor are needed in a group of 30 or 40 treated animals for the difference to be statistically significant. As the number of "spontaneous" tumors in the controls increases, the proportion of treated animals with that tumor needed for statistical significance increases rapidly if the groups are small. In practice, groups of 50 animals are a good compromise between sensitivity and the costs and effort in maintaining larger groups of animals. The marginal increase in sensitivity of 100 animals versus 50 animals is probably not sufficient to justify adding 50 animals to the group.

II. Species and Sex

There is little but economy and convenience to recommend one species over another for conduct of an in vivo carcinogenesis assay, although experience has taught that guinea pigs are peculiarly insusceptible to several types of carcinogen, for example, aromatic amines (MILLER et al. 1964) and polycyclic aromatic compounds (OBERLING et al. 1939). Rats are the most common species used routinely, and several strains have been employed, especially Fischer 344 and Sprague-Dawley. Mice are an equally commonly used species, also with several strains. In bioassays conducted by cutaneous application of an agent, mice are used almost exclusively, mainly because of the vast background of information arising from skin painting experiments in mice with polycyclic aromatic compounds.

Rats and mice both respond to carcinogen treatment by producing a great variety of tumors, depending on the nature of the carcinogenic agent. *N*-Nitroso compounds are the most broadly acting group of carcinogens, and the types of tumor commonly induced in rats by these compounds are listed in Table 2; the list is compiled from a review of PREUSSMANN and STEWART (1984) and from personal studies (LIJINSKY 1987a). Possibly a corollary to the sensitivity of so many organs of the rat to tumor induction is the high incidence of several types of "spontaneous" tumor in this species, of which the most common are also listed in Table 2. They are mainly tumors of endocrine and reproductive organs and are not usually increased in incidence or in time of appearance by treatment with a carcinogen. In mice, the common spontaneous tumors include those in liver and lung (often the organs in which carcinogens act), which increase in incidence as the result of carcinogen treatment. Weak carcinogenic effects in mice are sometimes difficult to detect because of the lack of sensitivity in groups containing small numbers, as discussed in the previous section.

Another species, less commonly used than rats and mice, is the Syrian golden hamster. This species has a low background incidence of "spontaneous" tumors (Table 3) but, in parallel, has low sensitivity to many types of carcinogens (e.g., aromatic amines, polycyclic aromatic compounds, many *N*-nitroso compounds). As has been mentioned, guinea pigs are not very sensitive to most classes of carcinogen and respond usually with tumors of the liver or no tumors at all (CARDY and LIJINSKY 1980).

It is difficult to generalize about the relative sensitivity of one sex compared with the other. To some types of carcinogen females appear to be more sensitive, to other carcinogens males are more sensitive. To most nitrosamines, male rats are more sensitive than female, but in hamsters, females are equally, or more, sensitive than males. To directly acting nitrosoalkylureas, female rats are more sensitive than male (and frequently develop different types of tumor), as is also the case in hamsters; hemangiosarcomas of the spleen are almost the only tumors induced in hamsters by nitrosoalkylureas (LIJINSKY et al. 1985).

To guard against the possibility that one sex is much more susceptible to the action of a particular carcinogen than the other, both sexes of the species should be used. This implies, of course, that a significantly increased incidence of tumors related to the treatment in only one sex of a species is adequate for identification of the treatment as carcinogenic. Analysis of a large number of bioassays conducted in both sexes of rats and mice has shown a high concordance between the two sexes and the two species in tumor responses, such as to suggest that a third species (for example, humans) would be more, rather than less, likely to respond similarly (HASEMAN and HUFF 1987).

III. Route of Administration

1. Oral

Most in vivo carcinogenesis assays are conducted by feeding the substance to animals mixed in food or dissolved in drinking water. This not only mimics human exposure to many foreign chemicals, as food additives or ingredients for example, but is the way of administering the maximum dose without undue acute toxic effects. Such gradually received doses of chemicals tend not to overload the detoxifying systems of the body, as happens with large, pulsed doses; hence the tumor responses are often greater than to the same total doses given in more concentrated form. When comparison has been made, treatment of rats with nitrosamines in drinking water has often induced tumors earlier than equimolar doses given by gavage (LIJINSKY 1987a); several nitrosodialkylureas, directly acting carcinogens, on the other hand, have been less effective when given to rats in drinking water than by gavage (LIJINSKY, unpublished data); different patterns of tumors are sometimes produced as a result of the two modes of oral administration. Treatment of animals with mixtures of feed containing substances that are volatile is obviously not completely safe to the animals or to the personnel conducting the study, although, apart from wastefulness, there is no objection to the procedure. In the case of volatile compounds, gavage or injection are preferred methods of treatment. Gavage and subcutaneous or intravenous injection have the advantage of precision of dosing, but they have the disadvantage of rapid exposure of the animal to a large amount of toxic compound, which can distort the animal's physiology, as well as providing results not directly applicable to human experience. Pulsed treatments, on the other hand, are often unavoidable, as in the case of hamsters, which do not drink enough and are messy drinkers.

In the case of water-soluble test compounds, administration in drinking water is simple and can be fairly quantitative in rats and mice. Frequently, drinking water solutions are offered ad libitum (as is food), so that quantification of the dose requires measurement of residual, unconsumed solution at the end of some period of time, often daily. This is time consuming. An additional drawback is that the animals might find some substances or higher doses unpalatable and drink less of them, making dose comparisons more difficult. We have found that offering smaller volumes than their normal daily consumption of water (we give 20 ml per day to each rat, 5 ml per day to each mouse) results in almost all of the solution being drunk. To compensate for any water deficiency the animals might incur, the test solution is given for 5 days a week, and water ad libitum on the remaining 2 days of each week. When, as is often the case, the solution is unpleasant to the animals, after 1 or 2 weeks of reluctant consumption on this regimen, the animals accommodate and consume the daily ration.

2. Skin Painting

Skin painting is a frequent means of testing a substance for carcinogenicity and is preferably conducted in mice. It is most appropriate for substances with which human skin comes into contact or for substances thought to be directly acting. The substance is dissolved in a solvent, such as acetone or toluene and applied in measured drops (25 or 50 μl) to the back of a mouse, once or twice a week, usually for a fixed number of weeks. The appearance of the tumors can be seen from the beginning, and a true latent period can be stated. The length of the latent period, the time between the beginning of treatment and the first observation of a tumor, is inversely related to the potency of the treatment, a more potent carcinogen or higher doses of a carcinogen resulting in a shorter latent period.

It is usual to use only female mice for skin painting studies, unless males are individually housed. Male mice fight, and the resulting skin lesions can seriously hamper observation of skin tumors and, in many cases, result in the removal of tumors that have appeared. Several strains of mouse have been, and are, used for skin painting studies, including Swiss, C3H, BALB/c and the more recently induced "Sencar" (*sen*sitive to *car*cinogens), selected by successive mating of the most sensitive mice of a group. The Sencar mice seem to be very sensitive to skin tumor induction by UV radiation or carcinogenic polycyclic aromatic compounds, but whether or not this is a truly genetic trait is not clear, since Sencar mice are less sensitive than Swiss or BALB/c mice to tumors of the liver and esophagus induced by systemic treatment with a nitrosamine, nitrosohexamethyleneimine (STRICKLAND and LIJINSKY, unpublished data).

Many skin tests in mice for carcinogenic activity have used the initiation-promotion protocol, described in detail by BERENBLUM and SHUBIK (1947), utilizing a single treatment with the test compound, followed by twice (or thrice) weekly application of a solution of a "promoting" agent, such as croton oil or the more refined phorbol diester TPA, for 10, 15, or 20 weeks. The disadvantage of this mechanistically interesting in vivo assay is that many compounds, which are quite effective carcinogens when applied chronically, are not good initiators and give poor tumor yields in the two-stage protocol; they include a number of partially hydrogenated polycyclic aromatic hydrocarbons and the potent alkylating agents nitrosomethylnitroguanidine (HECKER and LIJINSKY, unpublished data) and nitrosomethylurea (WAYNFORTH and MAGEE 1975). Chronic skin application of a substance seems preferable to initiation-promotion because it is less likely that a carcinogen will fail to induce tumors under the former conditions.

3. Inhalation

Testing of substances for carcinogenicity through inhalation exposure of rodents is not often employed, because there are few facilities in which it can be done, and it is very expensive, time consuming, and somewhat limited in application. It can only be used with substances that are volatile or can be converted to stable aerosols, and problems of safety are more serious. Although inhalation is the normal route of exposure of people to many carcinogens, this does not imply a need to test suspected compounds by this route, since most carcinogens act systemically. One severe limitation on the value of chronic inhalation studies, if the results are negative, is that such a result might be due simply to the failure to administer an adequate dose by inhalation to give rise to tumors within the short lifespan of laboratory rodents. As has been pointed out elsewhere, the short life of rodents limits their sensitivity to carcinogenesis. If a carcinogen is no more potent in humans than in, say rats, it would be necessary to deliver 25–35 times the dose that humans receive to compensate for the difference in lifespan alone, with the same incidence of tumors in the two species as the endpoint. There is no evidence that the time at which tumors appear following exposure to carcinogens is related to the lifespan of the species. Experiments in which a comparison has been possible show that the time-to-tumor induced by, for example, a similar dose of nitrosodiethylamine is similar across species varying from a few years to decades in longevity. A further exaggeration of the dose is needed in experimental animal studies to compensate for small groups of animals which must represent large groups of humans exposed to the same substance; both reduction in time-to-tumor and an increase in tumor incidence accompany an increase in dose of carcinogen, although the response is not necessarily linear. As previously mentioned, the incidence of tumors related to the treatment must reach a minimal value before statistical significance can be claimed. Because of these strictures, inhalation exposure is not optimal for detecting carcinogens since the dose cannot be exaggerated sufficiently. The tolerance of animals for many inhaled compounds is restricted by their irritant properties, which limits the concentration in the gas phase. Animals do not breathe a greater volume of air, proportionately, than humans, so the limitations of concentration can be a serious impediment. Several substances tested by inhalation have induced in rats, mice, or hamsters tumors of the nasal mucosa, which are an indication of carcinogenic activity but have no equivalent in humans.

4. Conclusions

In summary, oral administration and skin painting are the best modes for in vivo carcinogenesis assays. Intraperitoneal injection is not precise in localizing the compound, and differences in diffusibility of the test substance between one animal and another affect the outcome and cannot be controlled or quantified. Similarly, subcutaneous injection has these deficiencies and the additional handicap that large molecules will tend to remain at the injection site much longer than small molecules, which will impair the effectiveness of systemic carcinogens. In experiments with polynuclear hydrocarbons many years ago we discovered that several were quite effective carcinogens when painted on mouse skin but were ineffective by subcutaneous injection, in this case because the more soluble compounds did not remain at the injection site to induce fibrosarcomas (LIJINSKY et al. 1970). Inhalation exposures are useful but not practical for widespread use.

IV. Size of Dose and Dose Selection

1. Studies in Adults

Ideally, the dose of a carcinogen given to animals should be large enough to give rise to tumors in a high proportion of animals without causing toxicologic effects, other than the tumors, that influence the health or behavior of the animals. With such an outcome, it is difficult for any but the most sceptical critics to insist that the test substance does not present a carcinogenic risk to humans. In order to obtain the greatest chance of inducing a significant incidence of tumors in a small group of experimental animals (50), the maximum possible dose of the test substance should be administered but not so large a dose that overt damage to the health of the animals occurs in the short term. Conversely, if at this maximum dose there is no significant incidence of tumors induced in the animals, it can be reasonably assumed that the substance is not carcinogenic and will pose no increased carcinogenic risk in humans exposed to it, although that statement always requires some qualification in the form that species may differ in their response to any carcinogen.

The setting of a dose for a chronic bioassay must be based on the results of short-term studies, unless there is a lot of information about the substance already available. Some informed judgement can be used if the substance resembles in chemical structure another of known toxicity and carcinogenicity. If not, an acute toxicity test is conducted in small groups of animals (5) and an LD_{50} calculated, in males and females. A larger group of animals (10), 7–8 weeks old, is then given, twice or thrice a week, various proportions of the LD_{50}, differing by factors of two (i.e. 1/5, 1/10, 1/20, 1/40, 1/80 of the LD_{50}) for 13 or 26 weeks, during which time the animals are weighed every week, then killed, and examined histopathologically. That dose at which there is no substantial (greater than 10%) depression of weight gain compared with controls and in which there is minimal or zero injury detected is chosen as the dose for the chronic study. To ensure that there is no delayed cumulative toxicity which might lead to early death of the animals (and hence no tumor response), a second group of animals of both sexes is given chronically one-half of the chosen maximal tolerated dose or MTD, and the two are conducted in parallel with controls given no treatment or, in the case of gavage studies, treatment with the vehicle alone.

Corn oil is frequently used for gavage of substances that are not soluble in water. There have been suggestions that corn oil itself might induce tumors or elevate the incidence of some "spontaneous" tumors, but this has not occurred in my experience. A recent study by ANDERSON (1987) indicated an effect of chronic administration of corn oil on the intestinal epithelium of rats, but this was a consequence of quite massive doses of corn oil, far larger than are used in an in vivo carcinogenesis study.

Animals are treated for most of their lifespan, at least for 104 weeks, and then observed until death.

2. Multigeneration Studies

In some cases, the importance of a substance, commercial or natural, demands that the maximum sensitivity be built into a chronic toxicity test in animals. A way of achieving this is to expose animals in utero by treating pregnant females during the last third of pregnancy (in rats or mice), taking advantage of the great sensitivity of the fetus, and administering MTD. After birth, the infants are treated as soon as practical with the compound delivered at the MTD for most of their lifespan. In this way it was possible to demonstrate convincingly the carcinogenicity of the widely used sweetener saccharin (ARNOLD et al. 1980), and there are other examples.

There seems to be no particular advantage to exposing transplacentally fetuses in utero to a chemical and not continuing the exposure of the offspring after birth.

V. Conduct of the Experiment

The experimental animals should be of a well-maintained stock, bred in hygienic conditions so as to be free of unusual pathogens. Preferably, there should be a history of use of these animals in the facility so that unusual outcomes can be avoided, for example the appearance of unusual spontaneous tumors that might be mistaken to tumors related to the treatment. The animals at the start of the experiment should be of similar age (6–7 weeks) and weight, and placed randomly into experimental or control groups. Good animal husbandry is essential, and care should be taken to avoid spread of disease or contamination from the personnel to the animals, or from one animal to another. In order to minimize contamination with the suspect carcinogens, volatile materials should be contained and drinking water bottles filled and closed in a special room. Gavage and skin painting should be done in well-ventilated rooms.

Rats should be housed no more than five in a cage and mice, no more than ten. Food should be supplied ad libitum; a number of good quality feeds are available that are largely free of all but traces of such carcinogens as aflatoxins and nitrosamines. The feed should be autoclaved to prevent introduction of organisms. The use of synthetic or highly purified diets is not necessary. Indeed, in a recent study using purified synthetic diets the control animals survived significantly less well than animals given ordinary rat chow (LIJINSKY et al. 1989), suggesting that some important but unknown factors were missing from the synthetic diet. Deionized tap water is satisfactory for provision ad libitum to the animals or for dissolving test substances for drinking water treatment. In some facilities drinking water (not containing test substances) is acidified to suppress growth of *Pseudomonas* organisms.

Animals should be weighed periodically, to check that they remain healthy, and maintained on the treatment for 2 years, after which they are simply observed until death. Mortality checks should be conducted at least twice a day. In standard bioassays supported by the National Toxicology Program, animals are killed shortly after the end of the 2-year treatment. Although this provides better samples for histopathological examination, well-maintained rats and mice can survive almost to 3 years of age. In the time interval between the end of the 2-year treatment and the death of the animals, it is very possible that many small tumors, otherwise not detected, would grow to a sufficient size to be diagnosed and classified correctly. If the mortality checks are carried out conscientiously, there need be relatively little loss of tissues through autolysis. It is probable that the detection of a significantly elevated number of hepatocellular tumors in rats treated with sodium nitrite is due to allowing most of the animals, treated and controls, to live their maximum lifespan (LIJINSKY et al. 1983a). The need for maximum observation time in a carcinogenesis experiment is here illustrated, and there seems to be no justification for reducing the possibility of observing tumors, which is necessary to compensate for the short lifespan of laboratory animals compared with humans.

VI. Examination of Animals and Evaluation of Results

All animals at death or when killed moribund or at 130 weeks of the study are dissected, and all major organs and tissues, together with all lesions, are placed in formalin for fixation. Organs and tissues routinely removed include liver, lungs, brain, kidneys, heart, spleen, pancreas, bladder, ovaries, uterus, testes, entire gastrointestinal tract (esophagus, stomach, duodenum, ileum, jejunum, colon and cecum), thyroid, parathyroids, adrenal glands, prostate, spinal cord, and nasal cavity. Sections are prepared from the tissues embedded in paraffin and stained, and histopathological slides are prepared in the usual way for pathological examination. The diagnoses are reported in a standard fashion and the results tabulated.

A comparison of the numbers of each type of neoplasm, benign or malignant, and nonneoplastic lesions in the treated and control groups serves for statistical analysis of the differences in incidence. Those differences that are statistically significant are considered, in the absence of contrary evidence, to be related to the treatment. If the lesions are neoplasms, especially if malignant, they are evidence that the treatment is carcinogenic. The conclusion may be based on such a positive result in only one sex of one species at one dose level of the test substance, but this conclusion can be (and has been) disputed when the statistical significance of the increase in tumor incidence was marginal. The conviction of the outcome is increased if both species, both sexes, or both dose levels show a parallel response. However, it must be pointed out that there are many technical and experimental reasons for the failure of a carcinogen to be shown to induce tumors in a particular group of animals but much less likelihood that a substance that induces an increased incidence of neoplasms does so by accident. Therefore, a positive result is much more important than a negative result with the same substance.

There is no reason to attach more significance to one positive tumor outcome than to another, although the temptation is strong. For example, an increase in the incidence of benign liver neoplasms or of benign lung neoplasms (which are found in controls) might seem less impressive than a small incidence of esophageal neoplasms, which are virtually never seen in untreated animals. The point can be argued, but the original intent of the investigators who recommended and devised in vivo methods for testing substances for carcinogenic activity cannot be. That is, the detection of carcinogenic activity as a statistically significant increase in incidence of benign or malignant neoplasms is prima facie evidence of this activity and the starting point for consideration of what risk is presented to humans exposed to the substance.

D. Conclusions

It must surely be accepted that a substance which induces a significant incidence of tumors in a test on a smaller scale than 50 animals of each sex at two dose levels in two species must also be considered a carcinogen. However, the corollary is that, in a small scale assay or in one conducted at much below the MTD, a negative outcome – that is, a less than significant increase in the incidence of a tumor – is no assurance of the noncarcinogenicity of the treatment. The use of small numbers of animals, of doses that are too low, and the killing of the animals after too short a time for tumors to be detected are the principal reasons for discordance in the reports of testing of some substances, which are found to be carcinogenic in one experiment and not in another. Some examples are the reports in the classical paper of DRUCKREY, PREUSSMANN, et al. (1967) that nitrosodiphenylamine and nitrosomethyl-*n*-heptylamine are not carcinogenic, subsequently found to be incorrect (CARDY et al. 1979; LIJINSKY et al. 1983 b).

The finding of carcinogenic properties in a substance to which there is, or has been, human exposure leads to attempts to extrapolate the experimental results in order to calculate risks to humans. There have been many mathematical approaches to doing this, none remarkably successful. The main problem is that in small groups of animals high doses of carcinogen are needed to produce statistically significant tumor incidences. Humans are exposed to much smaller doses over much longer time periods than are the animals in our experiments. In few cases – perhaps none – is the carcinogenic substance the only one to which humans are exposed (exposure in utero to drugs, etc. might be an exception). Therefore, it is nonsense to speak of such an exposure as causing one case of cancer in 10^5 or 10^6 people. The shape of the dose-response curve in the experimental animals is known only at high concentrations. At lower concentrations, particularly those to which humans might be exposed, the shape is entirely unknown. Depending whether it is linear, concave, or convex, very different risk estimates will be made. Until we have a better understanding of the mechanisms of carcinogenesis, especially how many cells must be affected before progression to a visible tumor becomes inevitable, and how long that progression takes under a variety of circumstances, quantitative risk estimates should be very conservative, assuming the maximum effect of the smallest dose. Other essential information is an understanding of the pharmacology of carcinogens – information woefully lacking and not much sought at the present – and of the mechanisms by which the carcinogen induces tumors. That is, the nature of the interactions within cells that produce the essential – and presumably irreversible – transformation of the cells which are the targets of the carcinogen. This latter is often difficult to discover, because we measure interactions (for example with cellular DNA) in whole organs or in large masses of cells within an organ, most of which are not the stem cells that are presumed to be the only cells sufficiently undifferentiated as to be susceptible to progression to tumors. It might be some time before the positive result of an in vivo carcinogenesis assay enables us to say more than that the substance presents a carcinogenic risk to humans of unknown magnitude. Precise statements of the magnitude of the risk, even apparently learned ones, must be scrutinized and viewed with scepticism. This is particularly true if the estimates are made by those having a vested interest in whether they are high or low.

Acknowledgements. This manuscript was written mostly while I was a Visiting Scientist at the National Cancer Center Research Institute, Tokyo, Japan, sponsored by the Foundation for Promotion of Cancer Research, which I thank for its support. Research was sponsored by the National Cancer Institute, DHHS, under contract no. NO1-VCO-74101 with Bionetics Research, Inc. The contents of this publication do not necessarily reflect the views or policies of the Department of Health and Human Services, nor does mention of trade names, commercial products or organisations imply endorsements by the US Government.

References

Anderson RL (1987) Intestinal responses in the male rat to gavaged corn oil. Cancer Lett 36:55–63

Arnold DL, Moodie CA, Grice HC, Charbonneau SM, Stavric B, Collins BT, McGuire PF, Munro IC (1980) Long term toxicity of orthotoluenesulfonamide and saccharin in the rat. Toxicol Appl Pharmacol 52:113–152

Berenblum I, Shubik P (1947) A new quantitative approach to the study of the stages of chemical carcinogenesis in the mouse's skin. Br J Cancer 1:983–991

Boyland E (1957) The determination of carcinogenic activity. Acta Unio Intern Contra Cancrum 13:271–279

Butler WH, Greenblatt M, Lijinsky W (1969) Carcinogenesis in rats by aflatoxins B_1, G_1, and B_2. Cancer Res 29:2206–2211

Cardy RH, Lijinsky W (1980) Comparison of the carcinogenic effects of five nitrosamines in guinea pigs. Cancer Res 40:1879–1884

Cardy RH, Lijinsky W, Hildebrandt P (1979) Neoplastic and nonneoplastic urinary bladder lesions induced in Fischer 344 rats and B6C3F1 hybrid mice by *N*-nitrosodiphenylamine. Ecotoxicol Environ Safety 3:29–35

Carr CJ, Newberne PM, Oser BL, Rogers AE, Van Duuren B, Williams GM, Bernard B, Bieber MA, Enig M (1983) Report of the ad hoc working group on oil gavage in toxicology meeting, July 14–15, Arlington, VA. Nutrition Foundation, Washington, DC

Cook JW, Hewett CL, Hieger I (1933) The isolation of a cancer-producing hydrocarbon from coal tar, parts I, II, and III. J Chem Soc Trans 1:395–405

Druckrey H, Preussmann R, Schmähl D, Ivankovic S (1967) Organotrope carcinogene Wirkungen bei 65 verschiedenen *N*-Nitroso-Verbindungen an BD-Ratten. Z Krebsforsch 69:103–201

FDA (1971) Food and drug administration advisory committee on protocols for safety evaluations: panel on carcinogenesis report on cancer testing in the safety evaluation of food additives and pesticides. Toxicol Appl Pharmacol 20:419–438

Hartwell JW (1955–1986) Survey of compounds tested for carcinogenic activity. U.S. Public Health Service publication no. 149
Haseman JK, Huff JE (1987) Species correlation in long-term carcinogenicity studies. Cancer Lett 37:125–132
Haseman JK, Huff JE, Zeiger E, McConnell EE (1987) Comparative results of 327 chemical carcinogenicity studies. Environ Health Perspect 74:229–235
Hueper WC, Wiley FH, Wolfe HD (1938) Experimental production of bladder tumors in dogs by administration of beta-naphthylamine. J Ind Hyg Toxicol 20:85–91
International Agency for Research on Cancer (1973–1986) Information bulletin on the survey of chemicals being tested for carcinogenicity, numbers 1–12. IARC, Lyon
Lijinsky W (1984a) Chronic toxicity tests of pyrilamine maleate and methapyrilene hydrochloride in F344 rats. Food Chem Toxicol 22:27–30
Lijinsky W (1984b) Structure-activity relations in carcinogenesis by *N*-nitroso compounds. In: Rao TK et al. (eds) Genotoxicology of *N*-nitroso compounds. Plenum, New York, pp 189–231
Lijinsky W (1986a) A chronic toxicity study of benzyl chloride in F344 rats and (C57BL/6J × BALB/c)F_1 mice. JNCI 76:1231–1236
Lijinsky W (1986b) Rat and mouse forestomach tumors induced by chronic oral administration of styrene oxide. JNCI 77:471–476
Lijinsky W (1987a) Structure-activity relations in carcinogenesis by *N*-nitroso compounds. Cancer Metastasis Rev 6:301–356
Lijinsky W (1987b) Structural relations and dose-response studies in nitrosamine carcinogenesis. In: Mehlman MA (ed) Safety evaluation: toxicology, methods, concepts and risk assessment. Advances in modern environmental toxicology, Vol X pp 215–241, Princeton University Press
Lijinsky W (1988a) Nucleic acid alkylation by *N*-nitroso compounds related to organ-specific carcinogenesis. In: Politzer P, Roberts L (eds) Chemical carcinogens, activation mechanisms, structural and electronic factors, and reactivity. Elsevier, Amsterdam, pp 242–263
Lijinsky W (1988b) Nitrogen-containing alkylating carcinogens. Banbury report no. 31: new directions in qualitative and quantitative aspects in carcinogen risk assessment. Cold Spring Harbor, New York, pp 15–31
Lijinsky W (1988c) The importance of animal experiments in carcinogenesis research. Environ Molec Mutagen 11:307–314
Lijinsky W, Kovatch RM (1986) The effect of age on susceptibility of rats to carcinogenesis by two nitrosamines. Gann 77:1222–1226
Lijinsky W, Reuber MD (1983) Carcinogenicity of hydroxylated alkylnitrosoureas and of nitrosooxazolidones by mouse skin painting and by gavage in rats. Cancer Res 43:214–221
Lijinsky W, Reuber MD (1988) Tumors in Swiss mice following skin-painting with nitrosoalkylureas. J Cancer Res Clin Oncol 114:245–249
Lijinsky W, Taylor HW (1977) Nitrosamines and their precursors in food. Cold Spring Harbor symposium on the origins of human cancer. Book C, pp 1579–1590
Lijinsky W, Garcia H, Saffiotti U (1970) Structure-activity relationships among some polynuclear hydrocarbons and their hydrogenated derivatives. JNCI 44:641–649
Lijinsky W, Taylor HW, Keefer LK (1976) Reduction of rat liver carcinogenicity of nitrosomorpholine by alpha deuterium substitution. JNCI 57:1311–1313
Lijinsky W, Reuber MD, Blackwell BN (1980a) Liver tumors induced in rats by chronic oral administration of the common antihistaminic methapyrilene hydrochloride. Science 209:817–819
Lijinsky W, Reuber MD, Manning WB (1980b) Potent carcinogenicity of nitrosodiethanolamine in rats. Nature 288:309–310
Lijinsky W, Reuber MD, Riggs CW (1981) Dose-response studies in rats with nitrosodiethylamine. Cancer Res 41:4997–5003
Lijinsky W, Reuber MD, Davies TS, Saavedra JE, Riggs CW (1982a) Dose-response studies in carcinogenesis by nitrosomethyl-2-phenylethylamine in rats and the effect of deuterium. Food Cosmet Toxicol 20:393–399

Lijinsky W, Reuber MD, Davies TS, Riggs CW (1982b) Dose-response studies with nitrosoheptamethyleneimine and its alpha deuterium labeled derivative in F344 rats. JNCI 69:1127–1133

Lijinsky W, Kovatch RM, Riggs CW (1983a) Altered incidences of hepatic and hematopoietic neoplasms in F344 rats fed sodium nitrite. Carcinogenesis 4:1189–1191

Lijinsky W, Reuber MD, Singer GM (1983b) Induction of tumors of the esophagus in rats by nitrosomethylalkylamines. J Cancer Res Clin Oncol 106:171–175

Lijinsky W, Knutsen GM, Kovatch RM (1985) Carcinogenic effect of nitrosoalkylureas and nitrosoalkylcarbamates in Syrian hamsters. Cancer Res 45:542–545

Lijinsky W, Keefer LK, Saavedra JE, Hansen TH, Kovatch RM, Fiddler WE, Miller AT (1988) Carcinogenicity of cyclic nitrosamines containing sulfur in F344 rats. Food Chem Toxicol 26:3–7

Lijinsky W, Kovatch RM, Riggs CW (1987b) Carcinogenesis by nitrosodialkylamines and azoxyalkanes given by gavage to rats and hamsters. Cancer Res 47:3968–3972

Lijinsky W, Kovatch RM, Riggs CW, Walters PT (1988a) A dose-response carcinogenesis study of nitrosomorpholine in F344 rats. Cancer Res 48:2089–2095

Lijinsky W, Milner JA, Kovatch RM, Thomas BJ (1989) Lack of effect of selenium on induction of tumors of esophagus and bladder in rats by two nitrosamines. Toxicol Industr Health 5:63–72

Magee PN, Montesano R, Preussmann R (1976) *N*-Nitroso compounds and related carcinogens. American Chemical Society Monograph 173:491–625

McCann J, Choi E, Yamasaki E, Ames BN (1975) Detection of carcinogens as mutagens in the salmonella/microsome test. Proc Natl Acad Sci USA 72:5135–5139

Mico BA, Swagzdis JE, Hu Hs-W, Keefer LK, Oldfield NF, Garland WA (1985) Low dose in vivo pharmacokinetics and deuterium isotope effects in studies of *N*-nitrosodimethylamine in rats. Cancer Res 45:6280–6285

Miller JA, Miller EC (1953) The carcinogenic aminoazo dyes. Adv Cancer Res 1:339–396

Miller EC, Miller JA, Enomoto M (1964) The comparative carcinogenicities of 2-acetylaminofluorene and its *N*-hydroxy metabolite in mice, hamsters, and guinea pigs. Cancer Res 24:2018–2031

National Research Council (1959) Problems in the evaluation of carcinogenic hazard from 53 of food additives. National Academy of Sciences-National Research Council, publication 749

Oberling C, Guérin M, Guërin P (1939) Particularités évolutives des tumeurs produites avec le doses of benzopyrene. Bull Assoc Fr Cancer 28:198–213

Peto R, Gray R, Brantom P, Grasso P (1984) Nitrosamine carcinogenesis in 5120 rodents: chronic administration of sixteen different concentrations of NDEA, NDMA, NPYR and NPIP in the water of 4440 inbred rats, with parallel studies in NDEA alone of the effect of age of starting (3, 6, or 20 weeks) and of species (rats, mice or hamsters). In: O'Neill IK, Von Borstel RC, Miller CT, Long J, Bartsch H (eds) *N*-Nitroso compounds: occurrence, biological effects and relevance to human cancer. IARC scientific publications no. 57:627–665

Preussmann R, Stewart BW (1984) *N*-Nitroso compounds. In: Searle CE (ed) Chemical carcinogens. American Chemical Society Monograph no. 182:643–828

Reddy JK, Qureshi SA (1979) Tumorigenicity of the hypolipidemic peroxisome proliferator ethyl-δ-p-chlorophenoxyisobutyrate (clofibrate) in rats. Br J Cancer 40:476–482

Reddy JK, Azarnoff DL, Hignite CE (1980) Hypolipidaemic hepatic peroxisome proliferators form a novel class of chemical carcinogens. Nature 283:397–398

Rosenkranz HS, Klopman G (1987) Computer automated structure evaluation of the carcinogenicity of *N*-nitrosothiazolidine and *N*-nitrosothiazolidine-4-carboxylic acid. Food Chem Toxicol 25:253–256

Rous P (1911) A sarcoma of the fowl transmissible by an agent separable from the tumor cells. J Exp Med 13:397–411

Saffiotti U, Cefis F, Kolb LM (1968) A method for the experimental induction of bronchogenic carcinoma. Cancer Res 28:104–124

Schmähl D (1985) Critical remarks on the validity of promoting effects in human carcinogenesis. J Cancer Res Clin Oncol 109:260–262
Schmähl D, Habs M, Ivankovic S (1978) Carcinogenesis of *N*-nitrosodiethylamine (DENA) in chickens and domestic cats. Int J Cancer 22:552–557
Shubik P, Sicé J (1956) Chemical carcinogenesis as a chronic toxicity test. A review. Cancer Res 16:728–742
Solleveld HA, Haseman JK, McConnell EE (1984) Natural history of body weight gain, survival, and neoplasia in the F344 rat. JNCI 72:929–940
Swenberg JA, Kerns WD, Mitchell RI, Gralla EJ, Pavkov KL (1980) Induction of squamous cell carcinomas of the rat nasal cavity by inhalation exposure to formaldehyde vapor. Cancer Res 40:3398–3402
Tennant RW, Margolin BH, Shelby MD, Zeiger E, Haseman JK, Spalding J, Caspary W, Resnick M, Stasiewicz S, Anderson B, Minor R (1987) Prediction of chemical carcinogenicity from in vitro genetic toxicity assays. Science 236:933–941
USHEW (1971) Evaluation of environmental carcinogens, report to the Surgeon General, USPHS. U.S. Senate Hearings "Chemicals and the Future of Man", April 6 and 7, pp 180–198
Van Duuren BL, Katz C, Goldschmidt M, Frenkel K, Sivak A (1972) Carcinogenicity of halo-ethers II. Structure-activity relationships of analogs of bis(chloromethyl)ether. JNCI 48:1431–1439
Waynforth HB, Magee PN (1975) The effect of various doses and schedules of administration of *N*-methyl-*N*-nitrosourea, with and without croton oil promotion on skin papilloma production in BALB/c mice. Gann Monogr Cancer Res 17:439–448
Wilson RH, DeFeds F, Cox AJ (1941) The toxicity and carcinogenic activity of 2-acetaminofluorene. Cancer Res 1:595–608
Yamagiwa K, Ichikawa K (1915) Experimentelle Studie über die Pathogenese der Epithelialgeschwulste. Mitt Med Fak Tokio 15:295–344
Zeiger E (1987) Carcinogenicity of mutagens: Predictive capability of the *Salmonella* mutagenesis assay for rodent carcinogenicity. Cancer Res 47:1287–1296

CHAPTER 7

Transformation of Cells in Culture

M. A. KNOWLES

A. Introduction

Our understanding of the aetiology and development of neoplasia centres around our knowledge of the behaviour of mammalian cells and their transformed counterparts. Fundamental cancer research today aims to define precisely the molecular lesions which distinguish the "normal" from the "malignant" cell and to determine how these affect cellular behaviour in vivo. When these mechanisms are elucidated we will have the basis for a more rational design of diagnostic and prognostic tests and of therapeutic agents.

The development of cell culture systems for the study of transformation in vitro is one of the major landmarks in cancer research and has made a great contribution in progress towards this goal. In vitro systems with their rapidity, sensitivity, ease of quantitation and detailed control of experimental conditions offer several advantages over in vivo systems for the study of cell transformation. Powerful techniques such as somatic cell hybridisation and gene transfer and all the gamut of molecular techniques now available to clone and manipulate the structure of mammalian genes provide us with tools to investigate the genotype of normal and transformed cells. At last, we can ask fundamental questions about the molecular basis of carcinogenesis.

The possibilities which present themselves rely on three areas of past and present research:

1. The development of culture conditions suitable for the propagation or maintenance of various cell types in vitro.
2. The development of cell systems in which cells treated with various agents in vitro can be converted to the tumorigenic phenotype.
3. The development of techniques for molecular analysis.

In this review I will not deal with studies concerned with the design of optimal in vitro culture environments but rather concentrate on the cell systems which are available with emphasis on what has been learned from these about the molecular mechanisms of transformation. There is an enormous literature concerned with these topics, and I shall not present an exhaustive review of the profusion of phenomena which have been described but instead select what I believe to be key observations or culture systems of particular usefulness. More exhaustive reviews on particular topics will be referred to where appropriate.

Most of the available evidence suggests that tumours derive from a single cell (FIALKOW 1972). Subsequently, over long periods of time, the progeny of such single cells undergo a process of clonal evolution involving several steps. This

Table 1. Evidence that carcinogenesis is a multistage process

1. Mathematical models based on age-specific incidence curves in humans predict 2–7 independent events for most tumours
2. Histological evidence of distinct premalignant lesions preceding tumour development in many tissues
3. Carcinogensis studies in animals show distinct stages, e.g. initiation and promotion. Tumour incidences induced by fractionated doses of carcinogen are consistent with several discrete events
4. Individuals with inherited predispositions to specific cancers (e.g. familial retinoblastoma, Wilm's tumour, polyposis coli) can be shown to be hemizygous for deletions at specific genetic loci. Tumours arise when a second event (development of homozygosity at these loci) occurs
5. Cell transformation studies in vitro demonstrate a number of discrete sequential phenotypes during transformation. Studies with cloned oncogenes show that at least two cooperating oncogenes are required to transform normal rodent cells

development of tumours as a multistep process was pointed out by FOULDS (1969, 1975) and critically discussed by him and several others since (e.g. FARBER and CAMERON 1980; FARBER 1984). Evidence for such a multistep process comes from a variety of observations in humans and animals (Table 1). In particular, this includes the observations that a very long latent period following exposure to a carcinogen usually precedes the appearance of a tumour and that many cancers in humans and animals are preceded by what are variously termed preneoplastic or precancerous conditions recognisable by histopathology, cytopathology or cytogenetics. In several human genetic traits (e.g. familial retinoblastoma and polyposis coli) a specific germ line lesion is thought to represent one of the so-called stages in transformation, and a subsequent second somatic event leads to tumour development. Compelling evidence is also provided by mathematical models based on age-specific incidence curves for cancer in humans or dose-response curves for tumours induced in laboratory animals which predict a finite number of discrete events during the process (reviewed by FARBER and CAMERON 1980).

Although this concept of carcinogenesis as a multistage process is generally accepted, it must be emphasised that the nature of the steps is not clear. There is, of course, overwhelming evidence that somatic mutations play a major role in carcinogenesis, and some (perhaps all) of the steps are likely to involve a mutation of some kind. However, there is also the significant possibility that some could be non-mutational (i.e. epigenetic) events. The transformation process then may involve a combination of mutational and non-mutational events which may be different in different tumour, tissue or cell types. For most tumours, at present, no genetic lesions are known which correlate absolutely with tumour histology or cell type of origin. A few examples of common lesions do exist, e.g. in several tumour types up to a 40% incidence of *ras* gene mutation has been reported, although with no clear correlation with cell type or histopathology. An even higher incidence of a common genetic lesion has been reported recently in lung tumours of different histological types (KOK et al. 1987).

The analysis in vitro of such a complex and diverse process is obviously difficult, and definitive proofs are elusive. Attempts to elucidate mechanisms at the molecular level depend on detailed biological observations in a number of systems and on the availability of model systems in which steps or stages can be defined. It is on such multistage systems that I shall concentrate in this review.

As already stated, my aim is to discuss those studies which contribute to our understanding of the molecular mechanisms of carcinogenesis. The vast literature concerned with the development of test systems for potential carcinogens will be referred to only in this context. No lesser importance of such studies is implied by this omission, and their value as rapid, simple alternatives to animal tests with good predictive value must be stressed. Many contributions to our fundamental understanding of the process of transformation have come from such studies, and these will be discussed below. Several useful reviews on this subject are available (e.g. HEIDELBERGER et al. 1983).

A great deal has been learned from studies of mesenchymal cell transformation, and this will be discussed. However, my particular area of interest is mechanisms of epithelial cell transformation in vitro, and I will therefore use several epithelial systems as illustrations. Since more than 80% of human malignancies are carcinomas, I make no apology for this but rather state my belief that the study of epithelial cells, though in some cases extremely difficult technically, may teach us lessons about epithelial cancers which we cannot learn from mesenchymal cells.

B. Definition of Terms

A multitude of terms has developed to describe the phenomena observed during in vitro carcinogenesis. The term "transformation" has been used to describe any heritable change in the character of cultured cells. This usage led to some confusion in the earlier literature since the types of transformation described were varied, and many had no relevance to carcinogenesis. Because of the diversity of the heritable changes encompassed by this term and the associated term "reversion", it has been suggested that they be reserved for changes associated with the acquisition or loss of neoplastic potential, i.e. those cells which have been demonstrated to grow as malignant neoplasms in vivo (SANFORD 1974). This suggestion has been largely adopted in recent years. Other terms or descriptions are often added to qualify the type of transformation described, e.g. partial, full, morphological, proliferative, transformation to anchorage independence, etc., and many investigators prefer to use neoplastic or malignant transformation to ensure correct interpretation.

In a study of cell transformation in vitro, the terms "preneoplastic" or "partially transformed" are used to describe cells which, though not yet tumorigenic, show altered properties that appear to indicate that they are precursors of neoplastic cells. However, these terms are not necessarily synonymous with the in vivo term "precancer", and it is not yet known to what extent cells cultured from in vivo precancer resemble in vitro-induced preneoplastic cells. In recent years as it has become clear that one of the major changes involved in transformation is the acquisition of unlimited proliferative potential, a change which in many cases is not accompanied by tumorigenicity and other so-called "markers of transformation", a new term "immortalisation" has emerged. As will be seen later, the distinction between immortalisation and transformation is very clear-cut in many cell types, and this distinction becomes important in certain human cells which acquire certain markers of transformation (though not tumorigenicity) in the absence of immortalisation.

In vitro transformation can be induced by a variety of agents including chemicals, viruses, physical agents and cloned fragments of DNA. In addition, transformation may occur in the absence of any deliberate treatment, and it is then referred to as "spontaneous". In this chapter, I use the term "transformation" as a general term to describe heritable changes which are known to be correlated with the acquisition of tumorigenicity in vivo. Where cells are of known tumorigenicity, the terms "fully transformed" or "neoplastically transformed" are used.

C. Transformation of Rodent Mesenchymal Cells

Most systems for in vitro carcinogenesis have used mesenchymal cells or "fibroblasts" because of the ease with which such cells can be maintained in vitro. Over 40 years ago it was shown that normal mouse and rat mesenchymal cells, after a period of proliferation in culture, sometimes produce malignant tumours when re-implanted into syngeneic hosts (Gey 1941; Earle 1943; Earle and Nettleship 1943). The first clear demonstration of chemical transformation of cells in culture was made by Sachs and colleagues in the 1960s (Berwald and Sachs 1963, 1965). They treated primary or secondary cultures of Syrian hamster embryo (SHE) cells with polycyclic aromatic hydrocarbons and showed that whilst untreated cultures invariably had a limited lifespan, their treated counterparts were capable of continuous proliferation and ultimately produced progressively growing tumours upon re-inoculation into adult hamsters. No spontaneous transformation was observed in control SHE cultures. Since then, hamster embryo cells have been transformed in vitro by a variety of chemical carcinogens (e.g. Huberman et al. 1972; Kamahora and Kakunaga 1967; Sato and Kuroki 1966; Sanders and Burford 1967; Inui et al. 1972; Tsuda et al. 1973; Markovits et al. 1974; Kouri et al. 1975).

The absence of spontaneous transformation of hamster mesenchymal cells in vitro is in sharp contrast to the high frequencies reported with mouse and rat cells. There is a striking species-related tendency of cells to transform spontaneously, this being highest in mouse cells, which almost invariably transform, and lowest in chick and human cells, the latter always exhibiting a finite lifespan. The reasons for this are not clear, though in many cases (with the notable exception of chickens), parallels may be drawn between the species lifespan and the time taken for spontaneous transformation to occur in vitro, suggesting intrinsic molecular mechanisms perhaps related to in vivo ageing. It seems likely that the ease by which transformation can be induced or that spontaneous transformation occurs may be related to the number of molecular events required to transform the cells of different species, which is predicted by mathematical models to be higher in humans than rodents and higher in epithelial than mesenchymal cells.

Transformation of rodent mesenchymal cells in culture has been reproduced in many laboratories and with cells from a variety of origins. Only recently, however, has it been shown conclusively that as predicted by in vivo studies, more than one event is required for full transformation. The systems used fall into two groups, those using cells with a finite lifespan and those using "established" (immortal) cell lines. A number of quantitative and reproducible systems have been developed, and the field has been reviewed in detail on numerous occasions (e.g. HEIDELBERGER 1973, 1975; HEIDELBERGER et al. 1983; MISHRA and DI MAYORCA 1974; FREEMAN et al. 1975; CASTO and DIPAOLO 1975). Only a few examples are discussed below.

I. Cells with a Limited Lifespan

Following the pioneering studies of Sachs and co-workers with SHE cells, this system was adopted both by workers seeking a short-term test system for carcinogens and by those with interest in the fundamental mechanisms of carcinogenesis. A great deal is now known about transformation of these cells, and SHE cell transformation will therefore be used as an example of the transformation of cells with a finite lifespan.

Mass cultures of SHE cells almost invariably show a limited in vitro lifespan, whereas carcinogen-treated cultures are immortal and show characteristic morphological alterations described as criss-cross orientation or piling-up of cells. Assays were developed based on these morphological criteria in which cells were treated at clonal density and the resulting colonies scored on the basis of morphology (BERWALD and SACHS 1963, 1965). High incidences of morphological transformants were induced by chemical carcinogens within 7 days, and this has formed the basis of many short-term assays. Later, it was shown that these morphological transformants were not tumorigenic, so that the earlier results based on morphological criteria alone which had seemed inconsistent with a multistage model for carcinogenesis, on the contrary, pointed to more than one event. More recently, detailed studies of SHE transformation by BARRETT and co-workers have demonstrated that multiple phenotypic and genotypic steps are involved.

Induction of morphological transformation of SHE cells shows a linear increase in logarithm of frequency with the logarithm of dose. The slope of the line for transformation by many carcinogens is 1, suggesting a single-hit mechanism for induction of this phenotype (HUBERMAN and SACHS 1966; DIPAOLO et al. 1971; GART et al. 1979). BARRETT and colleagues (reviewed 1985) extended these studies to compare compounds classified as "genotoxic" and "epigenetic" (WEISBERGER and WILLIAMS 1981). All genotoxic compounds assessed apart from nickel chloride showed a linear response, as did all the epigenetic compounds which included asbestos and diethylstilbestrol. The latter two compounds gave linear dose-response curves on a log-log plot with slopes of ~ 0.3, which though lower than those for other compounds are still consistent with a one-hit mechanism for this step, with other factors, e.g. heterogeneity of target cells, influencing the dose response (see SWARTZ et al. 1982; BARRETT et al. 1983, 1984 for full discussion).

Morphological alteration in SHE cells must be interpreted as the change of a normal diploid cell to a preneoplastic variant since the altered cells have an increased probability of transformation to tumorigenicity. Interestingly, though it is generally accepted that somatic mutation represents the major mechanism of carcinogenesis, the frequency of morphological conversion in these assays is 20–100 times higher than measured frequencies of mutation at other loci measured at the same time and in the same cells. Indeed, morphological transformation by asbestos and diethylstilbestrol has been reported in the absence of detectable gene mutation. However, if the term "mutagenesis" is extended to include so-called chromosomal mutations (i.e. changes in chromosome number or structure), the relationship between transforming efficiency and mutagenicity is maintained for most chemicals (BARRETT et al. 1985). In the case of asbestos, it has been shown that non-random chromosome changes, particularly trisomy of chromosome 11 is present in at least 75% of immortal cell lines, which indicates that a genetic mechanism is involved. The presence of such non-random alterations suggests common lesions in transformants induced in this model system which presents the exciting possibility that the genes involved can be identified.

A number of modifications to the original clonal SHE system have been used including selection for transformants at confluence. Under these conditions three types of morphologically transformed foci can be identified in carcinogen-treated cultures (CASTO et al. 1977), and these are tumorigenic. It must be inferred therefore that these foci are not equivalent to the morphologically altered colonies scored in the colony assay but must have acquired a further event(s) as a result of selection at confluence.

In the colony assay, at least two steps in SHE 942!.3&ø2-!4)ø. by chemical carcinogens are predicted by the phenotypic changes observed, the first an immortalisation step, the second resulting in full transformation. It can be predicted, however, that the second step involves at least one "silent" event. For example, in contrast to the first morphological change, there is a delay of several passages before immortal SHE cells acquire in vitro markers of transformation or become tumorigenic. There is also considerable variation in the rate at which different immortal lines progress to anchorage-independent growth, though this is relatively constant for individual lines (BARRETT 1985). In some recent elegant experiments, BARRETT's group have presented evidence suggesting that one of the events which occurs during this latent period involves the loss of a tumour suppressor function by the immortal cells (KOI and BARRETT 1986). When normal diploid SHE cells were fused to highly tumorigenic benzo[*a*]pyrene-transformed cells, anchorage independence and tumorigenicity were suppressed in the resulting hybrids. Immortal cell lines at early passage also possessed this tumour suppressor function (sup+), but at later passages the function was lost (sup−). It was also shown that transfection of a v-H-*ras* plasmid or DNA from the tumorigenic cells into sup− immortal cells but not sup+ cells induces anchorage-independent growth, providing further evidence for a genotypic change between early and late passage immortal cells. Taken together, these data suggest that transformation of immortal SHE cells involves two steps; loss of a tumour suppressor function and activation of a dominantly acting oncogene. Analysis of carcinogen-induced tumorigenic cells for transforming genes using the NIH-3T3 transfection assay has yielded positive results in ~50% of the cell lines examined to date, and in at least some of these activated *ras* genes have been identified (J. C. BARRETT, personal communication).

Several other transformation systems based on normal diploid cells have been described, though to date none has yielded such informative mechanistic data as the SHE system. Newborn Syrian hamster dermal (SHD) cells have been used as an alternative to SHE cells for studies directed towards an analysis of the event(s) involved in immortalisation (NEWBOLD et al. 1982). These have the advantage that the cell population is homogenous and that untreated cultures have a very short in vitro lifespan and senesce after about 15 population doublings. Carcinogen-immortalised SHD cells are non-tumorigenic and progress to tumorigenicity and anchorage independence as SHE cells.

II. Established Cell Lines

The search for an ideal test system for carcinogens led many workers to study carcinogen-induced transformation of a range of permanent cell lines. Such studies have provided a wealth of information on chemical carcinogenicity and potentially can give insight into the molecular mechanisms of late stages in transformation. The marked propensity of rodent mesenchymal cells, especially mouse cells, to undergo spontaneous transformation in vitro has provided numerous immortal cell lines, many of which are non-tumorigenic but have acquired some properties associated with transformation, e.g. ability to clone and grow at low density, which makes them particularly suitable for use in in vitro assays. These lincs are considered preneoplastic since they show increased probability of full transformation compared with diploid cell strains. Indeed, some lines, e.g. Swiss and BALB/c-3T3 (TODARO and GREEN 1963; AARONSON and TODARO 1968) have a high frequency of spontaneous transformation, and rigorous culture schedules are required to prevent this. The most widely used cell lines have been those which exhibit density-dependent inhibition of growth and low or absent rates of spontaneous transformation. Of these, the C3H mouse prostate cell line described by CHEN and HEIDELBERGER (1969a–c) and the C3H10/T1/2 cell line (REZNIKOFF et al. 1973a) have been the most extensively studied.

It is clear that different mesenchymal cell lines are distinct genotypically and may represent different stages in transformation or progression. Thus, BHK cells transform in one step when treated with chemicals, apparently as a result of somatic mutation (BOUCK and DI MAYORCA 1976), which has been defined as a recessive lesion (BOUCK and DI MAYORCA 1982). The most likely explanation for a single-hit mutation to a recessive phenotype is that BHK cells are hemizygous at a suppressor locus. This has now been confirmed (TOLSMA et al. 1988), and this suppressor gene shown to be powerful enough at single copy number to suppress transformation in hybrids containing two pseudodiploid complements from the transformed cells. Interestingly, it has been shown that BHK transformation can also be induced apparently epigenetically by 5-azacytidine, and cell fusion experiments suggest that the same gene is involved (BOUCK et al. 1984). Spontaneous transformation of FOL-2 cells (CRAWFORD et al. 1983) has been shown to occur at a rate consistent with mutation at a single specific gene locus, and it might therefore be expected that chemical mutagens would induce transformation as a single-hit event as in BHK cells. However, treatment with mutagens does not increase the transformation rate over background (BARRETT et al. 1980), and it has been proposed that transformation of these cells may result from chromosome segregation involving a single gene locus (CRAWFORD et al. 1983). Transformation of BALB/c-3T3 cells is also thought to involve a single genetic event. These cells readily give rise to tumours when implanted into animals on a solid substrate (BOONE 1975; BOONE et al. 1976), a property shared by other cells requiring only a single further event for tumorigenicity.

Some cell lines appear to require more than one event for transformation, as described above for immortal SHE cells. In the original studies of C3H10/T1/2 by REZNIKOFF et al. (1973a, b), transformed foci of three morphological types (I, II, and III) could be scored 6 weeks after treatment, the latter two of which were tumorigenic. Dose-response curves for these experiments appeared linear on a log-log plot, though it was noted that the transformation frequency seemed to be related to cell density at the time of carcinogen treatment. It is now clear that more than one event is involved in this process. The first, termed "initiation", occurs in almost every cell in the treated population (MONDAL and HEIDELBERGER 1970; KENNEDY et al. 1980; FERNANDEZ et al. 1980; KENNEDY and LITTLE 1980) and has been suggested to involve the expression of a function in many or all of the treated cells. Expression of this function then increases the probability of a second (possibly mutagenic) step occurring when the cells are maintained at confluence.

More recently, it has been suggested that the initial event may resemble the SOS response of bacteria treated with mutagens (WALKER 1985) and may determine the frequency of subsequent rare genetic events (KENNEDY et al. 1984). In support of this hypothesis is the finding that certain protease inhibitors known to block the SOS response (MEYN et al. 1977) also reverse the initial step in C3H10/T1/2 transformation (KENNEDY and LITTLE 1978).

In contrast, the second step is a rare event ($P < 10^{-6}$) thought to occur at confluence since the number of cells at confluence in these density-arrested cells is constant regardless of the initial seeding density (KENNEDY et al. 1980). Interestingly, this probability cannot be significantly raised by treatment of cultures with mutagens such as X-rays and UV light as they approach confluence (KENNEDY et al. 1984), so that if this event is as suggested by its frequency a mutation, then it may be an unusual class of genetic event unaffected by mutagens of this type. BARRETT et al. (1984) on re-analysis of the data of KENNEDY et al. (1980) and FERNANDEZ et al. (1980) calculated a spontaneous rate of $1\text{–}6 \times 10^{-7}$ per cell per generation for this second event. This is similar to that calculated for the preneoplastic hamster cell line FOL+ (CRAWFORD et al. 1980), which also is unaffected by mutagens. "Initiated" C3H10/T1/2 cells and FOL+ cells therefore may require a similar event for transformation. An alternative hypothesis, suggested to explain the second event by HABER and colleagues (HABER et al. 1977), is a dose-dependent influence of carcinogen treatment on cell-cell interaction, which by modulating the proposed suppressive effects of normal cells on transformed cells will lead to an apparent alteration in focus incidence.

SAGER and co-workers have made detailed studies of the hamster fibroblast (CHEF) cell lines. Again, multiple steps in transformation of an immortal line are predicted (SMITH and SAGER 1982). A comparison of the subclones CHEF 16-2 and CHEF 18-1 (SAGER and KOVAC 1978) shows that CHEF 16-2 are highly tumorigenic and that CHEF 18-1, though immortal and with increased cloning efficiency, share many properties with normal cells. They are diploid, remain non-tumorigenic and anchorage dependent after many passages, and do not produce tumours when implanted attached to a solid substrate. Treatment with mutagens induces anchorage-independent clones and clones with reduced serum dependence at a much higher frequency than tumorigenicity (SMITH and SAGER 1982), and these have a higher probability than the parent line to undergo further change to tumorigenicity. Cells with both anchorage independence and reduced serum requirement are not tumorigenic, showing that these two in vitro phenotypes are not sufficient for tumorigenicity. Analysis of anchorage-independent mutants using cell hybridisation (MARSHALL and SAGER 1981) shows at least two complementation groups for this phenotype, and one has been mapped to chromosome 1.

Thus, immortal mesenchymal cells may require one or more further events for transformation to tumorigenicity and in some cases these events are predicted to be non-mutational. In C3H10/T1/2 cells there is general agreement that transformation is a 2-step process, that the first event is rapid and occurs in a high proportion of the treated population and that the second, rare event may represent an unusual genetic event (not increased in frequency by mutagens) or could be related to the size of the transformed colonies at confluence which allows them to overcome the suppressive effects of surrounding normal cells. In other cases in which a single step is predicted, this may involve mutation in a recessive gene (e.g. BHK), possible chromosomal segregation (FOL 2) or other mutational event. Taken together, these results demonstrate that the lesions involved in mesenchymal cell transformation are not all simple somatic mutations. They also suggest that immortal rodent cells represent a multitude of genotypes with the possibility that a number of combinations of genes or types of genes are involved.

III. Oncogenes and the Transformation of Rodent Mesenchymal Cells

The discovery of viral oncogenes, their identification as transduced cellular genes, and the finding that these same genes are altered in structure and/or function in human tumours has provided the key to understanding at least some of the somatic lesions involved in transformation. A recent count shows upwards of 60 oncogenes described to date from viral and other sources (for a detailed description, see Chap. 10, Part 2). Oncogene research has relied on in vitro assay systems from the start. Following the initial observations that several retroviruses contain transforming genes which are derived from cellular sequences, evidence was sought that these so-called cellular proto-oncogenes are involved in some way during human carcinogenesis. This evidence came most powerfully from the use of DNA-mediated gene transfer (transfection) into mammalian cells. Using the calcium phosphate-mediated transfection technique (GRAHAM and VAN DER EB 1973), high molecular weight DNA extracted from chemically transformed mouse cells (SHIH et al. 1979), a variety of human tumours and tumour-derived cell lines (SHIH et al. 1981), and carcinogen-induced animal tumours (BALMAIN and PRAGNELL 1983; SUKUMAR et al. 1983) was found to induce foci when introduced into the immortal mouse cell line NIH3T3. This biological assay has been used to identify and clone a number of oncogenes and has demonstrated that a range of single genes, when introduced into a clonal immortal rodent cell line can induce transformation. In the few years since these original studies, a vast number of experiments have been carried out in which cloned viral and cellular oncogenes have been introduced alone or in combination into a variety of cells. Such experiments have provided us with a great deal of information about the genetic lesions required to transform different cell types and have enabled detailed functional assays to be carried out on individual cloned oncogenes and their molecularly altered derivatives.

1. Introduction of Genetic Material into Mammalian Cells

The ability to introduce genetic material in a variety of forms (e.g. chromosomes, high molecular weight DNA, cloned genes in plasmids, bacteriophages, viruses and RNA) into cultured cells in an efficient and reproducible manner is essential for molecular studies of mammalian cells. The pivotal importance of such techniques demands that some discussion be given to the available methodologies. Although it has been possible to introduce DNA into cells for more than 10 years, the cellular processes involved are poorly understood, and most advances have been made empirically. Nevertheless, for most cell types, highly efficient gene transfer and expression can now be achieved. A number of methods for gene transfer which have been shown to be applicable to a range of cell types are listed in Table 2.

At present, the method most frequently used is the transfection technique described by GRAHAM and VAN DER EB (1973), in which DNA is introduced into cells following co-precipitation with calcium phosphate, in a manner reminiscent of bacterial transformation. Since it appears that particles of precipitate enter the cells by phagocytosis and that particle size and adherence to the cell surface are critical, several methods to enhance adsorption of precipitates to the cell membrane have been tried. These include agents known to enhance the adsorption of retroviruses to cells, such as DEAE-dextran, polybrene and poly-L-ornithine, which markedly increase transfection frequencies in certain cell types. DMSO, glycerol and polyethylene glycol have also been shown to enhance DNA uptake, presumably by affecting membrane fluidity. Following uptake of the DNA, it is assumed that fusion with lysosomes occurs, so that much of the internalised DNA may be degraded before it reaches the nucleus. Some lysosomotropic chemicals which raise lysosomal pH such as chloroquine, 3-methyl adenine, carbonyl cyanide, *p*-trifluoromethoxyphenyl hydrazone and NH_4Cl increase the efficiency of integration of transfected sequences, presumably by inactivating

Table 2. Methods to introduce DNA and RNA into mammalian cells

Method		References
Calcium phosphate co-precipitation		GRAHAM and VAN DER EB (1973); GRAHAM et al. (1980)
+glycerol "shock"		FROST and WILLIAMS (1979); PARKER and STARK (1979)
+DMSO		LEWIS et al. (1980); LOWY et al. (1978); STOW and WILKIE (1976)
+sodium butyrate		GORMAN and HOWARD (1983)
at 3% CO_2 (pH 6.95)		CHEN and OKAYAMA (1987)
Strontium phosphate co-precipitation		BRASH et al. (1987)
Polybrene + DMSO "shock"		KAWAI and NISHIZAWA (1984); MORGAN et al. (1986)
Poly-L-ornithine + DMSO "shock"		BOND and WOLD (1987)
DEAE dextran		VOGT (1967)
+PEG		GOPAL (1985)
Liposome fusion		FRALEY et al. (1980); SCHAEFFER-RIDDER et al. (1981); KANEDA et al. (1987); FELGNER et al. (1987)
Sendai virus envelope fusion		VOLSKY et al. (1984)
Protoplast fusion		RASSOULZADEGAN et al. (1982); YOAKUM et al. (1983)
Microinjection		SHEN et al. (1982); CAPECCHI (1980)
"Pricking"		YAMAMOTO and FURASAWA (1978); Yamamoto et al. (1982); KUDO et al. (1982)
Laser		KURATA et al. (1986)
Electroporation		KINOSITA and TSONG (1977a, b); Chu et al. (1987)
Infection:	retrovirus	SHIMOTONO and TEMIN (1981); WEI et al. (1981); TABIN et al. (1982); MILLER et al. (1983); CEPKO et al. (1984)
	SV 40	ASANO et al. (1985)

lysosomal enzymes (LUTHMAN and MAGNOSSON 1983; EGE et al. 1984). The events involved in transport of DNA into the nucleus, its integration into the chromatin and subsequent expression remain obscure. However, sodium butyrate, an agent known to alter chromatin structure, can increase the proportion of cells able to express foreign DNA and the level of transcription obtained.

Using the calcium phosphate precipitation method, a range of cell types can be transfected. However, some cultured primary cells lyse in the presence of calcium phosphate precipitates, and calcium ions induce several cell types (e.g. keratinocytes) to undergo squamous differentiation (HENNINGS et al. 1980). Strontium phosphate precipitation has been used to obtain efficient transfection in such cells.

Other methods (Table 2) are based on the fusion of membrane-packaged DNA to cells. A recent report using gangliosides to facilitate Sendai virus fusion of liposomes with cells claims efficiencies of 6×10^{-4}–2.5×10^{-3}, and a novel "lipofection" method based on liposomes prepared from a synthetic cationic lipid gives marked increases in efficiency compared with the calcium phosphate and DEAE-dextran techniques for some cell lines. Finally, various direct physical methods have been described, including microinjection, the so-called "pricking" method in which cells are pricked with a needle and DNA molecules dissolved in the medium can enter when the needle is withdrawn, a laser method based on a similar principle and electroporation, which creates pores in the plasma membrane by exposure of the cells to a pulsed electric field. This latter method has been shown to be capable of transforming >1% of viable cells to the stable expression of a marker.

Recently, defective retroviral vectors have been developed for the efficient transfer of functional foreign sequences into mammalian cells. This method takes advantage of the extremely efficient integration of such viruses into the host genome and gives very high frequencies of stable gene transfer (HWANG and GILBOA 1984).

2. Transformation of Immortal Fibroblasts by Oncogenes

The initial finding that the cell line NIH3T3 could be transformed in one step by DNA sequences from a proportion of human tumours and the subsequent identification and cloning of single human oncogenes from such transfectants demonstrated that a single oncogene can transform certain immortal mesenchymal cells. Thus NIH3T3 cells transform in response to a number of single activated cellular oncogenes, e.g. the three members of the *ras* gene family, N-*ras* (HALL et al. 1983), H-*ras* 1 and K-*ras* 2 (DER et al. 1982), *raf* (SHIMIZU et al. 1985), *neu*/HER2 (HUDZIAK et al. 1987), *met* (COOPER et al. 1984), *hst* (SAKAMOTO et al. 1986), *dbl* (EVA and AARONSON 1985), *lca* (OCHIYA et al. 1986), *trk* (MARTIN-ZANCA et al. 1986), *sis* (CLARKE et al. 1984) and a number of viral oncogenes (e.g. polyoma middle T antigen). This demonstrates an important general principle which is now emerging from several lines of study, that some phenotypic stages in transformation can be achieved by any one of a number of genetic lesions, that is, the oncogenes which can transform NIH3T3 form a complementation group. By inference, it seems unlikely that NIH3T3 has lesions in any of these genes but has a number of as yet uncharacterised genetic alterations. All the oncogenes identified using the NIH3T3 assay act in a dominant way to transform the cells.

NIH3T3 cells are not alone in their transformability by this group of oncogenes. Immortal hamster dermal fibroblasts (NEWBOLD and OVERELL 1983), C3H10/T1/2 cells (MANOHARAN et al. 1985), rat-1 cells (LAND et al. 1983), certain immortal SHE cells (KOI and BARRETT 1986) and many other immortal rodent cell lines are transformed by *ras* oncogenes. Immortalisation per se is not sufficient for transformability by *ras* genes, however. Several cell lines require a second introduced gene in addition to a *ras* gene for transformation. For example, the established cell line REF 52 behaves like certain primary cells with regard to transformation by an activated H-*ras* gene and requires a co-operating adenovirus E1A gene (FRANZA et al. 1986). Also, m5S, an immortal mouse fibroblast cell line is not transformed by v-H-*ras* (TSUNOKAWA et al. 1984), and EK-3, a derivative of NIH3T3, can be transformed by *myc* and *ras* in combination but not by *ras* alone (KATZ and CARTER 1986). In the SHE system, some immortal lines are transformable by *ras* and others not, and it appears that *ras*-

transformability is correlated with the loss of a tumour-suppressor function (Koi and Barrett 1986) (see Sect. C.I).

3. Transformation of Primary Cells

Transformation of primary cells requires at least two separate functions or stages (as discussed above), and these can be subserved by transfected oncogenes. The introduction of a *ras* oncogene into primary cells induces a morphological change and in some cases anchorage independence but not tumorigenicity, and the cells ultimately senesce (Land et al. 1983; Newbold and Overell 1983; Ruley 1983; Thomassen et al. 1985a). Similarly, other single oncogenes which can transform immortal cells do not transform primary cells. However, a combination of oncogenes can transform primary cells, and a number of studies have defined a range of oncogenes which can fulfil an establishment function and complement *ras* and other genes. Thus, adenovirus early region 1A (E1A) or portions of polyomavirus large T antigen can co-operate with H-*ras* or polyomavirus middle T antigen to transform baby rat kidney (BRK) cells (Ruley 1983), v-*myc* and c-*myc* can cooperate with *ras* (Land et al. 1983; Lee et al. 1985) as can a number of other genes, e.g. N-*myc* (Schwab et al. 1985; Yancopoulos et al. 1985), p53 (Parada et al. 1984; Eliyahu et al. 1984; Jenkins et al. 1984), c-*myb* and *ski* (Weinberg 1985).

These results have led to the adoption of several terminologies and classifications. Thus transformation is often divided into "early" and "late" stage events, the former being generally concerned with immortalisation, the latter with the acquisition of tumorigenicity. This has led to the classification of the genes concerned into two so-called complementation groups of "immortalising" and "transforming" genes (Weinberg 1985). The former include *myc*, N-*myc*, *myb*, p53, *ski*, *fos*, E1A, SV40 large T and polyomavirus large T, all of which have protein products which localise in the nucleus and are believed to be concerned with control of DNA replication and/or regulation of gene expression. The latter include the *ras* genes, *src*, *erb*B1, *erb*B2, *fms*, *fps*, *yes*, *raf*, *mos*, and *abl*, all of which show cytoplasmic or membrane localisation. The functions of these gene products vary but are related to events involved in growth factor stimulation and transduction of the resulting stimuli, e.g. some are altered forms of growth factor receptors (*erb*B1, *erb*B2, *fms*), others have tyrosine kinase activity (*src*), and others are related to G-proteins (*ras*). As with most attempts to define and classify newly described biological phenomena, these classifications are inadequate in various respects. Nevertheless, with the caveat that the suggestion of a temporal requirement for different functions implied by the "early" and "late" nomenclature may be dictated only by the in vitro environment, these definitions have proved useful for the purposes of discussion. Certainly, the two-stage, oncogene-mediated transformation model is consistent with studies of carcinogen-induced transformation described above.

There are, however, indications that the concept of a two-stage transformation process may be simplistic and that more detailed examination of transformation of primary cells will reveal other steps. In SHE cells, transfection with v-*myc* and v-Ha-*ras* appears to transform the cells (Thomassen et al. 1985a). However, v-*myc* alone does not immortalise these cells, and when v-*myc* + v-Ha-*ras* transfectant colonies are examined closely, it is found that initially the cells are not immortal but undergo crisis before immortal variants escape senescence and give rise to transformed lines (J. C. BARRETT, personal communication). Interestingly, it has been found that the transformed cell lines have a non-random loss of chromosome 15 (OSHIMURA et al. 1985). This lesion has now been associated more precisely with an event required for *myc* + *ras* transformation rather than the immortalising event. Following fusion of *myc* + *ras*-transformed cells with normal SHE cells, many hybrids senesced, indicating that immortality was recessive as in other cell types (see below). In hybrids which escaped senescence, expected numbers of chromosome 15 were found, indicating that this chromosome is not involved in immortalisation. However, these hybrids were suppressed for tumorigenicity and anchorage independence despite expression of the transfected genes. Loss of copies of chromosome 15 was associated with the emergence of transformed variants, which suggests that hamster chromosome 15 has a tumour-suppressor gene whose loss is essential for transformation by *myc* and *ras* (OSHIMURA et al. 1988).

The so-called immortalising genes differ in their ability to induce establishment of primary cells, and it is questionable whether single-step immortalisation does occur. However, polyomavirus large T antigen appears to immortalise cells as a one-step process (RASSOULZADEGAN et al. 1983). Similarly, adenovirus E1A induces immortalisation (HOUWELING et al. 1980; RULEY et al. 1984), apparently as a direct response to the introduced gene and not as the result of secondary events within the transfected cells. However, some recent results suggest that E1A extends the in vitro proliferative potential of baby rat kidney (BRK) cells but that further adaptive changes are required for establishment ($\equiv$ immortalisation) (ZERLER et al. 1986). The situation for *myc, myb,* and p53 is still less clear, and immortalisation by these genes may be even more dependent on additional events. Cells transfected with these genes have a markedly increased frequency of immortalisation compared with controls (LAND et al. 1986; JENKINS et al. 1984), but it appears that their ability to complement *ras* is a much more efficient function than their immortalisation function (LAND et al. 1986).

This classification of genes as immortalising or transforming depends to a certain extent on the assay used and the phenotypes scored. For example, an activated c-*myc* gene can transform immortal rat and mouse cell lines (KEATH et al. 1984) as can E1A, p53 and normal c-*myc* genes (KELEKAR and COLE 1986). In all these cases, transformation is not accompanied by a significant morphological change but results in a reduced requirement for serum growth factors, limited ability to grow in agar and tumorigenicity. Similarly, an activated c-H-*ras* has been shown to immortalise BRK cells (KELEKAR and COLE 1987). Clearly these two terms are not mutually exclusive, and there is overlap in function, but more useful definitions must await further information about the ways in which specific oncogenes subvert normal growth control pathways.

Early findings with retroviruses suggested that normal cells can be transformed by a single oncogene under certain conditions (HANAFUSA 1977; WEISS et al. 1982), and these results have always presented an apparent contradiction to the multistep theory of carcinogenesis and to the results described above. This possibility has been explored using cloned oncogenes in primary cells. Evidence has been presented that a *ras* oncogene acting alone can fully transform rat embryo fibroblasts (REFs) and Chinese hamster lung cells (SPANDIDOS and WILKIE 1984). Plasmids containing a mutant H-*ras* gene linked to transcriptional enhancers and a dominant selectable marker were transfected into early passage cells. Drug-resistant colonies were picked, passaged to assay for rescue of cells from senescence and plated in soft agar to assay for anchorage independence. A high proportion of transfectants were morphologically altered, immortal and anchorage independent, an apparently full transformation by high levels of a single oncogene product. Others have failed to confirm these findings. For example, LAND et al. (1986) were unable to induce foci in REF cells using several *ras* plasmic constructs including those of SPANDIDOS and WILKIE. Similarly, KELEKAR and COLE (1987) reported a low frequency of morphologically altered foci in BRK cells transfected with an activated c-H-*ras,* all of which senesced. In vivo experiments with transgenic mice containing various activated oncogenes provide a useful parallel to in vitro studies. Here, it has been shown that only a very small proportion of cells expressing these genes develop tumours, providing further evidence that single oncogenes are not sufficient for transformation.

In summary, transformation of rodent mesenchymal cells involves multiple phenotypic steps which can be correlated with underlying genetic or epigenetic lesions. Some of these can be defined as involving dominantly acting oncogenes and others, recessive oncogenes or tumour-suppressor genes. Using gene transfer techniques, cloned oncogenes have been used to induce transformation of a range of mesenchymal cells with at least two oncogenes required to transform primary cells. An increasing number of results support the theory that in many (perhaps all) cases, recessive lesions are required in addition to a combination of dominantly acting genes for transformation.

IV. Role of Immortalisation in Transformation

The molecular basis for immortalisation of cultured cells is not completely understood, and several aspects are still puzzling. The relationship of cellular lifespan in vitro in which the end of the proliferative lifespan is marked by crisis or senescence, to species lifespan in vivo has already been described (Sect. C). Culture lifespan can be extended in some cases by altering medium composition, particularly by replacing serum (e.g. ORLY et al. 1980; AMBESI-IMPIOMBATO et al. 1980; MASUI et al. 1986), but eventual senescence followed by the emergence of rare, immortalised variants is the rule. The nature of the observed crisis or senescence phenomenon is therefore of great interest. Recently, it was reported that the lifespan of mouse embryo cells could be extended indefinitely in a serum-free medium (LOO et al. 1987). In the absence of serum, no crisis or chromosomal aberrations were observed, and the cells were apparently immortal. Addition of serum to the cultures inhibited cell growth, and it was postulated that in serum-containing media, selection of cells which are unresponsive to putative serum or plasma inhibitors may contribute to the crisis phenomenon and be a pre-requisite for immortalisation.

The importance of immortalisation as a key event in carcinogenesis has been the subject for considerable debate. In vitro, immortalisation can be induced by carcinogens and facilitates full transformation (NEWBOLD et al. 1982), and in vivo, treatment with carcinogens leads to the induction of cell variants which can be isolated in vitro or transplanted as immortal cell lines. In addition, where cellular senescence occurs, it represents a dominant phenotype and can restrict tumorigenicity even in the presence of high expression levels of transforming oncogenes (O'BRIEN et al. 1986). However, the fact that immortalisation happens is no proof that it is an essential step. It can be argued that our in vitro preference for handling large numbers of proliferating cells will tend to select for this characteristic whether or not it is necessary for transformation, as will repeated transplantation of tissue in vivo. Indeed, it can be calculated that for human cells with approximately 50 potential population doublings, sufficient proliferative potential exists for the production of a large tumour mass of mortal cells. The inability to culture or transplant cells from many tumours suggests that this may be the case, though the marked increase in success rates for several cell types with improvements in media formulations in recent years implies that inadequate technique was responsible for many failures. Estimates of the number of clonogenic cells within tumours are in the range 0.001%–1% of the total cells (SALMON 1980).

Current models of tissue maintenance and tumour growth propose a spectrum of proliferative potential in the cells of a tissue, with stem cells or clonogenic cells capable of self-renewal at one end, giving rise to differentiating cells capable of limited clonal expansion and finally end cells which are terminally differentiated, non-dividing cells at the other end of the spectrum (BUICK and POLLACK 1984). Results indicate that the latter two compartments constitute the bulk of most tissues, and it is proposed that tumours arise as a consequence of lesions occurring in stem cells. Lesions in stem cells might for example result in a reduced commitment to differentiation and an increase in the population of clonally expanding cells or in an increased probability of stem cell renewal. Either could give rise to a tumour without exceeding the 50 potential divisions of a stem cell, so that immortalisation would not be essential.

Definitive proof of either hypothesis will no doubt be elusive, but it is clear that the study of this phenomenon will teach us a great deal about the control of cell proliferation and differentiation. It seems likely that the so-called immortalising genes (e.g. *myb*, *myc*, p53) are intimately involved in proliferative control in cells with a limited lifespan and could therefore effect loss of steady state proliferation kinetics in the type of cell differentiation hierarchy described above.

In terms of mechanisms of cellular immortalisation in vitro very little is known. As discussed above, several oncogenes appear to act in a dominant way to immortalise cells under certain conditions, and it will be important to determine how this happens and how these genes interact with the apparently recessive genes also implicated. The sequences of adenovirus E1A which are involved in immortalisation and co-operation with *ras* genes in the transformation of BRK cells and transcriptional activation have been analysed using a series of plasmids containing partial E1A coding sequences (ZERLER et al. 1986). E1A functions required for establishment and *ras* co-operation are linked and may be associated with a single biochemical activity but are not associated with functions required for transcriptional activation.

Somatic cell hybridisation studies indicate that immortality is a recessive phenotype in hybrids and that at least four complementation groups for immortality exist (PEREIRA-SMITH and SMITH 1988). This suggests that more than one type of immortality suppressor gene or senescence gene exists. Senescent cells produce a membrane-associated protein that inhibits initiation of DNA synthesis (PEREIRA-SMITH et al. 1985) and abundant, anti-proliferative mRNAs have been identified in non-proliferating normal human cells which may code for this and/or other anti-proliferative proteins (LUMPKIN et al. 1986), putative products of senescence genes.

Specific sequences which are growth inhibitory for HeLa cells have been detected in DNA from quiescent human fibroblasts using a gene transfer assay (PADMANABHAN et al. 1987). Interestingly, the sequences identified are highly represented in the mammalian genome. The increased activity in this assay of DNA from quiescent (compared with proliferating) cells may suggest a role for DNA modification in the regulation of these inhibitory sequences. The construction of cosmid libraries from these DNAs and of cDNA libraries from the human anti-proliferative mRNAs (KLEINSEK and SMITH 1987) have been reported, so that the relevant molecular clones should be available soon.

D. Transformation of Rodent Epithelial Cells

If our goal is to understand the molecular basis for human cancer, then epithelial cells must necessarily represent important targets for research. More than 80% of adult human cancers are carcinomas, and it is these solid neoplasms which present major therapeutic problems. The past 15–20 years have seen intense efforts to culture epithelial cells and to study their differentiation, proliferative control and transformation. Early attempts to study epithelial transformation were hampered by an inability to provide a suitable culture environment for specific cell types. Now these problems have been overcome, and a variety of epithelial cells can be serially propagated, many in serum-free media and at clonal densities. Since these cells represent the functional cells of the body tissues and as such show markedly different characteristics in vivo, it is not surprising that cells derived from different tissues have very different growth requirements and properties in vitro. Nevertheless, as will become apparent, there are broad similarities between transformation of cells from diverse sites which may indicate fundamental mechanisms common to all. Superimposed upon these are a multitude of differences which are probably related to the differentiation programme of specific cell types. This is in contrast to many studies on mesenchymal cells where in the majority of cases cells derived from diverse tissues of origin are thought to represent a similar cell type (FRANKS and WILSON 1977).

Evidence from a number of cultured epithelia points to a multistep process of transformation, and several phenotypic stages in transformation are recognised, which in a number of cell types are very similar. In cultures treated with carcinogens the first phenotypic alteration recognised is typically described as epithelial foci, altered foci, hyperplastic foci or enhanced growth variants, which are recognised because in general the normal epithelial cells do not divide more than a few times. From such foci, immortal cell lines can be isolated, and these progress to acquire various in vitro markers of transformation and ultimately tumorigenicity. This type of in vitro progression has been described in cultures derived from submandibular gland (KNOWLES and FRANKS 1977), epidermis (SLAGA et al. 1978), bladder (SUMMERHAYES and FRANKS 1979) and trachea (STEELE et al. 1977). Since recognition of altered epithelial foci is based on their survival and proliferation under conditions under which normal cells do not proliferate, this pattern has not been recognised in cell types which proliferate well in vitro. For example, rat liver cultures proliferate and spontaneously immortalise, often with no recognisable period of senescence (WILLIAMS 1976; WILLIAMS et al. 1971), and here transformation has not been divided easily into recognisable phenotypic stages.

To discuss epithelial transformation studies with no attention to cell type will not generate a coherent picture. I shall describe several individual cell systems, therefore, and then discuss their similarities and some general implications. This description is based partly on personal experience (with submandibular gland and bladder) and partly on the amount of information which is available on different tissue systems; it should be noted that several other epithelial cell types have been transformed in vitro.

I. Submandibular Gland

1. Phenotypic Stages

In mouse submandibular gland (SMG) cultures several phenotypic stages in epithelial transformation can be defined in primary cultures treated with carcinogens (KNOWLES and FRANKS 1977; WIGLEY 1979). Explants attach to the culture vessel, and there is a wave of epithelial cell proliferation giving rise to an epithelial outgrowth in which proliferation virtually ceases by 20–30 days (stage I). In some outgrowths ductal differentiation occurs, and in some carcinogen-treated cultures extensive hyperplastic duct systems develop from around 30 days onwards (stage II). Proliferating epithelial foci derived either from ducts or from flat epithelial areas can be identified from 60 to 70 days onwards at a frequency related to carcinogen dose (stage III). These foci are anchorage dependent and non-tumorigenic and initially grow extremely slowly. Many give rise to immortal, non-tumorigenic cell lines, and from these, anchorage-independent, tumorigenic variants arise spontaneously after several passages (KNOWLES and FRANKS 1978; FRANKS and KNOWLES 1978). The latter show distinct morphological alterations compared with the preneoplastic foci, and this morphological alteration was originally designated stage IV (KNOWLES and FRANKS 1977). It now seems clear that the relatively rapidly proliferating immortal cell lines (WIGLEY 1979) represent an additional stage between the slow-growing foci (stage III) initially identified and the morphologically altered cell lines (stage IV) (KNOWLES and FRANKS 1978).

It is difficult to make accurate measurements of the frequency of focus induction by different treatment regimes in this system. Focus incidence has been estimated on a per explant basis with a correction for carcinogen-induced cytotoxicity. Elaborate dose-response experiments have not been carried out, but is has been shown that higher doses and multiple carcinogen exposures induce more foci than single exposures. Based on counts of cells dissociated from floating explants at the time of carcinogen treatment and counts made after treatment to give a measure of cytotoxicity, it has been estimated that foci arise in control cultures at a frequency of 2×10^{-7} treated cells and in cultures treated with an optimum focus-inducing regime of multiple exposures to benzo[*a*]pyrene, at 1.25×10^{-5} (WIGLEY 1979).

These proliferating epithelial foci have been designated preneoplastic since they are non-tumorigenic but show increased propensity to progress to immortalisation and tumorigenicity. Indeed, it is tempting to speculate that parallels exist between these foci and the foci of squamous metaplasia which develop in hydrocarbon-treated SMG in vivo, preceding development of adenocarcinoma (WIGLEY and CARBONELL 1976).

A more careful analysis of the behaviour of individual foci suggests that more than one type of focus exists. When attempts are made to subculture foci, a high proportion can be passaged at least once (~75%), though carcinogen-induced foci give a significantly higher success rate than control foci (WIGLEY 1979). Subsequently, some foci degenerate, whilst others survive and proliferate to give rise to immortal cell lines (WIGLEY 1979; KNOWLES, unpublished observations). It may be postulated, therefore, that the foci scored initially are of at least two types, with the majority of spontaneous foci showing extended lifespan but no further spontaneous progression whilst carcinogen-induced foci show a significant probability of spontaneous immortalisation. This suggests that foci are not immortal ab initio and that an additional event(s) is required for immortalisation. Future analyses of focus heterogeneity will clearly be important.

2. Effect of a Tumour Promoter

Studies with the potent tumour promoter 12-*O*-tetradecanoylphorbol-13-acetate (TPA) have shown that as in the mouse two-stage skin papilloma induction model (reviewed by BOUTWELL et al. 1982) SMG foci are promoted by TPA treatment following initiation with carcinogen (KNOWLES 1979; WIGLEY 1983). Epithelial proliferation is stimulated, larger epithelial outgrowths develop in TPA-treated cultures, and following a sub-threshold dose of carcinogen which alone gives no foci, a high incidence of foci is induced. Cumulative incidence curves show that foci from TPA-treated cultures appear earlier (~50 days compared with 80–100) and grow faster than those induced by carcinogen alone. However, TPA does not affect immortalisation of foci or progression to tumorigenicity.

3. Analysis of DNA Content and Karyotypic Markers

Analysis of the DNA content of SMG epithelial cells at different stages of transformation shows that foci consist of 30%–60% tetraploid cells compared with 100% diploid cells in the primary outgrowth and in dissociated SMG tissue

(COWELL and WIGLEY 1980). Immortal preneoplastic cell lines are predominantly sub-tetraploid, and chromosome analyses of these early immortal variants show progressive changes with time and characteristic losses of chromosomes 1, 4, 7, 9 and 14 accompanying acquisition of tumorigenicity (COWELL 1981; COWELL and WIGLEY 1982). The consistent relationship between tumorigenicity and the loss of specific chromosomes or segments of chromosomes is particularly interesting since it suggests that recessive genetic lesions may be involved at this stage in transformation.

II. Bladder

1. Phenotypic Stages

Transitional epithelium derived from mouse and rat bladder has been used for in vitro transformation studies, and in both cases a multistage process has been described (HASHIMOTO and KITAGAWA 1974; SUMMERHAYES and FRANKS 1979; KNOWLES et al. 1986a; KNOWLES and JANI 1986). Much information is available on the induction of experimental tumours in rat and mouse bladder in vivo (HICKS and CHOWANIEC 1978), and results indicate that multiple steps are involved (HICKS 1980). We have carried out a number of comparative studies on organ-cultured bladder in vitro using direct-acting carcinogens and promoting agents and have shown that several changes resembling those seen in vivo can be induced (KNOWLES et al. 1985, 1986b). However, neither in vivo nor in vitro in organ culture has it so far been possible to identify in the organised tissue multiple phenotypic changes which might correspond to the predicted "stages" in transformation.

In order to approach the question of what these stages may represent in molecular terms, we are currently using an in vitro rat urothelial transformation model system based on mixed primary monolayer cultures similar to those described for mouse bladder by SUMMERHAYES and FRANKS (1979; KNOWLES et al. 1986a). In these, as in other epithelial systems, several distinct stages can be identified and studied independently.

Cultures are prepared from normal adult rat bladders by chopping to give a suspension of explants (0.5–1 mm^3) which, when plated, attach to the surface of the dishes within 4–7 days and produce outgrowths consisting predominantly of epithelial cells (Fig. 1). The proliferative index is high initially but falls to a low level by 3 weeks. The outgrowths then degenerate slowly from 30 to 40 days onwards, and most have been lost by 70–80 days.

The nitrosamide *N*-methyl-*N*-nitrosourea (MNU) has been used for most studies since it is a direct-acting carcinogen and much is known about its effects in the bladder in vivo. MNU treatment induces proliferating epithelial foci which can be identified against the background of quiescent and degenerating cells from approximately 45 days (Fig. 2). Estimations of focus incidence based on the number of surviving explants at the time of scoring show a clear dose-related response. Similar foci do arise in untreated cultures but at low frequency (KNOWLES and JANI 1986).

These foci are slow-growing, non-tumorigenic and anchorage dependent, but in contrast to the primary cells, they can be subcultured, with a much higher suc-

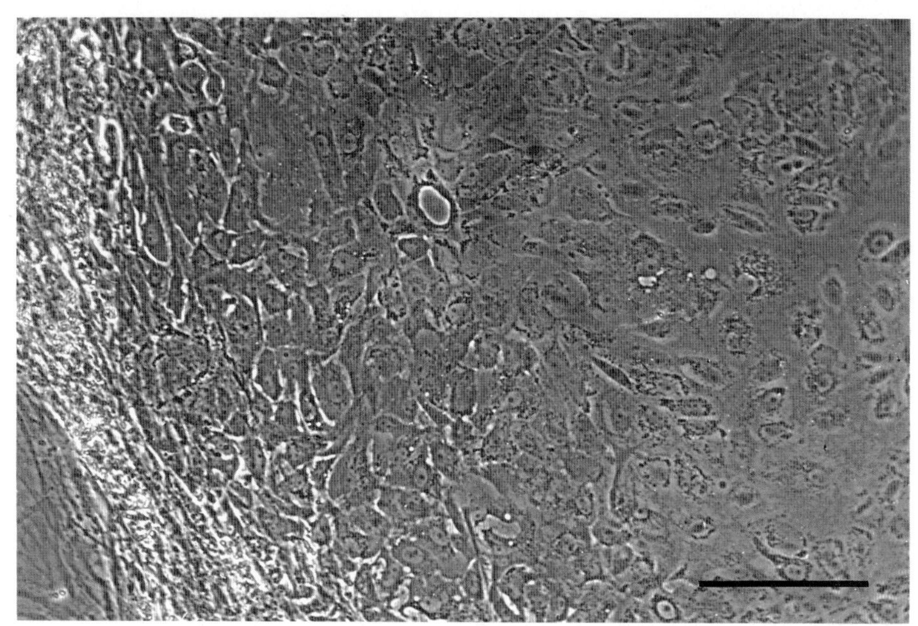

Fig. 1. Primary outgrowth from an untreated rat bladder explant 28 days in culture consisting predominantly of epithelial cells (*right*) with a rim of mesenchymal cells at the edge of the outgrowth (*left*). Phase contrast. *Bar,* 100 μm

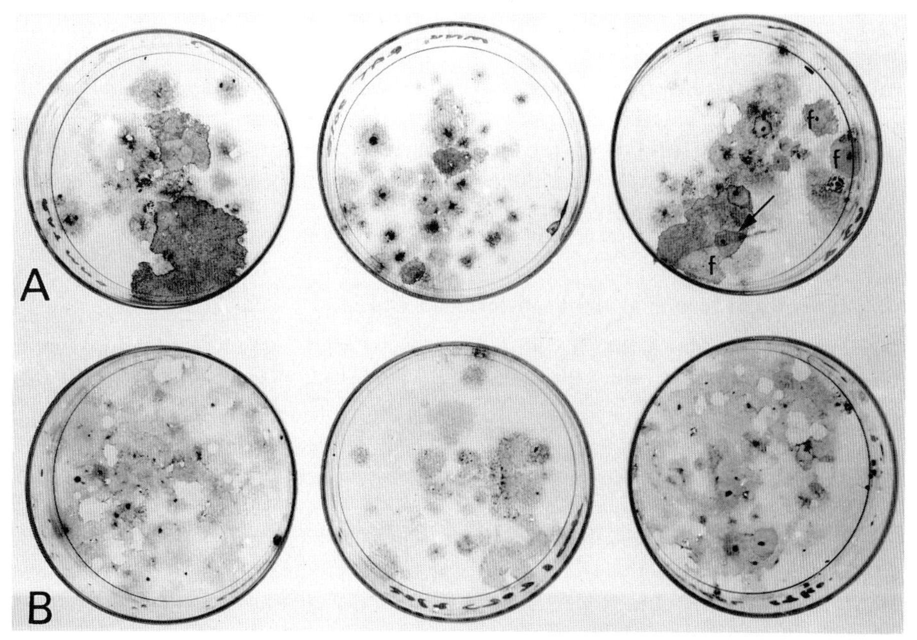

Fig. 2. Typical 65-day rat bladder primary cultures. **A** MNU-treated, **B** control. MNU-treated cultures contain several proliferating epithelical foci (*f*). In some of the foci, the limit of the original primary outgrowth is denoted by the presence of a dark rim of mesenchymal cells within the monolayer (*arrow*)

cess rate for MNU-induced foci (>70%) than for spontaneous foci. All foci can be defined therefore as having an extended lifespan compared with controls, though two types can be defined according to transferability, indicating, as in the SMG system, the possibility that more than one event or a different event is required for this phenotype. When transferred foci are maintained, those from control cultures almost invariably degenerate, usually by 120–150 days (Knowles and Jani 1986; Knowles, unpublished observations), and in a series of many experiments, only a single focus from a control culture has ever survived to give rise to an immortal cell line. In contrast, 40% (Knowles and Jani 1986) or more (L. Nicholson, personal communication) of MNU-induced, subculturable foci give rise to immortal cell lines. The requirement for an additional or different lesion for progression of the subculturable foci to immortalisation can therefore be predicted, so that immortalisation of the extended lifespan foci initially scored may require two further events.

2. Effect of Promoting Agents

Focus induction has been divided into two stages using an initiation-promotion schedule (Knowles and Jani 1986; Nicholson and Jani 1988). Following initiation with a sub-threshold dose of MNU, promotion with sodium saccharin gives rise to a significantly higher incidence of foci than either agent alone. The artificial sweetener sodium cyclamate, also a tumour promoter in the rat bladder in vivo, stimulates a very marked epithelial hyperplasia in urothelial primary cultures and a very high incidence of proliferating foci. It is not clear at present whether these are the result of spontaneous events with a finite probability related to cell proliferation or whether they are induced directly by the compound. Similar treatment with saccharin alone gives no significant increase in focus incidence above controls despite a transient early hyperplasia induced by this compound. In these cultures TPA induces significant numbers of foci in the absence of initiation with carcinogen.

3. Properties of Immortal and Transformed Cell Lines

A series of epithelial cell lines obtained from MNU− and MNU+ saccharin-treated cultures (designated RM1–RM6 and RMS1, RMS2E, respectively) and the single, control culture-derived line (RC1) have been assessed for anchorage independence and tumorigenicity. Anchorage independence is not a good marker for urothelial cell transformation. It should be noted, however, that tumour latent periods for all the tumorigenic cell lines (RM2 at high passage, RM3, RM4, RM6, RC1) were long (9 weeks at least), which may suggest that additional spontaneous events are required for tumorigenicity. Some of the cell lines (RM1, RM5, RMS1 and RMS2E) are stably immortal, and tumorigenic variants have not arisen even at high passage levels. No absolute in vitro correlates with tumorigenicity have been found. However, analysis of one cell line which transformed at high passage (RM2) for correlations of phenotype with anchorage independence and tumorigenicity shows that low passage, non-tumorigenic cells produce a TGF-β-like activity and that increased production of TGF-β and additional production of a TGF-α-like factor(s) is correlated with the development of anchorage independence (KNOWLES and EYDMANN, manuscript in preparation).

4. Role of Oncogenes and Suppressor Genes

As discussed in Sects. C.II and C.III a growing body of evidence suggests that the changes which occur during transformation involve at least two classes of gene; proto-oncogenes, which when activated appear to act as positive or dominant regulators of cell proliferation, and suppressor genes (≡ anti-oncogenes or emerogenes) which appear to exert a negative regulatory effect on aspects of the transformed phenotype, act in a recessive mode and therefore require the inactivation of both alleles during the transformation process (KNUDSON 1983, 1985; SAGER 1985, 1986).

In the rat bladder transformation system, we are attempting to classify stages in the transformation process according to the involvement of these two classes of gene. As discussed above, several "silent" steps may be involved in the generation of immortal cell lines, and at present it is not possible to classify foci except retrospectively according to their behaviour. Therefore, the studies have begun with immortal and transformed cell lines. These lines have been analysed for the presence of dominantly acting oncogenes using the NIH3T3 transfection assay. Using focus induction as an endpoint, all lines tested gave negative results. Some of the cell lines have also been analysed using NIH3T3 transfection followed by nude mouse assay (BOS et al. 1985), and no activated *ras* genes were identified using this technique. In contrast, a single fibroblast cell line established following MNU treatment of bladder primary cultures induced foci and was found to contain an activated K-*ras* with a G→A mutation at the second base in codon 12 (KNOWLES et al. 1987). This G→A transition is the most common MNU-induced point mutation and has been found at codon 12 of H-*ras* in 83% of rat mammary tumours induced in vivo by MNU (ZARBL et al. 1985). Using NIH3T3 transfection followed by nude mouse assay, DNA from a single immortal epithelial cell line gives rise to tumours, and DNA from these induces foci at high frequency in subsequent rounds of transfection. Hybridisations of DNA from these foci reveal no heterologous sequences related to the *ras* genes, *met, raf, sis, neu, ros,* and *mas*. Further studies of this putative activated oncogene are in progress. The absence of activated *ras* genes in the MNU-transformed urothelial cell lines may indicate that *ras* activation is not necessary for transformation of this cell type, that when induced as the initial step in transformation it confers no selective advantage on the cells or may even confer a selective disadvantage.

Cloned *ras* genes have now been introduced into immortal urothelial cell lines with some interesting results. When mutant H- and N-*ras* oncogenes were introduced by transfection into two immortal urothelial cell lines, the lines responded very differently to the transfected genes (KNOWLES et al., submitted). One of the lines (RMS1) was transformed apparently in a single step by H- and N-*ras* oncogenes to anchorage-independent growth and tumorigenicity. The other (RM1), after transfection with the same constructs, showed a distinct growth disadvantage, giving rise to G418-resistant colonies with significantly reduced diameter compared with controls. When these colonies were expanded, several episodes of "crisis" occurred before stable cell lines could be re-established. At this stage, however, the cells were tumorigenic. The initial poor growth of the *ras*-transfectants suggests that another event(s) is required following transfection for stable proliferation to occur, and we have obtained some further evidence for this. Early "unstable" G418-resistant colonies were pooled and injected into nude mice, and these have shown significantly longer tumour latent periods than stable cell lines.

Of particular interest is the transformation to anchorage independence of RMS1 by H- and N-*ras* proto-oncogenes. In contrast to rodent mesenchymal cells (e.g. NIH3T3), which require very high levels of expression of normal p21 from a large number of copies of the gene (50+) for transformation, the gene copy number in RMS1 transformants is relatively low. None of the stable RM1 or RMS1 mutant *ras* or proto-oncogene transfectants express high levels of p21. Indeed the human protein product is virtually undetectable in these cells by immunoprecipitation. All transfectants have a low copy number of the genes, and we have been unable to isolate transfectants with a high gene copy number since with increasing amounts of transfected plasmid, the number of colonies obtained declines, suggesting that these epithelial cells can only tolerate very small changes in p21 expression. These results indicate that in some epithelial cells very small changes in expression of *ras* proto-oncogenes may have a marked effect on the phenotype of the cell, and mutation of the gene may not be required for transformation. The difference in transformation of these two immortal lines by *ras* provides the basis for a subdivision of immortal lines into two classes, and other immortal lines are now being compared.

To examine the role of suppressor gene inactivation in rat urothelial transformation somatic cell hybridisation is being used. The ability of various immortal and transformed cell lines to suppress anchorage independence of fully transformed cell lines and of RMS1-*ras* transfectants has been examined. Results to date indicate that fully transformed RM2 cells are suppressed for anchorage independence by some immortal epithelial lines, and also by some tumorigenic cell lines. The latter result shows that fully transformed urothelial cells established following the same treatment regime may have different genotypes. We have also shown that RM2, RMS1-Ha1 (a mutant H-*ras* transformant of RMS1) and RMS2F (the fully transformed bladder mesenchymal cell line containing an activated K-*ras* gene) can be assigned to different complementation groups for suppression of anchorage independence based on their ability to be suppressed by a panel of immortal rodent mesenchymal and epithelial cell lines (KNOWLES et al., manuscript in preparation). These studies will enable us to classify all transformants according to the types of recessive lesion they have sustained and should provide the basis for a rational approach to the selective cloning of a series of individual suppressor genes.

III. Trachea

1. Phenotypic Stages

Primary cultures of rat trachea have been used in an extensive series of studies by NETTESHEIM, MARCHOK and co-workers. Initial experiments carried out using organ cultures treated with MNNG in vivo and "planted" and re-planted in culture vessels to give rise to epithelial outgrowths identified hyperplastic foci late in culture (120–140 days). From many such foci cell lines were established, several of which produce squamous cell carcinomas upon inoculation into immunosuppressed rats (STEELE et al. 1977). The induction of foci in rat trachea has subsequently been studied in great detail.

Tracheas treated in vivo and then dissociated and cultured in vitro give rise to cells with enhanced growth capacity, designated expanding foci, under conditions in which normal tracheal epithelial cells do not proliferate (TERZAGHI and NETTESHEIM 1979; TERZAGHI et al. 1982). This assay was subsequently used to analyse in vitro-treated tracheal cells (PAI et al. 1983; GRAY et al. 1983; THOMASSEN et al. 1983; STEELE et al. 1984). The assay (GRAY et al. 1983; THOMASSEN et al. 1983; NETTESHEIM and BARRETT 1985) is based on the finding that normal RTE cells do not divide more than a few times in standard tissue culture media without the addition of feeder cells or conditioned medium and a collagen-coated substratum. Suspensions of normal RTE cells plated at low density onto irradiated 3T3 feeder layers are exposed to carcinogen, and subsequently (4–7 days later) the 3T3 feeders are removed. Under these conditions, normal RTE cells stop proliferating and die. However, carcinogen-altered EG (extended growth) variants proliferate and can be scored 4–5 weeks after carcinogen exposure. At this stage, although the cells can often be subcultured, they are anchorage dependent and non-tumorigenic. It has been found more recently that removal of pyruvate from an enriched Waymouth's medium brings about death of normal cells but allows EG variants to prolifereate (MARCHOK et al. 1984).

Thus, the appearance of EG variants represents the first recognisable stage in RTE transformation. The frequency of EG variant induction is carcinogen dose-dependent and is estimated as $>2.6\%$ of colony-forming cells in the original population (THOMASSEN et al. 1983). A number of carcinogens have induced EG variants (NETTESHEIM and BARRETT 1984). For the induction of EG variants by MNNG it has been shown that a linear dose-response curve with a slope of 1 is obtained when log dose is plotted against log focus frequency, a result consistent with a one-hit mechanism for induction of this phenotype (Thomassen et al. 1983). This event can be compared, therefore, with the single-hit induction of morphologically altered foci in SHE cells by chemical carcinogens. Spontaneous EG variants are estimated to arise at a frequency of 0.02%–0.03%. A more recent estimate of the rate of spontaneous transformation based on studies in serum-free medium shows that it is a function of cell proliferation and is $7.5 \pm 4.1 \times 10^{-6}$ variants per cell generation (THOMASSEN 1986).

As in the SMG and bladder systems, tracheal EG variant cells seem to form a heterogeneous population. Only a proportion of EG variants are subculturable, and only a proportion of these become anchorage independent (TERZAGHI et al. 1982), suggesting that a series of events is involved in the establishment of anchorage-independent cell lines from foci. The rates of spontaneous generation and carcinogen induction of anchorage-independent variants have been quantitated recently (THOMASSEN et al. 1985b). The spontaneous rate was calculated as 0.5×10^{-4}–5.4×10^{-4} variants/cell per generation and induction by MNNG occurred with a frequency of approximately 10^{-3} variants/surviving cell. These rates and frequencies are similar to those for mutations at some known gene loci (e.g. STEGLICH and DE MARS 1982) and similar to those reported for some mesenchymal cells (e.g. BOUCK and DI MAYORCA 1976). Tumorigenicity follows anchorage-independence several passages later. DNA content estimations on freshly dissociated cells, early primary cultures and late primary cultures (day 40 and day 60) have shown an association between tetraploidy and the emergence of EG variants at around day 40. Increased aneuploidy and polyploidy occur with in vitro passage, and sub-populations in the sub-tetraploid and sub-octoploid ranges were detected several passages before neoplastic variants could be detected (VANDERLAAN et al. 1983).

2. Promotion and Inhibition

TPA treatment of rat tracheal cultures during the first few weeks following carcinogen treatment does not lead to an increased frequency of EG variants (NETTESHEIM et al. 1984) but does accelerate the appearance of anchorage-independent variants later in culture (STEELE et al. 1984). TPA also has a direct effect in inducing anchorage-independent variants from immortal anchorage-dependent cell lines but at low frequency, most immortal lines showing toxicity (NETTESHEIM et al. 1985), in contrast to the enhanced plating efficiency of RTE primary cells in the presence of TPA (MASS et al. 1984a).

In addition to experiments with promoting agents, the effects of the carcinogenesis "inhibitor" retinoic acid has been studied. A concentration-dependent inhibition of the induction of EG variants following MNNG treatment was observed (MASS et al. 1984b), and it has been shown that transformants become progressively more resistant to the growth inhibitory effects of the retinoid (FITZGERALD et al. 1986).

3. Role of Oncogenes

Recently, results of a study of the expression of cellular oncogenes in transformed tracheal cell lines was published (WALKER et al. 1987). Expression of 11 cellular oncogenes previously implicated in pulmonary or epithelial carcinogenesis was examined in five cell lines and normal RTE cells. Expression of N-*myc, abl, fes, erb*B, and *myb* was not detected in the transformed cells, *myc, fos, raf,* and K-*ras* were expressed at similar levels in normal and transformed cells, and H-*ras* was slightly but significantly over-expressed in the transformed cells. The oncogene *fms* was expressed at 5–19 times the normal level in three cell lines. No gene amplification or rearrangement was detected. Northern analyses showed an apparent 9.5-kb *fms*-related transcript in these cells compared with the normal 4-kb rat *fms* transcript. The precise role of this *fms*-related gene in rat tracheal transformation awaits further investigation[1].

[1] It has now been reported that these "*fms*-related" transcripts represent mouse retroviral sequences detected by viral pol sequences contained in the *fms* probe (WALKER et al. 1989. Cancer Res 49:625–628).

IV. Epidermis

Studies in mouse skin in vivo provided the first evidence that carcinogenesis could be divided into at least two stages, which were termed initiation and promotion. Many important concepts in carcinogenesis first proposed in this system have now been applied to other systems both in vivo and in vitro. A great deal is now known about epidermal carcinogenesis in vivo in the mouse, and the development of epidermal culture systems provides a unique opportunity to study carcinogenesis in parallel in vivo and in vitro. The comparisons made have in some instances shown reproducibility and predictability from in vivo to in vitro and vice versa, but some important differences are also apparent.

1. Induction of Foci Resistant to Calcium-Induced Differentiation

The first report of transformation of mouse epidermal cells in vitro was by FUSENIG et al. (1973), and since then a number of groups have used cultured epidermal cells as a model system for carcinogenesis studies. Most work has been done on newborn BALB/c mouse epidermis using a culture method based on that developed by YUSPA and co-workers (YUSPA and HARRIS 1974; YUSPA et al. 1976a, b, 1980a). Early reports showed that carcinogen treatment in vitro induces proliferating epithelial foci which can be recognised after 10–14 weeks in much the same way as described above for SMG, bladder and trachea (COLBURN et al. 1978; SLAGA et al. 1978). Many of these foci give rise to cell lines, some of which progress to tumorigenicity. Similar results have been reported with rat (INDO and MIYAJI 1979) and hamster epidermis (SUN et al. 1981; SINA et al. 1982).

More recently, this culture method has been modified using a medium containing reduced calcium levels (0.02–0.09 m*M*) to allow selective culture of cells with characteristics of epidermal basal cells (HENNINGS et al. 1980). Under these conditions, the cells do not stratify, have widened intercellular spaces and morphological, cell kinetic and protein markers characteristic of basal cells (YUSPA et al. 1981b) and can be grown at clonal density (YUSPA et al. 1981a). At higher calcium concentrations (1.2–1.4 m*M*) the cells stop dividing and terminally differentiate (HENNINGS et al. 1980, 1981). In contrast to normal cells, cultured malignant epidermal cells continue to proliferate when switched to high calcium medium (YUSPA et al. 1980b; KULESZ-MARTIN et al. 1980), and this has provided the key to exploring in detail the differences between normal and carcinogen-altered epidermal cells. It was shown that cultures from skin initiated by carcinogen in vivo yield cells resistant to calcium-induced inhibition of proliferation (YUSPA and MORGAN 1981). Similarly, cultures treated in vitro with chemical carcinogens give rise to foci which are able to proliferate in high calcium medium (KULESZ-MARTIN et al. 1980). The number of foci induced is proportional to carcinogen dose (KILKENNY et al. 1985), and these foci can give rise to cell lines, some of which become tumorigenic.

A number of results have been interpreted as showing a strong correlation between calcium-resistant focus formation in vitro and initiation in vivo (KILKENNY et al. 1985; KAWAMURA et al. 1985). These have been discussed in detail by YUSPA (1985) and include the finding of calcium-resistant foci both in initiated skin in vivo and in carcinogen-treated cultures in vitro, the persistence of focus-forming cells in vivo for many weeks after initiation, the existence of focus-forming cells in the epidermis of SENCAR mice which are sensitive to papilloma development by promotion alone, the correlation of focus number with strength and dose of initiating carcinogen in vivo and the non-tumorigenicity of the foci. The first detectable phenotypic change during epidermal transformation in vivo and in vitro is, therefore, and alteration in cell response to an external differentiation signal. Whether this is the key event in initiation, whether these altered cells are progenitors for papillomas and the precise molecular mechanism(s) involved remain to be elucidated. Indeed, as discussed below (Sect. IV.3) recent observations suggest that there are important molecular distinctions between calcium-resistant cells induced in vivo and in vitro.

Calcium-resistant foci induced in vitro are non-tumorigenic on first isolation, and cell strains isolated from such foci have been studied in some detail (e.g. YUSPA et al. 1980b; KULESZ-MARTIN et al. 1983). Both spontaneous and carcinogen-induced foci give rise to immortal cell lines, some of which become tumorigenic at later passages. No reliable correlates with tumorigenicity have been detected. The only predictable marker for preneoplastic and neoplastic cells from a variety of sources is the ability to grow in medium containing >0.1 mM Ca^{2+}. Growth in agar is a particularly poor marker in these keratinising cell lines (YUSPA et al. 1980b; KULESZ-MARTIN et al. 1983).

Although resistance to calcium-induced differentiation is closely correlated with the initiation event in vitro, it has proved possible to select cells which are immortalised but can still be "initiated" by carcinogens (KULESZ-MARTIN et al. 1985). Thus, aneuploid (sub-tetraploid) mouse epidermal cell lines have been established for use in an improved quantitative assay for carcinogen-induced alteration in differentiation. At present the properties of the Ca^{2+}-resistant foci have not been compared with those induced in normal epidermal cultures. It is descriptively difficult to use the term "initiation" to describe this event in previously immortalised cells. However, it is not difficult to accept the concept that the order in which different lesions are sustained during transformation may be different under certain conditions.

2. Effect of TPA

It has been proposed (YUSPA et al. 1981b, 1985) that in vivo papillomas arise from initiated cells by a process of selective clonal expansion. TPA induces pleiotropic responses in the epidermis involving both proliferative and differentiative events. YUSPA has proposed that basal cells induced to differentiate by TPA migrate into the upper layers soon after exposure. Thus, initiated cells with an altered differentiation program might proliferate and clonally expand in the basal layer to give rise ultimately to a benign tumour. When cultured basal cells are exposed to TPA they also respond in a heterogeneous manner (YUSPA et al. 1982), some differentiating and some remaining as proliferative foci which resist differentiation on subsequent exposure to TPA. Cell lines isolated from foci induced by carcinogens in vivo or in vitro show both resistance to TPA-induced and Ca^{2+}-induced differentiation (YUSPA et al. 1986).

3. Role of *ras* Oncogenes

Molecular analysis of changes associated with initiation and promotion has been facilitated by the ability to isolate cells in culture from initiated skin, papillomas and carcinomas. Great impetus was given to such studies by the findings of BALMAIN and colleagues (BALMAIN and PRAGNELL 1983; BALMAIN et al. 1984; BALMAIN 1985; BROWN et al. 1986; BALMAIN and BROWN 1988), who showed that more than 85% of DMBA-induced carcinomas and papillomas have an A:T→T:A transversion at the second base of codon 61 of c-H-*ras* (QUINTANILLA et al. 1986). It was also shown that mouse skin could be initiated by treatment with HaMSV and then promoted with TPA to develop papillomas. This latter result demonstrates that in vivo, a *ras* oncogene can substitute for treatment with an initiating carcinogen. Skin infected with HaMSV developed no tumours in the absence of promotion but when treated with TPA developed papillomas even if TPA treatment was delayed. Some of these benign tumours later progressed to carcinomas. If mutation in a *ras* gene is a critical lesion in initiation, as suggested by these results, then it might be expected that foci induced in vitro, which are phenotypically similar to cells initiated in vivo and considered by many to be biologically equivalent, would also carry this lesion. This does not appear to be the case. Cell lines derived from cells initiated in vitro have not given positive results in the NIH3T3 focus assay and show only a normal H-*ras* p21 protein in immunoprecipitation tests (A. BALMAIN, personal communication). Transfection of an activated human H-*ras* gene into one such line induces morphological changes and tumorigenicity, suggesting that these cells have sustained different genetic lesions from cells initiated in vivo. Similar results have also been reported in hamster epidermal lines (STORER et al. 1986).

Results obtained when cultured cells were infected with Kirsten or Harvey sarcomaviruses provide additional evidence that cells initiated by chemicals in vitro are not equivalent to cells initiated (or virus-infected) in vivo (YUSPA et al. 1983). Following virus infection cell proliferation was induced, but the cells were only partially blocked in their programme of terminal differentiation. These changes in the differentiation pattern of the virus-infected cells could be overcome by treatment with TPA exactly as might be expected of in vivo-initiated cells.

These results point to two different classes of initiating lesion in vivo and in vitro despite some apparent similarities in phenotype of the two cell populations. In vitro selection pressures may dictate which cells will survive, and these may differ from initiated cells in vivo (BALMAIN and BROWN 1988). For example, by analogy with experiments on rodent mesenchymal cells in vitro (Sect. C.III.3) a *ras* mutation alone would not be expected to immortalise, and cells containing only this lesion might be lost from the population. On the contrary, cells immortalised or with extended lifespan and/or with defects in terminal differentiation would be most likely to survive.

There is indeed evidence that culture conditions may determine which cell populations are selected. Several cell lines derived from pooled populations of DMBA+TPA-induced papillomas (YUSPA et al. 1986) do not contain activated c-H-*ras* genes (HARPER et al. 1986) despite the fact that most of the papillomas from which they were derived presumably contained such activations. This suggests that either the *ras* oncogene was lost during culture or that the cultured cells represent a different population from the papilloma cells. In many mouse strains, at least two classes of skin papillomas develop following initiation-promotion treatment (BURNS et al. 1976). Promoter-dependent tumours require repeated exposure to promoter for both expression and maintenance and regress on removal of the promoting stimulus. Autonomous papillomas, once established, require no further treatment with TPA and are more likely to progress to carcinomas. It is possible that the differences between carcinogen-induced foci in vitro (or cells isolated from papillomas) and virus-infected keratinocytes, are consistent with each representing a different class of initiated cell. In this case the virus-infected cells may represent the TPA-dependent class and the carcinogen-altered foci, the TPA-independent class. It can be predicted that the establishment of cell lines from papillomas in high calcium medium in the absence of TPA would select for promoter-independent cells which contain no activated *ras* gene (HARPER et al. 1986). Interestingly, it has been shown that by using different culture conditions cell lines can be established from papillomas which do express the mutated form of the H-*ras* protein (PERA and GORMAN 1984; A. BALMAIN, personal communication).

These results illustrate the need for extreme caution when comparing events during in vivo and in vitro carcinogenesis in the same tissue. It seems likely that unless culture conditions are carefully controlled, selection for immortalised cells is inevitable, and this may not represent the first lesion commonly seen in vivo. However, the differences observed may be in the particular sequence of events and not in the nature of the molecular events themselves. A more detailed comparison of stages in epidermal carcinogenesis in vivo and in vitro is given by BALMAIN and BROWN (1988).

V. Discussion

I have deliberately discussed several epithelial model systems at some length. During the past 10 years, great advances have been made in the development of these systems for studies of the multiple stages of carcinogenesis, and they offer several advantages over the mesenchymal systems described to date. Perhaps the most obvious of these is the clear definition of some of these so-called "stages". This allows comparison of distinct populations of cells predicted to differ from one another in only one or two molecular events.

In three of the systems described (SMG, bladder and trachea) the transformation process appears very similar (Fig. 3). The earliest phenotypic event in all is the appearance several weeks after carcinogen treatment of foci which show extended lifespan but none of the usual in vitro markers of transformation (e.g. anchorage independence) and are not tumorigenic. In all these systems four stages in transformation to tumorigenicity have been defined. This almost certainly represents a minimum. Indeed, it is likely that the long latent periods between certain of the observed phenotypes indicate the existence of additional phenotypically "silent" events.

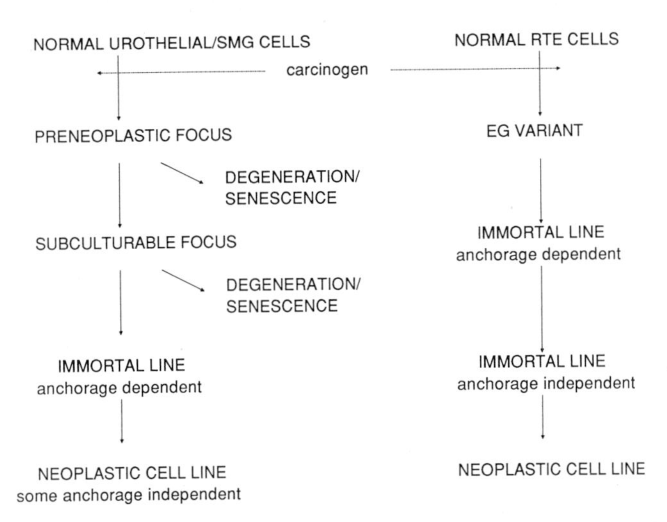

Fig. 3. Phenotypic stages recognised during in vitro transformation of urothelial, submandibular gland (*SMG*) and respiratory tract epithelial cells. *EG*, extended growth

Interestingly, in bladder and SMG at least two altered phenotypes can be identified before immortalisation occurs. As discussed in Sect. IV.3, the culture environment may influence the type and/or order of molecular lesions for which selection occurs. It is possible that in these primary cultures in which little epithelial proliferation occurs early in culture and subculture of the cells is not attempted, there is little selection pressure for immortalisation as an early event.

The nature of the molecular events underlying the observed phenotypic changes is only partially understood, but progress in this area promises to be rapid. In the tracheal system, the first and third events have been defined (Fig. 3) are likely to be single mutational events (based on dose-response profiles). It is possible that dominantly acting genes may be the targets, with point mutations or rearrangements activating the genes. Such dominantly acting genes have not been identified in these systems to date. The development of alternative assays to the NIH3T3 assay may be required to facilitate this, particularly in the bladder and epidermal culture systems in which *ras* genes do not seem to be involved.

The karyotypic changes described in SMG and trachea, viz. the appearance of polyploidy followed by specific chromosome losses, point to the involvement of recessive lesions at certain stages in the transformation process. There is evidence for similar events during epidermal transformation in vivo (BALMAIN and BROWN 1988), although no data for in vitro epidermal transformation have appeared. In the bladder, cell fusion experiments indicate that recessive lesions play an important role and that distinct genes are involved in different clonal populations with the same phenotype. A possible interpretation of these findings is that a multitude of combinations of genes collaborate together to generate each phenotype recognised. Whether this is the case will almost certainly be known within the next year or two, and some of the genes involved will have been cloned. Then it will be a short step to the assessment of the role played by these genes in human neoplasia.

E. Transformation of Human Cells

I. Differences Between Human and Rodent Cells

One of the major problems in experimental cancer research is the extrapolation of data obtained from animals to humans. Although animal models are required for studies of in vivo carcinogenesis, cultured cells provide a unique opportunity to study the process of human cell transformation and by comparison with rodent systems in vivo and in vitro should provide the basis for extrapolation to humans.

Despite the obvious advantages of using cultured human cells, progress has been much slower than with rodent cells. In vitro transformation of normal human cells has proved much more difficult than transformation of rodent cells (KAKUNAGA et al. 1983; DIPAOLO 1983). The reasons for this are not yet entirely clear, but several possibilities have been proposed.

As discussed above (Sect. C), normal human fibroblasts have a limited proliferative capacity in vitro and senesce after a mean of approximately 50 population doublings (HAYFLICK and MOOREHEAD 1961). Similarly, human epithelial cells have a finite lifespan, e.g. 20–50 population doublings for dermal keratinocytes (RHEINWALD and GREEN 1975). Spontaneous transformants never arise, and the karyotype remains diploid throughout the culture lifespan. Since the appearance of aneuploidy is clearly correlated with immortalisation in rodent cells (OSHIMURA and BARRETT 1986), the relative genetic stability of human cells may be of key significance. Indeed, chromosomal instability associated with the introduction of a cloned oncogene is thought to have contributed to transformation of certain cells (HARRIS 1987). This genetic stability may be related partly to extremely efficient DNA repair by human cells, though differences in transformation frequency between DNA repair-deficient human cells such as xeroderma pigmentosum cells and normal human and rodent cells are not great enough for this to represent a significant factor (MAHER et al. 1982).

The survival of a long-lived species such as man requires not only efficient repair mechanisms but also elaborate growth control mechanisms. If stringency of growth control is related to lifespan it must be expected that more steps or events are required to liberate human cells from these restraints. Mathematical models of the age distribution of cancer in humans support this hypothesis and predict upwards of five "steps" in the induction of most human tumours (COOK et al. 1969).

A relative inability to culture human cells compared with rodent cells has been proposed as a contributary factor. Certainly, culture conditions can affect cell phenotype, e.g. anchorage independence (MCCORMICK et al. 1985; SILINSKAS et al. 1981; PEEHL and STANBRIDGE 1981; KATOH et al. 1982), differentiation (MASUI et al. 1986) and in vitro lifespan (PACKER and FUEHR 1977; TODARO and GREEN 1964; GROVE and CRISTOFALO 1977), and such factors could inhibit the emergence of phenotypically altered human cells. Many human cells including epithelia (HARRIS 1987) can be cultured for long periods, and in general there is no difficulty in identifying phenotypic differences between normal and tumour-derived human cells. However, if as we suppose human cells are subject to a more complex hierarchy of growth controls than rodent cells, it is possible that rare variants may be more susceptible to phenotypic suppression by their normal neighbours (SAGER 1986). Such suppression could be caused for example by diffusible inhibitors or via intercellular communication. If this is the case, different assay systems may be necessary to detect transformants, particularly preneoplastic variants. In the final instance, transformed cells are tested using the available animal in vivo test systems such as athymic nude mice. Though useful in some instances, these do not appear to represent ideal hosts for many human cancers removed at surgery since the "take rate" is low.

II. Transformation of Human Mesenchymal Cells

Despite the difficulty of transforming human cells in vitro, some success has been achieved with both mesenchymal and epithelial cells. Early experiments succeeded in "transforming" phenotypically normal human fibroblasts cultured from tumours or from patients with clinical syndromes which predispose to cancer. IGEL and colleagues (1975; BENEDICT et al. 1975), in an extensive survey using a battery of carcinogens and 20 cell strains from normal patients and 50 from patients with tumours or genetic defects, induced transformation in only two, both from patients with von Recklinghausen's disease, a familial form of neoplasia (CROWE et al. 1956). Similarly, human osteosarcoma-derived cells (HOS; RHIM et al. 1975) were transformed at low frequency to tumorigenicity in nude mice by *N*-methyl-*N'*-nitro-*N*-nitrosoguanidine (MNNG) (RHIM et al. 1975, 1977), and cells from patients with xeroderma pigmentosum were transformed by UV irradiation (MAHER et al. 1982). KAKUNAGA (1978) reported the first transformation of normal adult human skin fibroblasts by the carcinogens 4-nitroquinoline 1-oxide and MNNG. Transformed cells which showed anchorage independence and tumorigenicity were isolated at a calculated frequency of $1–3.3 \times 10^{-7}$ after numerous subcultures. This frequency is several orders of magnitude lower than for example that obtained with Syrian hamster fibroblasts (BARRETT and Ts'O 1978). Other workers have attempted to transform human mesenchymal cells with a wide range of chemical carcinogens and/or irradiation, often using complex multiple treatment regimes. Few attempts have succeeded in inducing tumorigenicity. Other less stringent criteria for transformation have therefore been applied, and there is a wealth of reports of the induction of anchorage independence and/or morphological transformation by a wide range of agents. Some representative examples are shown in Table 3.

When these results are compared with similar experiments with rodent fibroblasts, some differences are immediately apparent. The most obvious is the induction of anchorage-independent growth in human cells apparently as an early step in transformation. In rodent cells, this phenotype is observed generally as a late-stage event (BARRETT and Ts'O 1978; FREEDMAN and SHIN 1974; JONES et al. 1976) and shows good correlation with tumorigenicity. In human cells it is an early event and can be induced by a wide range of chemical and physical agents and at high frequency (SILINSKAS et al. 1981) in the absence of tumorigenicity. These authors showed that anchorage-independent variants are induced with single-hit kinetics and with a similar dose-response curve to the induction of thioguanine resistance, a well-studied mutational marker, suggesting that

Table 3. In vitro transformation of human mesenchymal cells

Cell origin	Agent	Morphological alterations	Extended in vitro lifespan	Immortalisation	Anchorage independence	Tumorigenicity in nude mice	Karyotype	References
Lip fibroblasts	4NQO, MNNG	+	+	+	+	+	A	Kakunaga (1978)
Embryo fibroblasts	^{60}Co	+	+	+	+	–	A	Namba et al. (1985)
Foreskin fibroblasts	N-Ac-AAF	+	+	–	+	+	D/PD	Zimmerman and Little (1983)
	PS, β-PL, AFBl	+	+	NR	+	+	D	Milo and DiPaolo (1978)
	MNNG, 4NQO, EMS							Milo et al. (1981)
	UV	+	+	NR	+	+	NR	Maher et al. (1982)
	PS	+	+	NR	+	+	NR	Silinskas et al. (1981)
	SV40	+	+	+	+	–	A	Girardi et al. (1965)
Embryonic skin and muscle	UV	+	+	–	+	–	NR	Sutherland et al. (1980)
Skin mixed epithelial and fibroblasts cultures	MCA	–	–	–	+	–	A	Freeman et al. (1977)
Mesothelium	Asbestos	+	+	–	–	–	A	Lechner et al. (1985)
Endometrium	MNNG	+	+	+	+	–	NR	Dorman et al. (1983)
B lymphocytes	EBV		+	+	–	–	D	Henderson et al. (1977); Miller et al (1971)
T lymphocytes	EBV		+	+	–	–	D	Stevenson et al. (1986)

DMNA, dimethylnitrosamine; MNU, N-methyl-N-nitrosourea; AFBl, aflatoxin Bl; MNNG, N-methyl-N′-nitro-N-nitrosoguanidine; β-PL, β-propiolactone; UV, ultraviolet light; 4NQO, 4-nitroquinoline-l-oxide; BP, benzo[a]pyrene; N-Ac-AAF, N-acetoxy-2-(acetylamino) fluorene; PS, propane sultone; EMS, ethylmethanesulphonate; MCA, 3-methylcholanthrene; EBV, Epstein-Barr virus; NR, not recorded; A, aneuploid; D, diploid; PD, pseudodiploid; Ad 12, adenovirus 12; KiSV, Kirsten Sarcoma virus; CES, chick embryonic skin.

anchorage-independent colonies arise as the result of point mutation in a single gene. The observed phenotype probably reflects an altered response to growth factors since normal human fibroblasts can be induced to form colonies in agar in the presence of increased amounts of serum or certain growth factors (PEEHL and STANBRIDGE 1981). The molecular basis for this response is not yet known.

Anchorage independence has also been reported at high frequency in foreskin fibroblasts following transfection with human tumour cell DNA (SUTHERLAND and BENNETT 1984; SUTHERLAND et al. 1985). Such transfectants have an extended but not indefinite lifespan and are not tumorigenic. The nature of the transforming sequences is not yet defined.

There are several reports of the effects of single transfected oncogenes in human fibroblasts. None are transforming. For example, several groups have reported that following transfection with *ras* oncogenes human fibroblasts fail to show morphological alteration, focus formation or tumorigenicity (SAGER et al. 1983; SUTHERLAND et al. 1985), though a high frequency of anchorage-independent cells is induced (SUTHERLAND et al. 1985). It has been suggested that the lack of focus induction by *ras* genes may be the result of insufficient expression of the oncogene protein product in the transfectants, and a recent study has reported that as in rodent fibroblasts, morphological alteration and focus induction can be induced in the presence of high levels of p21 protein (HURLIN et al. 1987). These focus-forming cells were non-tumorigenic in nude mice and eventually senesced. Transfection of skin fibroblasts with v-*sis* also results in focus production, increased saturation density and decreased serum dependence with no increase in lifespan or tumorigenicity (FRY et al. 1986). The ability of these transfectants to grow in agar was not reported.

III. Transformation of Human Epithelial Cells

With the exception of SV40-induced transformation (see below), there are very few reports describing transformation of human epithelial cells. A number of examples are listed in Table 4. Historically, the inability to maintain epithelial cells in culture prevented major efforts to transform them in vitro. Recent improvements in epithelial culture technique have made transformation studies technically possible, but it appears that as with human mesenchymal cells, human epithelial cells do not transform readily in response to treatments which transform their rodent counterparts.

Some phenotypic changes have been reported in response to chemical carcinogens. In many cases, hyperplasia or epithelial atypia was observed in the treated tissues, most of which were treated in organ culture (CROCKER et al. 1973; LASNITZKI 1968; SHABAD et al. 1975; HAUGEN et al. 1982; EL-GERZAWI et al. 1982). In benzo[*a*]pyrene-treated, mammary, epithelial cell cultures morphological alterations and extended lifespan were induced, and two immortal lines were established. The immortal cells, which did not grow in agar, were confirmed as epithelial by keratin staining (STAMPFER and BARTLEY 1985). Anchorage-independent growth of fetal tracheal cells in response to diethylnitrosamine (DEN) and of foreskin epithelial cells in response to aflatoxin B1, MNNG, propane sultone, β-propiolactone and UV light have been reported (EMURA et al.

Table 4. In vitro transformation of human epithelial cells

Cell origin	Agent	Changes induced	Karyotype	Tumorigenicity	References
Adult pancreas	DMNA, MNU	Organ cultures show ductal hyperplasia and carcinoma	NR	+	Parsa et al. (1981a, b)
Fetal pancreas	MNU	Organ cultures show ductal hyperplasia and carcinoma	D/A	+	Parsa et al. (1984)
Foreskin keratinocytes	AFBl, MNNG PS, β-PL, UV	Anchorage independence, invasion into CES, extended lifespan	NR	–	Milo et al. (1981)
Epidermal keratinocytes	SV40	Immortalisation, anchorage independence	NR	–	Steinberg and Defendi (1983)
	SV40 ori⁻	Immortalisation (no crisis)	NR	+	Parkinson et al. (1983) Brown and Parkinson (1984)
	Ad12-SV40 + KiSV	Immortalisation, morphological changes, anchorage independence	A	+	Rhim et al. (1985)
	Ad12-SV40 + 4NQO/ MNNG	Immortalisation, morphological changes, anchorage independence	A	+	Rhim et al. (1986)
Mammary	BP	Immortalisation, morphological changes, anchorage independence	A	–	Stampfer and Bartley (1985)
	SV40	Immortalisation, morphological changes, anchorage independence	A	–	Chang et al. (1982)
Prostate	SV40	Extended lifespan, morphological alterations, reduced serum dependence	A	NR	Kaighn et al. (1980)
Ciliary epithelium	SV40	Extended lifespan, morphological alterations, reduced serum dependence	NR	NR	Coca-Prados and Wax (1986)
Urothelium	SV40	Immortalisation, morphological changes, reduced growth factor and extracellular matrix requirements. Loss of dependence on feeder cells for clonal growth	A	–	Christian et al. (1987)
Retinoblasts	Ad12 DNA	Immortalisation (no crisis)	A	–	Byrd et al. (1982)
Embryo kidney	Ad12	Immortalisation	PD/A	+	Whittaker et al. (1984)
Bronchus	v-H-*ras*	Immortalisation, anchorage independence, reduced growth factor requirement	A	+	Yoakum et al. (1985)

[a] For abbreviations, see footnote to Table 3.

1985; MILO et al. 1981). Neither cell type was tumorigenic, but invasiveness into cultured chick embryo skin was observed in both cases.

The only report of induction of tumorigenicity by chemical carcinogen treatment in cultured human epithelium is for fetal and adult pancreas (PARSA et al. 1981 a, b, 1984). The tissues were treated in organ culture, and ductal hyperplasia and atypia followed by appearances resembling carcinoma were observed during culture periods of up to 5 months. Cells derived from adult pancreas 10 weeks after treatment with dimethylnitrosamine (DMNA) and MNU produced nodules in nude mice which ranged in morphology from undifferentiated scirrhous carcinoma to well-differentiated papillary adenocarcinoma (PARSA et al. 1981 a). Cells derived from 4–5-month treated fetal pancreas were also tumorigenic, producing adenocarcinomas in which keratins and duct cell surface markers could be identified (PARSA et al. 1984).

The number of reports of human epithelial transformation by viruses or subgenomic viral fragments is much higher than for chemical or physical agents, with SV40 virus providing the most examples. Extensive reviews of human cell transformation by SV40 have appeared (SACK 1981; CHANG 1986), and only a brief discussion will be given here. As can be seen from Table 4 a number of different types of epithelial cell have been transformed by SV40. Human cells are termed semi-permissive for this virus since infection leads in some cases to stable integration of the virus and transformation and in others to lytic infection causing death of the cells and release of progeny virus. In recent years, many workers have used origin-defective (ori-) mutants to prevent lytic infection (SMALL et al. 1982).

Transformation by SV40 demonstrates several distinctive features. Some kind of morphological alteration is usually noted several weeks after infection. Commonly, foci of altered cells appear which can be shown to express large T antigen. Cells isolated from such foci usually show a reduced serum requirement, loss of dependence on feeder cells and an extended in vitro lifespan. However, the cells are not immortal, and almost invariably these altered populations of cells demonstrate a phenomenon termed "crisis". This was first described in SV40-transformed fibroblasts (GIRARDI et al. 1965) and is characterised by a stage of markedly increased cell loss. It is not clear whether crisis is equivalent to senescence, but as originally described crisis seems to be distinguished by the continued cycling of cells at the same time as widespread detachment and loss. However, many descriptions of crisis in the literature appear to refer to a phenomenon more akin to senescence in fibroblasts. The period of crisis may last for several weeks and either results in loss of the entire population or leaves a few rare variant colonies, which are progenitors of immortal cell lines. There are some reports of establishment of SV40-immortalised human epithelial cells in the apparent absence of an episode of crisis. This appears to be more common when SV40 ori-mutants are used, suggesting that viral replication may be involved. Nevertheless, some ori-infected cells do show crisis (CHANG 1986, and personal communication). The observation of crisis in some SV40-infected cultures but not others seems to indicate that some other event is required for immortalisation and that this can happen early or late in culture. There is considerable controversy over what happens at crisis, and one viewpoint is that it is related to the limited in vitro lifespan of normal cells. It can be argued that SV40-infected cells enjoy an extended lifespan compared with their uninfected counterparts by virtue of the properties imparted by the viral products, e.g. reduced growth factor requirements, and that with improved culture conditions uninfected cells might attain the same number of population doublings. Crisis then may represent the end of the in vitro lifespan but with a continued stimulus to divide provided by viral products, and at this stage, the first selective advantage for immortal variants will exist. In support of this hypothesis is the finding that DNA synthesis can be reinitiated in senescent WI-38 cells by SV40 infection (GORMAN and CRISTOFALO 1985).

Post-crisis, immortal epithelial cell lines are generally anchorage independent, and in SV40-transformed keratinocytes, alterations in control of terminal differentiation and keratin and fibronectin expression are recorded (DEFENDI et al. 1982; EDELMAN et al. 1985; HRONIS et al. 1984; STEINBERG and DEFENDI 1979; BANKS-SCHLEGEL and HOWLEY 1983; TAYLOR-PAPADIMITRIOU et al. 1982), but tumorigenicity is rarely observed (BROWN and PARKINSON 1984).

Human adenoviruses, which can morphologically transform rodent cells in vitro have been used to transform human embryonic kidney epithelium and retinoblasts (GRAHAM et al. 1977; BYRD et al. 1982; WHITTAKER et al. 1984). Something resembling SV40-induced crisis was recorded in Ad 12-transformed human embryo kidney cells (WHITTAKER et al. 1984). Interestingly, attempts to transform human fibroblasts in the same way have failed (WHITTAKER et al. 1984; TODARO and AARONSON 1969).

It is clear from these results with SV40 and adenovirus transformation that additional cellular events are required for transformation by these viruses. A similar conclusion can be drawn from experiments in which human bronchial epithelial cells were transfected with a single viral oncogene, v-H-*ras* (YOAKUM et al. 1985). What has been described as a "cascade" of events (HARRIS 1987) occurred in the transfected cells, giving rise eventually to a stable, morphologically altered and highly tumorigenic cell line (YOAKUM et al. 1985). It has been shown that these *ras*-transfected cells are genetically less stable than controls, and it has been proposed that this facilitates transformation (HARRIS 1987).

A combination of viral oncogenes were used by RHIM and colleagues (RHIM et al. 1985) to induce transformation of epidermal keratinocytes. Primary cultures infected with an Ad12-SV40 hybrid virus (SCHELL et al. 1966) produced actively growing foci with unlimited lifespan, and infection of these with Kirsten sarcomavirus induced morphological alterations, anchorage independence and tumorigenicity. In a subsequent study (RHIM et al. 1986) it was shown that this second stage could be effected by treatment with chemical carcinogens (4NQO or MNNG) though this did not appear to be mediated by activation of the endogenous human c-K-*ras* gene.

In conclusion, human cells and tissues are now beginning to make a significant contribution to our understanding of malignancy. The induction of transformation in human cells has proved extremely difficult but is now achieved with some regularity. However, all the available data suggest that many steps are required to transform human cells. As more partially and fully transformed cells become available (e.g. cells from cancer-prone individuals, cells from preneoplastic lesions, cells immortalised or transformed in vitro or isolated from tumours) the role of individual cloned genes in human cell transformation can be investigated more precisely.

F. Conclusions

A myriad of approaches is being used to unravel the molecular complexity of the transformation process. Cultured cells and tissues are the subjects of these various strategies as often as in vivo tissues. My aim has been to illustrate the range of tissue culture systems available for studies of cell transformation and to discuss those results which give useful mechanistic information.

The development of in vitro systems in which different stages in the transformation process can be identified together with methods for gene transfer have led to rapid advances in the identification and understanding of the individual genes involved in transformation. There is no doubt that cultured cells transformed in vitro provide extremely useful model systems which show clear parallels with results obtained with fresh tissue or cultures derived directly from tumours. The conclusion previously drawn from in vivo studies, that carcinogenesis is a multistep process, is borne out by in vitro findings with both mesenchymal and epithelial cells, and many genes implicated in the genesis of tumours in vivo have been shown to contribute to the transformed phenotype of cells in vitro.

As the mechanisms of transformation become clearer, it is apparent that proliferation of normal cells is regulated by a complex, balanced network of positive and negative regulatory signals. Already, a number of oncogenes have been identified as growth factors, growth factor receptors or part of the signal transduction mechanism, and when altered or "activated" during the transformation process, these appear to exert a dominant effect in the cell. It is becoming clear that a distinct class of genes, commonly termed suppressor genes, also play a role in carcinogenesis. These genes are thought to act in the normal cell to suppress proliferation (? and/or induce differentiation) and when inactivated during carcinogenesis lead to unregulated growth (? immortalisation). The involvement of both classes of gene during in vitro transformation has now been demonstrated, and with the imminent isolation and characterisation of suppressor genes from cultured cells, the next few years promise some extremely exciting in vitro experiments.

The genetic damage contributing to neoplasia in the majority of human tumours is at present unidentified. There is no doubt that some of the lesions involved are of a subtlety which will defy current methodology. Nor is there any doubt that new technologies and strategies will be developed to identify and characterise these lesions and that these will rely in great part on the use of cell culture models for transformation.

Acknowledgements. I thank Drs G. Currie, J. Denner and S. Chang for many helpful discussions and critical reading of the manuscript and Drs A. Balmain and J.C. Barrett for communicating their results in advance of publication. I am very grateful to Mrs. J. Marr for careful and patient typing of the references and tables.

References

Aaronson SA, Todaro GJ (1968) Development of 3T3-like lines from Balb/c mouse embryo cultures: transformation susceptibility to SV40. J Cell Physiol 72:141–148
Ambesi-Impiombato FS, Parks LA, Coon HG (1980) Culture of hormone-dependent functional epithelial cells from rat thyroids. Proc Natl Acad Sci USA 77:3455–3459
Asano M, Iwakura Y, Kawade Y (1985) SV40 vector with early gene replacement efficient in transducing exogenous DNA into mammalian cells. Nucleic Acids Res 13:8573–8586
Balmain A (1985) Transforming *ras* oncogenes and multistage carcinogenesis. Br J Cancer 51:1–7
Balmain A, Brown K (1988) Oncogene activation in chemical carcinogenesis. Adv Cancer Res 51:147–182
Balmain A, Pragnell I (1983) Mouse skin carcinomas induced in vivo by chemical carcinogens have a transforming Harvey-*ras* oncogene. Nature 303:72–74
Balmain A, Ramsden M, Bowden GT, Smith J (1984) Activation of the mouse cellular Harvey-*ras* gene in chemically induced benign skin papillomas. Nature 307:658–660
Banks-Schlegel SP, Howley PM (1983) Differentiation of human epidermal cells transformed by SV40. J Cell Biol 196:330–337
Barrett JC (1985) Cell culture models of multistep carcinogenesis. In: Likhachev A, Anisimov V, Montesano R (eds) Age-related factors in carcinogenesis. IARC Sci Publ no 58 (International Agency for Research on Cancer) Lyon, p 181
Barrett JC, Ts'O POP (1978) Evidence for the progressive nature of neoplastic transformation in vitro. Proc Natl Acad Sci USA 75:3761–3765
Barrett JC, Crawford BD, Ts'O POP (1980) The role of somatic mutation in a multistage model of carcinogenesis. In: Mishra V, Dunkel VC, Mehlamn M (eds) Mammalian cell transformation by chemical carcinogens. Senate Press, Princeton, p 467
Barrett JC, Thomassen DG, Hesterberg TW (1983) Role of gene and chromosomal mutations in cell transformation. Ann NY Acad Sci 407:291–300
Barrett JC, Hesterberg TW, Thomassen DG (1984) Use of cell transformation systems for carcinogenicity testing and mechanistic studies of carcinogenesis. Pharmacol Rev 36:53S–70S
Barrett JC, Hesterberg TW, Oshimura M, Tsutsui T (1985) Role of chemically induced mutagenic events in neoplastic transformation of Syrian hamster embryo cells. In: Barrett JC, Tennant RW (eds) Carcinogenesis, vol 9. Raven, New York, p 123
Benedict WF, Jones PA, Laug WE, Igel HJ, Freeman AE (1975) Characterisation of human cells transformed in vitro by urethane. Nature 256:322–324
Berwald Y, Sachs L (1963) In vitro cell transformation with chemical carcinogens. Nature 200:1182–1184
Berwald Y, Sachs L (1965) In vitro transformation of normal cells to tumor cells by carcinogenic hydrocarbons. JNCI 35:641–661
Bond VC, Wold B (1987) Poly-L-ornithine-mediated transformation of mammalian cells. Mol Cell Biol 7:2286–2293
Boone CW (1975) Malignant hemangioendotheliomas produced by subcutaneous inoculation of Balb/3T3 cells attached to glass beads. Science 188:68–70
Boone CW, Takeichi N, Paranjpe M, Gilden R (1976) Vasoformative sarcomas arising from BALB/3T3 cells attached to solid substrates. Cancer Res 36:1626–1633
Bos JL, Toksoz D, Marshall CJ, Verlaan-de Vries M, Veeneman GH, van der Eb AJ, van Boom JH, Janssen JWG, Steenvoorden ACM (1985) Amino-acid substitution at codon 13 of the N-*ras* oncogene in human acute myeloid leukaemia. Nature 315:726–730
Bos JL, Fearon ER, Hamilton SR, Verlaan-de Vries M, van Boom JH, van der Eb AJ, Vogelstein B (1987) Prevalence of *ras* gene mutations in human colorectal cancers. Nature 327:293–297
Bouck N, di Mayorca G (1976) Somatic mutation as the basis for malignant transformation of BHK cells by chemical carcinogens. Nature 264:722–727
Bouck N, di Mayorca G (1982) Chemical carcinogens transform BHK cells by inducing a recessive mutation. Mol Cell Biol 2:97–105

Bouck N, Kokkinakis D, Ostrowsky J (1984) Induction of a step in carcinogenesis that is normally associated with mutagenesis by nonmutagenic concentrations of 5-azacytidine. Mol Cell Biol 4:1231–1237

Boutwell RK, Verma AK, Ashendel CL, Astrup E (1982) Mouse skin: a useful model system for studying the mechanism of chemical carcinogenesis. In: Hecker E et al. (eds) Carcinogenesis, vol 7. Raven, New York, p1

Brash DE, Reddel RR, Quanrud M, Yang K, Farrell MP, Harris CC (1987) Strontium phosphate transfection of human cells in primary culture: stable expression of the simian virus 40 large T-antigen gene in primary human bronchial epithelial cells. Mol Cell Biol 7:2031–2034

Brown KW, Parkinson EK (1984) Extracellular matrix components produced by SV40-transformed human epidermal keratinocytes. Int J Cancer 33:257–263

Brown K, Quintanilla M, Ramsden M, Kerr IB, Young S, Balmain A (1986) v-*ras* genes from Harvey and BALB murine sarcoma viruses can act as initiators of two-stage mouse skin carcinogenesis. Cell 46:447–456

Buick RN, Pollack MN (1984) Perspectives on clonogenic tumor cells, stem cells, and oncogenes. Cancer Res 44:4909–4918

Burns FJ, Vanderlaan M, Sivak A, Albert RE (1976) Regression kinetics of mouse skin papillomas. Cancer Res 36:1422–1426

Byrd P, Brown KW, Gallimore PH (1982) Malignant transformation of human embryo retinoblasts by cloned adenovirus 12 DNA. Nature 298:69–71

Capecchi MR (1980) High efficiency transformation by direct microinjection of DNA into cultured mammalian cells. Cell 22:479–488

Casto BC, DiPaolo JA (1975) In vitro transformation: interaction of chemicals, viruses and irradiation. Bibl Haematologica 40:197–199

Casto BC, Janosko N, DiPaolo JA (1977) Development of a focus assay model for transformation of hamster cells in vitro by chemical carcinogens. Cancer Res 37:3508–3515

Cepko CL, Roberts BE, Mulligan RC (1984) Construction and applications of a highly transmissible murine retrovirus shuttle vector. Cell 37:1053–1062

Chang SE (1986) In vitro transformation of human epithelial cells. Biochim Biophys Acta 823:161–194

Chang SE, Keen J, Lane EB, Taylor-Papadimitriou J (1982) Establishment and characterization of SV40-transformed human breast epithclial cell lines. Cancer Res 42:2040–2053

Chen C, Okayama H (1987) High efficiency transformation of mammalian cells by plasmid DNA. Mol Cell Biol 7:2745–2752

Chen TT, Heidelberger C (1969a) Cultivation in vitro of cells derived from adult C3H mouse ventral prostate. JNCI 42:903–914

Chen TT, Heidelberger C (1969b) In vitro malignant transformation of cells derived from mouse prostate in the presence of 3-methylcholanthrene. JNCI 42:915–925

Chen TT, Heidelberger C (1969c) Quantitative studies on the malignant transformation of mouse prostate cells by carcinogenic hydrocarbons in vitro. Int J Cancer 4:166–178

Christian BJ, Loretz LJ, Oberley TD, Reznikoff CA (1987) Characterization of human uroepithelial cells immortalized in vitro by simian virus 40. Cancer Res 47:6066–6073

Chu G, Hayakawa H, Berg P (1987) Electroporation for the efficient transfection of mammalian cells with DNA. Nucleic Acids Res 15:1311–1326

Clarke MF, Westin E, Schmidt D, Josephs SF, Ratner L, Wong-Staal F, Gallo RC, Reitz MS Jr (1984) Transformation of NIH3T3 cells by a human c-*sis* cDNA clone. Nature 308:464–467

Coca-Prados M, Wax MB (1986) Transformation of human ciliary epithelial cells by simian virus 40: induction of cell proliferation and retention of β_2-adrenergic receptors. Proc Natl Acad Sci USA 83:8754–8758

Colburn NH, Vorder Bruegge WF, Bates JR, Gray RH, Rossen JD, Kelsey WH, Shimada T (1978) Correlation of anchorage-independent growth with tumorigenicity of chemically transformed mouse epidermal cells. Cancer Res 38:624–634

Cook PJ, Doll R, Fellingham SA (1969) A mathematical model for the age distribution of cancer in man. Int J Cancer 4:93–112

Cooper CS, Park M, Blair DG, Tainsky MA, Huebner K, Croce CM, Vande Woude GF (1984) Molecular cloning of a new transforming gene from a chemically transformed human cell line. Nature 311:29–33
Cowell JK (1981) Chromosome abnormalities associated with salivary gland epithelial cell lines transformed in vitro and in vivo with evidence of a role for genetic imbalance in transformation. Cancer Res 41:1508–1517
Cowell JK, Wigley CB (1980) Changes in DNA content during in vitro transformation of mouse salivary gland epithelium. JNCI 64:1443–1449
Cowell JK, Wigley CB (1982) Chromosome changes associated with the progression of cell lines from preneoplastic to tumorigenic phenotype during transformation of mouse salivary gland epithelium in vitro. JNCI 69:425–433
Crawford B, Klein L, Melville M, Morry D, Ts'O POP (1980) Somatic genetics of in vitro neoplastic transformation. Proc Am Assoc Cancer Res 21:127
Crawford B, Barrett JC, Ts'O POP (1983) Neoplastic conversion of preneoplastic syrian hamster cells: rate estimation by fluctuation analysis. Mol Cell Biol 3:931–945
Crocker TT, O'Donnell TV, Nunes LL (1973) Toxicity of benzo(a)pyrene and air pollution composite for adult human bronchial mucosa in organ culture. Cancer Res 33:88–93
Crowe FW, Schull WJ, Neel JV (1956) A clinical pathology and genetic study of multiple neurofibromatosis. CC Thomas, Springfield
Defendi V, Naimski P, Steinberg ML (1982) Human cells transformed by SV40 revisited: the epithelial cells. J Cell Physiol [Suppl] 2:131–140
Der CJ, Krontiris TG, Cooper GM (1982) Transforming genes of human bladder and lung carcinoma cell lines are homologous to the *ras* genes of Harvey and Kirsten sarcoma viruses. Proc Natl Acad Sci 79:3637–3640
DiPaolo JA (1983) Relative difficulties in transforming human and animal cells in vitro. JNCI 70:3–8
DiPaolo JA, Donovan PJ, Nelson RL (1971) In vitro transformation of hamster cells by polycyclic hydrocarbons: factors influencing the number of cells transformed. Nature New Biol 230:240–242
Dorman BH, Siegfried JM, Kaufman DG (1983) Alterations of human endometrial stromal cells produced by *N*-methyl-*N*′-nitro-*N*-nitrosoguanidine. Cancer Res 43:3348–3357
Earle WR (1943) Production of malignancy in vitro. IV. The mouse fibroblast cultures and changes in the living cells. JNCI 4:165–212
Earle WR, Nettleship A (1943) Production of malignancy in vitro. V. Results of injection of cultures into mice. JNCI 4:213–227
Edelman B, Steinberg ML, Defendi V (1984) Changes in fibronectin synthesis and binding distribution in SV40-transformed human keratinocytes. Int J Cancer 35:219–225
Ege T, Reisbig RR, Rogne S (1984) Enhancement of DNA-mediated gene transfer by inhibitors of autophagic-lysosomal function. Exptl Cell Res 155:9–16
El-Gerzawi S, Heatfield BM, Trump BF (1982) *N*-methyl-*N*-nitrosourea and saccharin: effects on epithelium of normal human urinary bladder in vitro. JNCI 69:577–583
Eliyahu D, Raz A, Gruss P, Givol D, Oren M (1984) Participation of p53 cellular tumour antigen in transformation of normal embryonic cells. Nature 312:646–649
Emura M, Mohr U, Kakunaga T, Hilfrich J (1985) Growth inhibition and transformation of a human fetal tracheal epithelial cell line by long-term exposure to diethylnitrosamine. Carcinogenesis 6:1079–1085
Eva A, Aaronson SA (1985) Isolation of a new human oncogene from a diffuse B-cell lymphoma. Nature 316:273–275
Farber E (1984) The multistep nature of cancer development. Cancer Res 44:4217–4223
Farber E, Cameron R (1980) The sequential analysis of cancer development. Adv Cancer Res 35:125–226
Felgner PL, Gadek TR, Holm M, Roman R, Chan HW, Wenz M, Northtop JP, Ringold GM, Danielson M (1987) Lipofection: a highly efficient, lipid-mediated DNA-transfection procedure. Proc Natl Acad Sci USA 84:7413–7417

Fernandez A, Mondal S, Heidelberger C (1980) Probabilistic view of the transformation of cultured C3H/10T1/2 mouse embryo fibroblasts by 3-methylcholanthrene. Proc Natl Acad Sci USA 77:7272–7276

Fialkow PJ (1972) Use of genetic markers to study cellular origin and development of tumors in human females. Adv Cancer Res 15:191–226

Fitzgerald DJ, Barrett JC, Nettesheim P (1986) Changing responsiveness to all-trans retinoic acid of rat tracheal epithelial cells at different stages in neoplastic transformation. Carcinogenesis 7:1715–1721

Foulds L (1969) Neoplastic development, vol 1. Academic, London

Foulds L (1975) Neoplastic development, vol 2. Academic, London

Fraley R, Subramani S, Berg P, Papahadjopoulos D (1980) Introduction of liposome-encapsulated SV40 DNA into cells. J Biol Chem 255:10431–10435

Franks LM, Knowles MA (1978) The structure of tumours derived from mouse submandibular gland epithelium transformed in vitro. Br J Cancer 37:240–247

Franks LM, Wilson PD (1977) Origin and ultrastructure of cells in vitro. Int Rev Cytol 48:55–139

Franza BR, Maruyama K, Garrels JI, Ruley HE (1986) In vitro establishment is not a sufficient prerequisite for transformation by activated *ras* oncogenes. Cell 44:409–418

Freedman VH, Shin S (1974) Cellular tumorigenicity in nude mice: correlation with cell growth in semi-solid medium. Cell 3:355–359

Freeman AE, Igel HJ, Price PJ (1975) Carcinogenesis in vitro. In vitro transformation of rat embryo cells: correlations with the known tumorigenic activation of chemicals in rodents. In Vitro 11:107–116

Freeman AE, Lake RS, Igel HJ, Gernand L, Pezzutti MR, Malone JM, Mark C, Benedict WF (1977) Heteroploid conversion of human skin cells by methylcholanthrene. Proc Natl Acad Sci USA 74:2451–2455

Frost E, Williams J (1978) Mapping temperature-sensitive and host-range mutations of adenovirus type 5 by marker rescue. Virology 91:39–50

Fry DG, Milam LD, Maher VM, McCormick JJ (1986) Transformation of diploid human fibroblasts by DNA transfection with the v-*sis* oncogene. J Cell Physiol 128:313–321

Fusenig NE, Samsel W, Thon W, Worst PKM (1973) Malignant transformation of epidermal cells in culture by DMBA. INSERM 19:219–228

Gart JJ, Di Paolo JA, Donovan PJ (1979) Mathematical models and the statistical analyses of cell transformation experiments. Cancer Res 39:5069–5075

Gey GO (1941) Cytological and cultural observations on transplantable rat sarcomata produced by inoculation of altered normal cells maintained in continuous culture. Cancer Res 1:737

Girardi AJ, Jensen FC, Koprowski H (1965) SV40-induced transformation of human diploid cells: crisis and recovery. J Cell Physiol 65:69–83

Gopal TV (1985) Gene transfer method for transient gene expression. Stable transformation and cotransformation of suspension cell cultures. Mol Cell Biol 5:1188–1190

Gorman CM, Howard BH (1983) Expression of recombinant plasmids in mammalian cells is enhanced by sodium butyrate. Nucleic Acids Res 11:7631–7648

Gorman SD, Cristofalo VJ (1985) Reinitiation of cellular DNA synthesis in BrdU-selected, nondividing, senescent WI-38 cells by simian virus 40 infection. J Cell Physiol 125:122–126

Graham FL, van der Eb AJ (1973) A new technique for the assay of infectivity of human adenovirus 5 DNA. Virology 62:456–467

Graham FL, Smiley J, Russell WC, Nairn R (1977) Characteristics of a human cell line transformed by DNA from human adenovirus type 5. J Gen Virol 36:59–72

Graham FL, Bacchetti S, McKinnon R (1980) Transformation of mammalian cells with DNA using the calcium technique. In: Baserga R, Croce C, Rovera G (eds) Introduction of macromolecules into viable mammalian cells. Liss, New York, p 3

Gray TE, Thomassen DG, Mass MJ, Barrett JC (1983) Quantitation of cell proliferation, colony formation, and carcinogen-induced cytotoxicity of rat tracheal epithelial cells grown in culture on 3T3 feeder layers. In Vitro 19:559–570

Grove GL, Cristofalo VJ (1977) Characterization of the cell cycle of cultured human diploid cells: effects of aging and hydrocortisone. J Cell Physiol 90:415–422

Haber DA, Fox DA, Dynan WS, Thilly WG (1977) Cell density dependence of focus formation in the C3H/10T1/2 transformation assay. Cancer Res 37:1644–1648

Hall A, Marshall CJ, Spurr NK, Weiss RA (1983) Identification of transforming gene in two human sarcoma cell lines as a new member of the *ras* gene family located on chromosome 1. Nature 303:396–400

Hanafusa H (1977) Cell transformation by RNA tumor viruses. Comp Virol 10:401–483

Harper JR, Roop DR, Yuspa SH (1986) Transfection of the EJ ras^{Ha} gene into keratinocytes derived from carcinogen-induced mouse papillomas causes malignant progression. Mol Cell Biol 6:3144–3149

Harris H (1987) Human tissues and cells in carcinogenesis research. Cancer Res 47:1–10

Hashimoto Y, Kitagawa HS (1974) In vitro neoplastic transformation of epithelial cells of rat urinary bladder by nitrosamines. Nature 252:497–499

Haugen A, Schafer P, Lechner JF, Stoner GD, Trump BF, Harris CC (1982) Cellular ingestion, toxic effects and lesions observed in human respiratory epithelium cultured with asbestos and glass fibers. Int J Cancer 30:265–272

Hayflick L, Moorehead PS (1961) The serial cultivation of human diploid cell strains. Exptl Cell Res 25:585–621

Heidelberger C (1973) Chemical oncogenesis in culture. Adv Cancer Res 18:317–366

Heidelberger C (1975) Chemical carcinogenesis. Ann Rev Biochem 44:79–121

Heidelberger C, Freeman AE, Pienta RJ, Sivak A, Bertram JA, Casto BC, Dunkel VC, Francis MC, Kakunaga T, Little JB, Schechtman LM (1983) Cell transformation by chemical agents: a review and analysis of the literature. Mutat Res 114:283–385

Henderson E, Miller G, Robinson J, Heston L (1977) Efficiency of transformation of lymphocytes by Epstein-Barr virus. Virology 76:152–763

Hennings H, Michael D, Creng C, Steinert P, Holbrook K, Yuspa SH (1980) Calcium regulation of growth and differentiation of mouse epidermal cells in culture. Cell 19:245–254

Hennings H, Steinert P, Buxman MM (1981) Calcium induction of transglutaminase and the formation of ε (γ-glutamyl) lysine cross-links in cultured mouse epidermal cells. Biochem Biophys Res Commun 102:739–745

Hicks RM (1980) Multistage carcinogenesis in the urinary bladder. Br Med Bull 36:39–46

Hicks RM, Chowaniec J (1978) Experimental induction, histology and ultrastructure of hyperplasia and neoplasia of the urinary bladder epithelium. Int Rev Exp Pathol 18:199–280

Houweling A, van der Elsen PJ, van der Eb AJ (1980) Partial transformation of primary rat cells by the leftmost 4.5% fragment of adenovirus 5 DNA. Virology 105:537–654

Hronis TS, Steinberg ML, Defendi V, Sun T-T (1984) Simple epithelial nature of some simian virus-40-transformed human epidermal keratinocytes. Cancer Res 44:5797–5804

Huberman E, Sachs L (1966) Cell susceptibility to transformation and cytotoxicity by the carcinogenic hydrocarbon benzo(*a*)pyrene. Proc Natl Acad Sci USA 56:1123–1129

Huberman E, Donovan PJ, DiPaolo JA (1972) Mutation and transformation of cultured mammalian cells by *N*-acetoxy-*N*-2-fluorenylacetamide. JNCI 48:837–840

Hudziak RM, Schlessinger J, Ullrich A (1987) Increased expression of the putative growth factor receptor $p185^{\mathrm{HER2}}$ causes transformation and tumorigenesis of NIH3T3 cells. Proc Natl Acad Sci USA 84:7159–7163

Hurlin PJ, Fry DG, Maher VM, McCormick JJ (1987) Morphological transformation, focus formation and anchorage independence in diploid human fibroblasts by expression of a transfected H-*ras* oncogene. Cancer Res 47:5752–5757

Hwang L-HS, Gilboa E (1984) Expression of genes introduced into cells by retroviral infection as more efficient than that of genes introduced into cells by DNA transfection. J Virol 50:417–424

Igel HJ, Freeman AE, Spiewak JE, Kleinfeld KL (1975) Carcinogenesis in vitro. II. Chemical transformation of diploid human cell cultures: a rare event. In Vitro 11:117–129

Indo K, Miyaji H (1979) Qualitative changes in the biologic characteristics of cultured fetal rat keratinizing epidermal cells during the process of malignant transformation after benzo(a)pyrene treatment. JNCI 63:1017–1027

Inui N, Takayama S, Sugimura T (1972) Neoplastic transformation and chromosomal aberrations induced by *N*-methyl-*N'*-nitro-*N*-nitrosoguanidine in hamster lung cells in tissue culture. JNCI 48:1409–1417

Jenkins JR, Rudge K, Currie GA (1984) Cellular immortalization by a cDNA clone encoding the transformation-associated phosphoprotein p53. Nature 312:651–653

Jones PA, Laug WE, Gardner A, Nye CA, Fink LM, Benedict WF (1976) In vitro correlates of transformation in C3H/10T1/2 clone 8 mouse cells. Cancer Res 36:2863–2867

Kaighn ME, Narayan KS, Ohnuki Y, Jones LW, Lechner JF (1980) Differential properties among clones of simian virus 40-transformed human epithelial cells. Carcinogenesis 1:635–645

Kakunaga T (1978) Neoplastic transformation of human diploid fibroblast cells by chemical carcinogens. Proc Natl Acad Sci USA 75:1334–1338

Kakunaga T, Crow JD, Hamada H, Hirakawa T, Leavitt J (1983) Mechanisms of neoplastic transformation in human cells. In: Harris CC, Autrup H (eds) Human carcinogenesis. Academic, New York, p 371

Kamahora J, Kakunaga T (1967) Malignant transformation of hamster embryonic cells in vitro by 4-nitroquinoline-1-oxide. Biken J 10:219–242

Kaneda Y, Uchida T, Kim J, Ishiura M, Okada Y (1987) The improved efficient method for introducing macromolecules into cells using HVJ (Sendai virus) liposomes with gangliosides. Exptl Cell Res 173:56–69

Katoh Y, Kazuo U, Shozo T (1982) Induction of anchorage-independent growth of normal fibroblasts by growth factors. Proc Jpn Acad 58 (B):83

Katz E, Carter BJ (1986) A mutant cell line derived from NIH/3T3 cells: two oncogenes required for in vitro transformation. JNCI 77:909–914

Kawai S, Nishizawa M (1984) New procedure for DNA transfection with polycation and dimethylsulfoxide. Mol Cell Biol 4:1172–1174

Kawamura H, Strickland JE, Yuspa SH (1985) Association of resistance to terminal differentiation with initiation of carcinogenesis in adult mouse epidermal cells. Cancer Res 45:2748–2752

Keath EJ, Caimi PG, Cole MD (1984) Fibroblast lines expressing activated c-*myc* oncogenes are tumorigenic in nude mice and syngeneic animals. Cell 39:339–348

Kelekar A, Cole MD (1986) Tumorigenicity of fibroblast lines expressing the adenovirus E1a, cellular p53, or normal c-*myc* genes. Mol Cell Biol 6:7–14

Kelekar A, Cole MD (1987) Immortalization by c-*myc*, H-*ras* and E1a oncogenes induces differential cellular gene expression and growth factor responses. Mol Cell Biol 7:3899–3907

Kennedy AR, Little JB (1978) Protease inhibitors suppress radiation-induced malignant transformation in vitro. Nature 276:825–826

Kennedy AR, Little JB (1980) Investigation of the mechanism for enhancement of radiation transformation in vitro by 12-*O*-tetradecanoylphorbol-13-acetate. Carcinogenesis 1:1039–1047

Kennedy AR, Fox M, Murphy G, Little JB (1980) Relationship between x-ray exposure and malignant transformation in C3H 10T 1/2 cells. Proc Natl Acad Sci USA 77:7262–7266

Kennedy AR, Cairns J, Little JB (1984) Timing of the steps in transformation of C3H 10T 1/2 cells by x-irradiation. Nature 307:85–86

Kilkenny AE, Morgan D, Spangler EF, Yuspa SH (1985) Correlation of initiating potency of skin carcinogens with potency to induce resistance to terminal differentiation in cultured mouse keratinocytes. Cancer Res 45:2219–2225

Kinosita K Jr, Tsong TY (1977a) Voltage-induced pore formation and hemolysis of human erythrocytes. Biochim Biophys Acta 471:227–242

Kinosita K Jr, Tsong TY (1977b) Formation and resealing of pores of controlled sizes in human erythrocyte membrane. Nature 268:438–441

Kleinsek DA, Smith JR (1987) Construction of a cDNA library from senescent human diploid fibroblast cells in culture. In Vitro 23:13A
Knowles MA (1979) Effects of the tumor-promoting agent 12-*O*-tetradecanoylphorbol-13-acetate on normal and "preneoplastic" mouse submandibular gland epithelial cells in vitro. JNCI 62:349–352
Knowles MA, Franks LM (1977) Stages in neoplastic transformation of adult epithelial cells by 7,12-dimethylbenz(*a*)anthracene in vitro. Cancer Res 37:3917–3924
Knowles MA, Franks LM (1978) Ultrastructure and biological markers of neoplastic change in adult mouse epithelial cells transformed in vitro. Br J Cancer 37:603–611
Knowles MA, Jani H (1986) Multistage transformation of cultured rat urothelium: the effects of *N*-methyl-*N*-nitrosourea, sodium saccharin, sodium cyclamate and 12-*O*-tetradecanoylphorbol-13-acetate. Carcinogenesis 7:2059–2065
Knowles MA, Jani H, Hicks RM, Berry RJ (1985) *N*-Methyl-*N*-nitrosourea induces dysplasia and cell surface markers of neoplasia in long-term rat bladder organ cultures. Carcinogenesis 6:1047–1054
Knowles MA, Summerhayes IC, Hicks RM (1986a) Carcinogenesis studies using cultured rat and mouse bladder. In: Webber MM, Sekely L (eds) In vitro models for cancer research. CRC Press, Boca Raton, p 127
Knowles MA, Jani H, Hicks RM (1986b) Induction of morphological changes in the urothelium of cultured adult rat bladder by sodium saccharin and sodium cyclamate. Carcinogenesis 7:767–774
Knowles MA, Edymann ME, Proctor A, Padua RA, Roberts J (1987) *N*-Methyl-*N*-nitrosourea-induced transformation of rat urothelial cells in vitro is not mediated by activation of *ras* oncogenes. Oncogene 1:143–148
Knudson AG Jr (1983) Model hereditary cancers of man. Prog Nucleic Acid Res Mol Biol 29:17–25
Knudson AG Jr (1985) Hereditary cancer oncogenes and antioncogenes. Cancer Res 45:1437–1443
Koi M, Barrett JC (1986) Loss of tumor-suppressive function during chemically induced neoplastic progression of Syrian hamster embryo cells. Proc Natl Acad Sci USA 83:5992–5996
Kok K, Osinga J, Carritt B, Davis MB, van der Hout AH, van der Veen AY, Landsvater RM, de Leij LFMH, Berendsen HH, Postmus PE, Poppema S, Buys CHCM (1987) Deletion of a DNA sequence at the chromosomal region 3p 21 in all major types of lung cancer. Nature 330:578–584
Kouri RE, Kurtz SA, Price PJ, Benedict WF (1975) 1-β-D-Arabinofuranosylcytosine-induced malignant transformation of hamster and rat cells in culture. Cancer Res 35:2413–2419
Kudo A, Yamamoto F, Furusawa M, Kuroiwa A, Natori S, Obinata M (1982) Structure of thymidine kinase gene introduced into mouse ltk$^-$ cells by a new injection method. Gene 19:11–19
Kulesz-Martin MF, Koehler B, Hennings H, Yuspa SH (1980) Quantitative assay for carcinogen altered differentiation in mouse epidermal cells. Carcinogenesis 1:995–1006
Kulesz-Martin M, Kilkenny AE, Holbrook KA, Digernes V, Yuspa SH (1983) Properties of carcinogen altered mouse epidermal cells resistant to calcium-induced terminal differentiation. Carcinogenesis 4:1367–1377
Kulesz-Martin MF, Yoshida MA, Prestine LA, Yuspa SH, Bertram JS (1985) Mouse cell clones for improved quantitation of carcinogen-induced altered differentiation. Carcinogenesis 6:1245–1254
Kurata S-I, Tsukakoshi M, Kasuya T, Ikawa Y (1986) The laser method for efficient introduction of foreign DNA into cultured cells. Exptl Cell Res 162:372–378
Land H, Parada LF, Weinberg RA (1983) Tumorigenic conversion of primary embryo fibroblasts requires at least two cooperating oncogenes. Nature 304:596–602
Land H, Chen AC, Morgenstern JP, Parada LF, Weinberg RA (1986) Behaviour of *myc* and *ras* oncogenes in transformation of rat embryo fibroblasts. Mol Cell Biol 6:1917–1925

Lasnitzki I (1968) The effect of a hydrocarbon-enriched fraction of cigarette smoke condensate on human fetal lung grown in vitro. Cancer Res 28:510–516

Lechner JF, Tokiwa T, LaVeck M, Benedict WF, Banks-Schlegel S, Yeager H Jr, Banerjee A, Harris CC (1985) Asbestos-associated chromosomal changes in human mesothelial cells. Proc Natl Acad Sci USA 82:3884–3888

Lee WMF, Schwab M, Westaway D, Varmus HE (1985) Augmented expression of normal c-*myc* is sufficient for cotransformation of rat embryo cells with a mutant *ras* gene. Mol Cell Biol 5:3345–3356

Lewis WH, Srinivasan PR, Stokoe N, Siminovitch L (1980) Parameters governing the transfer of the genes for thymidine kinase and dihydrofolate reductase into mouse cells using metaphase chromosomes or DNA. Somatic Cell Genet 6:333–348

Loo DT, Fuquay JI, Rawson CL, Barnes DW (1987) Extended culture of mouse embryo cells without senescence: inhibition by serum. Science 236:200–202

Lowy DR, Rands E, Scolnick EM (1978) Helper-independent transformation by unintegrated Harvey sarcoma virus DNA. J Virol 26:291–298

Lumpkin CK, McClung JK, Pereira-Smith OM, Smith JR (1986) Existence of high abundance antiproliferative mRNAs in senescent human diploid fibroblasts. Science 232:393–395

Luthman H, Magnosson G (1983) High efficiency polyoma DNA transfection of chloroquine treated cells. Nucleic Acids Res 11:1295–1308

Maher VM, Rowan LA, Silinskas KC, Kateley SA, McCormick JJ (1982) Frequency of UV-induced neoplastic transformation of diploid human fibroblasts is higher in xeroderma pigmentosum cells than in normal cells. Proc Natl Acad Sci USA 79:2613–2617

Manoharan TH, Burgess IA, Ho D, Newell CL, Fahl WE (1985) Integration of a mutant c-Ha-*ras* oncogene into C3H/10T1/2 cells and its relationship to tumorigenic transformation. Carcinogenesis 6:1295–1301

Marchok AC, Huang SF, Martin DH (1984) selection of carcinogen-altered rat tracheal epithelial cells preexposed to 7,12-dimethylbenz(*a*)anthracene by their loss of a need for pyruvate to survive in culture. Carcinogenesis 5:789–796

Markovits P, Coppey J, Papadopoulo D, Mazabraud A, Hubert-Habart M (1974) Transformation maligne de cellules d'embryon de hamster en culture par la dimethyl-7,10-benzo(*c*)acridine. Int J Cancer 14:215–225

Marshall CJ, Sager R (1981) Genetic analysis of tumorigenesis. IX. Suppression of anchorage independence in hybrids between transformed hamster cell lines. Somat Cell Genet 7:713–723

Martin-Zanca D, Hughes SH, Barbacid M (1986) A human oncogene formed by the fusion of truncated tropomyosin and protein tyrosine kinase sequences. Nature 319:743–748

Mass MJ, Nettesheim P, Gray TE, Barrett JC (1984a) The effects of 12-*O*-tetradecanoylphorbol-13-acetate and other tumor promoters on the colony formation of rat tracheal epithelial cells in culture. Carcinogenesis 5:1597–1601

Mass MJ, Nettesheim P, Beeman DK, Barrett JC (1984b) Inhibition of transformation of primary rat tracheal epithelial cells by retinoic acid. Cancer Res 44:5688–5691

Masui T, Wakefield LM, Lechner JF, Laveck MA, Sporn MB, Harris CC (1986) Type β transforming growth factor is the primary differentiation-inducing serum factor for normal human bronchial epithelial cells. Proc Natl Acad Sci USA 83:2438–2442

McCormick JJ, Kately-Kohler S, Maher VM (1985) Factors involved in quantitating induction of anchorage independence in diploid human fibroblasts by carcinogens. In: Barrett JC, Tennant RW (eds) Carcinogenesis – a comprehensive survey. Raven, New York, p 233

Meyn MS, Rossman T, Troll W (1977) A protease inhibitor blocks SOS functions in *Escherichia coli:* antipain prevents lambda repressor inactivation, ultraviolet mutagenesis and filamentous growth. Proc Natl Acad Sci USA 74:1152–1156

Miller DA, Jolly DJ, Friedmann T, Verma IM (1983) A transmissable retrovirus expressing human HPRT: gene transfer into cells obtained from humans deficient in HPRT. Proc Natl Acad Sci USA 80:4709–4713

Miller G, Lisco H, Kohn HT, Stitt D (1971) Establishment of cell lines from normal adult human blood leukocytes by exposure to Epstein-Barr virus and neutralization by human sera with Epstein-Barr virus antibody. Proc Soc Exp Biol Med 137:1459–1465

Milo GE, DiPaolo JA (1978) Neoplastic transformation of human diploid cells in vitro after chemical carcinogen treatment. Nature 275:130–132

Milo GE, Noyes I, Donahoe J, Weisbrode S (1981) Neoplastic transformation of human epithelial cells in vitro after exposure to chemical carcinogens. Cancer Res 41:5096–5102

Mishra NK, Di Mayorca G (1974) In vitro malignant transformation of cells by chemical carcinogens. Biochim Biophys Acta 355:205–219

Mondal S, Heidelberger C (1970) In vitro malignant transformation by methylcholanthrene of the progeny of single cells derived from C3H mouse prostate. Proc Natl Acad Sci USA 65:219–225

Morgan TL, Maher VM, McCormick JJ (1986) Optimal parameters for the polybrene-induced DNA transfection of diploid human fibroblasts. In Vitro Cell Develop Biol 22:317–319

Namba M, Nishitani K, Hyodoh F, Fukushima F, Kimoto T (1985) Neoplastic transformation of human diploid fibroblasts (KMST-6) by treatment with ^{60}Co gamma rays. Int J Cancer 35:275–280

Nettesheim P, Barrett JC (1984) Tracheal epithelial cell transformation: a model system for studies on neoplastic progression. In: Goldberg L (ed) CRC Crit Rev Toxicol vol 12. CRC Press, Boca Raton, p 215

Nettesheim P, Barrett JC (1985) In vitro transformation of rat tracheal epithelial cells: a model for the study of multistage carcinogenesis. In: Barrett JC, Tennant RW (eds) Carcinogenesis, vol 9. Raven, New York, p 283

Nettesheim P, Barrett JC, Mass MK, Steele V, Gray TE (1984) Studies on the action of tumor promoter and antipromoters on respiratory tract epithelium. In: Borzonsonyi M, Day NE, Lapis K, Yamasaki H (eds) Models, mechanisms and etiology of tumor promoters. IARC Sci Publ no 56, Lyon, p 109

Nettesheim P, Gray T, Barrett JC (1985) The toxic response of preneoplastic rat tracheal epithelial cells to 12-*O*-tetradecanoylphorbol-13-acetate. Carcinogenesis 6:1427–1434

Newbold RF, Overell RW (1983) Fibroblast immortality is a prerequisite for transformation by EJ c-Ha-*ras* oncogene. Nature 304:648–651

Newbold RF, Overell RW, Connell JR (1982) Induction of immortality is an early event in malignant transformation of mammalian cells by carcinogens. Nature 299:633–635

Nicholson L, Jani H (1988) Effects of sodium cyclamate and sodium saccharin on focus induction in explant cultures of rat bladder. Int J Cancer 42:295–298

O'Brien W, Stenman G, Sager R (1986) Suppression of tumor growth by senescence in virally transformed human fibroblasts. Proc Natl Acad Sci USA 83:8659–8663

Ochiya T, Fujiyama A, Fukushige S, Hatada I, Matsubara K (1986) Molecular cloning of an oncogene from a human hepatocellular carcinoma. Proc Natl Acad Sci USA 83:4993–4997

Orly J, Sato G, Erickson G (1980) Serum suppresses the expression of hormonally induced functions in cultured granulosa cells. Cell 20:817–827

Oshimura M, Barrett JC (1986) Chemically induced aneuploidy in mammalian cells: mechanisms and biological significance in cancer. Environment Mutag 8:129–159

Oshimura M, Gilmer TM, Barrett JC (1985) Nonrandom loss of chromosome 15 in Syrian hamster tumours induced by v-Ha-*ras* plus v-*myc* oncogenes. Nature 316:636–639

Oshimura M, Koi M, Ozawa N, Sugawara O, Lamb PW, Barrett JC (1988) Role of chromosome loss in *ras*/*myc*-induced Syrian hamster tumors. Cancer Res 48:1623–1632

Packer L, Fuehr K (1977) Low oxygen concentration extends the lifespan of cultured human diploid cells. Nature 267:423–425

Padmanabhan R, Howard TH, Howard BH (1987) Specific growth inhibitory sequences in genomic DNA from quiescent human embryo fibroblasts. Mol Cell Biol 7:1894–1899

Pai SB, Steele VE, Nettesheim P (1983) Neoplastic transformation of primary tracheal epithelial cell cultures. Carcinogenesis 4:369–374

Parada LF, Land H, Weinberg RA, Wolf D, Rotter V (1984) Cooperation between gene encoding p53 tumour antigen and *ras* in cellular transformation. Nature 312:649–651

Parker BA, Stark GR (1979) Regulation of simian virus 40 transcription: sensitive analysis of the RNA species present early in infections by virus or viral DNA. J Virol 31:360–369

Parkinson EK, Grabham P, Emmerson A (1983) A subpopulation of cultured human keratinocytes which is resistant to the induction of terminal differentiation-related changes by phorbol, 12-myristate, 13-acetate: evidence for an increase in the resistant population following transformation. Carcinogenesis 4:857–861

Parsa I, Marsh WH, Sutton AL (1981a) An in vitro model for human pancreas carcinogenesis. Cancer 47:1543–1551

Parsa I, Marsh WH, Sutton AL, Butt KMH (1981b) Effects of dimethylnitrosamine on organ-cultured adult human pancreas. Am J Pathol 102:403–411

Parsa I, Bloomfield RD, Foye CA, Sutton AL (1984) Methylnitrosourea-induced carcinoma in organ-cultured fetal human pancreas. Cancer Res 44:3530–3538

Peehl DM, Stanbridge EJ (1981) Anchorage-independent growth of normal human fibroblasts. Proc Natl Acad Sci USA 78:3053–3057

Pera MF, Gorman PA (1984) In vitro analysis of multistage epidermal carcinogenesis: development of indefinite renewal capacity and reduced growth factor requirements in colony-forming keratinocytes precedes malignant transformation. Carcinogenesis 5:671–682

Pereira-Smith OM, Smith JR (1988) Genetic analysis of indefinite division in human cells: identification of four complementation groups. Proc Natl Acad Sci USA 85:6042–6046

Pereira-Smith OM, Smith JR (1983) Evidence for the recessive nature of cellular immortality. Nature 221:964–966

Pereira-Smith OM, Fisher SF, Smith JR (1985) Senescent and quiescent cell inhibitors of DNA synthesis: membrane-associated proteins. Exp Cell Res 160:297–306

Quintanilla M, Brown K, Ramsden M, Balmain A (1986) Carcinogen-specific mutation and amplification of Ha-*ras* during mouse skin carcinogenesis. Nature 322:78–80

Rassoulzadegan M, Binetruy B, Cuzin F (1982) High frequency of gene transfer after fusion between bacteria and eukaryotic cells. Nature 295:257–259

Rassoulzadegan M, Naghashfar Z, Cowie A, Carr A, Grisoni M, Kamen R, Cuzin F (1983) Expression of the large T protein of polyomavirus promotes the establishment in culture of "normal" fibroblast cell lines. Proc Natl Acad Sci USA 80:4354–4358

Reznikoff CA, Brankow DW, Heidelberger C (1973a) Establishment and characterization of a cloned line of C3H mouse embryo cells sensitive to post confluence inhibition of division. Cancer Res 33:3231–3238

Reznikoff CA, Bertram JS, Brankow DW, Heidelberger C (1973b) Quantitative and qualitative studies of chemical transformation of cloned C3H mouse embryo cells sensitive to post confluence inhibition of cell division. Cancer Res 33:3239–3249

Rheinwald JG, Green H (1975) Serial cultivation of strains of human epidermal keratinocytes: the formation of keratinising colonies from single cells. Cell 6:331–343

Rhim JS, Kim CM, Arnstein P, Huebner RJ, Weisburger EK, Nelson-Rees WA (1975) Transformation of human osteosarcoma cells by a chemical carcinogen. JNCI 55:1291–1294

Rhim JS, Putman DL, Arnstein P, Huebner RJ, McAllister RM (1977) Characterization of human cells transformed in vitro by *N*-methyl-*N'*-nitro-*N*-nitrosoguanidine. Int J Cancer 19:505–510

Rhim JS, Jay G, Arnstein P, Price FM, Sanford KK, Aaronson SA (1985) Neoplastic transformation of human epidermal keratinocytes by AD12-SV40 and Kirsten sarcoma viruses. Science 227:1250–1252

Rhim JS, Fujita J, Arnstein P, Aaronson SA (1986) Neoplastic conversion of human keratinocytes by adenovirus 12-SV40 virus and chemical carcinogens. Science 232:385–388

Ruley HE (1983) Adenovirus early region 1A enables viral and cellular transforming genes to transform primary cells in culture. Nature 304:602–606

Ruley HE, Moomaw JF, Maruyama K (1984) Avian myelocytomatosis virus *myc* and adenovirus early region 1A promote the in vitro establishment of cultured primary cells. In: Vande Woude GF, Levine AJ, Topp WC, Watson JD (eds) Cancer cells 2. Cold Spring Harbor, New York, p 481

Ruley HE, Moomaw J, Chang C, Garrels JI, Furth M, Franza BR (1985) Multistep transformation of an established cell line by the adenovirus E1A and T24 Ha-*ras*-1 genes. Cancer Cells 3:257–264

Sack GH (1981) Human cell transformation by simian virus 40 – a review. In Vitro 17:1–19

Sager R (1985) Genetic suppression of tumor formation. Adv Cancer Res 44:43–68

Sager R (1986) Genetic suppression of tumour formation: a new frontier in cancer research. Cancer Res 46:1573–1580

Sager R, Kovac PE (1978) Genetic analysis of tumorigenesis. I. Expression of tumor-forming ability in hamster hybrid cell lines. Somat Cell Genet 4:375–392

Sager R, Tanaka K, Lau CC, Ebina Y, Anisowicz A (1983) Resistance of human cells to tumorigenesis induced by cloned transforming genes. Proc Natl Acad Sci USA 80:7601–7605

Sakamoto H, Mori M, Taira M, Yoshida T, Matsukawa S, Shimizu K, Sekiguchi M, Terada M, Sugimura T (1986) Transforming gene from human stomach cancers and a non-cancerous portion of stomach mucosa. Proc Natl Acad Sci USA 83:3997–4001

Salmon SE (1980) Cloning of human tumor stem cells. Liss, New York

Sanders FK, Burford BO (1967) Morphological conversion of cells in vitro by *N*-nitrosomethylurea. Nature 213:1171–1173

Sanford KK (1974) Biologic manifestations of oncogenesis in vitro: a critique. JNCI 53:1481–1485

Sato H, Kuroki T (1966) Malignization in vitro of hamster embryonic cells by chemical carcinogens. Proc Jpn Acad 42:1211–1216

Schaeffer-Ridder M, Wang Y, Hofschneider PH (1981) Liposomes as gene carriers: efficient transformation of mouse L cells by thymidine kinase gene. Science 215: 166–168

Schell K, Lane WT, Casey MJ, Huebner RJ (1966) Potentiation of oncogenicity of adenovirus type 12 grown in African green monkey kidney cell cultures preinfected with SV40 virus: persistence of both T antigens in the tumors and evidence for possible hybridization. Proc Natl Acad Sci USA 55:81–88

Schwab M, Varmus HE, Bishop JM (1985) Human N-*myc* gene contributes to neoplastic transformation of mammalian cells in culture. Nature 316:160–162

Shabad LM, Kolesnichenko TS, Golub NI (1975) The effect produced by some carcinogenic nitrosocompounds on organ cultures from human lung and kidney tissues. Int J Cancer 16:768–778

Shen YM, Hirshhorn RR, Mercer WE, Surmacz E, Tsutsui Y, Soprano K, Baserga R (1982) Gene transfer: DNA microinjection compared with DNA transfection with a very high efficiency. Mol Cell Biol 2:1145–1154

Shih C, Shilo B-Z, Goldfarb MP, Dannenberg A, Weinberg RA (1979) Passage of phenotypes of chemically transformed cells via transfection of DNA and chromatin. Proc Natl Acad Sci USA 76:5714–5718

Shih C, Padhy LC, Murray M, Weinberg RA (1981) Transforming genes of carcinomas and neuroblastomas introduced into mouse fibroblasts. Nature 290:261–264

Shimizu K, Yoshimichi N, Sekiguchi M, Hokamura K, Tanaka K (1985) Molecular cloning of an activated human oncogene, homologous to v-*raf*, from primary stomach cancer. Proc Natl Acad Sci USA 82:5641–5645

Shimotono K, Temin H (1981) Formation of infectious progeny virus after insertion of herpes TK gene into DNA of an avian retrovirus. Cell 26:67–77

Silinskas KC, Kateley SA, Tower JE, Maher VM, McCormick JJ (1981) Induction of anchorage independent growth in human fibroblasts by propane sultone. Cancer Res 41:1620–1627

Sina JF, Bradley MO, O'Brien TG (1982) Neoplastic transformation of syrian hamster epidermal cells in vitro. Cancer Res 42:4116–4123

Slaga TJ, Viaje A, Bracken WM, Buty SG, Miller DR, Fischer SM, Richter CK, Dumont JN (1978) In vitro transformation of epidermal cells from newborn mice. Cancer Res 38:2246–2252

Small MB, Gluzman Y, Ozer HL (1982) Enhanced transformation of human fibroblasts by origin-defective simian virus 40. Nature 296:671–672

Smith BL, Sager R (1982) Multistep origin of tumor-forming ability in Chinese hamster embryo fibroblast cells. Cancer Res 42:389–396

Spandidos DA, Wilkie NM (1984) Malignant transformation of early passage rodent cells by a single mutated human oncogene. Nature 310:469–475

Stampfer M, Bartley JC (1985) Induction of transformation and continuous cell lines from normal human mammary epithelial cells after exposure to benzo(α)pyrene. Proc Natl Acad Sci USA 82:2394–2398

Steele VE, Marchok AC, Nettesheim P (1977) Transformation of tracheal epithelium exposed in vitro to *N*-methyl-*N'*-nitro-*N*-nitrosoguanidine (MNNG). Int J Cancer 20:234–238

Steele VE, Beeman DK, Nettesheim P (1984) Enhanced induction of the anchorage-independent phenotype in initiated rat-tracheal epithelial cell cultures by the tumor promoter 12-*O*-tetradecanoylphorbol-13-acetate. Cancer Res 44:5068–5072

Steglich CS, De Mars R (1982) Mutations causing deficiency of APRT in fibroblasts cultured from humans heterozygous for mutant APRT alleles. Somatic Cell Genet 8:115–141

Steinberg ML, Defendi V (1979) Altered patterns of growth and differentiation in human keratinocytes infected by simian virus 40. Proc Natl Acad Sci USA 76:331–334

Steinberg ML, Defendi V (1983) Transformation and immortalization of human keratinocytes by SV40. J Invest Dermatol 81:131s–136s

Stevenson M, Volsky B, Hedenskog M, Volsky DJ (1986) Immortalization of human T lymphocytes after transfection of Epstein-Barr virus DNA. Science 233:980–984

Storer RD, Stein RB, Sina JF, DeLuca JG, Allen HL, Bradley MO (1986) Malignant transformation of a preneoplastic hamster epidermal cell line by the EJ c-Ha-*ras* oncogene. Cancer Res 46:1458–1464

Stow ND, Wilkie NM (1976) An improved technique for obtaining enhanced infectivity with herpes simplex type-1 DNA. J Gen Virol 33:447–458

Sukumar S, Notario V, Martin-Zanca D, Barbacid M (1983) Induction of mammary carcinomas in rats by nitroso-methylurea involves malignant activation of H-*ras*-1 locus by single point mutations. Nature 306:658–661

Summerhayes IC, Franks LM (1979) Effects of donor age on neoplastic transformation of adult mouse bladder epithelium in vitro. JNCI 62:1017–1023

Sun N-C, Sun CRY, Chao L, Fung W-P, Tennant RN, Hsie AW (1981) In vitro transformation of syrian hamster epidermal cells by *N*-methyl-*N'*-nitro-*N*-nitrosoguanidine. Cancer Res 41:1669–1676

Sutherland BM, Bennett PV (1984) Transformation of human cells by DNA transfection. Cancer Res 44:2769–2772

Sutherland BM, Cimino JS, Delihas N, Shih AG, Oliver RP (1980) Ultraviolet light-induced transformation of human cells to anchorage-independent growth. Cancer Res 40:1934–1939

Sutherland BM, Bennett PV, Freeman AG, Moore SP, Strickland PT (1985) Transformation of human cells by DNAs ineffective in transformation of NIH3T3 cells. Proc Natl Acad Sci USA 82:2399–2403

Swartz JB, Riddiough CR, Epstein SS (1982) Analyses of carcinogenesis dose-response relations with dichotomous data: implications for carcinogenic risk assessment. Teratogen Carcinogen Mutagen 2:179–204

Tabin CJ, Hoffman JW, Goff SP, Weinberg RA (1982) Adaptation of a retrovirus as a eukaryotic vector transmitting the herpes simplex virus thymidine kinase gene. Mol Cell Biol 2:426–436

Taylor-Papadimitriou J, Purkis P, Lane EB, McKay IA, Chang SE (1982) Effects of SV40 transformation on the cytoskeleton and behavioural properties of human keratinocytes. Cell Differ 11:169–180

Terzaghi M, Nettesheim P (1979) Dynamics of neoplastic development in carcinogen-exposed tracheal mucosa. Cancer Res 39:4003–4010
Terzaghi M, Nettesheim P, Riester L (1982) Effect of carcinogen dose on the dynamics of neoplastic development in rat tracheal epithelium. Cancer Res 42:4511–4518
Thomassen DG (1986) Role of spontaneous transformation in carcinogenesis: development of preneoplastic rat tracheal epithelial cells at a constant rate. Cancer Res 46:2344–2348
Thomassen DG, Gray TE, Moss MJ, Barrett JC (1983) High frequency of carcinogen-induced, early, preneoplastic changes in rat tracheal epithelial cells in culture. Cancer Res 43:5956–5963
Thomassen DG, Gilmer TM, Annab LA, Barrett JC (1985a) Evidence for multiple steps in neoplastic transformation of normal and preneoplastic syrian hamster embryo cells following transfection with Harvey murine sarcoma virus oncogene (v-Ha-*ras*). Cancer Res 45:726–732
Thomassen D, Nettesheim P, Gray TE, Barrett JC (1985b) Quantitation of the rate of spontaneous generation and carcinogen-induced frequency of anchorage-independent variants of rat tracheal epithelial cells in culture. Cancer Res 45:1516–1524
Todaro GJ, Aaronson SA (1969) Human cell strains susceptible to focus formation by human adenovirus type 12. Proc Natl Acad Sci USA 61:1272–1278
Todaro GJ, Green H (1963) Quantitative studies of the growth of mouse embryo cells in culture and their development into established cell lines. J Cell Biol 17:299–313
Todaro GJ, Green H (1964) Serum albumin supplemented medium for long term cultivation of mammalian fibroblast strains. Proc Soc Exp Biol Med 116:688–692
Tolsma SS, Thomas E, Bauer KD, Bouck N (1988) Genetic assessment of the strength of a cancer suppressor gene in hamster cells. Cancer Res 48:46–51
Tsuda H, Inui N, Takayama S (1973) In vitro transformation of newborn hamster cells by sodium nitrite. Biochem Biophys Res Commun 55:1117–1124
Tsunokawa Y, Esumi H, Sasaki MS, Mori M, Sakamoto H, Terada M, Sugimura T (1984) Integration of v-*ras*^{H} does not necessarily transform an immortalised murine cell line. Gann 75:732–736
Vanderlaan M, Steele V, Nettesheim P (1983) Increased DNA content as an early marker of transformation in carcinogen-exposed rat tracheal cell cultures. Carcinogenesis 4:721–727
Vogt PK (1967) DEAE-dextran: enhancement of cellular transformation induced by avian sarcoma viruses. Virology 33:175–177
Volsky DJ, Gross T, Sinangil F, Kuszynski C, Bartzatt R, Dambaugh T, Kieff E (1984) Expression of Epstein-Barr virus (EBV) DNA and cloned DNA fragments in human lymphocytes following Sendai virus envelope-mediated gene transfer. Proc Natl Acad Sci USA 81:5926–5930
Walker C, Nettesheim P, Barrett JC, Gilmer TM (1987) Expression of a *fms*-related oncogene in carcinogen-induced neoplastic epithelial cells. Proc Natl Acad Sci USA 84:1804–1808
Walker GC (1985) Inducible DNA repair systems. Ann Rev Biochem 54:425–457
Wei C-M, Gibson M, Spear PG, Scolnick EM (1981) Construction and isolation of a transmissible retrovirus containing the *src* gene of Harvey murine sarcoma virus and the thymidine kinase gene of herpes simplex virus type 1. J Virol 39:935–944
Weinberg RA (1985) The action of oncogenes in the cytoplasm and nucleus. Science 230:770–776
Weisberger JH, Williams GM (1981) Carcinogen testing: current problems and new approaches. Science 214:401–407
Weiss R, Teich N, Varmus H, Coffin J (eds) (1982) RNA tumor viruses. Cold Spring Harbor Laboratory, New York
Whittaker JL, Byrd PJ, Grand RJA, Gallimore PH (1984) Isolation and characterization of four adenovirus type 12-transformed human embryo kidney cell lines. Mol Cell Biol 4:110–116

Wigley CB (1979) Transformation in vitro of adult mouse salivary gland epithelium: a system for studies on mechanisms of initiation and promotion. In: Franks LM, Wigley CB (eds) Neoplastic transformation in differentiated epithelial cell systems in vitro. Academic, London

Wigley CB (1983) TPA affects early and late stages of chemically induced transformation in mouse submandibular salivary epithelial cells in vitro. Carcinogenesis 4:101–106

Wigley CB, Carbonell AW (1976) The target cell in chemical induction of carcinomas in mouse submandibular gland. Eur J Cancer 12:737–741

Williams GM (1976) Primary and long-term culture of adult rat liver epithelial cells. Methods Cell Biol 14:357–364

Williams GM, Weisburger EK, Weisburger JH (1971) Isolation and long-term cell culture of epithelial-like cells from rat liver. Exptl Cell Res 69:106–112

Yamamoto F, Furasawa M (1978) A simple microinjection technique not employing a micromanipulator. Exptl Cell Res 117:441–445

Yamamoto F, Furusawa M, Furusawa I, Obinata M (1982) The pricking method. A new efficient technique for mechanically introducing foreign DNA into the nuclei for culture cells. Exptl Cell Res 142:79–84

Yancopoulos GD, Nisen PD, Tesfaye A, Kohl NE, Goldfarb MP, Alt FW (1985) N-*myc* can cooperate with *ras* to transform normal cells in culture. Proc Natl Acad Sci USA 82:5455–5459

Yoakum GH, Korba BE, Lechner JF, Tokiwa T, Gazdar AF, Seeley T, Siegel M, Leeman L, Autrup H, Harris CC (1983) High-frequency transfection and cytopathology of the hepatitis B virus core antigen gene in human cells. Science 222:385–389

Yoakum GH, Lechner JF, Gabrielson EW, Korba BE, Malan-Shibley L, Willey JC, Valerio MG, Shamsuddin AM, Trump BF, Harris CC (1985) Transformation of human bronchial epithelial cells transfected by Harvey *ras* oncogene. Science 227:1174–1179

Yuspa SH (1985) Mechanisms of transformation and promotion of mouse epidermal cells. In: Barrett JC, Tennant RW (eds) Carcinogenesis, vol 9. Raven, New York, p 271

Yuspa SH, Harris CC (1974) Altered differentiation of mouse epidermal cells treated with retinyl acetate in vitro. Exptl Cell Res 86:95–105

Yuspa SH, Morgan DL (1981) Mouse skin cells resistant to terminal differentiation associated with initiation of carcinogcncsis. Naturc 293:72–74

Yuspa SH, Lichti U, Ben T, Patterson E, Hennings H, Slaga TJ, Colburn N, Kelsey W (1976a) Phorbol-ester tumor promoters stimulate DNA synthesis and ornithine decarboxylase activity in mouse epidermal cell cultures. Nature 262:402–404

Yuspa SH, Hennings H, Dermer P, Michael D (1976b) Dimethyl sulfoxide-induced enhancement of 7,12-dimethylbenz(a)anthracene metabolism and DNA binding in differentiating mouse epidermal cell cultures. Cancer Res 36:947–951

Yuspa SH, Hawley-Nelson P, Stanley JR, Hennings H (1980a) Epidermal cell culture. Transplant Proc [Suppl 1] 12:114–122

Yuspa SH, Hawley-Nelson P, Koehler B, Stanley JR (1980b) A survey of transformation markers in differentiating epidermal cell lines in culture. Cancer Res 40:4694–4703

Yuspa SH, Koehler B, Kulesz-Martin M, Hennings H (1981a) Clonal growth of mouse epidermal cells in medium with reduced calcium concentration. J Invest Dermatol 76:144–146

Yuspa SH, Hennings H, Lichti U (1981b) Initiator and promoter induced specific changes in epidermal function and biological potential. J Supramol Str Biochem 17:245–257

Yuspa SH, Ben T, Hennings H, Lichti U (1982) Divergent responses in epidermal basal cells exposed to the tumor promoter 12-*O*-tetradecanoylphorbol-13-acetate. Cancer Res 42:2344–2349

Yuspa SH, Vass W, Scolnick E (1983) Altered growth and differentiation of cultured mouse epidermal cells infected with oncogenic retrovirus: contrasting effects of viruses and chemicals. Cancer Res 43:6021–6030

Yuspa SH, Kilkenny AE, Stanley J, Lichti U (1985) Keratinocytes blocked in phorbol ester-responsive early stage of terminal differentiation by sarcoma viruses. Nature 314:459–462

Yuspa SH, Morgan D, Lichti U, Spangler EF, Michael D, Kilkenny A, Hennings H (1986) Cultivation and characterization of cells derived from mouse skin papillomas induced by an initiation-promotion protocol. Carcinogenesis 7:949–958

Zarbl H, Sukumar S, Arthur AV, Martin-Zanca D, Barbacid M (1985) Direct mutagenesis of Ha-*ras*-1 oncogene by *N*-nitroso-*N*-methylurea during initiation of mammary carcinogenesis in rats. Nature 315:382–386

Zerler B, Moran B, Maruyama K, Moomaw J, Grodzicker T, Ruley HE (1986) Adenovirus E1A coding sequences that enable *ras* and *pmt* oncogenes to transform cultured primary cells. Mol Cell Biol 6:887–899

Zimmerman RJ, Little JB (1983) Characteristics of human diploid fibroblasts transformed in vitro by chemical carcinogens. Cancer Res 43:2183–2189

Part IV.
Reactions of Carcinogens with DNA

CHAPTER 8

Metabolic Activation and DNA Adducts of Aromatic Amines and Nitroaromatic Hydrocarbons

F. A. BELAND and F. F. KADLUBAR

A. Introduction

The commercial production of aromatic amines and nitroaromatic hydrocarbons began in the middle 1800s following the synthesis of aniline and the aniline-based dyes, rosaniline and mauve. Thirty years later, REHN (1895) noted an increased incidence of urinary bladder cancer in German dyestuff workers, which LEICHTENSTERN (1898) suggested could be due to their exposure to naphthylamines. This initial report was rapidly followed by additional epidemiological studies until it became incontrovertible that exposure to 2-naphthylamine, benzidine, and 4-aminobiphenyl could lead to the induction of bladder cancer in humans (reviewed in PARKES and EVANS 1984).

Animal models for aromatic amine and nitroaromatic hydrocarbon carcinogenesis developed considerably later than the initial clinical observations. The carcinogenicity of 2-naphthylamine to the urinary bladder was not demonstrated until 1938 (HUEPER et al. 1938), followed by benzidine in 1950 (SPITZ et al. 1950) and 4-aminobiphenyl in 1954 (WALPOLE et al. 1954). These epidemiological and experimental studies led to curtailed industrial use of carcinogenic aromatic amines and their nitro analogues; nevertheless, significant human exposure to these compounds still occurs from a number of sources. Cigarette smoke, for instance, contains nanogram amounts of 2-naphthylamine and 4-aminobiphenyl (PATRIANAKOS and HOFFMANN 1979). These compounds, plus other primary aromatic amines that are present in microgram quantities, may account for the positive correlation between cigarette smoking and the incidence of bladder cancer in humans (WYNDER and GOLDSMITH 1977; WIGLE et al. 1980; MOOLGAVKAR and STEVENS 1981; MOMMSEN and AAGAARD 1983). During high-temperature cooking processes certain amino acids, such as tryptophan and glutamic acid along with creatinine and glucose, are pyrolyzed to extremely mutagenic heterocyclic aromatic amines. These derivatives are carcinogenic at a number of sites in experimental animals and may contribute to the etiology of human cancer (SUGIMURA 1986). Recently, a number of nitroaromatic hydrocarbons, resulting from a variety of combustion processes, have been detected in the environment. As with the heterocyclic aromatic amines, some of these agents are both very potent bacterial mutagens and animal carcinogens (TOKIWA and OHNISHI 1986). Although the importance of nitroaromatic hydrocarbons in the induction of human cancer is presently unknown, truck drivers exposed to diesel emission, a

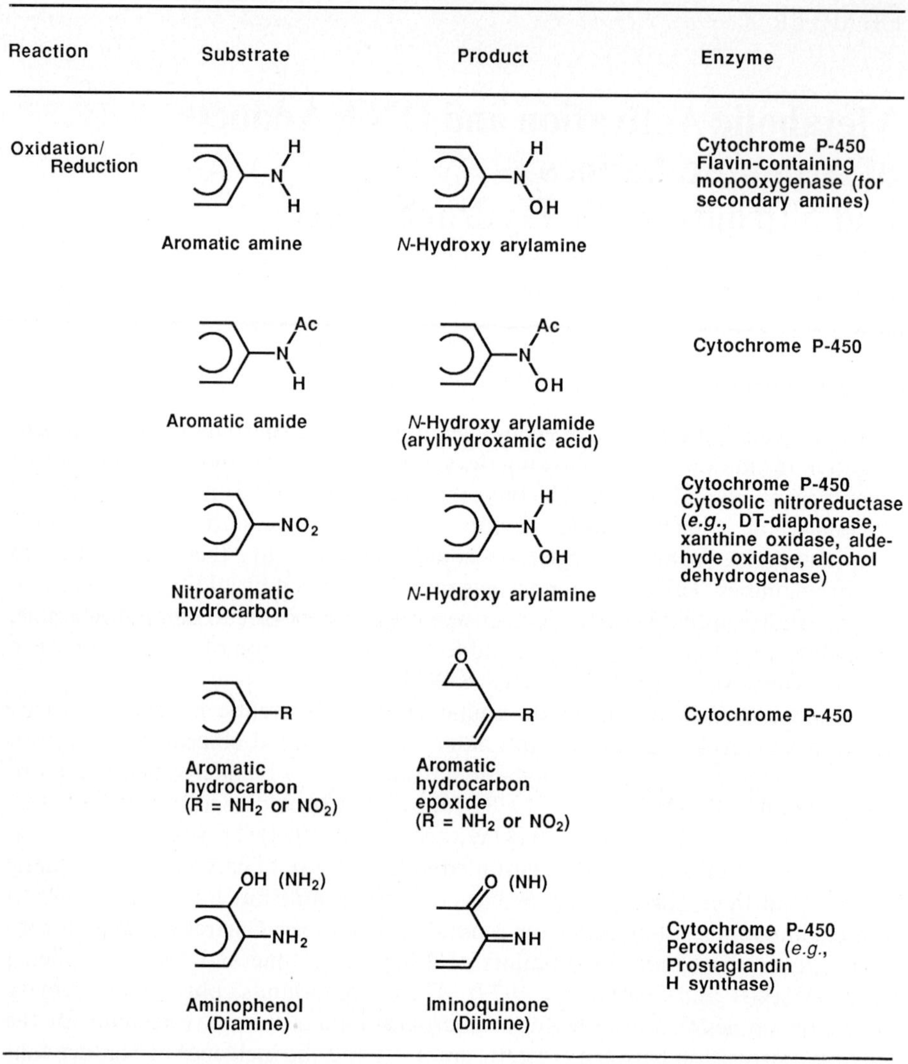

Reaction	Substrate	Product	Enzyme
Oxidation/ Reduction	Aromatic amine	*N*-Hydroxy arylamine	Cytochrome P-450 Flavin-containing monooxygenase (for secondary amines)
	Aromatic amide	*N*-Hydroxy arylamide (arylhydroxamic acid)	Cytochrome P-450
	Nitroaromatic hydrocarbon	*N*-Hydroxy arylamine	Cytochrome P-450 Cytosolic nitroreductase (*e.g.*, DT-diaphorase, xanthine oxidase, aldehyde oxidase, alcohol dehydrogenase)
	Aromatic hydrocarbon (R = NH_2 or NO_2)	Aromatic hydrocarbon epoxide (R = NH_2 or NO_2)	Cytochrome P-450
	Aminophenol (Diamine)	Iminoquinone (Diimine)	Cytochrome P-450 Peroxidases (*e.g.*, Prostaglandin H synthase)

Fig. 1. Metabolic activation pathways of aromatic amines and nitroaromatic hydrocarbons

particularly rich source of these compounds, are reported to be at increased risk for the development of urinary bladder cancer (SILVERMAN et al. 1983).

As with most chemical carcinogens, aromatic amines and nitroaromatic hydrocarbons need to be metabolized into reactive electrophiles in order to exert their carcinogenic effects. With aromatic amines and amides, this typically involves an initial *N*-oxidation to *N*-hydroxy arylamines and *N*-hydroxy arylamides (arylhydroxamic acids; Fig. 1), which in rat liver is mediated primarily by cytochrome P-450 isozymes *c* (BNF-B) and *d* (ISF-G) (reviewed in KADLUBAR and HAMMONS 1987). Arylamine *N*-oxidation is also catalyzed by the flavin-containing monooxygenase, but this appears to be restricted generally to

Reaction	Substrate	Product	Enzyme
Esterification	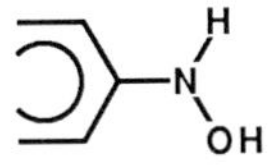 N-Hydroxy arylamine (Ar–N(H)–OH)	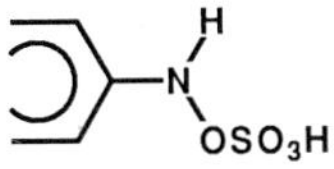 N-Sulfonyloxy arylamine (Ar–N(H)–OSO_3H)	Sulfotransferase
	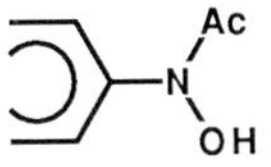 N-Hydroxy arylamide (Ar–N(Ac)–OH)	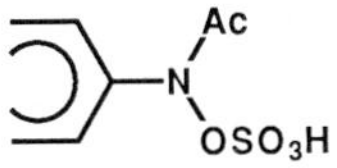 N-Sulfonyloxy arylamide (Ar–N(Ac)–OSO_3H)	Sulfotransferase
	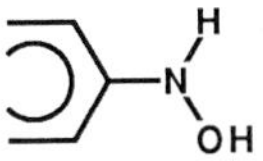 N-Hydroxy arylamine (Ar–N(H)–OH)	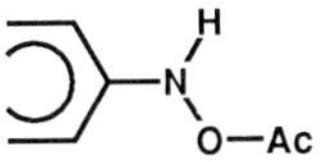 N-Acetoxy arylamine (Ar–N(H)–O—Ac)	Transacetylase
	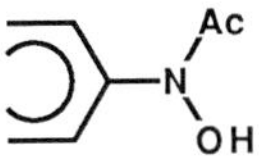 N-Hydroxy arylamide (Ar–N(Ac)–OH)	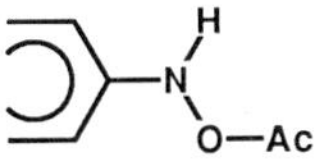 N-Acetoxy arylamine (Ar–N(H)–O—Ac)	Acyltransferase
	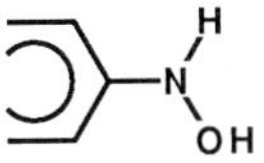 N-Hydroxy arylamine (Ar–N(H)–OH)	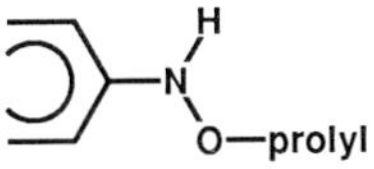 Amino acyl ester of N-hydroxy arylamine (Ar–N(H)–O—prolyl)	Amino acyl synthase

Fig. 1 (continued)

secondary amine substrates. The initial activation of nitroaromatic hydrocarbons is likewise through the formation of an *N*-hydroxy arylamine, a reduction catalyzed by both microsomal and cytosolic enzymes (KADLUBAR and HAMMONS 1987). Although microsomal nitroreduction has not been studied as extensively as amine oxidation, it appears to depend upon cytochrome P-450, in particular rat liver isozymes *c, d, b* (PB-B), and *e* (PB-D). Cytosolic nitroreductase activity is associated with a number of enzymes including DT-diaphorase, xanthine oxidase, aldehyde oxidase, and alcohol dehydrogenase. In addition to nitrogen oxidation and reduction reactions serving as activation pathways, certain aromatic amines and nitroaromatic hydrocarbons are converted into electrophilic derivatives through one of two general ring-oxidation pathways. The first is a cytochrome P-450-catalyzed epoxidation to metabolites analogous to those found with aromatic hydrocarbons (e.g. DJURIĆ et al. 1986a), while the second is through oxidation of phenolic arylamine metabolites to electrophilic iminoquinones (reviewed in KADLUBAR and BELAND 1985), which can be catalyzed by either cytochrome P-450 or by extrahepatic peroxidases (e.g. prostaglandin *H* synthase). *N*-Hydroxy arylamines, iminoquinones, and epoxide derivatives are directly electrophilic metabolites, while *N*-hydroxy arylamides require esterification before becoming capable of reacting with DNA (KADLUBAR and BELAND 1985). A number of different *N*-hydroxy arylamide esters have been described, and in some instances analogous esters exist for *N*-hydroxy arylamines, which further enhance their reactivity. These pathways are summarized in Fig. 1.

In the following review we will consider the metabolic pathways by which a representative series of aromatic amines and nitroaromatic hydrocarbons are activated. We will discuss the DNA adducts that result from this metabolic activation and, where possible, relate these adducts to their effects upon the structure of DNA and discuss their relevance to mutagenic and tumorigenic responses.

I. 1-Naphthylamine and 1-Nitronaphthalene

Although 1-naphthylamine was originally considered to be a human bladder carcinogen, the results of subsequent epidemiological studies coupled with the failure to demonstrate a carcinogenic response in animal models indicate that this is not the case (RADOMSKI 1979; RADOMSKI et al. 1980; PURCHASE et al. 1981). This lack of carcinogenicity appears to be due to the failure of 1-naphthylamine to be metabolized to a reactive electrophile. As noted in the introduction, the major activation pathway for most aromatic amines is a cytochrome P-450-catalyzed *N*-oxidation to an *N*-hydroxy arylamine. When the metabolism of 1-naphthylamine was studied using hepatic microsomes and purified hepatic cytochromes P-450 from rats, dogs, and humans (and also with the flavin-containing monooxygenase from porcine liver), *N*-oxidized metabolites were not detected (HAMMONS et al. 1985). Instead, the only metabolite found was the ring-oxidation product, 1-amino-2-naphthol (Fig. 2). This aminophenol has the potential to be oxidized to an electrophilic derivative, 1-imino-2-naphthoquinone. While the isomeric iminoquinone from 2-naphthylamine readily binds to DNA (YAMAZOE et al. 1985a), binding has not been investigated for the iminoquinone of 1-naphthylamine. Nevertheless, the low levels of DNA binding detected in vivo when 1-naphthylamine was administered to dogs (KADLUBAR et al. 1981a) suggest that either 1-amino-2-naphthol is effectively detoxified through conjugation, thus preventing the formation of substantial quantities of 1-imino-2-naphthoquinone, or it is not as electrophilic as the isomeric 2-imino-1-naphthoquinone.

Although 1-naphthylamine has not been found to be carcinogenic, its *N*-oxidized derivative, *N*-hydroxy-1-naphthylamine, is strongly tumorigenic (BELMAN et al. 1968; RADOMSKI et al. 1971; DOOLEY et al. 1984). *N*-Hydroxy-1-naphthylamine also readily binds to DNA, and as is observed with other *N*-

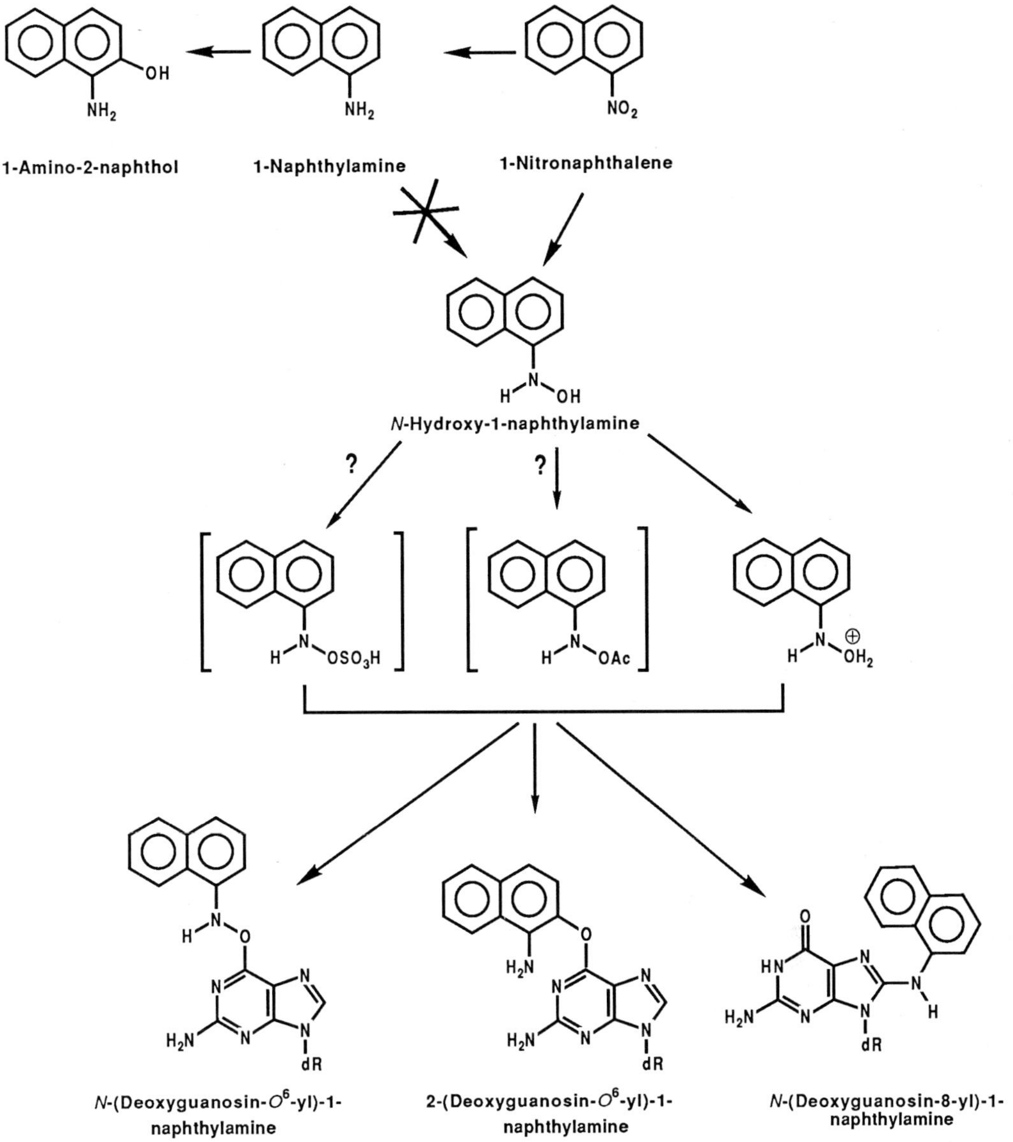

1-Amino-2-naphthol

1-Naphthylamine

1-Nitronaphthalene

N-Hydroxy-1-naphthylamine

N-(Deoxyguanosin-O^6-yl)-1-naphthylamine

2-(Deoxyguanosin-O^6-yl)-1-naphthylamine

N-(Deoxyguanosin-8-yl)-1-naphthylamine

Fig. 2. Metabolic activation pathways and DNA adducts of 1-naphthylamine and 1-nitronaphthalene

hydroxy arylamines the reaction is facilitated by acidic conditions, with 20-fold more binding observed at pH 5 than at pH 7 (KADLUBAR et al. 1977). This reaction results in the formation of two major DNA adducts through reaction of the aryl nitrogen and ortho carbon atoms at O^6 of deoxyguanosine (KADLUBAR et al. 1978). Evidence has also been presented that a minor adduct is formed by *N*-substitution at C8 of deoxyguanosine (MUROFUSHI et al. 1981; Fig. 2). Adduct formation at C8 of deoxyguanosine is typical for *N*-hydroxy arylamines, whereas reaction with the O^6 position, which is normally associated with S_N1-type reactions, thus far appears to be unique to *N*-hydroxy-1-naphthylamine.

The reaction of *N*-hydroxy arylamines with DNA is proposed to proceed through a protonated nitrenium ion pair (reviewed in KADLUBAR and BELAND 1985). Substitution of O^6 of deoxyguanosine would occur in compounds with greater nitrenium ion character. Interestingly, Hückel molecular orbital calculations indicate that the nitrenium ion of 1-naphthylamine has greater charge delocalization than the nitrenium ions of other arylamines (HAMMONS et al. 1985). This suggests that the protonated *N*-hydroxy derivative of 1-naphthylamine should react with more S_N1 character, which is consistent with the site of substitution observed in DNA. SCRIBNER and FISK (1978) arrived at a similar conclusion using Hückel molecular orbital-based polyelectronic perturbation theory.

The DNA adducts obtained in vivo in animals treated with *N*-hydroxy-1-naphthylamine are similar to those found in vitro, with *N*-(deoxyguanosin-O^6-yl)-1-naphthylamine being the major product (DOOLEY et al. 1984). This adduct is relatively persistent in target tissues (~30% loss in 1 week), which is consistent with both computational (BELAND 1978) and experimental (KADLUBAR et al. 1981 b) studies indicating that it should reside in the major groove of the DNA helix without causing any perturbations. *N*-Hydroxy-1-naphthylamine also binds to target tissue DNA 20-fold more than *N*-hydroxy-2-naphthylamine (DOOLEY et al. 1984). Since the same difference in reactivity is observed in vitro (KADLUBAR et al. 1977), adduct formation in vivo is likely to result from the direct reaction of the *N*-hydroxy arylamine with cellular DNA. Nonetheless, *N*-hydroxy-1-naphthylamine is a substrate for sulfotransferases (KADLUBAR et al. 1976 b), and although its substrate specificity for transacetylases has not been examined, these enzymes are present in cutaneous tissue (KAWAKUBO et al. 1988). Therefore, additional activation through the formation of *N*-sulfonyloxy or *N*-acetoxy esters cannot presently be excluded.

1-Nitronaphthalene is similar to 1-naphthylamine in that it has not been demonstrated to be carcinogenic in any animal model (NATIONAL CANCER INSTITUTE 1978). The reduction of 1-nitronaphthalene has been examined in vivo (JOHNSON and CORNISH 1978) and in anaerobic incubations with both microsomes and cytosol (POIRIER and WEISBURGER 1974; STERNSON 1975; TATSUMI et al. 1986). In each instance the major product was 1-naphthylamine; however, minor amounts of *N*-hydroxy-1-naphthylamine were detected in some of the in vitro incubations (STERNSON 1975; TATSUMI et al. 1986). This is consistent with the relatively low mutagenicity of 1-nitronaphthalene in bacterial mutagenesis assays (SCRIBNER et al. 1979 a) and suggests that only very low concentrations of *N*-hydroxy-1-naphthylamine are formed. The DNA adducts obtained from 1-nitronaphthalene have not been characterized; nevertheless, from bacterial mutagenesis assays using various strains of *Salmonella,* it can be inferred that they should be the same as those detected from *N*-hydroxy-1-naphthylamine. Thus, when 1-nitronaphthalene is assayed in *S. typhimurium* TA98NR, a strain deficient in nitroreductase activity, there is a 90% reduction in mutagenic activity compared with that observed in strain TA98 (ROSENKRANZ and MERMELSTEIN

1983). Since there is only a modest reduction in mutagenicity when 1-nitronaphthalene is tested in strain TA98/1,8-DNP$_6$ (ROSENKRANZ and MERMELSTEIN 1983), a strain deficient in transacetylase activity, it appears that the intermediate *N*-hydroxy arylamine is not further activated by *O*-acetylation, which is in accord with the in vivo data discussed earlier.

II. 2-Naphthylamine and 2-Nitronaphthalene

In contrast to 1-naphthylamine and its nitro analogue, 2-naphthylamine and 2-nitronaphthalene are carcinogenic in a number of laboratory animals (reviewed in FREDERICK and BELAND 1988). Furthermore, as previously noted, sufficient epidemiological data exist to conclude that 2-naphthylamine is a human urinary bladder carcinogen (IARC 1974).

2-Naphthylamine is metabolically activated through two general pathways: *N*-oxidation to *N*-hydroxy-2-naphthylamine and C-oxidation to 2-imino-1-naphthoquinone (Fig. 3). The first route is catalyzed by cytochrome P-450, and in the rat only one isozyme, *d*, appears to be capable of performing this *N*-oxidation (HAMMONS et al. 1985). *N*-Hydroxy-2-naphthylamine then undergoes acid-catalyzed reactions with DNA to yield three adducts (Fig. 3), two through substitution at C8 and N^2 of deoxyguanosine (50% and 30% of the total, respectively) and one through reaction with N^6 of deoxyadenosine (15%; KADLUBAR et al. 1980). Interestingly, the C8 deoxyguanosine adduct appears to occur as an 8,9-purine ring-opened structure, which is not typically found with arylamine adducts. *N*-Hydroxy-2-naphthylamine is a reasonable substrate for sulfotransferases (KADLUBAR et al. 1976b) but a poor substrate for transacetylases (FLAMMANG and KADLUBAR 1986); therefore, this *N*-hydroxy arylamine may be expected to be further activated through the formation of an *N*-sulfonyloxy but not an *N*-acetoxy ester.

When animals are treated with *N*-hydroxy-2-naphthylamine, the same adducts that have been detected in vitro are found in carcinogen-target tissues (DOOLEY et al. 1984). As noted in the section on 1-naphthylamine, the binding of *N*-hydroxy-2-naphthylamine to DNA in vitro (KADLUBAR et al. 1977) and in vivo (DOOLEY et al. 1984) is considerably less than that of *N*-hydroxy-1-naphthylamine, a difference that correlates with the extent of tumor induction by these two *N*-hydroxy compounds (DOOLEY et al. 1984).

DNA adducts indicative of *N*-hydroxy-2-naphthylamine formation have also been detected in the urinary bladders of dogs treated with 2-naphthylamine (KADLUBAR et al. 1981a). Since this target tissue does not contain detectable levels of cytochrome P-450 (WISE et al. 1984b), a number of investigations have been conducted to elucidate the mechanisms by which 2-naphthylamine and other aromatic amines are activated to bladder carcinogens. One hypothesis suggests that *N*-oxidation and *N*-glucuronidation occur in the liver (KADLUBAR et al. 1977; POUPKO et al. 1979). The resultant *N*-hydroxy *N*-glucuronide is then transported to the urinary bladder where it can undergo acid-catalyzed hydrolysis to release *N*-hydroxy-2-naphthylamine. In support of this hypothesis, a pharmacokinetic model based upon urinary pH and relative voiding interval

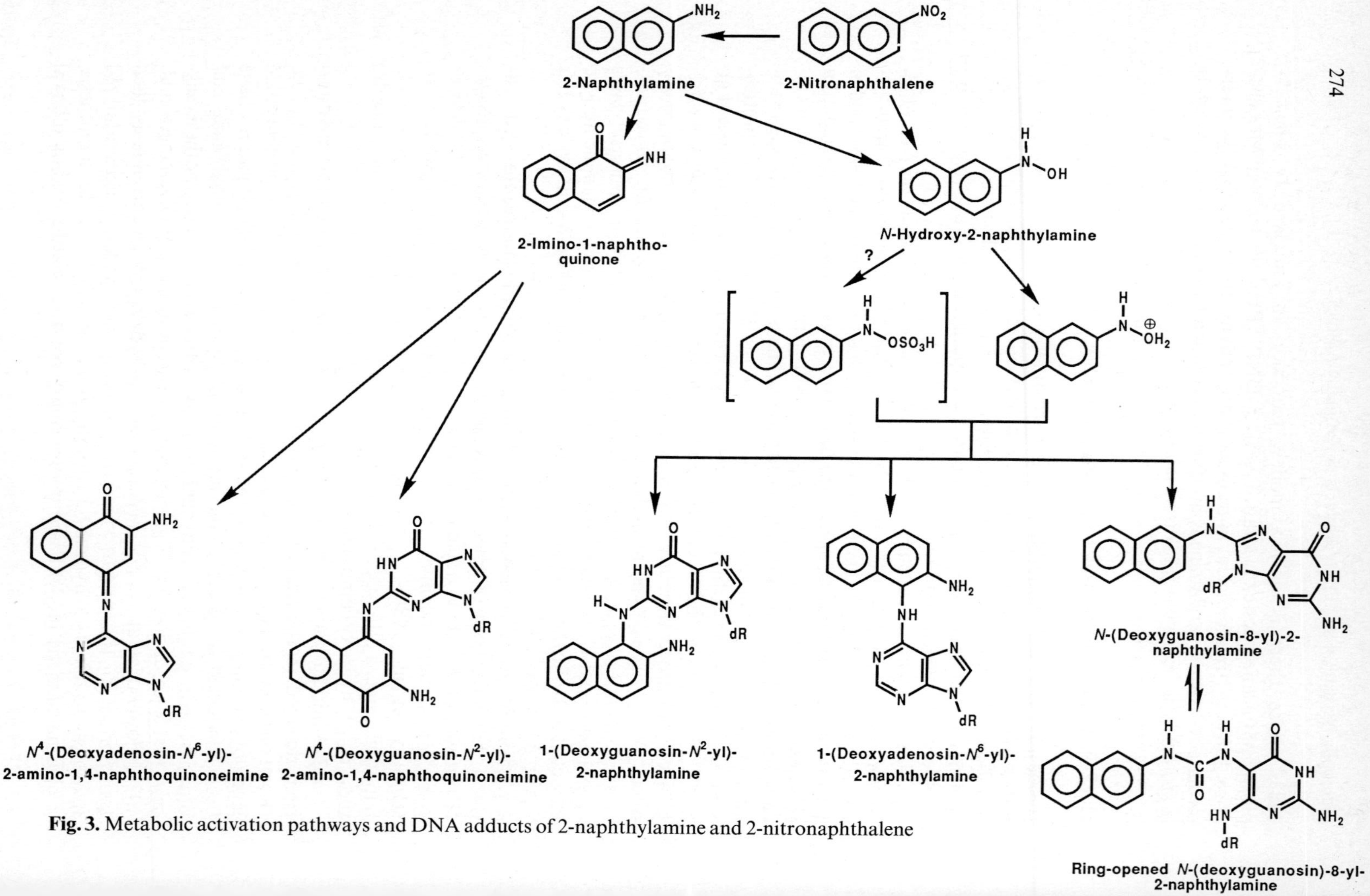

2-Naphthylamine

2-Nitronaphthalene

2-Imino-1-naphthoquinone

N-Hydroxy-2-naphthylamine

N^4-(Deoxyadenosin-N^6-yl)-2-amino-1,4-naphthoquinoneimine

N^4-(Deoxyguanosin-N^2-yl)-2-amino-1,4-naphthoquinoneimine

1-(Deoxyguanosin-N^2-yl)-2-naphthylamine

1-(Deoxyadenosin-N^6-yl)-2-naphthylamine

N-(Deoxyguanosin-8-yl)-2-naphthylamine

Ring-opened *N*-(deoxyguanosin)-8-yl-2-naphthylamine

Fig. 3. Metabolic activation pathways and DNA adducts of 2-naphthylamine and 2-nitronaphthalene

correctly predicts species susceptibility to bladder tumor induction by 2-naphthylamine (YOUNG and KADLUBAR 1982).

A second hypothesis suggests that reactive metabolites of 2-naphthylamine are generated by peroxidases, in particular prostaglandin *H* synthase (KADLUBAR et al. 1982b; MORTON et al. 1983; BOYD and ELING 1987). This enzyme is found in high concentrations in dog and human bladders (WISE et al. 1984b; KADLUBAR et al. 1988a) and has been shown to catalyze the C-oxidation of 2-naphthylamine to 2-imino-1-naphthoquinone (Fig. 3; KADLUBAR et al. 1982b; YAMAZOE et al. 1985a; BOYD and ELING 1987). This iminoquinone is electrophilic, and, upon reaction with DNA, two adducts are formed through substitution of N^2 of deoxyguanosine and N^6 of deoxyadenosine, respectively (YAMAZOE et al. 1985a). Approximately 20% of the radioactivity bound to DNA in the bladder epithelium, but not the liver, of a dog administered [^{3}H]2-naphthylamine coeluted with these markers (YAMAZOE et al. 1985a). This indicates that 2-imino-1-naphthoquinone is formed in vivo and implies that prostaglandin *H* synthase and/or other peroxidases may play a role in the activation of 2-naphthylamine. It should be noted, however, that 2-amino-1-naphthol readily oxidizes to 2-imino-1-naphthoquinone. Furthermore, 2-amino-1-naphthol and its *O*-sulfuric acid ester and *O*-glucuronide conjugate account for >90% of the urinary metabolites of 2-naphthylamine, while 2-naphthylamine and its *N*-glucuronide are relatively minor (~2%) metabolites (DEICHMANN and RADOMSKI 1969; also see KADLUBAR et al. 1981c). Presumably, the majority of the urinary 2-amino-1-naphthol results from hepatic (i.e., cytochrome P-450) ring oxidation (HAMMONS et al. 1985); thus, a role of hepatic metabolism in the formation of the imino-DNA adducts that are detected in the bladder cannot presently be excluded.

Prostaglandin *H* synthase has also been proposed to catalyze the *N*-oxidation of 2-naphthylamine (KADLUBAR et al. 1982b; YAMAZOE et al. 1985a), although this hypothesis has been questioned (BOYD and ELING 1987). DNA adducts coeluting with products derived from *N*-hydroxy-2-naphthylamine have been detected in incubations with prostaglandin *H* synthase (YAMAZOE et al. 1985a); however, the adduct derived from C8 of deoxyguanosine is a relatively minor adduct whereas it is the major adduct obtained in in vitro incubations with *N*-hydroxy-2-naphthylamine (KADLUBAR et al. 1980). These results in addition to others led BOYD and ELING (1987) to suggest that the additional DNA adducts resulting from prostaglandin *H* synthase metabolism were due to a free radical intermediate.

2-Nitronaphthalene has been studied less extensively than 2-naphthylamine, but available data suggest that they act through common intermediates. For example, both compounds exhibit similar target specificity by inducing bladder tumors in dogs and monkeys (reviewed in FREDERICK and BELAND 1988). In addition, 2-nitronaphthalene is readily reduced to 2-naphthylamine during in vitro incubations with hepatic cytosol and microsomes (POIRIER and WEISBURGER 1974); and the urinary metabolic profile of 2-nitronaphthalene is similar to that obtained from 2-naphthylamine (KADLUBAR et al. 1981c). Furthermore, the strain sensitivity in *Salmonella* mutagenesis assays indicates that nitroreduction is critical for the metabolic activation because there is a marked decrease in mutations in strain TA98NR compared with the normal tester strain TA98 (ROSENKRANZ and MERMELSTEIN 1983). Since the mutagenicity is not greatly decreased when 2-nitronaphthalene is assayed in strain TA98/1,8-DNP$_6$, the intermediate, *N*-hydroxy-2-naphthylamine, does not appear to be further activated by bacterial *O*-acetylation. This is consistent with results obtained with the hepatic transacetylases (FLAMMANG and KADLUBAR 1986).

III. 4-Aminobiphenyl and Derivatives

4-Aminobiphenyl is similar to 2-naphthylamine in that it is a bladder carcinogen in both dogs (WALPOLE et al. 1954; DEICHMANN et al. 1958 b) and humans (IARC 1972). As such, many of the principles discussed for 2-naphthylamine are applicable to this aromatic amine. For instance, as was observed with 2-naphthylamine, the *N*-oxidation of 4-aminobiphenyl is catalyzed to the greatest extent by rat liver cytochrome P-450*d* (KADLUBAR et al. 1988 a, c); however, in contrast to 2-naphthylamine, isozymes *a, b, c, e,* and *h* also show low but significant activity (MCMAHON et al. 1980; KAMATAKI et al. 1983; MASSON et al. 1983; KADLUBAR et al. 1988 a, c).

4-Aminobiphenyl may also be metabolized to *N*-hydroxy-4-aminobiphenyl by a noncytochrome P-450 pathway. Specifically, 4-aminobiphenyl is readily *N*-methylated by *S*-adenosylmethionine-dependent *N*-methyltransferases, and the resultant secondary amine is a substrate for the hepatic flavin-containing monooxygenase (ZIEGLER et al. 1988). Subsequent oxidation of the *N*-hydroxy-*N*-methyl-4-aminobiphenyl to an arylnitrone followed by hydrolysis will yield *N*-hydroxy-4-aminobiphenyl.

Although the *N*-glucuronidation of *N*-hydroxy-4-aminobiphenyl has long been regarded as the mechanism by which *N*-hydroxy-4-aminobiphenyl is transported to the urinary bladder (KADLUBAR et al. 1977; POUPKO et al. 1979), recent studies in the dog indicate that the unconjugated *N*-hydroxy-4-aminobiphenyl is the predominant form that enters the bladder lumen (KADLUBAR et al. 1988 b). For example, following a single treatment with 4-aminobiphenyl, *N*-hydroxy-4-aminobiphenyl accounts for about 1% of the administered dose, while its *N*-glucuronide is present in the urine at only about 0.3% of the dose. This is consistent with in vitro findings that, relative to other *N*-hydroxy arylamines, *N*-hydroxy-4-aminobiphenyl is a poor substrate for hepatic microsomal glucuronosyl transferases (KADLUBAR et al. 1977). Furthermore, the high levels of aromatic amine-hemoglobin adducts ($\sim$10% of the administered dose) that are found in vivo after 4-aminobiphenyl treatment (GREEN et al. 1984) are also consistent with high levels of free *N*-hydroxy-4-aminobiphenyl in the circulation.

The metabolically formed *N*-hydroxy-4-aminobiphenyl can undergo acid-catalyzed reactions with DNA (Fig. 4), which produce adducts substituted through C8 of deoxyguanosine and deoxyadenosine (70% and 15%, respectively) and N^2 of deoxyguanosine (5%; KADLUBAR et al. 1982 a; BELAND et al. 1983). Interestingly, the N^2 deoxyguanosine adduct is linked through the amine nitrogen to give a hydrazo structure; such an adduct has not been observed with any other aromatic amine carcinogen, with the exception of benzidine (YAMAZOE et al. 1988 b).

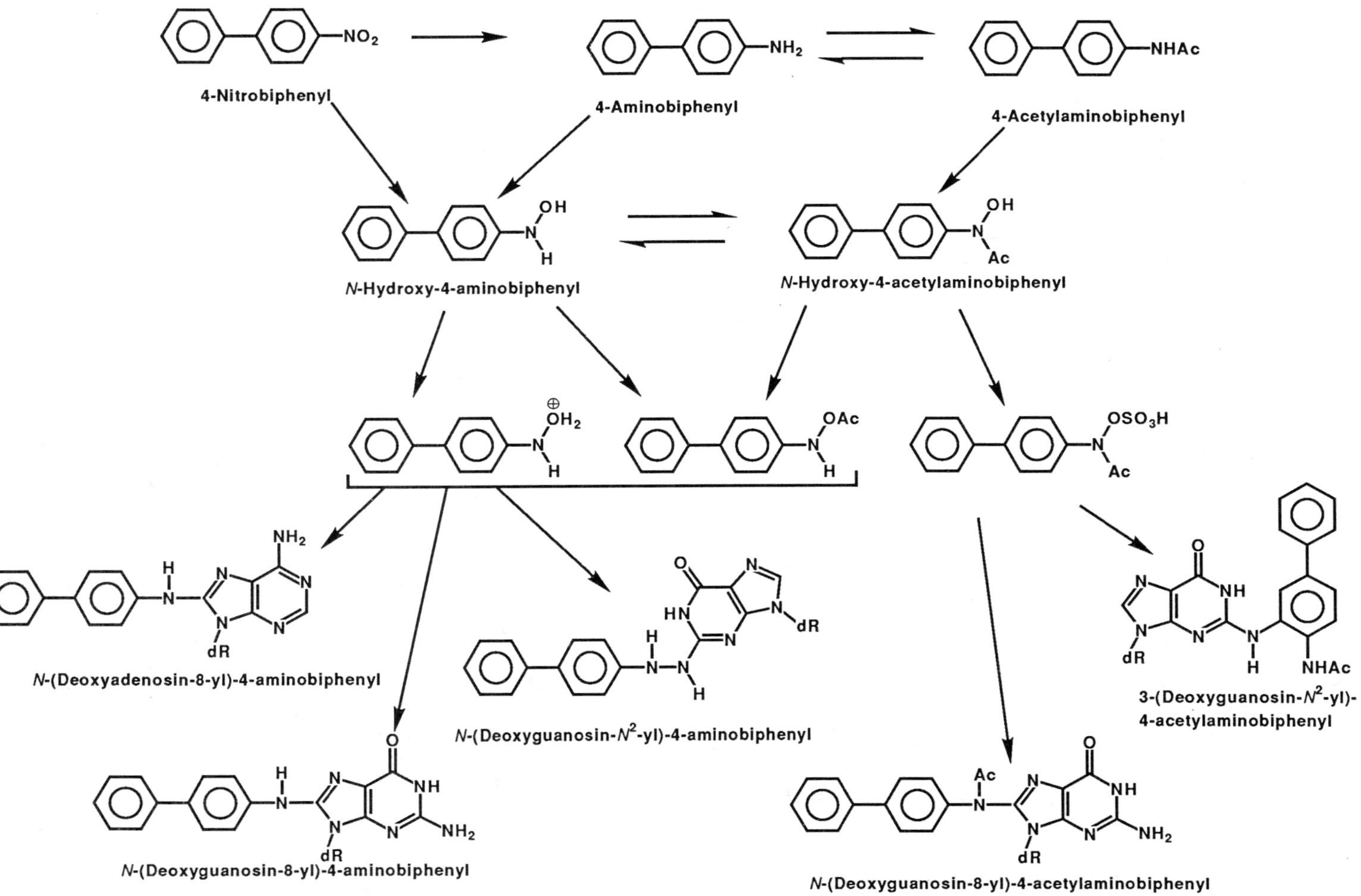

Fig. 4. Metabolic activation pathways and DNA adducts of 4-aminobiphenyl, 4-acetylaminobiphenyl, and 4-nitrobiphenyl

4-Aminobiphenyl is also a substrate for prostaglandin *H* synthase (KADLUBAR et al. 1982b; MORTON et al. 1983), and when incubations are conducted in the presence of DNA, one major and at least one minor adduct are formed (KADLUBAR et al. 1988a). Although the major adduct has not yet been characterized, the minor adduct coelutes with *N*-(deoxyguanosin-8-yl)-4-aminobiphenyl, which is consistent with the formation of *N*-oxidized metabolites during in vitro incubations with this peroxidase (KADLUBAR et al. 1982b). However, the adduct profile in the bladder epithelium of dogs treated with [^{3}H]4-aminobiphenyl is nearly the same as that found after in vitro incubations with *N*-hydroxy-4-aminobiphenyl (KADLUBAR et al. 1982a, 1988a; BELAND et al. 1983). Since only a small amount of radioactivity ($\sim$10%) from dog bladder DNA hydrolysates coelutes with the major prostaglandin *H* synthase adduct (KADLUBAR et al. 1988a), this implies that peroxidase-catalyzed activation of 4-aminobiphenyl plays only a minor role in this species. It should be noted, however, that with human bladder microsomes there is a significantly higher level of prostaglandin *H* synthase activity toward 4-aminobiphenyl and other aromatic amines (KADLUBAR et al. 1988a). Thus, the peroxidative activation pathway may be of greater relevance for aromatic amine-induced bladder tumors in humans as compared with experimental animals.

In addition to inducing urinary bladder tumors in dogs and humans, 4-aminobiphenyl and its amide derivative, 4-acetylaminobiphenyl, induce mammary gland tumors in rats (WALPOLE et al. 1952; MILLER et al. 1956), but, as opposed to the closely related arylamide, 2-acetylaminofluorene, 4-acetylaminobiphenyl is not hepatocarcinogenic in rats (SHIRAI et al. 1981b). This difference in tumorigenicity appears to be a function of the extent of DNA binding because, when the *N*-hydroxy arylamide derivatives are administered at equimolar doses, the binding of *N*-hydroxy-4-acetylaminobiphenyl is only 10%–25% that of *N*-hydroxy-2-acetylaminofluorene (KRIEK 1971; GUPTA and DIGHE 1984). At least three hepatic DNA adducts are formed in rats from *N*-hydroxy-4-acetylaminobiphenyl. The major hepatic adduct is *N*-(deoxyguanosin-8-yl)-4-aminobiphenyl (GUPTA and DIGHE 1984), which presumably arises from *N,O*-acyltransferase activation (SHIRAI et al. 1981a) or through deacetylation of the arylhydroxamic acid and possibly *O*-acetylation of the resultant *N*-hydroxy arylamine (FLAMMANG and KADLUBAR 1986). There are also two acetylated adducts formed through substitution at C8 and N^2 of deoxyguanosine (KRIEK and WESTRA 1979), which is consistent with sulfotransferase catalysis. The mammary gland DNA adducts have apparently not been examined; however, this tissue contains *N,O*-acyltransferase, and when the substrate specificities of *N*-hydroxy-4-acetylaminobiphenyl and *N*-hydroxy-2-acetylaminofluorene are compared, the extent of their nucleic acid binding parallels their relative tumorigenicities (KING et al. 1979; SHIRAI et al. 1981b). These data, together with the results obtained with *N*-hydroxy-2-acetylaminofluorene (ALLABEN et al. 1983), suggest that the predominant adduct is *N*-(deoxyguanosin-8-yl)-4-aminobiphenyl.

The persistence of the adducts derived from 4-aminobiphenyl and 4-acetylaminobiphenyl differs from species to species. Following a single dose of *N*-hydroxy-4-acetylaminobiphenyl to rats, there is >60% decrease in all hepatic DNA adducts within 9 days (GUPTA and DIGHE 1984). In contrast, in dogs treated with 4-aminobiphenyl, hepatic and bladder DNA adducts remain at constant levels for at least 1 week (BELAND et al. 1983). Oddly, the concentration of similar adducts derived from 2-acetylaminofluorene is reduced 80% in dog liver and bladder during the same period (BELAND et al. 1983). This difference in persistence could be due to conformational differences between the adducts; for example, SHAPIRO et al. (1986) have noted that during replication *N*-(deoxyguanosin-8-yl)-4-aminobiphenyl may exhibit less stacking with adjacent bases than *N*-(deoxyguanosin-8-yl)-2-aminofluorene. This conformational difference may reduce the recognition of *N*-(deoxyguanosin-8-yl)-4-aminobiphenyl by mammalian repair enzymes but might also account for its relatively low frameshift mutagenicity in *S. typhimurium* TA1538 as compared with *N*-(deoxyguanosin-8-yl)-2-aminofluorene (BELAND et al. 1983).

There have been comparatively fewer investigations of the DNA adducts formed by 4-aminobiphenyl derivatives. 4-Nitrobiphenyl is known to be a bladder carcinogen in dogs (DEICHMANN et al. 1958a), although less potent than 4-aminobiphenyl. Since both compounds produce adducts in dog bladder epithelium that are similar to those obtained from *N*-hydroxy-4-aminobiphenyl (BELAND et al. 1983), they probably act through a common intermediate. 4-Nitrobiphenyl also binds to bladder epithelium DNA at only 5% of the level observed with 4-aminobiphenyl (BELAND et al. 1983), which is consistent with the lower tumorigenicity of the nitro analogue.

As noted earlier, 4-acetylaminobiphenyl is a mammary, not a liver carcinogen in rats. Interestingly, 4′-fluoro-4-acetylaminobiphenyl, in addition to inducing mammary gland tumors, is a liver and kidney carcinogen in rodents (MATTHEWS and WALPOLE 1958). The binding of *N*-hydroxy-4-acetylaminobiphenyl and its 4′-fluoro derivative has been compared, and, while similar adducts are formed, the fluorinated analogue binds to a greater extent to liver and kidney DNA (KRIEK 1971; KRIEK and HENGEVELD 1977, 1978; KRIEK and WESTRA 1979). The reasons for this difference in binding are not known. It has been shown that the *N*-sulfonyloxy esters of these *N*-hydroxy arylamides bind to a similar extent in vitro (KRIEK and HENGEVELD 1978); however, the major adducts in vivo are non-acetylated and presumably do not arise from this type of reactive intermediate. It is possible that 4-acetylaminobiphenyl is detoxified more efficiently than 4′-fluoro-4-acetylaminobiphenyl, perhaps through hydroxylation and conjugation at the 4′-position.

In contrast to other aromatic amines, 3,2′-dimethyl-4-aminobiphenyl is primarily a colon carcinogen in rats (WALPOLE et al. 1952). As with aromatic amines carcinogenic to the bladder, 3,2′-dimethyl-4-aminobiphenyl has been proposed to be *N*-hydroxylated and *N*-glucuronidated in the liver and then transported in the bile to the intestine where the conjugate undergoes either enzymatic or acid hydrolysis to release *N*-hydroxy-3,2′-dimethyl-4-aminobiphenyl (KADLUBAR et al. 1981c; NUSSBAUM et al. 1983). In support of this hypothesis, *N*-hydroxy-3,2′-dimethyl-4-aminobiphenyl *N*-glucuronide has been detected as a major biliary metabolite following administration of 3,2′-dimethyl-4-aminobiphenyl (NUSSBAUM et al. 1983). In addition, *N*-hydroxy-3,2′-dimethyl-4-aminobiphenyl reacts readily with DNA in vitro (FLAMMANG et al. 1985) to give the same types of adducts found in vivo (WESTRA et al. 1985), with a C8

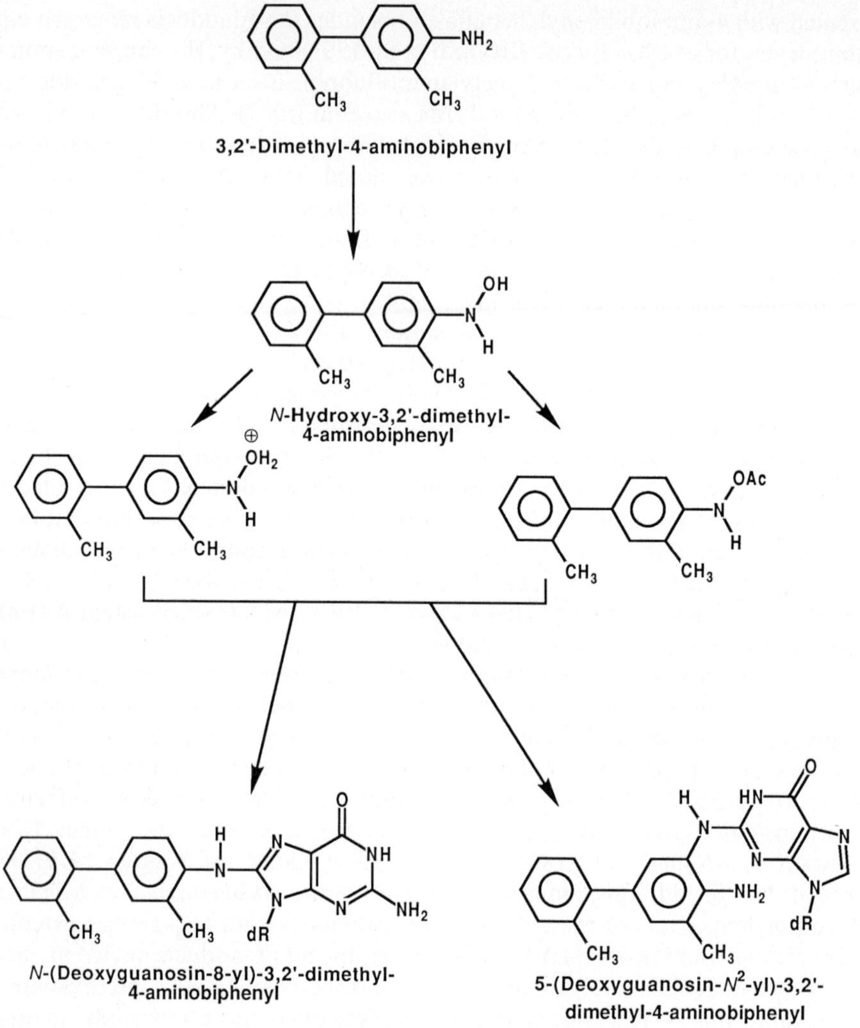

3,2'-Dimethyl-4-aminobiphenyl

N-Hydroxy-3,2'-dimethyl-4-aminobiphenyl

N-(Deoxyguanosin-8-yl)-3,2'-dimethyl-4-aminobiphenyl

5-(Deoxyguanosin-N^2-yl)-3,2'-dimethyl-4-aminobiphenyl

Fig. 5. Metabolic activation pathways and DNA adducts of 3,2′-dimethyl-4-aminobiphenyl

deoxyguanosine adduct being the major product (Fig. 5). 3,2′-Dimethyl-4-aminobiphenyl differs from a number of arylamine carcinogens in that its arylhydroxamic acid is not a substrate for either sulfotransferase or *N,O*-acyltransferase (FLAMMANG et al. 1985). The lack of metabolic activation by these pathways resulted in a search for alternate routes, which led to the discovery of a direct *O*-acetylation (but not *O*-sulfonation) of *N*-hydroxy-3,2′-dimethyl-4-aminobiphenyl (FLAMMANG et al. 1985). This pathway has subsequently been found to be important for a number of other *N*-hydroxy arylamines (SHINOHARA et al. 1985, 1986; DJURIĆ et al. 1985; FLAMMANG and KADLUBAR 1986).

Although 3,2′-dimethyl-4-aminobiphenyl is not a liver carcinogen, it binds to hepatic DNA to a greater extent than to intestinal DNA (WESTRA et al. 1985). The fact that 3,2′-dimethyl-4-aminobiphenyl induces intestinal tumors has been attributed to the higher rate of DNA synthesis in this tissue as compared with the liver, which would be expected to enhance the formation of mutagenic lesions through base mispairing and replication (WESTRA et al. 1985). It may also be significant that the *N*-oxidized metabolites of this aromatic amine are not substrates for sulfotransferase since there seems to be a positive relationship between this property and the ability of an aromatic amine to serve as a complete hepatocarcinogen in rats (RINGER and NORTON 1987).

IV. 2-Acetylaminofluorene, 2-Aminofluorene, and 2-Nitrofluorene

2-Acetylaminofluorene, which was initially intended to be used as an insecticide (WILSON et al. 1941), is unquestionably one of the most extensively studied chemical carcinogens. This aromatic amide and its amine and nitro derivatives induce tumors at a wide variety of sites including liver, urinary bladder, mammary gland, intestine, and forestomach (reviewed in GARNER et al. 1984). As with the compounds discussed previously, their initial activation normally involves the formation of *N*-hydroxy intermediates. The *N*-oxidation of 2-aminofluorene and 2-acetylaminofluorene is catalyzed by cytochrome P-450, and in rat liver this is primarily due to isozymes *c* and *d*, although a number of additional isozymes are also capable of *N*-oxidizing 2-aminofluorene (reviewed in KADLUBAR and HAMMONS 1987). The reduction of 2-nitrofluorene has been shown to be catalyzed by cytochrome P-450, in particular rat liver isozymes *b* and *d* (KAWANO et al. 1985), and is believed to give an *N*-hydroxy arylamine intermediate. Ring-oxidized derivatives are also formed during the in vivo metabolism of 2-nitrofluorene (MÖLLER et al. 1988). These metabolites retain their mutagenic activity and may form reactive *N*-hydroxy or iminoquinone derivatives (MÖLLER et al. 1988).

The metabolic intermediate, *N*-hydroxy-2-aminofluorene, will react directly with DNA (KRIEK 1965; KING and PHILLIPS 1969), but in contrast to the *N*-hydroxy arylamines discussed previously, only one adduct, *N*-(deoxyguanosin-8-yl)-2-aminofluorene (Fig. 6), appears to be formed (WESTRA and VISSER 1979; BELAND et al. 1980a). This is also the major DNA adduct obtained in vivo following the administration of 2-aminofluorene, 2-acetylaminofluorene, or their *N*-hydroxy derivatives (MEERMAN et al. 1981; VISSER and WESTRA 1981; BELAND et al. 1982, 1983; POIRIER et al. 1982, 1984, 1988; ALLABEN et al. 1983; GUPTA and DIGHE 1984; LAI et al. 1985, 1987; ELING et al. 1988); however, the reactive intermediate leading to this adduct can differ significantly from compound to compound and species to species. In rats treated with *N*-hydroxy-2-acetylaminofluorene, the *N*-(deoxyguanosin-8-yl)-2-aminofluorene that is formed in liver and mammary gland has been suggested to arise from an *N,O*-acyltransferase-catalyzed reaction (KING 1974; SHIRAI et al. 1981a; KING et al. 1979; ALLABEN et al. 1982, 1983; KING and GLOWINSKI 1983). In principle, this adduct could also arise from *N*-deacetylation to *N*-hydroxy-2-aminofluorene and its direct reaction with DNA (FREDERICK et al. 1982). However, incubation of rat

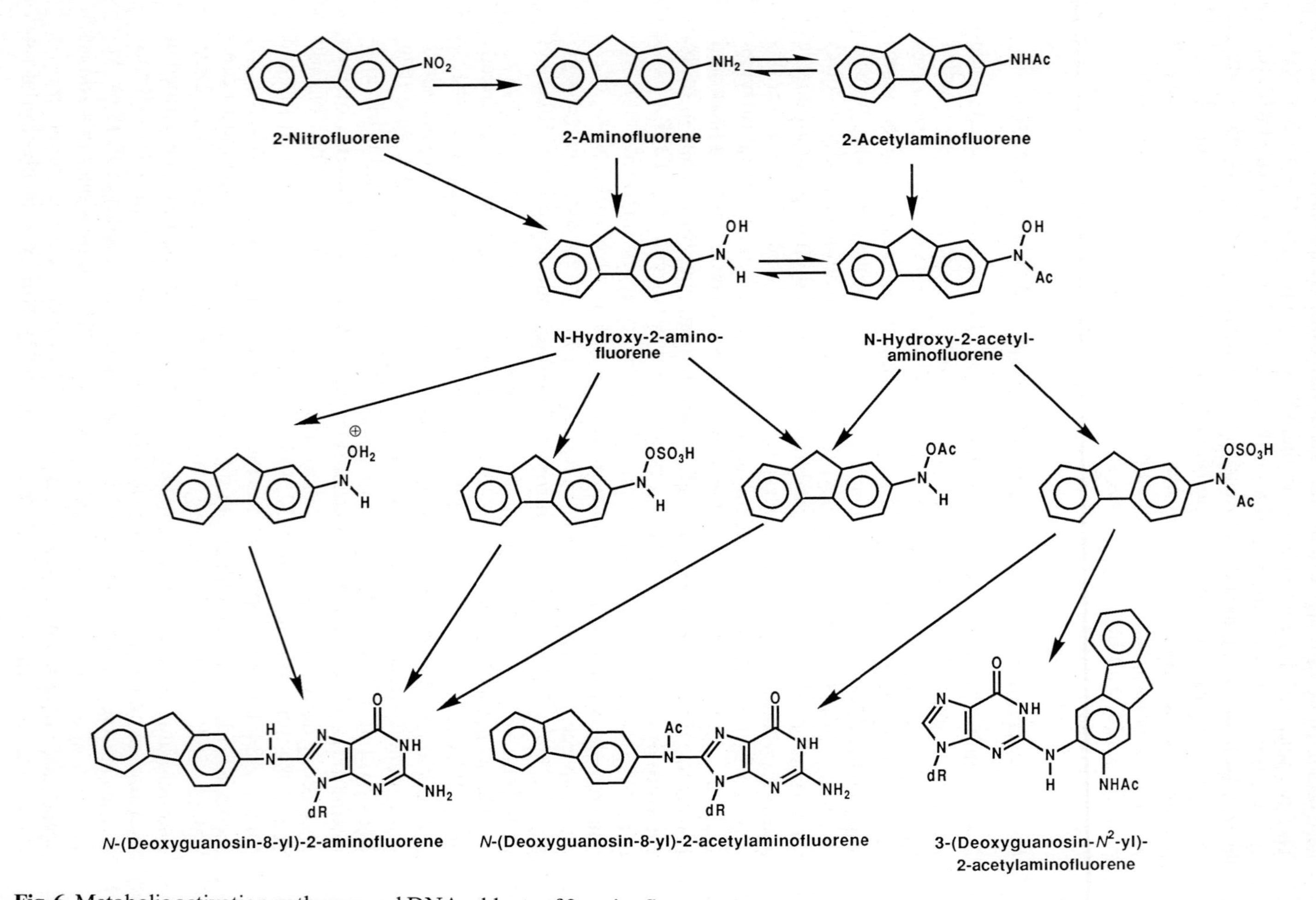

Fig. 6. Metabolic activation pathways and DNA adducts of 2-aminofluorene, 2-acetylaminofluorene, and 2-nitrofluorene

hepatocytes with *N*-hydroxy-2-acetylaminofluorene and the deacylase inhibitor, paraoxon, does not decrease the level of *N*-(deoxyguanosin-8-yl)-2-aminofluorene (HOWARD et al. 1981).

In mice a different metabolic activation pathway appears to predominate. This species has very low *N,O*-acyltransferase activity (KING and GLOWINSKI 1983), and recent studies indicate that *N*-hydroxy-2-acetylaminofluorene is deacetylated to *N*-hydroxy-2-aminofluorene, which is converted into a reactive *N*-sulfonyloxy ester (LAI et al. 1985, 1987).

An additional ester of *N*-hydroxy-2-aminofluorene that appears to be important for bacterial mutagenicity (MCCOY et al. 1982, 1983) and for DNA adduct formation in vivo is *N*-acetoxy-2-aminofluorene. Cytosols from a number of species and tissues are capable of catalyzing the *O*-acetylation of this *N*-hydroxy metabolite (FLAMMANG and KADLUBAR 1986; SHINOHARA et al. 1986), and studies with rabbits and mice whose transacetylase levels are genetically determined show that the rapid acetylator strains are much more susceptible to RNA and DNA damage (MCQUEEN et al. 1982; SHINOHARA et al. 1986) and to 2-aminofluorene-RNA and -DNA adduct formation (SHINOHARA et al. 1986; LEVY and WEBER 1988).

Two *N*-acetylated adducts, *N*-(deoxyguanosin-8-yl)-2-acetylaminofluorene and 3-(deoxyguanosin-N^2-yl)-2-acetylaminofluorene (Fig. 6), have also been detected in rat and mouse liver following the administration of 2-acetylaminofluorene and/or *N*-hydroxy-2-acetylaminofluorene (KRIEK 1972; WESTRA et al. 1976; BELAND et al. 1979, 1982; MEERMAN et al. 1981; VISSER and WESTRA 1981; POIRIER et al. 1982, 1984, 1988; GUPTA and DIGHE 1984; LAI et al. 1985, 1987). Results from experiments with the sulfotransferase inhibitor pentachlorophenol indicate that both of these adducts arise from *N*-sulfonyloxy-2-acetylaminofluorene (MEERMAN et al. 1981; LAI et al. 1985, 1987). In principle, they could also be formed from the *O*-acetylated intermediate, *N*-acetoxy-2-acetylaminofluorene, which is formed by the direct reaction of acetyl coenzyme A with *N*-hydroxy-2-acetylaminofluorene (LOTLIKAR and LUHA 1971); however, treatment of cells with this derivative generally results in the extensive formation of *N*-(deoxyguanosin-8-yl)-2-aminofluorene (MAHER et al. 1980; POIRIER et al. 1980; ARCE et al. 1987; HEFLICH et al. 1988), presumably as a result of the *N*-deacetylation of *N*-acetoxy-2-acetylaminofluorene to *N*-acetoxy-2-aminofluorene (HEFLICH et al. 1988). A number of minor adducts have also been reported both in vitro and in vivo, but none of these has been characterized extensively (KRIEK and REITSEMA 1971; HARVAN et al. 1977; GUPTA and DIGHE 1984).

Numerous studies have been conducted to determine the structural and biological consequences of these adducts. When incorporated into DNA, deoxyguanosine normally adopts an *anti* conformation about its glycosyl bond. In contrast, both experimental and theoretical studies indicate that, due to its *N*-acetyl moiety, *N*-(deoxyguanosin-8-yl)-2-acetylaminofluorene exists in a *syn* conformation (GRUNBERGER et al. 1970; NELSON et al. 1971; FUCHS and DAUNE 1972, 1974; LAVINE et al. 1974; FUCHS et al. 1976; LEFÉVRE et al. 1978; EVANS et al. 1980; LENG et al. 1980; BROYDE and HINGERTY 1982, 1985; EVANS and MILLER 1982; HINGERTY and BROYDE 1982, 1983; LIPKOWITZ et al. 1982; NEIDLE et al.

1984; DAUNE et al. 1985). This causes the 2-acetylaminofluorene to be inserted into the helix, which results in a denaturation of ~12 base pairs (FUCHS and DAUNE 1974), a process termed "base displacement" (LAVINE et al. 1974) or "insertion denaturation" (FUCHS 1975). A greater variety of conformations is available to *N*-(deoxyguanosin-8-yl)-2-aminofluorene, and several structures have been proposed including an outside binding model, a "kinked" complex, and an insertion model similar to that described for *N*-(deoxyguanosin-8-yl)-2-acetylaminofluorene (SPODHEIM-MAURIZOT et al. 1979, 1980; EVANS et al. 1980; LENG et al. 1980; LIPKOWITZ et al. 1982; BROYDE and HINGERTY 1983; HINGERTY and BROYDE 1986; VAN HOUTE et al. 1987). Any or all of these conformations may exist, but the net effect is that there is considerably less denaturation in DNA modified with *N*-(deoxyguanosin-8-yl)-2-aminofluorene as compared with *N*-(deoxyguanosin-8-yl)-2-acetylaminofluorene (KRIEK and SPELT 1979; MELCHIOR and BELAND 1984). There have not been any experimental studies conducted with 3-(deoxyguanosin-N^2-yl)-2-acetylaminofluorene, although a theoretical investigation has suggested that it can reside without steric hindrance in the minor groove of the double helix (BELAND 1978).

The conformational differences between these adducts appear to affect their biological properties. For example, whereas each of these adducts is chemically stable, *N*-(deoxyguanosin-8-yl)-2-acetylaminofluorene has a half-life in rat liver of approximately 7 days, while the other two adducts are relatively persistent (KRIEK 1972; WESTRA et al. 1976; VISSER and WESTRA 1981; POIRIER et al. 1982, 1984, 1988; BELAND et al. 1982; GUPTA and DIGHE 1984). This difference in persistence may result from *N*-(deoxyguanosin-8-yl)-2-acetylaminofluorene inducing structural distortions into the DNA, which could be a signal for DNA repair enzymes that cause excision of the adduct. The rapid removal of this adduct, coupled with the decrease in sulfotransferase activity upon continuous administration of 2-acetylaminofluorene (DEBAUN et al. 1970; JACKSON and IRVING 1972; RINGER et al. 1983; RINGER and NORTON 1987), increases the ratio of non-acetylated to acetylated adducts found in DNA. Thus, while *N*-(deoxyguanosin-8-yl)-2-aminofluorene represents approximately 60% of the total binding in rat liver DNA after a single dose of 2-acetylaminofluorene, it is essentially the only adduct detected in hepatic DNA after 2–3 weeks of feeding (POIRIER et al. 1982, 1988).

Polynucleotides modified with *N*-(deoxyguanosin-8-yl)-2-acetylaminofluorene are readily converted from the B-form normally found in DNA to the Z-form in which the modified guanine is in a *syn* conformation (SAGE and LENG 1980; SANTELLA et al. 1981 a, b). This conformational change, which presumably occurs to decrease steric interactions, will also take place in *N*-(deoxyguanosin-8-yl)-2-aminofluorene-modified polynucleotides; however, considerably higher salt or alcohol concentrations are required (SAGE and LENG 1980). Since this B→Z transition may decrease the rate of adduct repair (BOITEUX et al. 1985), it seems that the original structural distortion existing in B-DNA containing *N*-(deoxyguanosin-8-yl)-2-acetylaminofluorene may be the important determinant in adduct removal.

N-(Deoxyguanosin-8-yl)-2-acetylaminofluorene appears to be a more lethal lesion than *N*-(deoxyguanosin-8-yl)-2-aminofluorene as indicated by a decrease in plaque-forming frequency in *E. coli* transfected with $\phi\chi$174 DNA modified with these adducts (TANG et al. 1982). Furthermore, the excision of *N*-(deoxyguanosin-8-yl)-2-acetylaminofluorene in *E. coli* requires all three genes, A, B and C, of the *uvr* complex, while this is not the case with *N*-(deoxyguanosin-8-yl)-2-aminofluorene (TANG et al. 1982; BICHARA and FUCHS 1988). *N*-(Deoxyguanosin-8-yl)-2-acetylaminofluorene has also been shown to induce primarily frameshift mutations in the tetracycline resistance gene of *E. coli*, whereas the nonacetylated adduct causes mainly base substitution mutations of this gene of which the majority are G→T transversions (BICHARA and FUCHS 1985; FUCHS et al. 1988). The mutagenic potential of *N*-(deoxyguanosin-8-yl)-2-acetylaminofluorene does not appear to have been established in mammalian cells; however, the presence of *N*-(deoxyguanosin-8-yl)-2-aminofluorene has been correlated with the induction of mutations (presumably base substitutions) at the hypoxanthine-guanine phosphoribosyl transferase locus in human diploid fibroblasts and Chinese hamster ovary cells (MAHER et al. 1981; HEFLICH et al. 1986c, 1988; ARCE et al. 1987; CAROTHERS et al. 1988). Likewise, the concentration of *N*-(deoxyguanosin-8-yl)-2-aminofluorene has been related to the induction of frameshift mutations in *S. typhimurium* TA1538 (BERANEK et al. 1982; ARCE et al. 1987). Finally, as noted earlier, when mice are treated with a hepatocarcinogenic dose of *N*-hydroxy-2-acetylaminofluorene, the major adduct formed in liver DNA is *N*-(deoxyguanosin-8-yl)-2-aminofluorene (LAI et al. 1985, 1987). Examination of the mouse tumor DNA indicates a G→T transversion in the c-Ha-*ras* protooncogene (WISEMAN et al. 1986), the type of mutation predicted to result from this adduct (BELAND and KADLUBAR 1985).

In addition to being *N*-oxidized by cytochrome P-450, 2-aminofluorene but not 2-acetylaminofluorene is a substrate for prostaglandin *H* synthase and is metabolized through radical intermediates to 2-nitrofluorene, 2,2′-azobisfluorene, and 2-aminodifluorenylamine (BOYD et al. 1983; BOYD and ELING 1984). During this oxidation, metabolites are formed that bind to DNA to give a small amount of *N*-(deoxyguanosin-8-yl)-2-aminofluorene plus a number of other adducts that have not yet been characterized (KRAUSS and ELING 1985). Similar adducts, in addition to *N*-(deoxyguanosin-8-yl)-2-aminofluorene, were detected in the urinary bladder epithelium and the renal medulla of a dog administered 2-aminofluorene (ELING et al. 1988). The quantitative importance of peroxidase-catalyzed metabolism in the activation of 2-aminofluorene is still uncertain, but presumably it will depend upon a number of factors including the relative rates of hepatic *N*-acetylation and deacetylation, the cytochrome P-450 content of the tissue under consideration, the presence or absence of conjugating enzymes such as sulfotransferase, and in the case of the urinary bladder, the relative voiding interval.

V. Benzidine

Benzidine was initially shown to be a bladder carcinogen in humans and dogs (SPITZ et al. 1950; IARC 1982), and while it is carcinogenic in rats, mice, and hamsters (SPITZ et al. 1950; SAFFIOTTI et al. 1967; VESSELINOVITCH et al. 1975; MORTON et al. 1981; NELSON et al. 1982), the primary target tissue in these species is the liver. As with other aromatic amines, benzidine is metabolically activated through oxidation; however, as opposed to the aromatic amines considered previously, it is not a substrate for cytochrome P-450 (WISE et al. 1984b) or the flavin-containing monooxygenase (ZIEGLER et al. 1988). Instead, benzidine is oxidized by various extrahepatic peroxidases, including prostaglandin *H* synthase and lactoperoxidase, as well as by chloroperoxidase and horseradish peroxidase, through free radical intermediates to benzidine diimine, a reactive DNA-binding metabolite (ZENSER et al. 1979, 1980; RICE and KISSINGER 1982; KADLUBAR et al. 1982b; JOSEPHY et al. 1983a, b; MORTON et al. 1983; WISE et al. 1984a, b, 1985; TSURUTA et al. 1985; YAMAZOE et al. 1988b). Like 4-aminobiphenyl, benzidine is also an excellent substrate for hepatic *N*-methyltransferases, and *N*-methylbenzidine is oxidized by the flavin-containing monooxygenase (ZIEGLER et al. 1988). The presumed product, *N*-hydroxy-*N*-methylbenzidine, should readily decompose to benzidine diimine and result in DNA binding; however, the importance of this pathway in vivo is presently unknown.

In contrast to benzidine, the acetylated benzidine metabolites, *N*-acetylbenzidine and *N,N'*-diacetylbenzidine, are substrates for cytochrome P-450, which oxidizes them to *N*-hydroxy metabolites that bind to DNA either directly or after subsequent metabolism (MORTON et al. 1979, 1980; FREDERICK et al. 1985). These differences in the metabolism of benzidine and its acetylated derivatives probably contribute to the target organ specificity of this aromatic amine.

In rats administered single intraperitoneal doses of benzidine, *N*-acetylbenzidine, or *N,N'*-diacetylbenzidine, the relative order of hepatic DNA binding is *N*-acetylbenzidine > benzidine >> *N,N'*-diacetylbenzidine (MARTIN et al. 1982, 1983; KENNELLY et al. 1984). These results indicate that in rats a single *N*-acetylation is involved in the metabolic activation of benzidine, but that acetylation of the remaining amine function constitutes a detoxification. Examination of the hepatic DNA adducts from benzidine- and *N*-acetylbenzidine-treated rats supports this contention because a single adduct, *N*-(deoxyguanosin-8-yl)-*N'*-acetylbenzidine (Fig. 7), is detected with either compound.

N-(Deoxyguanosin-8-yl)-*N'*-acetylbenzidine can be formed from the acid-catalyzed reaction of *N*-hydroxy-*N'*-acetylbenzidine with DNA (MARTIN et al. 1982). In addition, results from in vitro experiments with rat liver preparations indicate that this adduct can also be formed by *O*-acetylation (but not *O*-sulfonation) of *N*-hydroxy-*N'*-acetylbenzidine and by the *N*-acetylation of *N*-hydroxy-*N'*-acetylbenzidine to give *N*-hydroxy-*N,N'*-diacetylbenzidine followed by an *N,O*-acyltransferase-catalyzed rearrangement to *N*-acetoxy-*N'*-acetylbenzidine (FREDERICK et al. 1985). Although each of these pathways may contribute to the metabolic activation of benzidine in rats, the data from in vitro experiments suggest that the latter pathway is the most important. This interpretation is consistent with the fact that benzidine and *N*-hydroxy-*N,N'*-diacetylbenzidine have comparable carcinogenicities that are greater than that of *N,N'*-diacetylbenzidine (MORTON et al. 1981).

As noted earlier, *N,N'*-diacetylbenzidine gives relatively low hepatic DNA binding when administered in vivo. Nevertheless, it gives rise to an adduct, *N*-

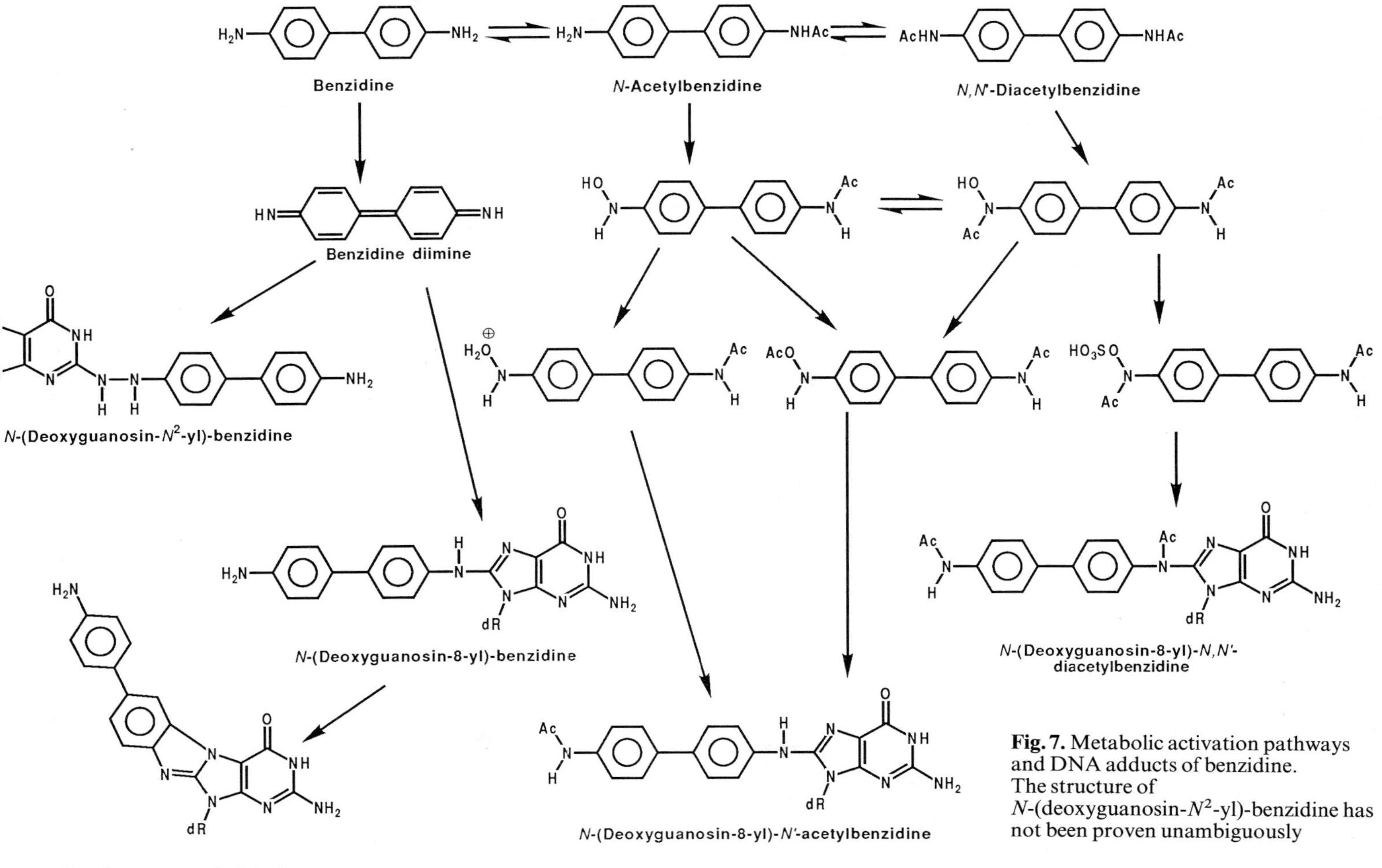

Fig. 7. Metabolic activation pathways and DNA adducts of benzidine. The structure of *N*-(deoxyguanosin-N^2-yl)-benzidine has not been proven unambiguously

(deoxyguanosin-8-yl)-*N,N'*-diacetylbenzidine (Fig. 7), that does not result from benzidine or *N*-acetylbenzidine (KENNELLY et al. 1984). This adduct presumably arises from the sulfotransferase-catalyzed activation of *N*-hydroxy-*N,N'*-diacetylbenzidine (MORTON et al. 1980), which can be formed from *N*-oxidation of *N,N'*-diacetylbenzidine (MORTON et al. 1979) or by *N*-deacetylation of *N,N'*-diacetylbenzidine followed by a sequential *N*-oxidation and *N*-acetylation (FREDERICK et al. 1985). Since *N,N'*-diacetylbenzidine is a rather poor substrate for rat liver cytochrome P-450 and *N*-deacetylases (FREDERICK et al. 1985), it is presently not known which of these two pathways is more important.

As mentioned previously, *N*-hydroxy-*N,N'*-diacetylbenzidine is regarded as the major proximate carcinogenic metabolite of benzidine and *N*-acetylbenzidine in rats (FREDERICK et al. 1985), and yet *N*-(deoxyguanosin-8-yl)-*N,N'*-diacetylbenzidine is not detected in rat liver in vivo following administration of these compounds (MARTIN et al. 1982, 1983; KENNELLY et al. 1984). The failure to form this adduct from benzidine or *N*-acetylbenzidine may be due to the efficient detoxification of *N*-hydroxy-*N,N'*-diacetylbenzidine and/or its sulfuric acid ester through glucuronide and/or glutathione conjugation (LYNN et al. 1984). Apparently, when rats are treated with *N,N'*-diacetylbenzidine, a greater concentration of *N*-hydroxy-*N,N'*-diacetylbenzidine is produced than occurs with benzidine or *N*-acetylbenzidine. This would allow the formation of *N*-(deoxyguanosin-8-yl)-*N,N'*-diacetylbenzidine in rats given *N,N'*-diacetylbenzidine but not the other two derivatives.

In mice given benzidine (MARTIN et al. 1982) and in hamsters treated with *N*-acetylbenzidine (KENNELLY et al. 1984), the only adduct detected in hepatic DNA is *N*-(deoxyguanosin-8-yl)-*N'*-acetylbenzidine. TALASKA et al. (1987) have reported the formation of two adducts in mice administered benzidine; however these adducts, which correlate with the induction of chromosomal aberrations, were not characterized nor was their relationship to one another established. In contrast to what is observed in rats, the results from in vitro experiments with mouse liver preparations suggest that *N*-(deoxyguanosin-8-yl)-*N'*-acetylbenzidine is formed through the direct reaction of *N*-hydroxy-*N'*-acetylbenzidine or its *O*-acetoxy derivative with DNA (FREDERICK et al. 1985). The metabolic activation pathways of benzidine in hamsters have not been elucidated.

The mutagenic potential of *N*-(deoxyguanosin-8-yl)-*N'*-acetylbenzidine has been assessed in both bacteria and in mammalian cells. In *S. typhimurium* TA1538, the concentration of *N*-(deoxyguanosin-8-yl)-*N'*-acetylbenzidine correlates with the induction of frameshift mutations (BELAND et al. 1983), and interestingly, this adduct seems to be more efficient at causing these mutations than the analogous DNA adducts from *N*-hydroxy-2-aminofluorene, *N*-hydroxy-4-aminobiphenyl, *N*-hydroxy-2-naphthylamine, 1-nitropyrene, and 1,8-dinitropyrene (BERANEK et al. 1982; BELAND et al. 1983; HOWARD et al. 1983; HEFLICH et al. 1985a; DJURIĆ et al. 1986b; ARCE et al. 1987). The reasons for the variation in mutagenic efficiency among these adducts are not known but may be due to conformational differences within the DNA, such as the relative ability to adopt a *syn* conformation about the glycosyl bond (LIPKOWITZ et al. 1982; BELAND et al. 1983; BROYDE and HINGERTY 1983; HINGERTY and BROYDE 1986; SHAPIRO et al. 1986). In Chinese hamster ovary cells, the concentration of *N*-(deoxyguanosin-8-yl)-*N'*-acetylbenzidine has been correlated with the induction of mutations at the hypoxanthine-guanine phosphoribosyl transferase locus and with the formation of sister chromatid exchanges (HEFLICH et al. 1986c). As was observed in *Salmonella*, *N*-(deoxyguanosin-8-yl)-*N'*-acetylbenzidine appears to be more efficient than *N*-(deoxyguanosin-8-yl)-2-aminofluorene or *N*-(deoxyguanosin-8-yl)-1-aminopyrene at inducing mutations in Chinese hamster ovary cells; however, it is slightly less efficient than *N*-(deoxyguanosin-8-yl)-2-aminofluorene in inducing sister chromatid exchanges.

Although benzidine is not a substrate for hepatic monooxygenases, it is readily oxidized by peroxidases to a radical cation and to benzidine diimine (JOSEPHY et al. 1983a, b; WISE et al. 1984a). When DNA is included in these incubations, extensive binding occurs, to a much greater extent than is observed with any other aromatic amine substrate (ZENSER et al. 1980; KADLUBAR et al. 1982b; WISE et al. 1984b; TSURUTA et al. 1985; YAMAZOE et al. 1988b). At least four DNA adducts result from the peroxidase-catalyzed activation of benzidine (YAMAZOE et al. 1988b). The major adduct arises from oxidation to benzidine diimine followed by reaction with deoxyguanosine to give *N*-(deoxyguanosin-8-yl)-benzidine (Fig. 7; YAMAZOE et al. 1986, 1988b). Subsequent oxidation of this adduct followed by reaction of the carbon *ortho* to the amine with N7 of guanine yields the second most prevalent adduct, *N*,3-(deoxyguanosin-7,8-yl)-benzidine (Fig. 7; YAMAZOE et al. 1988b). The remaining two adducts have not been formed in sufficient quantity to be characterized fully; nevertheless, one appears to be a degradation product of *N*-(deoxyguanosin-8-yl)-benzidine, while the other has been suggested to be *N*-(deoxyguanosin-N^2-yl)-benzidine (YAMAZOE et al. 1988b).

The importance of these adducts in vivo is presently uncertain. Substantial levels of DNA binding have been detected in in vitro experiments with extrahepatic tissues that are low in cytochrome P-450 but high in prostaglandin *H* synthase (e.g., dog bladder and kidney, human bladder) (WISE et al. 1984b; KADLUBAR et al. 1988c). Furthermore, benzidine but not *N*-acetylbenzidine or *N,N'*-diacetylbenzidine induces unscheduled DNA synthesis in cultured rabbit bladder (MCQUEEN et al. 1987). Binding has also been detected in urinary bladder DNA of dogs administered radiolabeled benzidine and *N*-acetylbenzidine; however, only small amounts of radioactivity are released upon hydrolysis of this DNA (BELAND et al. 1983; YAMAZOE et al. 1988b). This low extent of hydrolysis may be due to the facile oxidation of initial DNA adducts and subsequent intermolecular DNA-DNA crosslinking (FOURNEY et al. 1986), which in the case of *N*-(deoxyguanosin-8-yl)-benzidine has been shown to decrease the extent of DNA hydrolysis (YAMAZOE et al. 1988b).

VI. *N,N*-Dimethyl-4-aminoazobenzene and Its Demethylated Derivatives

N,N-Dimethyl-4-aminoazobenzene is a typical aromatic amine in that it is hepatocarcinogenic in mice (ANDERVONT and EDWARDS 1943) and rats (KINOSHITA 1937) and a urinary bladder carcinogen in dogs (NELSON and WOODward 1953). However, it differs from the compounds considered previously because it is not a primary amine or an amide. It is also intensely colored, a property that led to its use as a dye (WILLIAMS 1962) and facilitated its study in the days preceding radiolabeled carcinogens (MILLER and MILLER 1983). The metabolic activation of *N,N*-dimethyl-4-aminoazobenzene includes an *N*-demethylation to *N*-methyl-4-aminoazobenzene (Fig. 8) followed by an *N*-hydroxylation to give *N*-hydroxy-*N*-methyl-4-aminoazobenzene (KADLUBAR et al. 1976a). A subsequent *N*-demethylation of *N*-methyl-4-aminoazobenzene yields 4-aminoazobenzene, which is not carcinogenic in adult rats but is hepatocarcinogenic in weanling mice (MILLER et al. 1979; FUJII 1983; DELCLOS et al. 1984). The *N*-demethylation of *N,N*-dimethyl-4-aminoazobenzene and *N*-methyl-4-aminoazobenzene is catalyzed by cytochrome P-450, primarily isozyme *c* (LEVINE and LU 1982). Likewise, cytochrome P-450*c* *N*-hydroxylates *N*-methyl-4-aminoazobenzene (KIMURA et al. 1982, 1984), but this only occurs to a limited extent, with most of the oxidation being catalyzed by the flavin-containing monooxygenase (KADLUBAR et al. 1976a). In contrast, the primary aromatic amine, 4-aminoazobenzene, is *N*-hydroxylated almost exclusively by cytochrome P-450 (KADLUBAR et al. 1976a), in particular isozymes *c* and *d* (KIMURA et al. 1985). The oxidation of *N*-hydroxy-*N*-methyl-4-aminoazobenzene to *N*-methyl-*N*-(*p*-phenylazophenyl)-nitrone followed by a hydrolysis also leads to the formation of *N*-hydroxy-4-aminoazobenzene (Fig. 8; KADLUBAR et al. 1976a).

Primary *N*-hydroxy aromatic amines normally undergo acid-catalyzed reactions with DNA (KRIEK 1965; KADLUBAR and BELAND 1985). The acidic conditions allow protonation of the *N*-hydroxy group, which promotes nitrenium ion formation. This does not occur with *N*-hydroxy-*N*-methyl-4-aminoazobenzene because under acidic conditions the azo linkage rather than the hydroxy group appears to become protonated (CILENTO et al. 1956). Therefore, *N*-hydroxy-*N*-methyl-4-aminoazobenzene must be further metabolized in order to become bound to DNA; and, of the pathways investigated, only the cytosol-catalyzed formation of *N*-sulfonyloxy-*N*-methyl-4-aminoazobenzene seems to be important (KADLUBAR et al. 1976b). Although the synthesis of this *N*-sulfonyloxy ester has been reported (COLES et al. 1984), the DNA adducts that have been characterized from *N*-methyl-4-aminoazobenzene have been prepared from using the model ester, *N*-benzoyloxy-*N*-methyl-4-aminoazobenzene (LIN et al. 1975b; BELAND et al. 1980b; TARPLEY et al. 1980; TULLIS et al. 1981). These adducts are (Fig. 8): *N*-(deoxyguanosin-8-yl)-*N*-methyl-4-aminoazobenzene, 3-(deoxyguanosin-N^2-yl)-*N*-methyl-4-aminoazobenzene, and 3-(deoxyadenosin-N^6-yl)-*N*-methyl-4-aminoazobenzene. There is also evidence for the formation of *N*7-guanine adducts, which undergo rapid depurination (TARPLEY et al. 1982). In other experiments, *N*-methyl-*N*-(*p*-phenylazophenyl)-nitrone, which is the oxidation product of *N*-hydroxy-*N*-methyl-4-aminoazobenzene, has been shown to react under anhydrous conditions with double bonds in purines and pyrimidines (KADLUBAR et al. 1976a). It is not known whether similar reactions occur with DNA.

When rats are given single doses of *N*-methyl-4-aminoazobenzene, two hepatic DNA adducts, *N*-(deoxyguanosin-8-yl)-*N*-methyl-4-aminoazobenzene and 3-(deoxyguanosin-N^2-yl)-*N*-methyl-4-aminoazobenzene, are detected (LIN et

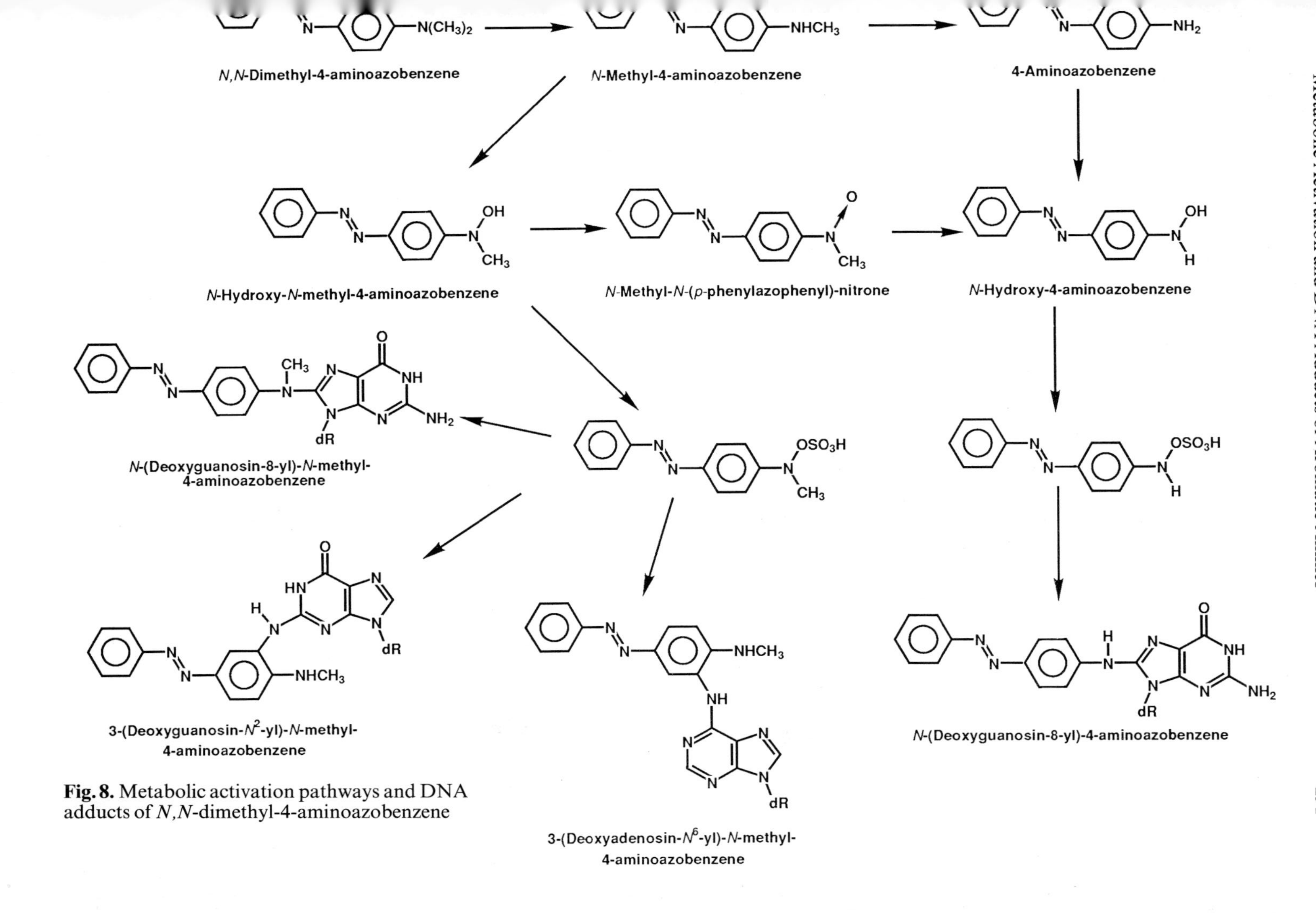

Fig. 8. Metabolic activation pathways and DNA adducts of *N,N*-dimethyl-4-aminoazobenzene

al. 1975a; BELAND et al. 1980b; TARPLEY et al. 1980). If additional doses are given, 3-(deoxyadenosin-N^6-yl)-*N*-methyl-4-aminoazobenzene is also found (TULLIS et al. 1981); and, during a carcinogenic feeding regimen, *N*-(deoxyguanosin-8-yl)-4-aminoazobenzene and an 8,9-purine ring-opened derivative of *N*-(deoxyguanosin-8-yl)-*N*-methyl-4-aminoazobenzene are also present (TULLIS et al. 1987).

After a single dose of *N*-methyl-4-aminoazobenzene, *N*-(deoxyguanosin-8-yl)-*N*-methyl-4-aminoazobenzene initially accounts for 70% of the binding; however, it is rapidly removed and cannot be detected 7 days after treatment (BELAND et al. 1980b). In contrast, 3-(deoxyguanosin-N^2-yl)-*N*-methyl-4-aminoazobenzene remains at a constant level for at least 2 weeks. Following multiple doses of *N*-methyl-4-aminoazobenzene, 3-(deoxyguanosin-N^2-yl)-*N*-methyl-4-aminoazobenzene becomes the major adduct accounting for approximately 40% of the total hepatic DNA binding, while *N*-(deoxyguanosin-8-yl)-*N*-methyl-4-aminoazobenzene and 3-(deoxyadenosin-N^6-yl)-*N*-methyl-4-aminoazobenzene each account for 20% (TULLIS et al. 1981). When *N*-methyl-4-aminoazobenzene is administered in the diet, 3-(deoxyguanosin-N^2-yl)-*N*-methyl-4-aminoazobenzene increases in a linear manner to become the major adduct after 5 weeks of feeding, whereas the concentrations of *N*-(deoxyguanosin-8-yl)-*N*-methyl-4-aminoazobenzene, 3-(deoxyadenosin-N^6-yl)-*N*-methyl-4-aminoazobenzene, and *N*-(deoxyguanosin-8-yl)-4-aminoazobenzene appear to reach a steady-state level after 1 week (TULLIS et al. 1987). If the *N*-methyl-4-aminoazobenzene diet is discontinued, the latter three adducts are rapidly removed from the hepatic DNA, so that the only persistent lesion is 3-(deoxyguanosin-N^2-yl)-*N*-methyl-4-aminoazobenzene.

Other investigators have shown that continued aromatic amine administration rapidly decreases the concentration of *N*-hydroxy-2-acetylaminofluorene sulfotransferase (DEBAUN et al. 1970; JACKSON and IRVING 1972; RINGER et al. 1983; RINGER and NORTON 1987). The kinetics of adduct formation and removal during *N*-methyl-4-aminoazobenzene feeding suggest that the sulfotransferase responsible for the sulfonation of *N*-hydroxy-*N*-methyl-4-aminoazobenzene differs from *N*-hydroxy-2-acetylaminofluorene sulfotransferase and is not depleted during continual carcinogen administration. This interpretation is consistent with the differences in pH optima and cofactor requirements observed in vitro when hepatic sulfotransferases are assayed using *N*-hydroxy-*N*-methyl-4-aminoazobenzene and *N*-hydroxy-2-acetylaminofluorene as substrates (KADLUBAR et al. 1976b).

The adduct profile in the livers of weanling mice administered a single dose of *N*-methyl-4-aminoazobenzene is similar to that observed in rats; *N*-(deoxyguanosin-8-yl)-*N*-methyl-4-aminoazobenzene is the major adduct, accounting for 45% of the initial binding while 3-(deoxyguanosin-N^2-yl)-*N*-methyl-4-aminoazobenzene is the second most prevalent adduct contributing an additional 10% (TARPLEY et al. 1980). At least 10 other adducts, some of which are due to *cis-trans* isomerism about the azo linkage, are also detected at levels of 5% or less. As was observed in rat liver, *N*-(deoxyguanosin-8-yl)-*N*-methyl-4-aminoazobenzene is removed from mouse liver DNA much more rapidly than 3-(deoxyguanosin-N^2-yl)-*N*-methyl-4-aminoazobenzene.

Following a single dose of 4-aminoazobenzene to weanling mice, one major adduct, *N*-(deoxyguanosin-8-yl)-4-aminoazobenzene, is detected (Fig. 8; DELCLOS et al. 1984). As with the adducts obtained form *N*-methyl-4-aminoazobenzene, this adduct seems to be formed from a sulfotransferase-catalyzed metabolite, *N*-sulfonyloxy-4-aminoazobenzene (DELCLOS et al. 1986). In contrast to *N*-(deoxyguanosin-8-yl)-*N*-methyl-4-aminoazobenzene, *N*-(deoxyguanosin-8-yl)-4-aminoazobenzene is relatively persistent with only a 60% loss seen in 21 days (DELCLOS et al. 1984). This persistence is similar to what is observed with other primary arylamine C8-substituted deoxyguanosine adducts (e.g., *N*-(deoxyguanosin-8-yl)-2-aminofluorene) and suggests that the methyl group of *N*-(deoxyguanosin-8-yl)-*N*-methyl-4-aminoazobenzene significantly perturbs the DNA helix, thus inducing excision repair. Nuclear magnetic resonance studies of the C8-substituted deoxyguanosine adducts of *N*-methyl-4-aminoazobenzene and 4-aminoazobenzene support this interpretation (DELCLOS et al. 1984).

N-(Deoxyguanosin-8-yl)-4-aminoazobenzene is also the major hepatic DNA adduct in weanling mice administered single doses of *N,N*-dimethyl-4-aminoazobenzene (DELCLOS et al. 1984). The failure to detect this adduct in an earlier investigation with *N*-methyl-4-aminoazobenzene (TARPLEY et al. 1980) appears to be due to differences in the chromatographic conditions between the two studies.

VII. 2-Acetylaminophenanthrene

The arylamide, 2-acetylaminophenanthrene, will induce leukemia and tumors of the mammary gland, ear duct, and small intestine in rats, but as with 4-acetylaminobiphenyl, it is not hepatocarcinogenic, even following promotion (MILLER et al. 1955, 1966; SCRIBNER and MOTTET 1981; SCRIBNER et al. 1983). This latter property has led to the use of 2-acetylaminophenanthrene in comparative studies to determine what factors are critical for the induction of hepatic tumors by aromatic amines. When administered at equimolar doses, 2-acetylaminophenanthrene and the hepatocarcinogen, 2-acetylaminofluorene, bind to hepatic DNA in rats to nearly the same extent (SCRIBNER and KOPONEN 1979; SCRIBNER et al. 1983). In addition, persistent adducts result from both compounds.

In order to characterize the 2-acetylaminophenanthrene adducts, SCRIBNER and NAIMY (1975) prepared the arylhydroxamic acid esters, *N*-acetoxy-2-acetylaminophenanthrene and *N*-sulfonyloxy-2-acetylaminophenanthrene, and reacted them with DNA to give *N*-(deoxyguanosin-8-yl)-2-acetylaminophenanthrene and 1-(deoxyadenosin-N^6-yl)-2-acetylaminophenanthrene as major products (Fig. 9). Subsequent work with ring- and acetyl-tritiated 2-acetylaminophenanthrene showed, however, that none of the hepatic DNA adducts retained the *N*-acetyl group (SCRIBNER and KOPONEN 1979). This contrasts with 2-acetylaminofluorene where ~25% of the initial hepatic DNA binding is in the form of acetylated adducts (e.g., *N*-(deoxyguanosin-8-yl)-2-acetylaminofluorene; KRIEK 1969). The lack of acetylated adduct formation from 2-acetylaminophenanthrene also parallels with the effect of this arylamide upon

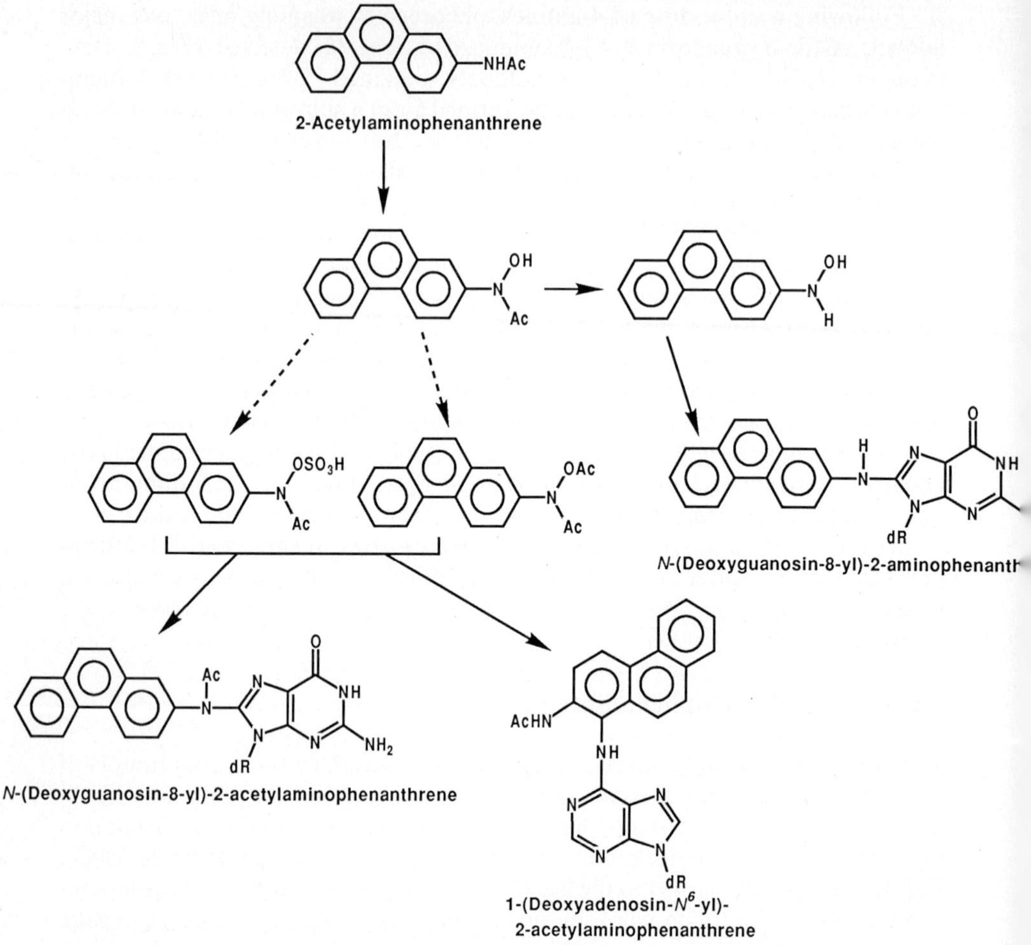

Fig. 9. Metabolic activation pathways and DNA adducts of 2-acetylaminophenanthrene. The *dashed lines* represent synthetic routes that have not been demonstrated to occur in vivo

hepatic sulfotransferase. Dietary administration of 2-acetylaminophenanthrene to rats causes only a slight (~15%) decrease in *N*-hydroxy-2-acetylaminofluorene sulfotransferase activity, whereas a 70% reduction in activity is observed with the hepatocarcinogen, 2-acetylaminofluorene (RINGER and NORTON 1987).

Recently, GUPTA and DIGHE (1984) reexamined the kinetics of adduct formation and removal with the *N*-hydroxy derivatives of 2-acetylaminophenanthrene, 4-acetylaminobiphenyl and 2-acetylaminofluorene, and obtained essentially the same relationships found earlier. In addition, by conducting reactions of *N*-hydroxy-2-aminophenanthrene with DNA and homopolymers, they were able to demonstrate that two deoxyguanosine adducts accounted for at least 70% of the total hepatic DNA binding. Subsequent studies have shown one of these adducts to be *N*-(deoxyguanosin-8-yl)-2-aminophenanthrene (Fig. 9; GUPTA et al. 1985). These adducts are also present in the carcinogen target tissues for 2-acetylaminophenanthrene, although at a much lower level than found in the liver (GUPTA et al. 1985, 1986).

VIII. 4-Acetylaminostilbene

4-Acetylaminostilbene is an aromatic amide that induces primarily ear duct (Zymbal's gland) and mammary gland tumors in rats (HADDOW et al. 1948; BALDWIN et al. 1963; ANDERSEN et al. 1964). It is only weakly hepatocarcinogenic; however, the liver tumor incidence can be increased if the carcinogen is given after partial hepatectomy and/or is followed by administration of a tumor promoter such as phenobarbital, 1,1,1-trichloro-2,2-bis(4-chlorophenyl)ethane, or diethylstilbestrol (HILPERT et al. 1983). 4-Acetylaminostilbene also synergistically increases the number of hepatic tumors induced by 2-acetylaminofluorene (KUCHLBAUER et al. 1985; RUTHSATZ and NEUMANN 1988).

As with other aromatic amides, 4-acetylaminostilbene appears to be metabolically activated by an initial *N*-hydroxylation, which produces *N*-hydroxy-4-acetylaminostilbene (Fig. 10). This arylhydroxamic acid has been detected as both a microsomal and urinary metabolite (ANDERSEN et al. 1964; BALDWIN and SMITH 1965) and is more carcinogenic that its parent amide (BALDWIN et al. 1963; ANDERSEN et al. 1964). For these reasons, much effort has gone into identifying nucleoside adducts formed from reactive *N*-hydroxy-4-acetylaminostilbene esters, in particular *N*-acetoxy-4-acetylaminostilbene. Compared with other arylhydroxamic acid esters, this has been difficult because the double bond in stilbene becomes a reactive center upon arylnitrenium ion formation. Thus, substitution can occur in this region in addition to sites normally reactive in aromatic amine carcinogens. Furthermore, reaction with the double bond results in the generation of two optically active centers that can lead to the formation of enantiomeric and diastereomeric adducts. To date, the only adducts that have been characterized from reactions with *N*-acetoxy-4-acetylaminostilbene result from addition to this double bond (SCRIBNER et al. 1978, 1979b; SCRIBNER and SCRIBNER 1979; FRANZ et al. 1986; FRANZ and NEUMANN 1987). These products are quite unusual because C8-guanine derivatives have not been reported. Instead, most of the adducts arise from reactions at exocyclic nitrogens and oxygens; this in some instances is followed by an attack on the purine ring nitrogens to give cyclic derivatives. These include: 1-(guanosin-N1-yl)-1-(4-acetylaminophenyl)-2-hydroxy-2-phenylethane; 1-(guanosin-O^6-yl)-1-(4-acetylaminophenyl)-2-hydroxy-2-phenylethane; 1-(adenosin-N^6-yl)-1-(4-acetylaminophenyl)-2-hydroxy-2-phenylethane; 1-(uridin-N3-yl)-1-(4-acetylaminophenyl)-2-hydroxy-2-phenylethane; 1,2-(adenosin-N1,N^6-yl)-1-(4-acetylaminophenyl)-2-phenylethane; (R,R)- and (S,S)-1,2-(guanosin-N^2,N3-yl)-1-(4-acetylaminophenyl)-2-phenylethane; and (R,R)- and (S,S)-1,2-(guanosin-N^2,N3-yl)-2-(4-acetylaminophenyl)-1-phenylethane. It should be noted that the uridine adduct results from reaction with cytidine followed by a deamination (SCRIBNER et al. 1978).

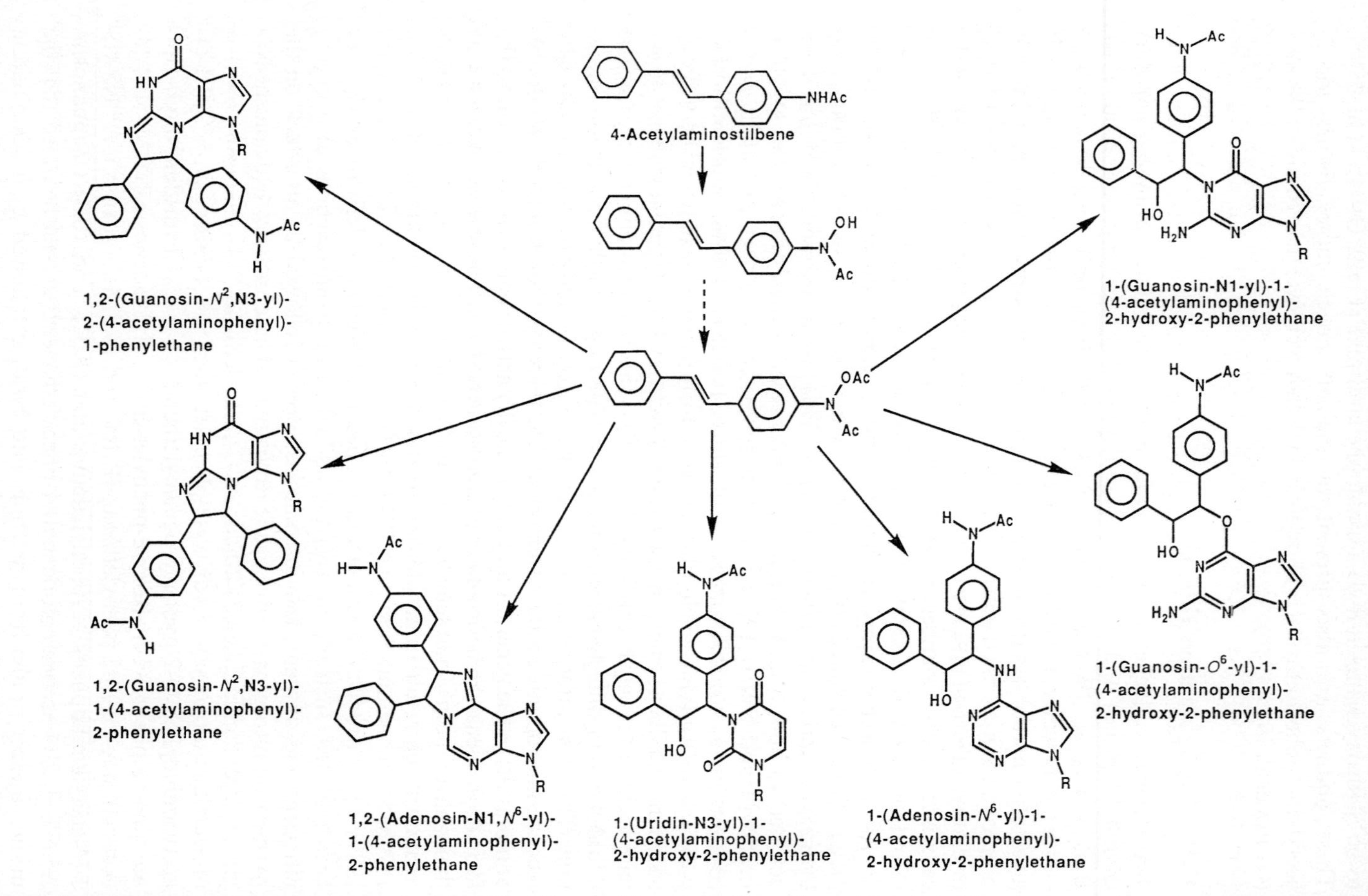

Fig. 10. Metabolic activation pathways and nucleoside adducts of 4-acetylaminostilbene. The *dashed line* represents a synthetic route that has not been demonstrated to occur in vivo

Upon administration of radiolabeled 4-acetylaminostilbene to rats, DNA binding is detected in both target and nontarget tissues with the relative order of binding being liver > kidney > Zymbal's gland $\simeq$ lung $\simeq$ glandular stomach $\simeq$ mammary gland (BAUR and NEUMANN 1980; NEUMANN 1981; HILPERT and NEUMANN 1985). Repeated treatment with radiolabeled 4-acetylaminostilbene results in an additive increase in DNA binding in the same relative order (HILPERT and NEUMANN 1985). The DNA adducts formed from 4-acetylaminostilbene in vivo have been examined only in the liver, kidney, lung, and glandular stomach, and some variation in the adduct profile is observed among these tissues (GAUGLER et al. 1979; GAUGLER and NEUMANN 1979; BAUR and NEUMANN 1980; NEUMANN 1981; HILPERT and NEUMANN 1985). More im portant, however, is the fact that the adducts characterized in vitro appear to make only a small contribution to the total binding found in vivo. This is perhaps not too surprising because as noted previously, the major adducts obtained from aromatic amides in vivo are nonacetylated rather than acetylated adducts. Interestingly, when hepatic DNA adducts of *N*-hydroxy-4-acetylaminostilbene are assayed by ^{32}P-postlabeling (REDDY et al. 1984), one major adduct is detected, and it has elution characteristics similar to other nonacetylated arylamine adducts [e.g., *N*-(deoxyguanosin-8-yl)-2-aminofluorene and *N*-(deoxyguanosin-8-yl)-2-aminophenanthrene]. This suggests that *N*-(deoxyguanosin-8-yl)-4-aminostilbene should be the major DNA adduct formed in vivo. The failure to detect this adduct when radiolabeled 4-acetylaminostilbene was administered may be due to the fact that the analyses have been conducted using Sephadex LH20 with a basic buffer that contained only a low percentage of alcohol. Alkaline conditions tend to destroy C8-substituted deoxyguanosine aromatic amine adducts (KRIEK and WESTRA 1980), and generally a higher percentage of alcohol is necessary to elute these adducts (BELAND et al. 1979, 1980a). In this regard, it is interesting to note that DNA adducts from *N*-acetoxy-2-acetylaminophenanthrene and *N*-acetoxy-4-acetylaminostilbene in human diploid fibroblasts have similar intrinsic cytotoxicities (MAHER et al. 1981). Although the adducts in this experiment were not characterized, treatment of human diploid fibroblasts with *N*-acetoxy-2-acetylaminofluorene gives only *N*-(deoxyguanosin-8-yl)-2-aminofluorene (MAHER et al. 1980).

IX. Nitropyrenes

Nitropyrenes, in particular 1-nitropyrene and 1,3-, 1,6-, and 1,8-dinitropyrene (Fig. 11), were originally discovered to be mutagenic components of xerographic toners (LÖFROTH et al. 1980; ROSENKRANZ et al. 1980) and of diesel engine exhaust (PEDERSON and SIAK 1981; SCHUETZLE et al. 1981; XU et al. 1982; SCHUETZLE 1983). Subsequently, they have been found as contaminants in a number of substances including coal fly ash (WEI et al. 1982), urban air particulates (RAMDAHL et al. 1982; SWEETMAN et al. 1982; TOKIWA et al. 1983; NIELSEN et al. 1984), kerosene heater emissions (TOKIWA et al. 1985), river sediment (SATO et al. 1985), and certain foods (KINOUCHI et al. 1986b; OHNISHI et al. 1986). All four of these nitropyrenes are usually detected in environmental samples, with 1-nitropyrene typically present in much greater concentrations than the

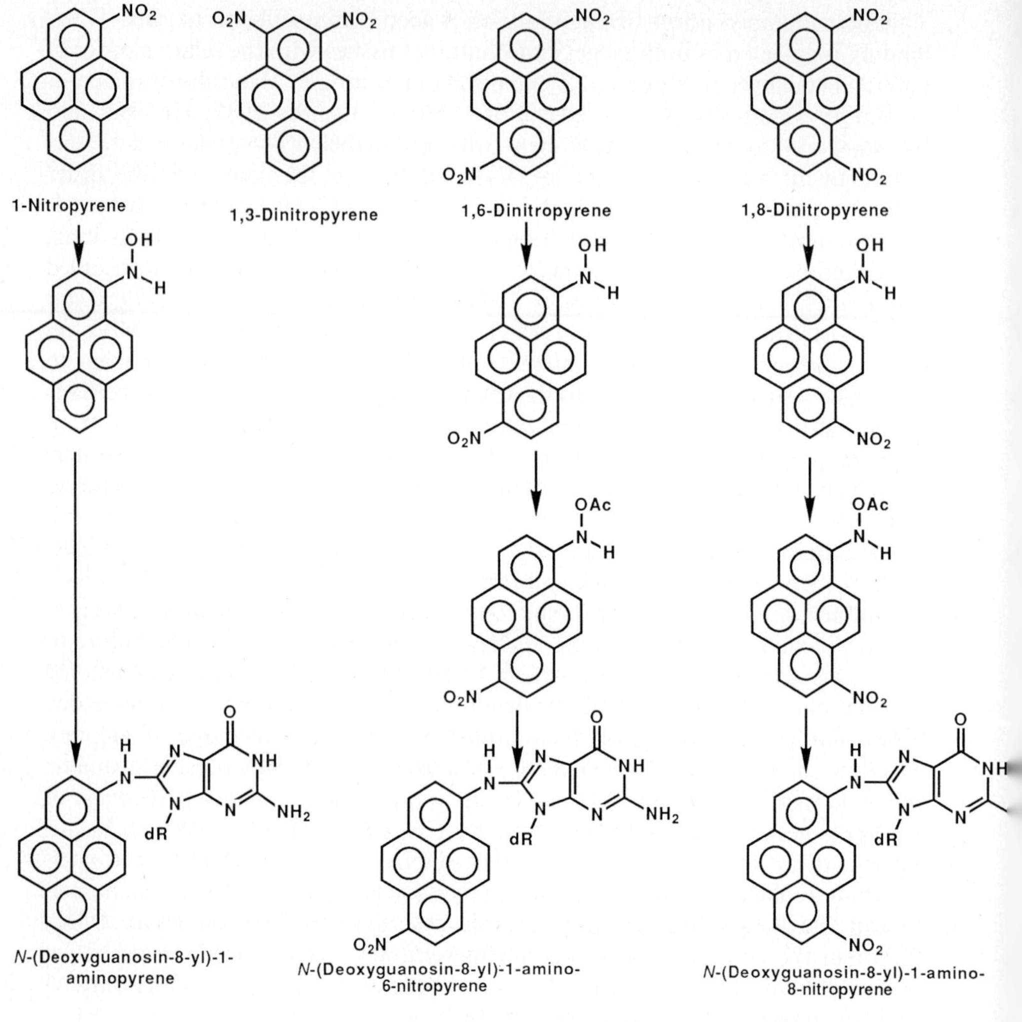

Fig. 11. Metabolic activation pathways and DNA adducts of 1-nitropyrene, 1,3-dinitropyrene, 1,6-dinitropyrene, and 1,8-dinitropyrene

dinitropyrenes. The dinitropyrenes, however, are extremely potent bacterial mutagens, which was the property that led to their initial discovery. As a result, in most samples the dinitropyrenes and 1-nitropyrene contribute equally to the observed bacterial mutagenicity. 1-Nitropyrene is tumorigenic at the injection site and in the mammary gland in rats (HIROSE et al. 1984), in the lungs of adult mice (EL-BAYOUMY et al. 1984), and in the livers of newborn mice (WISLOCKI et al. 1986). The dinitropyrenes, in particular 1,6- and 1,8-dinitropyrene, appear to be considerably more tumorigenic than 1-nitropyrene and, when tested in the same animal model, show the same organ specificity as 1-nitropyrene (NESNOW et al. 1984; OHGAKI et al. 1984, 1985; TOKIWA et al. 1984; TAKAYAMA et al. 1985; MAEDA et al. 1986; WISLOCKI et al. 1986; OTOFUJI et al. 1987; KING 1988).

An initial observation concerning nitropyrenes was their decreased bacterial mutagenicity in strains deficient in nitroreductases (LÖFROTH et al. 1980; ROSENKRANZ et al. 1980; MERMELSTEIN et al. 1981; McCOY et al. 1981). As a consequence, the major emphasis on the metabolic activation of these compounds has centered upon their nitroreduction to reactive polycyclic *N*-hydroxy arylamine derivatives. In rat liver, the microsomal nitroreduction of 1-nitropyrene appears to be catalyzed preferentially by cytochromes P-450 *c, d, b,* and *e* (SAITO et al. 1984). The isozyme specificity for nitroreduction of the dinitropyrenes has not been established; however, different isozymes may be involved because SKF-525A completely abolishes the microsomal reduction of 1,6-dinitropyrene but does not affect the nitroreduction of 1-nitropyrene (DJURIĆ et al. 1988). Nitroreduction is also catalyzed by the cytosolic enzymes xanthine oxidase, aldehyde oxidase, and other unknown NADPH- and NADH-dependent enzymes (HOWARD and BELAND 1982; HOWARD et al. 1983; TATSUMI et al. 1986; DJURIĆ et al. 1988).

As with other *N*-hydroxy arylamines, the intermediate nitroreduction product, *N*-hydroxy-1-aminopyrene, will undergo acid-catalyzed binding to DNA and gives a C8-substituted deoxyguanosine adduct, *N*-(deoxyguanosin-8-yl)-1-aminopyrene (Fig. 11), as the major product (HOWARD et al. 1983; HEFLICH et al. 1985b). A small amount of binding to deoxyadenosine has also been reported; however, these adducts have not been characterized (KINOUCHI and OHNISHI 1986). *N*-(Deoxyguanosin-8-yl)-1-aminopyrene is the major adduct detected in *S. typhimurium* TA1538 treated with 1-nitropyrene and 1-nitrosopyrene and has a mutagenic efficiency that is intermediate between *N*-(deoxyguanosin-8-yl)-*N'*-acetylbenzidine and *N*-(deoxyguanosin-8-yl)-2-aminofluorene (HOWARD et al. 1983; HEFLICH et al. 1985b). This adduct is also the major DNA adduct present in Chinese hamster ovary cells (HEFLICH et al. 1985a, 1986b, c) and in human diploid fibroblasts (BELAND et al. 1986; PATTON et al. 1986) incubated with 1-nitropyrene and/or 1-nitrosopyrene.

Metabolic reduction of dinitropyrenes leads to the formation of *N*-hydroxy-amino-mononitro derivatives (e.g., *N*-hydroxy-1-amino-8-nitropyrene). As with *N*-hydroxy-1-aminopyrene, these intermediates will react directly with DNA; in the case of 1,6-dinitropyrene, the major adduct is *N*-(deoxyguanosin-8-yl)-1-amino-6-nitropyrene (DJURIĆ et al. 1988), while 1,8-dinitropyrene gives *N*-(deoxyguanosin-8-yl)-1-amino-8-nitropyrene (HEFLICH et al. 1985a; ANDREWS et al. 1986; Fig. 11). DNA adducts from 1,3-dinitropyrene have not been characterized. The same adducts observed in vitro are found in *Salmonella* treated with 1,6- and 1,8-dinitropyrene (HEFLICH et al. 1985a; ANDREWS et al. 1986; DJURIĆ et al. 1986b). This observation indicates that the initial reduction of the dinitropyrenes to *N*-hydroxyamino-mononitro derivatives is an activation pathway but that reduction to diaminopyrenes is not associated with the mutagenic response. This interpretation is consistent with the results obtained when comparing the mutagenicity of aminonitropyrenes to dinitropyrenes (BRYANT et al. 1984; CERNIGLIA et al. 1988).

The mutagenic efficiencies of *N*-(deoxyguanosin-8-yl)-1-amino-8-nitropyrene and *N*-(deoxyguanosin-8-yl)-1-aminopyrene have been compared in *S. typhimurium* TA1538 and appear to be similar (HEFLICH et al. 1985a; DJURIĆ et al. 1986b). Therefore, the extreme mutagenicity of the dinitropyrenes is not due to a unique adduct but rather to their efficient metabolism to a DNA-binding intermediate. This difference in mutagenicity between 1-nitropyrene and the dinitropyrenes seems to be a result of the *N*-hydroxy amino intermediates of the dinitropyrenes (e.g., *N*-hydroxy-1-amino-8-nitropyrene) serving as substrates for transacetylases and forming reactive *N*-acetoxy derivatives (e.g., *N*-acetoxy-1-amino-8-nitropyrene). This was first demonstrated using transacetylase-deficient strains of *Salmonella* (MCCOY et al. 1983; ORR et al. 1985) and was subsequently shown to occur in eukaryotes (DJURIĆ et al. 1985). It is presently not known why 1-nitropyrene does not give similar *N*-acetoxy intermediates, but this may account for the decreased tumorigenicity of this compound compared with the dinitropyrenes.

N-(Deoxyguanosin-8-yl)-1-aminopyrene, along with other adducts, has been found in liver, kidney, and mammary gland DNA of rats given a single intraperitoneal dose of 1-nitropyrene (HASHIMOTO and SHUDO 1985; STANTON et al. 1985), although in a more recent study, similar results were not obtained (DJURIĆ et al. 1988). Binding to lung DNA has been reported in mice given a single intratracheal dose of 1-nitropyrene (MITCHELL 1985; HOWARD et al. 1986); however, *N*-(deoxyguanosin-8-yl)-1-aminopyrene accounted for only 20% of the total binding (MITCHELL 1988). The identification of 1-nitropyrene-DNA adducts is complicated by the fact that the major metabolic pathways in vivo are via ring oxidation rather than nitroreduction (EL-BAYOUMY and HECHT 1984; BALL et al. 1984b; BALL and KING 1985; HOWARD et al. 1985; KINOUCHI et al. 1986a). Some of these ring-oxidized metabolites are as mutagenic as 1-nitropyrene (EL-BAYOUMY and HECHT 1983; BALL et al. 1984a; BALL and KING 1985; FIFER et al. 1986) and will bind to DNA both directly and following nitroreduction (DJURIĆ et al. 1986a). Thus, a large number of adducts can potentially be formed, which is what is observed when incubations are conducted with rat liver microsomes (DJURIĆ et al. 1986a). Likewise, when rabbit lung epithelial cells are incubated with 1-nitropyrene, *N*-(deoxyguanosin-8-yl)-1-aminopyrene is detected but contributes only a small amount to the total binding (GALLAGHER et al. 1988). These unidentified adducts are also present in mouse lung after 1-nitropyrene treatment, and their properties suggest that they are derived from ring oxidation and subsequent nitroreduction pathways (MITCHELL 1988).

The only metabolites detected from dinitropyrenes in vitro and in vivo have been those resulting from nitroreduction (DJURIĆ et al. 1985, 1986c, 1988; HEFLICH et al. 1985a, 1986a). Similarly, when target tissues for dinitropyrene tumorigenesis have been examined, the only adducts detected are those previously identified from in vitro incubations (DELCLOS et al. 1987b; DJURIĆ et al. 1988).

X. 6-Aminochrysene and 6-Nitrochrysene

6-Aminochrysene is an aromatic amine that has been used as a chemotherapeutic agent in the treatment of splenomegaly, myeloid leukemia, and breast cancer (Buu-Hoi et al. 1962a, b, c; Groupe Européen 1967). Although relatively non-toxic, 6-aminochrysene is tumorigenic; newborn mice treated intraperitoneally develop liver and lung tumors (Roe et al. 1969), while liver, lung, and skin tumors arise in adult mice treated topically (Lambelin et al. 1975). 6-Nitrochrysene, which has been detected as an air pollutant (Garner et al. 1986), is similar to 6-aminochrysene in that it induces liver and lung tumors when administered to newborn mice (Busby et al. 1985; Wislocki et al. 1986). It also causes a high incidence of malignant lymphoma (Busby et al. 1985; Wislocki et al. 1986) and, of the polycyclic nitroaromatic hydrocarbons examined in the newborn mouse tumor assay, appears to be the most potent (Wislocki et al. 1986; Busby et al. 1988). In addition, 6-nitrochrysene is a skin tumor initiator in adult mice treated topically (El-Bayoumy et al. 1982) and will transform Syrian hamster embryo cells (DiPaolo et al. 1983).

6-Aminochrysene and its nitro derivative, 6-nitrochrysene, have considerably more hydrocarbon character than most of the aromatic amines considered previously. Therefore, as was observed with 1-nitropyrene, the metabolic activation of these compounds may involve oxidation/reduction of the amine/nitro group, ring oxidation, or a combination of the two. As with other *N*-hydroxy arylamines, *N*-hydroxy-6-aminochrysene will react with DNA under slightly acidic conditions. Three DNA adducts are formed, 5-(deoxyguanosin-N^2-yl)-6-aminochrysene, *N*-(deoxyguanosin-8-yl)-6-aminochrysene, and *N*-(deoxyinosin-8-yl)-6-aminochrysene, the last product apparently arising from the oxidative deamination of *N*-(deoxyadenosin-8-yl)-6-aminochrysene (Fig. 12; Delclos et al. 1987a). Two of these adducts, *N*-(deoxyguanosin-8-yl)-6-aminochrysene and *N*-(deoxyinosin-8-yl)-6-aminochrysene, have been detected in in vitro incubations of 6-nitrochrysene with hepatocytes isolated from untreated mice and rats and from phenobarbital-pretreated rats (Delclos et al. 1987a; Kadlubar et al. 1988c). This suggests that metabolic activation can occur through nitroreduction to *N*-hydroxy-6-aminochrysene. Likewise, the same adducts are formed when 6-aminochrysene is incubated with hepatic microsomes from untreated or phenobarbital-pretreated rats (Kadlubar et al. 1989), which suggests that 6-aminochrysene readily undergoes hepatic *N*-oxidation. In newborn mice, the metabolic activation pathway is strikingly different, as the major adduct detected in lung and liver DNA of animals treated with 6-nitrochrysene or 6-aminochrysene does not correspond to any of the adducts formed from *N*-hydroxy-6-aminochrysene (Delclos et al. 1987b). Instead, the major adduct is derived from the subsequent metabolism of 6-aminochrysene *trans*-1,2-dihydrodiol (Delclos et al. 1988). Although this adduct has not been characterized, preliminary spectral data (Kadlubar et al. 1988c) indicate that it arises from 6-aminochrysene *trans*-1,2-dihydrodiol-3,4-epoxide or a quinimine methide derivative analogous to that described by Hulbert and Grover (1983) for 9-hydroxychrysene-1,2-dihydrodiol-3,4-epoxide. The same adduct is formed when 6-aminochrysene *trans*-1,2-dihydrodiol is incubated with purified rat cytochrome

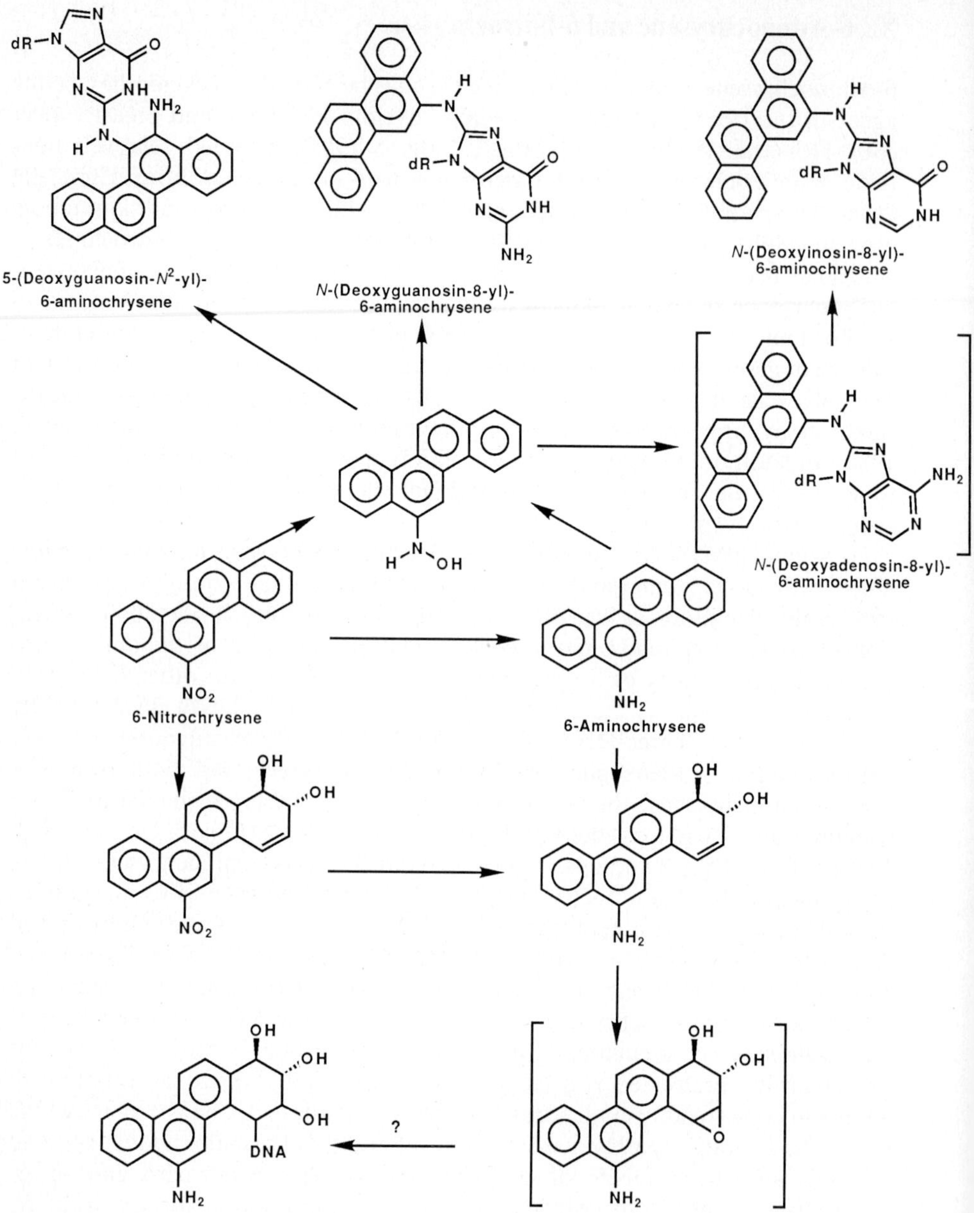

Fig. 12. Metabolic activation pathways and DNA adducts of 6-aminochrysene and 6-nitrochrysene

P-450 *c* but not isozymes *b* or *d* (KADLUBAR et al. 1988c). Cytochrome P-450 *c* is the major isozyme responsible for the oxidation of polycyclic aromatic hydrocarbons to reactive dihydrodiol epoxides (CONNEY 1982; KADLUBAR and HAMMONS 1987). This presumed dihydrodiol epoxide-derived adduct is also observed in in vitro incubations of 6-nitrochrysene with hepatocytes from mice and rats pretreated with Aroclor 1254 (KADLUBAR et al. 1988c). Therefore, it appears that the specific metabolic pathways by which 6-aminochrysene and 6-nitrochrysene are activated will be quite dependent on the particular cytochromes P-450 that are present.

XI. Heterocyclic Aromatic Amines

During the cooking of food, a series of complex heterocyclic amines can be formed by the pyrolysis of amino acids, creatinine, and sugars (SUGIMURA 1986). Representative examples of these compounds include 3-amino-1,4-dimethyl-5*H*-pyrido[4,3-*b*]indole (Trp-P-1); 3-amino-1-methyl-5*H*-pyrido[4,3-*b*]indole (Trp-P-2); 2-amino-6-methyldipyrido[1,2-*a*:3′,2′-*d*]imidazole (Glu-P-1); 2-amino-dipyrido[1,2-*a*:3′2′-*d*]imidazole (Glu-P-2); 2-amino-3-methylimidazo-[4,5-*f*]quinoline (IQ); 2-amino-3,4-dimethylimidazo[4,5-*f*]quinoline (MeIQ); and 2-amino-3,8-dimethylimidazo[4,5-*f*]quinoxaline (MeIQx) (Fig. 13). Some of these heterocyclic amines are extremely potent mutagens, especially in the *S. typhimurium* test system (reviewed in KATO 1986; FELTON et al. 1988). In addition, chronic studies in rodents indicate that these compounds are moderately potent carcinogens inducing tumors in a variety of tissues, including the liver, mammary, Zymbal's, and clitoral glands, skin, and intestine (SUGIMURA 1986). As such, they may contribute significantly to human cancers, especially tumors of the colon that have been strongly associated with dietary factors (BRUCE 1987).

As with other carcinogenic aromatic amines, hepatic *N*-oxidation is regarded as a necessary step in the metabolic activation of heterocyclic aromatic amines (KATO 1986; KATO and YAMAZOE 1987). Of the hepatic monooxygenases that have been thus far examined, rat liver cytochrome P-450 *d* has high catalytic activity for Glu-P-1 and IQ (KATO et al. 1983; YAMAZOE et al. 1983), while both isozymes *d* and *c* effectively *N*-oxidize Trp-P-2 and MeIQx (KAMATAKI et al. 1983; KATO et al. 1983; YAMAZOE et al. 1988a). Several heterocyclic aromatic amines are also substrates for prostaglandin *H* synthase, which converts them to DNA-bound products (PETRY et al. 1986; KADLUBAR et al. 1988a); however, the identity of the metabolites, the nature of the DNA adducts formed, and the role of this pathway in vivo are presently not known.

The *N*-hydroxy metabolites of Trp-P-2, Glu-P-1, and IQ are known to react directly with DNA (HASHIMOTO et al. 1980b; MITA et al. 1982; KATO and YAMAZOE 1987; SNYDERWINE et al. 1988), although in contrast to typical *N*-hydroxy arylamines the reaction is not necessarily facilitated by acidic conditions (MITA et al. 1982; SNYDERWINE et al. 1988). The reactivity of the *N*-hydroxy derivatives is greatly enhanced, however, by in situ generation of an *N*-acetoxy ester using acetic anhydride or ketene (HASHIMOTO et al. 1980a, b, 1982b; KATO and YAMAZOE 1987; SNYDERWINE et al. 1988), and for the *N*-hydroxy derivatives

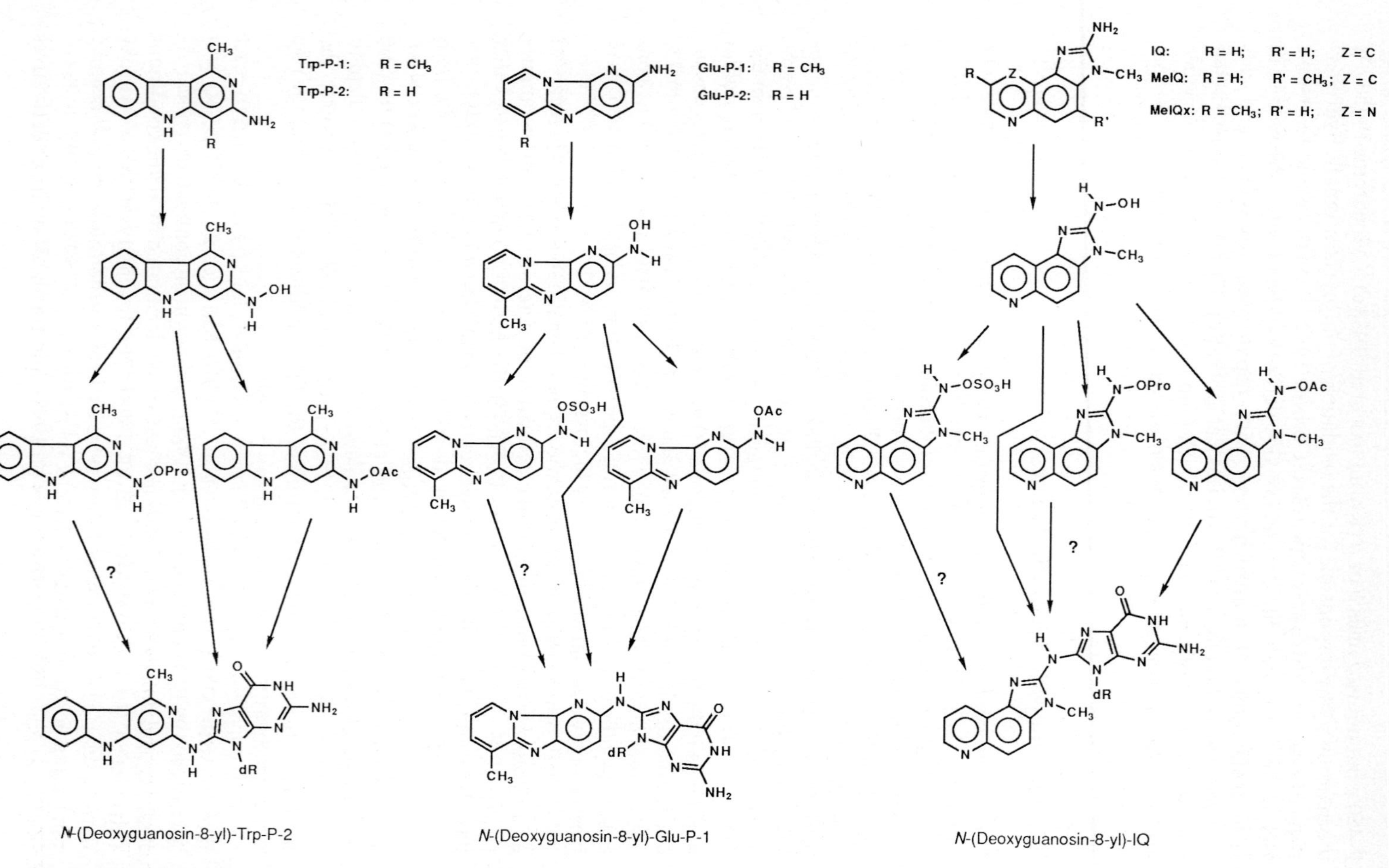

Fig. 13. Metabolic activation pathways and DNA adducts of heterocyclic aromatic amines

of Trp-P-2, Glu-P-1, and IQ, the major adducts formed with DNA in vitro are C8-deoxyguanosine-substituted products analogous to those found with other aromatic amine carcinogens (HASHIMOTO et al. 1980a, b, 1982b; SNYDERWINE et al. 1988; Fig. 13). Similar C8-deoxyguanosine adducts have been detected in the livers of rats treated with Trp-P-2 and Glu-P-1 (HASHIMOTO et al. 1982a).

The DNA adduct profiles from heterocyclic aromatic amines do not appear to have been examined in mutagenesis assays. Nevertheless, the mutations induced by IQ and MeIQ in *S. typhimurium* TA98 and TA1538 result from a 2-base CG deletion (FELTON et al. 1988), which is consistent with the formation of C8-deoxyguanosine adducts. Likewise, the mutagenic efficiency of IQ and Trp-P-2 DNA adducts in Chinese hamster ovary cells (BROOKMAN et al. 1985) is nearly the same as that observed with *N*-(deoxyguanosin-8-yl)-*N*′-acetylbenzidine (HEFLICH et al. 1986c). With a similar metabolic activation system and in the presence of DNA, Trp-P-2 is activated to give *N*-(deoxyguanosin-8-yl)-Trp-P-2 (HASHIMOTO et al. 1979).

As suggested by the in vitro DNA adduct results discussed above, enzymatic *O*-esterification of the *N*-hydroxy metabolites appears to be an important activation pathway for the heterocyclic aromatic amines. *N*-Hydroxy-Glu-P-1 is a good substrate for hepatic sulfotransferase (KATO and YAMAZOE 1987), and *N*-hydroxy-Trp-P-2 is readily activated by prolyl-tRNA synthetase (YAMAZOE et al. 1985b). Both of these *N*-hydroxy derivatives are converted to reactive esters by transacetylases (SHINOHARA et al. 1985), while *N*-hydroxy-IQ is metabolically activated by all three enzyme systems (SHINOHARA et al. 1986; KATO and YAMAZOE 1987). Of these pathways, the acetyl coenzyme A-dependent *O*-acetylation is likely to be of major importance in both mutagenesis and carcinogenesis. This enzyme system is present in the standard *S. typhimurium* tester strains, except TA98/1,8-DNP_6, and these heterocyclic amines show decreased mutagenicity in this strain (SAITO et al. 1985). Transacetylases are also widely distributed in carcinogen-target tissues of several species including humans (FLAMMANG and KADLUBAR 1986; SHINOHARA et al. 1986; FLAMMANG et al. 1988), and these enzymes exhibit a genetic polymorphism resulting in both rapid and slow acetylator individuals (WEBER 1987; HEIN 1988). In this regard, groups of patients with histories of colorectal cancer show higher proportions of rapid acetylator individuals (LANG et al. 1986; ILETT et al. 1987), and heterocyclic aromatic amines have been proposed as important etiologic agents for the induction of these tumors (KADLUBAR et al. 1988a).

B. Conclusions and Comments

In the nearly 100 years since REHN (1895) originally proposed that aromatic amines were responsible for urinary bladder cancer in humans, much effort has been expended in attempts to elucidate the mechanisms of action of this class of carcinogens. It is now quite clear that these compounds must be metabolized into electrophilic derivatives and that the initial step for aromatic amines with three or fewer rings is an enzymatic *N*-hydroxylation. This oxidation is catalyzed primarily by cytochrome P-450, and considerable isozyme specificity is observed. Similar studies on the substrate specificity of human cytochromes P-450 and their differential expression as a consequence of enzyme induction or genetic polymorphism should provide an assessment of interindividual differences in human susceptibility to aromatic amine carcinogenesis.

With aromatic amines containing more than three rings the initial oxidation may also involve an *N*-hydroxylation. Alternatively, these compounds can be activated by oxidative pathways traditionally associated with polycyclic aromatic hydrocarbons or by a combination of both pathways. Peroxidases, in particular prostaglandin *H* synthase, may also play a role in the initial oxidation of aromatic amines, especially in tissues containing low levels of cytochrome P-450. Although peroxidase-catalyzed activation of aromatic amines has been demonstrated in vitro, the data from in vivo experiments are presently equivocal as to the relative importance of this pathway.

The *N*-hydroxy arylamines resulting from the initial oxidation can undergo acid-catalyzed reactions with DNA. Subsequent activation can also occur through ester formation, in particular *O*-acetylation and *O*-sulfonation. Hepatic *O*-acetylation is catalyzed by acyltransferases, which also appear to be capable of catalyzing the *N*-acetylation of aromatic amines and the conversion of *N*-hydroxy arylamides to *N*-acetoxy arylamines (HEIN et al. 1987). Acyltransferases are polymorphic, and both rapid and slow phenotypes are observed in humans and in a variety of experimental animals. Because of this polymorphic distribution, acetylation is an important factor in the organ specificity observed with aromatic amines. For instance, if *N*-acetylation precedes *N*-oxidation, the concentration of *N*-hydroxy arylamine available for transport to the bladder should decrease. Thus, individuals with a rapid acetylator phenotype should be at a lower risk for bladder cancer from exposure to aromatic amines, which is what has been observed (reviewed in HEIN 1988). Likewise, the inability of dogs to *N*-acetylate aromatic amines is consistent with their susceptibility to bladder tumors. While acetylation appears to afford protection from bladder tumor induction, the opposite may be true for other tissues. Thus, a higher incidence of colon cancer has been found in low-risk individuals with a rapid acetylator phenotype. Furthermore, mice and rabbits with a rapid acetylator phenotype form substantially more hepatic DNA adducts from 2-aminofluorene than those with a slow acetylator phenotype. This difference may be due to the formation of greater concentrations of reactive *N*-acetoxy arylamine intermediates in rapid acetylator individuals.

Sulfotransferase-catalyzed sulfonation of *N*-hydroxy arylamines and *N*-hydroxy arylamides clearly leads to the formation of arylamine- and arylamide-DNA adducts. Nevertheless, the primary role of sulfotransferase, at least in rat liver, may not be in generating DNA adducts but rather in metabolizing arylamine derivatives to metabolites capable of eliciting a cytotoxic response that could serve as a "chemical partial hepatectomy", i.e., a promoting stimulus. This hypothesis is consistent with the observation that *N*-hydroxy arylamides that are not substrates for rat liver sulfotransferase, as indicated by the formation of arylamide-DNA adducts, are not hepatocarcinogenic in rats. It is also in accord with the finding that inhibiting sulfotransferase, in addition to decreasing the extent of arylamide-DNA adduct formation, abolishes the hepatotoxicity associated with *N*-hydroxy-2-acetylaminofluorene (MEERMAN and MULDER 1981). Furthermore, if this inhibition is followed by the administration of a tumor promoter, more preneoplastic foci are formed than in the absence of the inhibitor (MEERMAN 1985). This latter finding suggests that arylamide adducts are not required for initiation, and that electrophilic metabolites formed from sulfotransferase catalysis are cytotoxic to both initiated and normal cells. Sulfotransferase may have additional roles in carcinogenesis because the activity of this enzyme is diminished by a wide variety of hepatocarcinogens. Whether or not this is causally related to the hepatocarcinogenic response is presently unknown. It is also not known what role, if any, sulfotransferase plays in the etiology of aromatic amine-induced tumors in humans.

Aromatic amines and their nitro analogues form DNA adducts at C8, N^2, and O^6 of guanine and at C8 and N^6 of adenine. A number of cyclic adducts have also been reported, in particular those arising from 4-acetylaminostilbene. Nevertheless, in almost all instances the major adduct obtained is an arylamine derivative substituted at C8 of deoxyguanosine (e.g., *N*-(deoxyguanosin-8-yl)-2-aminofluorene). C8-deoxyguanosine-arylamine adducts have been correlated with the induction of mutations in bacteria and mammalian cells. They have also been shown to cause point mutations, of which the majority are G→T transversions. Precisely this type of mutation has been detected at codon 61 of the *ras* protooncogene during the induction of mouse liver tumors by *N*-hydroxy-2-acetylaminofluorene. Recent studies suggest that C8-deoxyguanosine-arylamine adducts are also formed in human tissues. First, individuals exposed to 4-aminobiphenyl in tobacco smoke form hemoglobin adducts indicative of the formation of *N*-hydroxy-4-aminobiphenyl (BRYANT et al. 1987). Second, DNA adducts detected by ^{32}P-postlabeling in exfoliated buccal cells from cigarette smokers have an enzymatic sensitivity consistent with C8-deoxyguanosine-arylamine adducts (GUPTA and EARLEY 1988). Third, immunoassays have indicated the presence of *N*-(deoxyguanosin-8-yl)-4-aminobiphenyl in human lung and urinary bladder DNA (KADLUBAR et al. 1988d). Finally, a point mutation at codon 61 of the *ras* protooncogene has been shown to occur in a proportion of human bladder cancers (FUJITA et al. 1985). Thus, C8-deoxyguanosine-arylamine adducts may well be involved in the etiology of aromatic amine-induced cancers in humans.

Acknowledgements: We thank Cynthia Hartwick and Roy Collins for help in preparing this manuscript.

References

Allaben WT, Weeks CE, Weis CC, Burger GT, King CM (1982) Rat mammary gland carcinogenesis after local injection of *N*-hydroxy-*N*-acyl-2-aminofluorenes: relationship to metabolic activation. Carcinogenesis 3:233–240

Allaben WT, Weis CC, Fullerton NF, Beland FA (1983) Formation and persistence of DNA adducts from the carcinogen *N*-hydroxy-2-acetylaminofluorene in rat mammary gland in vivo. Carcinogenesis 4:1067–1070

Andersen RA, Enomoto M, Miller EC, Miller JA (1964) Carcinogenesis and inhibition of the Walker 256 tumor in the rat by *trans*-4-acetylaminostilbene, its *N*-hydroxy metabolite, and related compounds. Cancer Res 24:128–143

Andervont HB, Edwards JE (1943) Carcinogenic action of two azo compounds in mice. JNCI 3:349–354
Andrews PJ, Quilliam MA, McCarry BE, Bryant DW, McCalla DR (1986) Identification of the DNA adduct formed by metabolism of 1,8-dinitropyrene in *Salmonella typhimurium*. Carcinogenesis 7:105–110
Arce GT, Cline DT Jr, Mead JE (1987) The ^{32}P-post-labeling method in quantitative DNA adduct dosimetry of 2-acetylaminofluorene-induced mutagenicity in Chinese hamster ovary cells and *Salmonella typhimurium* TA1538. Carcinogenesis 8:515–520
Baldwin RW, Smith WRD (1965) *N*-Hydroxylation in aminostilbene carcinogenesis. Br J Cancer 19:433–443
Baldwin RW, Smith WRD, Surtees SJ (1963) Carcinogenic action of *N*-hydroxy-4-acetylaminostilbene. Nature 199:613–614
Ball LM, King LC (1985) Metabolism, mutagenicity, and activation of 1-nitropyrene in vivo and in vitro. Environ Int 11:355–362
Ball LM, Kohan MJ, Claxton LD, Lewtas J (1984a) Mutagenicity of derivatives and metabolites of 1-nitropyrene: activation by rat liver S9 and bacterial enzymes. Mutat Res 138:113–125
Ball LM, Kohan MJ, Inmon JP, Claxton LD, Lewtas J (1984b) Metabolism of 1-nitro[^{14}C]pyrene in vivo in the rat and mutagenicity of urinary metabolites. Carcinogenesis 5:1557–1564
Baur H, Neumann H-G (1980) Correlation of nucleic acid binding by metabolites of *trans*-4-aminostilbene derivatives with tissue specific acute toxicity and carcinogenicity in rats. Carcinogenesis 1:877–885
Beland FA (1978) Computer-generated graphic models of the N^2-substituted deoxyguanosine adducts of 2-acetylaminofluorene and benzo[*a*]pyrene and the O^6-substituted deoxyguanosine adduct of 1-naphthylamine in the DNA double helix. Chem Biol Interact 22:329–339
Beland FA, Kadlubar FF (1985) Formation and persistence of arylamine DNA adducts in vivo. Environ Health Perspect 62:19–30
Beland FA, Dooley KL, Casciano DA (1979) Rapid isolation of carcinogen-bound DNA and RNA by hydroxyapatite chromatography. J Chromatogr 174:177–186
Beland FA, Allaben WT, Evans FE (1980a) Acyltransferase-mediated binding of *N*-hydroxyarylamides to nucleic acids. Cancer Res 40:834–840
Beland FA, Tullis DL, Kadlubar FF, Straub KM, Evans FE (1980b) Characterization of DNA adducts of the carcinogen *N*-methyl-4-aminoazobenzene in vitro and in vivo. Chem Biol Interact 31:1–17
Beland FA, Dooley KL, Jackson CD (1982) Persistence of DNA adducts in rat liver and kidney after multiple doses of the carcinogen *N*-hydroxy-2-acetylaminofluorene. Cancer Res 42:1348–1354
Beland FA, Beranek DT, Dooley KL, Heflich RH, Kadlubar FF (1983) Arylamine-DNA adducts in vitro and in vivo: their role in bacterial mutagenesis and urinary bladder carcinogenesis. Environ Health Perspect 49:125–134
Beland FA, Ribovich M, Howard PC, Heflich RH, Kurian P, Milo GE (1986) Cytotoxicity, cellular transformation and DNA adducts in normal human diploid fibroblasts exposed to 1-nitrosopyrene, a reduced derivative of the environmental contaminant, 1-nitropyrene. Carcinogenesis 7:1279–1283
Belman S, Troll W, Teebor G, Mukai F (1968) The carcinogenic and mutagenic properties of *N*-hydroxy-aminonaphthalenes. Cancer Res 28:535–542
Beranek DT, White GL, Heflich RH, Beland FA (1982) Aminofluorene-DNA adduct formation in *Salmonella typhimurium* exposed to the carcinogen *N*-hydroxy-2-acetylaminofluorene. Proc Natl Acad Sci USA 79:5175–5178
Bichara M, Fuchs RPP (1985) DNA binding and mutation spectra of the carcinogen *N*-2-aminofluorene in *Escherichia coli*. A correlation between the conformation of the premutagenic lesion and the mutation specificity. J Mol Biol 183:341–351
Bichara M, Fuchs RPP (1988) The uvrC gene product has no specific role in the repair of *N*-2-aminofluorene adducts. In: King CM, Romano LJ, Schuetzle D (eds) Carcinogenic and mutagenic responses to aromatic amines and nitroarenes. Elsevier, New York, pp 385–387

Boiteux S, de Oliveira RC, Laval J (1985) The *Escherichia coli* O^6-methylguanine-DNA methyltransferase does not repair promutagenic O^6-methylguanine residues when present in Z-DNA. J Biol Chem 260:8711–8715
Boyd JA, Eling TE (1984) Evidence for a one-electron mechanism of 2-aminofluorene oxidation by prostaglandin H synthase and horseradish peroxidase. J Biol Chem 259:13885–13896
Boyd JA, Eling TE (1987) Prostaglandin H synthase-catalyzed metabolism and DNA binding of 2-naphthylamine. Cancer Res 47:4007–4014
Boyd JA, Harvan DJ, Eling TE (1983) The oxidation of 2-aminofluorene by prostaglandin endoperoxide synthetase. Comparison with other peroxidases. J Biol Chem 258:8246–8254
Brookman KW, Salazar EP, Thompson LH (1985) Comparative mutagenic efficiencies of the DNA adducts from the cooked-food-related mutagens Trp-P-2 and IQ in CHO cells. Mutation Res 149:249–255
Broyde S, Hingerty BE (1982) DNA backbone conformation in AAF modified dCpdG: variable conformational modes of achieving base displacement. Chem Biol Interact 40:113–119
Broyde S, Hingerty B (1983) Conformation of 2-aminofluorene-modified DNA. Biopolymers 22:2423–2441
Broyde S, Hingerty B (1985) Base displacement in AAF-modified Z-DNA. Carcinogenesis 6:151–154
Bruce WR (1987) Recent hypotheses for the origin of colon cancer. Cancer Res 47:4237–4242
Bryant DW, McCalla DR, Lultschik P, Quilliam MA, McCarry BE (1984) Metabolism of 1,8-dinitropyrene by *Salmonella typhimurium*. Chem Biol Interact 49:351–368
Bryant MS, Skipper PL, Tannenbaum SR, Maclure M (1987) Hemoglobin adducts of 4-aminobiphenyl in smokers and nonsmokers. Cancer Res 47:602–608
Busby WF Jr, Garner RC, Chow FL, Martin CN, Stevens EK, Newberne PM, Wogan GN (1985) 6-Nitrochrysene is a potent tumorigen in newborn mice. Carcinogenesis 6:801–803
Busby WF Jr, Stevens EK, Kellenbach ER, Cornelisse J, Lugtenburg J (1988) Dose-response relationships of the tumorigenicity of cyclopenta[*cd*]pyrene, benzo[*a*]pyrene and 6-nitrochrysene in a newborn mouse lung adenoma bioassay. Carcinogenesis 9:741–746
Buu-Hoí NP, Bui-Quoc-Huong, Tran-lu-Y, (1962a) Activité chimiothérapeutique du 6-aminochrysène dans les splénomégalies. Bull Acad Natl Med (Paris) 146:313–317
Buu-Hoí NP, Bui-Quoc-Huong, Tran-lu-Y, Chau-Van-Tuong (1962b) Résultats préliminaires d'une tentative de chimiothérapie spécifique au moyen du 6-aminochryséne. Chemotherapia (Basel) 4:31–42
Buu-Hoí NP, Bui-Quoc-Huong, Tran-lu-Y (1962c) Action leucopéniante du 6-aminochryséne dans les leucémies myeloides chez l'homme. Chemotherapia (Basel) 4:413–418
Carothers AM, Urlaub G, Steigerwalt RW, Chasin LA, Grunberger D (1988) Spectrum of *N*-2-acetylaminofluorene-induced mutations in the dihydrofolate reductase gene of Chinese hamster ovary cells. In: King CM, Romano LJ, Schuetzle D (eds) Carcinogenic and mutagenic responses to aromatic amines and nitroarenes. Elsevier, New York, pp 337–349
Cerniglia CE, Lambert KJ, White GL, Heflich RH, Franklin W, Fifer EK, Beland FA (1988) Metabolism of 1,8-dinitropyrene by human, rhesus monkey, and rat intestinal microflora. Toxicity Assessment: An International Journal 3:147–159
Cilento G, Miller EC, Miller JA (1956) On the addition of protons to derivatives of 4-aminoazobenzene. J Am Chem Soc 78:1718–1722
Coles B, Ketterer B, Beland FA, Kadlubar FF (1984) Glutathione conjugate formation in the detoxification of ultimate and proximate carcinogens of *N*-methyl-4-aminoazobenzene. Carcinogenesis 5:917–920
Conney AH (1982) Induction of microsomal enzymes by foreign chemicals and carcinogenesis by polycyclic aromatic hydrocarbons: G.H.A. Clowes Memorial Lecture. Cancer Res 42:4875–4917

Daune MP, Westhof E, Koffel-Schwartz N, Fuchs RPP (1985) Covalent binding of a carcinogen as a probe for the dynamics of deoxyribonucleic acid. Biochemistry 24:2275–2284

DeBaun JR, Miller EC, Miller JA (1970) *N*-Hydroxy-2-acetylaminofluorene sulfotransferase: its probable role in carcinogenesis and in protein-(methion-*S*-yl) binding in rat liver. Cancer Res 30:577–595

Deichmann WB, Radomski JL (1969) Carcinogenicity and metabolism of aromatic amines in the dog. JNCI 43:263–269

Deichmann WB, MacDonald WM, Coplan MM, Woods FM, Anderson WAD (1958a) *Para* nitrobiphenyl, a new bladder carcinogen in the dog. Ind Med Surg 27:634–637

Deichmann WB, Radomski JL, Anderson WAD, Coplan MM, Woods FM (1958b) The carcinogenic action of *p*-aminobiphenyl in the dog. Ind Med Surg 27:25–26

Delclos KB, Tarpley WG, Miller EC, Miller JA (1984) 4-Aminoazobenzene and *N,N*-dimethyl-4-aminoazobenzene as equipotent hepatic carcinogens in male C57BL/6 × C3H/He F_1 mice and characterization of *N*-(deoxyguanosin-8-yl)-4-aminoazobenzene as the major persistent hepatic DNA-bound dye in these mice. Cancer Res 44:2540–2550

Delclos KB, Miller EC, Miller JA, Liem A (1986) Sulfuric acid esters as major ultimate electrophilic and hepatocarcinogenic metabolites of 4-aminoazobenzene and its *N*-methyl derivatives in infant male C57BL/6J × C3H/HeJ F_1 ($B6C3F_1$) mice. Carcinogenesis 7:277–287

Delclos KB, Miller DW, Lay JO Jr, Casciano DA, Walker RP, Fu PP, Kadlubar FF (1987a) Identification of *C*8-modified deoxyinosine and N^2- and *C*8-modified deoxyguanosine as major products of the in vitro reaction of *N*-hydroxy-6-aminochrysene with DNA and the formation of these adducts in isolated rat hepatocytes treated with 6-nitrochrysene and 6-aminochrysene. Carcinogenesis 8:1703–1709

Delclos KB, Walker RP, Dooley KL, Fu PP, Kadlubar FF (1987b) Carcinogen-DNA adduct formation in the lungs and livers of preweanling CD-1 male mice following administration of [^{3}H]-6-nitrochrysene, [^{3}H]-6-aminochrysene, and [^{3}H]-1,6-dinitropyrene. Cancer Res 47:6272–6277

Delclos KB, El-Bayoumy K, Hecht SS, Walker RP, Kadlubar FF (1988) Metabolism of the carcinogen [^{3}H]-δ-nitrochrysene in the preweanling mouse: identification of 6-aminochrysene-1,2-dihydrodiol as the probable proximate carcinogenic metabolite. Carcinogenesis 9:1875–1884

DiPaolo JA, DeMarinis AJ, Chow FL, Garner RC, Martin CN, Doniger J (1983) Nitration of carcinogenic and non-carcinogenic polycyclic aromatic hydrocarbons results in products able to induce transformation of Syrian hamster cells. Carcinogenesis 4:357–359

Djurić Z, Fifer EK, Beland FA (1985) Acetyl coenzyme A-dependent binding of carcinogenic and mutagenic dinitropyrenes to DNA. Carcinogenesis 6:941–944

Djurić Z, Fifer EK, Howard PC, Beland FA (1986a) Oxidative microsomal metabolism of 1-nitropyrene and DNA-binding of oxidized metabolites following nitroreduction. Carcinogenesis 7:1073–1079

Djurić Z, Heflich RH, Fifer EK, Beland FA (1986b) Metabolic activation of mutagenic and tumorigenic dinitropyrenes. In: Harris C (ed) Biochemical and molecular epidemiology of cancer. Liss, New York, pp 441–447

Djurić Z, Potter DW, Heflich RH, Beland FA (1986c) Aerobic and anaerobic reduction of nitrated pyrenes in vitro. Chem Biol Interact 59:309–324

Djurić Z, Fifer EK, Yamazoe Y, Beland FA (1988) DNA binding by 1-nitropyrene and 1,6-dinitropyrene in vitro and in vivo: effects of nitroreductase induction. Carcinogenesis 9:357–364

Dooley KL, Beland FA, Bucci TJ, Kadlubar FF (1984) Local carcinogenicity, rates of absorption, extent and persistence of macromolecular binding, and acute histopathological effects of *N*-hydroxy-1-naphthylamine and *N*-hydroxy-2-naphthylamine. Cancer Res 44:1172–1177

El-Bayoumy K, Hecht SS (1983) Identification and mutagenicity of metabolites of 1-nitropyrene formed by rat liver. Cancer Res 43:3132–3137
El-Bayoumy K, Hecht SS (1984) Metabolism of 1-nitro[*U*-4,5,9,10-^{14}C]pyrene in the F344 rat. Cancer Res 44:4317–4322
El-Bayoumy K, Hecht SS, Hoffmann D (1982) Comparative tumor initiating activity on mouse skin of 6-nitrobenzo[*a*]pyrene, 6-nitrochrysene, 3-nitroperylene, 1-nitropyrene and their parent hydrocarbons. Cancer Lett 16:333–337
El-Bayoumy K, Hecht SS, Sackl T, Stoner GD (1984) Tumorigenicity and metabolism of 1-nitropyrene in A/J mice. Carcinogenesis 5:1449–1452
Eling TE, Petry TW, Hughes MF, Krauss RS (1988) Aromatic amine metabolism catalyzed by prostaglandin H synthase. In: King CM, Romano LJ, Schuetzle D (eds) Carcinogenic and mutagenic responses to aromatic amines and nitroarenes. Elsevier, New York, pp 161–172
Evans FE, Miller DW (1982) Conformation and dynamics associated with the site of attachment of a carcinogen to a nucleotide. Biochem Biophys Res Commun 108:933–939
Evans FE, Miller DW, Beland FA (1980) Sensitivity of the conformation of deoxyguanosine to binding at the *C*-8 position by *N*-acetylated and unacetylated 2-aminofluorene. Carcinogenesis 1:955–959
Felton JS, Knize MG, Shen NH, Wu R, Becher G (1988) Mutagenic heterocyclic imidazoamines in cooked foods. In: King CM, Romano LJ, Schuetzle D (eds) Carcinogenic and mutagenic responses to aromatic amines and nitroarenes. Elsevier, New York, pp 73–85
Fifer EK, Howard PC, Heflich RH, Beland FA (1986) Synthesis and mutagenicity of 1-nitropyrene 4,5-oxide and 1-nitropyrene 9,10-oxide, microsomal metabolites of 1-nitropyrene. Mutagenesis 1:433–438
Flammang TJ, Kadlubar FF (1986) Acetyl coenzyme A-dependent metabolic activation of *N*-hydroxy-3,2′-dimethyl-4-aminobiphenyl and several carcinogenic *N*-hydroxy arylamines in relation to tissue and species differences, other acyl donors, and arylhydroxamic acid-dependent acyltransferases. Carcinogenesis 7:919–926
Flammang TJ, Westra JG, Kadlubar FF, Beland FA (1985) DNA adducts formed from the probable proximate carcinogen, *N*-hydroxy-3,2′-dimethyl-4-aminobiphenyl, by acid catalysis or *S*-acetyl coenzyme A-dependent enzymatic esterification. Carcinogenesis 6:251–258
Flammang TJ, Hein DW, Talaska G, Kadlubar FF (1988) *N*-Hydroxy-arylamine *O*-acetyltransferase and its relationship to aromatic amine *N*-acetyltransferase polymorphism in the inbred hamster and in human tissue cytosol. In: King CM, Romano LJ, Schuetzle D (eds) Carcinogenic and mutagenic responses to aromatic amines and nitroarenes. Elsevier, New York, pp 137–148
Fourney RM, O'Brien PJ, Davidson WS (1986) Peroxidase catalyzed aggregation of plasmid pBR322 DNA by benzidine metabolites in vitro. Carcinogenesis 7:1535–1542
Franz R, Neumann H-G (1987) Reaction of *trans*-4-*N*-acetoxy-*N*-acetylaminostilbene with guanosine, deoxyguanosine, RNA and DNA in vitro: predominant product is a cyclic N^2,*N*3-guanine adduct. Chem Biol Interact 62:143–155
Franz R, Schulten H-R, Neumann H-G (1986) Identification of nucleic acid adducts from *trans*-4-acetylaminostilbene. Chem Biol Interact 59:281–293
Frederick CB, Beland FA (1988) Comparison of the metabolism, biological activity, and chemical reactivity of 1-naphthylamine and 2-naphthylamine. In: Yang SK, Silverman BD (eds) Polycyclic aromatic hydrocarbon carcinogenesis: structure-activity relationships, vol II. CRC Press, Boca Raton, pp 111–139
Frederick CB, Mays JB, Ziegler DM, Guengerich FP, Kadlubar FF (1982) Cytochrome P-450- and flavin-containing monooxygenase-catalyzed formation of the carcinogen *N*-hydroxy-2-aminofluorene and its covalent binding to nuclear DNA. Cancer Res 42:2671–2677
Frederick CB, Weis CC, Flammang TJ, Martin CN, Kadlubar FF (1985) Hepatic *N*-oxidation, acetyl-transfer and DNA-binding of the acetylated metabolites of the carcinogen, benzidine. Carcinogenesis 6:959–965
Fuchs RPP (1975) In vitro recognition of carcinogen-induced local denaturation sites in native DNA by S_1 endonuclease from *Aspergillus oryzae*. Nature 257:151–152

Fuchs R, Daune M (1972) Physical studies on deoxyribonucleic acid after covalent binding of a carcinogen. Biochemistry 11:2659–2666

Fuchs RPP, Daune MP (1974) Dynamic structure of DNA modified with the carcinogen *N*-acetoxy-*N*-2-acetylaminofluorene. Biochemistry 13:4435–4440

Fuchs RPP, Lefevre J-F, Pouyet J, Daune MP (1976) Comparative orientation of the fluorene residue in native DNA modified by *N*-acetoxy-*N*-2-acetylaminofluorene and two 7-halogeno derivatives. Biochemistry 15:3347–3351

Fuchs RPP, Bichara M, Koffel-Schwartz N (1988) Molecular mechanisms involved in mutagenesis induced by *N*-2-aminofluorene derivatives. In: King CM, Romano LJ, Schuetzle D (eds) Carcinogenic and mutagenic responses to aromatic amines and nitroarenes. Elsevier, New York, pp 373–384

Fujii K (1983) Induction of tumors in transplacental or neonatal mice administered 3′-methyl-4-dimethylaminoazobenzene or 4-aminoazobenzene. Cancer Lett 17:321–325

Fujita J, Srivastava SK, Kraus MH, Rhim JS, Tronick SR, Aaronson SA (1985) Frequency of molecular alterations affecting *ras* protooncogenes in human urinary tract tumors. Proc Natl Acad Sci USA 82:3849–3853

Gallagher JE, Robertson IGC, Jackson MA, Dietrich AM, Ball LM, Lewtas J (1988) ^{32}P-Postlabeling analysis of DNA adducts of two nitrated polycyclic aromatic hydrocarbons in rabbit tracheal epithelial cells. In: King CM, Romano LJ, Schuetzle D (eds) Carcinogenic and mutagenic responses to aromatic amines and nitroarenes. Elsevier, New York, pp 277–281

Garner RC, Martin CN, Clayson DB (1984) Carcinogenic aromatic amines and related compounds. In: Searle CE (ed) Chemical carcinogens, vol 1. American Chemical Society, Washington DC, pp 175–276 (ACS Monograph 182)

Garner RC, Stanton CA, Martin CN, Chow FL, Thomas W, Hubner D, Herrmann R (1986) Bacterial mutagenicity and chemical analysis of polycyclic aromatic hydrocarbons and some nitro derivatives in environmental samples collected in West Germany. Environ Mutagen 8:109–117

Gaugler BJM, Neumann H-G (1979) The binding of metabolites formed from aminostilbene derivatives to nucleic acids in the liver of rats. Chem Biol Interact 24:355–372

Gaugler BJM, Neumann H-G, Scribner NK, Scribner JD (1979) Identification of some products from the reaction of *trans*-4-aminostilbene metabolites and nucleic acids in vivo. Chem Biol Interact 27:335–342

Green LC, Skipper PL, Turesky RJ, Bryant MS, Tannenbaum SR (1984) In vivo dosimetry of 4-aminobiphenyl in rats via a cysteine adduct in hemoglobin. Cancer Res 44:4254–4259

Groupe Européen du Cancer du Sein (1967) Induction par le 6-aminochrysène de rémission du cancer du sein en phase avancée chez la femme. Eur J Cancer 3:75–77

Grunberger D, Nelson JH, Cantor CR, Weinstein IB (1970) Coding and conformational properties of oligonucleotides modified with the carcinogen *N*-2-acetylaminofluorene. Proc Natl Acad Sci USA 66:488–494

Gupta RC, Dighe NR (1984) Formation and removal of DNA adducts in rat liver treated with *N*-hydroxy derivatives of 2-acetylaminofluorene, 4-acetylaminobiphenyl, and 2-acetylaminophenanthrene. Carcinogenesis 5:343–349

Gupta RC, Earley K (1988) ^{32}P-adduct assay: comparative recoveries of structurally diverse DNA adducts in the various enhancement procedures. Carcinogenesis 9:1687–1693

Gupta RC, Earley K, Fullerton NF, Beland FA (1985) Formation and removal of DNA adducts in rats administered multiple doses of 2-acetylaminophenanthrene. Proc Am Assoc Cancer Res 26:86

Gupta RC, Earley K, Fullerton NF, Beland FA (1986) Formation and removal of DNA adducts in target and nontarget tissues of female rats administered multiple doses of 2-acetylaminophenanthrene. Proc Am Assoc Cancer Res 27:107

Haddow A, Harris RJC, Kon GAR, Roe EMF (1948) The growth-inhibitory and carcinogenic properties of 4-aminostilbene and derivatives. Trans Roy Soc (London) A 241:147–195

Hammons GJ, Guengerich FP, Weis CC, Beland FA, Kadlubar FF (1985) Metabolic oxidation of carcinogenic arylamines by rat, dog, and human hepatic microsomes and by purified flavin-containing and cytochrome P-450 monooxygenases. Cancer Res 45:3578–3585
Harvan DJ, Hass JR, Lieberman MW (1977) Adduct formation between the carcinogen *N*-acetoxy-2-acetylaminofluorene and synthetic polydeoxyribonucleotides. Chem Biol Interact 17:203–210
Hashimoto Y, Shudo K (1985) Modification of nucleic acids with 1-nitropyrene in the rat: identification of the modified nucleic acid base. Gann 76:253–256
Hashimoto Y, Shudo K, Okamoto T (1979) Structural identification of a modified base in DNA covalently bound with mutagenic 3-amino-1-methyl-5*H*-pyrido[4,3-*b*]indole. Chem Pharm Bull 27:1058–1060
Hashimoto Y, Shudo K, Okamoto T (1980a) Metabolic activation of a mutagen, 2-amino-6-methyldipyrido[1,2-a:3′2′-d]imidazole. Identification of 2-hydroxyamino-6-methyldipyrido[1,2-a:3′2′-d]imidazole and its reaction with DNA. Biochem Biophys Res Commun 92:971–976
Hashimoto Y, Shudo K, Okamoto T (1980b) Activation of a mutagen, 3-amino-1-methyl-5*H*-pyrido[4,3-b]indole. Identification of 3-hydroxyamino-1-methyl-5*H*-pyrido[4,3-b]indole and its reaction with DNA. Biochem Biophys Res Commun 96:355–362
Hashimoto Y, Shudo K, Okamoto T (1982a) Modification of nucleic acids with mutacarcinogenic heteroaromatic amines in vivo. Identification of modified bases in DNA extracted from rats injected with 3-amino-1-methyl-5*H*-pyrido[4,3-*b*]indole and 2-amino-6-methyldipyrido[1,2-*a*:3′,2′-*d*]imidazole. Mutation Res 105:9–13
Hashimoto Y, Shudo K, Okamoto T (1982b) Modification of DNA with potent mutacarcinogenic 2-amino-6-methyldipyrido[1,2-*a*:3′,2′-*d*]imidazole isolated from a glutamic acid pyrolysate: structure of the modified nucleic acid base and initial chemical event caused by the mutagen. J Am Chem Soc 104:7636–7640
Heflich RH, Fifer EK, Djurić Z, Beland FA (1985a) DNA adduct formation and mutation induction by nitropyrenes in Salmonella and Chinese hamster ovary cells: relationships with nitroreduction and acetylation. Environ Health Perspect 62:135–143
Heflich RH, Howard PC, Beland FA (1985b) 1-Nitrosopyrene: an intermediate in the metabolic activation of 1-nitropyrene to a mutagen in *Salmonella typhimurium* TA1538. Mutation Res 149:25–32
Heflich RH, Djurić Z, Fifer EK, Cerniglia CE, Beland FA (1986a) Metabolism of dinitropyrenes to DNA-binding derivatives in vitro and in vivo. In: Ishinishi N, Koizumi A, McClellan RO, Stöber W (eds) Carcinogenic and mutagenic effects of diesel engine exhaust. Elsevier, Amsterdam, pp 185–197
Heflich RH, Fullerton NF, Beland FA (1986b) An examination of the weak mutagenic response of 1-nitropyrene in Chinese hamster ovary cells. Mutation Res 161:99–108
Heflich RH, Morris SM, Beranek DT, McGarrity LJ, Chen JJ, Beland FA (1986c) Relationships between the DNA adducts and the mutations and sister-chromatid exchanges produced in Chinese hamster ovary cells by *N*-hydroxy-2-aminofluorene, *N*-hydroxy-*N*′-acetylbenzidine and 1-nitrosopyrene. Mutagenesis 1:201–206
Heflich RH, Djurić Z, Zhuo Z, Fullerton NF, Casciano DA, Beland FA (1988) Metabolism of 2-acetylaminofluorene in the Chinese hamster ovary cell mutation assay. Environ Mol Mutagen 11:167–181
Hein DW (1988) Acetylator genotype and arylamine-induced carcinogenesis. Biochim Biophys Acta 948:37–66
Hein DW, Flammang TJ, Kirlin WG, Trinidad A, Ogolla F (1987) Acetylator genotype-dependent metabolic activation of carcinogenic *N*-hydroxyarylamines by *S*-acetyl coenzyme A-dependent enzymes of inbred hamster tissue cytosols: relationship to arylamine *N*-acetyltransferase. Carcinogenesis 8:1767–1774
Hilpert D, Neumann H-G (1985) Accumulation and elimination of macromolecular lesions in susceptible and non-susceptible rat tissues after repeated administration of *trans*-4-acetylaminostilbene. Chem Biol Interact 54:85–95
Hilpert D, Romen W, Neumann H-G (1983) The role of partial hepatectomy and of promoters in the formation of tumors in non-target tissues of *trans*-4-acetylaminostilbene in rats. Carcinogenesis 4:1519–1525

Hingerty B, Broyde S (1982) Conformation of the deoxydinucleoside monophosphate dCpdG modified at carbon 8 of guanine with 2-(acetylamino)fluorene. Biochemistry 21:3243–3252
Hingerty B, Broyde S (1983) AAF linked to the guanine amino group: a B-Z junction. Nucleic Acids Res 11:3241–3254
Hingerty BE, Broyde S (1986) Energy minimized structures of carcinogen-DNA adducts: 2-acetylaminofluorene and 2-aminofluorene. J Biomolecular Structure Dynamics 4:365–372
Hirose M, Lee M-S, Wang CY, King CM (1984) Induction of rat mammary gland tumors by 1-nitropyrene, a recently recognized environmental mutagen. Cancer Res 44: 1158–1162
Howard AJ, Mitchell CE, Dutcher JS, Henderson TR, McClellan RO (1986) Binding of nitropyrenes and benzo[*a*]pyrene to mouse lung deoxyribonucleic acid after pretreatment with inducing agents. Biochem Pharmacol 35:2129–2134
Howard PC, Beland FA (1982) Xanthine oxidase catalyzed binding of 1-nitropyrene to DNA. Biochem Biophys Res Commun 104:727–732
Howard PC, Beland FA, Casciano DA (1981) Quantitation of *N*-hydroxy-2-acetylaminofluorene mediated DNA adduct formation and the subsequent repair in primary rat hepatocyte cultures. J Supramol Struc Cell Biochem [Suppl] 5:197
Howard PC, Heflich RH, Evans FE, Beland FA (1983) Formation of DNA adducts in vitro and in *Salmonella typhimurium* upon metabolic reduction of the environmental mutagen 1-nitropyrene. Cancer Res 43:2052–2058
Howard PC, Flammang TJ, Beland FA (1985) Comparison of the in vitro and in vivo hepatic metabolism of the carcinogen 1-nitropyrene. Carcinogenesis 6:243–249
Hueper WC, Wiley FH, Wolfe HD, Ranta KE, Leming MF, Blood FR (1938) Experimental production of bladder tumors in dogs by administration of β-naphthylamine. J Ind Hyg Toxicol 20:46–84
Hulbert PB, Grover PL (1983) Chemical rearrangement of phenol-epoxide metabolites of polycyclic aromatic hydrocarbons to quinone-methides. Biochem Biophys Res Commun 117:129–134
IARC (1972) IARC monographs of the evaluation of carcinogenic risk of chemicals to man, vol 1. 4-Aminobiphenyl. International Agency for Research on Cancer. World Health Organization, Lyon, pp 74–79
IARC (1974) IARC monographs of the evaluation of carcinogenic risk of chemicals to man, vol 4. Some aromatic amines, hydrazine and related substances, *N*-nitroso compounds and miscellaneous alkylating agents. 2-Naphthylamine. International Agency for Research on Cancer. World Health Organization, Lyon, pp 97–111
IARC (1982) IARC monographs of the evaluation of carcinogenic risk of chemicals to man, vol 29. Some industrial chemicals and dyestuffs. Benzidine and its sulphate, hydrochloride and dihydrochloride. International Agency for Research on Cancer. World Health Organization, Lyon, pp 149–183
Ilett KF, David BM, Detchon P, Castleden WM, Kwa R (1987) Acetylation phenotype in colorectal carcinoma. Cancer Res 47:1466–1469
Jackson CD, Irving CC (1972) Sex differences in cell proliferation and *N*-hydroxy-2-acetylaminofluorene sulfotransferase levels in rat liver during 2-acetylaminofluorene administration. Cancer Res 32:1590–1594
Johnson DE, Cornish HH (1978) Metabolic conversion of 1- and 2-nitronaphthalene to 1- and 2-naphthylamine in the rat. Toxicol Appl Pharmacol 46:549–553
Josephy PD, Eling TE, Mason RP (1983a) An electron spin resonance study of the activation of benzidine by peroxidases. Mol Pharmacol 23:766–770
Josephy PD, Eling TE, Mason RP (1983b) Co-oxidation of benzidine by prostaglandin synthase and comparison with the action of horseradish peroxidase. J Biol Chem 258:5561–5569
Kadlubar FF, Beland FA (1985) Chemical properties of ultimate carcinogenic metabolites of arylamines and arylamides. In: Harvey RG (ed) Polycyclic hydrocarbons and carcinogenesis. American Chemical Society, Washington DC, pp 341–370 (ACS symposium series no. 283)

Kadlubar FF, Hammons GJ (1987) The role of cytochrome P-450 in the metabolism of chemical carcinogens. In: Guengerich FP (ed) Mammalian cytochromes P-450, vol II. CRC Press, Boca Raton, pp 81–130

Kadlubar FF, Miller JA, Miller EC (1976a) Microsomal *N*-oxidation of the hepatocarcinogen *N*-methyl-4-aminoazobenzene and the reactivity of *N*-hydroxy-*N*-methyl-4-aminoazobenzene. Cancer Res 36:1196–1206

Kadlubar FF, Miller JA, Miller EC (1976b) Hepatic metabolism of *N*-hydroxy-*N*-methyl-4-aminoazobenzene and other *N*-hydroxy arylamines to reactive sulfuric acid esters. Cancer Res 36:2350–2359

Kadlubar FF, Miller JA, Miller EC (1977) Hepatic microsomal *N*-glucuronidation and nucleic acid binding of *N*-hydroxy arylamines in relation to urinary bladder carcinogenesis. Cancer Res 37:805–814

Kadlubar FF, Miller JA, Miller EC (1978) Guanyl O^6-arylamination and O^6-arylation of DNA by the carcinogen *N*-hydroxy-1-naphthylamine. Cancer Res 38:3628–3638

Kadlubar FF, Unruh LE, Beland FA, Straub KM, Evans FE (1980) In vitro reaction of the carcinogen, *N*-hydroxy-2-naphthylamine, with DNA at the *C*-8 and N^2 atoms of guanine and at the N^6 atom of adenine. Carcinogenesis 1:139–150

Kadlubar FF, Anson JF, Dooley KL, Beland FA (1981a) Formation of urothelial and hepatic DNA adducts from the carcinogen 2-naphthylamine. Carcinogenesis 2:467–470

Kadlubar FF, Melchior WB Jr, Flammang TJ, Gagliano AG, Yoshida H, Geacintov NE (1981b) Structural consequences of modification of the oxygen atom of guanine in DNA by the carcinogen *N*-hydroxy-1-naphthylamine. Cancer Res 41: 2168–2174

Kadlubar FF, Unruh LE, Flammang TJ, Sparks D, Mitchum RK, Mulder GJ (1981c) Alteration of urinary levels of the carcinogen, *N*-hydroxy-2-naphthylamine, and its *N*-glucuronide in the rat by control of urinary pH, inhibition of metabolic sulfation, and changes in biliary excretion. Chem Biol Interact 33:129–147

Kadlubar FF, Beland FA, Beranek DT, Dooley KL, Heflich RH, Evans FE (1982a) Arylamine-DNA adduct formation in relation to urinary bladder carcinogenesis and *Salmonella typhimurium* mutagenesis. In: Sugimura T, Kondo S, Takebe H (eds) Environmental mutagens and carcinogens. Liss, New York, pp 385–396

Kadlubar FF, Frederick CB, Weis CC, Zenser TV (1982b) Prostaglandin endoperoxide synthetase-mediated metabolism of carcinogenic aromatic amines and their binding to DNA and protein. Biochem Biophys Res Commun 108:253–258

Kadlubar FF, Butler MA, Hayes BE, Beland FA, Guengerich FP (1988a) Role of microsomal cytochromes P-450 and prostaglandin *H* synthase in 4-aminobiphenyl-DNA adduct formation. In: Miners J, Birkett DJ, Drew R, McManus M (eds). Microsomes and drug oxidations. Taylor and Francis, London, pp 370–379

Kadlubar FF, Dooley KL, Benson RW, Roberts DW, Butler MA, Teitel CH, Bailey JR, Young JF (1988b) Pharmacokinetic model of aromatic amine-induced urinary bladder carcinogenesis in beagle dogs administered 4-aminobiphenyl. In: King CM, Romano LJ, Schuetzle D (eds) Carcinogenic and mutagenic responses to aromatic amines and nitroarenes. Elsevier, New York, pp 173–184

Kadlubar FF, Guengerich FP, Butler MA, Delclos KB (1988c) Cytochromes P-450 as determinants of susceptibility to carcinogenesis by aromatic amines and nitroaromatic hydrocarbons. In: Feo F, Pani P, Columbano A, Garcea R (eds) Chemical carcinogenesis. Models and mechanisms. Plenum Corp., New York, pp 17–23

Kadlubar FF, Talaska G, Lang NP, Benson RW, Roberts DW (1988d) Assessment of exposure and susceptibility to aromatic amine carcinogens. In: Bartsch H, Hemminki K, O'Neill IK (eds) Methods for detecting DNA damaging agents in humans: applications in cancer epidemiology and prevention. IARC scientific publications no. 89, Lyon, pp 166–174

Kadlubar FF, Butler MA, Culp SJ, Teitel CH, Fu PP, Shaikh AU, Delclos KB, Beland FA (1989) Metabolic activation of arylamine derivatives of carcinogenic polycyclic nitroaromatic hydrocarbons. In: Kato R, Estabrook RW, Cayen MN (eds) Xenobiotic metabolism and disposition. Taylor and Francis, London, pp 375–381

Kamataki T, Maeda K, Yamazoe Y, Matsuda N, Ishii K, Kato R (1983) A high-spin form of cytochrome P-450 highly purified from polychlorinated biphenyl-treated rats. Catalytic characterization and immunochemical quantitation in liver microsomes. Mol Pharmacol 24:146–155

Kato R (1986) Metabolic activation of mutagenic heterocyclic aromatic amines from protein pyrolysates. CRC Crit Rev Toxicol 16:307–348

Kato R, Yamazoe Y (1987) Metabolic activation and covalent binding to nucleic acids of carcinogenic heterocyclic amines from cooked foods and amino acid pyrolysates. Gann 78:297–311

Kato R, Kamataki T, Yamazoe Y (1983) *N*-Hydroxylation of carcinogenic and mutagenic aromatic amines. Environ Health Perspect 49:21–25

Kawakubo Y, Manabe S, Yamazoe Y, Nishikawa T, Kato R (1988) Properties of cutaneous acetyltransferase catalyzing *N*- and *O*-acetylation of carcinogenic arylamines and *N*-hydroxyarylamine. Biochem Pharmacol 37:265–270

Kawano S, Kamataki T, Maeda K, Kato R, Nakao T, Mizoguchi I (1985) Activation and inactivation of a variety of mutagenic compounds by the reconstituted system containing highly purified preparations of cytochrome *P*-450 from rat liver. Fundam Appl Toxicol 5:487–498

Kennelly JC, Beland FA, Kadlubar FF, Martin CN (1984) Binding of *N*-acetylbenzidine and *N,N'*-diacetylbenzidine to hepatic DNA of rat and hamster in vivo and in vitro. Carcinogenesis 5:407–412

Kimura T, Kodama M, Nagata C (1982) *N*-Hydroxylation enzymes of carcinogenic aminoazo dyes: possible involvement of cytochrome P-448. Gann 73:55–62

Kimura T, Kodama M, Nagata C (1984) Role of cytochrome P-450 and flavin-containing monooxygenase in the *N*-hydroxylation of *N*-methyl-4-aminoazobenzene in rat liver: analysis with purified enzymes and antibodies. Gann 75:895–904

Kimura T, Kodama M, Nagata C, Kamataki T, Kato R (1985) Comparative study on the metabolism of *N*-methyl-4-aminoazobenzene by two forms of cytochrome P-448. Biochem Pharmacol 34:3375–3377

King CM (1974) Mechanism of reaction, tissue distribution, and inhibition of arylhydroxamic acid acyltransferase. Cancer Res 34:1503–1515

King CM (1988) Metabolism and biological effects of nitropyrene and related compounds. Health Effects Institute, Cambridge, pp 1–22 (Health Effects Institute research report no. 16)

King CM, Glowinski IB (1983) Acetylation, deacetylation and acyltransfer. Environ Health Persp 49:43–50

King CM, Phillips B (1969) *N*-Hydroxy-2-fluorenylacetamide. Reaction of the carcinogen with guanosine, ribonucleic acid, deoxyribonucleic acid, and protein following enzymatic deacetylation or esterification. J Biol Chem 244:6209–6216

King CM, Traub NR, Lortz ZM, Thissen MR (1979) Metabolic activation of arylhydroxamic acids by *N-O*-acyltransferase of rat mammary gland. Cancer Res 39:3369–3372

Kinoshita R (1937) The cancerogenic chemical substances. Trans Soc Pathol Jpn 27:665–725

Kinouchi T, Ohnishi Y (1986) Metabolic activation of 1-nitropyrene and 1,6-dinitropyrene by nitroreductases from *Bacteroides fragilis* and distribution of nitroreductase activity in rats. Microbiol Immunol 30:979–992

Kinouchi T, Morotomi M, Mutai M, Fifer EK, Beland FA, Ohnishi Y (1986a) Metabolism of 1-nitropyrene in germ-free and conventional rats. Gann 77:356–369

Kinouchi T, Tsutsui H, Ohnishi Y (1986b) Detection of 1-nitropyrene in yakitori (grilled chicken). Mutation Res 171:105–113

Krauss RS, Eling TE (1985) Formation of unique arylamine:DNA adducts from 2-aminofluorene activated by prostaglandin H synthase. Cancer Res 45:1680–1686

Kriek E (1965) On the interaction of *N*-2-fluorenylhydroxylamine with nucleic acids in vitro. Biochem Biophys Res Commun 20:793–799

Kriek E (1969) On the mechanism of action of carcinogenic aromatic amines. I. Binding of 2-acetylaminofluorene and *N*-hydroxy-2-acetylaminofluorene to rat-liver nucleic acids in vivo. Chem Biol Interact 1:3–17

Kriek E (1971) On the mechanism of action of carcinogenic aromatic amines. II. Binding of *N*-hydroxy-*N*-acetyl-4-aminobiphenyl to rat-liver nucleic acids in vivo. Chem Biol Interact 3:19–28

Kriek E (1972) Persistent binding of a new reaction product of the carcinogen *N*-hydroxy-*N*-2-acetylaminofluorene with guanine in rat liver DNA in vivo. Cancer Res 32:2042–2048

Kriek E, Hengeveld GM (1977) Interaction of the carcinogen *N*-4-(4′-fluorobiphenyl)acetamide (4′F-4BAA) with DNA in liver and kidney of the rat. Colloques internationaux du CNRS 256:115–118

Kriek E, Hengeveld GM (1978) Reaction products of the carcinogen *N*-hydroxy-4-acetylamino-4′-fluorobiphenyl with DNA in liver and kidney of the rat. Chem Biol Interact 21:179–201

Kriek E, Reitsema J (1971) Interaction of the carcinogen *N*-acetoxy-*N*-2-acetylaminofluorene with polyadenylic acid: dependence of reactivity on conformation. Chem Biol Interact 3:397–400

Kriek E, Spelt CE (1979) Differential excision from DNA of the C-8 deoxyguanosine reaction products of *N*-hydroxy-2-aminofluorene and *N*-acetoxy-*N*-acetyl-2-aminofluorene by endonuclease S_1 from *Aspergillus oryzae*. Cancer Lett 7:147–154

Kriek E, Westra JG (1979) Metabolic activation of aromatic amines and amides and interactions with nucleic acids. In: Grover PL (ed) Chemical carcinogens and DNA, vol II. CRC Press, Boca Raton, pp 1–28

Kriek E, Westra JG (1980) Structural identification of the pyrimidine derivatives formed from *N*-(deoxyguanosin-8-yl)-2-aminofluorene in aqueous solution at alkaline pH. Carcinogenesis 1:459–468

Kuchlbauer J, Romen W, Neumann H-G (1985) Syncarcinogenic effects on the initiation of rat liver tumors by *trans*-4-acetylaminostilbene and 2-acetylaminofluorene. Carcinogenesis 6:1337–1342

Lai C-C, Miller JA, Miller EC, Liem A (1985) Sulfoöxy-2-aminofluorene is the major ultimate electrophilic and carcinogenic metabolite of *N*-hydroxy-2-acetylaminofluorene in the livers of infant male C57BL/6J × C3H/HeJ F_1 (B6C3F_1) mice. Carcinogenesis 6:1037–1045

Lai C-C, Miller EC, Miller JA, Liem A (1987) Initiation of hepatocarcinogenesis in infant male B6C3F_1 mice by *N*-hydroxy-2-aminofluorene or *N*-hydroxy-2-acetylaminofluorene depends primarily on metabolism to *N*-sulfooxy-2-aminofluorene and formation of DNA-(deoxyguanosin-8-yl)-2-aminofluorene adducts. Carcinogenesis 8:471–478

Lambelin G, Roba J, Roncucci R, Parmentier R (1975) Carcinogenicity of 6-aminochrysene in mice. Eur J Cancer 11:327–334

Lang NP, Chu DZJ, Hunter CF, Kendall DC, Flammang TJ, Kadlubar FF (1986) Role of aromatic amine acetyltransferase in human colorectal cancer. Arch Surg 121: 1259–1261

Lavine AF, Fink LM, Weinstein IB, Grunberger D (1974) Effect of *N*-2-acetylaminofluorene modification on the conformation of nucleic acids. Cancer Res 34:319–327

Lefévre J-F, Fuchs RPP, Daune MP (1978) Comparative studies on the 7-iodo and 7-fluoro derivatives of *N*-acetoxy-*N*-2-acetylaminofluorene: binding sites on DNA and conformational change of modified deoxytrinucleotides. Biochemistry 17:2561–2567

Leichtenstern O (1898) Über Harnblasenentzündung und Harnblasengeschwulste bei Arbeitern in Farbfabriken. Deut Med Wochenschr 24:709–713

Leng M, Ptak M, Rio P (1980) Conformation of acetylaminofluorene and aminofluorene modified guanosine and guanosine derivatives. Biochem Biophys Res Commun 96:1095–1102

Levine WG, Lu AYH (1982) Role of isozymes of cytochrome P-450 in the metabolism of *N,N*-dimethyl-4-aminoazobenzene in the rat. Drug Metab Dispos 10:102–109

Levy GN, Weber WW (1988) HPLC analysis of ^{32}P-postlabeled DNA-2-aminofluorene adducts. In: King CM, Romano LJ, Schuetzle D (eds) Carcinogenic and mutagenic responses to aromatic amines and nitroarenes. Elsevier, New York, pp 283–287

Lin J-K, Miller JA, Miller EC (1975a) Structures of hepatic nucleic acid-bound dyes in rats given the carcinogen *N*-methyl-4-aminoazobenzene. Cancer Res 35:844–850
Lin J-K, Schmall B, Sharpe ID, Miura I, Miller JA, Miller EC (1975b) *N*-Substitution of carbon 8 in guanosine and deoxyguanosine by the carcinogen *N*-benzoyloxy-*N*-methyl-4-aminoazobenzene in vitro. Cancer Res 35:832–843
Lipkowitz KB, Chevalier T, Widdifield M, Beland FA (1982) Force field conformational analysis of aminofluorene and acetylaminofluorene substituted deoxyguanosine. Chem Biol Interact 40:57–76
Löfroth G, Hefner E, Alfheim I, Møller M (1980) Mutagenic activity in photocopies. Science 209:1037–1039
Lotlikar PD, Luha L (1971) Acetylation of the carcinogen *N*-hydroxy-2-acetylaminofluorene by acetyl coenzyme A to form a reactive ester. Mol Pharmacol 7:381–388
Lynn RK, Garvie-Gould CT, Milam DF, Scott KF, Eastman CL, Ilias AM, Rodgers RM (1984) Disposition of the aromatic amine, benzidine, in the rat: characterization of mutagenic urinary and biliary metabolites. Tox Appl Pharmacol 72:1–14
Maeda T, Izumi K, Otsuka H, Manabe Y, Kinouchi T, Ohnishi Y (1986) Induction of squamous cell carcinoma in the rat lung by 1,6-dinitropyrene. JNCI 76:693–701
Maher VM, Hazard RM, Beland FA, Corner R, Mendrala AL, Levinson JW, Heflich RH, McCormick JJ (1980) Excision of the deacetylated *C*-8-guanine DNA adduct by human fibroblasts correlates with decreased cytotoxicity and mutagenicity. Proc Am Assoc Cancer Res 21:71
Maher VM, Heflich RH, McCormick JJ (1981) Repair of DNA damage induced in human fibroblasts by *N*-substituted aryl compounds. Natl Cancer Inst Monogr 58:217–222
Martin CN, Beland FA, Roth RW, Kadlubar FF (1982) Covalent binding of benzidine and *N*-acetylbenzidine to DNA at the *C*-8 atom of deoxyguanosine in vivo and in vitro. Cancer Res 42:2678–2686
Martin CN, Beland FA, Kennelly JC, Kadlubar FF (1983) Binding of benzidine, *N*-acetylbenzidine, *N,N'*-diacetylbenzidine and Direct Blue 6 to rat liver DNA. Environ Health Perspect 49:101–106
Masson HA, Ioannides C, Gorrod JW, Gibson GG (1983) The role of highly purified cytochrome P-450 isozymes in the activation of 4-aminobiphenyl to mutagenic products in the Ames test. Carcinogenesis 4:1583–1586
Matthews JJ, Walpole AL (1958) Tumours of the liver and kidney induced in Wistar rats with 4'-fluoro-4-aminodiphenyl. Br J Cancer 12:234–241
McCoy EC, Rosenkranz HS, Mermelstein R (1981) Evidence for the existence of a family of bacterial nitroreductases capable of activating nitrated polycyclics to mutagens. Environ Mutagenesis 3:421–427
McCoy EC, McCoy GD, Rosenkranz HS (1982) Esterification of arylhydroxylamines: evidence for a specific gene product in mutagenesis. Biochim Biophys Res Commun 108:1362–1367
McCoy EC, Anders M, Rosenkranz HS (1983) The basis of the insensitivity of *Salmonella typhimurium* strain TA98/1,8-DNP$_6$ to the mutagenic action of nitroarenes. Mutation Res 121:17–23
McMahon RE, Turner JC, Whitaker GW (1980) The *N*-hydroxylation and ring-hydroxylation of 4-aminobiphenyl in vitro by hepatic mono-oxygenases from rat, mouse, hamster, rabbit and guinea-pig. Xenobiotica 10:469–481
McQueen CA, Maslansky CJ, Glowinski IB, Crescenzi SB, Weber WW, Williams GM (1982) Relationship between the genetically determined acetylator phenotype and DNA damage induced by hydralazine and 2-aminofluorene in cultured rabbit hepatocytes. Proc Natl Acad Sci USA 79:1269–1272
McQueen CA, Way BM, Williams GM (1987) Biotransformation of aromatic amines to DNA-damaging products by urinary bladder organ cultures. Carcinogenesis 8:401–404
Meerman JHN (1985) The initiation of γ-glutamyltranspeptidase positive foci in the rat liver by *N*-hydroxy-2-acetylaminofluorene. The effect of the sulfation inhibitor pentachlorophenol. Carcinogenesis 6:893–897
Meerman JHN, Mulder GJ (1981) Prevention of the hepatotoxic action of *N*-hydroxy-2-acetylaminofluorene in the rat by inhibition of *N-O*-sulfation by pentachlorophenol. Life Sciences 28:2361–2365

Meerman JHN, Beland FA, Mulder GJ (1981) Role of sulfation in the formation of DNA adducts from *N*-hydroxy-2-acetylaminofluorene in rat liver in vivo. Inhibition of *N*-acetylated aminofluorene adduct formation by pentachlorophenol. Carcinogenesis 2:413–416
Melchior WB Jr, Beland FA (1984) Preferential reaction of the carcinogen *N*-acetoxy-2-acetylaminofluorene with satellite DNA. Chem Biol Interact 49:177–187
Mermelstein R, Kiriazides DK, Butler M, McCoy EC, Rosenkranz HS (1981) The ex traordinary mutagenicity of nitropyrenes in bacteria. Mutation Res 89:187–196
Miller EC, Sandin RB, Miller JA, Rusch HP (1956) The carcinogenicity of compounds related to 2-acetylaminofluorene. III. Aminobiphenyl and benzidine derivatives. Cancer Res 16:525–534
Miller EC, Lotlikar PD, Pitot HC, Fletcher TL, Miller JA (1966) *N*-Hydroxy metabolites of 2-acetylaminophenanthrene and 7-fluoro-2-acetylaminofluorene as proximate carcinogens in the rat. Cancer Res 26:2239–2247
Miller EC, Kadlubar FF, Miller JA, Pitot HC, Drinkwater NR (1979) The *N*-hydroxy metabolites of *N*-methyl-4-aminoazobenzene and related dyes as proximate carcinogens in the rat and mouse. Cancer Res 39:3411–3418
Miller JA, Miller EC (1983) Some historical aspects of *N*-aryl carcinogens and their metabolic activation. Environ Health Perspect 49:3–12
Miller JA, Sandin RB, Miller EC, Rusch HP (1955) The carcinogenicity of compounds related to 2-acetylaminofluorene. II. Variations in the bridges and the 2-substituent. Cancer Res 15:188–199
Mita S, Yamazoe Y, Kamataki T, Kato R (1982) Effects of ascorbic acid on the non-enzymatic binding to DNA and the mutagenicity of *N*-hydroxylated metabolite of a tryptophan-pyrolysis product. Biochem Biophys Res Commun 105:1396–1401
Mitchell CE (1985) Effect of aryl hydrocarbon hydroxylase induction on the in vivo covalent binding of 1-nitropyrene, benzo[*a*]pyrene, 2-aminoanthracene, and phenanthridone to mouse lung deoxyribonucleic acid. Biochem Pharmacol 34:545–551
Mitchell CE (1988) Formation of DNA adducts in mouse tissues after intratracheal instillation of 1-nitropyrene. Carcinogenesis 9:857–860
Möller L, Corrie M, Midtvedt T, Rafter J, Gustafsson J-Å (1988) The role of the intestinal microflora in the formation of mutagenic metabolites from the carcinogenic air pollutant 2-nitrofluorene. Carcinogenesis 9:823–830
Mommsen S, Aagaard J (1983) Tobacco as a risk factor in bladder cancer. Carcinogenesis 4:335–338
Moolgavkar SH, Stevens RG (1981) Smoking and cancers of bladder and pancreas: risks and temporal trends. JNCI 67:15–23
Morton KC, King CM, Baetcke KP (1979) Metabolism of benzidine to *N*-hydroxy-*N,N′*-diacetylbenzidine and subsequent nucleic acid binding and mutagenicity. Cancer Res 39:3107–3113
Morton KC, Beland FA, Evans FE, Fullerton NF, Kadlubar FF (1980) Metabolic activation of *N*-hydroxy-*N,N′*-diacetylbenzidine by hepatic sulfotransferase. Cancer Res 40:751–757
Morton KC, Wang CY, Garner CD, Shirai T (1981) Carcinogenicity of benzidine, *N,N′*-diacetylbenzidine, and *N*-hydroxy-*N,N′*-diacetylbenzidine for female CD rats. Carcinogenesis 2:747–752
Morton KC, King CM, Vaught JB, Wang CY, Lee M-S, Marnett LJ (1983) Prostaglandin H synthase-mediated reaction of carcinogenic arylamines with tRNA and homopolyribonucleotides. Biochem Biophys Res Commun 111:96–103
Murofushi Y, Hashimoto Y, Shudo K, Okamoto T (1981) Reaction of 1-naphthylhydroxylamine with calf thymus deoxyribonucleic acid. Isolation and synthesis of *N*-(guanin-C^8-yl)-1-naphthylamine. Chem Pharm Bull 29:2730–2732
National Cancer Institute (1978) Bioassay of 1-nitronaphthalene for possible carcinogenicity. Carcinogenesis tech rep ser no 64, pp 1–47
Neidle S, Kuroda R, Broyde S, Hingerty BE, Levine RA, Miller DW, Evans FE (1984) Studies on the conformation and dynamics of the *C*8-substituted guanine adduct of the carcinogen acetylaminofluorene: model for a possible Z-DNA modified structure. Nucleic Acids Res 12:8219–8233

Nelson AA, Woodard G (1953) Tumors of the urinary bladder, gall bladder, and liver in dogs fed *o*-aminoazotoluene or *p*-dimethylaminoazobenzene. JNCI 13:1497–1509
Nelson CJ, Baetcke KP, Frith CH, Kodell RL, Schieferstein G (1982) The influence of sex, dose, time, and cross on neoplasia in mice given benzidine dihydrochloride. Toxicol Appl Pharmacol 64:171–186
Nelson JH, Grunberger D, Cantor CR, Weinstein IB (1971) Modification of ribonucleic acid by chemical carcinogens. IV. Circular dichroism and proton magnetic resonance studies of oligonucleotides modified with *N*-2-acetylaminofluorene. J Mol Biol 62:331–346
Nesnow S, Triplett LL, Slaga TJ (1984) Tumor initiating activities of 1-nitropyrene and its nitrated products in SENCAR mice. Cancer Lett 23:1–8
Neumann H-G (1981) Significance of metabolic activation and binding to nucleic acids of aminostilbene derivatives in vivo. Natl Cancer Inst Monogr 58:165–171
Nielsen T, Seitz B, Ramdahl T (1984) Occurrence of nitro-PAH in the atmosphere in a rural area. Atmospheric Environ 18:2159–2165
Nussbaum M, Fiala ES, Kulkarni B, El-Bayoumy K, Weisburger JH (1983) In vivo metabolism of 3,2′-dimethyl-4-aminobiphenyl (DMAB) bearing on its organotropism in the Syrian golden hamster and the F344 rat. Environ Health Perspect 49:223–231
Ohgaki H, Negishi C, Wakabayashi K, Kusama K, Sato S, Sugimura T (1984) Induction of sarcomas in rats by subcutaneous injection of dinitropyrenes. Carcinogenesis 5:583–585
Ohgaki H, Hasegawa H, Kato T, Negishi C, Sato S, Sugimura T (1985) Absence of carcinogenicity of 1-nitropyrene, correction of previous results, and new demonstration of carcinogenicity of 1,6-dinitropyrene in rats. Cancer Lett 25:239–245
Ohnishi Y, Kinouchi T, Tsutsui H, Uejima M, Nishifuji K (1986) Mutagenic nitropyrenes in foods. In: Hayashi Y, Nagao M, Sugimura T, Takayama S, Tomatis L, Wattenberg LW, Wogan GN (eds) Diet, nutrition and cancer. Japan Sci Soc Press, Tokyo, pp 107–118
Orr JC, Bryant DW, McCalla DR, Quilliam MA (1985) Dinitropyrene-resistant *Salmonella typhimurium* are deficient in an acetyl-CoA acetyltransferase. Chem Biol Interact 54:281–288
Otofuji T, Horikawa K, Maeda T, Sano N, Izumi K, Otsuka H, Tokiwa H (1987) Tumorigenicity test of 1,3- and 1,8-dinitropyrene in BALB/c mice. JNCI 79:185–188
Parkes HG, Evans AEJ (1984) Epidemiology of aromatic amine cancers. In: Searle CE (ed) Chemical carcinogens, vol 1. American Chemical Society, Washington DC, pp 277–301 (ACS Monograph 182)
Patrianakos C, Hoffmann D (1979) Chemical studies on tobacco smoke. LXIV. On the analysis of aromatic amines in cigarette smoke. J Anal Toxicol 3:150–154
Patton JD, Maher VM, McCormick JJ (1986) Cytotoxic and mutagenic effects of 1-nitropyrene and 1-nitrosopyrene in diploid human fibroblasts. Carcinogenesis 7:89–93
Pederson TC, Siak J-S (1981) The role of nitroaromatic compounds in the direct-acting mutagenicity of diesel particle extracts. J Appl Toxicol 1:54–60
Petry TW, Krauss RS, Eling TE (1986) Prostaglandin H synthase-mediated bioactivation of the amino acid pyrolysate product Trp P-2. Carcinogenesis 7:1397–1400
Poirier LA, Weisburger JH (1974) Enzymic reduction of carcinogenic aromatic nitro compounds by rat and mouse liver fractions. Biochem Pharmacol 23:661–669
Poirier MC, Williams GM, Yuspa SH (1980) Effect of culture conditions, cell type, and species of origin on the distribution of acetylated and deacetylated deoxyguanosine *C*-8 adducts of *N*-acetoxy-2-acetylaminofluorene. Mol Pharmacol 18:581–587
Poirier MC, True B, Laishes BA (1982) Formation and removal of (guan-8-yl)DNA-2-acetylaminofluorene adducts in liver and kidney of male rats given dietary 2-acetylaminofluorene. Cancer Res 42:1317–1321
Poirier MC, Hunt JM, True B, Laishes BA, Young JF, Beland FA (1984) DNA adduct formation, removal and persistence in rat liver during one month feeding of 2-acetylaminofluorene. Carcinogenesis 5:1591–1596
Poirier MC, Fullerton NF, Beland FA (1988) DNA adduct formation and removal during chronic dietary administration of 2-acetylaminofluorene. In: King CM, Romano LJ, Schuetzle D (eds) Carcinogenic and mutagenic responses to aromatic amines and nitroarenes. Elsevier, New York, pp 321–328

Poupko JM, Hearn WL, Radomski JL (1979) *N*-Glucuronidation of *N*-hydroxy aromatic amines: a mechanism for their transport and bladder-specific carcinogenicity. Toxicol Appl Pharmacol 50:479–484
Purchase IFH, Kalinowski AE, Ishmael J, Wilson J, Gore CW, Chart IS (1981) Lifetime carcinogenicity study of 1- and 2-naphthylamine in dogs. Br J Cancer 44:892–901
Radomski JL (1979) The primary aromatic amines: their biological properties and structure-activity relationships. Ann Rev Pharmacol Toxicol 19:129–157
Radomski JL, Brill E, Deichmann WB, Glass EM (1971) Carcinogenicity testing of *N*-hydroxy and other oxidation and decomposition products of 1- and 2-naphthylamine. Cancer Res 31:1461–1467
Radomski JL, Deichmann WB, Altman NH, Radomski T (1980) Failure of pure 1-naphthylamine to induce bladder tumors in dogs. Cancer Res 40:3537–3539
Ramdahl T, Becher G, Bjørseth A (1982) Nitrated polycyclic aromatic hydrocarbons in urban air particles. Environ Sci Technol 16:861–865
Reddy MV, Gupta RC, Randerath E, Randerath K (1984) ^{32}P-Postlabeling test for covalent DNA binding of chemicals in vivo: application to a variety of aromatic carcinogens and methylating agents. Carcinogenesis 5:231–243
Rehn L (1895) Blasengeschwülste bei Fuchsin-Arbeitern. Arch Klin Chir 50:588–600. Reprinted in translation: Shimkin MB (1980) Some classics of experimental oncology: 50 selections, 1775–1965. NIH publication no. 80–2150, pp 44–51
Rice JR, Kissinger PT (1982) Cooxidation of benzidine by horseradish peroxidase and subsequent formation of possible thioether conjugates of benzidine. Biochem Biophys Res Commun 104:1312–1318
Ringer DP, Norton TR (1987) Further characterization of the ability of hepatocarcinogens to lower rat liver aryl sulfotransferase activity. Carcinogenesis 8:1749–1752
Ringer DP, Kampschmidt K, King RL Jr, Jackson S, Kizer DE (1983) Rapid decrease in *N*-hydroxy-2-acetylaminofluorene sulfotransferase activity of liver cytosols from rats fed carcinogen. Biochem Pharmacol 32:315–319
Roe FJC, Carter RL, Adamthwaite S (1969) Induction of liver and lung tumours in mice by 6-aminochrysene administered during the first 3 days of life. Nature 221:1063–1064
Rosenkranz HS, Mermelstein R (1983) Mutagenicity and genotoxicity of nitroarenes. All nitro-containing chemicals were not created equal. Mutation Res 114:217–267
Rosenkranz HS, McCoy EC, Sanders DR, Butler M, Kiriazides DK, Mermelstein R (1980) Nitropyrenes: isolation, identification, and reduction of mutagenic impurities in carbon black and toners. Science 209:1039–1043
Ruthsatz M, Neumann H-G (1988) Synergistic effects on the initiation of rat liver tumors by *trans*-4-acetylaminostilbene and 2-acetylaminofluorene, studied at the level of DNA adduct formation. Carcinogenesis 9:265–269
Saffiotti U, Cefis F, Montesano R, Sellakumar AR (1967) Induction of bladder cancer in hamsters fed aromatic amines. In: Lampe KF, Penalver RA, Soto A (eds) Bladder cancer, a symposium. Aesculapius, Birmingham, pp 129–135
Sage E, Leng M (1980) Conformation of poly(dG-dC).poly (dG-dC) modified by the carcinogens *N*-acetoxy-*N*-acetyl-2-aminofluorene and *N*-hydroxy-*N*-2-aminofluorene. Proc Natl Acad Sci USA 77:4597–4601
Saito K, Kamataki T, Kato R (1984) Participation of cytochrome P-450 in reductive metabolism of 1-nitropyrene by rat liver microsomes. Cancer Res 44:3169–3173
Saito K, Shinohara A, Kamataki T, Kato R (1985) Metabolic activation of mutagenic *N*-hydroxyarylamines by *O*-acetyltransferase in *Salmonella typhimurium* TA98. Arch Biochem Biophys 239:286–295
Santella RM, Grunberger D, Broyde S, Hingerty BE (1981 a) Z-DNA conformation of *N*-2-acetylaminofluorene modified poly(dG-dC).poly(dG-dC) determined by reactivity with anti cytidine antibodies and minimized potential energy calculations. Nucleic Acids Res 9:5459–5467
Santella RM, Grunberger D, Weinstein IB, Rich A (1981 b) Induction of the Z conformation in poly(dG-dC).poly(dG-dC) by binding of *N*-2-acetylaminofluorene to guanine residues. Proc Natl Acad Sci USA 78:1451–1455
Sato T, Kato K, Ose Y, Nagase H, Ishikawa T (1985) Nitroarenes in Suimon River sediment. Mutation Res 157:135–143

Schuetzle D (1983) Sampling of vehicle emissions for chemical analysis and biological testing. Environ Health Persp 47:65–80
Schuetzle D, Lee FS-C, Prater TJ, Tejada SB (1981) The identification of polynuclear aromatic hydrocarbon (PAH) derivatives in mutagenic fractions of diesel particulate extracts. Intern J Environ Anal Chem 9:93–144
Scribner JD, Fisk SR (1978) Reproduction of major reactions of aromatic carcinogens with guanosine, using HMO-based polyelectronic perturbation theory. Tetrahedron Lett 4759–4762
Scribner JD, Koponen G (1979) Binding of the carcinogen 2-acetamidophenanthrene to rat liver nucleic acids: lack of correlation with carcinogenic activity, and failure of the hydroxamic acid ester model for in vivo activation. Chem Biol Interact 15:201–209
Scribner JD, Mottet NK (1981) DDT acceleration of mammary gland tumors induced in the male Sprague-Dawley rat by 2-acetamidophenanthrene. Carcinogenesis 2:1235–1239
Scribner JD, Naimy NK (1975) Adducts between the carcinogen 2-acetamidophenanthrene and adenine and guanine of DNA. Cancer Res 35:1416–1421
Scribner NK, Scribner JD (1979) Reactions of the carcinogen *N*-acetoxy-4-acetamidostilbene with polynucleotides in vitro. Chem Biol Interact 26:47–55
Scribner JD, Smith DL, McCloskey JA (1978) Deamination of 1-methylcytosine by the carcinogen *N*-acetoxy-4-acetamidostilbene: implications for hydrocarbon carcinogenesis. J Org Chem 43:2085–2087
Scribner JD, Fisk SR, Scribner NK (1979a) Mechanisms of action of carcinogenic aromatic amines: an investigation using mutagenesis in bacteria. Chem Biol Interact 26:11–25
Scribner NK, Scribner JD, Smith DL, Schram KH, McCloskey JA (1979b) Reactions of the carcinogen *N*-acetoxy-4-acetamidostilbene with nucleosides. Chem Biol Interact 26:27–46
Scribner JD, Woodworth B, Koponen G, Holmes EH (1983) Use of 2-acetamidophenanthrene and 2-acetamidofluorene in investigations of mechanisms of hepatocarcinogenesis. Environ Health Perspect 49:81–86
Shapiro R, Underwood GR, Zawadzka H, Broyde S, Hingerty BE (1986) Conformation of d(CpG) modified by the carcinogen 4-aminobiphenyl: a combined experimental and theoretical analysis. Biochemistry 25:2198–2205
Shinohara A, Saito K, Yamazoe Y, Kamataki T, Kato R (1985) DNA binding of *N*-hydroxy-Trp-P-2 and *N*-hydroxy-Glu-P-1 by acetyl-CoA dependent enzyme in mammalian liver cytosol. Carcinogenesis 6:305–307
Shinohara A, Saito K, Yamazoe Y, Kamataki T, Kato R (1986) Acetyl coenzyme A dependent activation of *N*-hydroxy derivatives of carcinogenic arylamines: mechanism of activation, species difference, tissue distribution, and acetyl donor specificity. Cancer Res 46:4362–4367
Shirai T, Fysh JM, Lee M-S, Vaught JB, King CM (1981a) Relationship of metabolic activation of *N*-hydroxy-*N*-acylarylamines to biological response in the liver and mammary gland of the female CD rat. Cancer Res 41:4346–4353
Shirai T, Lee M-S, Wang CY, King CM (1981b) Effects of partial hepatectomy and dietary phenobarbital on liver and mammary tumorigenesis by two *N*-hydroxy-*N*-acylaminobiphenyls in female CD rats. Cancer Res 41:2450–2456
Silverman DT, Hoover RN, Albert S, Graff KM (1983) Occupation and cancer of the lower urinary tract in Detroit. JNCI 70:237–245
Snyderwine EG, Roller PP, Adamson RH, Sato S, Thorgeirsson SS (1988) Reaction of *N*-hydroxylamine and *N*-acetoxy derivatives of 2-amino-3-methylimidazolo[4,5-*f*]quinoline with DNA. Synthesis and identification of *N*-(deoxyguanosin-8-yl)-IQ. Carcinogenesis 9:1061–1065
Spitz S, Maguigan WH, Dobriner K (1950) The carcinogenic action of benzidine. Cancer 3:789–804
Spodheim-Maurizot M, Saint-Ruf G, Leng M (1979) Conformational changes induced in DNA by in vitro reaction with *N*-hydroxy-*N*-2-aminofluorene. Nucleic Acids Res 6:1683–1694
Spodheim-Maurizot M, Saint-Ruf G, Leng M (1980) Antibodies to *N*-hydroxy-2-aminofluorene modified DNA as probes in the study of DNA reacted with derivatives of 2-acetylaminofluorene. Carcinogenesis 1:807–812

Stanton CA, Chow FL, Phillips DH, Grover PL, Garner RC, Martin CN (1985) Evidence for *N*-(deoxyguanosin-8-yl)-1-aminopyrene as a major DNA adduct in female rats treated with 1-nitropyrene. Carcinogenesis 6:535–538

Sternson LA (1975) Detection of arylhydroxylamines as intermediates in the metabolic reduction of nitro compounds. Experientia 31:268–270

Sugimura T (1986) Past, present, and future of mutagens in cooked foods. Environ Health Perspect 67:5–10

Sweetman JA, Karasek FW, Schuetzle D (1982) Decomposition of nitropyrene during gas chromatographic-mass spectrometric analysis of air particulate and fly-ash samples. J Chromatography 247:245–254

Takayama S, Ishikawa T, Nakajima H, Sato S (1985) Lung carcinoma induction in Syrian golden hamsters by intratracheal instillation of 1,6-dinitropyrene. Gann 76:457–461

Talaska G, Au WW, Ward JB Jr, Randerath K, Legator MS (1987) The correlation between DNA adducts and chromosomal aberrations in the target organ of benzidine exposed, partially-hepatectomized mice. Carcinogenesis 8:1899–1905

Tang M-S, Lieberman MW, King CM (1982) *uvr* Genes function differently in repair of acetylaminofluorene and aminofluorene DNA adducts. Nature 299:646–648

Tarpley WG, Miller JA, Miller EC (1980) Adducts from the reaction of *N*-benzoyloxy-*N*-methyl-4-aminoazobenzene with deoxyguanosine or DNA in vitro and from hepatic DNA of mice treated with *N*-methyl- or *N,N*-dimethyl-4-aminoazobenzene. Cancer Res 40:2493–2499

Tarpley WG, Miller JA, Miller EC (1982) Rapid release of carcinogen-guanine adducts from DNA after reaction with *N*-acetoxy-2-acetylaminofluorene or *N*-benzoyloxy-*N*-methyl-4-aminoazobenzene. Carcinogenesis 3:81–88

Tatsumi K, Kitamura S, Narai N (1986) Reductive metabolism of aromatic nitro compounds including carcinogens by rabbit liver preparations. Cancer Res 46:1089–1093

Tokiwa H, Ohnishi Y (1986) Mutagenicity and carcinogenicity of nitroarenes and their sources in the environment. CRC Crit Rev Toxicol 17:23–60

Tokiwa H, Kitamori S, Nakagawa R, Horikawa K, Matamala L (1983) Demonstration of a powerful mutagenic dinitropyrene in airborne particulate matter. Mutation Res 121:107–116

Tokiwa H, Otofuji T, Horikawa K, Kitamori S, Otsuka H, Manabe Y, Kinouchi T, Ohnishi Y (1984) 1,6-Dinitropyrene: mutagenicity in *Salmonella* and carcinogenicity in BALB/c mice. JNCI 73:1359–1363

Tokiwa H, Nakagawa R, Horikawa K (1985) Mutagenic/carcinogenic agents in indoor pollutants; the dinitropyrenes generated by kerosene heaters and fuel gas and liquefied petroleum gas burners. Mutation Res 157:39–47

Tsuruta Y, Josephy PD, Rahimtula AD, O'Brien PJ (1985) Peroxidase-catalyzed benzidine binding to DNA and other macromolecules. Chem Biol Interact 54:143–158

Tullis DL, Straub KM, Kadlubar FF (1981) A comparison of the carcinogen-DNA adducts formed in rat liver in vivo after administration of single or multiple doses of *N*-methyl-4-aminoazobenzene. Chem Biol Interact 38:15–27

Tullis DL, Dooley KL, Miller DW, Baetcke KP, Kadlubar FF (1987) Characterization and properties of the DNA adducts formed from *N*-methyl-4-aminoazobenzene in rats during a carcinogenic treatment regimen. Carcinogenesis 8:577–583

van Houte LPA, Bokma JT, Lutgerink JT, Westra JG, Retél J, van Grondelle R, Blok J (1987) An optical study of the conformation of the aminofluorene – DNA complex. Carcinogenesis 8:759–766

Vesselinovitch SD, Rao KVN, Mihailovich N (1975) Factors modulating benzidine carcinogenicity bioassay. Cancer Res 35:2814–2819

Visser A, Westra JG (1981) Partial persistency of 2-aminofluorene and *N*-acetyl-2-aminofluorene in rat liver DNA. Carcinogenesis 2:737–740

Walpole AL, Williams MHC, Roberts DC (1952) The carcinogenic action of 4-aminodiphenyl and 3:2′-dimethyl-4-aminodiphenyl. Br J Ind Med 9:255–263

Walpole AL, Williams MHC, Roberts DC (1954) Tumours of the urinary bladder in dogs after ingestion of 4-aminodiphenyl. Br J Ind Med 11:105–109

Weber WW (1987) The acetylator genes and drug response. Oxford University Press, New York, pp 1–248

Wei C-i, Raabe OG, Rosenblatt LS (1982) Microbial detection of mutagenic nitro-organic compounds in filtrates of coal fly ash. Environ Mutagenesis 4:249–258

Westra JG, Visser A (1979) Quantitative analysis of *N*-(guanin-8-yl)-*N*-acetyl-2-aminofluorene and *N*-(guanin-8-yl)-2-aminofluorene in modified DNA by hydrolysis in trifluoroacetic acid and high pressure liquid chromatography. Cancer Lett 8:155–162

Westra JG, Kriek E, Hittenhausen H (1976) Identification of the persistently bound form of the carcinogen *N*-acetyl-2-aminofluorene to rat liver DNA in vivo. Chem Biol Interact 15:149–164

Westra JG, Flammang TJ, Fullerton NF, Beland FA, Weis CC, Kadlubar FF (1985) Formation of DNA adducts in vivo in rat liver and intestinal epithelium after administration of the carcinogen 3,2′-dimethyl-4-aminobiphenyl and its hydroxamic acid. Carcinogenesis 6:37–44

Wigle DT, Mao Y, Grace M (1980) Relative importance of smoking as a risk factor for selected cancers. Can J Public Health 71:269–275

Williams MHC (1962) Environmental and industrial bladder cancer. Preventive measures. Acta Unio Int Contra Cancrum 18:676–683

Wilson RH, DeEds F, Cox AJ Jr (1941) The toxicity and carcinogenic activity of 2-acetaminofluorene. Cancer Res 1:595–608

Wise RW, Zenser TV, Davis BB (1984a) Characterization of benzidinediimine: a product of peroxidase metabolism of benzidine. Carcinogenesis 5:1499–1503

Wise RW, Zenser TV, Kadlubar FF, Davis BB (1984b) Metabolic activation of carcinogenic aromatic amines by dog bladder and kidney prostaglandin H synthase. Cancer Res 44:1893–1897

Wise RW, Zenser TV, Davis BB (1985) Prostaglandin H synthase oxidation of benzidine and *o*-dianisidine: reduction and conjugation of activated amines by thiols. Carcinogenesis 6:579–583

Wiseman RW, Stowers SJ, Miller EC, Anderson MW, Miller JA (1986) Activating mutations of the c-Ha-*ras* protooncogene in chemically induced hepatomas of the male B6C3 F_1 mouse. Proc Natl Acad Sci USA 83:5825–5829

Wislocki PG, Bagan ES, Lu AYH, Dooley KL, Fu PP, Han-Hsu H, Beland FA, Kadlubar FF (1986) Tumorigenicity of nitrated derivatives of pyrene, benz[*a*]anthracene, chrysene and benzo[*a*]pyrene in the newborn mouse assay. Carcinogenesis 7:1317–1322

Wynder EL, Goldsmith R (1977) The epidemiology of bladder cancer. A second look. Cancer 40:1246–1268

Xu XB, Nachtman JP, Jin ZL, Wei ET, Rappaport SM, Burlingame AL (1982) Isolation and identification of mutagenic nitro-PAH in diesel-exhaust particulates. Anal Chim Acta 136:163–174

Yamazoe Y, Shimada M, Kamataki T, Kato R (1983) Microsomal activation of 2-amino-3-methylimidazo[4,5-*f*]quinoline, a pyrolysate of sardine and beef extracts, to a mutagenic intermediate. Cancer Res 43:5768–5774

Yamazoe Y, Miller DW, Weis CC, Dooley KL, Zenser TV, Beland FA, Kadlubar FF (1985a) DNA adducts formed by *ring*-oxidation of the carcinogen 2-naphthylamine with prostaglandin *H* synthase in vitro and in the dog urothelium in vivo. Carcinogenesis 6:1379–1387

Yamazoe Y, Shimada M, Shinohara A, Saito K, Kamataki T, Kato R (1985b) Catalysis of the covalent binding of 3-hydroxyamino-1-methyl-5H-pyrido-[4,3-*b*]indole to DNA by a L-proline- and adenosine triphosphate-dependent enzyme in rat hepatic cytosol. Cancer Res 45:2495–2500

Yamazoe Y, Roth RW, Kadlubar FF (1986) Reactivity of benzidine diimine with DNA to form *N*-(deoxyguanosin-8-yl)-benzidine. Carcinogenesis 7:179–182

Yamazoe Y, Abu-Zeid M, Manabe S, Toyama S, Kato R (1988a) Metabolic activation of a protein pyrolysate promutagen 2-amino-3,8-dimethylimidazo[4,5-*f*]-quinoxaline by rat liver microsomes and purified cytochrome P-450. Carcinogenesis 9:105–109

Yamazoe Y, Zenser TV, Miller DW, Kadlubar FF (1988b) Mechanism of formation and structural characterization of DNA adducts derived from peroxidative activation of benzidine. Carcinogenesis 9:1635–1641

Young JF, Kadlubar FF (1982) A pharmacokinetic model to predict exposure of the bladder epithelium to urinary *N*-hydroxyarylamine carcinogens as a function of urine pH, voiding interval, and resorption. Drug Metab Dispos 10:641–644
Zenser TV, Mattammal MB, Davis BB (1979) Cooxidation of benzidine by renal medullary prostaglandin cyclooxygenase. J Pharmacol Exp Ther 211:460–464
Zenser TV, Mattammal MB, Armbrecht HJ, Davis BB (1980) Benzidine binding to nucleic acids mediated by the peroxidative activity of prostaglandin endoperoxide synthetase. Cancer Res 40:2839–2845
Ziegler DM, Ansher SS, Nagata T, Kadlubar FF, Jakoby WB (1988) *N*-Methylation: potential mechanism for metabolic activation of carcinogenic primary arylamines. Proc Natl Acad Sci USA 2514–2517

CHAPTER 9

Polycyclic Aromatic Hydrocarbons: Metabolism, Activation and Tumour Initiation

M. HALL and P. L. GROVER

A. Introduction

Polycyclic aromatic hydrocarbons (PAH) are formed as products of the incomplete pyrolysis of organic materials and are present in considerable quantities in fossil fuel from which they are released by a variety of combustion processes (see GUERIN 1978). Sources of environmental PAH therefore include, in the wider sense, power plants, domestic heating systems, petrol and diesel engines, refuse burning and various industrial activities, whilst tobacco smoke provides a more localized source of supply. Each of these sources of PAH produces a mixture containing between 100 and 300 different individual hydrocarbons, and the estimated total annual emission in the USA of just one of them, benzo[*a*]pyrene (BaP), is some 1200 tons (GRIMMER 1983). Since only micrograms of this hydrocarbon are required to initiate tumours on mouse skin, it would be surprising if the human population was not placed at increased risk of developing cancer as a result of pollution of the environment on this scale by PAH.

Two aspects of PAH research have particularly intrigued scientists in the years since the initial isolation of a pure carcinogenic hydrocarbon from coal tar in the 1930s (COOK et al. 1933). The first has been the very marked structure-activity relationships that exist within this class of chemical carcinogen (see YANG and SILVERMAN 1988), as exemplified by two pairs of isomeric PAH, dibenz[*a,c*]- and dibenz[*a,h*]anthracene and benzo[*e*]- and benzo[*a*]pyrene (for formulae, see tables). In each case the former isomer is almost inactive whilst the latter compound is a potent carcinogen. The second aspect of interest has been the mechanisms by which the carcinogenic PAH initiate tumours.

As might be expected for compounds that are formed at very high temperatures, the polycyclic hydrocarbons are, in chemical terms, relatively inert. However, PAH are also lipid soluble and would tend to accumulate in organisms which come into contact with them, unless they can be metabolised to more water-soluble derivatives that can be excreted. Paradoxically it is now known that if PAH were not metabolised, they would not be carcinogenic.

The rapidly expanding literature on all aspects of polycyclic hydrocarbon research makes a comprehensive review of the area impossible in the space available here. Indeed some 10 years ago three volumes were required to cover the subject adequately (GELBOIN and TS'O 1978), and in the most recent series whole volumes are being given over to each family of PAH (OSBORNE and CROSBY 1987; COOMBS and BHATT 1987). In what follows, therefore, an attempt has been made to cover just four particular areas of PAH research relating to (a) the general mechanisms by which they are metabolised, (b) the pathways by which different PAH are activated by metabolism, (c) certain stereoselective aspects of metabolic activation and (d) the basis for tissue and species susceptibility to their carcinogenic effects.

B. Metabolism

It is now clear that the formation of epoxides is the initial step in the metabolism of unsubstituted polycyclic hydrocarbons and that this involves the addition of one atom of oxygen across a double bond. The reaction is usually catalysed by the microsomal cytochrome P-450-linked mono-oxygenases, a group of inducible enzymes whose properties have been extensively examined (see ESTERBROOK et al. 1978; NEBERT and GONZALEZ 1987). The formation of epoxides at some particularly olefinic double bonds can also be catalysed by prostaglandin *H* synthase (MARNETT et al. 1979).

Following the proposal (BOYLAND 1950) that epoxides are universal intermediates in the oxidative metabolism of aromatic double bonds, a great deal of work was carried out on the metabolism of hydrocarbons ranging from naphthalene to BaP (see SIMS and GROVER 1974). The results that were obtained, when considered together with those of many earlier studies, were entirely consistent with this prediction. Such work was facilitated by the increasing availability of synthetic epoxides that could be used in metabolism studies (NEWMAN and BLUM 1964), but because of the reactivity of epoxides and the ease with which they can be further metabolised, direct evidence for the existence of epoxides as metabolites was difficult to obtain. Most of the evidence was eventually gained from experiments in which a radioactive hydrocarbon was used as substrate and the epoxide metabolite was trapped by adding the unlabelled synthetic epoxide. The involvement of epoxides as intermediates in the metabolism of PAH has previously been reviewed (see JERINA and DALY 1974; SIMS and GROVER 1974).

Once formed, epoxides can be hydrated to yield diols, they can isomerise to phenols, or they can become conjugated with glutathione; each of these different types of product can then be further metabolised. The categories of metabolites formed by these various metabolic steps are considered below: in almost all cases detoxication products that can be more readily excreted are formed although in some rare, but important, instances, biologically active metabolites result.

I. Diols

Diols are formed by the hydration of epoxides in an enzyme-catalysed reaction involving epoxide hydrolases (EH). These are mainly microsomal activities, although a cytoplasmic form is known; their occurrence and properties have been well reviewed (see OESCH 1973; GUENTHER and OESCH 1981). The EH, which are widespread in mammalian tissues, are inducible, but they are not induced to nearly the same extent as the mono-oxygenases (GLATT et al. 1984).

Almost all the double bonds in a polycyclic hydrocarbon can be oxidised to yield epoxides, but the extent of diol formation at any particular position appears to depend partly on the stability of the epoxide (i.e. the rate at which it rearranges to give a phenol, see below) and partly on its ability to act as a substrate for either the epoxide hydrolases or the glutathione transferases. With BaP, for example (Fig. 1), the 9,10-epoxide yields both the 9-phenol and the 9,10-diol as metabolites. In contrast, oxidation at the 2,3-position yields only the 3-phenol; the corresponding 2,3-diol is not detected as a metabolite even though there is good evidence that an epoxide intermediate is formed (YANG et al. 1977). Oxidation at the K-region 4,5-bond of BaP in intact cells or tissues yields predominantly glutathione conjugates and their derivatives because K-region epoxides are good substrates for the cytoplasmic glutathione transferases (see below). In incubations with microsomal preparations, however, the 4,5-diol can be readily detected as a metabolite since a glutathione conjugating system is absent. The diols originally detected as metabolites of naphthalene, anthracene and phenanthrene in early in vivo experiments were found to have the *trans* configuration; subsequent studies have shown that this is true for almost all diol metabolites of PAH. *Trans*-diols normally adopt the quasi-diequatorial conformation, but if they are formed adjacent to a bay-region in the molecule or to an alkyl substituent, they may be forced to adopt a quasi-diaxial conformation.

Because diols possess asymmetric centres they can exist in enantiomeric forms, and currently a great deal of attention is being paid to the stereoselective

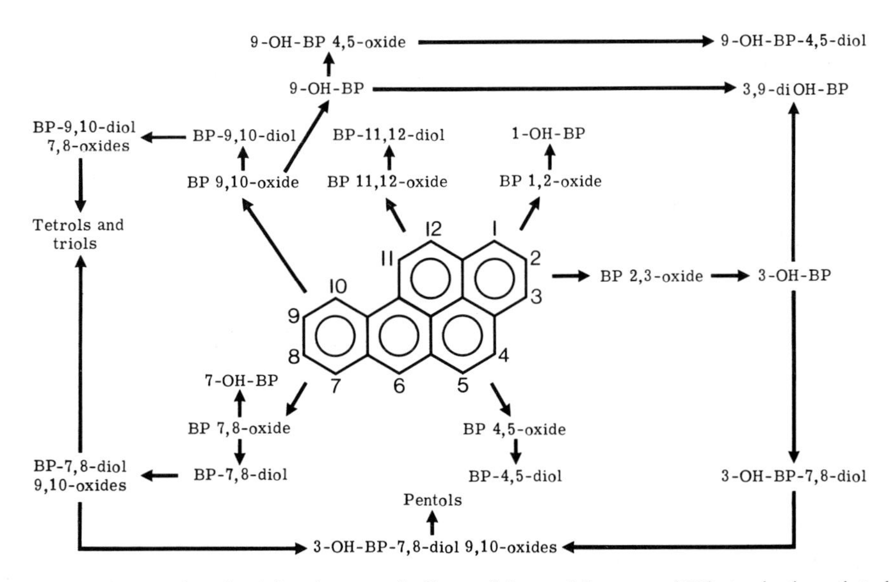

Fig. 1. Pathways involved in the metabolism of benzo[*a*]pyrene (*BP*) to hydroxylated derivatives. The enzymes concerned in catalysing the formation of epoxides or their hydration to diols and in conjugation reactions are mentioned in the text. For simplicity, the diagram does not include the epoxides presumed to be intermediate in the conversion of 3-OH-BP to 3,9-diOH-BP or 3-OH-BP-7,8-diol, the phenols that may arise from isomerisation of the K-region BP-4,5- and -11,12-oxides, or the reactions involved in the conjugation of epoxides of various types with glutathione or the conjugation of phenols or dihydrodiols with sulphuric or glucosiduronic acids

metabolism of PAH because it has been shown that some enantiomers of diol-epoxides possess much greater biological activity than others (see Sect. D).

II. Phenols

Although there are other possible mechanisms for the formation of phenols, it is now known that most phenolic metabolites arise from the isomerisation of epoxides. With BaP, for example, the 3-, 7- and 9-hydroxy derivatives are almost certainly formed from the 2,3-, 7,8- and 9,10-oxides, respectively (Fig. 1). It is worth noting that each epoxide rearranges spontaneously to form predominantly one phenol (FU et al. 1978) and that the related 2-, 8- and 10-hydroxy metabolites of BaP are not formed to any appreciable extent.

The amount of any particular phenol that is formed as a metabolite from an epoxide will therefore depend on factors including (a) the rate of oxidation of the hydrocarbon to the precursor epoxide, (b) the relative stability of the epoxide, i.e. the rate at which it isomerises, (c) the direction of isomerisation and (d) the rates at which the epoxide is removed by conjugation or by hydration to the related diol. For example, whilst several different phenols are found as metabolites of BaP, the 4- and 5-hydroxy derivatives are not normally seen, partly because the 4,5-oxide is more stable than, say, the 2,3-oxide, but also because the 4,5-oxide can be readily converted into a diol or a glutathione conjugate.

III. Conjugation with Glutathione

Mercapturic acids were first detected in the urine of animals fed naphthalene (BOURNE and YOUNG 1934) or anthracene (BOYLAND and LEVI 1936) and since then our knowledge of how such metabolites arise has increased considerably. What follows is an extremely brief outline of glutathione conjugation in so far as it impinges on PAH metabolism (for more detailed reviews, see BOYLAND and CHASSEAUD 1969; ARIAS and JAKOBY 1976; CHASSEAUD 1979; MANNERVIK 1985).

An enzyme that catalysed conjugations with glutathione was first detected in rat liver (BOOTH et al. 1961; COMBES and STAKELUM 1961). It has since been shown that the glutathione transferases consist of a family of dimeric isoenzymes whose individual members are composed of various combinations of two monomers (MANNERVIK and JENSSON 1982), and the nomenclature has now been standardised (JAKOBY et al. 1984). The glutathione transferases, which are widely distributed throughout the animal kingdom, are mainly cytosolic enzymes, and, in liver, they can comprise up to 10% of the soluble proteins present. Although the enzymes are specific for glutathione, a variety of epoxides, including PAH epoxides, can act as second substrates for the transferases. The conjugation with glutathione of simple epoxides (GROVER 1974), of diol-epoxides (COOPER et al. 1980a; JERNSTRÖM et al. 1985) and of triol-epoxides (HODGSON et al. 1986) derived from PAH can be catalysed by the transferases, but different epoxides may differ appreciably in their abilities to act as second substrates.

Prior to their excretion in urine, glutathione conjugates are converted in vivo into acetylcysteine derivatives, which are also commonly called mercapturic acids, by a series of enzyme-catalysed steps. Studies on the in vivo metabolism of benz[*a*]anthracene (BA) in rats have shown that acetylcysteine derivatives, in addition to being excreted in urine, also appear in the bile together with the corresponding glutathione, cysteinylglycine and cysteine derivatives of the hydrocarbon. At present the conjugation of PAH epoxides with glutathione is regarded as a true detoxication reaction.

IV. Conjugation with Glucuronic and Sulphuric Acids

Epoxides that are not conjugated with glutathione can be converted, as mentioned above, into phenols and diols. These hydroxylated metabolites are often not sufficiently polar to be excreted as such, and they are, therefore, conjugated with glucuronic or sulphuric acids to facilitate excretion. Glucuronic and sulphuric acid conjugates of PAH metabolites have been detected in urine, bile and faeces, and these conjugation reactions obviously form an important pathway for hydrocarbon elimination. The enzymes that catalyse glucuronide formation are the UDP glucuronyl-transferases which are located in the endoplasmic reticulum and which use UDP-glucuronic acid as an activated form of glucuronic acid. The transferases are present in almost all tissues, and studies have revealed that there are two types of enzyme that show different ranges of second substrate specificity. The enzyme catalysing the conjugation of hydroxylated hydrocarbon metabolites, such as 3-hydroxyBaP, with UDP-glucuronic acid is of the type that matures in foetal development. The different types of enzyme are differentially induced by, for example, 3-methylcholanthrene (3-MC) and phenobarbital (PB), and the evidence indicates that the form most active with hydrocarbon metabolites is that induced by 3-MC. For reviews of glucuronic acid conjugation see TEAGUE (1954), DUTTON and BURCHELL (1977), and NEMOTO (1978, 1981).

Glucuronic acid conjugates are generally regarded as detoxication products. However, it should be noted that enzymic hydrolysis of 3-hydroxyBaP glucuronide yields a DNA-binding product (KINOSHITA and GELBOIN 1978) and that the hydrolysis by bacterial glucuronidase in the gut of BaP conjugates, which are known to undergo an enterohepatic circulation (CHIPMAN et al. 1981), could conceivably contribute to the initiation of colon cancer (RENWICK and DRASAR 1976).

The enzymes responsible for catalysing conjugations with sulphate are the sulphotransferases which are cytosolic enzymes that use 3-phosphoadenosine-5-phosphosulphate as an activated form of sulphate: various facets of sulphate conjugation have been reviewed (see for example DODGSON 1977; NEMOTO 1981).

Sulphate esters are formed as metabolites of PAH in a variety of biological situations. In rabbits and rats treated with phenanthrene, for example, four phenols and two diols were excreted as urinary sulphates (SIMS 1962), and an extensive range of hydroxylated BaP metabolites were conjugated with sulphate when incubated with rat liver cytosol (NEMOTO et al. 1978). Tissues and cells in culture such as human colon (AUTRUP 1979) and hamster trachea (MOORE and COHEN 1978) will also form sulphate esters of hydroxylated BaP derivatives. Although much more water soluble than the parent hydrocarbons, sulphates are not as polar as either glucuronides or glutathione conjugates, and they are extractable from aqueous media into solvents (COHEN et al. 1976, 1977). Sulphate esters of PAH derivatives have not been shown to be unequivocally implicated in metabolic activation of PAH in vivo: however, with other classes of chemical carcinogens such as the aromatic amines (see Chap. 8) sulphate conjugation certainly leads to the formation of DNA-reactive species.

V. Hydroxylated Products – Further Metabolism

Hydroxylated derivatives of PAH may still be relatively non-polar, and a variety of additional oxidation/hydroxylation reactions are known to occur. Overall, there is evidence for the conversion of phenols to phenol-epoxides and hence to diphenols and triols, for the conversion of diols to vicinal diol-epoxides and tetrols and to triols, and for the metabolism of triols to triol-epoxides and pentols. These further oxidation reactions will be illustrated by reference to BaP and chrysene.

With BaP (see Fig. 1), the 7,8-diol is further metabolised through oxidation of the isolated olefinic 9,10-double bond to the 7,8-diol 9,10-epoxide (SIMS et al. 1974), which is important in metabolic activation (see below), and to the corresponding tetrols. Surprisingly, the reactive 7,8-diol 9,10-epoxides have recently been reported to be subject to further hydroxylation in microsomal preparations, and the products are believed to be triol-epoxides possessing a phenolic OH-group in either the 1- or 3-position (JERNSTRÖM et al. 1984). In contrast to the 7,8-diol, the 9,10-diol derivative yields on further metabolism a triol rather than a diol-epoxide, and this triol is also thought to have the additional phenolic OH-group in the 1- or 3-positions (THAKKER et al. 1978; MOORE and COHEN 1978). 9-HydroxyBaP, which can be involved in metabolic activation in some situations, yields the related 4,5-oxide and 4,5-diol as well as two products tentatively identified as the 4,9- and 5,9-diphenols when incubated with rat liver preparations (JERNSTRÖM et al. 1978) through metabolism at the K-region 4,5-bond. Recent work on the further metabolism of 3-hydroxyBaP in rats has shown that conjugates of the 3-hydroxyBaP 7,8-diol and the 3,5-diphenol are excreted in bile (RIBEIRO et al. 1985), although again it is not clear in what order the conjugation and/or further metabolism reactions occur. Some but by no means all of the pathways for the further metabolism of hydroxylated derivatives of BaP are given in Fig. 1.

Examples of the routes by which hydroxylated derivatives of chrysene are further metabolised are shown in Fig. 2. The 1,2-diol (III) is known to be metabolised to triols that have not been unequivocally identified but that are thought to have the phenolic OH-group in the 7,8,9,10-ring (NORDQVIST et al. 1981; JACOB et al. 1982). In addition, the 1,2-diol yields another triol, the 9-hydroxy derivative (IV), that can also be formed from the corresponding phenol (II) (HODGSON et al. 1985b). The 1,2-diol and the 9-hydroxy 1,2-diol can both yield epoxides (V and VI) that are involved, in mouse skin, in reactions with DNA (HODGSON et al. 1983; see Sect. C) and the epoxides can presumably be hydrated to the corresponding tetrols and pentols (VII and VIII). Since 9-hydroxychrysene can yield the 1,2-diol, a diphenol might also be formed through rearrangement of the intermediate phenol-epoxide, but this has not so far been described.

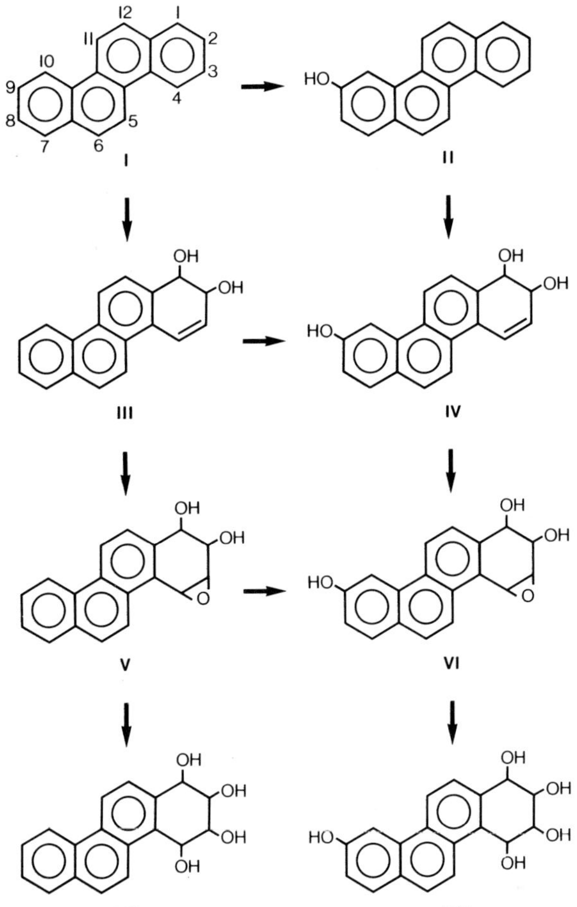

Fig. 2. Pathways involved in the metabolism of chrysene to diol- and triol-epoxides. Epoxides presumed to be involved as intermediates in the metabolism of I→II, I→III, II→IV, III→IV and V→VI are not shown

VI. Mechanisms Involved in Metabolic Activation

Whilst detailed information on the mechanisms and pathways involved in hydrocarbon metabolism and excretion was being accumulated in earlier years, parallel studies aimed at discovering how carcinogenic PAH might initiate tumours were also in progress. The initial detection, in mouse skin, of the covalent reaction of BaP with protein (MILLER 1951) was followed by studies showing that radioactively labelled dibenz[*a,h*]anthracene became covalently bound to DNA in this tissue (HEIDELBERGER and DAVENPORT 1961) and that the extent of binding of particular PAH to DNA in mouse skin could be related to their carcinogenic potencies (BROOKES and LAWLEY 1964). Once it had been shown that the covalent reaction of hydrocarbons with both protein and DNA could be mediated in vitro by microsomal enzymes (GROVER and SIMS 1968), the search for electrophilic metabolites intensified. Simple K-region epoxides came under suspicion because they possessed many of the properties thought relevant (see SIMS and GROVER 1974), but it was soon shown that the nucleic acid adducts formed when such epoxides reacted with DNA were not the same as those formed in tissues or cells treated with the parent hydrocarbon (BAIRD et al. 1973, 1975). The report by BORGEN et al. (1973), that the further metabolism of the 7,8-diol of BaP by microsomal mono-oxygenases in the presence of DNA gives rise to more covalent reaction with the nucleic acid than occurs with BaP itself, led rapidly to the identification of the BaP-7,8-diol 9,10-oxides as the vicinal diol-epoxides responsible for the covalent reactions that occur when embryo cells in culture (SIMS et al. 1974) or mouse skin (DAUDEL et al. 1975) are treated with the hydrocarbon. Since then work carried out in many different laboratories has confirmed that the general mechanism by which the PAH are activated is through vicinal diol-epoxide formation and that, in most cases, the diol-epoxides involved are of the bay-region type (JERINA and DALY 1977), i.e. the epoxide oxygen is located in a "bay" (BARTLE and JONES 1967) that is formed in a molecule by the presence of an angular benzo-ring. The detailed studies concerned have been reviewed elsewhere (see GELBOIN 1980; SIMS and GROVER 1981; CONNEY 1982; COOPER et al. 1983), and, most recently and comprehensively, by DIPPLE et al. (1984). It should perhaps be mentioned that this generalisation is based on examination of one or more aspects of the metabolic activation of many different PAH ranging from phenanthrene through the benzopyrenes to the dibenzanthracenes and that there are some exceptions to the general rule (see Sect. C below).

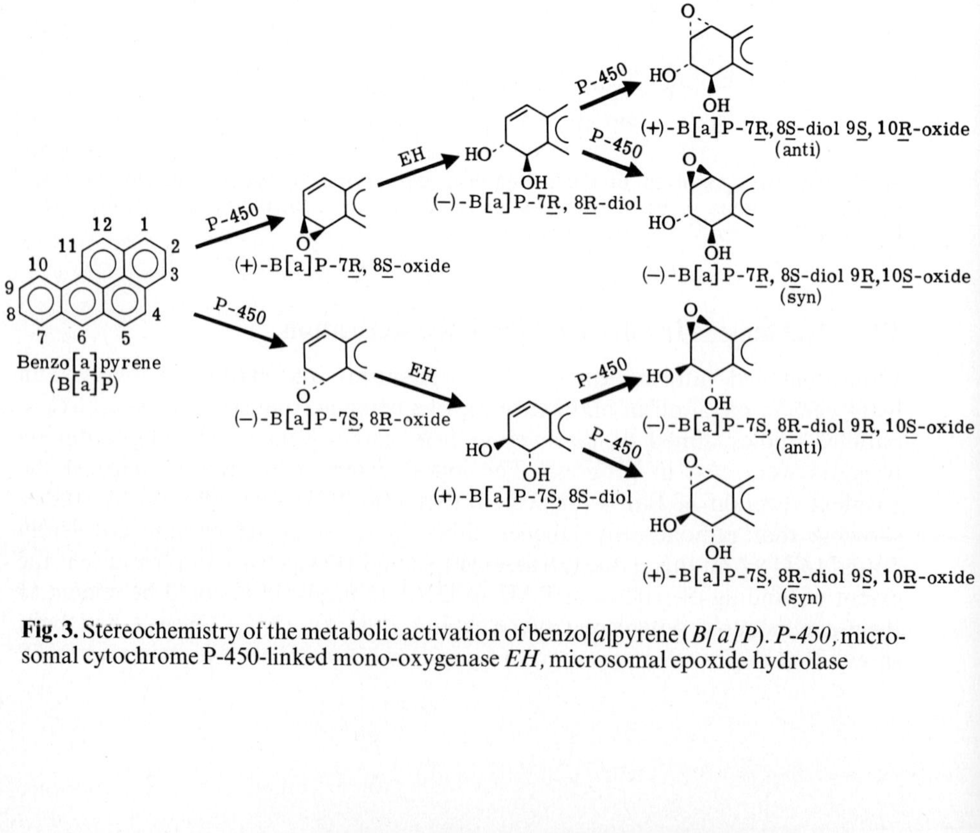

Fig. 3. Stereochemistry of the metabolic activation of benzo[*a*]pyrene (*B[a]P*). *P-450*, microsomal cytochrome P-450-linked mono-oxygenase *EH*, microsomal epoxide hydrolase

Soon after the identification of the first diol-epoxides as metabolites of BaP (Sims et al. 1974) and BA (Booth and Sims 1974), it was realised (a) that diol-epoxides could exist as either the *syn*- or *anti*-forms [which have, in the case of the BaP-7,8-diol 9,10-oxides (see Fig. 3) the epoxide oxygen on either the same (*syn*) or on the opposite (*anti*) face of the ring as the hydroxyl group in the 7-position (Hulbert 1975; Yagi et al. 1975)] and (b) that these isomers would differ markedly in chemical reactivity. In addition there are two enantiomers of both the *syn*- and the *anti*-diol-epoxides, and this aspect of their stereochemistry assumed increasing importance once it was shown (a) that the mono-oxygenases involved in the formation of the epoxides and diol-epoxides are stereoselective and form predominantly one of each pair of enantiomers and (b) that individual enantiomers may differ markedly in biological activity (see Sect. D below). The detailed investigation of the metabolic activation of a single PAH could therefore involve the synthesis and resolution of four stereoisomers of each diol-epoxide (and with BA, for example, there are four possible vicinal diol-epoxides) before tests for mutagenicity and/or tumour-initiating activity could be carried out and before DNA adducts could be completely characterised. This serves to illustrate the amount of effort that has been required to reach the position where it is now possible to state that the general mechanism of hydrocarbon activation is through the formation of bay-region vicinal diol-epoxides.

C. Pathways of Activation

A great deal of progress has been made in recent years in the identification of the metabolic pathways by which various PAH are activated to the species (i.e. the ultimate carcinogens) that are involved in tumour initiation, but the data that have been gathered are by no means complete. Although it is generally accepted (a) that polycyclic hydrocarbons are activated through conversion to electrophilic epoxides and (b) that in the well-researched examples almost all these epoxides are diol-epoxides of the bay-region type (see Sect. B), there are, of course, exceptions to be found in such a structurally diverse group of compounds. Table 1 attempts to collate the evidence available for those PAH whose pathways of metabolic activation have been examined; it includes only those hydrocarbons in which the activity of putative proximate or ultimate carcinogens have been tested in one or more biological systems or where partial or complete characterisation of DNA adducts has shed light on the likely activation pathway. In some instances in which the literature is very extensive (e.g. with BaP), review articles have been cited and in others reviews have been used to cover the older literature more concisely. Comparative data on the biological activities shown by PAH derivatives in systems in which they have been tested, for example as mutagens or tumour-initiating agents, have clearly been of value for the identification of activation pathways and the ultimate carcinogenic metabolites. There is no doubt, however, that the best evidence is provided by the isolation and characterisation of nucleic acid adducts from a susceptible tissue that has been treated with the parent hydrocarbon. With many of the PAH listed in Table 1 information of this type is not yet available, and the degree of certainty with which activation pathways can be correctly identified varies greatly from compound to compound.

Table 1. Pathways involved in the metabolic activation of polycyclic aromatic hydrocarbons

Hydrocarbon	Derivatives showing highest levels of biological activity	Putative ultimate carcinogens	References
Aceanthrylene 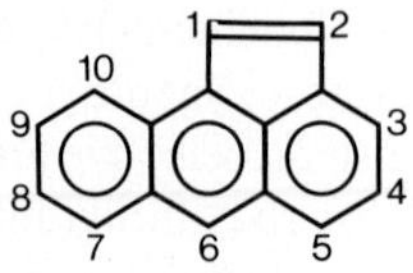	–	1,2-Oxide[h]	NESNOW et al. (1989)
Benzo[*j*]aceanthrylene 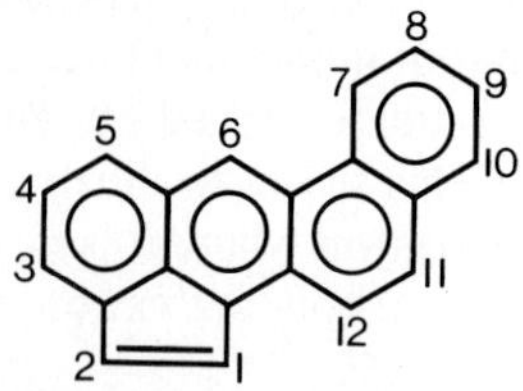	–	? 1,2-Oxide[c]	BARTCZAK et al. (1987); NESNOW et al. (1988)
Benzo[*l*]aceanthrylene 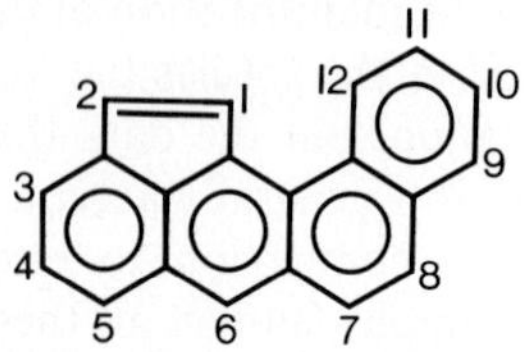	–	? 1,2-Oxide[c,d]	NESNOW et al. (1984); BARTCZAK et al. (1987); NESNOW et al. (1988)
Benz[*c*]acridine 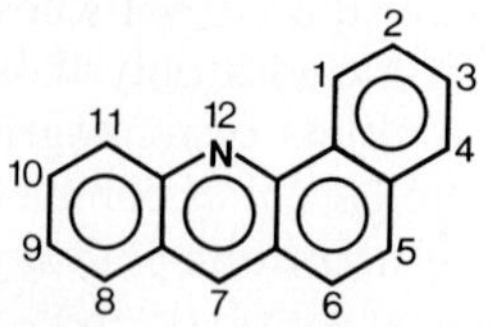	3,4-Diol[a,f,g]	3,4-Diol 1,2-oxide[c,d,f,g]	LEVIN et al. (1983); WOOD et al. (1983a); CHANG et al. (1984)

[a] Mutagenic to *S. typhimurium* with metabolic activation.
[b] Mutagenic to V79 Chinese hamster cells with metabolic activation.
[c] Direct-acting mutagen in *S. typhimurium*.
[d] Direct-acting mutagen in V79 Chinese hamster cells.
[e] Transforms cells in culture.
[f] Tumour initiator in mouse skin.
[g] Induces tumours in newborn mice.
[h] DNA adducts characterised.
[i] Not actually detected as a metabolite; activation may therefore occur via a different pathway.
[j] Although the 4,5-diol is the most active derivative so far tested, there is some evidence that adducts arise from the 9,10-diol.

Table 1 (continued)

Hydrocarbon	Derivatives showing highest levels of biological activity	Putative ultimate carcinogens	References
Benz[*a*]anthracene	3,4-Diol[a, b, f, g]	3,4-Diol 1,2-oxide[c, d, f, g, h]	SIMS and GROVER (1981) (review)
	8,9-Diol[a]	8,9-Diol 10,11-oxide[e, h]	CONNEY (1982) (review); WOOD et al. (1983b)
Benzo[*b*]fluoranthene 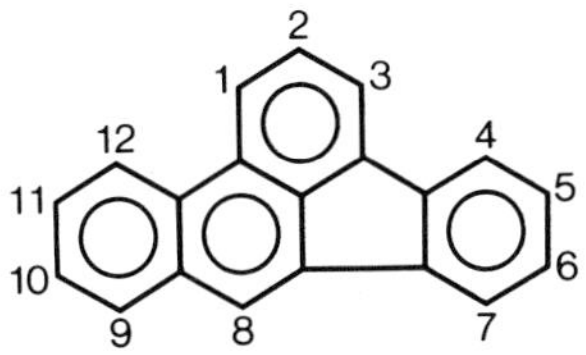	9,10-Diol[a, f, i]	? 9,10-Diol 11,12-oxide	GEDDIE et al. (1987)
Benzo[*j*]aceanthrylene 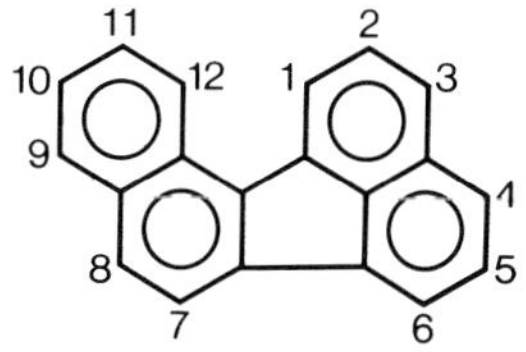	? 9,10-Diol[f, j]	? 9,10-Diol 11,12-oxide[h]	RICE et al. (1987b); WEYAND et al. (1987)
Benzo[*c*]phenanthrene 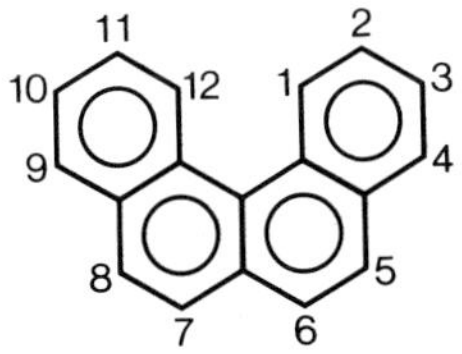	3,4-Diol[a, b, f, g]	3,4-Diol 1,2-oxide[e, d, f, g, h]	CONNEY (1982) (review); LEVIN et al. (1986); AGARWAL et al. (1987); DIPPLE et al. (1987); PREUSS-SCHWARTZ et al. (1987)
Benzo[*a*]pyrene 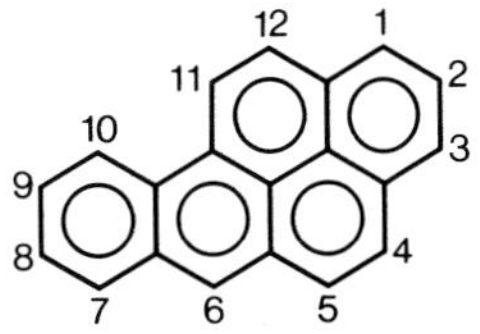	7,8-Diol[a, b, e, f]	7,8-Diol 9,10-oxide[c, d, g, h]	COOPER et al. (1983); OSBORNE and CROSBY (1987) (reviews)

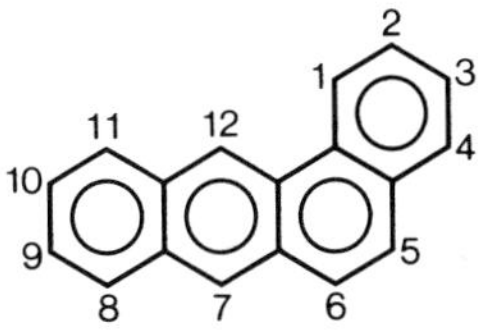

Table 1 (continued)

Hydrocarbon	Derivatives showing highest levels of biological activity	Putative ultimate carcinogens	References
Benzo[*e*]pyrene 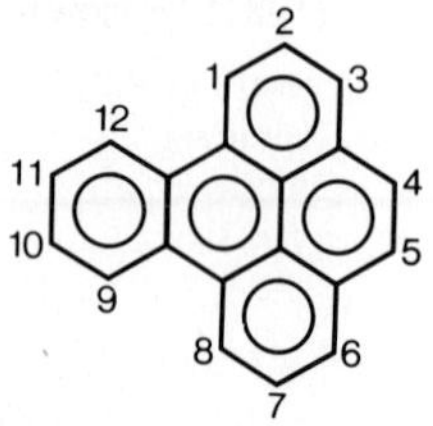	9,10-Diol[f]	? 9,10-Diol 11,12-oxide[g]	OSBORNE and CROSBY (1987) (review)
Benzo[*f*]quinoline 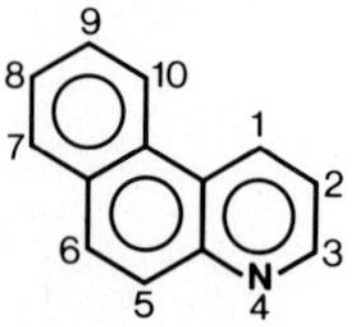	7,8-Diol[a]	? 7,8-Diol 9,10-oxide[c]	KUMAR et al. (1987)
Benzo[*h*]quinoline 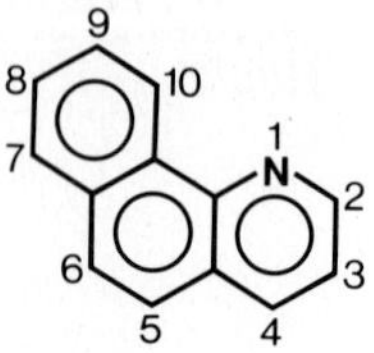	7,8-Diol[a]	? 7,8-Diol 9,10-oxide	KUMAR et al. (1987)
Chrysene 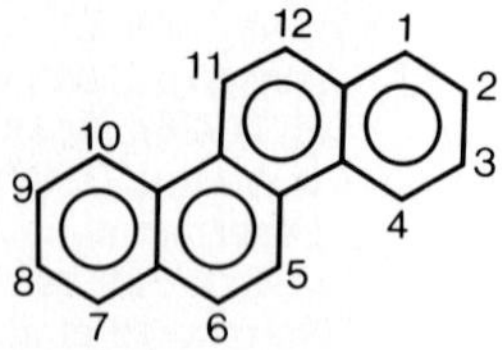	1,2-Diol[a, b, f]	1,2-Diol 3,4-oxide[c, d, e, h]	CONNEY (1982) (review)
	9-Hydroxy 1,2-diol[a, b]	9-Hydroxy-1,2-diol 3.4-oxide[c, d, e]	HODGSON et al. (1983); GLATT et al. (1986)
Cyclopenta[*c,d*]pyrene 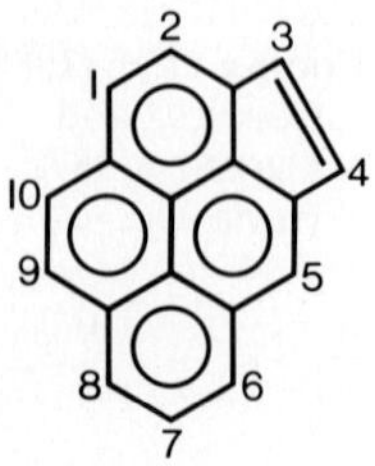	–	? 3,4-Oxide[c, d, e]	GOLD and EISENSTADT (1980); GOLD et al. (1980)

Table 1 (continued)

Hydrocarbon	Derivatives showing highest levels of biological activity	Putative ultimate carcinogens	References
15,16-Dihydro-11-methylcyclopenta[*a*]phenanthren-17-one 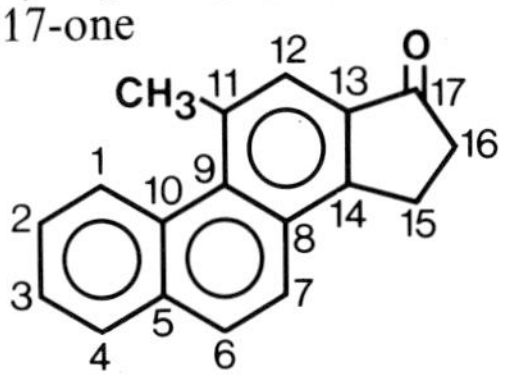	3,4-Diol[a,f]	3,4-Diol 1,2-oxide[h]	COOMBS and BHATT (1987) (review)
15,16-Dihydro 1,11-methanocyclopenta[*a*]phenanthrene-17-one 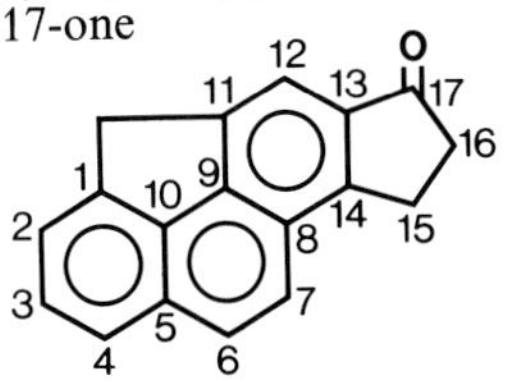	3,4-Diol[a]	? 3,4-Diol 1,2-oxide	COOMBS and BHATT (1987) (review)
Dibenz[*a,h*]acridine 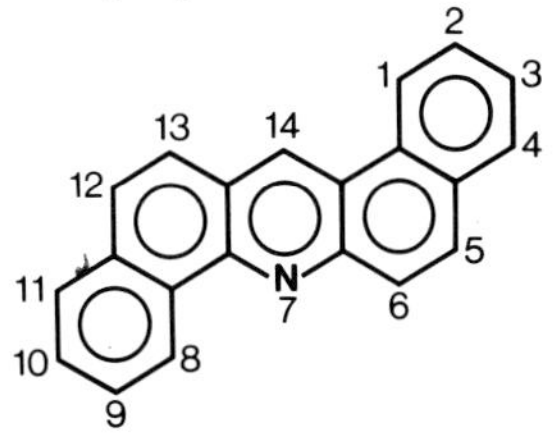	10,11-Diol[a]	10,11-Diol 8,9-oxide[c,d]	STEWARD et al. (1987); CHANG et al. (1987)
Dibenz[*c,h*]acridine 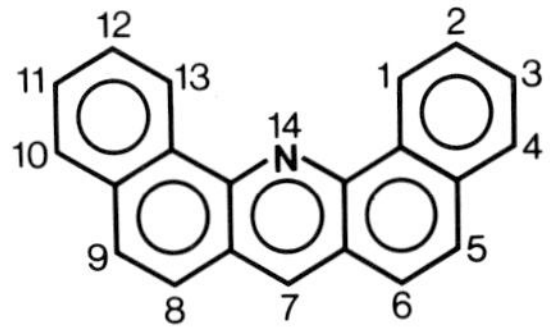	3,4-Diol[a]	3,4-Diol 1,2-oxide[c,d]	THAKKER et al. (1985b); WOOD et al. (1986)
Dibenz[*a,c*]anthracene 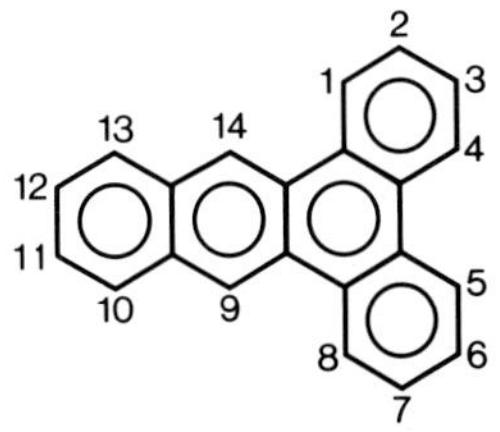	10,11-Diol[a]	? 10,11-Diol 12,13-oxide	SIMS and GROVER (1981) (review)

Table 1 (continued)

Hydrocarbon	Derivatives showing highest levels of biological activity	Putative ultimate carcinogens	References
Dibenz[*a,h*]anthracene 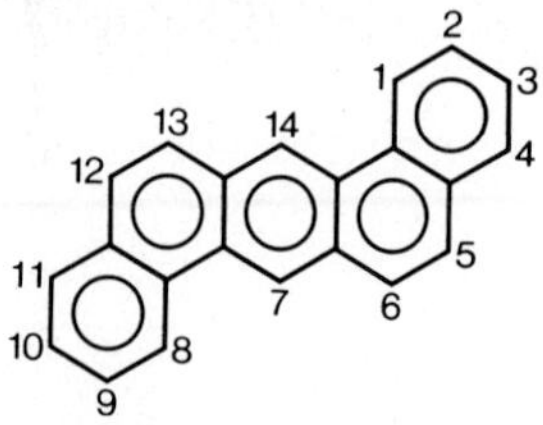	3,4-Diol[a,f,g]	? 3,4-Diol 1,2-oxide	CONNEY (1982) (review)
7H-Dibenz[*c,g*]carbazole 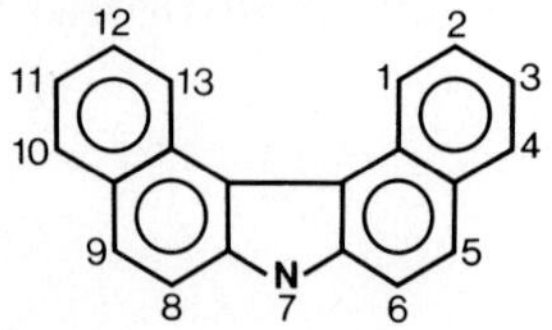	3-Hydroxy[a,h]	?	SCHURDAK et al. (1987); SCHOENY and WARSHAWSKY (1987)
Dibenz[*a,e*]fluoranthene 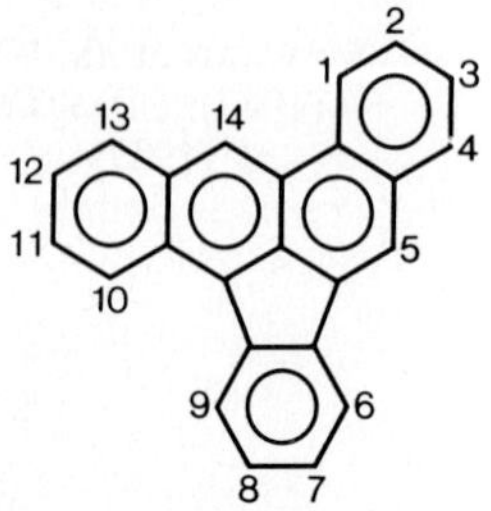	12,13-Diol[a,f] 3,4-Diol[a,f]	12,13-Diol 10,11-oxide[h] 3,4-Diol 1,2-oxide[h]	PERIN-ROUSSEL et al. (1983, 1984); SAGUEM et al. (1983a,b); ZAJDELA et al. (1987)
Dibenz[*a,h*]pyrene 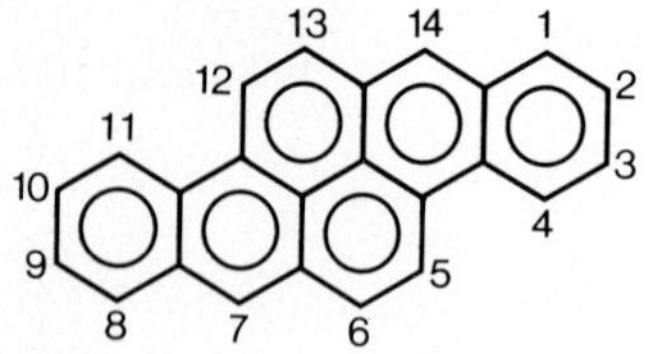	1,2-Diol[f,g]	? 1,2-Diol 3,4-oxide[g]	CHANG et al. (1982)
Dibenz[*a,i*]pyrene 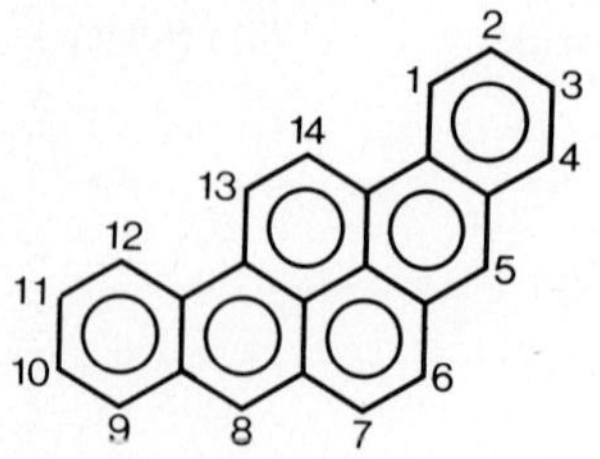	3,4-Diol[f,g]	? 3,4-Diol 1,2-oxide[g]	CHANG et al. (1982)

Table 1 (continued)

Hydrocarbon	Derivatives showing highest levels of biological activity	Putative ultimate carcinogens	References
7,12-Dimethylbenz[*a*]-anthracene 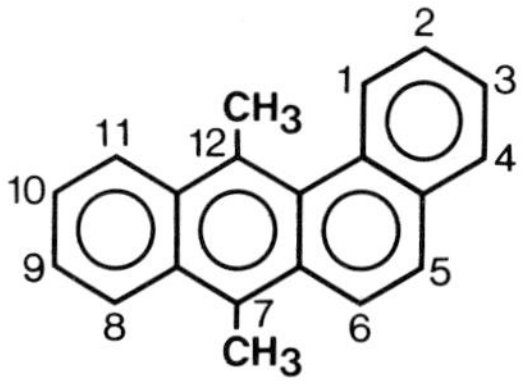	3,4-Diol[a,b,e,f]	3,4-Diol 1,2-oxide[h]	SIMS and GROVER (1981) (review); CONNEY (1982) (review); SAWICKI et al. (1983); DIPPLE et al. (1984) (review)
7-Ethylbenz[*a*]anthracene 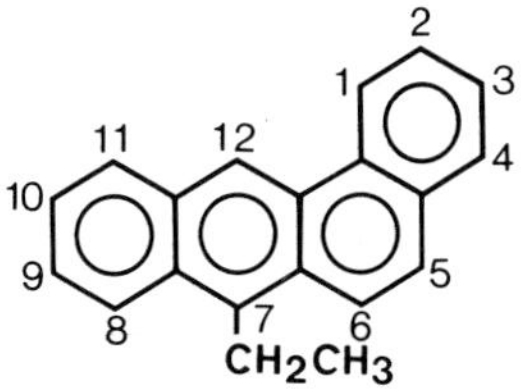	3,4-Diol[a]	? 3,4-Diol 1,2-oxide[c,h]	MCKAY (1987); MCKAY et al. (1988); GLATT et al. (1989)
Fluoranthene 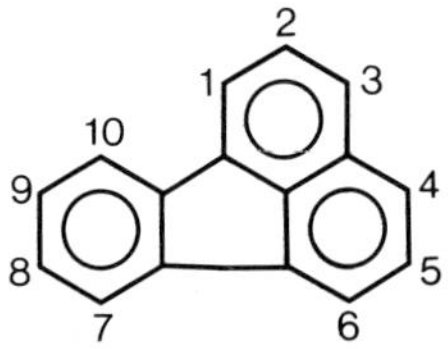	2,3-Diol[a]	2,3-Diol 1,10b-oxide[h]	LAVOIE et al. (1982); RASTETTER et al. (1982); BABSON et al. (1986)
Indeno[*c*,*d*]pyrene 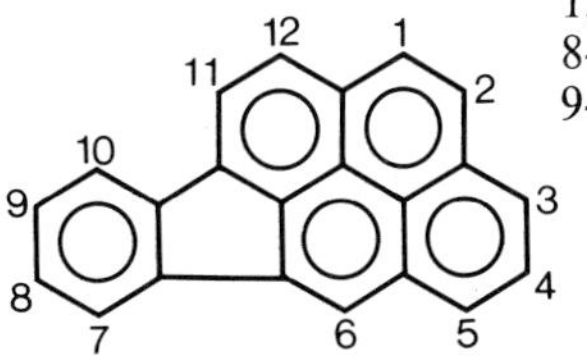	1,2-Oxide[c,f] 1,2-Diol[f] 8-Hydroxy[a] 9-Hydroxy[a]	?	RICE et al. (1985); RICE et al. (1986)
7-Methylbenz[*c*]acridine 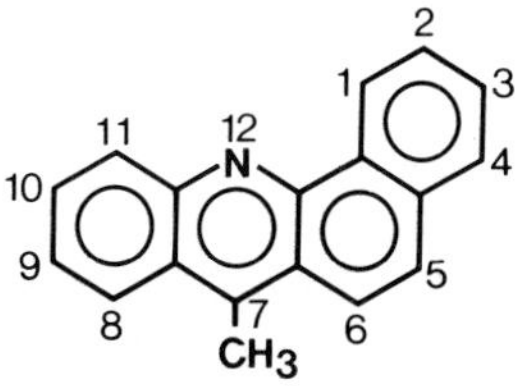	3,4-Diol[a]	3,4-Diol 1,2-oxide[c,d]	GILL et al. (1986)

Table 1 (continued)

Hydrocarbon	Derivatives showing highest levels of biological activity	Putative ultimate carcinogens	References
7-Methylbenz[*a*]anthracene 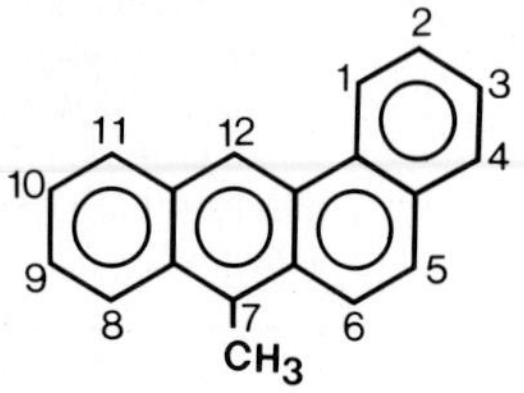	3,4-Diol[a,b,e,f]	3,4-Diol 1,2-oxide[c,h]	SIMS and GROVER (1981) (review); MCKAY (1987); MCKAY et al. (1988); GLATT et al. (1989)
3-Methylcholanthrene 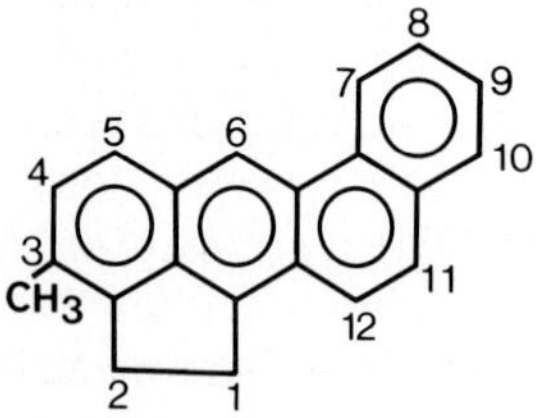	9,10-Diol[a,e,f]	? 9,10-Diol 7,8-oxide[f,h] ? 3-Hydroxymethyl-9,10-diol 7,8-oxide	SIMS and GROVER (1981) (review); CONNEY (1982) (review); DIGIOVANNI et al. (1985); OSBORNE et al. (1986)
5-Methylchrysene 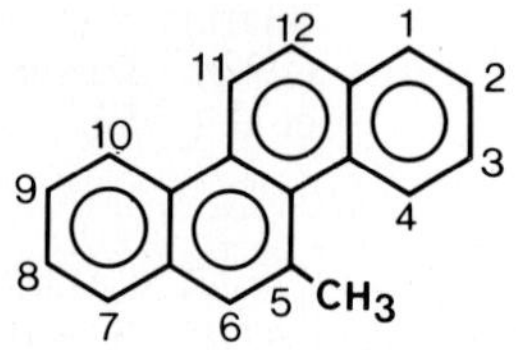	1,2-Diol[a,f]	1,2-Diol 3,4-oxide[d,g,h]	HECHT et al. (1986); BROOKES et al. (1986); REARDON et al. (1987); HECHT et al. (1987)

I. 7,12-Dimethylbenz[*a*]anthracene

With 7,12-dimethylbenz[*a*]anthracene (DMBA), all the available evidence points to a single pathway of activation in which the 3,4-diol is further metabolised to the bay-region 3,4-diol 1,2-oxides. Extensive investigations have led to the partial characterisation of adducts formed by reactions of these diol-epoxides with guanine and adenine moieties in the DNA in cells and tissues that have been treated with the parent hydrocarbon (SAWICKI et al. 1983; DIPPLE et al. 1984). More complete characterisation of adducts has been hampered, until recently, by difficulties in synthesising the relevant diol-epoxides (LEE and HARVEY 1986).

II. Benzo[*a*]pyrene

The most comprehensive data on mechanisms of activation have been obtained for BaP (see COOPER et al. 1983; OSBORNE and CROSBY 1987). It is quite clear that, in almost all in vivo situations, activation proceeds along a single pathway via the 7,8-diol to the 7,8-diol 9,10-oxides; the resulting DNA and RNA adducts have been well characterised (see SINGER and GRUNBERGER 1983). Here, however, secondary activation pathways have been noted in rat mammary tissue (PHILLIPS et al. 1985) and in mouse skin (VIGNY et al. 1980) which, in the latter case, appears to involve 9-hydroxyBaP and its conversion to the related 9-hydroxyBaP 4,5-oxide.

III. Dibenzo[*a,e*]fluoranthene

Dibenzo[*a,e*]fluoranthene is a non-alternant hydrocarbon that is both an environmental pollutant and a potent carcinogen. In this case all the evidence suggests that there are two parallel pathways of activation. Characterisation of hydrocarbon-nucleic acid adducts has shown that activation proceeds both via the conventional pathway involving the bay-region 3,4-diol 1,2-oxides and via the pseudo-bay-region 12,13-diol 10,11-oxides (PERIN-ROUSSEL et al. 1984). Although differences in the comparative biological activities of the precursor diols have been found (ZAJDELA et al. 1987), the data support the conclusions of the adduct characterisation studies.

IV. Dibenz[*a,h*]anthracene

With dibenz[*a,h*]anthracene, another potent carcinogen, the principle activation pathway has not been unequivocally established. The 3,4-diol has been found to be the most mutagenic diol in a microsomally mediated *Salmonella typhimurium* test system and is a tumour initiator in newborn mice and on mouse skin (see CONNEY 1982). It is therefore reasonable to suggest that the activation of dibenz[*a,h*]anthracene proceeds via the 3,4-diol 1,2-oxides, but it should be borne in mind that the synthetic *anti*-isomer of this bay-region diol-epoxide was found to be inactive when tested as a tumour initiator on mouse skin (SLAGA et al. 1980) and that no nucleic acid adducts appear to have been characterised to date.

V. Benz[*c*]acridine

A somewhat similar situation exists with benz[*c*]acridine. Here a great deal of circumstantial evidence (LEVIN et al. 1983; WOOD et al. 1983a; CHANG et al. 1984), derived from the determination of the biological activities of synthetic compounds in a variety of test systems, implicates the bay-region diol-epoxide pathway, but nucleic acid adducts formed in a target tissue have apparently yet to be isolated and characterised.

VI. Benz[*a*]anthracene and Chrysene

The pitfalls attendant upon the over-interpretation or extrapolation of available data can be illustrated by reference to BA and chrysene, PAH that are weak tumour-initiating agents but that are relatively abundant in tobacco smoke and in the environment as a whole (see GRIMMER 1983). In both cases data obtained from experiments in which the activities of synthetic diols and diol-epoxides as mutagens and tumour-initiating agents were examined (see SIMS and GROVER 1981; CONNEY 1982) indicated that the bay-region diol-epoxides were the most biologically active compounds. It was therefore tempting to assume that both BA and chrysene were activated solely by conversion to the bay-region diol-epoxides. However, when the nucleic acid adducts formed from BA in hamster embryo cells or in mouse skin were isolated, evidence was obtained indicating that a second activation pathway involving the non-bay-region diol-epoxide, BA-8,9-diol 10,11-oxide existed in addition to the predicted pathway that involved the bay-region 3,4-diol 1,2-oxide (COOPER et al. 1980 b).

With chrysene, the second activation pathway appears to involve, as the ultimate reactive species, a diol-epoxide of the bay-region type that also possesses a phenolic hydroxyl group in the 9-position (HODGSON et al. 1983). This triol-epoxide, 9-hydroxychrysene 1,2-diol 3,4-oxide, has been synthesised (SEIDEL et al. 1989) and found to react with DNA in vitro (PHILLIPS et al. 1987) and to be mutagenic in bacterial and mammalian cells (GLATT et al. 1986). Although the intermediate triol, 9-hydroxychrysene 1,2-diol, has been detected as a microsomal metabolite (HODGSON et al. 1985 a), it is not yet clear which of the pathways shown in Fig. 2 result in the formation of triol-epoxide-nucleic acid adducts in in vivo situations, since recent work has shown that the *anti*-isomer of the bay-region diol-epoxide can itself be further metabolised to the triol-epoxide (HODGSON et al. 1986; HALL et al. 1988). In this case the diol-epoxide and triol-epoxide adducts formed in, for example, mouse skin that has been treated with chrysene also require further characterisation.

From the foregoing it is clear that in the majority of PAH for which DNA adducts have been examined, metabolic activation involves diol-epoxides of the bay-region type. With several other PAH, circumstantial evidence favours the involvement of the same type of reactive species. There does not appear, however, to be any a priori reason why epoxides of other types should not also be involved in reactions with DNA in vivo, and it is hoped that sufficient examples have been mentioned above to demonstrate that this is, in fact, the case. The relatively small number of hydrocarbons for which DNA adducts have been characterised makes it very likely that other exceptions to the bay-region diol-epoxide generalisation will be found.

D. Stereochemistry of Activation Pathways

A large number of pharmacologically important substances, including a variety of drugs, have for some time been known to exist as pairs of enantiomers, two compounds related to one another as non-superimposable mirror images. Each member of a given pair differs physically from its antipode in the direction by which it rotates the plane of polarised light (+ or −) and also in the absolute configuration of groups around a chiral centre (*R* or *S*). Biologically, however, two members of a pair of enantiomers may exhibit large differences both in terms of their activity and in the details of their metabolism (ARIËNS 1984; TRAGER and JONES 1987).

PAH may also be included in this category of compounds. Although the parent hydrocarbons themselves are not enantiomeric, the metabolism of PAH in a variety of systems has been found to be stereoselective, and the resulting optically active derivatives have been demonstrated to possess quite different biological properties (see CONNEY 1982; THAKKER et al. 1985a). Over 50 years ago BOYLAND and LEVI (1935) observed that the anthracene 1,2-diol excreted in the urine of rats and rabbits which had been fed anthracene in their diet was optically active and that, in addition, the predominant enantiomer was different in the two species. More recently, in vitro studies on the activation of PAH to reactive derivatives, generally using either microsomes or reconstituted enzyme systems isolated from rat liver, have tended to concentrate on BaP as a substrate. These investigations have led to the elucidation of a stereochemical scheme for the metabolic activation of BaP by microsomal cytochrome P-450-linked monooxygenases and EH, as shown in Fig. 3. Similar schemes have been worked out for the oxidation of BaP away from the 7,8-position, for instance at the K-region, and this also applies to other PAH. However, discussion of these is outside the scope of the present review. DIPPLE et al. (1984), THAKKER et al. (1985a) and YANG et al. (1985) provide useful sources of references for these other examples.

I. Benzo[*a*]pyrene

It can be seen from Fig. 3 that both BaP-7,8-oxide and BaP-7,8-diol exist as pairs of enantiomers, while the 7,8-diol 9,10-oxide, generally regarded as the ultimate carcinogenic form of BaP (see Sect. C), is produced as two diastereomers, each comprising two enantiomers. [For a more detailed discussion of these stereochemical relationships and related nomenclature, see DIPPLE et al. (1984).] Of these four, the (+)-*anti*-BaP-7,8-diol 9,10-oxide has been found to possess greater biological activity than the other three in most systems in which they have been tested (see CONNEY 1982; COOPER et al. 1983; DIPPLE et al. 1984), and adducts derived from the covalent binding of (±)-*anti*-BaP-7,8-diol 9,10-oxide to DNA have been identified as the major DNA adducts extracted following application of BaP to a variety of tissues and cells maintained in culture (WESTON et al. 1983 and references therein), although this does not apply to all tissues studied (PHILLIPS et al. 1985; MOORE et al. 1987). In addition, the (+)-*anti*-enantiomer of the diol-epoxide is the major enantiomer formed from the metabolism of BaP and metabolic BaP-7,8-diol by rat liver microsomes prepared from 3-MC-induced animals and by isolated rat liver cytochrome P-450*c*, the major 3-MC inducible form of the enzyme [for cytochrome P-450 nomenclature, refer to WOLF (1986) and GUENGERICH (1987)]. In fact the majority of studies performed on the stereoselectivity of BaP activation in a variety of systems have indicated that the

Table 2. Stereoselectivity of the metabolism of polycyclic aromatic hydrocarbons (PAH) to diol derivatives: instances where this varies from the situation generally observed with 3-MC-induced rat liver microsomes or with rat liver cytochrome P-450c in a reconstituted system[a, g]

PAH	Metabolising system	Enantiomeric composition of diol product (%)		References
Anthracene		(−)-1*R*,2*R*-Diol	(+)-1,*S*,2*S*-Diol	
	Rat liver cytochrome P-450*b*+epoxide hydrolase (EH)	39[b, e]	61	VAN BLADEREN et al. (1985)
	Rat liver microsomes (RLM), untreated	80[c]	20	VON TUNGELN and FU (1986)
	RLM, phenobarbital (PB)-induced	65–71[b, e]	35–30	VAN BLADEREN et al. (1985);
		88[c]	12	VON TUNGELN and FU (1986)
	Rabbit liver and skin microsomes, untreated	34[c]	66	HALL and GROVER (1987)
9,10-Dimethyl-anthracene	RLM, untreated	63[c]	37	VON TUNGELN and FU (1986)
	RLM, PB-induced	65[c]	35	VON TUNGELN and FU (1986)
Benz[*a*]anthracene		(−)-3*R*,4*R*-Diol	(+)-3*S*,4*S*-Diol	
	RLM, untreated	83[c]	17	YANG et al. (1985)
	RLM, PB-induced	91[c]	9	YANG et al. (1985)
		69[b]	31	THAKKER et al. (1979)
7,12-Dimethylbenz[*a*]-anthracene	RLM, untreated	57[c]	43	YANG et al. (1985)
	RLM, PB-induced	62[c]	38	YANG et al. (1985)
	RLM, 3-methylcholanthrene (3-MC)-induced	64[c]	36	YANG et al. (1985)
7-Chlorobenz[*a*]-anthracene	RLM, untreated	60[c]	40	FU et al. (1985)
	RLM, PB-induced	60[c]	40	FU et al. (1985)
Benzo[*j*]-fluoranthene		?4*R*,5*R*-Diol	?4*S*,5*S*-Diol	
	Rat liver S-9 fraction, polychlorinated biphenyl (PCB)-induced	60[c]	40	RICE et al. (1987a)
	Mouse skin in vivo	80[c]	20	RICE et al. (1987b)
		?9*R*,10*R*-Diol	?9*S*,10*S*-Diol	
	Rat liver S-9 fraction, PCB-induced	73[b]	27	RICE et al. (1987a)
	Mouse skin in vivo	85[c]	15	RICE et al. (1987b)
Benzo[*c*]phenanthrene		(−)-3*R*,4*R*-Diol	(+)-3*S*,4*S*-Diol	
	RLM, untreated	89[b]	11	ITTAH et al. (1983);
		64–71[c, d, e]	36–29	MUSHTAQ and YANG (1987)
	RLM, PB-induced	20[b]	80	ITTAH et al. (1983);
		46–54[c, d, e]	54–46	MUSHTAQ and YANG (1987)

phenanthrene	RLM, PB-induced	86[b]	14	PRASAD et al. (1988)
Benzo[a]pyrene		(−)-7*R*,8*R*-Diol	(+)-7*S*,8*S*-Diol	
	Human skin in vitro	50, 83[c,f]	50, 17	HALL and GROVER (1988)
	Human liver microsomes	64, 75, 78[c,f]	36, 25, 22	HALL et al. (unpublished data)
7-Methylbenzo[a]-pyrene	RLM, benz[a]anthracene-induced	55[b]	45	KINOSHITA et al. (1982)
	RLM, PCB-induced	20[d]	80	CHIU et al. (1983)
Chrysene		(−)-1*R*,2*R*-Diol	(+)-1*S*,2*S*-Diol	
	RLM, untreated	76[b]	24	NORDQVIST et al. (1981);
		51[c]	49	WEEMS et al. (1986)
	RLM, PB-induced	55[b]	45	NORDQVIST et al. (1981);
		41[c]	59	WEEMS et al. (1986)
	Human skin in vitro	81, 88[c,f]	19, 12	WESTON et al. (1985)
Dibenz[a,h]anthracene		?3*R*,4*R*-Diol	?3*S*,4*S*-Diol	
	RLM, 3-MC-induced	80[b]	20	NORDQVIST et al. (1979)
Naphthalene		(−)-1*R*,2*R*-Diol	(+)-1*S*,2*S*-Diol	
	Rat liver cytochrome P-450*b*+EH	66–61[b,e]	39–34	VAN BLADEREN et al. (1985)
	RLM, PB-induced	66–50[b,e]	50–34	VAN BLADEREN et al. (1985)
Triphenylene		(−)-1*R*,2*R*-Diol	(+)-1*S*,2*S*-Diol	
	Rat liver cytochrome P-450*c*+EH	85[b]	15	THAKKER et al. (1988)
	RLM, untreated	78	22	THAKKER et al. (1985c)
	RLM, PB-induced	70	30	THAKKER et al. (1985c);
		84[b]	16	THAKKER et al. (1988)

[a] For present purposes this is taken as being formation of PAH diol with $\geq 90\%$ *R*,*R* absolute configuration.
[b] Determined by resolution of derivatised radioactive metabolic product on high performance liquid chromatography (HPLC).
[c] Determined by direct resolution of radioactive metabolic product on chiral HPLC.
[d] Determined by circular dichroism spectroscopy of metabolic product.
[e] Enantiomeric composition dependent upon amount of microsomal protein or purified enzyme included in the assay system.
[f] Samples from separate individuals.
[g] Where structural formulae are not shown, see Table 1.

major pathway of primary metabolism results in a predominance (≧90%) of the (+)-7*R*,8*S*-oxide and of the (−)-7*R*,8*R*-diol, whilst secondary metabolism of the diol yields mainly (+)-*anti*-7,8-diol 9,10-oxide from the *R,R* enantiomer and mainly (+)-*syn*-7,8-diol 9,10-oxide from the *S,S* enantiomer (see CONNEY 1982; COOPER et al. 1983; DIPPLE et al. 1984; YANG et al. 1985). Such studies have led to proposals for the stereochemistry of the binding site and the mechanism of action of individual enzymes involved in the activation pathway, such as cytochrome P-450*c* and EH (LEVIN et al. 1980; ARMSTRONG et al. 1981; JERINA et al. 1982; ARMSTRONG 1987; KADLUBAR and HAMMONS 1987).

II. Primary Metabolism

Through expediency, investigations into the stereoselectivity of activation of other PAH have often referred to the observations made on BaP as a standard. In the majority of cases, and particularly, but not exclusively, where the metabolising system used has been either 3-MC-induced rat liver microsomes or rat liver cytochrome P-450*c*, a similar pattern has been observed to that described above. Thus, primary metabolism of the following PAH, in addition to BaP, has led to a detection of oxides having mainly (≧90%), *R,S* absolute configuration and/or diols having mainly (≧90%) *R,R* absolute configuration: anthracene (VAN BLADEREN et al. 1984, 1985; VON TUNGELN and FU 1986) and its methylated (VON TUNGELN and FU 1986) and halogenated (FU et al. 1986) derivatives; BA (THAKKER et al. 1979; YANG et al. 1985) and its methylated (CHOU and YANG 1979; YANG 1982; YANG et al. 1982; FU et al. 1983; YANG and FU 1984; YANG et al. 1984; YANG et al. 1985) and halogenated (FU and YANG 1983; CHIU et al. 1984; FU et al. 1985) derivatives; benzo[*c*]phenanthrene (BcPh) (ITTAH et al. 1983; MUSHTAQ and YANG 1987; VAN BLADEREN et al. 1987; YANG et al. 1987) and 6-fluoro-BcPh (PRASAD et al. 1988); halogenated derivatives of BaP (FU and YANG 1982; BUHLER et al. 1983; CHOU and FU 1984; THAKKER et al. 1984; FU et al. 1986); chrysene (NORDQVIST et al. 1981; WESTON et al. 1985; WEEMS et al. 1986) and its methylated derivatives (AMIN et al. 1987); 15,16-dihydrocyclopenta[*a*]phenanthren-17-one and its methylated derivatives (HADFIELD et al. 1984; COOMBS and BHATT 1987); dibenz[*c,h*]acridine (THAKKER et al. 1985b); phenanthrene (NORDQVIST et al. 1981); and triphenylene (THAKKER et al. 1985c, 1988). It should be noted that not all the PAH listed here have, as yet, been shown to possess biological activity (see DIPPLE et al. 1984). However, they are included for the sake of comparison.

This point also applies to the compounds listed in Table 2, which presents exceptions to the general rule outlined above. As may be seen from this table, most of these "atypical" cases have arisen from metabolism by systems other than 3-MC-induced rat liver microsomes, in particular either untreated or PB-induced rat liver microsomes, although this last observation may be biased by the preponderance of studies made with variously induced rat liver microsomes compared with those employing extrahepatic tissues. It is known that treatment of animals with various xenobiotics leads to discriminatory induction of certain forms of cytochrome P-450 to levels higher than those found in the constitutive (untreated) state (see WOLF 1986; GUENGERICH 1987). For instance, while cytochrome P-450*c* is the major 3-MC-inducible form of the enzyme in rat liver, forms *b* and *e* predominate in livers from PB-treated animals. Thus the "unusual" stereoselective profiles noted in Table 2 for either untreated or PB-induced rat liver microsomes could be due to a difference in the steric arrangement of the binding site of the predominant form(s) of cytochrome P-450 present in these microsomes compared with that of cytochrome P-450*c*. For some PAH, such as naphthalene or anthracene (VAN BLADEREN et al. 1985), the stereoselectivity of EH as well as the P-450/EH ratio may also be important in determining the stereoselectivity of the oxide and diol products.

Compared with other PAH, the stereoselectivity of the primary metabolism of BaP at the 7,8-position is relatively invariable and seems little affected either by the tissue, strain or species used as the source of the microsomes, or by the enzyme induction regime employed or by the form of cytochrome P-450 used in a reconstituted system (YANG et al. 1985; HALL and GROVER 1986, 1987; HALL et al. 1987). The only exceptions noted so far have been with samples obtained from a few human individuals (Table 2), in whom differences in genetic make-up as well as exposure to environmental or therapeutic agents may contribute to the cytochrome P-450 profile.

For some PAH the stereoselectivity of diol formation has been investigated by different laboratories and different results have been obtained. In some cases, for example BaP, these differences appear to be subtle (YANG et al. 1985), whilst in others, for example BcPh, they are more pronounced (MUSHTAQ and YANG 1987). Such discrepancies are possibly attributable to differences in the strain of rat used as a source of tissue, in the amount of microsomal protein included in the incubation mixture and/or to the method employed for assessing the enantiomeric composition of the product. A detailed study of the stereoselective formation of BcPh-3,4-diol by MUSHTAQ and YANG (1987) has tended to preclude the first two possibilities, at least in the case of this hydrocarbon, but the last has not yet been examined.

III. Secondary Metabolism

Table 3 summarises the data available for the stereoselectivity exhibited in the secondary metabolism of PAH diols to diol-epoxides. Although it is recognised that these metabolites can exist as two pairs of enantiomers (Fig. 3), most data have been reported only for the two diastereomers (*anti* and *syn*), and so this is how they are presented here. Quantitation of the diol-epoxides in all cases was achieved by measurement of their tetrol hydrolysis products, which are relatively stable and can be separated by high performance liquid chromatography (HPLC). In an additional study on the metabolism of BcPh in rodent embryo cell cultures, >85% of the adducts formed between BcPh-3,4-diol 1,2-oxide and DNA were found to result from the *anti*-diastereomer having *R,S*-diol *S,R*-oxide absolute configuration, with the remainder arising from the *syn* form having *S,R*-diol *R,S*-oxide absolute configuration. This, together with the data presented in Table 3, indicate that, as with primary metabolism, the secondary metabolism of PAH generally follows the pattern determined for BaP by rat liver microsomes. Again, though, there appear to be exceptions to this, for instance in

Table 3. Stereoselectivity of the metabolism of polycyclic aromatic hydrocarbon (PAH) diols to bay-region diol-epoxides

PAH	Metabolising system	Substrate	Diastereomeric composition of diol-epoxide product (%)		References
			anti-	*syn-*	
Benz[*a*]anthracene	Rat liver microsomes (RLM), untreated, phenobarbital (PB)-, 3-methylcholanthrene (3-MC)-induced [a,b]	(±)-3,4-Diol	~100	—[c]	THAKKER et al (1982)
		(−)-3*R*,4*R*-Diol	~100	—	
		(+)-3*S*,4*S*-Diol	—	—	
	Ram seminal vesicle (RSV) microsomes, untreated	(±)-3,4-Diol	68	32	DIX et al. (1986)
Benzo[*c*]phenanthrene	RLM, untreated	(±)-3,4-Diol	41	59	THAKKER et al. (1986a)
		(−)-3*R*,4*R*-Diol	72	28	
		(+)-3*S*,4*S*-Diol	0	100	
	RLM, PB-induced	(±)-3,4-Diol	38	62	
		(−)-3*R*,4*R*-Diol	44	56	
		(+)-3*S*,4*S*-Diol	6	94	
	RLM, 3-MC-induced [a,b]	(±)-3,4-Diol	27	73	
		(−)-3*R*,4*R*-Diol	77	23	
		(+)-3*S*,4*S*-Diol	0	100	
	Mouse liver microsomes, untreated	(±)-3,4-Diol	28	72	
		(−)-3*R*,4*R*-Diol	45	55	
		(+)-3*S*,4*S*-Diol	30	70	
Benzo[*a*]pyrene	Majority of systems examined, including RLM	(±)-7,8-Diol	Major	minor	COOPER et al. (1983) (review); THAKKER et al. (1985a) (review)
		(−)-7*R*,8*R*-Diol	Major	minor	
		(+)-7*S*,8*S*-Diol	Minor	major	
	Rabbit liver cytochrome P-450 LM_2	(−)-7*R*,8*R*-Diol	~ 50	~ 50	DEUTSCH et al. (1978, 1979)
		(+)-7*S*,8*S*-Diol	~ 50	~ 50	
	Rabbit liver cytochrome P-450 LM_7	(−)-7*R*,8*R*-Diol	24	76	DEUTSCH et al. (1978)
	Peroxyl radical-dependent oxidation	(−)-7*R*,8*R*-Diol	Major	minor	MARNETT (1987) (review); GOWER and WILLS (1987)
		(+)-7*S*,8*S*-Diol	Major	minor	
6-Fluoro-benzo[*a*]-pyrene	RLM, 3-MC-induced[a]	(−)-7*R*,8*R*-Diol	89	11	THAKKER et al. (1984)

[a] Similar composition obtained with rat liver cytochrome P-450*c* ± epoxide hydrolase.
[b] Minor metabolites (<10%); major metabolites are bis-dihydrodiols.
[c] Not detected.
[d] Composition dependent upon amount of microsomal protein included in the assay system.

Table 3 (continued)

PAH	Metabolising system	Substrate	Diastereomeric composition of diol-epoxide product (%)		References
			anti-	*syn-*	
Chrysene					
	RLM, untreated	(−)-1*R*,2*R*-Diol	84–90[d]	16–10	Vyas et al. (1982a)
		(+)-1*S*,2*S*-Diol	31	69	
	RLM, PB-induced	(−)-1*R*,2*R*-Diol	86–90[d]	14–10	
		(+)-1*S*,2*S*-Diol	41	59	
	RLM, 3-MC-induced[a]	Metabolic (90% 1*R*, 2*R*)	79	21	Nordqvist et al. (1981)
		(−)-1*R*,2*R*-Diol	95–97[d]	5–3	Vyas et al. (1982a)
		(+)-1*S*,2*S*-Diol	14	86	Vyas et al. (1982a)
5-Methyl-chrysene, 6-Methyl-chrysene	Mouse skin in vivo	(−)-1*R*,2*R*-Diol	> 90	< 10	Hecht et al. (1987)
Naphthalene					
	RLM	(−)-1*R*,2*R*-Diol	83–89	17–11	Schaeffer et al. (1987)
		(+)-1*S*,2*S*-Diol	65–77	35–23	
Phenanthrene					
	RLM, untreated	(±)-1,2-Diol	57	43	Vyas et al. (1982b)
		(−)-1*R*,2*R*-Diol	68	32	
		(+)-1*S*,2*S*-Diol	37	63	
	RLM, PB-induced	(±)-1,2-Diol	57	43	
		(−)-1*R*,2*R*-Diol	77	23	
		(+)-1*S*,2*S*-Diol	32	68	
	RLM, 3-MC-induced	(±)-1,2-Diol	60	40	
		Metabolic (97% 1*R*,2*R*)	75	25	Nordqvist et al. (1981)
		(−)-1*R*,2*R*-Diol	85	15	Vyas et al. (1982b)
		(+)-1*S*,2*S*-Diol	15	85	
Triphenylene					
	RLM, untreated	(−)-1*R*,2*R*-Diol	Minor	major	Thakker et al. (1986b)
		(+)-1*S*,2*S*-Diol	Minor	major	
	RLM, PB-induced	(−)-1*R*,2*R*-Diol	Minor	major	
		(+)-1*S*,2*S*-Diol	Minor	major	
	RLM, 3-MC-induced[a]	(−)-1*R*,2*R*-Diol	Major	minor	
		(+)-1*S*,2*S*-Diol	Minor	major	

the further metabolism of the *R,R*-diol enantiomers of BcPh and triphenylene by microsomes other than those obtained from 3-MC-induced rat liver and also in the oxidation of BaP-7,8-diol by two forms of rabbit liver cytochrome P-450 (Table 3).

A second microsomal enzyme system distinct from cytochrome P-450 is also capable of oxidizing PAH diols to diol-epoxides but is not capable of the initial oxidation of the parent hydrocarbon. This involves prostaglandin *H* synthase or another activity that will generate peroxyl radicals. Although only three PAH diols have as yet been tested with such systems, the stereoselectivity of epoxidation by this peroxyl radical-dependent mechanism has been found to be different to that exhibited by cytochrome P-450-dependent systems in that the *S,S*-enantiomer is metabolised largely to the *anti*-diol-epoxide (MARNETT 1987; Table 3). Such activity has been demonstrated in a number of tissues including ram seminal vesicles, mouse keratinocytes and rat intestine (DIX et al. 1986; ELING et al. 1986; GOWER and WILLS 1987) and may be important in the activation of carcinogenic PAH in tissues where cytochrome P-450 activities are low. These might include uninduced mouse skin (MARNETT 1987) and, speculatively, bone marrow. However, much work remains to be done to assess the importance of this particular reaction to PAH metabolism in vivo.

E. Factors Governing Susceptibility to PAH-Induced Tumorigenesis

Examples of variable responses within the human population to exposure to similar levels of mixtures of PAH in the form of, for instance, tobacco smoke or industrial waste have been known for many years. The biochemical basis for such interindividual differences is, however, still largely obscure. Studies on PAH activation utilising cultured human tissues and cells (HARRIS et al. 1982; AUTRUP and HARRIS 1983) or organs (e.g. HALL and GROVER 1988) have as yet progressed little beyond the stage of recording phenomena without providing an actual rationale for these observations. Many of the conclusions drawn have of necessity been based on extrapolations from animal studies. Although the interindividual differences found in humans are generally larger when compared with inbred strains of animals, available evidence indicates that such extrapolation is usually valid (HARRIS 1987; see MOORE et al. 1987 for an exception). Indeed, it has been used to provide assessments of the carcinogenic risk of a variety of PAH to humans (IARC 1973, 1983).

The study of PAH-induced carcinogenesis in animal species, although not providing direct parallels with interindividual variations observed in humans, has presented examples of differences in susceptibility which have proved useful in examining the underlying biochemical mechanisms associated with these. Such differences have been found between different species, different strains of the same species, different tissues of the same animal and have also been shown to be influenced by the age and sex of the animal.

One of the most well-researched of these is the observation that mouse skin is susceptible to PAH-induced carcinogenesis while rat liver, ironically since it has been used as a source of microsomes for most metabolic studies (see Sect. D), is relatively resistant. It should be pointed out, however, that often only one species is chosen for a particular type of testing study, e.g. the mouse in the case of skin tumorigenesis, and therefore similar observations on other species tend to be limited. A further example is the finding that certain strains of mice exhibit differential sensitivity to PAH (KOURI et al. 1973). Indeed one strain (SENCAR) has been derived specifically for its susceptibility to the induction of skin tumours by chemical carcinogens (DIGIOVANNI et al. 1980). Such differential susceptibility could be due to one or more underlying factors, several of which have been examined more or less thoroughly. These include levels of enzymes and the extent of their induction, regio- and stereoselective metabolism of PAH, extent of binding of PAH metabolites to DNA, persistence of these DNA adducts, and sensitivity to promoting agents. Anatomical details may also be important, as noted by BOUTWELL et al. (1981) in the case of skin tumorigenesis, but will not be considered here.

I. Species and Strain Differences

Twenty years ago an observation was made which has provoked much work on the basis of differential susceptibility to PAH. NEBERT and GELBOIN (1969) found that certain strains of mice could be classified as being "responsive" to induction of so-called aryl hydrocarbon hydroxylase (AHH) activity in liver by PAH, whereas others were non-responsive. This variation in induction has been extensively studied in the mouse and has been ascribed to the expression, or otherwise, of a cytosolic protein, molecular weight 95K, which is capable of binding PAH within the cell (POLAND and GLOVER 1988; FERNANDEZ et al. 1988), an event postulated to result in expression of a number of genes encoding cytochrome P-450 and other enzyme activities (NEBERT and JENSEN 1979; NEBERT and GONZALEZ 1987). Thus it could be predicted that responsive strains would demonstrate increased susceptibility to PAH-induced tumorigenesis compared with equivalent non-responsive animals. In general this has been found to be the case, but the correlation is not necessarily a strict one (KINOSHITA and GELBOIN 1972). For instance, which PAH is used as tumour initiator (KOURI et al. 1973; NEBERT and JENSEN 1979), the route of its administration (LEGRAVEREND et al. 1980) or the tissue being examined (NEBERT and GELBOIN 1969; SEIFRIED et al. 1977; OKEY et al. 1979) may all be important determining factors. Again, there are particular strains (SEIFRIED et al. 1977) and species (BICKERS et al. 1983) that may be defined as being responsive, and yet they are resistant to hydrocarbon-induced tumours.

1. Metabolism

Studies on the metabolism of PAH by tissues of various mouse strains have produced somewhat conflicting results, possibly as a consequence of different strains being used in different studies. For instance, the metabolism of 3-MC by liver (BURKI et al. 1973) and of DMBA by epidermis (DIGIOVANNI et al. 1980) was found to be similar when responsive or non-responsive strains of mice were compared. On the other hand an increased metabolism of hydrocarbon to its proximate or ultimate carcinogen, especially following induction, and a correlation with susceptibility to PAH in systems derived from responsive compared with non-responsive strains have been noted for BaP metabolism in liver (HOLDER et al. 1975), lung (SEIFRIED et al. 1977) and skin (LEGRAVEREND et al. 1980; BICKERS et al. 1983).

In a parallel study (BICKERS et al. 1983) in which the activation of BaP by epidermal microsomes of a responsive, susceptible strain of mouse was compared with that of a responsive, non-susceptible strain of rat, it was concluded that neither the formation of BaP-7,8-diol nor the patterns of BaP metabolism were reliable indicators of susceptibility to PAH in rodent skin. Other studies in which the activation of PAH by mouse and rat tissues has been compared do indicate an interspecies variation in the extent of and the regio- and stereoselectivity of metabolism. Proportionally more diols were extracted following incubation of BaP with rat liver microsomes than with mouse liver microsomes (HOLDER et al. 1975); 7,8-benzoflavone enhanced the mutagenic activity of benzo[*e*]pyrene-9,10-diol and also its conversion to the bay-region diol-epoxide to a greater extent by mouse than by rat liver microsomes (THAKKER et al. 1981); a smaller percentage of BcPh-3,4-diol was activated to diol-epoxide by mouse than by rat liver microsomes and with a lower stereoselectivity (THAKKER et al. 1986a), although these were reported as being only "modest differences". CAMUS et al. (1980) made the interesting observation that although detectable AHH activity was considerably lower in mouse skin when compared with rat liver, the activation of BaP-7,8-diol to mutagenic derivatives by microsomes from the two tissues was similar. A direct comparison of the metabolism of BaP in mouse and rat skin both in vivo and in vitro (WESTON et al. 1982a) indicated that approximately threefold more diols were extracted from the former than the latter. In both cases the 7,8-diol was the major metabolite. Rat skin also showed formation of an 11,12-diol which mouse skin did not (WESTON et al. 1982b). A similar study with chrysene, using rodent skin incubated in vitro, demonstrated a 10-fold increase in extractable diols, a slightly higher proportion of 1,2-diol and a higher stereoselectivity of 1,2-diol formation in the mouse compared with the rat (WESTON et al. 1985).

2. Formation and Persistence of DNA Adducts

Such increased formation of the proximate carcinogen in mouse compared with rat skin could account for the higher levels of binding of the PAH to DNA noted in these studies (WESTON et al. 1982a, 1985). A number of other studies have also demonstrated this species difference in binding (ASHURST and COHEN 1981a; BAER-DUBOWSKA and ALEXANDROV 1981; BAER-DUBOWSKA et al. 1981). Similar work has been performed using rodent embryo cell cultures and a similar trend observed, with both BaP (SEBTI et al. 1985) and BcPh (PREUSS-SCHWARTZ et al. 1987). Differential binding of BaP and DMBA has also been noted in rat compared with human mammary cell cultures (MOORE et al. 1987). In general in these studies the particular PAH has been applied as one dose and the DNA extracted and analyzed at one time point (approximately 24 h) following application.

In addition, adducts appear to persist for longer in mouse compared with rat skin, possibly because of differences in the DNA repair system. This is so whether the initial levels of PAH-DNA adducts are greater in rat skin following application of BaP or BaP-4,5-oxide (ALEXANDROV et al. 1983; ROJAS and ALEXANDROV 1986a) or greater in mouse skin following application of BaP-7,8-diol or *anti*-BaP-7,8-diol 9,10-oxide (ROJAS and ALEXANDROV 1986b). In these reports, initial adducts were found to be qualitatively similar in both species, in contrast to the findings of WESTON et al. (1982a). The major adducts persisting 3 weeks after topical application of the hydrocarbon derivative to mouse skin were deoxyadenosine adducts in the case of BaP-4,5-oxide (ROJAS and ALEXANDROV 1986a) and *anti*-BaP-7,8-diol 9,10-oxide-deoxyguanosine adducts in the case of BaP-7,8-diol and *anti*-BaP-7,8-diol 9,10-oxide (ROJAS and ALEXANDROV 1986b).

Similar studies on the binding and rate of disappearance of PAH-DNA adducts in various strains of mice have found no significant qualitative or quantitative differences between responsive and non-responsive strains. This appears to be so irrespective of the PAH used, which to date include BaP (PHILLIPS et al. 1978; LEGRAVEREND et al. 1980; ASHURST and COHEN 1981b; SEBTI et al. 1985), DMBA (GOSHMAN and HEIDELBERGER 1967; PHILLIPS et al. 1978; MORSE et al. 1987a), 3-MC (PHILLIPS et al. 1978) and 15,16-dihydro-11-methylcyclopenta[*a*]phenanthren-17-one (ABBOTT 1983). Again, only a single application of hydrocarbon was made in each case. Unlike the situation in comparisons of mouse and rat skin, persistence of individual DNA adducts has not been studied in different mouse strains.

II. Tissue Differences

So far in this discussion the possible basis for variability in susceptibility to PAH-induced tumorigenesis has only been considered for different species and strains. Similar studies which have attempted to rationalise variations in sensitivity of different tissues are fewer in number. Both NEBERT and GELBOIN (1969) and BURKI et al. (1973) measured basal and induced AHH levels in a variety of tissues of mice and other species. In all cases liver was found to contain the highest activity, while the extent of induction varied in the other tissues. This activity might not, however, be a true reflection of the PAH-metabolising capacity of a given tissue. As mentioned above, results obtained by CAMUS et al. (1980) indicate that the AHH of mouse skin may be more efficient in the secondary metabolism of BaP than that of rat liver. A variation in the efficiency of a given tissue in the primary metabolism of a hydrocarbon to its proximate carcinogen, as compared with the secondary metabolism of this derivative to the ultimate carcinogen could provide an explanation for the observations of BICKERS et al. (1983) and also of LANGENBACH and NESNOW (1983). In the latter case a study on the metabolism of BaP to mutagenic derivatives by cultures of various rat tissues led to the conclusion that the mutagenic response is not directly correlated with formation of the 7,8-diol.

Studies on the metabolism of BaP by different tissues have found that lung microsomes produced proportionally more diols than did the liver microsomes of three mouse strains (SEIFRIED et al. 1977), while the stereoselective metabolism was similar but not identical in microsomes prepared from epidermis, lung and liver of an outbred strain of mouse (HALL and GROVER 1986). Two studies have investigated PAH-DNA binding in different tissues. Following topical application of BaP to the backs of rats and mice in vivo, more than twice as many adducts were detected in the epidermis than in the dermis of the skin (ROJAS et al. 1986). A comparison of the binding of the leukaemogenic PAH, DMBA and 7,8,12-trimethylbenz[*a*]anthracene (TMBA) to DNA of the liver, spleen and bone marrow of female rats indicated that there was no correlation between overall or specific adduct formation and tissue susceptibility (FALZON et al. 1987).

One factor which my be of importance in determining the apparent susceptibility of a given tissue to a particular PAH is the route of administration of the hydrocarbon. For instance, the levels of DMBA and of DMBA-DNA adducts in the epidermis of mice were greater following topical application than following oral dosing, while the reverse was true for three internal organs investigated (MORSE et al. 1987a). Interestingly, in studies in which complex mixtures of PAH (viz. crude coal tar, cigarette smoke condensate or juniper wood tar) were applied topically to the skin of mice in vivo, higher levels of hydrocarbon-DNA adducts were in general found in internal organs, including lung, than in the skin itself (MUKHTAR et al. 1986; RANDERATH et al. 1988; SCHOKET et al. 1989), although this was not always the case (SCHOKET et al. 1988). When the distribution of DNA adducts formed after treatment with crude coal tar was compared with that resulting from application of BaP, a higher proportion was detected in the skin compared with the lung for the latter (MUKHTAR et al. 1986). This might imply that an inhibitor of epidermal PAH metabolism exists in the complex mixtures, which allows a more widespread distribution of PAH within the body. Obviously this is of great potential importance when considering the effects of PAH in humans, since exposure is almost always to complex mixtures of hydrocarbons.

III. Influence of Sex and Age

Two additional influences on susceptibility, which may also provide explanations for some of the inconclusive data mentioned above, are the sex and age of the animals under investigation (HENNINGS et al. 1981). A comparison of the results from separate studies in which a similar amount of BaP was applied topically to the skin of either female (BAER-DUBOWSKA and ALEXANDROV 1981) or male (ALEXANDROV et al. 1983; ROJAS et al. 1986) Swiss mice in vivo indicates that the formation of hydrocarbon-DNA adducts was greater over a similar period of time in the former than in the latter. MORSE et al. (1987b) investigated this directly by comparing DMBA-DNA adduct formation and disappearance in the epidermis of male and female SENCAR mice. They found that at all times (3–48 h) following treatment, more adducts were present in male than in female animals, although qualitatively the adducts were similar. In the same study the extent of DMBA-DNA binding in the epidermis of male mice was found to be dependent upon the age of the animal. A similar phenomenon was observed for DMBA-DNA binding in the liver and mammary tissue of female Sprague-Dawley rats (JANSS and BEN 1978), but not the haematopoietic organs of female Long-Evans rats (FALZON et al. 1987). DMBA was found to be active, when administered orally, in inducing mammary tumours in young female Sprague-Dawley rats, but older females (>100 days) were resistant (HUGGINS et al. 1961). Although this resistance was initially thought to be due to a change, with age or hormonal status, in the sensitivity of the mammary glands to DMBA, it was subsequently shown that the mammary glands of old rats remained susceptible providing that small doses of DMBA were applied directly to the mammary fat pads (SINHA and DAO 1980).

AHH activity has been found to be greater in the livers of female compared with male mice (BURKI et al. 1973) but to show the opposite sex dependency in rat liver (KAMATAKI et al. 1986). The level of this activity, the extent of its inducibility and the degree of associated BaP metabolism in rat skin have all been shown to increase as a function of age (MUKHTAR and BICKERS 1983). Expression of other enzymes, for instance epoxide hydrolase (MEIJER et al. 1987) or specific forms of cytochrome P-450 (KAMATAKI et al. 1986; GUENGERICH 1987; MCCLENNAN-GREEN et al. 1987; BARROSO et al. 1988) may also be dependent upon the sex or age of the animal, as well as on the species, strain or tissue (ROBERTSON et al. 1986; MEIJER et al. 1987). The metabolism of PAH by these forms has not necessarily been well-characterised as yet (MCCLENNAN-GREEN et al. 1987). Unfortunately even less is known about the enzymology of the DNA repair system in relation to the factors discussed above.

In conclusion, the studies which have been assessed here generally indicate that differences between species in terms of PAH activation, DNA binding and enzymology are greater than those observed between strains of the same animal. Since each of these factors may contribute to individual sensitivity to PAH-induced tumorigenesis, the basis of interspecies differences in susceptibility may thus be more readily accounted for. A further influence on this susceptibility could arise from a differential response to various promoting agents, and this has been found to be so for different species (SHUBIK 1950) and for different mouse strains (HENNINGS et al. 1981; REINERS et al. 1983, 1984). Less rigorous investigations have been made into the basis of variability in tissue susceptibility. Indeed all these studies appear to suffer from a lack of standardisation in terms of approach, especially in regard to the species, strains and tissues examined and the sex and age of the animals used. Until a comprehensive, wide-ranging study is carried out which makes allowance for these variables, the underlying causes of observed differences in susceptibility to PAH-induced tumorigenesis will remain a matter for speculation.

F. Concluding Remarks

In the preceding pages an attempt has been made to provide a very brief outline of hydrocarbon metabolism and then to review some specific aspects of PAH activation. Considerable progress has recently been made in the identification of the metabolic pathways by which individual PAH are activated and in studies on the stereospecificities of the enzymes concerned in these processes. However, it should be quite clear from Sect. E that there is at present no satisfactory scientific explanation for the variation in tissue and species susceptibility observed in carcinogenicity tests with PAH.

In addition several studies strongly suggest that there is no simple relationship between the extent of covalent reaction of PAH metabolites with DNA and tissue susceptibility within rodent species (EASTMAN et al. 1978; BOROUJERDI et al. 1981; ADRIAENSSENS et al. 1983; DUNN 1983). For example, in mice and rabbits treated with different doses of BaP that were administered by different routes, BaP-DNA adducts were found 24 h later to be present in all the target and non-target tissues examined at "surprisingly similar" levels (STOWERS and ANDERSON 1984). Moreover these authors also drew attention to the fact that the extent of DNA adduct formation did not correlate with tissue variations in cytochrome P-450 levels: whilst there was only a twofold difference in BaP-DNA adduct levels between, for example, brain and liver, there is known to be a 400-fold difference in the abilities of microsomes prepared from these tissues to metabolise BaP (ROUET et al. 1981). In other experiments in which the formation and persistence of DNA adducts in mouse skin that had been treated with either the (+)- or (−)-enantiomers of *anti*-BaP-7,8-diol 9,10-oxide were examined, the differences in the amounts of adducts formed were not considered to be sufficient to account for the known differences in the biological activities in this tissue of the two diol-epoxide enantiomers studied (PELLING et al. 1984).

Similar problems were encountered in attempts to establish a simple relationship between DNA modification and tumour susceptibility in rat mammary gland. Here the covalent binding of DMBA to the DNA of mammary tissue and liver was compared using the sensitive Sprague-Dawley and the resistant Long-Evans strains (DANIEL and JOYCE 1984). These authors found (a) that the levels of DMBA-DNA adducts were higher in the non-target organ (the liver) than in the target organ (mammary gland) for both strains of rat and (b) that the DNA adduct levels were higher in both these organs in the resistant strain than they were in the susceptible rats. The resistant strain did provide some evidence, however, for adduct removal that was not apparent in the susceptible Sprague-Dawley rats.

Although Long-Evans rats are resistant to mammary tumour induction by DMBA, they are susceptible to the leukaemogenic effects of this hydrocarbon. When the covalent binding of DMBA and of TMBA, another leukaemogenic PAH, to target (spleen and bone marrow) and non-target (liver) organs were compared, again no direct correlation between either the amount or the nature of the PAH-DNA adducts formed and the susceptibility of the organ was found (FALZON et al. 1987).

The absence of a simple correlation between the extent of DNA modification and tissue susceptibility to carcinogenesis is not confined to PAH but also exists within the *N*-nitroso compounds. In a detailed review LIJINSKY (1988) has drawn attention to "anomalous" results such as (a) the apparent lack of DNA alkylation in vivo following treatment with certain carcinogenic cyclic and acyclic nitrosamines and (b) the alkylation of hepatic DNA following the administration of *N*-nitroso compounds that did not give rise to tumours. This author came to the conclusion that factors other than DNA alkylation must be of great importance in tumour induction.

There may be a variety of reasons why, in these examples of an apparent lack of a simple correlation between DNA adduct formation and tumour induction, the involvement of DNA modifications in the process of tumour initiation is not, in fact, precluded. Differences in adduct persistence in general have already been alluded to above, but it has to be said that a very rapid removal of adducts would be required to explain the apparent absence of alkylated bases reported following the treatment of animals with cyclic *N*-nitroso compounds (LIJINSKY 1988) unless another mechanism, such as the introduction of strand breaks, was involved.

Differences in adduct formation and persistence between initiated and uninitiated cells within the same tissue would be much more difficult to detect. It is conceivable that the reason why only a few tumours develop in, say, carcinogen-treated mouse skin is that only a few cells become modified in the manner required to effect initiation. This is an old objection to studies on carcinogen-DNA interactions and one that has been difficult to refute since high levels of overall DNA modification or extensive reactions with "hot spots" in a very few cells would have been masked by "noise" from the mass of uninitiated cells. The new and extremely sensitive methods of DNA adduct detection that are coming into more widespread use, such as radio-immune assays, synchronised fluorescence and ^{32}P-postlabelling (see Chap. 13) and that are especially applicable to work on polycyclic hydrocarbons, may enable this particular problem to be tackled in the near future.

Acknowledgements. Work carried out in our laboratory was supported in part by grants from the Cancer Research Campaign and the Medical Research Council and, in part, by PHS grant CA21959, awarded by the National Cancer Institute, DHHS. We wish to thank Drs G. M. Holder, D. M. Jerina, E. J. LaVoie, S. Nesnow, O. Perin-Roussel, J. E. Rice and D. Warshawsky for providing us with data prior to publication.

References

Abbott PJ (1983) Strain-specific tumourigenesis in mouse skin induced by the carcinogen, 15,16-dihydro-11-methylcyclopenta[*a*]phenanthren-17-one, and its relation to DNA adduct formation and persistence. Cancer Res 43:2261–2266

Adriaessens PI, White CM, Anderson MW (1983) Dose-response relationships for the binding of benzo[*a*]pyrene metabolites in DNA and protein in lung, liver and forestomach of control and butylated hydroxyanisole-treated mice. Cancer Res 43:3712–3719

Agarwal SK, Sayer JM, Yeh HJC, Pannell LK, Hilton BD, Pigott MA, Dipple A, Yagi H, Jerina DM (1987) Chemical characterization of DNA adducts derived from the configurationally isomeric benzo[*c*]phenanthrene-3,4-diol 1,2-epoxides. J Am Chem Soc 109:2497–2504

Alexandrov K, Rojas M, Bourgeois Y, Chouroulinkov I (1983) The persistence of benzo[*a*]pyrene diol-epoxide deoxyguanosine adduct in mouse skin and its disappearance in rat skin. Carcinogenesis 4:1655–1657

Amin S, Huie K, Balanikas G, Hecht SS, Pataki J, Harvey RG (1987) High stereoselectivity in mouse skin metabolic activation of methylchrysenes to tumorigenic dihydrodiols. Cancer Res 47:3613–3617

Arias IM, Jakoby WB (1976) Glutathione: metabolism and function. Raven, New York

Ariëns EJ (1984) Stereochemistry, a basis for sophisticated nonsense in pharmacokinetics and clinical pharmacology. Eur J Clin Pharmacol 26:663–668

Armstrong RN (1987) Enzyme-catalyzed detoxication reactions: mechanisms and stereochemistry. CRC Crit Rev Biochem 22:39–88

Armstrong RN, Kedzierski B, Levin W, Jerina DM (1981) Enantioselectivity of microsomal epoxide hydrolase toward arene oxide substrates. J Biol Chem 256:4726–4733
Ashurst SW, Cohen GM (1981a) The formation and persistence of benzo[*a*]pyrene metabolite-deoxyribonucleoside adducts in rat skin in vivo. Int J Cancer 28:387–392
Ashurst SW, Cohen GM (1981b) In vivo formation of benzo[*a*]pyrene diol epoxide-deoxyadenosine adducts in the skin of mice susceptible to benzo[*a*]pyrene-induced carcinogenesis. Int J Cancer 27:357–364
Autrup H (1979) Separation of water-soluble metabolites of benzo[*a*]pyrene formed by cultured human colon. Biochem Pharmacol 28:1727–1730
Autrup H, Harris CC (1983) Metabolism of chemical carcinogens by human tissues. In: Harris CC, Autrup H (eds) Human carcinogenesis. Academic Press, New York, pp 169–194
Babson JR, Russo-Rodriguez SE, Rastetter WR, Wogan GN (1986) In vitro DNA-binding of microsomally activated fluoranthene: evidence that the major product is a fluoranthene N^2-deoxyguanosine adduct. Carcinogenesis 7:859–865
Baer-Dubowska W, Alexandrov K (1981) The binding of benzo[*a*]pyrene to mouse and rat skin DNA. Cancer Lett 13:47–52
Baer-Dubowska W, Frayssinet C, Alexandrov K (1981) Formation of covalent deoxyribonucleic acid benzo[*a*]pyrene-4,5-epoxide adduct in mouse and rat skin. Cancer Lett 14:125–129
Baird WM, Dipple A, Grover PL, Sims P, Brookes P (1973) Studies on the formation of hydrocarbon-deoxyribonucleoside products by the binding of derivatives of 7-methylbenz[*a*]anthracene to DNA in aqueous solution and in mouse embryo cells in culture. Cancer Res 33:2386–2392
Baird WM, Harvey RG, Brookes P (1975) Comparison of the cellular DNA-bound products of benzo[*a*]pyrene with the products formed by reaction of benzo[*a*]pyrene 4,5-oxide with DNA. Cancer Res 35:54–57
Barroso M, Dargouge O, Lechner MC (1988) Expression of a constitutive form of cytochrome P450 during rat-liver development and sexual maturation. Eur J Biochem 172:363–369
Bartczak AW, Sangaiah S, Ball LM, Warren SH, Gold A (1987) Synthesis and bacterial mutagenicity of the cyclopenta oxides of the four cyclopenta-fused isomers of benzanthracene. Mutagenesis 2:101–105
Bartle KD, Jones DW (1967) Proton chemical shifts in tri-phenylene. Trans Faraday Soc 63:2868–2873
Bickers DR, Mukhtar H, Yang SK (1983) Cutaneous metabolism of benzo[*a*]pyrene: comparative studies in C57BL/6N and DBA/2N mice and neonatal Sprague-Dawley rats. Chem Biol Interact 43:263–270
Booth J, Sims P (1974) 8,9-Dihydro-8,9-dihydroxybenz[*a*]anthracene 10,11-oxide: a new type of polycyclic aromatic hydrocarbon metabolite. FEBS Lett 47:30–33
Booth J, Boyland E, Sims P (1961) An enzyme from rat-liver catalysing conjugations with glutathione. Biochem J 79:516–524
Borgen A, Darvey H, Castagnoli N, Crocker TT, Rasmussen RE, Wang I-Y (1973) Metabolic conversion of benzo[*a*]pyrene by Syrian hamster liver microsomes and binding of metabolites to deoxyribonucleic acid. J Med Chem 16:502–506
Boroujerdi M, Kung HC, Wilson AGE, Anderson MW (1981) Metabolism and DNA binding of benzo[*a*]pyrene in vivo in the rat. Cancer Res 41:951–957
Bourne MC, Young L (1934) Metabolism of naphthalene in rabbits. Biochem J 28:803–808
Boutwell RK, Urbach F, Carpenter G (1981) Chemical carcinogenesis. Experimental models. In: Laerum OD, Iversen OH (eds) UICC technical reports series, vol 63. Biology of skin cancer (excluding melanomas). UICC, Geneva, pp 109–123
Boyland E (1950) The biological significance of metabolism of polycyclic compounds. Biochem Soc Symp 5:40–54
Boyland E, Chasseaud LF (1969) The role of glutathione and glutathione *S*-transferases in mercapturic acid biosynthesis. Adv Enzymol 32:173–217
Boyland E, Levi AA (1935) Metabolism of polycyclic compounds. I. Production of dihydroxydihydroanthracene from anthracene. Biochem J 29:2697–2683

Boyland E, Levi AA (1936) Metabolism of polycyclic compounds; anthrylmercapturic acid. Biochem J 30:1225–1227

Brookes P, Lawley PD (1964) Evidence for the binding of polynuclear aromatic hydrocarbons to the nucleic acids of mouse skin: relation between carcinogenic power of hydrocarbons and their binding to deoxyribonucleic acid. Nature 202:781–784

Brookes P, Ellis MV, Pataki J, Harvey RG (1986) Mutation in mammalian cells by isomers of 5-methylchrysene. Carcinogenesis 7:463–466

Buhler DR, Unlu F, Thakker DR, Slaga TJ, Conney AH, Wood AW, Chang RL, Levin W, Jerina DM (1983) Effect of a 6-fluoro substituent on the metabolism and biological activity of benzo[*a*]pyrene. Cancer Res 43:1541–1549

Burki K, Liebelt AG, Bresnick E (1973) Induction of aryl hydrocarbon hydroxylase in mouse tissues from a high and low cancer strain and their F_1 hybrids. JNCI 50:369–380

Camus AM, Pyerin WG, Grover PL, Sims P, Malaveille C, Bartsch H (1980) Mutagenicity of benzo[*a*]pyrene 7,8-dihydrodiol and 7,12-dimethylbenz[*a*]anthracene 3,4-dihydrodiol in *S. typhimurium* mediated by microsomes from rat liver and mouse skin. Chem Biol Interact 32:257–265

Chang RL, Levin W, Wood AW, Lehr RE, Kumar S, Yagi H, Jerina DM, Conney AH (1982) Tumorigenicity of bay-region diol-epoxides and other benzo-ring derivatives of dibenzo[*a,h*]pyrene and dibenzo[*a,i*]pyrene on mouse skin and newborn mice. Cancer Res 42:25–29

Chang RL, Levin W, Wood AW, Kumar S, Yagi H, Jerina DM, Lehr RE, Conney AH (1984) Tumorigenicity of dihydrodiols and diol-epoxides of benz[*c*]acridine in newborn mice. Cancer Res 44:5161–5164

Chang RL, Levin W, Katz M, Conney AH, Jerina DM, Agarwal N, Sikka H, Kumar S, Wood AW (1987) Mutagenicity of two pairs of bay-region diol epoxide diastereomers derived from dibenz[*a,h*]acridine. Proc Am Assoc Cancer Res 28:110

Chasseaud LF (1979) The role of glutathione and glutathione *S*-transferases in the metabolism of chemical carcinogens and other electrophilic agents. Adv Cancer Res 29:175–274

Chipman JK, Frost GS, Hirom PC, Millburn P (1981) Biliary excretion, systemic availability and reactivity of metabolites following intraportal infusion of [^{3}H]benzo[*a*]pyrene in the rat. Carcinogenesis 2:741–745

Chiu P-L, Weems HB, Wong TK, Fu PP, Yang SK (1983) Stereoselective metabolism of benzo[*a*]pyrene and 7-methylbenzo[*a*]pyrene by liver microsomes from Sprague-Dawley rats pretreated with polychlorinated biphenyls. Chem Biol Interact 44:155–166

Chiu P-L, Fu PP, Yang SK (1984) Stereoselectivity of rat liver microsomal enzymes in the metabolism of 7-fluorobenz[*a*]anthracene and mutagenicity of metabolites. Cancer Res 44:562–570

Chou MW, Fu PP (1984) Stereoselective metabolism of 8- and 9-fluorobenzo[*a*]pyrene by rat liver microsomes: absolute configurations of *trans*-dihydrodiol metabolites. J Toxicol Environ Health 14:211–223

Chou MW, Yang SK (1979) Combined reversed-phase and normal-phase high-performance liquid chromatography in the purification and identification of 7,12-dimethylbenz[*a*]anthracene metabolites. J Chromatogr 185:635–654

Cohen GM, Haws SM, Moore BP, Bridges JW (1976) Benzo[*a*]pyren-3-yl hydrogen sulphate, a major ethyl acetate-extractable metabolite of benzo[*a*]pyrene in human, hamster and rat lung cultures. Biochem Pharmacol 25:2561–2570

Cohen GM, Moore BP, Bridges JW (1977) Organic solvent soluble sulphate ester conjugates of monohydroxybenzo[*a*]pyrene. Biochem Pharmacol 26:551–553

Combes B, Stakelum GS (1961) A liver enzyme that conjugates sulfobromophthalein sodium with glutathione. J Clin Invest 40:981–988

Conney AH (1982) Induction of microsomal enzymes by foreign chemicals and carcinogenesis by polycyclic aromatic hydrocarbons: GHA Clowes Memorial Lecture. Cancer Res 42:4875–4917

Cook JW, Hewett CL, Hieger I (1933) The isolation of a cancer producing hydrocarbon from coal tar. J Chem Soc 396–405

Coombs MM, Bhatt TS (1987) Cyclopenta[*a*]phenanthrenes. Polycyclic aromatic compounds structurally related to steroids. Cambridge University Press, Cambridge
Cooper CS, Hewer A, Ribeiro O, Grover PL, Sims P (1980 a) The enzyme-catalysed conversion of *anti*-benzo[*a*]pyrene-7,8-diol 9,10-oxide into a glutathione conjugate. Carcinogenesis 1:1075–1080
Cooper CS, Ribeiro O, Hewer A, Walsh C, Pal K, Grover PL, Sims P (1980 b) The involvement of a "bay-region" and a non-bay-region diol-epoxide in the metabolic activation of benz[*a*]anthracene in mouse skin and in hamster embryo cells. Carcinogenesis 1:233–243
Cooper CS, Grover PL, Sims P (1983) The metabolism and activation of benzo[*a*]pyrene. Prog Drug Metab 7:295–396
Daniel F, Joyce NJ (1984) 7,12-Dimethylbenz[*a*]anthracene-DNA adducts in Sprague-Dawley and Long-Evans female rats: the relationship of DNA adducts to mammary cancer. Carcinogenesis 5:1021–1026
Daudel P, Duquesne M, Vigny P, Grover PL, Sims P (1975) Fluorescence spectral evidence that benzo[*a*]pyrene-DNA products in mouse skin arise from diol-epoxides. FEBS Lett 57:250–253
De Baun JR, Smith JYR, Miller EC, Miller JA (1970) Reactivity in vivo of the carcinogen *N*-hydroxy-2-acetylaminofluorene: increase by sulfate ion. Science 167:184–186
Deutsch J, Leutz JC, Yang SK, Gelboin HV, Chiang YL, Vatsis KP, Coon MJ (1978) Regio- and stereoselectivity of various forms of purified cytochrome P-450 in the metabolism of benzo[*a*]pyrene and (−)-*trans*-7,8-dihydroxy-7,8-dihydrobenzo[*a*]pyrene as shown by product formation and binding to DNA. Proc Natl Acad Sci USA 75:3123–3127
Deutsch J, Vatsis KP, Coon MJ, Leutz JC, Gelboin HV (1979) Catalytic activity and stereoselectivity of purified forms of rabbit liver microsomal cytochrome P-450 in the oxygenation of the (−)- and (+)-enantiomers of *trans*-7,8-dihydroxy-7,8-dihydrobenzo[*a*]pyrene. Mol Pharmacol 16:1011–1018
DiGiovanni J, Slaga TJ, Boutwell RK (1980) Comparison of the tumour-initiating activity of 7,12-dimethylbenz[*a*]anthracene and benzo[*a*]pyrene in female SENCAR and CD-1 mice. Carcinogenesis 1:381–389
DiGiovanni J, Diamond L, Pritchett WP, Fisher EP, Harvey RG (1985) Tumor-initiating activity of the 9,10-dihydrodiol and 9,10-dihydrodiol-7,8-epoxide of 3-methylcholanthrene in Sencar mice. Cancer Lett 28:223–228
Dipple A, Moschel RC, Bigger CAH (1984) Polynuclear aromatic carcinogens. In: Searle CE (ed) Chemical carcinogens, vol 1, 2nd edn. American Chemical Society, Washington DC, pp 41–163
Dipple A, Pigott MA, Agarwal SK, Yagi H, Sayer JM, Jerina DM (1987) Optically active benzo[*c*]phenanthrene diol epoxides bind extensively to adenine in DNA. Nature 327:535–536
Dix TA, Buck JR, Marnett LJ (1986) Hydroperoxide-dependent epoxidation of 3,4-dihydrobenzo[*a*]anthracene by ram seminal vesicle microsomes and by hematin. Biochem Biophys Res Commun 140:181–187
Dodgson KS (1977) Conjugation with sulphate. In: Parke DV, Smith RL (eds) Drug metabolism – from microbe to man. Taylor and Francis, London, pp 91–104
Dunn BP (1983) Wide-range linear dose response curve for DNA binding of orally administered benzo[*a*]pyrene in mice. Cancer Res 43:2654–2658
Dutton GJ, Burchell B (1977) Newer aspects of glucuronidation. Prog Drug Metab 2:1–70
Eastman A, Sweetenham J, Bresnick E (1978) Comparison of in vivo and in vitro binding of polycyclic hydrocarbons to DNA. Chem Biol Interact 23:345–353
Eling T, Curtis J, Battista J, Marnett LJ (1986) Oxidation of (+)-7,8-dihydroxy-7,8-dihydrobenzo[*a*]pyrene by mouse keratinocytes: evidence for peroxyl radical- and monooxygenase-dependent metabolism. Carcinogenesis 7:1957–1963
Esterbrook RW, Werringloer J, Capdevila J, Prough RA (1978) The role of cytochrome P-450 and the microsomal electron transport system: the oxidative metabolism of benzo[*a*]pyrene. In: Gelboin HV, Ts'o POP (eds) Polycyclic hydrocarbons and cancer, vol 1. Academic Press, New York, pp 285–319

Falzon M, Vu VT, Roller PP, Thorgeirsson SS (1987) Relationship between 7,12-dimethyl- and 7,8,12-trimethylbenz[*a*]anthracene DNA adduct formation in hematopoietic organs and leukemogenic effects. Cancer Lett 37:41–49

Fernandez N, Roy M, Lesca P (1988) Binding characteristics of *Ah* receptors from rats and mice before and after separation from hepatic cytosols. 7-Hydroxyellipticine as a competitive antagonist of cytochrome P-450 induction. Eur J Biochem 172:585–592

Fu PP, Yang SK (1982) Stereoselective metabolism of 6-bromobenzo[*a*]pyrene by rat liver microsomes: absolute configuration of *trans*-dihydrodiol metabolites. Biochem Biophys Res Commun 109:927–934

Fu PP, Yang SK (1983) Stereoselective metabolism of 7-bromobenz[*a*]anthracene by rat liver microsomes: absolute configurations of *trans*-dihydrodiol metabolites. Carcinogenesis 4:979–984

Fu PP, Harvey RG, Beland FA (1978) Molecular orbital theoretical prediction of the isomeric products formed from reactions of arene oxide and related metabolites of polycyclic aromatic hydrocarbons. Tetrahedron 34:857–866

Fu PP, Cerniglia CE, Chou MW, Yang SK (1983) Differences in the stereoselective metabolism of 7-methylbenz[*a*]anthracene and 7-hydroxymethylbenz[*a*]anthracene by rat liver microsomes and by the filamentous fungus *Cunninghamella elegans*. In: Cooke M, Dennis AJ (eds) Polynuclear aromatic hydrocarbons: formation, metabolism and measurement. Battelle Press, Columbus, pp 531–543

Fu PP, Von Tungeln LS, Chou MW (1985) Stereoselective metabolism of 7-chlorobenz[*a*]anthracene by rat liver microsomes. Absolute configuration and optical purities of *trans*-dihydrodiol metabolites. Mol Pharmacol 28:62–71

Fu PP, Chou MW, Von Tungeln LS, Unruh LE, Yang SK (1986) Metabolism of halogenated polycyclic aromatic hydrocarbons. In: Cooke M, Dennis AJ (eds) Polynuclear aromatic hydrocarbons: chemistry, characterization and carcinogenesis. Battelle Press, Columbus, pp 321–332

Geddie JE, Amin S, Huie K, Hecht SS (1987) Formation and tumorigenicity of benzo[*b*]fluoranthene metabolites in mouse epidermis. Carcinogenesis 8:1579–1584

Gelboin HV (1980) Benzo[*a*]pyrene metabolism, activation and carcinogenesis: role and regulation of mixed-function oxidases and related enzymes. Physiol Rev 50:1107–1166

Gelboin HV, Ts'o POP (1978) Polycyclic hydrocarbons and cancer. Academic Press, New York

Gill JH, Bonin AM, Podobna E, Baker RSU, Duke CC, Rosario CA, Ryan AJ, Holder GM (1986) 7-Methylbenz[*c*]acridine: mutagenicity of some of its metabolites and derivatives, and the identification of *trans*-7-methylbenz[*c*]acridine-3,4-dihydrodiol as a microsomal metabolite. Carcinogenesis 7:23–31

Glatt HR, Mertes I, Wolfel J, Oesch F (1984) Epoxide hydrolases in laboratory animals and in man. In: Greim H, Jung R, Kramer M, Marquardt H, Oesch F (eds) Biochemical basis of chemical carcinogenesis. Raven, New York, pp 107–121

Glatt H, Seidel A, Bochnitschek W, Marquardt H, Marquardt H, Hodgson RM, Grover PL, Oesch F (1986) Mutagenic and cell-transforming activities of triol-epoxides as compared to other chrysene metabolites. Cancer Res 46:4556–4565

Glatt H, Harvey RG, Phillips DH, Hewer A, Grover PL (1989) Influence of the alkyl substituent on mutagenicity and covalent DNA binding of bay region diol-epoxides of 7-methyl and 7-ethylbenz[*a*]anthracene in Salmonella and V79 Chinese hamster cells. Cancer Res 49:1778–1782

Gold A, Eisenstadt E (1980) Metabolic activation of cyclopenta[*cd*]pyrene to 3,4-epoxycyclopenta[*cd*]pyrene by rat liver microsomes. Cancer Res 40:3940–3944

Gold A, Nesnow S, Moore M, Garland H, Curtis G, Howard B, Graham D, Eisenstadt E (1980) Mutagenesis and morphological transformation of mammalian cells by a non-bay-region polycyclic cyclopenta[*cd*]pyrene and its 3,4-oxide. Cancer Res 40:4482–4484

Goshman LM, Heidelberger C (1967) Binding of tritium-labeled polycyclic hydrocarbons to DNA of mouse skin. Cancer Res 27:1678–1688

Gower JD, Wills ED (1987) The oxidation of benzo[*a*]pyrene-7,8-dihydrodiol mediated by lipid peroxidation in the rat intestine and the effect of dietary lipids. Chem Biol Interact 63:63–74

Grimmer G (1983) Environmental carcinogens: polycyclic aromatic hydrocarbons. CRC Press, Boca Raton

Grover PL (1974) K-Region epoxides of polycyclic hydrocarbons: formation and further metabolism by rat-lung preparations. Biochem Pharmacol 23:333–343

Grover PL, Sims P (1968) Enzyme-catalysed reactions of polycyclic hydrocarbons with deoxyribonucleic acid and protein in vitro. Biochem J 110:159–160

Guengerich FP (1987) Cytochrome P-450 enzymes and drug metabolism. Prog Drug Metab 10:1–54

Guenther TM, Oesch F (1981) Microsomal epoxide hydrolase and its role in polycyclic aromatic hydrocarbon transformation. In: Gelboin HV, Ts'o POP (eds) Polycyclic hydrocarbons and cancer, vol 3. Academic Press, New York, pp 183–212

Guerin MR (1978) Energy sources of polycyclic aromatic hydrocarbons. In: Gelboin HV, Ts'o POP (eds) Polycyclic hydrocarbons and cancer, vol 1. Academic Press, New York, pp 3–42

Hadfield ST, Abbott PJ, Coombs MM, Drake AF (1984) The effect of methyl substituents on the in vitro metabolism of cyclopenta[*a*]phenanthren-17-ones: implications for biological activity. Carcinogenesis 5:1395–1399

Hall M, Grover PL (1986) Effects of inducers on the regio- and stereoselective metabolism of benzo[*a*]pyrene by mouse tissue microsomes. Chem Biol Interact 59:265–280

Hall M, Grover PL (1987) Differential stereoselectivity in the metabolism of benzo[*a*]pyrene and anthracene by rabbit epidermal and hepatic microsomes. Cancer Lett 38:57–64

Hall M, Grover PL (1988) Stereoselective aspects of the metabolic activation of benzo[*a*]pyrene by human skin in vitro. Chem Biol Interact 64:281–296

Hall M, Parker DK, Christou M, Jefcoate CR, Grover PL (1987) The stereoselective formation of benzo[*a*]pyrene dihydrodiols by purified rat liver cytochromes P-450. In: Benford D, Gibson GG, Bridges JW (eds) Drug metabolism – from molecules to man. Taylor and Francis, London, pp 411–414

Hall M, Parker DK, Hewer AJ, Phillips DH, Grover PL (1988) Further metabolism of diol-epoxides of chrysene and dibenz[*a,c*]anthracene to DNA binding species as evidenced by ^{32}P-postlabelling analysis. Carcinogenesis 9:865–868

Harris CC (1987) Human tissues and cells in carcinogenesis research. Cancer Res 47:1–10

Harris CC, Trump BF, Grafstrom R, Autrup H (1982) Differences in metabolism of chemical carcinogens in cultured human epithelial tissues and cells. J Cell Biochem 18:285–294

Hecht SS, Melikian AA, Amin S (1986) Methylchrysenes as probes for the mechanism of metabolic activation of carcinogenic methylated polynuclear aromatic hydrocarbons. Acc Chem Res 19:174–180

Hecht SS, Amin S, Huie K, Melikian AA, Harvey RG (1987) Enhancing effect of a bay region methyl group on tumorigenicity in newborn mice and mouse skin of enantiomeric bay region diol epoxides formed stereoselectively from methylchrysenes in mouse epidermis. Cancer Res 47:5310–5315

Heidelberger C, Davenport GR (1961) Local functional components of carcinogenesis. Acta Unio Int Cancr 17:55–63

Hennings H, Devor D, Wenk ML, Slaga TJ, Former B, Colburn NH, Bowden GT, Elgjo K, Yuspa SH (1981) Comparison of two-stage epidermal carcinogenesis initiated by 7,12-dimethylbenz[*a*]anthracene or *N*-methyl-*N'*-nitro-*N*-nitrosoguanidine in newborn and adult SENCAR and BALB/c mice. Cancer Res 41:773–779

Hodgson RM, Weston A, Grover PL (1983) Metabolic activation of chrysene in mouse skin: evidence for the involvement of a triol-epoxide. Carcinogenesis 4:1639–1643

Hodgson RM, Seidel A, Bochnitschek W, Glatt HR, Oesch F, Grover PL (1985a) The formation of 9-hydroxy-1,2-diol as an intermediate in the metabolic activation of chrysene. Carcinogenesis 6:135–139

Hodgson RM, Weston A, Seidel A, Bochnitschek W, Glatt HR, Oesch F, Grover PL (1985b) Metabolism of chrysene to triols and a triol-epoxide in mouse skin and rat liver preparations. In: Cooke M, Dennis AJ (eds) Polynuclear aromatic hydrocarbons: chemistry, characterization and carcinogenesis. Battelle Press, Columbus, pp 387–399

Hodgson RM, Seidel A, Bochnitschek W, Glatt HR, Oesch F, Grover PL (1986) Metabolism of the bay-region diol-epoxide of chrysene to a triol-epoxide and the enzyme-catalysed conjugation of these epoxides with glutathione. Carcinogenesis 7:2095–2098

Holder GM, Yagi H, Jerina DM, Levin W, Lu AYH, Conney AH (1975) Metabolism of benzo[*a*]pyrene. Effect of substrate concentration and 3-methylcholanthrene pretreatment on hepatic metabolism by microsomes from rats and mice. Arch Biochem Biophys 170:557–566

Huggins C, Grand LC, Brillantes FP (1961) Mammary cancer induced by a single feeding of polynuclear hydrocarbons and its suppression. Nature 189:204–207

Hulbert PB (1975) Carbonium ion as ultimate carcinogen of polycyclic aromatic hydrocarbons. Nature 256:146–148

IARC Monographs on the evaluation of the carcinogenic risk of chemicals to man (1973) Volume 3. Certain polycyclic aromatic hydrocarbons and heterocyclic compounds. IARC, Lyon

IARC Monographs on the evaluation of the carcinogenic risk of chemicals to man (1983) Volume 32. Polynuclear aromatic compounds, part 1, chemical, environmental and experimental data. IARC, Lyon

Ittah Y, Thakker DR, Levin W, Croisy-Delcey M, Ryan DE, Thomas PE, Conney AH, Jerina DM (1983) Metabolism of benzo[*c*]phenanthrene by rat liver microsomes and by a purified monooxygenase system reconstituted with different isozymes of cytochrome P-450. Chem Biol Interact 45:15–28

Jacob J, Schmoldt A, Grimmer G (1982) Formation of carcinogenic and inactive chrysene metabolites by rat liver microsomes of various monooxygenase activities. Arch Toxicol 51:255–265

Jakoby WB, Ketterer B, Mannervik B (1984) Glutathione transferases: nomenclature. Biochem Pharmacol 33:2539–2540

Janss DH, Ben TL (1978) Age-related modification of 7,12-dimethylbenz[*a*]anthracene binding to rat mammary gland DNA. JNCI 60:173–177

Jerina DM, Daly JW (1974) Arene oxides: a new aspect of drug metabolism. Science 185:573–582

Jerina DM, Daly JW (1977) Oxidation at carbon. In: Parke DV, Smith RL (eds) Drug metabolism – from microbc to man. Taylor & Francis, London, pp 13–32

Jerina DM, Michaud DP, Feldmann RJ, Armstrong RN, Vyas KP, Thakker DR, Yagi H, Thomas PE, Ryan DE, Levin W (1982) Stereochemical modeling of the catalytic site of cytochrome P-450c. In: Sato R, Kato R (eds) Microsomes, drug oxidations, and drug toxicity. Wiley, New York, pp 195–201

Jernström B, Vadi H, Orrhenius S (1978) Formation of DNA binding products from isolated benzo[*a*]pyrene metabolites in rat liver nuclei. Chem Biol Interact 20:311–321

Jernström B, Dock L, Martinez M (1984) Metabolic activation of benzo[*a*]pyrene-7,8-dihydrodiol and benzo[*a*]pyrene-7,8-diol-9,10-epoxide to protein-binding products and the inhibitory effect of glutathione and cysteine. Carcinogenesis 5:199–204

Jernström B, Martinez M, Meyer DJ, Ketterer B (1985) Glutathione conjugation of the carcinogenic and mutagenic electrophile $(\pm)$-7β,8α-dihydroxy-9α,10α-oxy-7,8,9,10-tetrahydrobenzo[*a*]pyrene catalysed by purified rat liver glutathione transferases. Carcinogenesis 6:85–89

Kadlubar FF, Hammons GJ (1987) The role of cytochrome P-450 in the metabolism of chemical carcinogens. In: Guengerich FP (ed) Mammalian cytochromes P-450, vol II. CRC Press, Boca Raton, pp 81–130

Kamataki T, Maeda K, Shimada M, Kato R (1986) Effects of phenobarbital, 3-methylcholanthrene and polychlorinated biphenyls on sex-specific forms of cytochrome P-450 in liver microsomes of rats. J Biochem 99:841–845

Kinoshita N, Gelboin HV (1972) The role of aryl hydrocarbon hydroxylase in 7,12-dimethylbenz[*a*]anthracene skin tumorigenesis: on the mechanism of 7,8-benzoflavone inhibition of tumorigenesis. Cancer Res 32:1329–1339

Kinoshita N, Gelboin HV (1978) β-Glucuronidase catalysed hydrolysis of benzo[*a*]pyrene-3-glucuronide and binding to DNA. Science 199:307–309

Kinoshita T, Konieczny M, Santella R, Jeffrey AM (1982) Metabolism and covalent binding to DNA of 7-methylbenzo[*a*]pyrene. Cancer Res 42:4032–4038

Kouri RE, Salerno RA, Whitmire CE (1973) Relationships between aryl hydrocarbon hydroxylase inducibility and sensitivity to chemically induced subcutaneous sarcomas in various strains of mice. JNCI 50:363–368

Kumar S, Dubey SK, Sikka H, Geddie NG, Dzech A, LaVoie EJ (1987) Mutagenic activity of dihydrodiols, epoxides, and dihydrodiol epoxides of benzo[*f*]quinoline and benzo[*h*]quinoline. Proc Am Assoc Cancer Res 28:111

Langenbach R, Nesnow S (1983) Cell-mediated mutagenesis, an approach to studying organ specificity of chemical carcinogens. In: Langenbach R, Nesnow S, Rice JM (eds) Organ and species specificity in chemical carcinogenesis. Plenum, New York, pp 377–389

LaVoie EJ, Hecht SS, Bedenko V, Hoffman D (1982) Identification of the mutagenic metabolites of fluoranthene, 2-methylfluoranthene, and 3-methylfluoranthene. Carcinogenesis 3:841–846

Lee H, Harvey RG (1986) Synthesis of the active diol-epoxide metabolites of the potent carcinogenic hydrocarbon 7,12-dimethylbenz[*a*]anthracene. J Org Chem 51:3502–3507

Legraverend C, Mansour B, Nebert DW, Holland JM (1980) Genetic differences in benzo[*a*]pyrene-initiated tumorigenesis in mouse skin. Pharmacology 20:242–255

Levin W, Buening MK, Wood AW, Chang RL, Kedzierski B, Thakker DR, Boyd DR, Gadaginamath GS, Armstrong RN, Yagi H, Karle JM, Slaga TJ, Jerina DM, Conney AH (1980) An enantiomeric interaction in the metabolism and tumorigenicity of (+)- and (−)-benzo[*a*]pyrene 7,8-oxide. J Biol Chem 255:9067–9074

Levin W, Wood AW, Chang RL, Kumar S, Yagi H, Jerina DM, Lehr RE, Conney AH (1983) Tumor-initiating activity of benz[*c*]acridine and twelve of its derivatives on mouse skin. Cancer Res 43:4625–4628

Levin W, Chang RL, Wood AW, Thakker DR, Yagi H, Jerina DM, Conney AH (1986) Tumorigenicity of optical isomers of the diastereomeric bay-region 3,4-diol 1,2-epoxides of benzo[*c*]phenanthrene in murine tumor models. Cancer Res 46:2257–2261

Lijinsky W (1988) Nucleic acid alkylation by *N*-nitroso compounds related to organ-specific carcinogenesis. In: Politzer P, Martin FJ (eds) Bioactive molecules, vol 5. Chemical carcinogens. Activation mechanisms, structural and electronic factors, and reactivity. Elsevier, Amsterdam, pp 242–263

Mannervik B (1985) The isozymes of glutathione transferase. Adv Enzymol 57:357–417

Mannervik B, Jensson H (1982) Binary combination of four protein subunits with different catalytic specificities explain the relationship between six basic glutathione *S*-transferases in rat liver cytosol. J Biol Chem 257:9909–9912

Marnett LJ (1987) Peroxyl free radicals: potential mediators of tumor initiation and promotion. Carcinogenesis 8:1365–1373

Marnett LJ, Johnson JT, Bienkowski MJ (1979) Arachidonic acid-dependent metabolism of 7,8-dihydroxy-7,8-dihydrobenzo[*a*]pyrene by ram seminal vesicles. FEBS Lett 106:13–16

McClennan-Green P, Waxman DJ, Caveness M, Goldstein JA (1987) Phenotypic differences in expression of cytochrome P-450 but not its mRNA in outbred male Sprague-Dawley rats. Arch Biochem Biophys 253:13–25

McKay S (1987) Metabolism and activation of 7-ethyl and 7-methylbenz[*a*]anthracene. PhD Thesis, University of London

McKay S, Phillips DH, Hewer AJ, Grover PL (1988) Metabolic activation of 7-ethyl- and 7-methylbenz[*a*]anthracene in mouse skin. Carcinogenesis 9:141–145

Meijer J, Lundqvist G, DePierre JW (1987) Comparison of the sex and subcellular distributions, catalytic and immunochemical reactivities of hepatic epoxide hydrolases in seven mammalian species. Eur J Biochem 167:269–279

Miller EC (1951) Studies on the formation of protein-bound derivatives of 3,4-benzpyrene in the epidermal fraction of mouse skin. Cancer Res 11:100–108

Moore BP, Cohen GM (1978) Metabolism of benzo[*a*]pyrene and its major metabolites to ethyl acetate-soluble and water-soluble metabolites by cultured rodent trachea. Cancer Res 38:3066–3075

Moore CJ, Pruess-Schwartz D, Mauthe RJ, Gould MN, Baird WM (1987) Interspecies differences in the major DNA adducts formed from benzo[*a*]pyrene but not 7,12-dimethylbenz[*a*]anthracene in rat and human mammary cell cultures. Cancer Res 47:4402–4406

Morse MA, Baird WM, Carlson GP (1987a) Distribution, covalent binding, and DNA adduct formation of 7,12-dimethylbenz[*a*]anthracene in SENCAR and BALB/c mice following topical and oral administration. Cancer Res 47:4571–4575

Morse MA, Baird WM, Carlson GP (1987b) Effects of sex and age on DMBA:DNA binding in epidermis of SENCAR mice following topical administration of dimethylbenz[*a*]anthracene. Cancer Lett 37:25–31

Mukhtar H, Bickers DR (1983) Age-related changes in benzo[*a*]pyrene metabolism and epoxide-metabolizing enzyme activities in rat skin. Drug Metab Dispos 11:562–567

Mukhtar H, Asokan P, Das M, Santella RM, Bickers DR (1986) Benzo[*a*]pyrene diol epoxide-1-DNA adduct formation in the epidermis and lung of SENCAR mice following topical application of crude coal tar. Cancer Lett 33:287–294

Mushtaq M, Yang SK (1987) Stereoselective metabolism of benzo[*c*]phenanthrene to the procarcinogenic *trans*-3,4-dihydrodiol. Carcinogenesis 8:705–709

Nebert DW, Gelboin HV (1969) The in vivo and in vitro induction of aryl hydrocarbon hydroxylase in mammalian cells of different species, tissues, strains, and developmental and hormonal states. Arch Biochem Biophys 134:76–89

Nebert DW, Gonzalez FJ (1987) P-450 genes: structure, evolution and regulation. Ann Rev Biochem 56:945–993

Nebert DW, Jensen NM (1979) The *Ah* locus: genetic regulation of the metabolism of carcinogens, drugs, and other environmental chemicals by cytochrome P-450-mediated monooxygenases. CRC Crit Rev Biochem 6:401–437

Nemoto N (1978) Glucuronidation in the metabolism of benzo[*a*]pyrene. In: Aitio A (ed) Conjugation reactions in drug biotransformation. Elsevier, Amsterdam, pp 17–27

Nemoto N (1981) Glutathione, glucuronide and sulphate transferase in polycyclic aromatic hydrocarbon metabolism. In: Gelboin HV, Ts'o POP (eds) Polycyclic hydrocarbons and cancer, vol 3. Academic Press, New York, pp 213–258

Nemoto N, Takayama S, Gelboin HV (1978) Sulphate conjugation of benzo[*a*]pyrene metabolites and derivatives. Chem Biol Interact 23:19–30

Nesnow S, Leavitt S, Easterling R, Watts R, Toncy SH, Claxton L, Sangaiah R, Toney GE, Wiley J, Fraher P, Gold A (1984) Mutagenicity of cyclopenta-fused isomers of benz[*a*]anthracene in bacterial and rodent cells and identification of the major rat microsomal metabolites. Cancer Res 44:4993–5003

Nesnow S, Easterling RE, Ellis S, Watts R, Ross J (1988) Metabolism of benz[*j*]aceanthrylene (cholanthrylene) and benz[*l*]aceanthrylene by induced rat liver S9. Cancer Lett 39:19–27

Nesnow S, Ross J, Mohapatra N, Briant BJ, Samgaiah R, Gold A, Gupta R (1989) Genotoxicity and identification of the major DNA-adducts of aceanthrylene. Proc 11th symp polynuclear aromatic hydrocarbons (in press)

Newman MS, Blum S (1964) A new cyclization reaction leading to epoxides of aromatic hydrocarbons. J Am Chem Soc 86:5598–5600

Nordqvist M, Thakker DR, Levin W, Yagi H, Ryan DE, Thomas PE, Conney AH, Jerina DM (1979) The highly tumorigenic 3,4-dihydrodiol is a principal metabolite formed from dibenz[*a,h*]anthracene by liver enzymes. Mol Pharmacol 16:643–655

Nordqvist M, Thakker DR, Vyas KP, Yagi H, Levin W, Ryan DE, Thomas PE, Conney AH, Jerina DM (1981) Metabolism of chrysene and phenanthrene to bay-region diol epoxides by rat liver enzymes. Mol Pharmacol 19:168–178

Oesch F (1973) Mammalian epoxide hydrases; inducible enzymes catalysing the inactivation of carcinogenic and cytotoxic metabolites derived from aromatic and olefinic compounds. Xenobiotica 3:305–340

Okey AB, Bondy GP, Mason ME, Kahl GF, Eisen HJ, Guenthner TM, Nebert DW (1979) Regulatory gene product of the *Ah* locus. Characterization of the cytosolic inducer-receptor complex and evidence for its nuclear translocation. J Biol Chem 254:11636–11648

Osborne MR, Crosby NT (1987) Benzopyrenes. Cambridge University Press, Cambridge
Osborne MR, Brookes P, Lee H, Harvey RG (1986) The reaction of a 3-methylcholanthrene diol epoxide with DNA in relation to the binding of 3-methylcholanthrene in the DNA of mammalian cells. Carcinogenesis 7:1345–1350
Pelling JC, Slaga TJ, DiGiovanni J (1984) Formation and persistence of DNA, RNA and protein adducts in mouse skin exposed to pure optical enantiomers of 7β,8α-dihydroxy-9α,10α-epoxy-7,8,9,10-tetrahydrobenzo[*a*]pyrene in vivo. Cancer Res 44:1081–1086
Perin-Roussel O, Saguem S, Ekert B, Zajdela F (1983) Binding to DNA of bay-region and pseudo-bay-region diol-epoxides of dibenzo[*a,e*]fluoranthene and comparison with adducts obtained with dibenzo[*a,e*]fluoranthene or its dihydrodiol in the presence of microsomes. Carcinogenesis 4:27–32
Perin-Roussel O, Croisy A, Ekert B, Zajdela F (1984) The metabolic activation of dibenzo[*a,e*]fluoranthene in vitro. Evidence that its bay-region and pseudo-bay-region diol-epoxides react preferentially with guanosine. Cancer Lett 22:289–298
Phillips DH, Grover PL, Sims P (1978) The covalent binding of polycyclic hydrocarbons to DNA in the skin of mice of different strains. Int J Cancer 22:487–494
Phillips DH, Hewer A, Grover PL (1985) Aberrant activation of benzo[*a*]pyrene in cultured rat mammary cells in vitro and following direct application to rat mammary glands in vivo. Cancer Res 45:4167–4174
Phillips DH, Hewer A, Grover PL (1987) Formation of DNA adducts in mouse skin treated with metabolites of chrysene. Cancer Lett 35:207–214
Poland A, Glover E (1988) Ca^{2+}-dependent proteolysis of the Ah receptor. Arch Biochem Biophys 261:103–111
Prasad GKB, Mirsadeghi S, Boehlert C, Byrd RA, Thakker DR (1988) Oxidative metabolism of the carcinogen 6-fluorobenzo[*c*]phenanthrene. Effect of a K-region fluoro substituent on the regioselectivity of cytochromes P-450 in liver microsomes from control and induced rats. J Biol Chem 263:3676–3683
Preuss-Schwartz D, Baird WM, Yagi H, Jerina DM, Pigott MA, Dipple A (1987) Stereochemical specificity in the metabolic activation of benzo[*c*]phenanthrene to metabolites that covalently bind to DNA in rodent embryo cell cultures. Cancer Res 47:4032–4037
Randerath E, Mittal D, Randerath K (1988) Tissue distribution of covalent DNA damage in mice treated dermally with cigarette "tar": preference for lung and heart DNA. Carcinogenesis 9:75–80
Rastetter WH, Nachbar RB, Russo-Rodriguez S, Wattley RV, Thilly WG, Andon BM, Jorgensen WL, Ibrahim M (1982) Fluoranthene: synthesis and mutagenicity of four diol epoxides. J Org Chem 47:4873–4878
Reardon DB, Prakash AS, Hilton BD, Roman JM, Pataki J, Harvey RG, Dipple A (1987) Characterisation of 5-methylchrysene-1,2-dihydrodiol-3,4-epoxide-DNA adducts. Carcinogenesis 8:1317–1322
Reiners J, Davidson K, Nelson K, Mamrack M, Slaga T (1983) Skin tumor promotion: a comparative study of several stocks and strains of mice. In: Langenbach R, Nesnow S, Rice JM (eds) Organ and species specificity in chemical carcinogenesis. Plenum, New York, pp 173–188
Reiners JJ, Nesnow S, Slaga TJ (1984) Murine susceptibility to two-stage carcinogenesis is influenced by the agent used for promotion. Carcinogenesis 5:301–307
Renwick AG, Drasar BS (1976) Environmental carcinogens and large bowel cancer. Nature 263:234–235
Ribeiro O, Kirkby CA, Hirom PC, Millburn P (1985) Secondary metabolites of benzo[*a*]-pyrene: 3-hydroxy-*trans*-7,8-dihydro-7,8-dihydroxybenzo[*a*]pyrene, a biliary metabolite of 3-hydroxybenzo[*a*]pyrene in the rat. Carcinogenesis 6:1507–1511
Rice JE, Coleman DT, Hosted TJ, LaVoie EJ, McCaustland DJ, Wiley JC (1985) Identification of mutagenic metabolites of ideno[1,2,3-*cd*]pyrene formed in vitro with rat liver enzymes. Cancer Res 45:5421–5425
Rice JE, Hosted TJ, DeFloria MC, LaVoie EJ, Fischer DL, Wiley JC (1986) Tumor-initiating activity of major in vivo metabolites of indeno[1,2,3-*cd*]pyrene on mouse skin. Carcinogenesis 7:1761–1764

Rice JE, Geddie NG, LaVoie EJ (1987a) Identification of metabolites of benzo[*j*]fluoranthene formed in vitro in rat liver homogenate. Chem Biol Interact 63:227–237

Rice JE, Weyand EH, Geddie NG, DeFloria MC, LaVoie EJ (1987b) Identification of tumorigenic metabolites of benzo[*j*]fluoranthene formed in vivo in mouse skin. Cancer Res 47:6166–6170

Robertson IGC, Jensson H, Mannervik B, Jernström B (1986) Glutathione transferases in rat lung: the presence of transferase 7-7, highly efficient in the conjugation of glutathione with the carcinogenic (+)-7β,8α-dihydroxy-9α,10α-oxy-7,8,9,10-tetrahydrobenzo[*a*]pyrene. Carcinogenesis 7:295–299

Rojas M, Alexandrov K (1986a) In vivo formation and persistence of DNA and protein adducts in mouse and rat skin exposed to ($\pm$)-benzo[*a*]pyrene-4,5-oxide. Carcinogenesis 7:235–240

Rojas M, Alexandrov K (1986b) In vivo formation and persistence of DNA adducts in mouse and rat skin exposed to ($\pm$)-*trans*-7,8-dihydroxy-7,8-dihydrobenzo[*a*]pyrene and ($\pm$)-7β,8α-dihydroxy-9α,10α-epoxy-7,8,9,10-tetrahydrobenzo[*a*]pyrene. Carcinogenesis 7:1553–1560

Rojas M, Baer-Dubowska W, Alexandrov K (1986) Comparison of benzo[*a*]pyrene-DNA adduct levels in mouse and rat epidermis and dermis. Cancer Lett 30:35–39

Rouet P, Alexandrov K, Markovits P, Frayssinet C, Dansette PM (1981) Metabolism of benzo[*a*]pyrene by brain microsomes of fetal and adult rats and mice. Induction by 5,6-benzoflavone, comparison with liver and lung microsomal activities. Carcinogenesis 2:919–926

Saguem S, Mispelter J, Perin-Roussel O, Lhoste JM, Zajdela F (1983a) Multi-step metabolism of the carcinogen dibenzo[*a,e*]fluoranthene. I. Identification of the metabolites from rat microsomes. Carcinogenesis 4:827–835

Saguem S, Perin-Roussel O, Mispelter J, Lhoste JM, Zajdela F (1983b) Multi-step metabolism of the carcinogen dibenzo[*a,e*]fluoranthene. II. Metabolism pathways. Carcinogenesis 4:837–842

Sawicki JT, Moschel RC, Dipple A (1983) Involvement of both *syn* and *anti*-dihydrodiol-epoxides in the binding of 7,12-dimethylbenz[*a*]anthracene to DNA in mouse embryo cell cultures. Cancer Res 43:3213–3218

Schaeffer V, Mpanza Z, Thakker D (1987) Novel stereoselectivity of rat liver cytochromes P-450 in the metabolism of (1S,2S)- and (1R,2R)-dihydrodiols of naphthalene to diol epoxides. Fed Proc 46:865

Schoeny R, Warshawsky D (1987) Mutagenicity of 7H-dibenzo[*c,g*]carbazole and metabolites in *Salmonella typhimurium*. Mutation Res 188:275–286

Schoket B, Hewer A, Grover PL, Phillips DH (1988) Covalent binding of components of coal-tar, creosote and bitumen to the DNA of the skin and lungs of mice following topical application. Carcinogenesis 9:1253–1258

Schoket B, Horkay I, Kosa A, Paldeak L, Hewer A, Grover PL, Phillips DH (1989) Formation of DNA adducts in the skin of psoriasis patients, in human skin in organ culture and in mouse skin and lung following topical application of coal-tar and juniper tar. J Invest Dermatol (in press)

Schurdak ME, Stong DB, Warshawsky D, Randerath K (1987) ^{32}P-postlabeling analysis of DNA adduction in mice by synthetic metabolites of the environmental carcinogen, 7H-dibenzo[*c,g*]carbazole: chromatographic evidence for 3-hydroxy-7H-dibenzo-[*c,g*]carbazole being a proximate genotoxicant in liver but not skin. Carcinogenesis 8:591–597

Sebti SM, Preuss-Schwartz D, Baird WM (1985) Species- and length of exposure-dependent differences in the benzo[*a*]pyrene: DNA adducts formed in embryo cell cultures from mice, rats, and hamsters. Cancer Res 45:1594–1600

Seidel A, Bochnitschek W, Glatt HR, Oesch F, Hodgson RM, Grover PL (1989) Activated metabolites of chrysene: synthesis of 9-hydroxychrysene-1,2-diol and its bay-region *syn* and *anti* triol-epoxides. Proc 11th int symp polynuclear aromatic hydrocarbons (in press)

Seifried HE, Birkett DJ, Levin W, Lu AYH, Conney AH, Jerina DM (1977) Metabolism of benzo[*a*]pyrene. Effect of 3-methylcholanthrene pretreatment on metabolism by microsomes from lungs of genetically "responsive" and "nonresponsive" mice. Arch Biochem Biophys 178:256–263

Shubik P (1950) Studies on the promoting phase in the stages of carcinogenesis in mice, rats, rabbits, and guinea pigs. Cancer Res 10:13–17

Sims P (1962) Metabolism of polycyclic compounds. 19. The metabolism of phenanthrene in rabbits and rats: phenols and sulphuric esters. Biochem J 84:558–563

Sims P, Grover PL (1974) Epoxides in polycyclic aromatic hydrocarbon metabolism and carcinogenesis. Adv Cancer Res 20:165–274

Sims P, Grover PL (1981) Involvement of dihydrodiols and diol-epoxides in the metabolic activation of polycyclic hydrocarbons other than benzo[*a*]pyrene. In: Gelboin HV, Ts'o POP (eds) Polycyclic hydrocarbons and cancer, vol 3. Academic Press, New York, pp 117–181

Sims P, Grover PL, Swaisland A, Pal K, Hewer A (1974) Metabolic activation of benzo[*a*]pyrene proceeds by a diol-epoxide. Nature 252:326–328

Singer B, Grunberger D (1983) Molecular biology of mutagens and carcinogens. Plenum, New York

Sinha DK, Dao TL (1980) Induction of mammary tumors in aging rats by 7,12-dimethylbenz[*a*]anthracene: role of DNA synthesis during carcinogenesis. JNCI 64:519–521

Slaga TJ, Gleason GL, Mills G, Ewald L, Fu PP, Lee HM, Harvey RG (1980) Comparison of the skin-tumour-initiating activities of dihydrodiols and diol-epoxides of various polycyclic aromatic hydrocarbons. Cancer Res 40:1981–1984

Steward AR, Kumar C, Sikka HC (1987) Metabolism of dibenz[*a,h*]acridine by rat liver microsomes. Carcinogenesis 8:1043–1050

Stowers SJ, Anderson MW (1984) Ubiquitous binding of benzo[*a*]pyrene metabolites to DNA and protein in tissues of the mouse and rabbit. Chem Biol Interact 51:151–166

Teague RS (1954) The conjugates of β-glucuronic acid of animal origin. Adv Carbohydrate Chem 9:186–246

Thakker DR, Yagi H, Lehr RE, Levin W, Buening M, Lu AYH, Chang RL, Wood AW, Conney AH, Jerina DM (1978) Metabolism of *trans*-9,10-dihydroxy-9,10-dihydrobenzo[*a*]pyrene occurs primarily by arylhydroxylation rather than formation of a diol epoxide. Mol Pharmacol 14:502–513

Thakker DR, Levin W, Yagi H, Turujman S, Kapadia D, Conney AH, Jerina DM (1979) Absolute stereochemistry of the *trans*-dihydrodiols formed from benzo[*a*]anthracene by rat liver microsomes. Chem Biol Interact 27:145–161

Thakker DR, Levin W, Buening M, Yagi H, Lehr RE, Wood AW, Conney AH, Jerina DM (1981) Species-specific enhancement by 7,8-benzoflavone of hepatic microsomal metabolism of benzo[*e*]pyrene 9,10-dihydrodiol to bay-region diol epoxides. Cancer Res 41:1389–1396

Thakker DR, Levin W, Yagi H, Tada M, Ryan DE, Thomas PE, Conney AH, Jerina DM (1982) Stereoselective metabolism of the (+)- and (−)-enantiomers of *trans*-3,4-dihydroxy-3,4-dihydrobenz[*a*]anthracene by rat liver microsomes and by a purified and reconstituted cytochrome P-450 system. J Biol Chem 257:5103–5110

Thakker DR, Yagi H, Sayer JM, Kapur U, Levin W, Chang RL, Wood AW, Conney AH, Jerina DM (1984) Effects of a 6-fluoro substituent on the metabolism of benzo[*a*]pyrene 7,8-dihydrodiol to bay-region diol epoxides by rat liver enzymes. J Biol Chem 259:11249–11256

Thakker DR, Yagi H, Levin W, Wood AW, Conney AH, Jerina DM (1985a) Polycyclic aromatic hydrocarbons: metabolic activation to ultimate carcinogens. In: Anders MW (ed) Bioactivation of foreign compounds. Academic Press, New York, pp 177–242

Thakker DR, Shirai N, Levin W, Ryan DE, Thomas PE, Lehr RE, Conney AH, Jerina DM (1985b) Metabolism of dibenz[*c,h*]acridine by rat liver microsomes and by cytochrome P-450c with and without epoxide hydrolase. Proc Am Assoc Cancer Res 26:114

Thakker DR, Boehlert C, Levin W, Yagi H, Jerina DM (1985c) Metabolism of triphenylene by liver microsomes from control, phenobarbital (PB)-treated and 3-methylcholanthrene (MC)-treated rats. Proc Am Assoc Cancer Res 26:113

Thakker DR, Levin W, Yagi H, Yeh HJC, Ryan DE, Thomas PE, Conney AH, Jerina DM (1986a) Stereoselective metabolism of the (+)-(*S,S*)- and (−)-(*R,R*)-enantiomers of *trans*-3,4-dihydroxy-3,4-dihydrobenzo[*c*]phenanthrene by rat and mouse liver microsomes and by a purified and reconstituted cytochrome P-450 system. J Biol Chem 261:5404–5413

Thakker DR, Boehlert C, Levin W, Conney AH, Ryan DE, Thomas PE, Yagi H, Jerina DM (1986b) Novel stereoselectivity in the metabolism of the enantiomeric trans 1,2-dihydrodiols of triphenylene to diol epoxides by hepatic cytochromes P-450. Proc Am Assoc Cancer Res 27:116

Thakker DR, Boehlert C, Mirsadeghi S, Levin W, Ryan DE, Thomas PE, Yagi H, Pannell LK, Sayer JM, Jerina DM (1988) Differential stereoselectivity on metabolism of triphenylene by cytochromes P-450 in liver microsomes from 3-methylcholanthrene- and phenobarbital-treated rats. J Biol Chem 263:98–105

Trager WF, Jones JP (1987) Stereochemical considerations in drug metabolism. Prog Drug Metab 10:55–83

van Bladeren PJ, Vyas KP, Sayer JM, Ryan DE, Thomas PE, Levin W, Jerina DM (1984) Stereoselectivity of cytochrome P-450*c* in the formation of naphthalene and anthracene 1,2-oxides. J Biol Chem 259:8966–8973

van Bladeren PJ, Sayer JM, Ryan DE, Thomas PE, Levin W, Jerina DM (1985) Differential stereoselectivity of cytochromes P-450*b* and P-450*c* in the formation of naphthalene and anthracene 1,2-oxides. J Biol Chem 260:10226–10235

van Bladeren PJ, Balani SK, Sayer JM, Thakker DR, Boyd DR, Ryan DE, Thomas PE, Levin W, Jerina DM (1987) Stereoselective formation of benzo[*c*]phenanthrene (+)-(3S,4R) and (+)-(5S,6R)-oxides by cytochrome P-450*c* in a highly purified and reconstituted system. Biochem Biophys Res Commun 145:160–167

Vigny P, Ginot YM, Kindts M, Cooper CS, Grover PL, Sims P (1980) Fluorescence spectral evidence that benzo[*a*]pyrene is activated by metabolism in mouse skin to a diol-epoxide and a phenol-epoxide. Carcinogenesis 1:945–950

Von Tungeln LS, Fu PP (1986) Stereoselective metabolism of 9-methyl, 9-hydroxymethyl- and 9,10-dimethylanthracenes: absolute configurations and optical purities of *trans*-dihydrodiol metabolites. Carcinogenesis 7:1135–1141

Vyas KP, Levin W, Yagi H, Thakker DR, Ryan DE, Thomas PE, Conney AH, Jerina DM (1982a) Stereoselective metabolism of the (+)- and (−)-enantiomers of *trans*-1,2-dihydroxy-1,2-dihydrochrysene to bay-region 1,2-diol-3,4-epoxide diastereomers by rat liver enzymes. Mol Pharmacol 22:182–189

Vyas KP, Thakker DR, Levin W, Yagi H, Conney AH, Jerina DM (1982b) Stereoselective metabolism of the optical isomers of *trans*-1,2-dihydroxy-1,2-dihydrophenanthrene to bay-region diol epoxides by rat liver microsomes. Chem Biol Interact 38:203–213

Weems HB, Fu PP, Yang SK (1986) Stereoselective metabolism of chrysene by rat liver microsomes. Direct separation of diol enantiomers by chiral stationary phase hplc. Carcinogenesis 7:1221–1230

Weston A, Grover PL, Sims P (1982a) Metabolism and activation of benzo[*a*]pyrene by mouse and rat skin in short-term organ culture and in vivo. Chem Biol Interact 42:233–250

Weston A, Grover PL, Sims P (1982b) Formation of the 11,12-diol as a metabolite of benzo[*a*]pyrene by rat skin in vivo. Biochem Biophys Res Commun 105:935–941

Weston A, Grover PL, Sims P (1983) Metabolic activation of benzo[*a*]pyrene in human skin maintained in short-term organ culture. Chem Biol Interact 45:359–371

Weston A, Hodgson RM, Hewer AJ, Kuroda R, Grover PL (1985) Comparative studies of the metabolic activation of chrysene in rodent and human skin. Chem Biol Interact 54:223–242

Weyand EH, Rice JE, Hussain N, LaVoie EJ (1987) Detection of DNA adducts of tumorigenic nonalternant polycyclic aromatic hydrocarbons by ^{32}P-postlabeling. Proc Am Assoc Cancer Res 28:102

Wolf CR (1986) Cytochrome P-450s: polymorphic multigene families involved in carcinogen activation. Trends Genet 2:209–214

Wood AW, Chang RL, Levin W, Ryan DE, Thomas PE, Lehr RE, Kumar S, Schaefer-Ridder M, Engelhardt U, Yagi H, Jerina DM, Conney AH (1983a) Mutagenicity of diol-epoxides and tetrahydroepoxides of benz[*a*]acridine and benz[*c*]acridine in bacteria and in mammalian cells. Cancer Res 43:1656–1662

Wood AW, Chang RL, Levin W, Yagi H, Thakker DR, van Bladeren PJ, Jerina DM, Conney AH (1983b) Mutagenicity of the enantiomers of the diastereomeric bay-region benz[*a*]anthracene 3,4-diol-1,2-epoxides in bacterial and mammalian cells. Cancer Res 43:5821–5825

Wood AW, Chang RL, Levin W, Kumar S, Shirai N, Jerina DM, Lehr RE, Conney AH (1986) Bacterial and mammalian cell mutagenicity of four optically active bay-region 3,4-diol-1,2-epoxides and other derivatives of the nitrogen heterocycle dibenz-[*c,h*]acridine. Cancer Res 46:2760–2766

Yagi H, Hernandez O, Jerina DM (1975) Synthesis of (+/−)-7 beta,8-alpha-dihydroxy-9-beta,10 beta-epoxy-7,8,9,10-tetrahydrobenzo[*a*]pyrene, a potential metabolite of the carcinogen benzo[*a*]pyrene with stereochemistry related to the antileukemic triptolides. J Am Chem Soc 97:6881–6883

Yang CS, Silverman BD (1988) Polycyclic aromatic hydrocarbon carcinogenesis: structure-activity relationships. CRC Press, Boca Raton

Yang SK (1982) The absolute stereochemistry of the major *trans*-dihydrodiol enantiomers formed from 11-methylbenz[*a*]anthracene by rat liver microsomes. Drug Metab Dispos 10:205–211

Yang SK, Fu PP (1984) Stereoselective metabolism of 7-methylbenz[*a*]anthracene: absolute configuration of five dihydrodiol metabolites and the effect of dihydrodiol conformation on circular dichroism spectra. Chem Biol Interact 49:71–88

Yang SK, Roller PP, Fu PP, Harvey RG, Gelboin HV (1977) Evidence for a 2,3-epoxide as an intermediate in the microsomal metabolism of benzo[*a*]pyrene to 3-hydroxybenzo[*a*]pyrene. Biochem Biophys Res Commun 77:1176–1182

Yang SK, Chou MW, Fu PP, Wislocki PG, Lu AYH (1982) Epoxidation reactions catalyzed by rat liver cytochromes P-450 and P-448 occur at different faces of the 8,9-double bond of 8-methylbenz[*a*]anthracene. Proc Natl Acad Sci USA 79:6802–6806

Yang SK, Chou MW, Evans FE, Fu PP (1984) Metabolism of 8-hydroxy-methylbenz[*a*]anthracene by rat liver microsomes. Stereochemistry of dihydrodiol metabolites and the effect of enzyme induction. Drug Metab Dispos 12:403–413

Yang SK, Mushtaq M, Chiu P-L (1985) Stereoselective metabolism and activations of polycyclic aromatic hydrocarbons. In: Harvey RG (ed) ACS symposium series no 283. Polycyclic hydrocarbons and carcinogenesis. American Chemical Society, Washington DC, pp 19–34

Yang SK, Mushtaq M, Weems HB (1987) Stereoselective formation and hydration of benzo[*c*]phenanthrene 3,4- and 5,6-epoxide enantiomers by rat liver microsomal enzymes. Arch Biochem Biophys 255:48–63

Zajdela F, Perin-Roussel O, Saguem S (1987) Marked differences between mutagenicity in *Salmonella* and tumour-initiating activities of dibenz[*a,e*]fluoranthene proximate metabolites; initiation inhibiting activity of norharman. Carcinogenesis 8:461–464

CHAPTER 10

Interactions of Fungal and Plant Toxins with DNA: Aflatoxins, Sterigmatocystin, Safrole, Cycasin, and Pyrrolizidine Alkaloids

J. D. GROOPMAN and L. G. CAIN

A. Introduction

Since the discovery of the aflatoxins in 1960, much effort has been made by research laboratories to investigate the association between exposure to naturally occurring carcinogens and long-term adverse health effects in people. In the case of aflatoxin B1, these health consequences range from acute hepatic liver toxicities to liver cancer. During the ensuing years, the vast majority of the mechanistic biochemical studies inquiring into the mode of action of naturally occurring chemical carcinogens have been carried out using aflatoxin B1. In fact, the aflatoxins are among the few ubiquitous and structurally identified environmental carcinogens for which quantitative estimates of human exposure have been systematically sought and risk assessments attempted. However, in the past 10 years compounds such as sterigmatocystin, cycasin, the family of pyrrolizidine alkaloids, and especially safrole and estragole have been more extensively probed. Stimulus for this research has undoubtedly been generated by the hypothesis that prevention of dietary exposure to naturally occurring plant and fungal carcinogens will improve the general health status of a population. Since it is almost axiomatic that the development of human cancer can be modulated by many factors both biological and chemical in nature, and because initiation, promotion, and progression-like events are required prior to the clinical diagnosis of a tumor, no one agent can be responsible for, or present at all the critical stages during the growth of a tumor. Therefore, the systematic investigation of the biological consequences of exposure to dietary carcinogens will help to develop appropriate cancer prevention strategies.

The purposes of this review are to discuss the current status of knowledge of the occurrence of plant and fungal toxins, their interactions with DNA and other macromolecular targets, their biological potencies, and where available, the evidence for epidemiological association of dietary exposure to such compounds with human disease states, such as cancer. A number of recent reviews are available on the compounds to be discussed (BUSBY and WOGAN 1984; ZEDECK 1984; GROOPMAN et al. 1986; MILLER and MILLER 1983). In addition, an excellent review of the general area of carcinogen-DNA adducts has recently been published (JEFFREY 1985). This chapter will focus upon the literature reports of interactions between naturally occurring carcinogens and DNA since 1982. There has been a vast increase in our knowledge of the biology of chemical carcinogens in the past few years because of the great advances in immunochemical techniques, analytical chemistry, and molecular biology. It will be clear that aflatoxin B1 and safrole are the compounds most studied to date. However, the scientific reports concerning these agents serve as examples of the information base which needs to be generated if we are to begin to understand the mechanism of action of naturally occurring carcinogens.

B. Aflatoxins

I. Occurrence

Human populations are exposed to aflatoxins as a result of the consumption of commodities that have been directly contaminated by the fungal strains *Aspergillus flavus* and *A. parasiticus* during growth, harvest, or storage. In general, diets can contain aflatoxin B1 (AFB1) and aflatoxin B2 (AFB2) in concentration ratios of 1.0 to 0.1, and when all four aflatoxins occur [AFB1, AFB2, aflatoxin G1 (AFG1), and aflatoxin G2 (AFG2)], a proportion of 1.0:0.1:0.3:0.03, respectively, exists. An extensive list of the grains and foodstuffs that have been found to be contaminated with aflatoxins can be found in BUSBY and WOGAN (1984) and includes corn, peanuts, milo, sorghum, copra, and rice. While contamination by the molds may be universal within a given geographical area, the levels or final concentrations of aflatoxins in the grain product can vary from less than 1 μg/kg (1 ppb) to greater than 12000 μg/kg (12 ppm). This problem is compounded by the unequal distribution of the mold metabolite, aflatoxin, within a lot of grain. For example, in many peanut lots only one peanut in 10000 may contain aflatoxin, but the level within a single peanut may be up to several hundred micrograms (CAMPBELL et al. 1986); thus, contamination of an entire shipment will occur once it has been blended, ground, and processed. It is for these reasons that the accurate measurement of human consumption of aflatoxin through sampling foodstuffs is difficult.

The present guidelines for permissible levels of aflatoxin contamination of agricultural commodities in the USA is 20 μg total aflatoxins/kg (20 ppb) (STOLOFF 1980). The United States Food and Drug Administration (FDA) has also set a practical action guideline of 0.5 μg aflatoxin M1 (AFM1)/liter (0.5 ppb) for fluid milk. Recently, AFM1 was shown to be about tenfold less carcinogenic than AFB1 (CULLEN et al. 1987), however, it is still a potent agent. In recent years, many people have advocated much lower tolerances for aflatoxin contamination of foodstuffs.

II. Aflatoxin Chemistry, Metabolism, DNA and Protein Adduct Formation

The aflatoxins are highly substituted coumarins containing a fused dihydrofurofuran moiety. AFB1 and AFB2 were named because of their strong blue fluorescence under ultraviolet light, whereas AFG1 and AFG2 fluoresced greenish-yellow. These properties permitted the very rapid development of screening methods for grains and commodities. The B toxins are characterized by

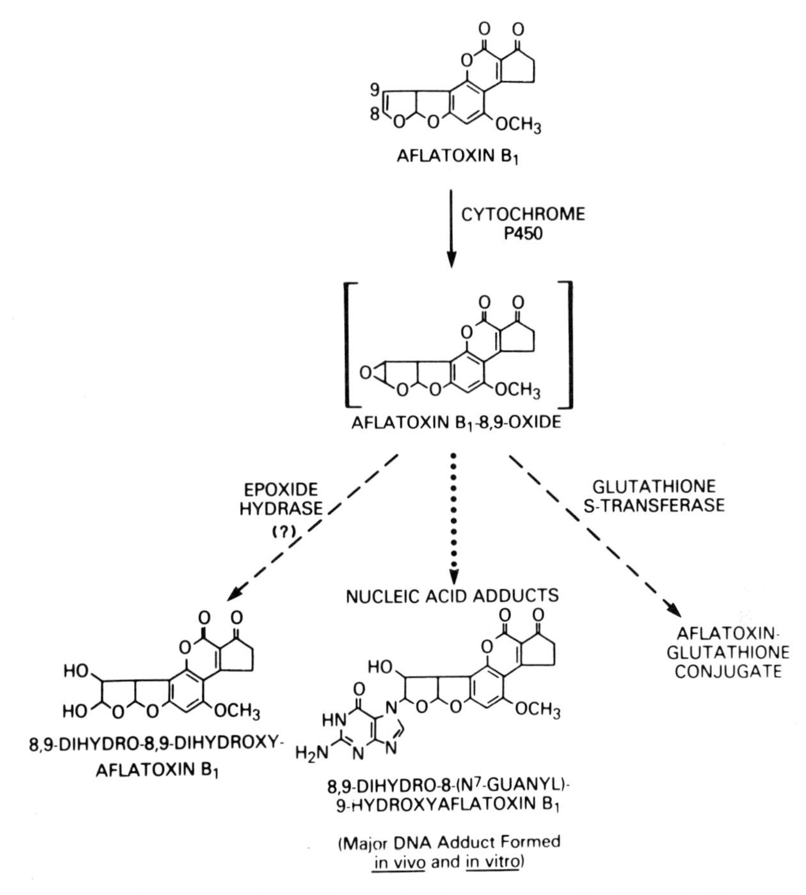

Fig. 1. Some metabolic activation and detoxification pathways for aflatoxin B1

the fusion of a cyclopentenone ring to the lactone ring of the coumarin structure whereas the G toxins contain an additional fused lactone ring. AFB1 and to a lesser extent AFG1 were responsible for the biological potency of aflatoxin-contaminated meals and crude fractions derived from toxigenic *A. flavus* cultures. These two toxins possess an unsaturated bond at the 2,3 position (the 8,9 position according to IUPAC nomenclature) on the terminal furan ring (see Fig. 1). AFB2 and AFG2 are essentially biologically inactive unless these agents are first metabolically oxidized to AFB1 and AFG1 in vivo.

The aflatoxins are primarily metabolized in animals by the microsomal mixed function oxygenase system, a complex organization of cytochrome-coupled, O_2- and NADPH-dependent enzymes localized mainly on the endoplasmic reticulum of liver cells but also present in kidney, lungs, skin, and other organs. These enzymes catalyze the oxidative metabolism of AFB1, resulting in the formation of various hydroxylated derivatives, as well as an unstable, highly reactive epoxide metabolite. Detoxification of AFB1 is accomplished by enzymatic conjugation of the hydroxylated metabolites with sulfate or glucuronic acid to form water-

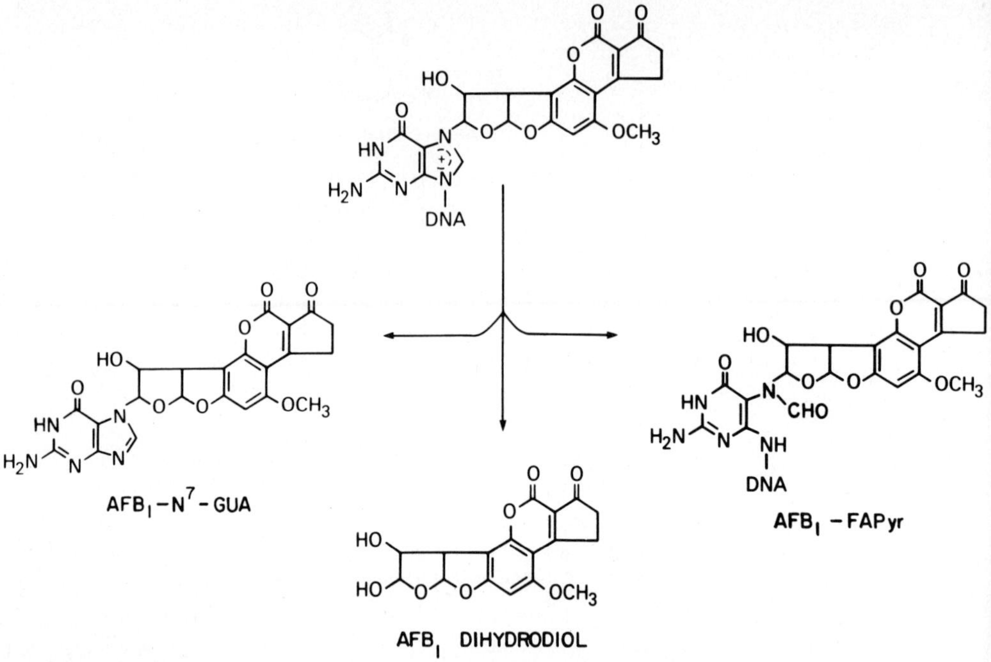

Fig. 2. Summary of aflatoxin B1-DNA adducts which occur in vivo

soluble sulfate or glucuronide esters that are excreted in urine or bile. An alternative route for removal of AFB1 from the organism involves the enzyme-catalyzed reaction of the epoxide metabolite with glutathione and its subsequent excretion in the bile. Some of the known detoxification pathways of AFB1 metabolism have been summarized in Fig. 1.

During the course of AFB1 metabolism, the reactive electrophilic epoxide can covalently react with various nucleophilic centers in cellular macromolecules such as DNA, RNA, and protein. The consequences of this "activation" reaction may be opposite to those of detoxification and potentially pose a biological hazard to the cell or organism and constitute a putative mechanism by which many compounds, including AFB1, exert toxic, carcinogenic, and genotoxic effects (BUSBY and WOGAN 1984).

The first AFB1-DNA adduct was identified by ESSIGMANN et al. (1977) as 2,3-dihydro-2-(N7-guanyl)-3-hydroxy-AFB1 (AFB1-N7-Gua) (Fig. 2), the major product liberated from DNA modified in vitro by incubation with AFB1 and a rat liver microsomal activation system. Its presence was subsequently confirmed in vivo (CROY et al. 1978). The binding of AFB1 residues to DNA in vivo was essentially a linear function of dose at a given time after treatment. A modification level of 125–1100 AFB1 residues/10^7 nucleotides was observed in rat liver 2 h after i.p. dosing with 0.125–1.0 mg AFB1/kg (CROY et al. 1978). Initial binding levels in DNA have been observed to fall rapidly within hours after AFB1 treatment (GROOPMAN et al. 1980). For example, maximum modification of rat liver DNA (1250 residues/10^7 nucleotides) was noted no later than 30 min after a 1 mg

AFB1/kg dose but declined to a level of 160 residues/10^7 nucleotides 36 h after treatment, giving an apparent half-life of AFB1 binding to DNA of approximately 12 h (GROOPMAN et al. 1980).

A number of other components, in addition to AFB1-N7-Gua, including AFB-dihydrodiol (Fig. 2), were isolated from nucleic acid hydrolysates activated in vivo and in vitro with AFB1. These adducts, designated I and IV by LIN et al. (1977), are apparently related to AFB1-N7-Gua by a precursor-product relationship. Thus, when AFB1-N7-Gua was treated under mildly alkaline conditions (pH 9.6), it was converted to these two other adducts. Furthermore, when both I and IV were subjected to additional acid hydrolysis, AFB1 dihydrodiol was formed as the major product, along with small amounts of AFB1-N7-Gua. Low levels of I were detected when IV was hydrolyzed and vice versa. On the basis of these results and of spectral data, I has been putatively identified as 2,3-dihydro-2-(N^5-formyl-2,5,6-triamino-4-oxopyrimidin-N^5-yl)-3-hydroxy AFB1 (AFB1-FAPyr; Fig. 2), a formamidopyrimidine derivative of AFB1-N7-Gua which contains an opened imidazole ring (LIN et al. 1977; HERTZOG et al. 1982). This proposed structure has recently been verified. A ring-closed structure, 2,3-dihydro-2-(8,9-dihydro-8-hydroxy-guan-7-yl)-3-hydroxy-AFB1, is proposed for IV although structural confirmation has not been obtained (LIN et al. 1977). HERTZOG et al. (1982) have disputed this structure, proposing that IV is a ring-opened isomer of I (AFB1-FAPyr) instead.

Recently, the reaction of AFB1 with nuclear DNA (nDNA) and mitochondrial DNA (mtDNA) was investigated (SHAMSUDDIN et al. 1987). This group used monoclonal antibodies directed against AFB1-modified guanosine and ultrastructural immunocytochemistry to localize the AFB1-guanosine adducts. Morphometric analysis of the electron micrographs demonstrated that localization of the AFB1-guanosine adducts is several-fold greater in mitochondria than in the nuclei. However, analytical chemical analysis found that only 77% of the covalent binding is detectable in nucleic acids of the mitochondria relative to the nuclear fractions. In RNA-free preparations, the extent of covalent modification of circular mtDNA was less than 50% of that of nDNA. These data indicate that 67% of AFB1 binding to mitochondrial nucleic acids is associated primarily with mtRNA. Nonetheless, the subcellular partitioning of aflatoxin-DNA adducts between mtDNA and nDNA may have biological importance. At the present time, it appears that between 95%–98% of the aflatoxin residues bound to total cellular DNA have been accounted for by chemical structural analysis and DNA adduct localization.

The investigation of the interactions and biological consequences of AFB1 with DNA has been an intensive area of study. Since DNA adduct formation is probably a prerequisite for the initiation of carcinogenesis by AFB1, the efforts of many investigators to study AFB1-DNA adduct formation is well justified on mechanistic grounds. Despite the suggestions of many years ago, protein-carcinogen interactions, sometimes called an epigenetic mechanism of initiation, has not been a research area favored by many laboratories. One aspect of protein-aflatoxin binding which was intensively studied was covalent nuclear protein interactions. GROOPMAN et al. (1980) described the covalent adduction of AFB1 with rat liver histones and noted the extensive modification of histone H1 by activated AFB1. A subtle observation reported in this paper was the direct linear correspondence between histone and DNA adduction. Indeed, histone binding following administration of a single dose of AFB1 was dose dependent over a 16-fold range and followed DNA binding in a precise manner. The obvious interpretation of these data is that the same metabolic activation pathway required for DNA adduct formation is used for protein adduct binding.

The interest in developing serum screening methods in order to assess human exposure to dietary aflatoxins has rejuvenated investigations into the mechanisms of aflatoxin binding to proteins in general and serum proteins in particular, with specific attention being paid to albumin. SABBIONI et al. (1987) have recently elucidated the structure of the major aflatoxin-albumin adduct found in vivo. The adduct appears to have formed by the metabolic activation of aflatoxin to the 8,9-epoxy-aflatoxin with subsequent chemical conversion to the dihydrodiol and sequential oxidation to the dialdehyde followed by condensation with the epsilon amino group of lysine. This adduct is a Schiff base which undergoes Amadori rearrangement to an alpha-amino ketone. This protein adduct has a completely modified aflatoxin structure retaining only the coumarin and cyclopentenone rings of the parent compound. WILD et al. (1987) examined the occurrence of albumin adducts in rat serum during chronic dosing with AFB1. While not measuring the individual protein adduct and assuming that the radioactivity measurements made reflect the adduct identified by SABBIONI et al. (1987), it appears that the albumin adduct accumulates during chronic administration and that a relationship between intake and albumin binding exists.

The significance of the protein adduct work as it relates to biological monitoring is that these adducts will represent the integrated level of aflatoxin exposure received over many previous weeks. The average half-life of albumin in humans is about 20 days. Therefore, an accumulated dose of aflatoxin will be present in albumin long after dietary exposure has ceased. This is a property not found for the DNA adducts because the half-life of the AFB1-N7-Gua adduct is short, about 12 h, and it is then excised and rapidly excreted in the urine. Taken together, the combination of measurements of aflatoxin-serum protein adducts and DNA adduct excretion in urine offers the promise of a method for determining both recent and long-term exposure to aflatoxin.

III. Experimental Animal Models for Aflatoxin Carcinogenesis

1. Animal Models, Dietary Antioxidants, and DNA Adduct Formation

The carcinogenic potency of AFB1 has been well established in many species of animals, including rodents, nonhuman primates, and fish (BUSBY and WOGAN 1984). The liver is the primary target organ affected and in which the toxin induces a high incidence of hepatocellular carcinomas. Variables such as animal species and strain, dose, route of administration, and dietary factors have all been investigated. A number of studies have also found that significant numbers of tumors have been induced at sites other than the liver.

Most of the published information on AFB1 carcinogenicity has been obtained from studies in rats, which are highly susceptible to the toxin. There has, however, been an increasing amount of literature in recent years dealing with the carcinogenic responses of the rainbow trout (an even more sensitive species than the rat) and the monkey (possibly a more appropriate model for human risk assessment). Such experiments have often examined dose-response characteristics and the influences of such parameters as route of administration, size and frequency of dose, sex, age, and strain of the test animal. Effects of various modifying factors on carcinogenic responses have been evaluated, including diet, hormonal status, liver injury, microsomal enzyme activity, and concurrent exposure to other carcinogens. Several studies have examined the potency and structure-activity relationships of aflatoxin congeners, structural analogues, and metabolites as inducers of liver tumors.

This information on the high potency of AFB1 provided the impetus to study the metabolism and DNA adduct formation reactions of AFB1 in order to begin to understand the underlying molecular mechanisms of how this compound initiates these processes.

GROOPMAN and KENSLER (1987) have recently discussed the following rat animal model based upon dietary antioxidant manipulations. It has been found that phenobarbital and B-napthoflavone are potent inhibitors of AFB1 carcinogenesis in rats (MCLEAN and MARSHALL 1971; GURTOO et al. 1985). Both these agents induce cytochrome P-450 isozymes that accelerate phase I metabolism of AFB1 to hydroxylated products which are considerably less active than either AFB1 or its 8,9-epoxide (WONG and HSEIH 1976; GURTOO et al. 1975). Presumably, these inductions serve to alter the balance between metabolic activation and detoxication of aflatoxin. Dietary antioxidants also inhibit AFB1 hepatocarcinogenesis when fed simultaneously with the carcinogen (CABRAL and NEAL 1983; WILLIAMS et al. 1986). In this instance the protective effects may arise from enhanced carcinogen inactivation through selective induction of phase II detoxication pathways which facilitate the clearance of activated metabolites through conjugation reactions (DELONG et al. 1983).

CABRAL and NEAL (1983) have demonstrated that the concurrent feeding of the commercial antioxidant ethoxyquin to rats fed aflatoxin-contaminated chow dramatically protects against the hepatocarcinogenic action of aflatoxin. Along similar lines, KENSLER et al. (1986) developed a more refined exposure protocol for assessing the mechanisms of chemoprotection by ethoxyquin and other antioxidants and to validate the possible use of adduct dosimetry in risk assessment. In this protocol, male F344 rats are placed on a purified diet of the AIN-76A formulation supplemented with 0.4% ethoxyquin. Beginning 1 week later, animals are dosed with 250 µg AFB1/kg body weight p.o. for 5 days a week for 2 weeks. One week after cessation of aflatoxin dosing, rats are turned to the basal diet. With this protocol, ethoxyquin supplementation reduces the number and volume of presumptive preneoplastic hepatic lesions (gamma glutamyl transpeptidase-positive foci) observed at 4 months by >95% when compared with rats maintained on the basal (unsupplemented) diet throughout the experimental period.

The induction of phase II enzymes such as the glutathione S-transferases (GSTs) by ethoxyquin or other dietary antioxidants is a prominent biochemical effect of antioxidant treatment in the rat. After 1 day on a semipurified diet supplemented with 0.4% ethoxyquin, the specific activities of GSTs were significantly elevated by about 1.5-fold. Maximal induction of four- to fivefold was observed after 1 week on the antioxidant-containing diet and persisted throughout the feeding period. Removal of ethoxyquin from the diet resulted in a rapid diminution of GST activities such that basal levels were reached within 10 days. Rats fed ethoxyquin at levels as low as 0.05% showed significant elevations of enzyme activities after 2 weeks on the diet. These findings indicate the transient nature of a dietary alteration which could have a significant impact on the short-term initiation phase of carcinogenesis. Treatment of rats maintained on antioxidant-supplemented diets results in large increases in the biliary elimination of AFB1-glutathione conjugates as well as greatly diminished levels of AFB1 modification of hepatic DNA following single or repetitive exposures to this carcinogen (KENSLER et al. 1985, 1986). We have observed a striking correlation between the degree of induction of hepatic GST, by structurally distinct antioxidants and the degree of chemoprotection as judged by reduced AFB1-N7-Gua levels in rat liver DNA.

It is well established that a single dose of AFB1 is not an efficient carcinogenic regimen in rats; however, a dosing regimen of small repeated doses can induce a high incidence of hepatocellular carcinomas (BUSBY and WOGAN 1984). Therefore, the effect of ethoxyquin on the kinetics of aflatoxin-DNA adduct formation and removal was examined in rats treated in the multiple-dosing protocol described for the gamma glutamyl transpeptidase-positive foci studies. The time-course for the formation and removal of total aflatoxin-DNA adducts in the liver in rats receiving p.o. injections of 250 μg AFB1/kg on each of days 8–12 and 15–19 is shown in Fig. 3. Maximal binding levels were achieved following the second dose and binding following the next three doses remained at a plateau level of about 140 pmol aflatoxin equivalents bound per mg DNA. Overall binding declined after the end of the first treatment period; however, resumption of AFB1 treatment produced only minor elevation of binding levels as the cycle of adduct formation and removal was renewed. This 50% diminution of aflatoxin-DNA binding during the second cycle presumably results from AFB1-induced alterations in cytochrome P-450-mediated AFB1 activation (CROY and WOGAN 1981; KENSLER et al. 1986). Total DNA adduct levels dropped fivefold in the 1st week following cessation of dosing and continued to decline at a comparable rate over the next 4 months to a level of 100 fmol aflatoxin equivalents bound per mg DNA at 133 days. Inclusion of ethoxyquin in the diet, beginning 1 week prior to and extending to 1 week beyond dosing with AFB1, produced a dissimilar pattern of effects and yielded substantially lower binding levels during the early time period. At 2 h after the first AFB1 dose, approximately 18-fold less binding was observed in the ethoxyquin-treated animals. By day 2 the difference declined to sixfold and was about 3.5-fold throughout the second dosing cycle. Remarkably, the difference in binding levels diminished during the post-dosing period such that binding levels in control and ethoxyquin rats were indistinguishable at days 106 and 133.

Liquid chromatographic analysis of hydrolyzed DNA from the livers of these animals revealed no remarkable qualitative differences in the adduct profile induced by ethoxyquin treatment at any time point. Aflatoxin-DNA adducts iso-

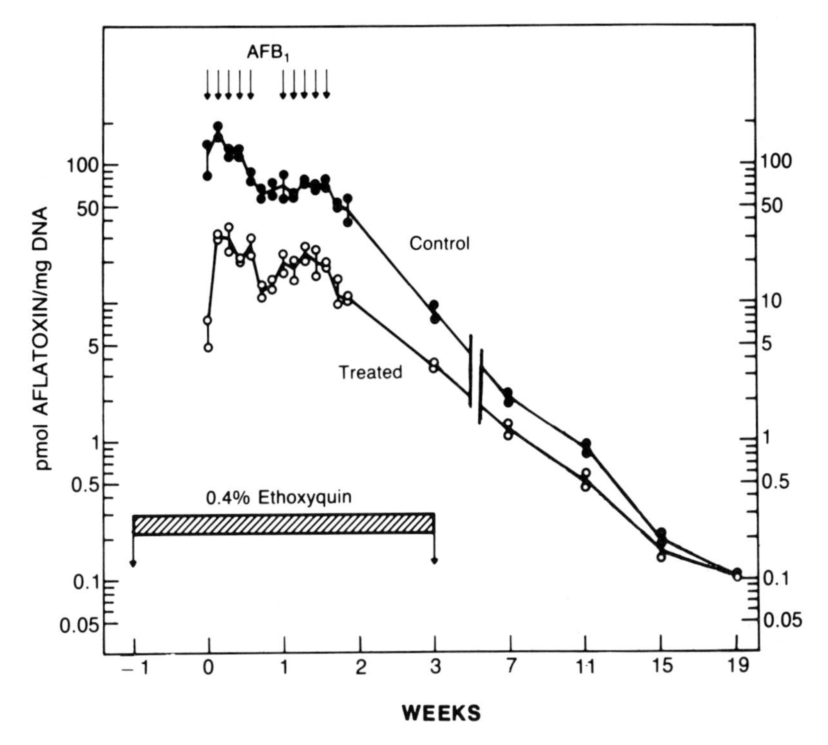

Fig. 3. Effect of ethoxyquin on aflatoxin-DNA adduct formation in rat liver

lated from control rat liver DNA showed that 80% of the DNA adducts are found in the form of AFB1-N7-Gua or decomposition products of this major adduct; namely, the formamido-pyrimidine derivatives [AFB1-N7-FAPyr (major) and AFB1-N7-FAPyr (minor)] and 8,9-dihydrodiol; however, ethoxyquin treatment reduced the amount of the AFB1-N7-Gua adduct by greater than 95%. The relative concentrations of the ring-opened formamido-pyrimidine adducts were also decreased to a comparable degree. When integrated across the 4-month time frame of the experiment, ethoxyquin treatment reduced the accumulation of AFB1-N7-Gua, AFB1-N7-FAPyr (major), and AFB1-N7-FAPyr (minor) adducts by 77%, 71%, and 76%, respectively. However, the temporal patterns for the different adducts were quite distinct. The levels of the two formamido-pyrimidine adducts remained constant over the 2-week dosing period, approximately 40 and 10 pmol bound/mg DNA for the major and minor derivatives, respectively. AFB1-N7-FAPyr (major) was the only adduct detectable after day 49, and ethoxyquin treatment had no effect on levels of this adduct at these late time points (days 106 and 133). By contrast, although the dominant species on the first day of dosing, levels of the AFB1-N7-Gua adduct decreased rapidly after day 1 such that levels during the second dosing cycle were only one-fifth to one-third those observed during the first cycle. No AFB1-N7-Gua adduct was detectable after day 21, indicating that this adduct is rapidly removed from DNA by chemical and/or enzymatic processes.

When these DNA adduct data are considered in the context of the quantitative 2- and 3-dimensional analyses of gamma glutamyl transpeptidase-positive lesions in the livers of rats treated with an identical antioxidant/aflatoxin exposure protocol, it is apparent that a strong relationship exists between the initial amount of DNA modification (AFB1-N7-Gua) in target tissue by aflatoxin and its pathologic effect. However, levels of aflatoxin-DNA adducts (i.e., the formamido-pyrimidine derivatives) at later time points did not appear to be related to the greatly diminished neoplastic outcome in the ethoxyquin-treated animals. These experiments indicate that dramatic alterations in the formation of specific adducts can result in a change in the carcinogenic outcome; they also serve to underscore the difficulties associated with exposure dosimetry assigned by the more simplistic approaches of monitoring total adduct levels.

MONROE and EATON (1987) also examined the effects of antioxidants on the mechanisms of action of aflatoxin. They conducted a comparative study of the effects of dietary butylated hydroxyanisole (BHA) on the hepatic in vivo DNA binding and in vitro biotransformation of AFB in the rat (AFB-susceptible species) and mouse (AFB-resistant species). Rats and mice were fed a control diet or an identical diet containing 0.75% BHA for 10 days. On the 11th day, one-half of the control and BHA-treated animals were given [^{3}H]AFB (0.25 mg/kg in dimethyl sulfoxide) via intraperitoneal injection. Animals were killed 2 h later and the level of covalent binding of AFB to hepatic DNA determined. BHA treatment resulted in a decrease in in vivo hepatic AFB-DNA adduct formation in mice to 68% of control, but in rats treatment decreased AFB-DNA binding to 18% of control. Furthermore, hepatic AFB-DNA binding in control mice was only 1.2% of that measured in control rats. The rate of in vitro activation of AFB to the epoxide was 3.4-fold greater in control mice relative to control rats. BHA pretreatment increased the activation of AFB in mice 3.3-fold but had no effect on oxidative metabolism in rats. Control mice had 52 times greater GST activity toward the AFB-epoxide but only 2.6 times greater GST activity toward 1-chloro-2,4-dinitrobenzene (CDNB), compared with control rats. In mice, BHA did not significantly increase GST activity toward the AFB-epoxide, but increased GST activity toward CDNB 3.1-fold. In rats, BHA increased GST activity toward the AFB-epoxide and CDNB by 3.2- and 2.1-fold, respectively. Epoxide hydrolase activity toward *p*-nitrostyrene oxide in mice was only 52% of the activity in rats. BHA increased epoxide hydrolase activity 3.8- and 2.5-fold in mice and rats, respectively. These data indicate that mice have high levels of an AFB-epoxide-specific GST activity relative to that of the rat. The rate of formation of the AFB-epoxide and the activity of epoxide hydrolase appear to be relatively unimportant under conditions of high GST activity, whereas elevated GST activity, and thus inactivation of the AFB-epoxide, appears to be the critical component in species- and BHA-induced differences in AFB-DNA adduct formation and, presumably, AFB hepatocarcinogenicity.

MANDEL et al. (1987) examined the effect of dietary administration of 0.5% ethoxyquin on the in vivo induction of enzymes and on AFB1-DNA binding in liver of male Fischer F344 rats. Ethoxyquin increases microsomal cytochrome P-450s, in particular those isozymes classes as phenobarbital inducible, and the in vitro rate of metabolism of AFB1. The formation of the metabolites aflatoxin

M1 and Q1 was enhanced to a greater extent than was the formation of the active metabolite, AFB1-8,9 epoxide (assessed by the level of AFB1-8,9-dihydrodiol). Prolonged feeding with ethoxyquin was accompanied eventually by a reduction in the initially elevated cytochrome P-450 content, but this was not reflected in any significant decrease in the rate of AFB1 metabolism in vitro. Ethoxyquin increased the GST activity of the liver cytosol fractions as assessed with the model substrate 1-chloro-2,4-dinitrobenzene. Reduced in vivo binding of [^{3}H]AFB1 to DNA of the liver and kidney was found to result from ethoxyquin treatment. It was concluded that the reduced hepatocarcinogenesis which results from feeding ethoxyquin simultaneously with AFB1 is due to the reduction in DNA adduct formation, which in turn is due, at least in part, to increased detoxifying metabolism in the microsomal, cytosolic, and plasma membrane compartments of the liver cells.

The use of dietary antioxidants as modifiers of aflatoxin biochemistry has also been extended to nonmammalian species. The trout is a highly susceptible species for aflatoxin toxicity and carcinogenesis. Some of the literature has been reviewed by BUSBY and WOGAN (1984). Since fish do not produce glutathione conjugates, the biochemical studies in trout reveal other pathways of toxication and detoxification of AFB1.

Recently, GOEGER et al. (1986) explored the mechanisms of action of indole-3-carbinol (I3C), a component of cruciferous vegetables, which has been shown to inhibit AFB1 carcinogenesis in trout. The purpose of their study was to examine the effect of I3C on AFB1 metabolism and hepatic DNA adduct formation in vivo and in vitro. When fed at 0.2%, I3C produced a 70% reduction in average in vivo hepatic DNA binding of injected AFB1 over a 21-day period compared with controls. A 24-h distribution study of injected tritiated AFB1 in I3C-treated fish showed less total radioactivity in thc blood and liver at all times examined compared with controls. These reductions were due primarily to reduced levels of AFB1 bound to red blood cell DNA, reduced plasma levels of the primary metabolite aflatoxicol (AFL), and decreased levels of AFB1 and polar metabolites present in the liver of I3C-treated fish. In contrast to blood, total radioactivity was significantly elevated in the bile of I3C-treated fish resulting from a sevenfold increase in AFL-M1 glucuronide levels over controls. No difference was observed in the concentration of AFL glucuronide, the primary conjugate present in control fish. There was also no difference in total radioactivity remaining in the carcass of I3C-treated or control fish. These findings indicate that I3C inhibition of AFB1 hepatocarcinogenesis in trout involves substantial changes in the pharmacokinetics of carcinogen distribution, metabolism, and elimination, leading to significantly reduced initial hepatic nDNA damage in vivo.

Taken together, these mammalian and nonmammalian models have helped to define better the relative importance of aflatoxin-DNA adduct formation and removal in the adverse biology induced by this carcinogen.

2. Interactions with Cellular Oncogenes

In the last few years, research on the role of cellular oncogenes in cancer has expanded at a very rapid rate. Some of this literature and the potential role of these genes are reviewed and discussed by BISHOP (1985) and WEINBERG (1985). Clearly the interactions of AFB1 with cellular oncogenes may provide very important data that will further our understanding of the mechanisms of action of this carcinogen.

The initial research in this area was done by MCMAHON et al. (1986) who treated weanling male Fischer rats with 40 intraperitoneal injections of AFB1 (25 μg per animal per day) over a 2-month period. This chronic treatment regimen resulted in the sequential formation of hyperplastic foci, preneoplastic nodules, and hepatocellular carcinomas in all of the animals treated. This is consistent with previously published data (BUSBY and WOGAN 1984). The presence of transforming DNA sequences was detected by formation of anchorage-independent foci after transfection of tumor-derived DNA into NIH 3T3 mouse fibroblasts. Transfection of genomic DNA isolated from individual tumors from eight animals resulted in specific transforming activities ranging from 0.05 to 0.2 foci per microgram of DNA. Primary transfectant DNAs were analyzed by Southern blot hybridization with DNA probes homologous to c-Ha-*ras,* c-Ki-*ras,* and N-*ras* oncogenes. A highly amplified c-Ki-*ras* oncogene of rat origin was detected in transformants derived from tumors in two of the eight animals tested. There was no evidence to suggest the presence of c-Ha-*ras* or N-*ras* sequences in any of the transformants. Analysis of primary liver tumor DNA showed no Ki-*ras* DNA amplification when compared with control liver DNA samples. Increased levels of c-Ki-*ras* p21 proteins were detected in 3T3 transformants containing activated rat c-Ki-*ras* genes. The presence of c-Ki-*ras* sequences of rat origin capable of inducing transformed foci can be taken as evidence that the c-Ki-*ras* gene has been activated in the primary liver tumors.

This initial study was followed up by MCMAHON et al. (1987). Activated c-Ki-*ras* genes in liver tumors from AFB1-treated rats were analyzed to determine the nature of their activation by characterization of two c-Ki-*ras* alleles present in tumor-derived NIH 3T3 mouse transformants. Using selective hybridization of synthetic oligonucleotides to transformant DNA, they found that a single G:C→A:T base transition in either the first or second position of the 12th codon was associated with activation of the gene. Such mutations would lead to amino acid substitutions of aspartate or serine for glycine in the mutant proteins. To confirm these findings, these researchers applied a technique for direct sequence analysis of a 90-base pair region of the rat c-Ki-*ras* gene produced by primer-directed enzymatic amplification. Findings produced by this approach, which provides a convenient method to characterize mutations in multiple alleles without the necessity to clone individual genes, confirmed the presence and identity of the 12th codon mutations in the activated oncogene, as initially determined by the oligonucleotide hybridization technique.

While it is still too early to understand fully the implications of the oncogene work in terms of AFB1 biochemistry, this research points to the importance of integrating a chemical knowledge of the covalent interactions of chemical carcinogens with DNA with the results of biological experiments.

IV. Affinity Chromatography for Aflatoxin-DNA Adducts and Other Metabolite Isolation from Biological Samples

Immunological methods for the detection of low molecular weight substances have been extensively used over the past 30 years. Numerous antibodies and antisera have been generated because thousands of chemicals are antigenic. The development of monoclonal antibody technology during the last decade now permits the isolation of monoclonal antibodies having a unique specificity. Simply stated, monoclonal antibodies are specific for a single epitope. The nature of the epitope is sometimes trivialized by the tendency to depict epitopes two-dimensionally on paper rather than by considering the three-dimensional structure that the antibody must recognize to permit binding. Since even small, perhaps even seemingly insignificant changes can affect the recognition of a chemical agent by monoclonal antibodies, it is very important before starting to produce monoclonal antibodies to consider the antigen to be synthesized, the screening methods to be used, and the overall goal for the production of the antibody. The selection of these parameters will dictate whether the antibody will be useful for in vitro studies using artificial conditions or for real samples.

Numerous monoclonal antibodies which recognize aflatoxins, using antigens ranging from aflatoxin-modified DNA to aflatoxin-adducted proteins, have been produced. These antibodies are being used in conjunction with other chemical and analytical techniques as noninvasive screening methodologies to monitor human exposure to these environmentally occurring mycotoxins. These methods depend upon the ability to quantify aflatoxin and its metabolites, including DNA adducts, in readily accessible compartments, such as serum and urine (GROOPMAN et al. 1982, 1984, 1985).

Efforts were initiated by GROOPMAN and coworkers to use the high affinity of the monoclonal antibodies to produce a preparative monoclonal antibody affinity column in order to isolate aflatoxins from complex biological fluids. Similar methodologies were also developed by Dr. T-T SUN at the Beijing Cancer Institute (SUN and CHU 1984).

Urine samples from rats treated with [^{14}C]AFB1 fractionate on a high-affinity monoclonal antibody affinity column were analyzed by competitive RIA and the results compared with the amount of aflatoxin determined from the radioactivity data. A precise correspondence was obtained, indicating that the majority of the aflatoxin derivatives in rat urine would be recognized by the monoclonal antibody. These data also demonstrated that the competitive RIA has the requisite sensitivity for the determination of the aflatoxin content in biological samples.

Two adult male Fisher 344 rats were each injected with 1 mg [^{14}C]AFB1 per kg body weight and their urine collected for 20 h, at which time 10%–12% of the radiolabel had been excreted into the urine. Aliquots (100 μl) of urine from each rat, containing 290 and 310 ng of [^{14}C]AFB1 equivalents, respectively, were diluted with 1.9 ml of PBS and applied to the antibody affinity column, and 65% of the applied ^{14}C bound to the affinity matrix. The eluate containing the initially unretained aflatoxin moieties was recycled back through the column, but this second passage failed to result in further binding. These data indicate that the unretained aflatoxins are not immunologically recognized by the antibody and thus consistent with the specificity reported for this monoclonal antibody. The retained aflatoxins were eluted from the column with 50% DMSO-PBS and analyzed by analytical reversed-phase HPLC.

The HPLC chromatograms of the UV and radioactivity aflatoxin profiles in the rat urine revealed that the predominant metabolite was AFM1, accounting for between 41% and 50% of the recovered ^{14}C. AFP1 and AFB1 were also detected but together account for less than 10% of the radioactivity and/or UV absorbance. The AFB1-N7-Gua adduct was a major metabolite and comprised 16% of the applied radioactivity. The level of AFB1-N7-Gua in the urine corresponded to the amount calculated from the pharmacokinetic data of BENNETT et al. (1981). The overall recovery of the radioactive aflatoxins applied to the HPLC column was greater than 95%. The unretained material from the affinity column (PBS washes) was analyzed by preparative HPLC procedures. All of the radiolabeled aflatoxin from this fraction chromatographed as unretained polar derivatives of AFB1. These data show that the major metabolites isolated from the urine of rats treated with AFB1 are AFM1, AFP1, and AFB1-N7-Gua. It appears highly probable that, by using other monoclonal antibodies with different specificities, the entire complement of urinary aflatoxin derivatives, including the oxidative conjugates, could be quantitatively recovered (GROOPMAN et al. 1985).

These initial experiments demonstrate the efficacy of using the monoclonal antibody affinity technique as a preparative tool to isolate the aflatoxins from exposed animals. The next concern addressed by animal studies is whether the levels of DNA adduct excreted into the urine correspond to the levels of initial DNA adduct formation within the rat liver. Six adult male Fisher 344 rats were each intubated p.o. with 0.25 mg [^{14}C]AFB1 per kg body weight. Three of the rats had been maintained for 1 week prior to dosing on a diet containing 0.4% ethoxyquin while the other three rats were maintained on an AIN-76A diet, as previously described in Sect. B.III.1. After 24 h the rats were killed, and the DNA was isolated from the livers. The urine and feces excreted by the rats over the 24-h period were also collected and analyzed. In accordance with previous data (KENSLER et al. 1985, 1986) the rats maintained on the ethoxyquin diet had a greater than 90% reduction in AFB-DNA adduct formation. The urine from each of the rats within the ethoxyquin and control dietary groups contained about 20%, while the feces accounted for about 60% of the administered [^{14}C]AFB1. Therefore, there were no apparent differences between ethoxyquin and control rats in the total amounts of aflatoxin excreted into urine and feces. The excretion of the major aflatoxin DNA adduct into the urine of rats maintained on the ethoxyquin diet was reduced by 66% compared with that in the urine of control diet animals. This finding qualitatively corresponds with the reduction in DNA adduct levels in the livers of the ethoxyquin-diet animals. However, the reduction in DNA adduct excretion is not quantitatively the same. This may be due to the contribution of DNA adducts formed in other organs whose level is not as dramatically reduced by ethoxyquin as those in the liver, as discussed in a previous paper from our laboratories (KENSLER et al. 1986). Another contributing factor to this difference may be excised (or turned over) aflatoxin RNA adducts.

These excretion data indicate that a general correlation does exist between levels of aflatoxin DNA adducts excreted into urine and initial levels of binding to DNA in a target organ. While further and more extensive investigations need to be performed to determine in more detail the kinetics of the excretion patterns in multiply or chronically dosed animals, our findings strongly suggest that measurement of AFB1-N7-Gua adducts in urine is a valid and quantitative indicator of recent exposure to aflatoxin and of the future development of neoplasia.

V. Human Liver Cancer and Aflatoxin: Epidemiology and Exposure Monitoring

Primary liver cancer is one of the leading causes of cancer mortality in Asia and Africa. For example, in the People's Republic of China, liver cancer accounts for 120000 deaths per year and is the third leading cause of cancer mortality in males, behind cancer of the esophagus and stomach (as reported by the NATIONAL CANCER OFFICE OF THE MINISTRY OF PUBLIC HEALTH, P.R.C. 1980). In parts of western Africa, liver cancer mortality can be up to 400 cases per 100000 per year, and it has been reported that, on an island outside of Shanghai, liver cancer incidence is 3000 cases per 100000 per year. In contrast, liver cancer incidence in the United States is about 0.5 cases per 100000 per year. Clearly, liver cancer incidence varies worldwide by at least 1000- to 10000-fold. Several epidemiological studies were conducted during the late 1960s and early 1970s in order to obtain information on the relationship between estimated dietary intake of aflatoxin and the incidence of primary human liver cancer in different parts of the world. Data from these studies showed that aflatoxin ingestion varied over a range of values from 3 to 222 ng/kg body weight per day (reviewed in GROOPMAN et al. 1986). Estimated liver cancer incidence values extended from a minimum of 2.0 to a maximum of 35.0 cases/100000 population per year. There was a positive association between high intakes of aflatoxin and high incidence rates of liver cancer. The association was most apparent in connection with incidence rates for adult men. The incidence of liver cancer in many of these studies was a linear function of the log of dietary aflatoxin intake (BUSHBY and WOGAN 1984; LINSELL and PEERS 1977).

This information provides a strong motivation for further investigations of the circumstantial relationship between aflatoxin ingestion and liver cancer incidence. The association between dietary aflatoxin and liver cancer, together with the extensive animal data on aflatoxin carcinogenicity, are sufficient to support the hypothesis that exposure to the carcinogen is associated with an elevated risk of this form of cancer. Further investigations are warranted in order to produce effective means by which to monitor and control the contamination of foods by aflatoxins.

Several recent epidemiologic studies have been published on the correspondence of aflatoxin exposure and liver cancer in Africa. In one of these reports, BULATAO-JAYME et al. (1982) compared the dietary intakes of 90 confirmed primary liver cancer patients against 90 age and sex-matched controls. By using dietary recall, the frequency and amounts of food items consumed were converted into units of aflatoxin load per day, using a standardized Philippine table of aflatoxin values for these items. Of the total subjects' aflatoxin load, 51.2% came from cassava, 20.3% from corn, 6.8% from peanuts, and 5.8% from sweet potato. The mean aflatoxin load per day of the patients was found to be 440% that of the controls. Dietary aflatoxin loads and alcohol intakes were subjectively allocated into heavy and light exposure groups. The comparison of patients versus controls generated a relative risk (RR) of developing primary liver cancer from ingesting contaminated foods. The following foods were found to be statistically significant in order of rank: cassava, peanuts, sweet potato, corn, and alcohol. Boiled rice, which has negligible aflatoxin content, gave no difference in risk. These researchers combined aflatoxin load and alcohol intake and determined a synergistic and statistically significant effect on RR with aflatoxin exposure and alcohol intake. While large alcohol consumption combined with light aflatoxin intake gives a RR of 3.9, a large aflatoxin consumption with light alcohol use extends the RR to 17.5. In people with heavy aflatoxin and alcohol exposure, the RR is 35.0. All of these RR are compared against light aflatoxin and alcohol use, whose RR was arbitrarily set to 1.0. This study helped to establish a possible relationship in humans between aflatoxin ingestion and the development of liver cancer when the effect of alcohol is assumed as a variable. These findings indicate a direct effect of alcohol upon aflatoxin consumption, especially among heavy drinkers, as a probable synergistic factor in liver cancer development.

VAN RENSBURG and his collaborators (1985) performed estimations of the incidence of hepatocellular carcinoma for the period 1968–1974 in the province of Inhambane, Mozambique. Taken together with rates observed in South Africa among mineworkers from the same province, these data indicate very high levels of liver cancer incidence in certain districts of Inhambane. Exceptionally high incidence levels in adolescents and young adults are not sustained at older ages, suggesting the existence of a subgroup of highly susceptible individuals. This is a striking association and indicates that public health intervention in lowering the exposure of children and adolescents to dietary aflatoxin can have a direct impact upon limiting the onset of liver cancer in young adults (20–30 years old). These investigators also noted a sharp decline in hepatocellular cancer incidence during the period of study. One probable hypothesis for this finding is a lowering of the aflatoxin levels in the diet. Concurrently with the studies of liver cancer incidence, 2183 samples of prepared food were randomly collected from six districts of Inhambane as well as from Manhica-Magude, a region of lower hepatocellular carcinoma incidence to the south. A further 623 samples were taken during 1976–1977 in Transkei, much further south, where an even lower incidence of liver cancer has been recorded. The mean aflatoxin dietary intake values calculated from these samples were significantly related to the cancer rates. Furthermore, data on AFB1 contamination of prepared food from five different countries showed a highly significant relationship with crude hepatocellular carcinoma rates.

A major confounding variable in any of these studies is the role of chronic hepatitis B virus (HBV) infection. There is evidence that chronic HBV infection may be a prerequisite for the development of virtually all cases of liver cancer, and, given the merely moderate prevalence of carrier status that has been observed in some high incidence regions, it is likely that an interaction between HBV and aflatoxin is responsible for the exceptionally high rates evident in parts of Africa and Asia. Various indications from Mozambique suggest that aflatoxin exposure may have a late stage effect on the development of liver cancer in HBV carriers (VAN RENSBURG et al. 1985). HBV carriers may be inherently predisposed to the initiating effects of aflatoxin. Variations in aflatoxin levels in foodstuffs would then account for the geographically varied distributions of liver cancer. On this basis, the incidence of hepatocellular carcinoma may be limited by the proportion of hepatitis surface antigen (HBSAg) carriers in the population. Until prospective studies are performed or appropriate animal models developed, the role of the interaction of these two potent carcinogens in predisposing people to tumor initiation and development will remain controversial. Fortunately, through the use of available vaccines and food storage conditions, both factors can be limited and, presumably, cancer rates lowered in the absence of these mechanistic experiments.

PEERS et al. (1987) published an epidemiological study conducted in Swaziland. The data collected were assessed for the relationship between aflatoxin exposure, HBV infection, and the incidence of liver cell carcinoma, the most commonly occurring malignancy among males in Swaziland. The levels of aflatoxin intake were evaluated in dietary samples from households across the country and crop samples taken from representative farms. The prevalence of HBV markers was estimated from the serum of blood donors, and liver cancer incidence was recorded for the years 1979–1983 through a national system of cancer registration.

Across four broad geographic regions, there was a more than fivefold variation in the estimated daily intake of aflatoxin, ranging from 3.1 to 17.5 μg. The proportion of HBV-exposed individuals was very high (86% in men) but varied relatively little by geographic region; the prevalence of carriers of the surface antigen was 23% in men and varied from 21% to 28%. Liver cancer incidence varied over a fivefold range and was strongly associated with estimated levels of aflatoxin. In an analysis involving ten smaller subregions, aflatoxin exposure emerged as a more important determinant of the variation in liver cancer incidence than the prevalence of HBV infection. These researchers also concluded that aflatoxin estimates from crop samples appeared to be a reasonable surrogate for dietary measurements. A comparison with dietary aflatoxin levels measured in an earlier survey in Swaziland suggested that programs aimed at reducing contamination levels had had some success. This again provides circumstantial evidence for the utility of concerted programs aimed at lowering dietary aflatoxin exposure.

AUTRUP, HARRIS and their associates have pioneered the application of synchronous fluorescence spectroscopy (SFS) to the analysis of aflatoxin-DNA adducts in urine. Synchronous fluorescence spectroscopy relies upon the sensitive and specific measurement of physico-chemical properties of chemical compounds. By monitoring differences in excitation and emission energies for a specific agent, very sensitive quantitative analyses can be performed. In AUTRUP et al. (1983) preliminary data on urine samples collected in Murang'a district,

Kenya, were found to be contaminated with 2,3-dihydro-2-(N7-guanyl)-3-hydroxyaflatoxin B1 (AFB-GuaI). Using high pressure liquid chromatographic methods, 6 of 81 samples had a detectable level of a compound whose fluorescence spectrum was identical to that of chemically synthesized AFB-GuaI as confirmed by photon-counting fluorescence spectrophotometry. This work was followed up by a more extensive study published by AUTRUP et al. (1987). In this study, conducted over the period from 1981–1984, more than a thousand urine samples from all over Kenya were analyzed. Of all tested individuals, 12.6% were positive for aflatoxin exposure as indicated by the urinary excretion of AFB1-guanine. Assuming no annual and/or seasonal variation, a regional variation in the exposure was observed. The highest rate of aflatoxin exposure was found in the Western Highlands and Central Province. The incidence of hepatitis infection nationwide as measured by the presence of the surface antigens was 10.6%, but a wide regional variation was observed. In this recent study a moderate degree of correlation between the exposure to aflatoxin and liver cancer was observed.

WILD et al. (1986) have used highly sensitive immunoassays to quantitate aflatoxins in human body fluids. Using an enzyme-linked immunosorbent assay (ELISA) they showed the ability to quantitate AFB1 over the range 0.01 ng/ml to 10 ng/ml. The assay system has been validated by using human urine samples spiked with AFB1 over this concentration range. To apply the methodology to human samples from presumptive exposed populations, 29 urine samples from the Philippines were analyzed and found to contain a range of levels from 0 to 4.25 ng/ml AFB1 equivalent, with a mean of 0.875 ng/ml. This compared with a mean of 0.666 ng/ml AFB1 equivalent in samples from France. Taken together, these data indicate that rapid and sensitive ELISAs can be used as a screening method to determine exposure groups. The advantage of this method compared with food analysis data is that it can be used to determine metabolism of aflatoxins.

Another example of the application of ELISA methods to human urine analysis is found in ZHU et al. (1987). They analyzed a total of 252 urine samples from 32 households in Fushui county of the Guangxi autonomous region of the People's Republic of China. A good correlation between total dietary AFB intake and total AFM excretion in human urine was observed during a 3-day study, and a linear regression equation of AFB consumed compared with excretion of AFM1 could be generated. Between 1.2% and 2.2% of dietary AFB1 was found to be present as AFM1 in human urine. A good correlation was also observed between the AFB concentration in corn and the AFM1 concentration in human urine. This is the same province in which YEH et al. (1985) reported the annual mortality rate of hepatocellular carcinoma at 15 cases per 100000 per year. YEH et al. (1985) also found marked variation in different countries, ranging from 5 to 55 cases per 100000 per year. The incidence of liver cancer was found to correlate well with the severity and extent of AFB1 contamination of foodstuffs. The percentage of AFM1 excreted into the urine of the people living in Guangxi province was similar to data collected by CAMPBELL et al. (1970) from the urine of Filipinos ingesting peanut butter heavily contaminated with approximately 500 μg AFB1/kg. It was estimated that 1%–4% of the ingested aflatoxin was excreted at this metabolite.

Finally, SUN et al. (1986) have employed monoclonal antibody affinity column and HPLC techniques in the analysis of human urine samples for aflatoxin content. These researchers were among the first to use this new technique, and they reported the measurement and quantitation of AFM1 in people dietarily exposed to this carcinogen. Their data provide support for the use of AFM1 as a dosimeter for recent exposure to the carcinogen.

Many of the epidemiological studies cited are of sufficient size and power to indicate that AFB1 in the diet is associated with human liver cancer. In VAN RENSBURG et al. (1985) an intriguing figure is depicted indicating a log-log relationship between aflatoxin in the diet and liver cancer incidence in a number of African countries. Further, data in this article demonstrate that there is an earlier onset of liver cancer in people living in these countries, suggesting a strong environmental, perhaps dietary, component in disease initiation. Unfortunately, none of the epidemiological studies to date provide a causal relationship between aflatoxin in the diet and liver cancer for the simple reason that no epidemiological study can prove causality, only association. In Africa and Asia, where both aflatoxin and hepatitis B are so prevalent, any association between these two powerful biological modifiers and a disease might be expected. Fortunately (or perhaps unfortunately), animal model data provide irrefutable information on the potency of aflatoxin as a liver carcinogen. Animal models for hepatitis B are fewer, limited to the duck and woodchuck, but in the next few years long-term AFB1 feeding and hepatitis B studies now underway using these models may reveal much useful data concerning the interactions of these two agents. Given the status of the epidemiological data we can, in the absence of information from prospective studies, use the technologies which have been developed for human exposure monitoring. The criteria for the exposure monitoring methods are that they are fast, accurate, sensitive, and can be carried out in large enough numbers to facilitate the eventual prospective epidemiological studies.

Dr. Groopman, Dr. Gerald Wogan (MIT) and Dr. Chen Jun-shi (Chinese Academiy of Preventive Medicine) have been studying people living in Guangxi Province, People's Republic of China, for aflatoxin exposure. These studies were started 5 years ago and are now beginning to yield the requisite data required to explore both dietary intake of the parent compound, AFB1, and the urinary output of aflatoxin metabolites in the same person. These pharmacokinetic data are essential for the conduct of the assessments which will address the question of the relationship between aflatoxin exposure and liver cancer.

People exposed to AFB1 from dietary sources were identified for pilot studies by collaborators in Beijing under the leadership of Dr. Chen Jun-shi (GROOPMAN et al. 1985, 1986). These urine samples were used to gain preliminary evidence of the applicability of the monoclonal antibody affinity column technique and HPLC analysis procedures for monitoring individuals for exposure to aflatoxins. For the initial study, 20 individuals were selected and two 25-ml aliquots of urine obtained from a morning voiding for each individual. The intake of AFB1 from the diet, primarily corn contaminated with AFB1 from 20 to 200 ppb (μg/kg), from the previous day (24 h) was calculated. The exposures ranged from 13.4 to 87.5 μg AFB1. Competitive RIA of the samples eluted from the monoclonal antibody column demonstrated that the aflatoxin concentration in the collected

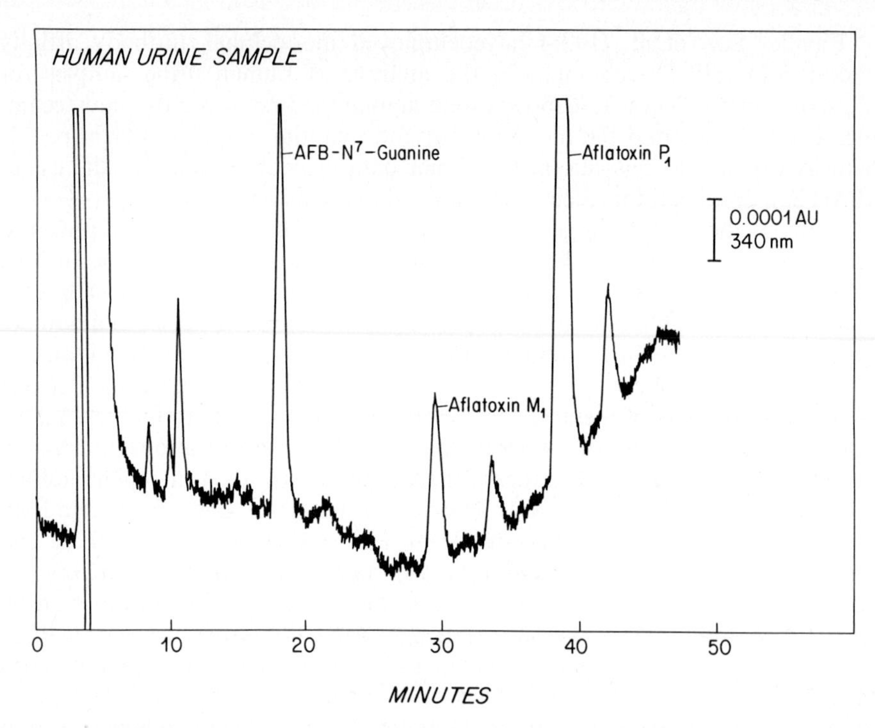

Fig. 4. High-performance liquid chromatography profile of ultraviolet absorbance from a human urine sample preparatively isolated by monoclonal antibody affinity chromatography

urine was in the 0.1–10 ng/ml range. These data were calculated using a linear extrapolation of the RIA data to standard curves generated using AFB1.

Urine samples from four individuals who had been exposed to the highest level (87.5 μg) the previous day were prepared with the antibody affinity column and then measured by analytical HPLC (Fig. 4). HPLC analysis demonstrated the presence of the major AFB1-DNA adduct, AFB1-N7-Gua, at levels representing between 7 and 10 ng of the adduct. These data indicate that the monoclonal antibody columns, coupled with HPLC, can quantify aflatoxin-DNA adducts in human urine samples obtained from environmentally exposed people.

The experience from the first China samples stimulated a more extensive study in Guangxi Province in 1985. In order to provide information about the relationship between dose and excretion of AFB1 and its adducts in chronically exposed people, the following protocol was developed. The diets of 30 males and 12 females, ages ranging from 25 to 64 years, were monitored for 1 week and total aflatoxin intake determined for each day. Urine was obtained in two 12-h fractions for three consecutive days during the 1 week period. These urine samples were obtained only after dietary aflatoxin levels had been measured for at least 3 consecutive days. Therefore, the urine collections were initiated on the 4th day of the protocol. These samples have also been analyzed by another analytical

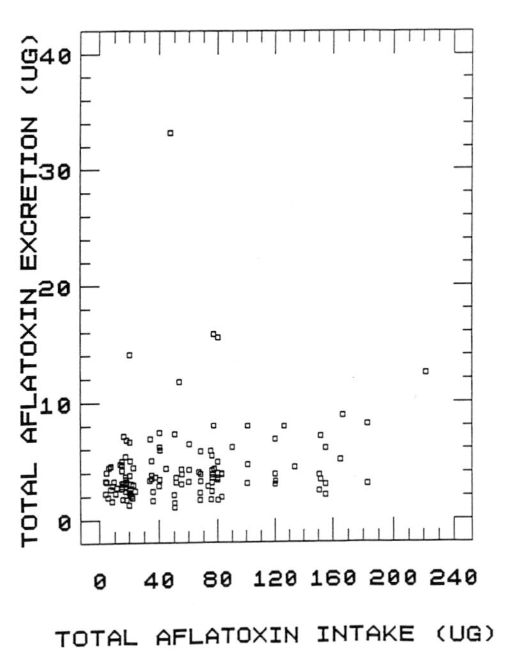

Fig. 5. Scatterplot, for aflatoxin, of total male and female intake compared with excretion data

method (ZHU et al. 1987). The average male intake of AFB1 was 48.4 μg per day, for a total exposure over the 7-day period of 276.8 μg. The average female daily intake was 92.4 μg per day. Immunoassays were performed on aliquots of the 12-h urine samples following clean-up of the samples by C18 Sep-Pak and monoclonal antibody affinity chromatography.

Total AFB excretion for each 12-h sample period was calculated by multiplying the urine volume by the concentration of AFB determined in the aliquot of urine. Figure 5 depicts a scatterplot comparison of aflatoxin intake with aflatoxin metabolite excretion. All of the male and female data were combined for this analysis. The aflatoxin intake data represents the total integrated ingestion by an individual for the day prior to urine collection and during the 3 days of urine collection. The excretion data are the composite of all aflatoxin metabolites excreted into the urine during the 3 days of urine sampling. These data reveal that, despite a 20-fold range of AFB1 intake, the amount of aflatoxin excreted generally varied only over a threefold range. This indicates that urinary excretion of AFB is a saturable process. We also performed HPLC analysis of the urine samples for AFM1, AFP1, and the major AFB-DNA adducts. Taken together, it appears that urine is a valid compartment to sample people for aflatoxin exposure, but more data must be collected in order to develope a risk model for people.

C. Sterigmatocystin

Sterigmatocystin (ST) is a potent toxin and hepatocarcinogen produced by the mold *Aspergillus versicolor*. This agent was reviewed by WOGAN and BUSBY (1980). Its structure is very similar to that of AFB1 (Fig. 6). ST retains the furofuran moiety and the unsaturated double bond of the aflatoxin ring structure. Therefore, it seems reasonable to assume that the mechanism of its activation to a DNA binding species would be similar and would be through epoxidation at that double bond. In the absence of extensive metabolism reports, ESSIGMANN and co-workers performed similar DNA adduct characterization methods to those employed for AFB1 (ESSIGMANN et al. 1977).

ST was covalently bound in vitro to calf thymus DNA by incubation in the presence of phenobarbital-induced rat liver microsomes (ESSIGMANN et al. 1979). Acid hydrolysis of ST-modified DNA liberated a major guanine-containing adduct, present in DNA at an estimated level of 1 ST residue per 100–150 nucleotides. The adduct was isolated by HPLC, and spectral and chemical data identified the adduct as 1,2-dihydro-2-N7-guanyl)-1-hydroxyST. The structure and stereochemistry of this adduct indicate that the exo-ST-1,2-oxide is the metabolite that reacts with DNA, and the quantitative yield of adduct indicates that this metabolite is a major product of the in vitro metabolism of ST. Therefore, the mechanism of activation is identical to AFB1.

This work was extended by REDDY et al. (1985) who employed the ^{32}P-postlabeling method to detect the in vitro and in vivo modification of DNA by ST. ST-modified DNA was initially incubated under buffered alkaline conditions in order to convert unstable ST-N7-guanine moieties into the stable, putative ST-formamido-pyrimidine derivatives. DNA was subsequently digested with micrococcal nuclease and spleen phosphodiesterase and the resulting ST-modified nucleotides purified by reverse-phase thin-layer chromatography (TLC). These adducts had been labeled at the 5′-position by incubation with [gamma-^{32}P]ATP and T4 polynucleotide kinase. Quantitation of excised TLC fractions indicated that ST-DNA moieties could be detected with a sensitivity of 1 ST adduct in 3–5 $\times 10^7$ nucleotides.

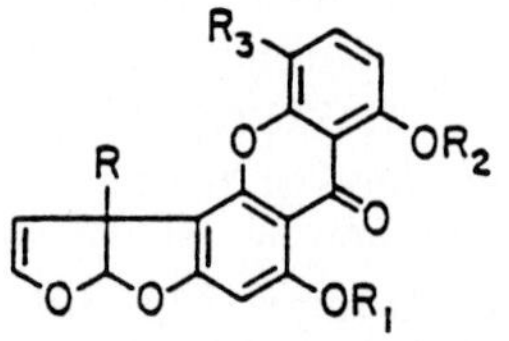

	R	R_1	R_2	R_3
Sterigmatocystin	H	CH_3	H	H
Aspertoxin	OH	CH_3	CH_3	H
O-Methylsterigmatocystin	H	CH_3	CH_3	H

Fig. 6. Structure of sterigmatocystin and related compounds

Using the post-labeling method a dose-dependent formation of ST-DNA adducts was detected in the liver of male Fischer 344 rats over a 27-fold range of ST administered (0.33–9 mg/kg). In addition, ST-DNA adducts formed in rats given a 9 mg/kg dose were found to persist for up to 105 days after treatment at a level of 0.5% of the 2-h value. Loss of these adducts from liver DNA was observed to exhibit a triphasic profile: rapid loss during the first 24 h ($t_{1/2}=12$ h), followed by a slower decline from 1 to 14 days post-dosing ($t_{1/2}=7$ days), and an extremely slow decline from days to 105 post-treatment ($t_{1/2}=109$ days). Therefore, this analytical approach to the study of mycotoxin-DNA interactions permits the quantitative description of DNA modification in ST-treated animals.

The question of the role of ST exposure in human health remains unanswered. There are no epidemiological studies reported that have attempted to associated exposure to this mycotoxin with any disease outcome. However, because the identity of the major DNA adduct is known and since the post-labelling method can be applied to the detection of ST-DNA lesions, one would anticipate that in the next few years investigations to detect the occurrence of ST-DNA lesions in human samples will begin.

D. Safrole and Related Compounds

I. Carcinogenicity

Safrole (4-allyl-1,2-methylenedioxybenzene) is a component of many spice flavors, including star anise oil, sassafras oil, oil of mace, and cinnamon leaf oil. Because of the biological potency and this agent and its congeners, there is concern about the potential impact of this group of compounds on human health. Safrole had been found to produce liver tumors in rats (reviewed by Miller and Miller 1983). Systematic structure-activity studies on the carcinogenicity of this class of compounds were performed by the Miller's laboratory. This research group examined 23 naturally occurring and synthetic alkenylbenzene derivatives structurally related to safrole for their hepatocarcinogenicity in mice. Estragole (1-allyl-4-methoxybenzene) and its proximate carcinogenic metabolite 1′-hydroxyestragole induced hepatic tumors on administration for 12 months in the diet of female CD-1 mice. Eugenol (1-allyl-4-hydroxy-3-methoxybenzene) and anethole (*trans*-4-methoxy-1-propenylbenzene) were inactive in this assay. Methyleugenol (1-ally-3,4-dimethoxybenzene) and its 1′-hydroxy metabolite had activities similar to those of estragole and its 1′-hydroxy metabolite for the induction of hepatic tumors in male B6C3F1 mice treated prior to weaning; 1-allyl-1′-hydroxy-4-methoxynaphthalene was somewhat less active. At the levels tested, myristicin (1-allyl-5-methoxy-3,4-methylenedioxybenzene), elemicin (1-allyl-3,4,5-trimethoxybenzene) and its 1′-hydroxy metabolite, dill apiol (1-allyl-2,3-dimethoxy-4,5-methylenedioxybenzene), parsley apiol (1-allyl-2,5-dimethoxy-3,4-methylenedioxybenzene), 1′-hydroxyallybenzene, 3′-hydroxyanethole, and benzyl and anisyl alcohols failed to initiate hepatic tumors on administration to male mice prior to weaning. The acetylenic derivative 1′-hydroxy-2′,3′-dehydroestragole was much more potent than either 1′-hydroxysafrole or 1′-hydroxyestragole when administered to preweanling mice. The 2′,3′-oxides of safrole, estragole, eugenol, and 1′-hydroxysafrole, which are metabolites of these alkenylbenzenes, had little or no activity in this test. The 2′,3′-oxides of safrole and estragole and their 1′-hydroxy derivatives likewise had little or no activity for the induction of lung adenomas in female A/J mice or for the induction of tumors on repetitive subcutaneous (s.c.) injection into male Fischer rats. However, the 2′,3′-oxides of safrole, estragole, eugenol, 1′-hydroxysafrole, and 1′-hydroxyestragole, when administered topically to female CD-1 mice at relatively high doses, initiated benign skin tumors that could be promoted with croton oil.

This work was extended by WISEMAN et al. (1987), who found that preweanling male C3H/HeJ mice were more susceptible than male C57BL/6J mice or females of either strain to liver tumor induction by 1′-hydroxyestragole (1′-hydroxy-1-allyl-4-methoxybenzene) and 1′-hydroxysafrole (1′-hydroxy-1-allyl-3,4-methylenedioxybenzene). Male C57BL/6J × C3H/HeJ F1 mice given a single dose of 1′-hydroxyestragole at 12 days of age developed approximately twice as many hepatomas per liver as did those given the same dose per g body weight at 1 day of age. The acetylenic compounds 1′-hydroxy-2′,3′-dehydroestragole and 1′-hydroxy-2′,3′-dehydrosafrole were the most potent derivatives studied; they were five- and tenfold more potent (based on the average numbers of hepatomas per liver) than the corresponding allylic benzene derivatives. 1′-Acetoxyestragole and 1′-acetoxysafrole had activities similar to those of their respective 1′-hydroxy derivatives; estragole derivatives were consistently two- to threefold more potent than the related safrole derivatives. 1′-Hydroxyelemicin (1′-hydroxy-1-allyl-3,4,5-trimethoxybenzene), its acetic acid ester 1′-oxoestragole, and 3′-bromo-*trans*-anethole (3′-bromo-1-*trans*-propenyl-4-methoxybenzene) each had very weak, but statistically significant hepatocarcinogenic activity. The propenylic derivatives *cis*-anethole, *trans*-isosafrole, 1:1 *cis,trans*-isosafrole, 3′-hydroxy-*trans*-anethole, piperine, and *trans*-cinnamaldehyde showed no hepatocarcinogenic activity at the levels examined. In contrast, the propenylic derivatives *cis*- and *trans*-asarone (1-propenyl-2,4,5-trimethoxybenzene) were each active; the hepatocarcinogenicities of the asarones were not inhibited by prior administration of pentachlorophenol, a sulfotransferase inhibitor that abolished the hepatocarcinogenicity of estragole under the same conditions. Furthermore, precocene II (6,7-dimethoxy-2,2-dimethyl-2*H*-1-benzopyran), a cyclic propenylic plant metabolite and asarone analogue, showed strong hepatocarcinogenic activity similar to that of 1′-hydroxy-2′,3′-dehydroestragole and 1′-hydroxy-2′,3′-dehydrosafrole; precocene I (the 7-methoxy analogue of precocene II) was less active than precocene II but more active than *cis*-asarone.

Taken together, this group of naturally occurring carcinogens may pose a potential human health hazard. Further investigation, similar to the AFB1 studies already described, into the mechanism of action of safrole and related compounds is warranted.

II. DNA Adduct Formation and Oncogene Interactions

The metabolism and DNA binding properties of safrole and estragole has been extensively investigated by the MILLERS. The metabolic activation of safrole to its

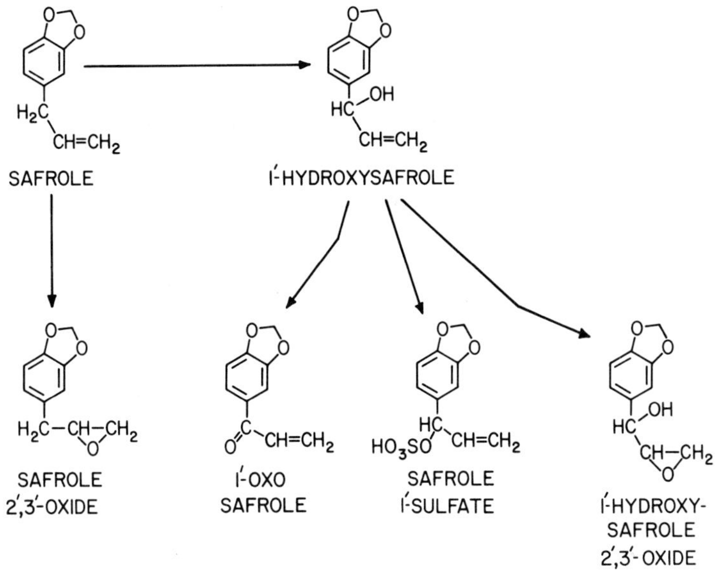

SAFROLE → 1′-HYDROXYSAFROLE

SAFROLE 2′,3′-OXIDE

1′-OXO SAFROLE

SAFROLE 1′-SULFATE

1′-HYDROXY-SAFROLE 2′,3′-OXIDE

Fig. 7. Structures of safrole and its proximate carcinogenic forms

proximate carcinogenic form, 1′-hydroxysafrole (Fig. 7) has been characterized (Miller and Miller 1983). The number of DNA adducts produced by this agent proved to be much more extensive than for AFB1. These data are summarized in the following discussion.

Phillips et al. (1981) administered 1′-[2′-3′-[^{3}H]-hydroxysafrole, a proximate carcinogen, to adult female mice, and this resulted in the formation of DNA-, ribosomal RNA-, and protein-bound adducts in the liver that reached maximum levels within 24 h post-dosing. The levels of all three macromolecule-bound adducts decreased rapidly between 1 and 3 days after injection, at which time the amounts of the DNA-bound adducts essentially plateaued at approximately 15% of the maximum level. Deoxyribonucleoside adducts obtained from the in vivo hepatic DNA were compared by HPLC with those formed by in vitro reaction of deoxyguanosine and deoxyadenosine with 1′-acetoxysafrole, 1′-hydroxysafrole-2′,3′-oxide, and 1′-oxosafrole. These data showed that the four in vivo adducts studied were derived from an ester of 1′-hydroxysafrole. Three of the four in vivo adducts comigrated with adducts formed by reaction of 1′-acetoxysafrole with deoxyguanosine; the fourth adduct comigrated with the major product of the reaction of this ester with deoxyadenosine. Two of the four DNA adducts were characterized as N2-(*trans*-isosafrol-3′-yl)deoxyguanosine and as N6-(*trans*-isosafrol-3′-yl)deoxyadenosine.

Wiseman et al. (1985) extended this research using the model electrophilic and carcinogenic esters 1′-acetoxysafrole or 1′-acetoxyestragole and DNA adducts formed in vivo in the hepatic DNA of 12-day-old male C57BL/6 × C3H/He F1 (B6C3F1) mice treated with 1′-hydroxysafrole or 1′-acetoxysafrole. They confirmed the identity of the previously described DNA adducts and further resolved these compounds into diastereomers. The proposed structures for each diastereomer were confirmed by nuclear magnetic resonance and circular dichroism spectroscopy. Two new adducts were isolated from the in vitro reaction mixture and from an analysis of their pKas and the loss of ^{3}H from [8^{3}H]deoxyguanosine; they were deduced to be 8-(*trans*-isosafrol-3′-yl)- and 8-(*trans*-isoestragol-3′-yl)deoxyguanosine, respectively. Other adducts were characterized in a similar manner as 7-(*trans*-isosafrol-3′-yl)- and 7-(*trans*-isoestragol-3′-yl)guanine, respectively. HPLC of hydrolysates of the hepatic DNA of male 12-day-old B6C3F1 mice killed 9 h after a single dose (0.1 μmol/g body weight) of [2′,3′-^{3}H]-1′-hydroxysafrole showed that the adducts S-Ia, S-Ib, S-II, S-IV [identified by PHILLIPS et al. (1981) as N6-(*trans*-isosafrol-3′-yl)deoxyadenosine] and the C-8 and N-7 guanine adducts were present at average levels of 3.5, 7.0, 24.4, 2.9, 1.2, and 3.6 pmol/mg DNA, respectively. Similar levels of these adducts were found in the hepatic DNA after administration of the same dose of [2′,3′-^{3}H]-1′-acetoxysafrole under identical conditions.

RANDERATH et al. (1984) studied the binding of a series of alkenylbenzenes to liver DNA of adult female CD1 mice, isolated 24 h after i.p. administration of nonradioactive test compound (2 or 10 mg/mouse), by a modified ^{32}P-postlabelling assay. The known hepatocarcinogens, safrole, estragole, and methyleugenol, exhibited the strongest binding to mouse liver DNA (1 adduct in 10000–15000 DNA nucleotides or 200–300 pmol adduct/mg DNA after administration of a 10 mg dose), while several related compounds, which have not been shown thus far to be carcinogenic in rodent bioassays, bound to mouse liver DNA at 3- to 200-fold lower levels. The latter compounds included allylbenzene, anethole, myristicin, parsley apiol, dill apiol, and elemicin. Eugenol did not bind. Low binding to mouse liver DNA was also observed for the weak hepatocarcinogen isosafrole. Two main ^{32}P-labelled adducts, which appeared to be guanine derivatives, were detected for each of the binding chemicals on TLC. The loss of safrole adducts from liver DNA was biphasic: a rapid loss during the 1st week ($t_{1/2}=3$ days) was followed by a much slower decline for up to 20 weeks ($t_{1/2}=2.5$ months). Adducts formed by reaction of 1′-acetoxysafrole, a model ultimate carcinogen, with mouse liver DNA in vitro were chromatographically identical to safrole-DNA adducts formed in vivo.

In a subsequent manuscript (PHILLIPS et al. 1984), a series of nine alkenylbenzenes were reported to have been administered to preweanling male mice. Male C57Bl × C3H/He F_1 mice were injected with 0.25, 0.5, 1.0, and 3.0 μmol of a compound on days 1, 8, 15, and 22 after birth, respectively. Groups of mice were killed and their liver DNA isolated on days 23, 29, and 43, and analyzed by a modified ^{32}P-postlabelling procedure. The highest levels of adducts were detected with methyleugenol (72.7 pmol/mg DNA), estragole (30.0 pmol/mg DNA), and safrole (17.5 pmol/mg DNA). After correction for liver growth, it was estimated that most of these adducts were still present at 43 days. Significant levels of DNA binding by myristicin (7.8 pmol/mg DNA) and elemicin (3.7 pmol/mg DNA) were also found, but in the former case the adducts were less persistent. Only low levels of adducts were detected with anethole, dill apiol, and parsley apiol (<1.4 pmol/mg DNA); no DNA binding was detected with eugenol. Thus, all but one of the alkenylbenzenes studied became bound to newborn mouse liver DNA, but the levels, and the persistence, of adducts formed by the carcinogenic compounds were greater.

FENNELL et al. (1986) compared the levels of DNA adducts formed in mouse liver after administration of the hepatocarcinogen [1′-^{3}H]1′-hydroxy-2′,3′-dehydroestragole determined by analysis of the ^{3}H-containing adducts and by ^{32}P-postlabelling analysis. They observed that the two diastereomers of N^2-(dehydroestragol-1′-yl)-deoxyguanosine were the only adducts detected by use of the tritiated carcinogen. Similarly, the unresolved diastereomers of N^2-(dehydroestragol-1′-yl)-deoxyguanosine-3′,5′-diphosphate were the only adducts detected by the postlabelling procedure. Analysis by ^{32}P-postlabelling of defined mixtures of the normal deoxynucleoside-3′-phosphates and synthetic N^2-(dehydroestragol-1′-yl)-deoxyguanosine-3′-phosphate showed that recovery of the labelled adduct was about 60% of that of the normal nucleotides. Likewise, the levels of the adduct in the hepatic DNA from mice treated with 1′-hydroxydehydroestragole, as determined by ^{32}P-postlabelling, were generally 60%–80% of those obtained by analysis for the tritiated adducts. Since 1′-oxodehydroestragole-deoxyadenosine adducts, the major products obtained upon reaction of 1′-oxodehydroestragole with DNA in vitro, were not detected by ^{32}P-postlabelling in the hepatic DNA from mice treated with 1′-hydroxydehydroestragole, these data provide further evidence that the covalent binding of 1′-hydroxydehydroestragole to liver DNA in vivo does not involve the 1′-oxo derivative.

Finally, the investigation of the activation of cellular oncogenes by chemical carcinogens has been studied using estragole. WISEMAN et al. (1986) found activated c-Ha-*ras* protooncogenes in hepatomas initiated by *N*-hydroxy-2-acetylaminofluorene, vinyl carbamate, or 1′-hydroxy-2′,3′-dehydroestragole. Southern analysis of NIH 3T3 cells transformed by DNA from 24 of these hepatomas revealed amplified and/or rearranged restriction fragments homologous to a Ha-*ras* probe. The other tumor contained an activated Ki-*ras* gene. Immunoprecipitation and electrophoretic analysis of p21 *ras* proteins in NIH 3T3 transformants derived from a majority of the hepatomas suggested that the activating mutations were localized in the 61st codon of the c-Ha-*ras* gene. Creation of a new *Xba* I restriction site by an A:T→T:A transversion at the second position of codon 61 was detected in DNA from primary tumors and NIH 3T3 cells transformed by DNA from 6 of 7 vinyl carbamate- and 5 of 10 1′-hydroxy-2′,3′-dehydroestragole-induced hepatomas. Selective oligonucleotide hybridization demonstrated a C:G→A:T transversion at the first position of the 61st codon in NIH 3T3 transformants derived from 7 of 7 *N*-hydroxy-2-acetylaminofluorene-indued hepatomas. By the same criterion, an A:T→G:C transition at the second position of codon 61 was the activating mutation in 1 of 7 vinyl carbamate- and 5 of 10 1′-hydroxy-2′,3′-dehydroestragole-induced tumors. Thus, c-Ha-*ras* activation is apparently an early event in B6C3F1 mouse hepatocarcinogenesis that results directly from reaction of ultimate chemical carcinogens with this gene in vivo.

In summary, most of the probable DNA adducts formed by safrole and estragole have been identified, and it appears that these adducts have the potential to activate cellular oncogenes following their formation at specific target sites. While data exists about the carcinogenic potency of these agents, at the present time little epidermiological data exist to indicate what the risk is for people dietarily exposed to safrole and its associated compounds.

E. Cycasin

Cycasin is a member of a family of azoxyglycosides produced by cycads. The DNA adduct formation and metabolism of this agent has recently been extensively reviewed by MORGAN and HOFFMAN (1983), HOFFMAN and MORGAN (1983), and ZEDECK (1984). Cycasin is carcinogenic only following deglucosylation to release its principal metabolite, methylazoxymethanol (MAM). Methylazoxymethanol is clearly responsible for DNA adduct formation. SHANK and MAGEE (1967) characterized the principal DNA adduct formed in vivo by MAM as 7-methylguanine. While historically cycasin has been associated with human toxicities and an extensive literature exists on the preparation of foods to detoxify the cycasin, very little recent literature exists on human exposure.

F. Pyrrolizidine Alkaloids

The pyrrolizidine alkaloids are a family of compounds produced by many different plants. Over 100 structurally related compounds are known and at least one-third of these agents have been found to be carcinogenic. The chemistry and metabolism of these compounds have been reviewed by BUSHBY and WOGAN (1984), ROBINS (1982), and SCHOENTAL (1982). Structures of some of the major pyrrolozidine alkaloids and metabolic pathways are found in Figs. 8 and 9. The metabolism of the pyrrolizidine alkaloids to DNA binding species requires oxidation to a pyrrole. Despite the large number of reports on pyrrolizidine alkaloid in the literature, there are relatively few papers on the structural identification of pyrrolizidine alkaloid-DNA adducts. ROBERTSON (1982) characterized one of the first DNA adducts from the carcinogenic pyrrolizidine alkaloid, dehydroretronecine. This agent was reacted with deoxyguanosine at pH 7.4 in vitro to yield two major adducts, which were isolated by TLC and HPLC. The DNA adducts were identified as derivatives with a bond between the C-7 of dehydrosupinidine and the N2 position of deoxyguanosine. Mass spectral fragmentation patterns and infrared and ultraviolet absorbance spectra were also consistent with N2 substitution. Circular dichroism spectra established the identities of each of the adducts as 7-(deoxyguanosin-N^2-yl)dehydrosupinidine, demonstrating that the reactive electrophile derived from protonated dehydroretronecine readily alkylates the N^2-position of deoxyguanosine at C-7 in an SN1 reaction to yield a racemic mixture of products.

A general binding profile of a pyrrolizidine alkaloid to macromolecules was reported by CANDRIAN et al. (1985). This group prepared retronecine-labelled [^{3}H]seneciphylline ([^{3}H]SPH) and [^{3}H]senecionine ([^{3}H]SON) biosynthetically with seedlings of *Senecio vulgaris* L. using [2,3-^{3}H]putrescine as the precursor. Rats of both sexes were treated with the labelled pyrrolizidine alkaloids and

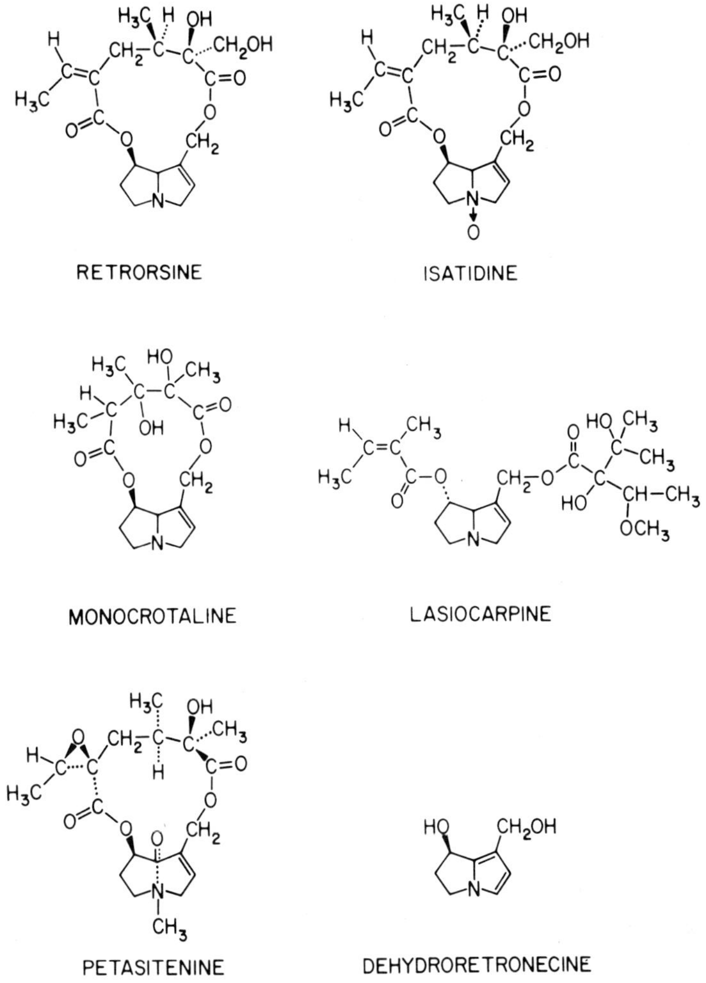

RETRORSINE

ISATIDINE

MONOCROTALINE

LASIOCARPINE

PETASITENINE

DEHYDRORETRONECINE

Fig. 8. Structures of the carcinogenic pyrrolizidine alkaloids

killed after 6 h or 4–5 days. DNA and proteins were isolated from liver, lungs, and kidneys, and covalent binding of the alkaloids to DNA was determined. A covalent binding index (CBI, µmol alkaloid bound per mol nucleotides/mmol alkaloid administered per kg body wt) of 210 ± 12 was found for the liver from SON-treated females, whereas binding to liver DNA of males was fourfold lower. The DNA damage determined 6 h after treatment persisted during the following 4 days. Administration of [^{3}H]SPH to female and male rats resulted in a CBI of 69 ± 7 and $73-92$, respectively, for the liver DNA. Furthermore, they found binding of both alkaloids to DNA of the lungs and kidneys in male and female rats.

PETRY et al. (1984) examined hepatic DNA damage induced by the pyrrolizidine alkaloid monocrotaline following i.p. administration to adult male

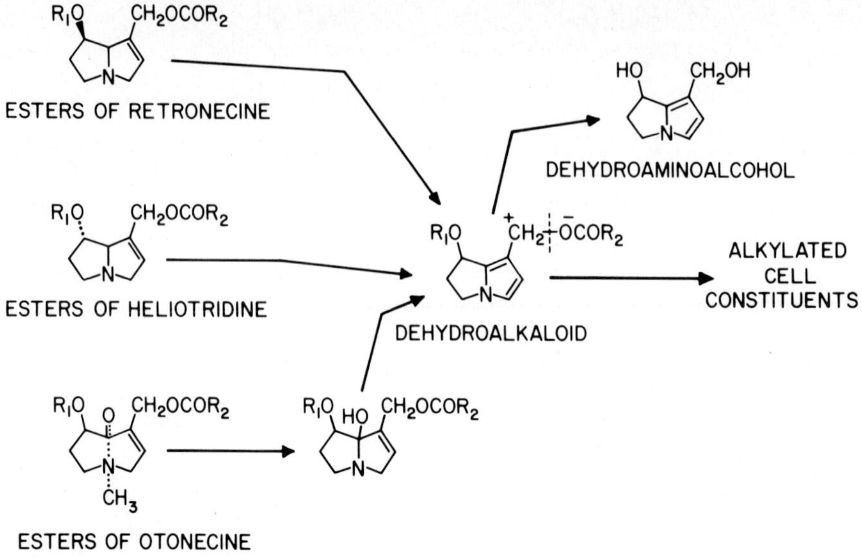

Fig. 9. Proposed metabolic activation pathways for pyrrolizidine alkaloids

Sprague-Dawley rats. Animals were treated with various doses up to 5 mg/kg. Hepatic nuclei were isolated 4 h after treatment for DNA damage assessment, as characterized by the alkaline elution technique. Alkaline elution is a highly sensitive method used to determine the cross-linking of DNA, DNA strand breaks, and protein-DNA cross-links. A mixture of DNA-DNA interstrand cross-links and DNA-protein cross-links was induced. Following an intraperitoneal injection of monocrotaline at 30 mg/kg, DNA-DNA interstrand cross-linking reached a maximum within 12 h or less and thereafter decreased over a protracted period of time. By 96 h post-administration, the calculated cross-linking factor was no longer statistically different from zero. No evidence for the induction of DNA single-strand breaks was observed, although the presence of small numbers of DNA single-strand breaks could have been masked by the overwhelming predominance of DNA cross-links.

Taken together, the evidence indicates that some pyrrolizidine alkaloids form single DNA adducts while others have the ability to induce DNA cross-links. Clearly, this is a more complex chemical outcome than was described for any of the other agents in this chapter. More research needs to be done to determine which of the pyrrolizidine alkaloid-DNA interactions might be associated with its hepatotoxic and hepatocarcinogenic effects.

G. Summary

The potential association between exposure to naturally occurring carcinogens and detrimental human health effects have spurred many inquiries into the mechanism of action of mycotoxins, safrole, the pyrrolizidine alkaloids, and cycasin. The vast majority of research reports in this field over the past 20 years has involved the aflatoxins. Nearly 4000 papers exist in the literature on aflatoxins compared with less than 1000 reports for cycasin, safrole, sterigmatocystin, and the pyrrolizidine alkaloids combined. However, the predominance of the aflatoxin literature should not be construed to mean that the other agents are less important in human disease. Of all the naturally occurring carcinogens studied, aflatoxins are the only group of natural carcinogens for which systematic regulatory guidelines and testing procedures are required. This, in combination with a number of epidemiological studies, has associated the exposure status of people to AFB1 as being important in the etiology of liver cancer. In turn, because the epidemiological studies have relied upon the criteria of presumptive intake data rather than relying upon quantitative analyses of aflatoxin-DNA adduct and metabolite content obtained by monitoring biological fluids from exposed people, the field of developing individual biological monitoring procedures has been greatly encouraged. The information to be obtained by monitoring exposed individuals for specific DNA adducts and metabolites will define the pharmacokinetics of the metabolism naturally occurring carcinogens in humans, thereby facilitating risk assessments. Therefore, the aflatoxin and safrole studies discussed in this chapter can serve as models to further the acquisition of mechanistic data about the mode of action of many other carcinogens. This increase in knowledge will help in obtaining the resources necessary to protect the food supply, whenever possible, from contamination by naturally occurring carcinogens as a readily obtainable and important goal for protecting the public's health in high-exposure regions of the world.

Acknowledgement: This work was supported by grants from the USPHS PO1ES00597 and CA39416.

References

Autrup H, Bradley KA, Shamsuddin AKM, Wakhisi J, Wasunna A (1983) Detection of putative adduct with fluorescence characteristics identical to 2,3-dihydro-2-(7′-guanyl)-3-hydroxyaflatoxin B1 in human urine collected in Murang'a district, Kenya. Carcinogenesis 4(9):1193–1195

Autrup H, Seremet T, Wakhisi J, Wasunna A (1987) Aflatoxin exposure measured by urinary excretion of aflatoxin B_1-guanine adduct and hepatitis B virus infection in areas with different liver cancer incidence in Kenya. Cancer Res 47:3430–3433

Bennett RA, Essigmann JM, Wogan GN (1981) Excretion of an aflatoxin-guanine adduct in urine of aflatoxin B1-treated rats. Cancer Res 41:650–654

Bishop JM (1985) Viral oncogenes. Cell 42:23–38

Bulatao-Jayme J, Almero EM, Castro CA, Jardeleza TR, Salamat L (1982) A case-control dietary study of primary liver cancer risk from aflatoxin exposure. Int J Epidemiol 11:112–119

Busby WF, Wogan GN (1984) Aflatoxins. In: Searle CE (ed) Chemical carcinogens, 2nd edn. American Chemical Society, Washington DC, pp 945–1136

Cabral JRP, Neal GE (1983) The inhibitory effects of ethoxyquin on the carcinogenic action of aflatoxin B1 in rats. Cancer Lett 9:125–132

Campbell AA, Whitaker TB, Pohland AE, Dickens JW, Park DL (1986) Sampling, sample preparation, and sampling plans for foodstuffs for mycotoxion analysis. Pure and Applied Chemistry 58:305–314

Campbell TC, Caedo JP, Bulatao-Jayme J, Salamat L, Engel RW (1970) Aflatoxin M_1 in human urine. Nature 227:403–404

Candrian U, Luthy J, Schlatter C (1985) In vivo covalent binding of retronecine-labelled [^{3}H] senecionine to DNA of rat liver, lung and kidney. Chem Biol Interact 54:57–69
Croy RG, Wogan GN (1981) Temporal patterns of covalent DNA adducts in rat liver after single and multiple doses of aflatoxin B1. Cancer Res 41:197–203
Croy RG, Essigmann JM, Reinhold VN, Wogan GN (1978) Identification of the principle aflatoxin B1-DNA adduct formed in vivo in rat liver. Proc Natl Acad Sci USA 75:1745–1749
Cullen JM, Ruebner BH, Hseih LS, Hyde DM, Hseih DP (1987) Carcinogenicity of dietary aflatoxin M1 in male Fischer rats compared to aflatoxin B1. Cancer Res 47:1913–1917
DeLong MJ, Prochaska HJ, Talalay P (1983) Substituted phenols as inducers of enzymes which inactivate electrophilic compounds. In: McBrien DC, Slater TF (eds) Protective agents in human and experimental cancer. Academic, London, pp 175–196
Donahue PR, Essigmann JM, Wogan GN (1982) Alfatoxin-DNA adducts: detection in urine as a dosimeter of exposure. Banbury Report 13:221–229
Essigmann JM, Croy RG, Nadzan AM, Busby WF Jr, Reinhold VN, Buchi G, Wogan GN (1977) Structural identification of the major DNA adduct formed by aflatoxin B1 in vitro. Proc Natl Acad Sci USA 74:1870–1874
Essigmann JM, Barker LJ, Fowler KW, Francisco MA, Reinhold VN, Wogan GN (1979) Sterigmatocystin-DNA interactions: identification of a major adduct formed after metabolic activation in vitro. Proc Natl Acad Sci USA 76:179–183
Fennell TR, Wiseman RW, Miller JA, Miller EC (1985) Major role of hepatic sulfotransferase activity in the metabolic activation, DNA adduct formation, and carcinogenicity of 1′-hydroxy-2′,3′-dehydroestragole in infant male C57BL/6J × C3H/HeJ F1 mice. Cancer Res 45:5310–5320
Fennell TR, Juhl U, Miller EC, Miller JA (1986) Identification and quantitation of hepatic DNA adducts formed in B6C3F1 mice from 1′-hydroxy-2′,3′-dehydroestragole: comparison of the adducts detected with the 1′-3H-labelled carcinogen and by 32P-postlabelling. Carcinogenesis 7:1881–1887
Goeger DE, Shelton DW, Hendricks JD, Bailey GS (1986) Mechanisms of anti-carcinogenesis by indole-3-carbinol: effect on the distribution and metabolism of aflatoxin B1 in rainbow trout. Carcinogenesis 7:2025–2031
Groopman JD, Kensler TW (1987) The use of monoclonal antibody affinity columns for assessing DNA damage and repair following exposure to aflatoxin B1. Pharmacol Ther 34:321–334
Groopman JD, Busby WF Jr, Wogan GN (1980) Nuclear distribution of aflatoxin B1 and its interaction with histones in rat liver in vivo. Cancer Res 40:4343–4351
Groopman JD, Haugen A, Goodrich GR, Harris CC (1982) Quantitation of aflatoxin B1 modified DNA using monoclonal antibodies. Cancer Res 42:3120–3124
Groopman JD, Trudel LJ, Donahue PR, Rothstein A, Wogan GN (1984) High affinity monoclonal antibodies for aflatoxins and their application to solid phase immunoassay. Proc Natl Acad Sci USA 81:7728–7731
Groopman JD, Donahue PR, Zhu J, Chen J, Wogan GN (1985) Aflatoxin metabolism in humans: detection of metabolites and nucleic acid adducts in urine by affinity chromatography. Proc Natl Acad Sci USA 82:6492–6497
Groopman JD, Busby WF, Donahue PR, Wogan GN (1986) Aflatoxins as risk factors for liver cancer: an application of monoclonal antibodies to monitor human exposure. In: Harris CC (ed) Biochemical and molecular epidemiology of cancer. Liss, New York, pp 233–256
Groopman JD, Donahue PR, Zhu J, Chen J, Wogan GN (1987) Temporal patterns of aflatoxin metabolites in urine of people living in Guangxi Province, P.R.C. Proc Am Assoc Cancer Res 28:36
Gurtoo HL, Dahms RP, Paigen B (1975) Metabolic activation of aflatoxins related to their mutagenicity. Biochem Biophys Res Comm 61:735–742
Gurtoo HL, Koser PL, Bansal SK, Fox HW, Sharma SD, Mulhern AI, Pavelic ZP (1985) Inhibition of aflatoxin B1-hepatocarcinogenesis in rats by B-napthoflavone. Carcinogenesis 6:675–678

Haugen A, Groopman JD, Hsu IC, Goodrich GR, Wogan GN, Harris CC (1981) Monoclonal antibody to aflatoxin B1 modified DNA detected by enzyme immunoassay. Proc Natl Acad Sci USA 78:4124–4127
Hertzog PJ, Lindsay Smith JR, Garner RC (1982) Production of monoclonal antibodies to guanine imidazole ring-opened aflatoxin B1-DNA, the persistent DNA adduct in vivo. Carcinogenesis 3:723–725
Hoffmann GR, Morgan RW (1984) Review: putative mutagens and carcinogens in foods. V. Cycad azoxyglycosides. Environ Mutagen 6:103–116
Jeffrey AM (1985) DNA modification by chemical carcinogens. Pharmacol Ther 28:237–272
Kensler TW, Enger PA, Trush MA, Bueding E, Groopman JD (1985) Modification of aflatoxin B1 binding to DNA in vivo in rats fed phenolic antioxidants, ethoxyquin and a dithiothione. Carcinogenesis 6:759–763
Kensler TW, Egner PA, Davidson NE, Roebuck BD, Pikul A, Groopman JD (1986) Modulation of aflatoxin metabolism, aflatoxin N7-guanine formation and hepatic tumorigenesis in rats fed ethoxyquin: role of induction of glutathione S-transferases. Cancer Res 46:3924–3931
Lin JK, Miller JA, Miller EC (1977) 2,3-Dihydro-2(guan-7-yl)-3-hydroxy-aflatoxin B1, a major acid hydrolysis product of aflatoxin B1-DNA or -ribosomal RNA adducts formed in hepatic microsome mediated reactions in rat liver in vivo. Cancer Res 37:4430–4438
Linsell CA, Peers FG (1977) Aflatoxin and liver cancer. Trans R Soc Trop Med Hyg 71:471–473
Mandel HG, Manson MM, Judah DJ, Simpson JL, Green JA, Forrester LM, Wolf CR, Neal GE (1987) Metabolic basis for the protective effect of the antioxidant ethoxyquin on aflatoxin B1 hepatocarcinogenesis in the rat. Cancer Res 47:5218–5223
McLean AEM, Marshall A (1971) Reduced carcinogenic effects of aflatoxin in rats given phenobarbitone. Br J Exp Pathol 52:322–329
McMahon G, Hanson L, Lee JJ, Wogan GN (1986) Identification of an activated c-Ki-*ras* oncogene in rat liver tumors induced by aflatoxin B1. Proc Natl Acad Sci USA 83:9418–9422
McMahon G, Davis E, Wogan GN (1987) Characterization of c-Ki-*ras* oncogene alleles by direct sequencing of enzymatically amplified DNA from carcinogen-induced tumors. Proc Natl Acad Sci USA 84:4974–4978
Miller EC, Miller JA (1986) Carcinogens and mutagens that may occur in foods. Cancer 58:1795–1803
Miller EC, Swanson AB, Phillips DH, Fletcher TL, Liem A, Miller JA (1983) Structure-activity studies of the carcinogenicities in the mouse and rat of some naturally occurring and synthetic alkenylbenzene derivatives related to safrole and estragole. Cancer Res 43:1124–1134
Miller JA, Miller EC (1983) The metabolic activation and nucleic acid adducts of naturally-occurring carcinogens: recent results with ethyl carbamate and the spice flavors safrole and estragole. Br J Cancer 48:1–15
Morgan RW, Hoffmann GR (1983) Cycasin and its mutagenic metabolites. Mutat Res 114:19–58
National Cancer Office of the Ministry of Public Health, P.R.C. (1980) Studies on mortality rates of cancer in China. People's Publishing House, Beijing
Peers F, Bosch X, Kaldor J, Linsell A, Pluijmen M (1987) Aflatoxin exposure, hepatitis B virus infection and liver cancer in Swaziland. Int J Cancer 39:545–553
Pestka JJ, Chu FS (1982) Reactivity of aflatoxin B2a antibody with aflatoxin B1-modified DNA and related metabolites. Appl Environ Microbiol 44:1159–1165
Petry TW, Bowden GT, Huxtable RJ, Sipes IG (1984) Characterization of hepatic DNA damage induced in rats by the pyrrolizidine alkaloid monocrotaline. Cancer Res 44:1505–1509
Phillips DH, Miller JA, Miller EC, Adams B (1981) N2 atom of guanine and N6 atom of adenine residues as sites for covalent binding of metabolically activated 1′-hydroxysafrole to mouse liver DNA in vivo. Cancer Res 41:2664–2671

Phillips DH, Reddy MV, Randerath K (1984) ^{32}P-Postlabelling analysis of DNA adducts formed in the livers of animals treated with safrole, estragole and other naturally-occurring alkenylbenzenes. II. Newborn male $B6C3F_1$ mice. Carcinogenesis 5:1623–1628

Randerath K, Haglund RE, Phillips DH, Reddy MV (1984) ^{32}P-Post-labelling analysis of DNA adducts formed in the livers of animals treated with safrole, estragole and other naturally-occurring alkenyl-benzenes. I. Adult female CD-1 mice. Carcinogenesis 5:1613–1622

Reddy MV, Irvin TR, Randerath K (1985) Formation and persistence of sterigmatocystin-DNA adducts in rat liver determined via 32P-postlabeling analysis. Mutat Res 152:85–96

Robertson KA (1982) Alkylation of N2 in deoxyguanosine by dehydroretronecine, a carcinogenic metabolite of the pyrrolizidine alkaloid monocrotaline. Cancer Res 42:8–14

Robins DJ (1982) The pyrrolizidine alkaloids. Fortschr Chem Org Naturst 41:115–203

Sabbioni G, Skipper P, Buchi G, Tannenbaum SR (1987) Isolation and characterization of the major serum albumin adduct formed by aflatoxin B1 in vivo in rats. Carcinogenesis 8:819–824

Schoental R (1982) Health hazards of pyrrolizidine alkaloids: a short review. Toxicol Lett 10:323–326

Shamsuddin AM, Harris CC, Hinzman MJ (1987) Localization of aflatoxin B1-nucleic acid adducts in mitochondria and nuclei. Carcinogenesis 8:109–114

Shank RC, Magee PN (1967) Similarities between the biochemical actions of cycasin and dimethylnitrosamine. Biochem J 105:521–527

Stoloff L (1980) Aflatoxin control – past and present. J Assoc Off Anal Chem 63:1067–1073

Sun TT, Chu YY (1984) Carcinogenesis and prevention strategy of liver cancer in areas of prevalence. J Cell Physiol [Suppl i] 3:39–44

Sun TT, Chu YR, Hsia CC, Wei YP, Wu SM (1986) Strategies and current trends of etiologic prevention of liver cancer. In: Harris CC (ed) Biochemical and molecular epidemiology. Liss, New York, pp 283–292

Van Rensburg SJ, Cook-Mozaffari P, Van Schalkwyk DJ, Van der Watt JJ, Vincent TJ, Purchase IF (1985) Hepatocellular carcinoma and dietary aflatoxin in Mozambique and Transkei. Br J Cancer 51:713–726

Weinberg RA (1985) The action of oncogenes in the cytoplasm and nucleus. Science 230:770–776

Wild CP, Umbenhauer D, Chapot B, Montesano R (1986) Monitoring of individual human exposure to aflatoxins (AF) and *N*-nitrosamines (NNO) by immunoassays. J Cell Biochem 30:171–179

Wild CP, Garner RG, Montesano R, Tursi F (1986) Aflatoxin B1 binding to plasma albumin and liver DNA upon chronic administration to rats. Carcinogenesis 7:853–858

Williams GM, Tanaka T, Maeura Y (1986) Dose-dependent inhibition of aflatoxin B1 induced hepatocarcinogenesis by the phenolic antioxidants, butylated hydroxytoluene and butylated hydroxyanisole. Carcinogenesis 7:1043–1050

Wiseman RW, Fennell TR, Miller JA, Miller EC (1985) Further characterization of the DNA adducts formed by electrophilic esters of the hepatocarcinogens 1′-hydroxysafrole and 1′-hydroxyestragole in vitro and in mouse liver in vivo, including new adducts at C-8 and N-7 of guanine residues. Cancer Res 45:3096–3105

Wiseman RW, Stowers SJ, Miller EC, Anderson MW, Miller JA (1986) Activating mutations of the c-Ha-*ras* protooncogene in chemically induced hepatomas of the male B6C3 F1 mouse. Proc Natl Acad Sci USA 83:5825–5829

Wiseman RW, Miller EC, Miller JA, Liem A (1987) Structure-activity studies of the hepatocarcinogenicities of alkenylbenzene derivatives related to estragole and safrole on administration to preweanling male C57BL/6J × C3H/HeJ F1 mice. Cancer Res 47:2275–2283

Wogan GN (1976) The induction of liver cancer by chemicals. In: Linsell DA, Warwick GP (eds) Liver cell cancer. Elsevier, New York, pp 121–150

Wong JJ, Hsieh DPH (1976) Mutagenicity of aflatoxins related to their metabolism and carcinogenic potential. Proc Natl Acad Sci USA 73:2241–2244

Yeh FS, Mo CC, Yen RC (1985) Risk factors for hepatocellular carcinoma in Guangxi, People's Republic of China. Nat Cancer Inst Monogr 69:47–48

Zedeck MS (1984) Hydrazine derivatives, azo and azoxy compounds, and methylazoxymethanol and cycasin. In: Searle CE (ed) Chemical carcinogens, 2nd edn. American Chemical Society, New York, pp 915–944

Zhu JQ, Zhang LS, Hu X, Xiao Y, Chen JS, Xu YC, Fremy J, Chu FS (1987) Correlation of dietary aflatoxin B_1 levels with excretion of aflatoxin M_1 in human urine. Cancer Res 47:1848–1852

CHAPTER 11

N-Nitroso Compounds

P. D. LAWLEY

A. Introduction: Development of Importance of *N*-Nitroso Compounds for Carcinogenesis Studies

I. Historical Origins

The *N*-nitroso compounds fall into the category of carcinogens discovered through proposed industrial use (BARNES 1974). In the early 1950s the then Director of the Medical Research Council's Toxicology Research Unit, the late Dr. J.M. BARNES, was asked by Dr. H. SWAFFIELD to investigate possible toxic hazards from the proposed use of a presumed relatively biologically inert, water-miscible, organic solvent, *N*-nitrosodimethylamine, $(CH_3)_2N.NO$ (often referred to as dimethylnitrosamine; the systematic nomenclature is *N*-methyl-*N*-nitroso-methanamine). The stimulus was the finding by Dr. SWAFFIELD that two men who had worked with this "new solvent" in a pilot plant had contracted cirrhosis of the liver.

Administration of this nitrosamine in the diet to rats was found to cause liver damage (centrilobular necrosis) (BARNES and MAGEE 1954) and, perhaps more unexpected, was the induction, with prolonged dosage, of hepatocellular carcinoma (MAGEE and BARNES 1956).

It was immediately evident that metabolic conversion to highly reactive intermediates was involved (MAGEE 1956), the liver being the main site of both decomposition and toxic action. As the earliest suggestion, diazomethane was proposed as a methylating agent, derived through α-oxidation of the nitrosamine (ROSE 1958), with concomitant liberation of the hydroxymethylating agent formaldehyde.

The designation of the proposed methylating molecular species as diazomethane was probably not intended to imply necessarily the basic form of this compound (as generally encountered by organic chemists in ethereal solution), since in neutral aqueous media its relatively high basicity [pKa in tetrahydrofuran-H_2O, 60:40 (v/v) = 10; MCGARRITY and SMYTH 1980] would result in its predominant occurrence as the conjugate acid $CH_3N_2^+$, the methyldiazonium ion (MCGARRITY and COX 1983) (Fig. 1).

Subsequent studies, extending to over 300 *N*-nitroso compounds (PREUSSMANN and STEWART 1984), have devoted much attention to the consequences of the in vivo reaction of this type of alkylating species (see LIJINSKY 1988). Their importance as determinants of the biological (mainly carcinogenic) effects of *N*-nitroso compounds is well established, but, perhaps not surprisingly,

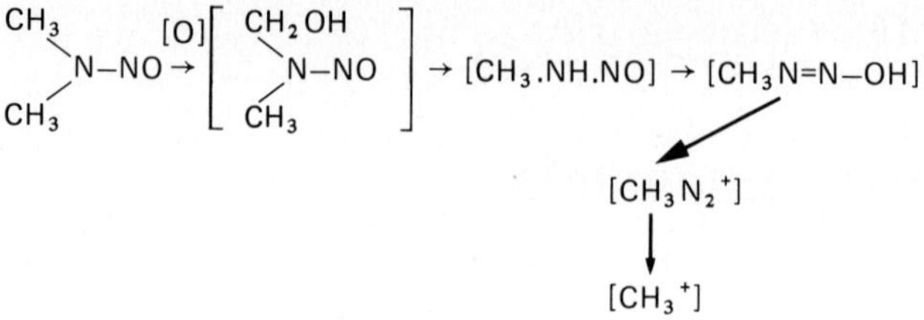

$$\begin{matrix} CH_3 \\ \quad \diagdown \\ \quad N{-}NO \\ \quad \diagup \\ CH_3 \end{matrix} \xrightarrow{[O]} \left[\begin{matrix} CH_2OH \\ \quad \diagdown \\ \quad N{-}NO \\ \quad \diagup \\ CH_3 \end{matrix} \right] \rightarrow [CH_3.NH.NO] \rightarrow [CH_3N{=}N{-}OH] \rightarrow [CH_3N_2^+] \rightarrow [CH_3^+]$$

N-Nitrosodimethylamine

$$CH_3N(NO)CONH_2 \xrightarrow{OH^-} [CH_3.NH.NO] + NCO^- + H^+$$

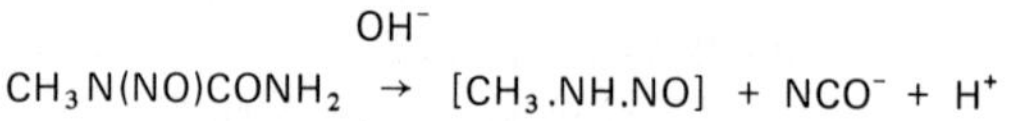

N-Methyl-*N*-nitrosourea

Fig. 1. Schematic basis of metabolic activation of *N*-nitrosodimethylamine or of alkali-catalysed hydrolysis of *N*-methyl-*N*-nitrosourea

there remain important, unresolved aspects of structure-activity relationships not accounted for in terms of known alkylation reactions (LIJINSKY 1987, 1988). This probably reflects the importance of alkylation for the initiation of cancer, whereas promotion and progression to malignancy are determined by other factors and do not necessarily reflect the genotoxicity of carcinogens (see e.g. BUTTERWORTH and SLAGA 1987). The importance of the non-mutagenic action of *N*-nitroso compounds would clearly predominate in instances in which initiation was through so-called spontaneous mutation.

II. Development of the Concept that *N*-Nitroso Compounds are Alkylating Carcinogens

Initiation of cancer is now widely accepted to result from induction of mutations in target (stem) cells. For chemical carcinogenesis it has emerged that the most potent initiating carcinogens are mutagens capable of causing proto-oncogene-activating base substitution through chemical modification of bases in DNA, the principal such reactions being alkylation or aralkylation. This concept has received important support from studies with *N*-nitroso compounds (ZARBL et al. 1985; for review, BARBACID 1987).

As noted, these began in the 1950s, over 20 years after the isolation of the first pure chemical carcinogens, which were polycyclic aromatic hydrocarbons (KENNAWAY and HIEGER 1930). The now commonplace identification of initiating carcinogens and mutagens was far from immediate (cf. LAWLEY 1989). The first generally recognised chemical mutagen, the alkylating agent mustard gas [di-(2-chloroethyl) sulphide] (AUERBACH and ROBSON 1946), was not a particularly potent carcinogen in tests available at the time (HESTON 1950), although it did emerge subsequently as one of the best-established carcinogens in humans (WADA et al. 1968). Aromatic hydrocarbons correspondingly responded weakly, if at all, to conventional tests for mutagenic action, and the stimulus to reinvestigate their potential in this respect probably emerged from studies showing their covalent reaction with DNA of a target tissue, mouse skin, in positive association with their carcinogenic potency (BROOKES and LAWLEY 1964).

The first identified reaction of alkylating agents with DNA (LAWLEY and WALLICK 1957; REINER and ZAMENHOF 1957) showed predominance of the N-7 atom of guanine in nucleophilic reactivity, although this was unexpected in view of the higher basicity (and expected nucleophilicity) of the cytosine and adenine moieties of the corresponding nucleotides, these proving to be less reactive in DNA. The availability of ^{35}S-labelled mustard gas enabled in vivo alkylation of DNA to be demonstrated (BROOKES and LAWLEY 1960), and the use of ^{14}C-labelled *N*-nitrosodimethylamine soon afterwards gave analogous experimental establishment of the concept of this chemical type of carcinogen as an in vivo alkylating agent (MAGEE and FARBER 1962).

In view of the WATSON and CRICK (1953) model for DNA replication and mutagenesis through anomalous base-pairing, it was suggested that the ionized form of 7-alkylguanine in DNA, which was deduced on physicochemical grounds to be more likely to exist at neutral pH than that of guanine, could cause miscoding (LAWLEY and BROOKES 1961) (Fig. 2). Studies with *N*-nitroso compounds, showing them to be much more potent mutagens (LOVELESS and HAMPTON 1969) and carcinogens (SWANN and MAGEE 1968) than conventional chemical methylating agents such as dimethyl sulphate or methyl methanesulphonate, effectively discounted this theory. However, the discovery (LOVELESS 1969) that the *N*-nitroso compounds could be distinguished by their ability to react more extensively at the extranuclear O-6 atom of deoxyguanosine provided a satisfying alternative model for miscoding in which transient ionized form was "fixed" by methylation.

Failure to find *O*-alkylated bases in DNA was essentially due to the use of too drastic conditions for hydrolysis in acid solution of DNA to yield bases. Use of milder acid (often 0.1 *M* HCl, 70° C, 0.5 h) proved convenient for the liberation of O^6-alkylguanines (LAWLEY and THATCHER 1970).

The potentially miscoding base structurally complementary to O^6-methylguanine, viz. O^4-methylthymine, appears to be even more unstable (LAWLEY et al. 1973) and requires enzymic degradation of DNA for its liberation (as the deoxyribonucloside O^4-methylthymidine). The search for this in methylated DNA led to the discovery of a further group of DNA *O*-alkylation products, the phosphotriesters (LAWLEY 1973; SWENSON et al. 1976).

Methods for the analytical separation of the various alkylation products have been developed through chromatographic procedures, first using cation-exchangers or Sephadex G-10 (reviewed LAWLEY 1976), later HPLC (FREI et al. 1978; BERANEK et al. 1980; WARREN 1984). Detection was through absorption of eluted markers coincident with radioactivity from isotopically labelled alkyl groups in *N*-nitroso compounds, or deuterium labelling and mass spectrometry (LIJINSKY et al. 1968) (which showed that methylation of DNA by *N*-nitrosodimethylamine transferred the CD_3 group intact, contraindicating involvement of CD_2N_2).

More recently, immunoassay (WILD et al. 1986; PARSA et al. 1987) and ^{32}P-postlabelling [REDDY et al. 1984; GUPTA 1987 (the Randerath method)] have been

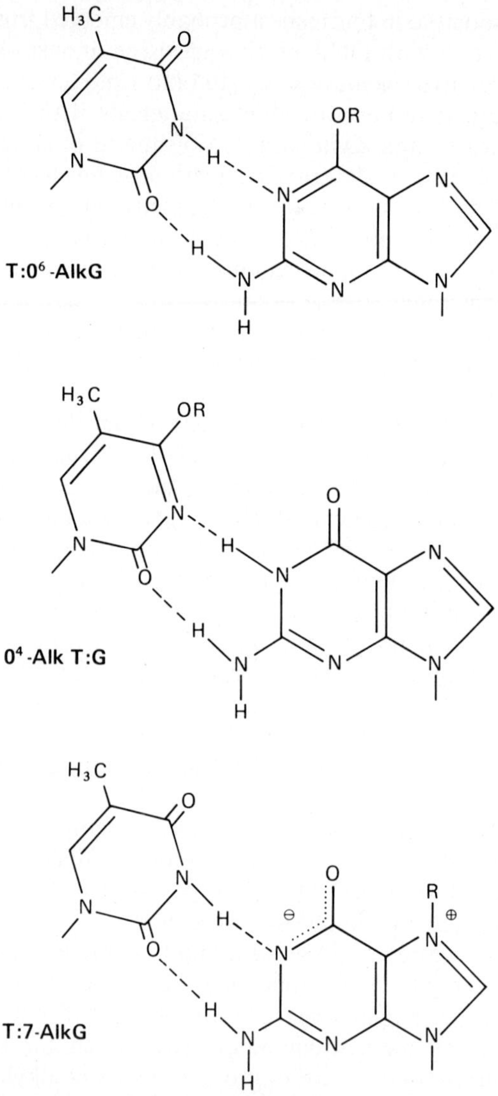

T:0^6-AlkG

0^4-Alk T:G

T:7-AlkG

Fig. 2. Suggested miscodings of alkylated bases in DNA. These were derived from studies comparing highly carcinogenic and mutagenic *N*-nitroso compounds (which alkylate extranuclear *O*-6 of guanine in DNA and extranuclear O-4 of thymine through alkyldiazonium ions, see Fig. 1) with S_N2 agent such as methyl methanesulphonate, which are much less reactive towards these atoms, e.g. the ratio O-6:N-7 alkylation of DNA guanine is 0.11 for *N*-methyl-*N*-nitrosourea and 0.004 for methyl methanesulphonate (LAWLEY and SHAH 1972). The comparatively low mutagenicity and carcinogenicity of methyl methanesulphonate ruled out the significance of miscoding by ionized 7-alkylguanines.

The base-pairing between O^4-alkylthymine and guanine requires the anti-conformation of the alkyl group (BRENNAN et al. 1986), and this has been contraindicated by studies with oligonucleotides (LI et al. 1987). O^4-Ethylguanine is deduced to miscode in mutagenesis induced by *N*-ethyl-*N*-nitrosourea in *Escherichia coli* (RICHARDSON et al. 1987) introduced to obviate the use of isotopically pre-labelled *N*-nitroso compounds, thus potentiating the detection of their reactions in humans. This was first achieved through fluorometry (HERRON and SHANK 1980), although the extent of DNA methylation in this case was comparatively high.

With regard to proteins, alkylation of ring-*N* atoms of histidine and the *S*-atom of cysteine are well-established reactions, potentially useful for monitoring human exposure to alkylating agents through studies with haemoglobin (OSTERMAN-GOLKAR et al. 1976; BAILEY et al. 1981). As with DNA phosphate, alkylation of protein carboxylate is expected but more difficult to demonstrate experimentally (KIM et al. 1977).

Other predicted reactions of metabolised *N*-nitrosamines include those of the aldehydes liberated through α-oxidation, such as formaldehyde from *N*-nitrosodimethylamine. The relative instability of products such as *N*-hydroxymethylated derivatives appear to preclude their isolation from DNA, but secondary reactions that "fix" the alkylations permit this, e.g. through cross-linking via a methylene bridge, or through cyclisation (Fig. 3). The last has been suggested to be particularly important for the action of cyclic nitrosamines (HECHT et al. 1982).

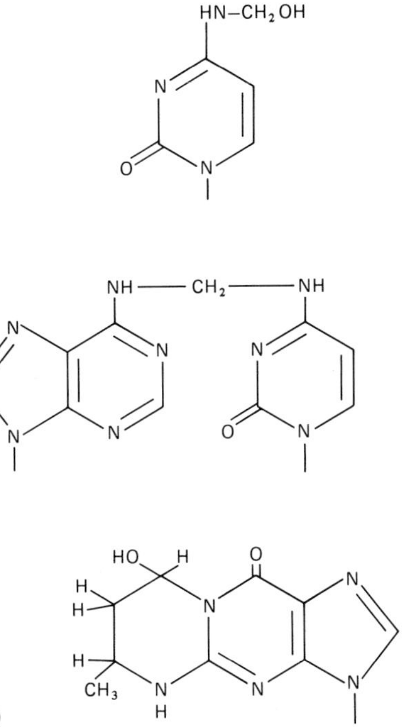

Fig. 3. Products from reactions of DNA bases with aldehydes. *I*, Methylol derivative from cytosine and formaldehyde; comparatively unstable (FELDMAN 1973). *II*, Cross-linked, stabilised product from reaction of DNA with formaldehyde involving secondary reaction of a derivative such as (I) to form a methylene bridge (CHAW et al. 1980). *III*, Product from reaction of guanine residue with crotonaldehyde (CH_3-CH=CH-CHO), a metabolite of *N*-nitrosopyrrolidine (HECHT et al. 1982)

III. *N*-Nitroso Compounds as Mutagenic Carcinogens

The ability of *N*-nitrosodimethylamine to methylate DNA in vivo (MAGEE and FARBER 1962) suggests its activity as a mutagen. This was first demonstrated as induction of X-linked recessive lethals in *Drosophila melanogaster* (PASTERNAK 1962). Evidently this organism can convert the nitrosamine to DNA-reactive metabolites, but the chemical nature of this damage does not appear to be documented (for a more recent study, see WOODRUFF et al. 1984).

Other test systems showing such mutagenic activity were investigated soon afterwards, with the realisation that conventional bacteria and yeasts gave negative results ascribed to their inability to metabolise the carcinogen. In vitro hydroxylation in an enzyme-free system (MALLING 1966) was shown to activate *N*-nitrosodimethylamines and *N*-nitrosodiethylamines as mutagens for *Neurospora*; a mouse liver homogenate activated *N*-nitrosodimethylamine to cause reversion mutations in *Salmonella typhimurium* (MALLING 1971).

These findings initiated a prodigious amount of work devoted to detection of carcinogens through their ability to cause mutations (mainly reversions from auxotrophy in bacteria) as the basis of "rapid screening" tests, of which the best known are those due to AMES and co-workers (e.g. 1973, 1975). Of course these tests were confined to the detection of mutagenic carcinogens (broadly equated with initiators of cancer) and would not be expected to detect tumour promoters that are non-mutagenic (see e.g. BUTTERWORTH and SLAGA 1987).

If the hope was that these and other short-term tests would obviate the need for the use of animals in conventional carcinogenesis tests, such hope proved to be short-lived. Extensive collaborative investigations of the quantitative correlation between potency of chemicals as mutagens in these tests, and as carcinogens, showed only poor positive correlations. Thus in a survey of tests for validation of bacterial mutagenicity tests of 180 chemicals including some *N*-nitroso compounds, the rank orders for mutagenicity and carcinogenicity were sufficiently different (BARTSCH et al. 1980) to lead to the conclusion that "no correlation could be made between quantitative aspects" of these parameters.

In another extensive collaborative study, supposedly potent carcinogens and non-carcinogens were compared (DE SERRES and ASHBY 1981). The *N*-nitroso compounds were represented by *N*-nitrosomorpholine (carcinogen); *N*-nitrosodiphenylamine and di-(*N*-nitroso)pentamethylene tetramine (Fig. 4) (classified

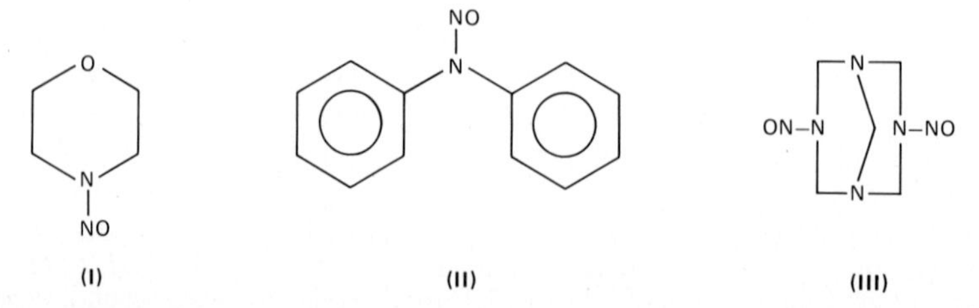

Fig. 4. Formulae of *N*-nitroso compounds used in a collaborative study relating short-term tests and carcinogenicity. *I*, *N*-nitrosomorpholine; *II*, *N*-nitrosodiphenylamine; *III*, di-(*N*-nitroso)pentamethylene tetramine

for the purposes of this study as non-carcinogens). The results for these compounds, in line with those throughout the collaborative study, were somewhat equivocal. The carcinogen was reported to give a positive response in 59% of tests; the supposed non-carcinogens gave 33% positive. In the case of *N*-nitrosodiphenylamine, this latter result was not perhaps too surprising, since the evaluation by the International Agency for Research on Cancer (IARC 1982a) concluded that "there is limited evidence for the carcinogenicity in experimental animals", but the principal data, showing the ability to induce bladder cancer in rodents, were not available when the quoted study was planned.

Ashby (1981) commented that the positive data for di-*N*-nitrosopentamethylene tetramine were "possibly consistent with the spontaneous release of formaldehyde in certain biological systems"; in particular, mutagenicity in *Drosophila* was found for both this (supposedly) non-carcinogenic *N*-nitroso compound and formaldehyde. The general problem of this interpretation was also mentioned – formaldehyde is metabolised through formate to doubtless inert products. The ability to remove this hydroxymethylating carcinogen might vary considerably from tissue to tissue [although there is no doubt of its carcinogenicity, notably its ability, following inhalation exposure, to induce nasopharyngeal carcinoma in the rat (IARC 1982b)]; a covalent reaction of inhaled ^{14}C-labelled formaldehyde with DNA in the target tissue, the respiratory mucosa, was detected (Casanova-Schmitz et al. 1984). The evaluation of di-*N*-nitrosopentamethylene tetramine as a non-carcinogen (IARC 1976) may have been correct, nevertheless, according to Ashby (1981). This interesting case of what might be termed a "borderline" non-carcinogen thus raises the apparently as yet unsolved question as to whether formaldehyde can mediate in the carcinogenic action of *N*-nitroso compounds.

The general problem also appears to remain that *N*-nitroso compounds, particularly those activated through metabolism, are difficult to detect in vitro (Ashby 1981). A possible reason is that the systems used for activation in vitro do not always mimic in vivo activation (see e.g. Masson et al. 1983). This might be checked by comparative quantitative measurements of in vivo and in vitro alkylation of supposedly critical receptors such as DNA, but this has not apparently been done for in vitro systems, in contrast to the extensive in vivo data (see e.g. Lijinsky 1988). The expected high reactivity and concomitant short half-life of the supposedly activated intermediates imply that the site of activation must be sufficiently near, and accessible, to the target of alkylation. Presumably this is true in vivo for cells of tissues in which DNA alkylation can be demonstrated; generally non-alkylated tissues are thought not to activate the carcinogen. In instances in which in vitro systems fail to detect induced mutations by a carcinogen, such as with *N*-nitrosodiethanolamine (Lijinsky 1988), in contrast to the observed alkylation of DNA in the target organ, rat liver, it may be that the activated intermediate transported from the site of metabolism to DNA is too unstable to reach this target in the in vitro system; the relatively extensive alkylation of soluble protein in comparison with that of DNA in liver may also reflect this.

Such considerations do not apply to most *N*-alkyl-*N*-nitrosoureas, which generally appear to be activated to alkylating agents through an alkali-catalysed hydrolytic mechanism (see Fig. 1). [An exception is *N*-methyl-*N*-nitroso-*N'*,*N'*-

diethylurea, which appears to require metabolic activation (LIJINSKY 1986).] Thus *N*-methyl-*N*-nitrosourea transfers the intact methyl group to a variety of *N*- and *O*-atom sites in DNA (LAWLEY and SHAH 1973). The supposed intermediates in this case are too unstable to be isolated, but the transient existence of alkyldiazonium ions can be demonstrated in extremely acidic solutions (MCGARRITY and COX 1983). The half-life of *N*-methyl-*N*-nitrosourea in neutral aqueous solution at 37° C is about 15 min, but it is quite stable in unbuffered water or saline, since the acid that is liberated by its initial hydrolysis rapidly suffices to inhibit further decomposition. Therefore this compound can be used as a potent water-soluble carcinogen that rapidly methylates throughout the body following injection in unbuffered saline (e.g. FREI et al. 1978). On the basis of

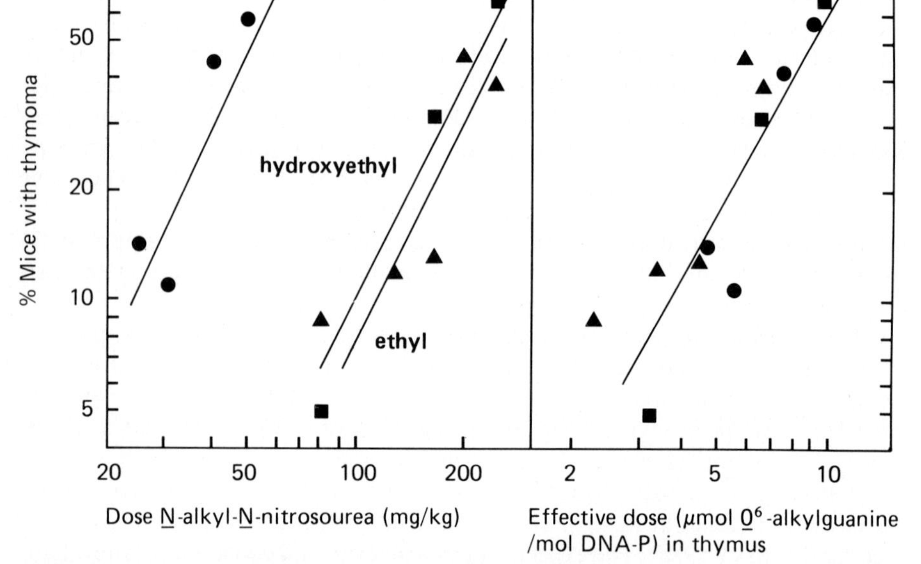

Fig. 5. The O-6 atom of guanine in thymus DNA as the significant target of *N*-alkyl-*N*-nitrosourea in the induction of thymic lymphoma in C57BL/Cbi mice. The proportion of mice with thymoma up to 250 days after a single injection of *N*-alkyl-*N*-nitrosourea is shown (*left*) (FREI et al. 1978; SWENSON et al. 1979, 1986). The effective dose to give 50% thymoma, correlating these three carcinogens (and also consistent with data for the weaker alkylating carcinogens methyl and ethyl methanesulphonate), is about 8 μmol O^6-alkylguanine per mol DNA-P, or about 9×10^4 such alkylations per genome (*right*). The dose dependence suggests a requirement for 2–3 such specific "hits" to induce thymoma. From the various proportions of alkylation products in DNA given by the carcinogens, it can be deduced that O^6-alkylation of guanine is the predominant tumour-initiating reaction, but the possibility that a proportion of tumours might result from other reactions, notably O^4-alkylation of thymine, cannot be ruled out (see the text). So far the principal base substitution associated with transforming oncogene activation detected in murine thymomas is GC→AT in the second position of codon 12 of Ki-*ras* (WARREN et al. 1987; W. WARREN, personal communication; instances of this mutation in N-*ras* have been reported by GUERRERO et al. 1986)

yield of tumours (thymic lymphoma in mice following a single injection by the intraperitoneal route), it must be considered as the most potent carcinogen per unit single dose so far found (FREI et al. 1978). Furthermore, the yields of tumours from five alkylating carcinogens covering a spectrum of reactivity showed that these were positively and quantitatively correlated with extent of DNA alkylation at O-6 of guanine in the assumed target organ, the thymus, but not with alkylation at any other site (FREI et al. 1978; SWENSON et al. 1986) (Fig. 5).

The implication of this type of alkylation as a quantitative determinant of carcinogenesis agrees with the concept that base-substitution mutations can initiate cancer. In this case the type of mutation expected is the GC→AT transition (see Fig. 2), and such mutations in *N*-methyl-*N*-nitrosourea-induced thymomas have recently been detected [so far, in about 23% of the tumours, at the second position of codon 12 in the Ki-*ras* gene in AKR mice (WARREN et al. 1987; W. WARREN, personal communication)].

It must be admitted, however, that this quantitative correlation is not sufficiently accurate to rule out minor contributions to total initiation of tumours through other mutagenic mechanisms, as for example TA→CG transitions (so far not detected) through miscoding of O^4-alkylthymidine (Fig. 2). From studies using polydeoxyribonucleotide templates containing various alkylated bases and DNA polymerase I, SAFFHILL and co-workers (ABBOTT and SAFFHILL 1979; SAFFHILL 1985) found that O^6-alkylation of guanine and O^4-alkylation of thymine were much more likely to cause base-substitution mutations through miscoding than other carcinogen-induced base modifications (Table 1). In assessing the relative miscoding effect per unit dose of carcinogen on this basis, it must be remembered that the ethylating agents generally give considerably less overall alkylation of target organ DNA than the methylating analogues. For example, in the case of mouse thymus DNA as a target of *N*-alkyl-*N*-nitrosoureas the O^6-

Table 1. Miscoding bases in DNA alkylated by *N*-nitroso-methyl or -ethyl compounds: relative occurrence and miscoding efficiencies

Alkylated base	Percentage of all products in alkylated DNA[a]	Miscoding efficiency		
		Observed per molecule[b]	Per alkylated DNA (rel.)	Estimated per unit dose in vivo (rel.)[c]
O^6-Methylguanine	7.5	up to 1.0[d]	1.0	1.0
O^6-Ethylguanine	7.9	up to 1.0[e]	1.0	0.18
O^4-Methylthymine	0.7	0.83	0.08	0.08
O^4-Ethylthymine[f]	1.0	0.81	0.10	0.02

[a] BERANEK et al. (1980); LAWLEY and SHAH (1973).
[b] ABBOTT and SAFFHILL (1979); SAFFHILL (1985); refers to DNA polymerase I in vitro.
[c] Refers to DNA of thymus of mice 1 h after injection (i.p.) of *N*-alkyl-*N*-nitrosoureas (FREI et al. 1978).
[d] Decreases as concentration of dCTP in incubation mixture increases.
[e] Assumed same as methyl.
[f] O^4-*iso*propylthymine also miscodes in this system (SINGER et al. 1986).

alkylation of guanine was 19 μmol/mol DNA-P for methylation, by a 1 mmol/kg dose of *N*-methyl-*N*-nitrosourea, as against 3.4 for ethylation, in agreement with the observed higher equicarcinogenic doses for the latter (see Fig. 5).

The mutagenic action of *N*-nitrosodialkylamines is expected to parallel that of *N*-alkyl-*N*-nitrosoureas reacting through the same alkyldiazonium ions (Fig. 1) in so far as it is ascribed to such alkylation, but other possible mutagenic metabolites have been considered. Thus, the mutagen nitrite can result from enzymic oxidative denitrosation of *N*-nitrosodialkylamines (APPEL and GRAF 1982). This acts through deamination of cytosine to uracil, or, more effectively, of 5-methylcytosine to thymine, since the latter presumably cannot be removed from DNA by repair enzymes (COULONDRE et al. 1978). Available evidence contraindicates nitrite as a predominant mediator of nitrosamine-induced mutation. Comparative studies of the mutagenic effectiveness of *N*-nitrosodimethylamine in rat or hamster cells show a quantitative, positive correlation with the ability of cellular microsomes to demethylate the carcinogen, i.e. to liberate the alkyldiazonium ion, but no such correlation is found with the ability to denitrosate, liberating nitrite (YOO et al. 1987).

Relatively recently, a proposed alternative mechanism for O^6-alkylguanine-induced mutation invokes deamination of complementary cytosine (WILLIAMS and SHAW 1987) rather than the direct miscoding by O^6-alkylguanine as proposed to occur in the in vitro studies (ABBOTT and SAFFHILL 1979). This would therefore also be expected to be most effective at the so-called hot spots where alkylated guanine is complementary to 5-methylcytosine, deaminated to the non-reparable thymine, i.e. in CG sequences of mammalian DNA [or in the second position of CCAGG sequences in certain strains of *Escherichia coli* (COULONDRE et al. 1978)].

The pattern of base-pair sites at which GC→AT transitions are preferentially induced by the methylating carcinogen *N*-methyl-*N*′-nitrosoguanidine is not in agreement with this prediction (COULONDRE and MILLER 1977). This mutagen, although methylating through the same intermediates as *N*-methyl-*N*-nitrosourea, hydrolyses spontaneously much more slowly at neutral pH, and its rapid decomposition in vivo, with concomitant methylating ability, results from reaction with thiol groups rather than with hydroxyl ions (Fig. 6a); the methylation pathway is not exclusive, since alternative attack by thiols on the nitroso group yields cystine, possibly potentiating reactions mediated by intermediate free radicals. *N*-Alkyl-*N*-nitrosourethanes are similarly activated by thiol reaction (Fig. 6b). The alkylating mutagen ethyl methanesulphonate gives a spectrum of mutations almost identical with that of the nitroso compound, e.g. more than 99% GC→AT transitions, consistent with almost complete predominance of alkylation of O-6 guanine as the cause of these mutations.

A disadvantage of the study by COULONDRE and MILLER (1977) is that the mutations were limited to those inducing or reverting the "stop" codons, amber and ochre. RICHARDSON et al. (1987) determined the base sequences of a spectrum of mutants induced (by *N*-methyl- or *N*-ethyl-*N*-nitrosoureas) in the plasmid pSV2gpt in *Escherichia coli*, thus permitting a wider variety of codons to be mutant sites. While confirming the overall conclusion of the previous study (Table 2) that the predominant base substitution was the GC→AT transition,

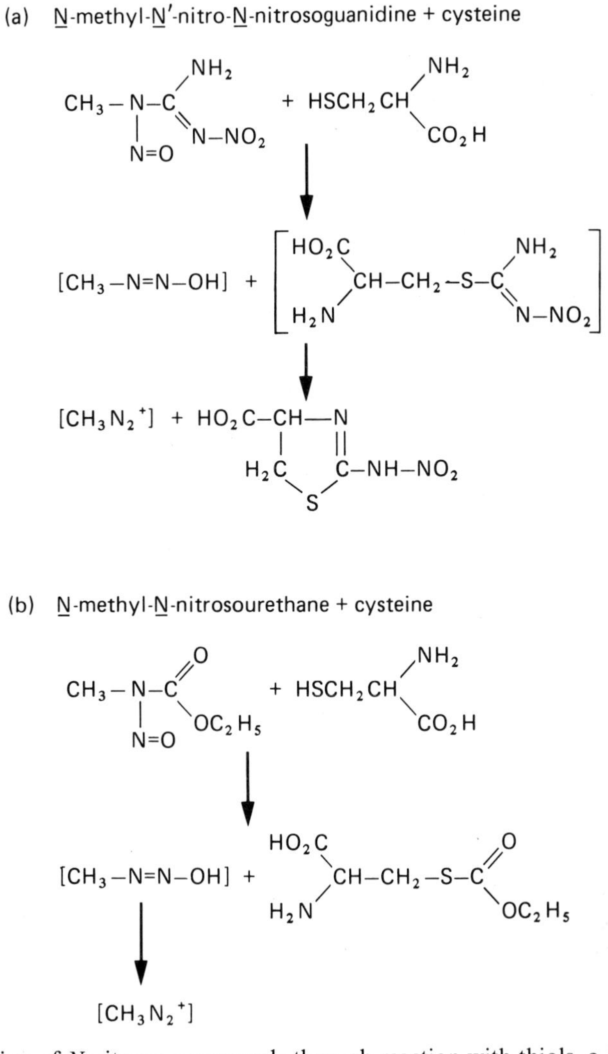

(a) N-methyl-N′-nitro-N-nitrosoguanidine + cysteine

$CH_3-N(N{=}O)-C(NH_2){=}N-NO_2$ + $HSCH_2CH(NH_2)CO_2H$

↓

$[CH_3-N{=}N-OH]$ + $[HO_2C(H_2N)CH-CH_2-S-C(NH_2){=}N-NO_2]$

↓

$[CH_3N_2^+]$ + $HO_2C-CH-N{=}C(-NH-NO_2)-S-CH_2$ (ring: $HO_2C-CH-N$, H_2C, $C-NH-NO_2$, S)

(b) N-methyl-N-nitrosourethane + cysteine

$CH_3-N(N{=}O)-C({=}O)OC_2H_5$ + $HSCH_2CH(NH_2)CO_2H$

↓

$[CH_3-N{=}N-OH]$ + $HO_2C(H_2N)CH-CH_2-S-C({=}O)OC_2H_5$

↓

$[CH_3N_2^+]$

Fig. 6a, b. Activation of *N*-nitroso compounds through reaction with thiols. **a** *N*-Methyl-*N*′-nitro-*N*-nitrosoguanidine: the preferred nitrimine structure (RICE et al. 1984) is shown; for thiol reactions see LAWLEY and THATCHER (1970) and SCHULZ and MCCALLA (1969); not all reactions lead to methylation, since the nitroso group is attacked by cysteine with the possibly concomitant generation of free radicals. **b** *N*-Methyl-N-nitrosourethane: the principal reaction pathway is shown (SCHOENTAL and RIVE 1965)

they also found a remarkable preference for mutation to occur at the second guanine in sequences GG(A or T); this, as noted later, is a predominant site of base-substitution-activating oncogenes.

Both studies agreed in finding a very low (or zero) proportion of AT→GC transitions among mutations induced by the methylating nitroso compounds, but as expected from the higher extent of ethylation at O-4 of thymine in DNA

Table 2. Base substitutions by *N*-nitroso compounds in *Escherichia coli*

Mutagen	Gene	Mutation (frequency)	Remarks	References
N-methyl-*N'*-nitro-*N*-nitroso-guanidine	*lac* I	GC→AT transitions (502/518) TGG (trp)→TAG (amber) CAG (gln)→TAG (amber) CAA (gln)→TAA (ochre) AT→GC transitions (1/518) (reversions from amber or ochre)	Study limited to induction of, or reversion from, amber and ochre codons	COULONDRE and MILLER (1977)
N-Methyl-*N*-nitrosourea	pSVgpt plasmid	GC→AT transitions (39/39)	Predominantly (85%) at 5'-GG-(A or T)-3' sequences	RICHARDSON et al. (1987)
N-Ethyl-*N*-nitrosourea		GC→AT transitions (24/33)	Predominantly (71%) at 5'-GG-(A or T)-3'-sequences	
		AT→GC transitions (7/33) GC→CG transversions (1/33) AT→CG transversions (1/33)		

relative to that at O-6 of guanine (reflecting the slower rate of removal by repair of the alkylthymine) these were more frequent for *N*-ethyl-*N*-nitrosourea (RICHARDSON et al. 1987). Differential repair (of O^6-methylguanine) may also contribute to the preferential mutation at specific guanine base sites (such as the second position of GGT or GGA sequences) (TOPAL et al. 1986). Observations that ethylation induces AT→GC transitions are consistent with the results of studies with ethylated polydeoxyribonucleotide templates (see Table 1).

As previously mentioned, other reaction pathways that could conceivably result in mutations include participation of aldehydes generated by oxidative metabolism of *N*-nitroso-*N*-alkylamines or cyclic *N*-nitrosamines, which react at extranuclear amino groups of DNA bases (FELDMAN 1973).

IV. *N*-Nitroso Compounds as Mutagenic Activators of Oncogenes

It is now possible to determine the nature of base-substitution mutations (often regarded, with less precision, as synonymous with the previously used term "point" mutations) induced by *N*-nitroso compounds. Then it can be argued what is the most likely cause of such specific changes in DNA structure. The most intensively studied mutations are those associated with activation of oncogenes, as first demonstrated by BARBACID and co-workers (REDDY et al. 1982).

The rationale previously outlined showed that the significant targets of alkylating carcinogens in mice are O-6 atoms of guanine. From the observed dose and time dependence of tumorigenesis (FREI and LAWLEY 1980) it was deduced that the process requires at least two and possibly three critical alkylations and a further time-dependent event. The requirement for multiple alkylations implies that, despite the extensive alkylation associated with the mean tumorigenic dose (i.e. that reducing survival of non-tumour-bearing mice to 0.37) covering a spectrum of values for strains of mice with various susceptibilities (FREI 1980) (from around 60000 O^6-alkylations of guanine per genome for the relatively sensitive RFM females), the available critical sites would be small in number, probably less than ten, because alkylations are expected to occur fairly randomly (CHANG et al. 1979; OSBORNE 1984).

It now seems reasonably clear that this expectation is fulfilled for at least one of the critical alkylations. The number of specific base sites in DNA at which oncogene-activating substitutions have been found to occur is indeed limited, and in the already classic case of mammary tumours induced by a single dose of *N*-methyl-*N*-nitrosourea in Buf/N rats (ZARBL et al. 1985) striking specificity was found as 48 out of 58 tumours showed activated oncogenes, and in all 48 adenine replaced guanine as the second base of codon 12 in the first exon of the Ha-*ras* gene. Thus, activation resulted from a GC→AT transition, which appears to be in good agreement with the expected molecular mechanism and also with the observed predominant type of mutation induced by methylation in *E. coli* [with *N*-methyl-*N'*-nitro-*N*-nitrosoguanidine, more than 98% (COULONDRE and MILLER 1977), the remainder being AT→CG transversions, although it should be noted that, in the system used, AT→GC transitions could not be scored] (see Table 2).

Some other studies of oncogene-activating mutations in tumours induced by *N*-nitroso compounds (Table 3) report base substitutions that do not accord with the principal expected GC→AT transition as predicted on chemical grounds.

Thus TA→AT transversions have been found (BARGMANN et al. 1986) in four cell lines derived from neuroblastomas of BDIX rats treated neonatally with *N*-ethyl-*N*-nitrosourea (SCHUBERT et al. 1974); this type of transversion was also found in mutant β-globin genes of offspring of mice treated with the same carcinogen (LEWIS et al. 1985).

In organs in which repair removal of O^6-alkylguanines is extensive (such as the liver of rodents), the less easily removed O^4-alkylthymines should predominate as promutagenic groups inducing TA→CG transitions (as observed for *E. coli*, see Table 2). The only report so far of this type of mutation in oncogene activation is in mouse liver carcinomas induced by *N*-nitrosodiethylamine (STOWERS et al. 1988); it was pointed out that, in the strain of mouse used, spontaneous oncogene activation occurs, so that it was uncertain whether the carcinogen had directly caused these mutations.

The paucity of data so far implicating TA→CG transitions in oncogene activation appears to contradict the prediction from in vitro miscoding of O^4-alkylthymine with guanine (SAFFHILL 1985) but may be held to support the view of LI et al. (1987) that these alkylated bases do not form stable hydrogen bonds with guanine, contrary to the deductions of BRENNAN et al. (1986) on physicochemical grounds and of RICHARDSON et al. (1987) from studies on

Table 3. Base substitutions activating proto-oncogenes in tumours induced by *N*-nitroso compounds

Carcinogen	Animal	Tumour	Oncogene	Codon no. (position)	Type of mutation[a]		No. observed / No. examined	References
N-Methyl-*N*-nitrosourea	Rat (Buf/N)	Mammary	Ha-*ras*	12 (2)	GGA (gly) → GGA (glu)	GC → AT transition	48/58	Zarbl et al. (1985)
	Mouse (AKR/RF)	Thymoma	N-*ras*	12 (2)	GGT (gly) → GAT (asp)	GC → AT transition	1/5	Guerrero et al. (1986)
	Mouse (AKR)	Thymoma	Ki-*ras*	12 (2)	GGT (gly) → GAT (asp)	GC → AT transition	9/37	Warren et al. (1987); W. Warren (personal communication)
	Mouse (AKR/RF)	Thymoma	N-*ras*	61 (1)	CAA (leu) → AAA (lys)	CG → AT transversion	1/5	Guerrero et al. (1986)
	Rat (F344)	Bladder cell line[b]	Ki-*ras*	12 (2)	GGT (gly) → GAT (asp)	GC → AT transition	1/1	Knowles et al. (1987)
N-Ethyl-*N*-nitrosourea	Rat (BDLX)	Neuro/glioblastoma cell lines	*neu*	664 (2)	GTG (val) → GAG (glu)	TA → AT transversion	4/6	Bargmann et al. (1986)
	Rat (F344/NCr)	Schwannoma	*neu*	664 (2)	GTG (val) → GAG (glu)	TA → AT transversion	11/13	Perantoni et al. (1987)
N-Methyl-*N'*-nitro-*N*-nitrosoguanidine	Guinea pig	Transformed cell line	N-*ras*	61 (3)	CAA (gln) → CAT (his)	AT → TA transversion	1/1	Doniger et al. (1987)
N-Nitrosodiethylamine	Mouse ($B6C3F_1$)	Liver adenoma	Ha-*ras*	61 (1)	CAA (gln) → AAA (lys)	CG → AT transversion	4/22	Stowers et al.[c] (1988)
				61 (2)	CAA (gln) → CTA (leu)	AT → TA transversion	4/22	
		Liver carcinoma	Ha-*ras*	61 (1)	CAA (gln) → AAA (lys)	CG → AT transversion	3/11	
				61 (2)	CAA (gln) → CGA (arg)	AT → GC transition	3/11	

[a] The sequences shown are codons in DNA, the direction being 5′-GGT- etc. corresponding to 5′-GGU etc. transcribed codons in mRNA.
[b] Non-epithelial cell line; 4/4 tumorigenic epithelial cell lines did not contain activated *ras*.
[c] 1/28 rat liver tumours induced by *N*-nitrosodiethylamine showed transforming DNA but did not contain activated *ras*.

mutagenesis in *E. coli*. A further discrepancy between the type of observed oncogene-activating mutation and the predicted predominant base substitution is the CG→AT transversion in the N-*ras* oncogene, found for a mouse thymoma induced by *N*-methyl-*N*-nitrosourea (GUERRERO et al. 1986).

In the case of the activation of N-*ras* in guinea pig embryo cell lines transformed in vitro (DONIGER et al. 1987), the AT→TA transversion, at position 3 of codon 61, was found in five instances of transformation by four different carcinogens, *N*-methyl-*N'*-nitro-*N*-nitrosoguanidine, *N*-nitroso-diethylamine, benzo[*a*]pyrene or 3-methylcholanthrene. It was considered that this transversion is unlikely to be induced directly by the two types of carcinogen (alkylating or aralkylating) since their predominant reactions were with guanine rather than adenine [the predominant mutagenic mechanism in *E. coli* reported for the reactive metabolite of benzo[*a*]pyrene being GC→TA (EISENSTADT et al. 1982)].

Although this suggestion is correct with respect to the major promutagenic reactions of these carcinogens, there is evidence for alternative mechanisms. Thus, with the diol epoxide metabolite of benzo[*a*]pyrene, 18% of base substitutions induced in *E. coli* were AT→TA transversions, presumably due to aralkylation of adenine at N-6 (DIPPLE et al. 1984).

N-Nitroso compounds alkylate DNA at a variety of sites, including N-1, N-3 and N-7 of adenine (FREI et al. 1978; BERANEK et al. 1980). The cytotoxic action of alkylation of adenine is indicated by rapid enzymic removal of 3-methyladenine (LAWLEY and WARREN 1976; KARRAN et al. 1980), and this type of DNA repair appears to occur at much the same rate in different types of cell, in contrast to the marked variations in ability to remove the promutagenic base O^6-methylguanine (e.g. see for human lymphocytes from various individuals, LAWLEY et al. 1986).

Furthermore, alkylation of adenine at N-1 blocks hydrogen-bonding in DNA essential for Watson-Crick base pairing, and 1-methyladenine was found to persist in DNA of mice methylated by *N*-methyl-*N*-nitrosourea (FREI et al. 1978).

Chemical modifications of the DNA template that appear to impair the action of DNA polymerase (but do not correspond with any plausible mechanism for miscoding such as those shown in Fig. 2) have been proposed as sources of mutation, and evidence has been found and summarised, notably by STRAUSS and co-workers (STRAUSS et al. 1982, 1986). The general deduction was that purines (more particularly adenine) were preferentially inserted opposite non-instructional bases, and therefore, inactivated pyrimidines would tend to cause transitions, inactivated purines, transversions; evidence in support was documented.

With regard to the non-bulky substituents introduced into DNA by *N*-nitroso compounds transferring methyl and ethyl groups, the role of potentially cytotoxic DNA polymerase-blocking reactions, which are induced by weak carcinogens such as methyl methanesulphonate, is as yet unclear. Suggestions that methylation at Watson-Crick hydrogen-bonding sites (notably N-3 of cytosine, N-3 or O-2 of thymine) are sources of mutation are associated mainly with SINGER and co-workers (e.g. reviewed by SINGER and KUSMIEREK 1982). Miscoding by the corresponding alkylated bases was detected in vitro using RNA polymerases, but corresponding studies using DNA polymerase showed, at most, weak effects. For example O^2-methylthymine showed less than one-tenth of the mutagenic efficiency of O^4-methylthymine for miscoding with guanine (SAFFHILL 1985); mispairing with thymine (such as to give TA→AT transversion) could not be detected in this system.

The overall conclusion remains, therefore, that miscoding alkylated bases, notably O^6-alkylguanines, are likely to be the predominant sources of induced mutation activating oncogenes, provided of course that they are not removed by repair from the DNA template before it is used for replication (LAWLEY and ORR 1970; CRADDOCK and HENDERSON 1984; SWENBERG et al. 1987; for a general review, see FRIEDBERG 1985).

In view of the variety of alkyl groups introduced into DNA by the numerous *N*-nitroso compounds used as carcinogens, it should be emphasised that the chemical nature of the group influences not only the rate of repair but also the type of repair system. This was first noted by WARREN and LAWLEY (1980), comparing removal of O^6-ethyl- and O^6-methylguanines from *E. coli*. They reasoned that if methylation damage was removed by repair mechanisms different from that found for removal of "bulky" adducts associated with cytotoxicity, e.g. alkylation-induced cross-links (LAWLEY and BROOKES 1968), as the size of the alkyl group increased, so would the mode of repair, with perhaps both types of mechanism operating, but in different proportions for different alkyl substituents in DNA. In the case quoted, it was deduced that O^6-ethylguanine is removed mainly by the relatively non-specific excision repair system that removes pyrimidine dimers or cross-links from DNA and also other bulky adducts such as aralkylated bases (see e.g. ROBERTS 1978), whereas the removal of O^6-methylguanine depends on a different, more specific and inducible enzyme.

With mammalian cells the same principle applies, but there are quantitative differences between various cell types. The specific removal of O^6-alkylguanines from either most mammalian cells or *E. coli* was subsequently found to be, strictly speaking, non-enzymic, since it involves transfer of the alkyl groups from DNA to a cysteinyl residue of an alkyl acceptor protein which is thereby inactivated (OLSSON and LINDAHL 1980; reviewed by YAROSH 1985); despite this, the methyltransferase terminology persists. A principal difference between the originally discovered *E. coli* and the mammalian quasi-enzymes so far studied is that the latter are constitutive (although differing quantitatively from one type of cell to another), whereas the bacteria have a low constitutive level with rapid inducibility through protein synthesis in response to damage; the mammalian system is therefore "saturated" when extent of alkylation of DNA is increased to equal the number of molecules of quasi-enzymes available (in human lymphoid cells, the most proficient so far found, about 100000 molecules per cell) (LAWLEY et al. 1986).

In the liver of rats, pretreatment of the animals with various cytotoxic and carcinogenic agents [including *N*-nitrosodimethylamine and its ethyl, propyl and butyl analogues (MARGISON 1982)] can enhance (by about fivefold) repair of *N*-nitrosodimethylamine (but not apparently *N*-methyl-*N*-nitrosourea-)-induced O^6-methylguanine. This effect (which does not occur in hamsters) has been referred to as "induced" repair, but the mechanism apparently remains obscure (see CRADDOCK et al. 1982; MONTESANO et al. 1983; RENARD and VERLY 1983).

The so-called alkyltransferase type repair functions less effectively as the size of the O^6-alkyl substituent on DNA guanine increases through methyl, ethyl, *n*-propyl and *n*-butyl, with *iso*-propyl, *iso*-butyl and 2-hydroxymethyl being increasingly more difficult to remove (MORIMOTO et al. 1985).

Removal of O^6-alkylguanines from DNA by excision repair should be recognized in human cells by impaired repair proficiencies of xeroderma pigmentosum patients who lack the ability to remove UV-induced pyrimidine dimers or bulkier chemical adducts. This does not appear to be the case for O^6-methylguanine but has been reported for O^6-ethylguanine (MAHER et al. 1986) in human fibroblasts. Furthermore, evidence has been obtained (BOYLE et al. 1986) for removal of O^6-*n*-butylguanine in Chinese hamster and human cells by excision repair. It seems likely therefore that O^6-alkylguanine can be removed by excision of oligonucleotides containing the adduct and that this becomes more prominent the larger the alkyl substituent. It may be noted here that *N*-nitroso-*N*-aralkylureas may be expected to aralkylate DNA at the N-2 atom of guanine, i.e. the site of attack by ultimate carcinogenic metabolites of polycyclic aromatic hydrocarbons; MOSCHEL et al. (1980) showed that *N*-nitroso-*N*-benzylurea aralkylates at N-2, N-7 and O-6 of guanosine; probably N^2-aralkylguanine in DNA would also be repaired through excision.

Of particular interest is the finding that formation of guanine-cytosine interstrand cross-links in DNA inducible by chloroethylating drugs such as 1,3-bis(2-chloroethyl)-1-nitrosourea (BCNU) can be largely prevented through the action of the alkyltransferase type of repair (LUDLUM et al. 1986b) (Fig. 7). This contrasts with guanine-guanine cross-linking by nitrogen and sulphur mustards, which was found to respond to excision repair in *E. coli* (LAWLEY and BROOKES 1968) and mammalian cells (for review, ROBERTS 1978; FRIEDBERG 1985); moreover, the O^6-alkylguanine induced by mustard gas is, at most, weakly responsive to alkyltransferase action (LUDLUM et al. 1986a), thus reinforcing the difference between the chloroethylating *N*-nitroso compounds and the mustards.

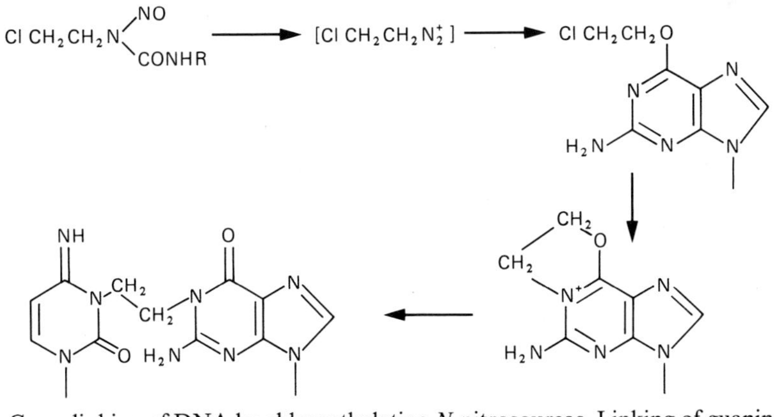

Fig. 7. Cross-linking of DNA by chloroethylating *N*-nitrosoureas. Linking of guanine and cytosine bases is shown; this is prevented by repair dealkylation of O^6-chloroethylguanine; alternative reactions are possible (see Ludlum 1986). In semustine (MeCCNU), *R* = *trans*-*cyclo*hexyl, —⬡—CH_3

Notable differences between the bacterial and mammalian repair of *O*-alkylated DNA products are that O^4-alkylthymines and phosphotriesters are less efficiently removed by the alkyltransfer mechanism than are O^6-alkylguanines in mammalian cells (see e.g. YAROSH et al. 1985). For continuous treatment of animals with alkylating *N*-nitroso compounds, therefore, O^4-alkylthymines tend to persist and, as the dose is increased, to accumulate in DNA, as for example in liver of rats fed *N*-nitrosodiethylamine (SWENBERG et al. 1984, 1987). This could therefore favour activation of oncogenes through alkylation of thymine rather than of guanine, more especially in target organs that are proficient in removal of O^6-alkylguanines. So far no activation of oncogenes through the TA→CG transition, as predicted to result from the in vitro studies (Table 1), appears to have been reported.

In mice and rats, the liver is the most proficient organ for repair of O^6-alkylguanines (O'CONNOR et al. 1973; FREI et al. 1978), brain the least proficient (GOTH and RAJEWSKI 1974a, b), with other organs intermediate (CRADDOCK and HENDERSON 1984). It might be expected, therefore, that not only would the repair factor influence organotropism of the various *N*-nitroso compounds (see later) but might also determine the nature of oncogene activations. So far little evidence is available to enable assessment of this question; somewhat surprisingly few reports on oncogene activation by base substitution in liver of rodents by *N*-nitroso compounds appear to be available. FUNATO et al. (1987) found that only 1 of 18 liver tumours induced in F344 rats by *N*-nitroso-dibutylamine gave DNA capable of transforming NIH 3T3 cells, which contained activated N-*ras*, but the mode of activation was not determined. They consider that activation through point mutation may be more characteristic of tumours obtained by single treatments with carcinogens, rather than by continuous treatment as is often required to induce liver tumours. STOWERS et al. (1988) reported activation of Ha-*ras* in tumours induced in the liver of B6C3F$_1$ mice by *N*-nitrosodiethylamine, but as they note this strain is subject to extensive spontaneous hepatoma initiation, and the pattern of mutations (see Table 2) may not reflect direct action of the carcinogen. In rats, only 1 of 28 liver tumours showed oncogene activation, as detected through transfection by tumour DNA, and no *ras* activation was found.

Current relevant data on mouse liver tumours induced by several other carcinogens, and also those supposedly of spontaneous occurrence have been reviewed (REYNOLDS et al. 1987). It is of interest that one of the carcinogens was an aldehyde, furfural, which has been regarded as non-mutagenic on the basis of rapid-screening; despite this, furfural-induced hepatocarcinomas and adenoma in C57BL X C3H (F1) mice showed oncogene activation of Ha-*ras*, including one AT→GC transition at codon 61(2), five CG→AT transversions at 61(1), GC→TA transversions at 13(2) and 117(3), and a GC→CG transversion at 13(1). The question (as yet unsolved) was raised whether these are mutations of spontaneous origin or due to direct action of the aldehyde. This is of some relevance to the action of *N*-nitrosodialkylamines as hepatocarcinogens, since they generate aldehydes through metabolic activation as well as alkylating agents, and these react at a variety of sites in DNA, notably extranuclear amino groups (FELDMAN 1973).

It should be emphasised here that activation of oncogenes through base-substitution mutations is not the only mechanism through which mutagenic carcinogens exert their effects. One aspect of their genotoxic action is to cause breakage of chromosomes detectable as sister chromatid exchange and chromosomal aberrations. As might be expected from the chemistry of DNA alkylation, these effects are not so obviously associated quantitatively with specific sites of alkylation as is base substitution. Per unit dose, the comparatively weak *O*-alkylator and weak carcinogen methyl methanesulphonate was somewhat less active as an inducer of sister chromatid exchange (about two- to threefold) in Chinese hamster V79 cells in vitro than was *N*-methyl-*N*-nitrosourea, but more so (by about the same factor) than *N*-ethyl-*N*-nitrosourea (SWENSON et al. 1980). From the known extents of alkylation by these agents at various sites in DNA, it was deduced that alkylation of adenine at N-3 (as previously noted, thought to be a powerful source of cytotoxic action) could correlate with the ability to induce chromosome breakage, although from a comparison of *N*-methyl-*N*-nitrosourea and dimethyl sulphate it was hypothesised that the major methylation at N-7 and other minor methylations might also contribute (CONNELL and MEDCALF 1982). Similarly, in the mouse in vivo, chromosome damage in bone marrow was caused by the weak carcinogen methyl methane sulphonate as well as by the very much more potent *N*-methyl-*N*-nitrosourea, although the latter did induce more than twice as many breaks and exchanges at equitoxic doses (FREI and VENITT 1975).

Chromosomal abnormalities in tumours are often (but not invariably) observed and appear generally (but again not always) (NOWELL 1986) to be associated with later stages of carcinogenesis (see e.g. SANDBERG 1980). Methylation-induced murine thymomas, in common with those induced by X-irradiation or virus, often but not invariably exhibit trisomy of chromosome 15, less so when the carcinogen (*N*-methyl-*N*-nitrosourea) was injected into adult, as opposed to neonatal, mice (CHAN et al. 1981).

A more detailed analysis of a chromosomal rearrangement thought to be caused by an *N*-nitroso compound, the methylating carcinogen *N*-methyl-*N′*-nitro-*N*-nitrosoguanidine, is that associated with the activation of the *met* oncogene in a human osteogenic sarcoma cell line (PARK et al. 1986; TEMPEST et al. 1986). Prolonged treatment of the non-tumorigenic cell line with the carcinogen yielded a cell line tumorigenic in nude mice, in which the DNA was (like carcinogen-activated, *ras*-gene-containing DNAs) able to cause transformation of NIH 3T3 cells through transfection. The activation of the oncogene in this instance was found to be due to fusion between 5′-sequences from the *tpr* locus in chromosome 1- and 3′-sequences from the *met* proto-oncogene on chromosome 7. This is the first report of oncogene activation via chromosome rearrangement following chemical carcinogen treatment of human cells in culture and which is therefore analogous to the type of oncogene activation associated with the well-known Philadelphia chromosome translocation in chronic myeloid leukaemia (see NOWELL 1986).

A further mechanism of possible significance in carcinogenesis by *N*-nitroso compounds is the deletion of oncogenes, as reported in one instance for a murine thymoma induced by *N*-methyl-*N*-nitrosourea (GUERRERO et al. 1985);

mutational activation of the N-*ras* oncogene was accompanied by loss of the normal allele; this does not generally occur as a concomitant of oncogene activation (Bos et al. 1984).

There may appear at the outset to be some discrepancy between the assumed random attack of alkylating carcinogens throughout DNA and the specific nature of the mutagenic or chromosome-breaking events associated with activation of oncogenes. It would be particularly difficult, for example, to envisage that the limited number of single base sites in *ras* oncogenes at which activating base substitutions occur should be in any way specifically susceptible to alkylation. The general assumption is that these hot spots reflect the requirement for specific amino acid substitutions to result from the mutations, in order to cause appropriate conformational change in the protein products of these genes, which can then act in pseudo-dominant fashion in transformation of cells from the normal to the tumorigenic state (e.g. see CLANTON et al. 1987; BARGMANN et al. 1986; SHIH and WEEKS 1984).

With regard to chromosome breakage, it has been proposed from time to time that there are fragile sites on chromosomes which are somewhat specifically susecptible to various types of carcinogens. Thus, YUNIS et al. (1987) commented that the fragility of sites on chromosomes of human lymphocytes susceptible to the action of dimethyl sulphate or benzo[*a*]pyrene probably reflects a relatively high guanine content in their DNA. When cellular DNA is fractionated in various ways, after treatment of animals with carcinogens, there appear from some earlier studies to be no very marked preferences for reaction with specific regions of DNA (cf. CHANG et al. 1979 for unique or repetitive sequences of rat liver DNA and *N*-methyl-*N*-nitrosourea). Using isolated rat liver chromatin and *N*-ethyl- or *N*-methyl-*N*-nitrosoureas, MARUSHIGE and MARUSHIGE (1983) found identical alkylation of nucleosomal core and linker DNAs.

However, another study using this type of fractionation following in vivo treatment of rats found linker DNA to be more extensively methylated than core by *N*-nitrosodimethylamine (RAMANATHAN et al. 1976). HeLa cells treated with *N*-methyl-*N*-nitrosourea give a similar result (HELLER and GOLDTHWAIT 1983), core DNA is shielded from methylation, i.e. it was 12% of the extent found for free DNA; with linker DNA this shielding is less (67%), but linker DNA is less accessible to the *N*-glycosylase repair enzyme that removes 3-methyladenine.

Removal by repair of methylpurines from DNA was also investigated by RYAN et al. (1986), who fractionated nuclei of liver of rats treated with *N*-nitrosodimethylamine to obtain transcriptionally active chromatin and a nuclear matrix fraction involved in semi-conservative replication of DNA. They found that promutagenic O^6-methylguanine tended to persist more in the DNA of the nuclear matrix; the presumed cytotoxic lesion 3-methyladenine was very rapidly removed from the DNA of all fractions.

Studies on the alkylation of synthetic oligodeoxyribonucleotides have supported the concept that the base sequence influences the reactivity of guanine and adenine towards *N*-methyl-*N*-nitrosourea (BRISCOE and COTTER 1985). For example, formation of promutagenic O^6-methylguanine was relatively highest in GGG sequences, lowest in TGT; no obvious correlation with the sequences associated with specific hot spot bases in oncogenes has emerged so far. However,

TOPAL et al. (1986) presented evidence that base sequence can influence repair removal of O^6-methylguanine and suggested that guanine in position 2 of codon 12 in *ras* may be subject to steric hindrance to such repair. As previously noted (see Table 2) mutations induced in a plasmid in *E. coli* preferentially occur at a site homologous with that of *ras* codon 12(2); whether the same phenomenon would occur in eukaryotic cells is as yet not known.

In summary, therefore, the present consensus favours fairly random chemical attack by alkylating carcinogens on cellular DNA, with some tendency for promutagenic alkylations to persist in nuclear matrix DNA but to be more rapidly removed from transcriptionally active DNA; possibly this could cause variations in the probability of oncogene activation in different types of target cells. From studies of excision repair, HANAWALT and co-workers (e.g. BOHR et al. 1986) proposed that transcribed regions of DNA coding for essential enzymes are preferentially repaired. While specificity in sites of base substitution that activate oncogenes may result from the requirement for specific amino acid changes in the protein products, it is remarkable that the same specificity in sites of mutation-inducing alkylation (or of removal of promutagenic bases through DNA repair) has emerged from studies with *E. coli*.

B. Organotropism in Carcinogenesis by *N*-Nitroso Compounds

I. Role of *N*-Nitroso Compounds in Multistage Carcinogenesis

The first demonstration of carcinogenesis by an *N*-nitroso compound, viz. hepatocarcinogenesis in rats, involved continuous treatment with *N*-nitrosodimethylamine. This requirement for multiple treatments has generally (but not always) been found for carcinogenesis by chemicals, in the sense that the resultant tumours have characteristics generally denoted as malignant, i.e. capable of invasion of surrounding tissue and of metastasis.

The classical system for investigation of carcinogenesis by chemicals in general, leading to the discovery of the first pure chemical carcinogens, polycyclic aromatic hydrocarbons (as previously noted), involved application of carcinogens to the skin of mice. Although malignant carcinomas could be induced by continuous application, it became clear that the process of carcinogenesis occurred in discrete stages, the first visible manifestation being the production of benign (non-invasive) tumours (papillomas). In order to obtain even these in relatively high yields from a single application of hydrocarbons, it was necessary to use multiple treatments with a promoting agent, the first of which to be discovered was croton oil (BERENBLUM 1941). MOTTRAM (1945) first established the protocol for classic two-stage carcinogenesis by showing that further treatment of the induced papillomas with hydrocarbon enhanced the ultimate yield of malignant cancers.

It is now widely believed that this basic mechanism is general for carcinogenesis. For example, MOOLGAVKAR and KNUDSON (1981) have derived a mathematical treatment for the time dependence of carcinogenesis. The first critical event is an initiating mutation in a stem cell. Promotion favours proliferation of initiated cells rather than normal and may thus result in the formation of a benign intermediate tumour. The enhanced number of initiated cells is envisaged to make probable a second mutagenic event (or perhaps some quasi-mutation such as a chromosomal rearrangement) in an initiated cell; this confers malignancy. Various lines of evidence indicate that malignancy is accompanied by genetical instability (NOWELL 1986; NICOLSON 1987). Many cancers thus show a diversity of cell types, although there can now be little doubt that these originate from a single initiated cell through clonal expansion (FIALKOW 1979). The predominant features of time dependence of the incidence of human cancer can be accounted for quantitatively in terms of the two-stage theory as outlined above (MOOLGAVKAR and KNUDSON 1981).

An important corollary of this theory is that while cancer is initiated by a mutation, this of itself is not sufficient to cause even benign tumours, without promotion. Promoters are not necessarily mutagens; in a classical instance, human breast cancer has been deduced from combined epidemiological and biochemical studies to be promoted by oestrogens (BULBROOK et al. 1984). Whereas a certain (admittedly low) level of so-called spontaneous mutation is inevitable, the promotional factor may vary between individuals, and this, in algebraic terms, has a relatively large quantitative effect on the ultimate tumour yield from a given extent of initiation.

With regard to dose-response relationships, this theory implies that initiation, in so far as it results from an activating mutation, should be a single-hit irreversible process, and evidence is available for this (e.g. for mouse skin and benzo[*a*]pyrene, with exogenous promotion, see ALBERT and BURNS 1977). For exogenous, non-genotoxic promoting agents, available evidence favours the existence of no-effect threshold. This qualitative difference between mutagenic, carcinogenic and non-mutagenic promoters obviously affects assessment of their likely carcinogenic risk (see e.g. WILLIAMS 1987, for a recent survey).

For a so-called complete carcinogen inducing malignant tumours often by multiple or continuous treatments, it is generally found that the yield of tumours is dependent on a power of dose of carcinogen greater than unity (often 2–3), however, theoretically and in practice (see PETO et al. 1984) as the dose decreases, this power of dose dependence also decreases, tending to a value of unity. For the case quoted (induction of liver tumours in rats fed *N*-nitroso-dimethylamine or *N*-nitroso-diethylamine) the implication is that these acted as mutagens, and at low doses their effect is additive to a spontaneous yield of tumours, caused by a mutagenic action analogous to that caused by the applied carcinogen; this could account for the observed dose dependence.

These preliminary remarks serve to illustrate further the potential complexity of the process of complete carcinogenesis by *N*-nitroso compounds. In common with other classes of chemical carcinogen, it is not possible to classify these compounds as purely initiators, although there can be little doubt that, as mutagens, they are eminently capable of tumour initiation.

This has been well illustrated in several instances (Table 4). In the classic mouse skin system, initiation by a single application of *N*-methyl-*N*-nitrosourea followed by promotion with croton oil (WAYNFORTH and MAGEE 1975) shows a dose dependence approximating to the single-hit type. *N*-Nitrosoureas may also

Table 4. *N*-Nitroso compounds as initiators

Initiator	Target organ	Promoters	References
N-Methyl-*N*-nitrosourea	Mouse skin	Croton oil (phorbol ester)	WAYNFORTH and MAGEE (1975)
N-Nitroso-dimethylamine	Rat liver	Partial hepatectomy	CRADDOCK (1971)
N-Nitroso-diethylamine	Rat liver	Phenobarbital, dioxin, etc.	PITOT et al. (1987)
N-Nitroso-*n*-butyl- -*n*-(4-hydroxybutyl)amine	Rat bladder	Sodium salts, urine, etc.	COHEN et al. (1987)

possess some promoting ability, in so far as they can carbamoylate phosphotidylethanolamine in cell membranes (YANO et al. 1987) through the isocyanate liberated on hydrolysis (Fig. 1).

Single doses of *N*-methyl-*N*-nitrosourea can induce multiple mammary tumours in 50-day-old female rats (e.g. THOMPSON and MEEKER 1983). Although often denoted as adenocarcinomas, these tumours rarely metastasise (even when induced by three doses of carcinogen, given at 0, 31, and 83 days), and most were considered benign by standard human criteria (WILLIAMS et al. 1981).

The first promotional mechanism revealed in liver carcinogenesis by *N*-nitroso compounds was cell division, induced about 24 h after partial hepatectomy; a single dose of *N*-nitrosodimethylamine given during this period of DNA synthesis is thus potentiated to induce cancer, presumably because replication of template DNA containing initiating O^6-methylguanine occurs more frequently than in quiescent liver, where repair of this promutagenic methylpurine is more extensive (CRADDOCK 1971; CRADDOCK and HENDERSON 1984).

The earliest precancerous stage in liver carcinogenesis that can be detected histologically is the induction of hyperplastic nodules (POPPER et al. 1960). Using a single dose of *N*-nitrosodiethylamine as initiator, the number of these nodules was proportional to dose at lower doses (indicating a single-hit process), levelling off at around 50 mg/kg, when toxicity of the carcinogen became evident (SCHERER and EMMELOT 1975). This discovery led to the development of protocols for the study of the stepwise nature of carcinogenesis in the liver of rats (reviewed by FARBER 1984; PITOT et al. 1987) and the establishment of *N*-nitroso compounds as initiators and of a variety of non-mutagenic chemicals as promoters.

A single dose of *N*-nitrosodiethylamine can be hepatocarcinogenic without exogenous promotion (in contrast to *N*-nitrosodimethylamine) (CRADDOCK 1975). Complete carcinogenesis by this *N*-nitroso compound implies that it can of itself effect the three principal stages of initiation, promotion and progression that are necessary. Of these, the first and third could reflect the persistent ability to induce mutation; as previously noted, promutagenic ethylated bases are known to be less easily removed through repair than are their methyl analogues, but whether this would suffice to account for the second irreversible mutational event proposed as necessary to confer malignancy on promoted, initiated foci remains unknown. It may be thought more likely that this should be classified as of spontaneous origin.

The principal factor of importance for *N*-nitroso compounds in general is the evident potential of these mutagens to act also as promoters. This action implies their ability to permit growth of initiated foci selectively, in turn suggesting that these are less susceptible to the cytotoxic action of the carcinogen than the surrounding normal cells. Not all initiated foci persist when challenged with appropriate cytotoxic treatment (Solt and Farber 1976); malignant cells appear to originate from these resistant foci. Therefore, in complete hepatocarcinogenesis by continuous treatment with *N*-nitroso compounds that require metabolic activation, an essential difference, among several enzymic changes which have been noted as specific to hyperplastic nodules, is a relative lack of ability to effect such activating metabolism; as noted by Farber (1984), this concept of the importance of differential cytotoxicity in carcinogenesis can be traced back to Haddow (1938).

In rat kidney, single doses of *N*-nitrosodimethylamine can induce mesenchymal tumours if animals are fed a protein-free diet for 3 days prior to dosing (Swann and McLean 1968). Putative preneoplastic foci are induced with single-hit dose dependence, while complete tumorigenesis is a multi-hit phenomenon (Driver et al. 1987); immunosurveillance was suggested as a host defence mechanism opposing promotion.

In the urinary bladder, *N*-nitroso compounds can also be initiators or complete carcinogens, by local application. As an example of the latter, Mohr et al. (1978) found that instillation of a single dose of 2 mg *N*-methyl-*N*-nitrosourea into the bladder of rats (195 g body weight) gave a bladder tumour incidence of 57% compared with 2% in controls; malignant invasive carcinoma was induced.

Despite this extensive complete carcinogenesis, which was accompanied by severe damage to the urothelium throughout the urinary tract, it has proved possible to use *N*-nitroso compounds as initiators (most effectively, *N*-nitroso-*n*-butyl-*n*-(4-hydroxybutyl)amine) in order to study the effects of promoters, in the first instance saccharin (Hicks et al. 1975; for review, IARC 1980; Cohen et al. 1987). It should be noted that although saccharin was classified as a promoter for rats, no evidence implicating it as a human carcinogen was found (IARC 1980). It appears to act as a weak, non-specific, hyperosmotic mitogen in bladder urothelium, like other sodium salts including sodium chloride (Shibata et al. 1986). Rat urine of itself appears to contain promoters (Oyasu et al. 1981). Thus, although the classic concepts of initiation and promotion apply to the action of *N*-nitroso compounds as bladder carcinogens, the target tissue is evidently highly susceptible to endogenous promoting factors, following initiator-induced tissue damage.

The brain of the rat is one of the most intensively studied organs with respect to tumour induction by *N*-nitroso compounds (for a recent review, see Lantos 1986); this follows from the finding by Swann and Magee (1968) that even the relatively weak mutagen methyl methanesulphonate could induce brain tumours in rats. Not surprisingly, therefore, *N*-nitroso-*N*-alkylureas (particularly the ethylurea), which are much more effective mutagens and which alkylate brain DNA systemically, have proved valuable in studies of brain tumorigenesis.

The evident relative lack of proficiency in repair of promutagenic alkylated bases in DNA, as found for rodent as compared with those of human cells, is most notable for the brain; this led to the view that the persistence of O^6-alkylguanine in alkylated brain DNA could account for the susceptibility of this organ to carcinogenesis by *N*-alkyl-*N*-nitrosoureas (GOTH and RAJEWSKY 1974a, b). However, more detailed studies of the persistence of O^6-ethylguanine in specific cell types within the brain, using a sensitive immunohistochemical method (HEYTING et al. 1983), while showing definite differences between cell types, were considered not to support the concept, although it was pointed out that the target cells within the broader classifications of brain cells have not yet been precisely defined. LANTOS (1986) concluded that mitotically active cells of the subependymal plate were the most susceptible targets of *N*-ethyl-*N*-nitrosourea, probably the stem cells from which most gliomas originate; the level of O^6-ethylation of guanine in DNA decreases in these cells with a half-life of about a week (HEYTING et al. 1983), possibly mainly reflecting the rate of cell division. Also as previously noted, O^4-ethylthymine might persist for longer than O^6-ethylguanine if repair does in fact occur.

It seems likely, therefore, that the replication of DNA on ethylated templates might account for the initiation of gliomas, as proposed for ethylation-induced liver cancer; a further similarity is the occurrence of multiple foci of early cell proliferations in ethylated brain.

In view of the likely importance of such proliferation, it is not perhaps surprising that for neurogenic tumour induction by single doses of *N*-ethyl-*N*-nitrosourea of around 40 mg (0.34 mmol)/kg body weight to rats, early postnatal injections or transplacental treatments are much more effective (about 30-fold) than are injections into young adults (DRUCKREY et al. 1966; reviewed by DRUCKREY 1975; KLEIHUES et al. 1976; RICE and WARD 1982). This applies also to other species examined, but it should be noted that mice and monkeys were found to be much less susceptible than rats; for these species the neurogenic tumour yields for transplacental treatments are low, about the same as for adult rats (a few percent), and the adult mouse and monkey give yields of less than 1% (RICE and WARD 1982).

In the mouse, single systemic treatments with *N*-methyl- or *N*-ethyl-*N*-nitrosoureas affect the lymphoid system, giving lymphocytic thymic lymphomas in high yields. Again, the susceptibility is maximal for young mice (Table 5). The data illustrate that initiation of mice systemically by a given extent of methyla-

Table 5. Induction of thymic lymphoma in mice by single intraperitoneal injection of *N*-methyl-*N*-nitrosourea 30 mg (0.291 mmol)/kg body wt. Percentage of mice with thymoma up to 250 days after injection is shown to illustrate strain and age dependence

Strain	Age at injection (weeks)				References
	3	4	8–10	18	
C3Hf/D	38	–	18	–	TERRACINI et al. (1976);
CBA (f)	–	25	0	–	J.V. FREI, P.M. FRY,
C57BL (f)	–	40	13	4	G. HARRIS, and P.D. LAWLEY
RFM (f)	–	–	55	–	(unpublished data); HARRIS et al. (1983)

–, Not determined.

tion (over a very short period of about 1 h) leads to a wide spectrum of thymoma yields, dependent on the strain of mouse and age at treatment.

This specific organotropism for thymus in mice is not found for other species; for example, analogous single methylation of young adult rats gave tumours mainly of the stomach, intestine and kidney (SWANN and MAGEE 1968). With Syrian golden hamsters, a single intraperitoneal injection of *N*-methyl- or *N*-ethylnitrosourea (two-thirds of the acute 50% lethal dose) gave mainly squamous cell carcinomas of the forestomach (LIKHACHEV et al. 1983). Feeding *N*-methyl-*N*-nitrosourea to monkeys required comparatively high doses (totalling over 50 g) to obtain tumours, and these were essentially localised to the oesophagus (SIEBER and ADAMSON 1979).

The high specific susceptibility of mice to induction of thymoma is further well illustrated by the effects if skin painting (with phorbol ester promotion) of *N*-methyl-*N*-nitrosourea on mice of the C57BL strain (G. HARRIS and P. D. LAWLEY, unpublished data): thymoma is readily induced but no skin tumours, reflecting the lack of promotion by phorbol ester in this strain, despite the higher extent of methylation of DNA in skin as compared with thymus when skin painting is used.

In summary, therefore, of the data relevant to *N*-nitroso compounds as initiators, it is not immediately obvious why exceptionally high susceptibilities to tumour induction should occur (more especially in the rat or mouse) in specific tissues. Systemic action of the highly water-soluble, and fairly evenly distributed *N*-alkyl-*N*-nitrosoureas causes correspondingly evenly distributed alkylation of DNA in all organs, which is virtually complete in less than 1 h after injection. In certain types of stem cells (not yet identified), notably in rat brain or mouse lymphoid system, the process of carcinogenesis is thus induced highly effectively. Lack of DNA repair and a comparatively high rate of cell proliferation are observed in numerous other tissues. Therefore, these factors do not of themselves suffice for the stages of promotion or progression necessary for carcinogenesis; the apparent specific promotional factors operating in the hypersensitive tissues have yet to be identified.

Some indications of what may be termed, in the broad sense, promotional factors are beginning to emerge from studies with inbred strains of mice showing different susceptibilities to *N*-nitroso carcinogens. The higher yields of lung tumours induced by feeding *N*-nitrosodimethylamine to male mice of the GRS/A, as opposed to the C3Hf/A, strain despite equal extents of lung DNA methylation were positively associated with a higher proliferate response of target alveolar cells to a given extent of carcinogen-induced damage (DE MUNTER et al. 1979). The possibility that some immunological mechanism can oppose methylation-induced tumour initiation has also been raised by the finding that H-2 haplotype can influence susceptibility (DEN ENGELSE et al. 1981).

The imprint of initiation by an *N*-nitroso compound is well illustrated by experiments in which fetal brain of rat offspring was sampled 2 days after transplacental treatment with *N*-ethyl-*N*-nitrosourea. The cells were grown in tissue culture media and tested for their ability to produce tumours after subcutaneous injection into syngeneic animals; glial tumours were obtained (ROSCOE and CLAISSE 1976). Thus, the processes of promotion and progression necessary for carcinogenesis could occur in vitro in conventional growth media, presumably after induction of an initiating mutation by the carcinogen; however, it should be noted that spontaneous transformation of brain cells not treated with the carcinogen was also found, presumably reflecting spontaneous initiating mutations.

As previously noted (Table 2) base-substitution mutations activating oncogenes in *N*-nitroso compound-induced tumours have been specified for rat mammary tumours (which some authors classify as non-malignant) and murine thymomas. In the case of mouse skin papillomas, activation of this type is associated with tumour initiation: QUINTANILLA et al. (1986) have identified GC→TA transversions in codon 61 of Ha-*ras* induced by the polycyclic hydrocarbon 7,12-dimethylbenz[*a*]anthracene and also reported an as yet unspecified activation of this gene by *N*-methyl-*N*′-nitro-*N*-nitrosoguanidine. The weight of opinion from the small amount of data as yet available therefore supports the view that tumour initiation by *N*-nitroso compounds can result from activating mutations through induction of base substitutions. It should be emphasised, however, that not all base substitutions cause structural mutations such as those activating oncogenes; the majority of mutations of this type detected in bacteria convert amino acid codons to stop codons (nonsense mutations) and could therefore delete action of genes; inactivation of so-called anti-oncogenes has been found in certain human tumours, notably retinoblastoma (for review, KNUDSON 1985). Therefore, the failure to find oncogene activation in all cases of tumorigenesis (Table 2) is perhaps not unexpected.

The nature of the postulated second mutations involved in conversion of benign to malignant tumours according to the two-stage theory remains unknown. It should be noted, however, that HENNINGS et al. (1983) used *N*-methyl-*N*′-nitro-*N*-nitrosoguanidine and O'CONNELL et al. (1986) used *N*-ethyl-*N*-nitrosourea to induce the second irreversible stage in the mouse skin system, i.e. they showed that this initiator would also enhance the rate of conversion of papillomas into carcinomas, as first demonstrated by MOTTRAM (1945) using benzo[*a*]pyrene. This induction of progression may therefore be due to a second base substitution, but, as previously noted, *N*-nitroso compounds can also transform cells through chromosomal rearrangement. In complete carcinogenesis by a single dose of initiator, it appears that the second mutagenic event is manifested only after a considerable latent period, long after the initiating carcinogen has been eliminated from the body. Presumably the postulated second mutation, if induced by the initiator, has remained dormant during the latent period or has been produced by a so-called spontaneous mechanism (perhaps due to an endogenous mutagen) during this latent period. In the latter case the specific type of mutation would not necessarily reflect the mode of action of the initiator.

II. Structure-Activity Relationships for *N*-Nitroso Compounds

Some obvious broad features of structure-activity relationships have already been mentioned. The first is that *N*-nitroso compounds may be directly acting or may require metabolic activation (as illustrated in Figs. 1 and 6).

The two modes of spontaneous activation, through hydrolysis or through reaction with thiols, have a further obvious effect. The *N*-alkyl-*N*-nitrosoureas

Table 6. Distribution of DNA alkylation in mice or rats after injection of *N*-nitroso compounds. Ratio of extent of alkylation at N-7 to that found in liver is shown; extent in liver quoted as μmol 7-alkylguanine/mol DNA-P per unit dose of 1 mmol/kg

	C57BL/Cbi mouse (i.p.) (lh)[a]			Albino rat (i.v.) (4h)[b]
	N-methyl-*N*-nitrosourea	*N*-methyl-*N'*-nitro-*N*-nitrosoguanidine	*N*-ethyl-*N*-nitrosourea	*N*-nitroso-dimethylamine
Liver	483	118	13	5246
Relative extent:				
Bone marrow	0.38	< 0.06	0.5	–
Brain	0.66	0.10	0.5	–
Kidney	0.60	0.44	–	0.12
Lung	0.52	0.67	0.7	0.067
Small bowel	0.72	1.92	0.8	0.002
Spleen	0.51	1.42	0.8	–
Thymus	0.43	0.10	0.5	–

–, Not determined.
[a] LAWLEY (1984).
[b] Swann and Magee (1968).

(Fig. 1) are generally water soluble, evenly distributed throughout the body, and alkylate correspondingly (Table 6). On the other hand, thiol-activated compounds such as *N*-methyl-*N'*-nitro-*N*-nitrosoguanidine tend to be more localised in their effects near the site of application; thus injection by the intraperitoneal route does not result in appreciable effects on the lymphoid system (FREI and LAWLEY 1976). With *N*-nitrosodimethylamine, the distribution of methylation reflects the predominance of liver as the organ in which most metabolic activation occurs (SWANN and MAGEE 1968).

LIJINSKY (1987) has reviewed his extensive studies of the carcinogenicity of *N*-nitroso compounds by oral administration, chiefly to rats. The predominant site for thiol-activated compounds is the forestomach; for the nitrosoalkylureas a noticeably wider spectrum of sites was found.

These studies reported data for 46 *N*-nitrosoalkylamides and amidines and for 123 *N*-nitrosodialkylamines or cyclic nitrosamines expected to required metabolic activation. PREUSSMANN and STEWART (1984) reported on 100 directly acting and 232 metabolically activated compounds, administered mainly orally but sometimes by other routes.

Clearly the scope of the present article permits only a brief attempt to pick out salient points of the structure-activity relationships found. These are further complicated by effects of the various animal species, routes of administration and dose rates used.

From the previous sections, an expression for tumour yield could in principle be derived from quantitative knowledge of three main factors (F_1, F_2, F_3) multiplicatively determining the stages of carcinogenesis. The first, expressing the probability of tumour initiation, would multiply the number of target stem cells at risk by a probability of initiating mutation. As already suggested, one type of mutation, that activates oncogenes through base substitution, has been indicated to be important because of the nature of the specific chemical reactions of *N*-nitroso compounds with DNA that are quantitatively and positively associated with carcinogenesis, notably O^6-alkylation of guanine.

As discussed elsewhere (LAWLEY 1984) the rate of mutation at a given base-pair site in DNA will depend on the measurable quantities of the extent of alkylation of DNA and the proportion of alkylations thought to induce miscoding, together with a repair factor, expressing removal of such groups from the template before its replication. As noted, this repair is not likely to be very efficient in most tissues of rodents but is believed to be most significant for alkylated DNA in the liver. Human cells have been found generally to be highly proficient in the removal of O^6-methylguanine from DNA, but it should be noted that sub-populations of cells within a given organ may be relatively deficient, e.g. some parenchymal cells of the human pancreas (PARSA et al. 1987); also some individuals may show less repair proficiency than others (e.g. for lymphocytes, see LAWLEY et al. 1986).

A principal unknown quantity in assessing initiation is the number of cells at risk; if this is constant for a given site and tumour type in a given inbred strain of animal, it has proved possible to relate structure and activity for three *N*-alkyl-*N*-nitrosoureas, as inducers of murine thymoma, through the principles outlined, since F_1 will depend mainly on O^6-alkylation of DNA guanine, while F_2 (the promotional factor) and F_3 are constant (see Fig. 5).

When mice of different ages and strains are employed (see Table 5), tumour yield can vary widely for a given dose of carcinogen. The most reasonable assumption is that this reflects differences either in F_1, due perhaps to different numbers of target cells per animal, or in F_2, due to (endogenous) promotional factors. No way of deciding the relative importance of these possibilities appears obvious as yet; either could, for example, be higher in younger animals. Perhaps for animals of a given age, the endogenous promotion could differ from strain to strain.

The third factor, F_3, refers to tumour progression or the rate of occurrence of a second rare, irreversible and probably mutational process in a supposed clone of intermediate cells derived from an initiated cell through clonal expansion. This area is the least explored of the three considered: the basic mechanisms involved could be manifold (see e.g. NICOLSON 1987) and could vary considerably according to the type of tumour.

For instances in which a single initiating dose of carcinogen suffices for complete carcinogenesis, the F_2 and F_3 factors, being predominantly endogenous since the initiating carcinogen will generally be rapidly removed from the animal, may well be approximately constant from once carcinogen to another.

However, as already discussed, few organs respond to such a regimen of carcinogenesis, and these probably possess high endogenous promotability following initial genotoxic damage by the initiator. Also, this damage may differ in its effect with the nature of the initiator sufficiently to influence subsequent parts of the carcinogenic process. Nevertheless, the problems of relating carcinogenic action to the properties of the carcinogen will obviously be expected to multiply when continuous treatment of animals is used, since all three factors may well be dependent on the nature of the carcinogen, whether mutagenic or non-genotoxic promoting actions are involved. These preliminary considerations therefore serve to underline the complexity of the problem of interpreting results of continuous feeding of *N*-nitroso compounds.

Few studies have been carried out in which the levels of DNA alkylation in target tissues have been monitored under these circumstances. As already mentioned, the importance of O^4-ethylation of thymine, as opposed to O^6-alkylation of guanine, in DNA in ethylation-induced liver carcinogenesis in rats has been indicated by this approach. The availability of methods in which the need for radioactive labelling is obviated should permit more extensive studies of this type.

III. Role of Metabolic Activation in Structure-Activity Relationships

Much attention has been devoted to the consideration of whether or not initiation of tumours by *N*-nitroso compounds is always the result of promutagenic alkylation of DNA; LIJINSKY (1988) has reviewed several instances in which no DNA adducts could be detected in target organs, notably following treatment with cyclic nitrosamines. In general, the participation of DNA alkylation will depend in the nature of the metabolic oxidation of the nitrosamines (see e.g. reviews by KRÜGER 1972; PREUSSMANN and STEWART 1984; DIPPLE et al. 1987; O'NEILL et al. 1984).

Earlier work showed the predominance of α-hydroxylation (see Fig. 1), which accounted for the methylation and alkylation of nucleic acids in vivo (mainly in liver of rats) by *N*-nitrosomethylalkylamines in which the alkyl chains were not cyclic and not hydroxylated. This follows from the instability of *N*-nitrosoalkyl-α-hydroxyalkylamines, yielding reactive alkyldiazonium ions and aldehydes; supporting evidence is that *N*-nitroso-α-acetoxymethylmethylamine is a potent carcinogen, both locally acting and systemic (lung, heart and kidney tumours) (HABS et al. 1978).

As the length of the alkyl chain increases, effects due to the nature of the reaction products result even when the unaltered alkyl group is transferred to DNA. As already noted, the rate of reaction of promutagenic O^6-alkylguanine in DNA with the alkyl acceptor repair protein (alkyltransferase) decreases, and the butyl analogue responds to excision repair. Also, for propylation, evidence consistent with the participation of the S_N1 mechanism for this alkylation has been found, since *N*-nitroso-di-*n*-propylamine yields a proportion of propylation products as *iso*-propyl isomers (PARK et al. 1980; MORIMOTO et al. 1983) (Fig. 8). As expected, this occurs more extensively when the less nucleophilic *O*-atom, as opposed to the *N*-atom, sites are alkylated (SCRIBNER and FORD 1982). These authors favour a modified S_N2 mechanism in which the transition complex of the propyldiazonium ion and O-6 of guanine is "looser" than with N-7, permitting more facile rearrangement in reaction with the former. As noted, O^6-isopropylguanine is expected to persist longer in DNA than the *n*-propyl isomer.

In some instances, α-oxidation of *N*-nitrosoalkylamines to liberate alkyldiazonium ions is obviated by the structure of the alkyl groups. This is the case with *N*-nitrosomethyl-*tert*-butylamine (Fig. 9, I), which could give *tert*-butyl-

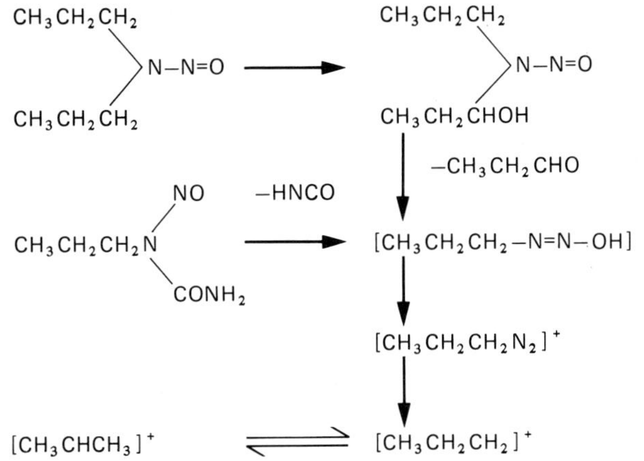

$(CH_3CH_2CH_2)_2N{-}N{=}O \rightarrow (CH_3CH_2CH_2)(CH_3CH_2CHOH)N{-}N{=}O$

$\downarrow -CH_3CH_2CHO$

$CH_3CH_2CH_2N(NO)CONH_2 \xrightarrow{-HNCO} [CH_3CH_2CH_2{-}N{=}N{-}OH]$

$\downarrow$

$[CH_3CH_2CH_2N_2]^+$

$\downarrow$

$[CH_3CHCH_3]^+ \rightleftharpoons [CH_3CH_2CH_2]^+$

Fig. 8. Scheme for alkylation by *N*-nitroso-di-*n*-propylamine or by *N*-nitroso-*N*-*n*-propylurea

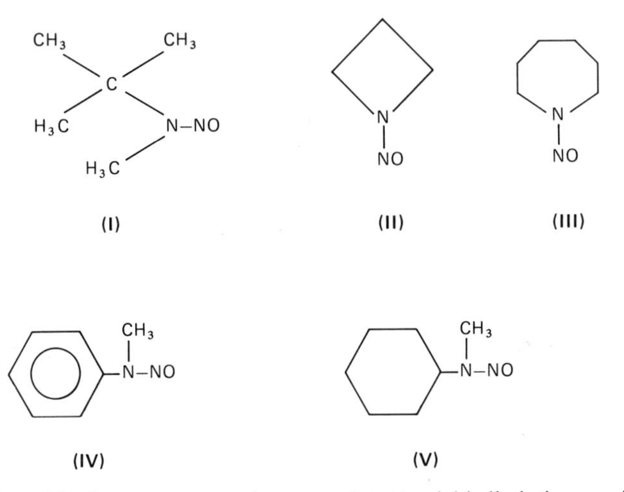

Fig. 9. Formulae of *N*-nitroso compounds reported not to yield alkylating species

amine but no ion of this type, and which proved to be neither hepatotoxic nor carcinogenic at doses up to 1000 mg/kg in the rat (HEATH 1962).

A similar failure to undergo metabolism to yield the alkylating molecular species has been claimed for some *N*-nitrosamines which are carcinogenic, notably certain cyclic nitrosamines. Thus LIJINSKY and ROSS (1969) reported that tritiated *N*-nitrosoazetidine (Fig. 9, II), -hexamethyleneimine (III), and -methylaniline (-methylphenylamine) (IV) showed no evidence of alkylation of DNA in the liver of rats, but *N*-nitrosomethylcyclohexylamine (V) did methylate. Subsequently, evidence apparently contrary to this was found (PEGG and LIJINSKY 1984) that (III) and *N*-nitroso-morpholine and -pyrrolidine partially inhibit the removal of O^6-methylguanine from methylated DNA in the liver of rats, implying that these cyclic nitrosamines do in fact effect promutagenic O^6-alkylation of DNA guanine, presumably through a hitherto undetectable but appropriate metabolic activation.

This method has also been applied by CRADDOCK and HENDERSON (1986) to confirm that *N*-nitrosomethylbenzylamine causes O^6-methylation of guanine in DNA of the target organ, the oesophagus, in the rat (FONG et al. 1979). The sequence of events during rapid carcinogenesis, following eight twice-weekly injections, has been deduced (CRADDOCK and DRIVER 1987) to involve initiation, presumably mainly through this reaction, potentiated by the saturation of the repair system. Initiated cells are resistant to the toxic action of the carcinogen, perhaps due to a reduced ability to metabolise it, as proposed to be important in liver carcinogenesis by *N*-nitrosoamines. This may be specific to the oesophagus, accounting for its particular susceptibility to this carcinogen.

However, *N*-nitrosomethylphenylamine (IV) did not deplete the alkylacceptor protein in the oesophagus (CRADDOCK and HENDERSON 1986), despite its being a potent carcinogen in this organ and presumably therefore does not alkylate O-6 of guanine. α-Oxidation would liberate the phenyldiazonium ion; no evidence as yet implicates this (as a possible arylating agent) in carcinogenesis (DIPPLE et al. 1987), but *N*-nitrosomethylphenylamine has been shown to be mutagenic in a modification of the Ames test (ARAKI et al. 1984) [in which *N*-nitrosodiphenylamine (Fig. 4) showed no positive response despite being a (somewhat weak) carcinogen, as previously mentioned]. KOEPKE et al. (1987) have suggested that its mechanism of mutagenesis (and therefore of tumour initiation) involves reaction of the phenyldiazonium ion with amino groups of DNA bases to yield unstable triazenes that decompose, thus causing deamination.

KRÜGER (1971, 1972) first raised the question whether oxidations other than α-oxidation could activate *N*-nitrosamines to ultimate carcinogens; by analogy with the metabolic degradation of fatty acids, involving oxidation β- to the carboxyl group, he pointed out that β-oxidation could well occur and found evidence suggesting this using *N*-nitrosodi-*n*-butylamines or -propylamines.

A further analogy between the effects of the *N*-nitroso group and the carboxyl group is the facilitation of the base-catalysed exchange of hydrogen atoms in the α-position to either group (KEEFER and FODOR 1970). As noted by LIJINSKY (1986, 1987), this carbanion formation is not usually considered important for carcinogenesis but does facilitate preparation of deuterated *N*-nitroso compounds, useful to deduce the significance of metabolic pathways.

Since the C-D bond is stronger than C-H, processes dependent on C-D bond cleavage will be slower than those for the non-deuterated compound. Thus the finding that $(CD_3)_2NNO$ was a weaker carcinogen than $(CH_3)_2NNO$ supports the generally held view that α-oxidation is a rate-limiting step in carcinogenesis (KEEFER et al. 1973), whereas the carcinogenicity of α-CD_2-labelled *N*-nitroso-di(2-oxopropyl)amine (given orally to rats or Syrian hamsters) is the same as that of the non-deuterated compound, contraindicating the importance of α-oxidation in this case (LIJINSKY 1987).

For N-nitroso-2,6-dimethylmorpholine (Fig. 10) carcinogenicity to the rat oesophagus was enhanced by β-deuteration; the opposite was found for car-

α-deuterated β-deuterated

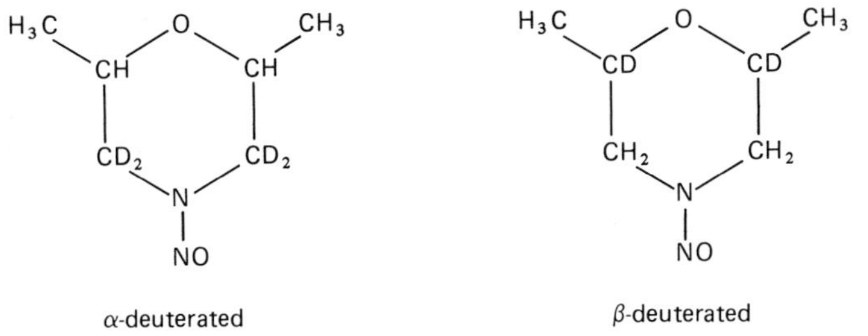

Fig. 10. Formulae of α- and β-deuterated *N*-nitroso-2,6-dimethylmorpholine

cinogenicity in hamsters, in which different target organs (pancreatic duct, liver, lung and nasal cavity) are affected (LIJINSKY 1986). This indicates the importance of β-oxidation for the hamster and α-oxidation for the rat; α-oxidation was also implicated in activation of *N*-nitrosomorpholine. The putative alkylating species derived from these oxidations are as yet not known (LIJINSKY 1986). STEWART et al. (1974) tentatively identified 7-(2-hydroxyethyl)guanine among other products in DNA from the liver of rats treated with *N*-nitrosomorpholine.

Organotropism can itself be influenced by deuteration. Thus *N*-nitrosomethylethylamine, at low doses, selectively gives liver tumours in rats, whereas the β-deuterated nitrosamine also gives oesophageal carcinomas (LIJINSKY et al. 1982). This suggests that reduced β-oxidation in rat liver causes increased circulation of the carcinogen to other organs, of which the oesophagus is the most susceptible.

WHITE et al. (1983) were the first to report evidence suggesting that β-oxidation of the ethyl group in *N*-nitrosodiethylamine occurs in mice; VON HOFE et al. (1986) found 2-hydroxylation of DNA in the liver of rats given *N*-nitrosomethylethylamine or *N*-nitrosodiethylamine, but the extents are limited compared with those from ethylation. Hydroxyethylation presumably depends here on the subsequent α-hydroxylation of the β-hydroxylated nitrosoamines.

However, it has been reported that *N*-nitrosomethyl-2-hydroxyethylamine (an important environmental nitrosamine) is not very extensively metabolised in rats (PREUSSMANN et al. 1978), and mechanisms other than α-oxidation have been suggested for its activation (Fig. 11). MICHEJDA et al. (1979) suggested sulphate conjugation. LOEPPKY et al. (1984) favoured β-nitrosaminoaldehydes as reactive intermediates, capable either of direct reaction with amino groups in DNA thus causing deamination of DNA bases (the latter is a well-recognised promutagenic process) or of generating alkyldiazonium ions.

A particularly important instance of organotropism associated with β-hydroxylation is the activity of 2,6-dimethyl-*N*-nitrosomorpholine in hamster pancreas. POUR et al. (1979) found that β-oxidised derivatives of *N*-nitroso-di-*n*-propylamine were proximate metabolites of *N*-nitroso-2,6-dimethylmorpholine and attributed the organotropism for hamster pancreas to the structural similarity between the cyclic form of *N*-nitroso(2-hydroxypropyl)(2-oxopropyl)amine and hexose sugars (Fig. 12), which could potentiate transport of this carcinogen into the target stem cells of pancreatic ductular tissue. The β-

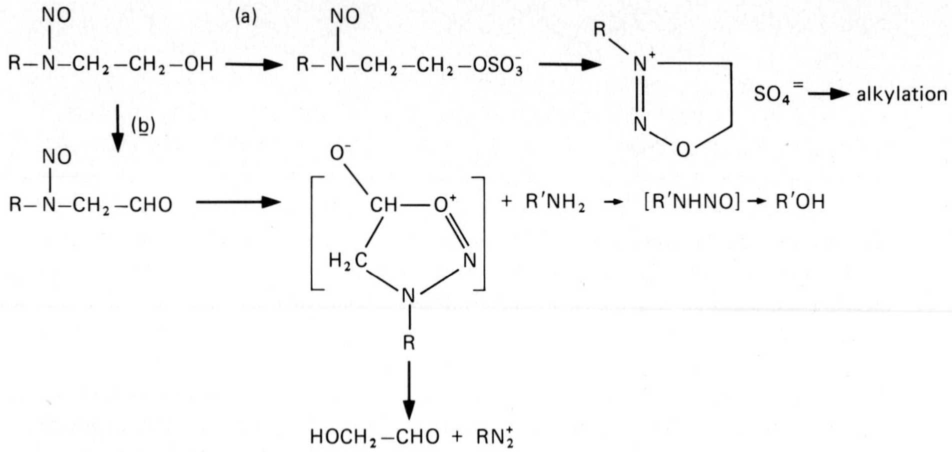

$$R{-}N(NO){-}CH_2{-}CH_2{-}OH \xrightarrow{(a)} R{-}N(NO){-}CH_2{-}CH_2{-}OSO_3^- \longrightarrow [R{-}N^+{=}N{-}O \text{ ring}]\; SO_4^{=} \longrightarrow \text{alkylation}$$

(b)

$$R{-}N(NO){-}CH_2{-}CHO \longrightarrow [O^-{-}CH{-}O^+{=}N{-}N(R){-}CH_2] + R'NH_2 \rightarrow [R'NHNO] \rightarrow R'OH$$

$$\downarrow$$

$$HOCH_2{-}CHO + RN_2^+$$

Fig. 11. Suggested activations of β-hydroxylated *N*-nitrosoethylamines: (*a*) MICHEJDA et al. (1979); (*b*) LOEPPKY et al. (1984)

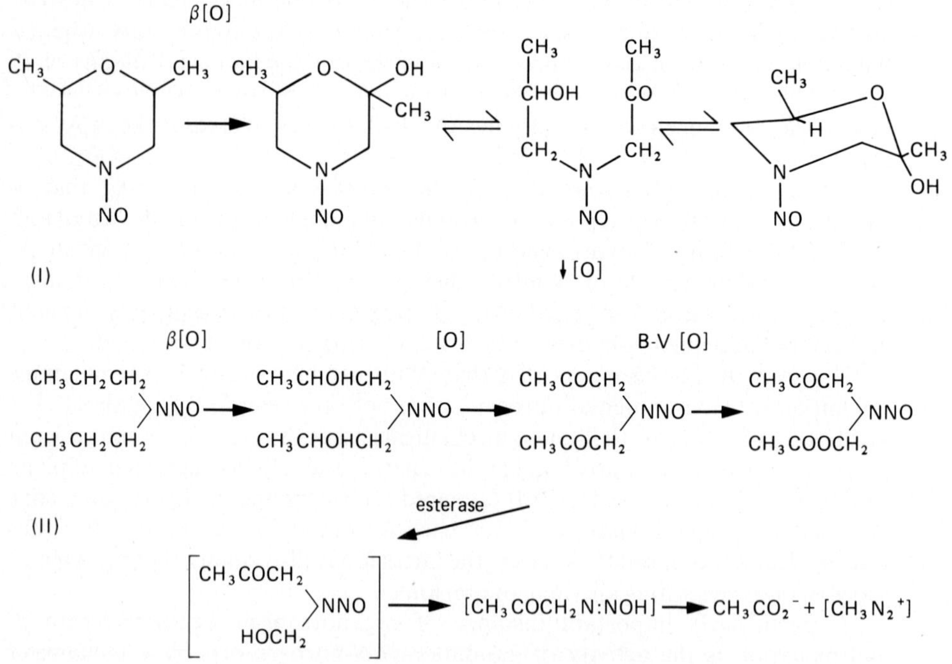

(I)

β[O]

CH_3, O, CH_3, N, NO $\longrightarrow$ CH_3, O, OH, CH_3, N, NO $\rightleftharpoons$ CH_3–CHOH–CH_2, CH_3–CO–CH_2, N, NO $\rightleftharpoons$ CH_3, H, O, CH_3, OH, N, NO

$\downarrow$[O]

(II)

β[O] [O] B-V [O]

$$(CH_3CH_2CH_2)_2NNO \longrightarrow (CH_3CHOHCH_2)_2NNO \longrightarrow (CH_3COCH_2)_2NNO \longrightarrow CH_3COCH_2(CH_3COOCH_2)NNO$$

esterase

$$[CH_3COCH_2(HOCH_2)NNO] \longrightarrow [CH_3COCH_2N{:}NOH] \longrightarrow CH_3CO_2^- + [CH_3N_2^+]$$

Fig. 12. Methylation through β-oxidation of *N*-nitrosamines (POUR et al. 1979; LEUNG and ARCHER 1984, 1985; DIPPLE et al. 1987). *I, N*-nitroso-2,6-dimethylmorpholine; *II, N*-nitroso-di-*n*-propylamine; *B-V [O]*, Baeyer-Villiger oxidation

oxidation-derived proximate carcinogens have been shown to methylate DNA; a suggested mechanism is shown in Fig. 12 (LEUNG and ARCHER 1984).

Bladder organotropism by systemic action was first reported for *N*-nitroso-*n*-butyl-*n*-4-hydroxybutylamine (DRUCKREY et al. 1964). *N*-Nitrosodi-*n*-butylamine and *N*-nitrosomethyl-*n*-alkylamines, in which the alkyl chains have 8, 10, 12 and 14 C atoms, also systemically but less specifically induce bladder tumours (LIJINSKY et al. 1981). DRUCKREY et al. (1967) attributed the specific action on the bladder to a metabolite formed in the liver but transported to the bladder and there converted to a proximate carcinogen. The detailed mechanisms involved are as yet unknown, but evidence implicates ω-oxidation of the *n*-butyl chain for *N*-nitroso-di-*n*-butylamine (i.e. giving the 4-hydroxy and -carboxy derivatives) (OKADA et al. 1976; SUZUKI and OKADA 1980) and degradative oxidation, through loss of 2-carbon fragments, of longer alkyl chains to give *N*-nitroso-methyl-*N*-(3-carboxypropyl)amine for the longer chain nitrosamines (SINGER et al. 1981) (Fig. 13), i.e. as also derived from ω-oxidation of the *n*-butyl chain.

Oxidations at the β- and γ-positions of the longer alkyl chains were associated with carcinogenesis in organs other than bladder (liver, lung, oesophagus) (SUZUKI and OKADA 1980).

It is clearly of interest to determine the effect of this metabolism on the nature of the ultimate DNA alkylation in vivo, and relevant studies have begun (VON HOFE et al. 1987). *N*-Nitrosomethyl-alkylamines with alkyl chain lengths C1-C12 have been assessed for their ability to induce O^6-methylguanine in DNA of tis-

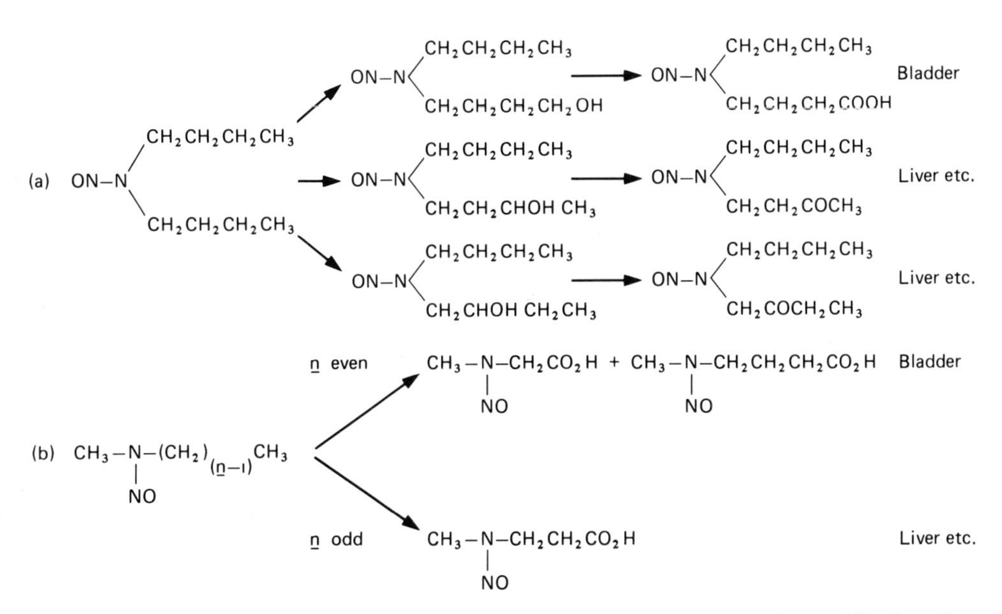

Fig. 13 a, b. Organotropism dependent on metabolism in rats: **a** *N*-nitroso-(di-*n*-butyl)-amine (SUZUKI and OKADA 1980); **b** *N*-nitrosomethyl-*n*-alkylamines (LIJINSKY et al. 1981). *Bladder* denotes association of specific metabolites with systemic induction of cancer in the urinary bladder (**a**) or together with other organs, liver, lung, oesophagus (**b**); alkyl chains with odd number of CH_2 groups are not bladder carcinogens

sues of the rat after single oral doses. Increasing the chain length decreases the yield of this promutagenic base; in the oesophagus, methylation parallels carcinogenic potency. For higher homologues, however, DNA methylation does not account for the complex pattern of tissue specificity; although methylation of urinary bladder DNA was detected with some long-chain nitrosamines, the extent was considered probably too limited to account for their carcinogenic action, and the existence of other alkylation products not detectable by the methods used was considered possible.

These examples serve to illustrate some aspects of the complexity of the correlations between organotropism and metabolism for *N*-nitroso compounds. For more comprehensive recent reviews see LIJINSKY (1987, 1988).

C. *N*-Nitroso Compounds and Human Cancer

I. Introduction

Studies of the carcinogenic action of *N*-nitroso compounds, as already noted, were initially stimulated by a report of suspected human hepatotoxicity attributed to industrial exposure to *N*-nitrosodimethylamine. This was not the first implication of this compound in human toxicology, since FREUND (1937) had documented accidental poisoning from its laboratory use, and chronic industrial exposure was reported by HAMILTON and HARDY (1949) to cause cirrhosis of the liver (see also KIMBROUGH 1982). FUSSGAENGER and DITSCHUNEIT (1980) (see also FLEIG et al. 1982) reported in some detail on a case of a women lethally poisoned by consuming a total dose of about 25 mg/kg over a period of about 2.5 years; this dose is much less than that of *N*-nitrosodiethylamine required to cause hepatocarcinoma in monkeys (totalling over 1400 mg/kg; KELLY et al. 1966). As already mentioned, another case of *N*-nitrosodimethylamine poisoning enabled the demonstration of characteristic, metabolically mediated methylation of DNA in liver (HERRON and SHANK 1980).

There is thus no doubt that metabolic activation of *N*-nitroso compounds can occur in humans. The discovery of their carcinogenic action in 1957 has since stimulated interest in reducing the levels encountered in various industries, such as the manufacture of rubber, leather, amines and rocket fuels, and in synthetic cutting fluids used in metal-working (see PREUSSMANN and EISENBRAND 1984).

The question then arose whether these compounds, like the polycyclic aromatic hydrocarbons, are an ubiquitous environmental carcinogenic hazard, and this became generally accepted (MAGEE 1982; PREUSSMANN and EISENBRAND 1984). As with the hydrocarbons, it remains difficult to implicate *N*-nitroso compounds in the causation of specific cancers (PRESTON-MARTIN 1987). A notable exception came from the use of chloroethylnitrosoureas in the chemotherapy of gastrointestinal cancer, in which treatment with semustine [methyl-CCNU, 1-(2-chloroethyl-3-(4-methylcyclohexyl)-1-nitrosourea], a cross-linking agent for DNA (see Fig. 7), was found to cause a 12.4-fold risk enhancement for non-lymphocytic leukaemia and preleukaemia compared with previous treatments (BOICE et al. 1983).

II. Environmental and Endogenous Formation of *N*-Nitroso Compounds

The generation of *N*-nitrosodialkylamines through nitrosation of secondary amines was first proposed as an environmental source of carcinogens by DRUCKREY and PREUSSMANN (1962), referring to nitrogen oxides in tobacco smoke as nitrosating agents. It is now well recognised that carcinogenic *N*-nitrosamines occur both in tobacco smoke and in unburnt tobacco (HECHT et al. 1987).

The food preservative nitrite is a classical reagent for nitrosation (acting as N_2O_3 in acidic solution) and is present in human saliva, derived from reduction of nitrate by oral microflora. The first report of the environmental occurrence of an *N*-nitrosodialkylamine was the finding by ENDER et al. (1964) that a toxic sample of fish meal contained 100 µg/kg (ppm) of *N*-nitrosodimethylamine. This led to the development of analytical methods for the determination of *N*-nitroso compounds and to a prodigious number of estimates of the amounts present in various environmental sources, reviewed by IARC at 2-yearly intervals since 1976 (for the most recent, see BARTSCH et al. 1987; PREUSSMANN and EISENBRAND 1984).

Endogenous nitrosation of secondary amines was first shown to occur in rats by SANDER and BÜRKLE (1969); feeding of rats with nitrite together with morpholine or *N*-methylbenzylamine induced liver and lung, or oesophageal cancer, respectively. Failure of previous attempts to induce cancer with nitrite plus diethylamine was attributed to the higher basicity of the latter; the reaction is essentially that between the non-protonated base (R_2NH) and N_2O_3, generated from acidic HNO_2; the pH dependence favours nitrosation of less basic amines in vivo; optimal pH values are around 3–4. Amides are more weakly basic than amines, and rates of nitrosation increase as pH decreases. Tertiary amines can be nitrosated through dealkylation in weakly acidic media at much lower rates than secondary amines (HEIN 1963).

CHALLIS and co-workers (CHALLIS and KYRTOPOULOS 1979; CHALLIS et al. 1987) have drawn attention to the possibility that nitrosation under non-acidic conditions could be effected by nitrogen oxides, thus generating carcinogenic diazopeptides (BRAMBILLA et al. 1972) in blood, although these would be unstable at normal gastric pH; under the latter conditions *N*-nitrosopeptides are the minor product (Fig. 14).

Much attention has been devoted to the catalysis of nitrosation by bacteria (SANDER 1968; HILL 1986) at pH values above the purely chemical optimum of 3–4, thus potentiating formation of *N*-nitroso compounds in any tissue in which such bacteria are present, apart from the spontaneous reactions in the acid stomach; furthermore, bacterially mediated nitrosation could occur in the achlorhydric stomach. This latter condition is associated with intestinal metaplasia, a precancerous condition prevalent in a population in Colombia at very high risk from gastric cancer; therefore, it was suggested that achlorhydria increases endogenous *N*-nitroso carcinogens (CORREA et al. 1975). However, bacteria can also reduce nitrate-nitrite levels in tissues, e.g. in the human intestine (SAUL et al. 1981), and the occurrence of *N*-nitroso compounds in the colon and faeces remains controversial (reviewed by HILL 1986).

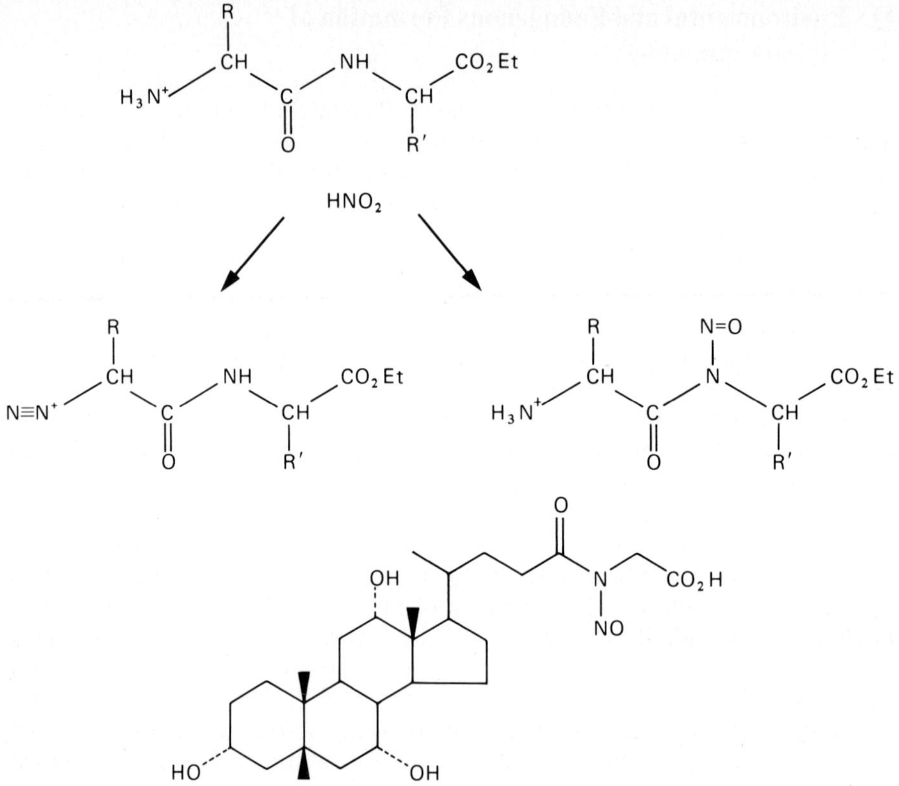

Fig. 14. Nitrosation of peptides or of bile acid conjugates

In order to measure in vivo nitrosation, OHSHIMA and BARTSCH (1981) introduced a method based on the assay of urinary *N*-nitrosoproline (a non-carcinogen which is rapidly nitrosated and of which more than 80% is excreted unchanged within 24 h); model experiments showed that the amounts excreted increase linearly with amounts of proline ingested and exponentially with amount of nitrate (the in vitro rate of nitrosation is proportional to the concentration of proline and the square of the concentration of nitrite). Urinary *N*-nitrosoproline levels are higher in smokers; increased nitrosation has been attributed to smoking-enhanced salivary thiocyanate (LADD et al. 1984), a known catalyst for this process (BOYLAND et al. 1971).

HALL et al. (1987) applied the urinary *N*-nitrosoproline method together with direct measurements of *N*-nitroso compounds in gastric juice. The influence of gastric pH was studied by comparing hypoacidic patients, after partial gastrectomy or with pernicious anaemia, with controls; levels of *N*-nitroso compounds are less around neutral pH than around pH 1–2, and results from the *N*-nitrosoproline urinary assays are in agreement. This evidence therefore fails to support implication of achlorhydria-induced, enhanced *N*-nitrosation as a risk factor in gastric cancer.

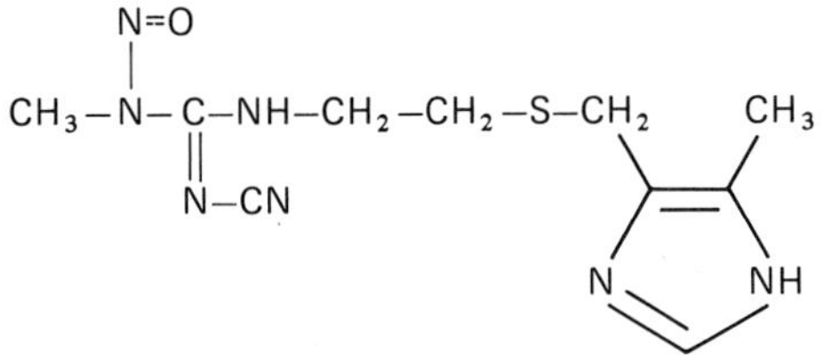

Fig. 15. *N*-Nitrosocimetidine

Much attention has been devoted to the possibility that the very widely used H_2-receptor blocking drug, cimetidine, could be nitrosated in vivo to yield a methylating analogue of the gastric carcinogen *N*-methyl-*N'*-nitro-*N*-nitrosoguanidine (Fig. 15). This was suggested by Prof. A. B. FOSTER, when ELDER et al. (1979) reported gastric carcinoma in three patients treated with the drug for dyspepsia. Neither cimetidine nor its *N*-nitroso derivative were carcinogenic when fed to rats (LIJINSKY and REUBER 1984), although *N*-nitrosocimetidine did prove to be a methylating agent in rats or hamsters (markedly less effectively per unit dose than *N*-methyl-*N*-nitrosourea) when administered by intravenous injection (JENSEN et al. 1987). This relatively weak methylating activity was ascribed to extensive denitrosation by thiol groups of haemoglobin. The current consensus appears to regard cimetidine treatment as not a significant carcinogenic hazard to humans, since it was not mentioned in BARTSCH et al. (1987).

Despite the apparent lack of evidence implicating *N*-nitroso compounds as causative agents for gastric cancer, it is not ruled out that such compounds might initiate cancer, for example in children by causing precancerous hyperplasia (CRESPI et al. 1987); this would be more in line with the general concept that *N*-nitroso compounds act as initiating mutagens rather than at later promotional stages in the carcinogenic process.

The same consideration may apply to the aetiology of colorectal cancer, as it has proved difficult to establish a positive correlation between the amounts of faecal mutagens (including *N*-nitroso compounds) and cancer risk (reviewed HILL 1986; THOMPSON and HILL 1987), although there is strong evidence implicating non-mutagens (bile acids) as promoters. Of interest in this connection is the demonstration by SHUKER et al. (1981) that naturally occurring bile acid conjugates can be nitrosated to yield mutagenic and carcinogenic derivatives which give carboxymethylated purines in DNA after intragastric administration to rats; urinary extraction of 7-carboxymethylguanine could be used to monitor this in vivo alkylation (SHUKER et al. 1987 b).

III. In Vivo Alkylation as a Measure of Human Exposure

As discussed previously, initiation of cancer by *N*-nitroso compounds is thought to result from alkyldiazonium-mediated alkylation of DNA in target stem cells; possibly aldehyde metabolites could also initiate, and in some cases deamination of DNA bases could be induced. Clearly the implication of these compounds as human carcinogens would be strengthened by demonstration of such reactions.

Methods have been developed for appropriate assays. Some idea of the required sensitivity can be deduced from the extents of alkylation of DNA in target organs of animals. From the data of PEGG and PERRY (1981) the accepted level of *N*-nitrosodimethylamine ingested by humans (about 0.02 mg/kg; PREUSSMANN and EISENBRAND 1984) would maximally produce about 10 methylations at O-6 of guanine in genomic DNA of liver (1.2×10^{10} nucleotides, i.e. about 1 nmol/mol DNA-P); this estimate assumes that 16 times the extent of reaction would occur in humans as in rats at the same dose (mg/kg), on the grounds that dose per unit surface area is a more relevant basis for the calculation of species equivalence.

Generally, human cells are more proficient than those of rodents in the removal of promutagenic alkylations. It seems unlikely therefore that humans have organs hypersensitive to complete systemic carcinogenesis by *N*-nitroso compounds, like the brain of neonatal rats or the thymus of young mice. Nevertheless, any residual alkylation of DNA after repair would constitute a potential source of tumour initiation. Individuals deficient in the repair of promutagenic alkylated bases would be expected to be more susceptible to such initiation. Studies of fibroblasts from representative patients with inherited predispositions to colon cancer (GARDNER's syndrome and familial polyposis coli) showed no general association with deficiency in the repair of O^6-methylguanine by methyl transfer (except for one case of GARDNER's syndrome) (MAHER et al. 1986).

Even if accurate measurements of DNA alkylation were made, correlation with expected cancer risk would not yet be possible. PETO et al. (1984) carried out one of the most comprehensive studies of liver cancer induction in rats by *N*-nitrosodi-methylamines and -ethylamines, but no determinations of alkylation of DNA appear to have been made. Even if they were available, these authors considered that the dose-response relationships for humans "might easily be a few orders of magnitude different in either direction" from those for rats, since factors subsequent to initiation in the process of carcinogenesis might well differ significantly for the two species.

Nevertheless, in view of the extensive chemical knowledge of the alkylation of DNA by this group of carcinogens, this so-called molecular dosimetry is expected to provide valuable data directed towards the goal of correlating promutagenic alkylation and biological effect, including carcinogenesis.

The principal methods available are through immunoassays (for *O*-alkylated bases principally) and by RANDERATH's ^{32}P-postlabelling procedures. Indirect indications of in vivo alkylation of DNA are obtained by monitoring alkylpurines excreted in urine, most usefully 3-alkyladenines (SHUKER et al. 1987a) which are rapidly removed from alkylated DNA by repair, are excreted unchanged, and are not naturally occurring purines (HANSKI and LAWLEY 1985). Alkylation of haemoglobin (FARMER et al. 1986), particularly the N-terminal valine, for which sensitive methods based on gas chromatography and mass spectrometry are available, is also expected to be proportional to DNA alkylation.

One of the most compelling implications of *N*-nitroso compounds in human carcinogenesis (CRADDOCK 1983) is the occurrence of *N*-nitrosonornicotine (NNN) and 4-(*N*-methyl-*N*-nitrosamino)-1-(3-pyridyl)-3-butanone (NNK) as

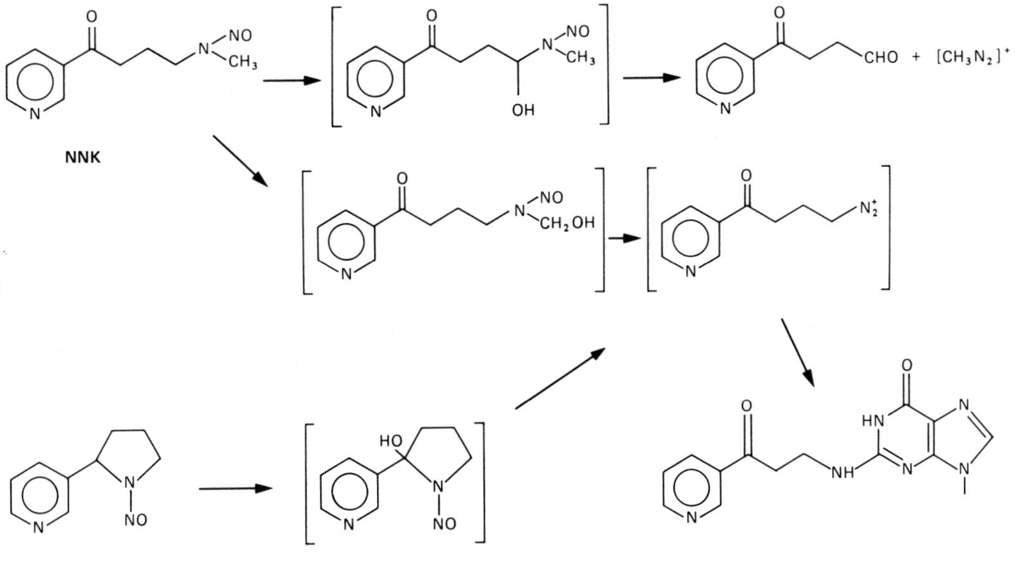

Fig. 16. Metabolic activations of tobacco-specific *N*-nitrosamines. *NNK*, 4-(*N*-methyl-*N*-nitrosamino)-1-(3-pyridyl)-1-butanone; *NNN, N*-nitrosonornicotine

major carcinogenic constituents of tobacco smoke and unburnt tobacco, particularly in snuff (HECHT et al. 1977; 1987).

From the known metabolic pathways, NNK methylates DNA through $CH_3N_2^+$, and NNK and NNN react at N-2 of guanine in DNA through 4-(carbethoxynitrosamino)-1-(3-pyridyl)-1-butanone (Fig. 16) (HECHT et al. 1987). Methylation of DNA has been implicated in carcinogenesis by NNK in rats, the most recent detailed study showing that induction of promutagenic adducts and cell proliferation secondary to toxicity are required for induction of nasal tumours (BELINSKY et al. 1987), i.e. a mechanism in line with the initiation, promotion and progression of malignant tumours at other sites such as liver and oesophagus as previously discussed.

However, no data yet available have provided direct evidence that DNA methylation is associated with humans at high risk from cancer of the respiratory tract due to smoking or the use of snuff (HECHT et al. 1987), possibly because of lack of sensitivity of the method (immunoassay) employed (about 1 mol O^6-methylguanine per 10^6 DNA bases).

Higher sensitivity of detection of adducts (1 per 10^9) has been attained in studies of DNA from human oral mucosal cells using ^{32}P-postlabelling (DUNN and STICH 1986), but although this is in principle applicable to methylation products (REDDY et al. 1984) these have yet to be studied. No adducts specific to high-risk groups (tobacco chewers or smokers) were found.

So far the most successful study using this approach has been with DNA from oesophageal tissues, from patients who underwent surgery for cancer of the oesophagus in Linxian county (China), an area in which foodstuffs are consumed that contain relatively high levels of N-nitroso-dimethyl, -diethyl-, and -methylbenzylamines (UMBENHAUER et al. 1985; WILD et al. 1987). Radioimmunoassay of O^6-alkyldeoxyguanosine isolated chromatographically from DNA digests was used with a sensitivity of around 8 nmol/mol DNA-P. Control oesophageal DNA, from Europeans believed not to have been exposed to abnormally high levels of nitrosamines, showed O^6-methylation up to about 15 nmol/mol DNA-P, whereas in some of the Chinese samples values of up to about 160 nmol/mol DNA-P were reported. Detectable methylation was therefore found, despite the ability of the oesophagus, as with other human tissues, to remove the promutagenic base through the so-called methyltransferase; it is not known whether the residual O^6-methylguanine was present in cells or regions of DNA not accessible to this quasi-enzyme (WILD et al. 1987); ethylation of DNA has not so far been detected.

In summary, the possible association between exposure to *N*-nitroso compounds and alkylation of DNA in human target tissues requires further study using methods of appropriate sensitivity. So far the most sensitive method appears to be ^{32}P-postlabelling; adducts have been found in DNA of rats by this method, at levels of around 1 nmol/mol DNA-P for animals at 10 months of age, much less at 1 month; these adducts are as yet unidentified and are not methylation products (RANDERATH et al. 1986) but are at about the level expected from average environmental exposures to *N*-nitrosodimethylamine (without taking repair into account).

IV. Activating Mutations in Human Tumour Oncogenes Possibly Consistent with Induction by *N*-Nitroso Compounds

From the concepts previously outlined, *N*-nitroso compounds are envisaged to initiate cancer mainly through GC→AT transition mutations, particularly at the second base-pair of codon 12 in *ras* proto-oncogenes. This mutation has been reported in human tumour DNA, most frequently in the Ki-*ras* oncogene (Table 7). In five instances of colon carcinoma, four indicated initiation of the precancerous adenomatous polyps through this mutation; for the other, it was possible that the mutation was associated with progression (BOS et al. 1987).

Perhaps surprisingly, studies with lung carcinomas have shown *ras* activation so far only in adenocarcinomas (generally regarded as less likely to be induced by cigarette smoking) rather than in classic smoking-induced cancers, such as small cell carcinoma (RODENHUIS et al. 1987); nevertheless in five instances in which *ras* activation was found, the patients were heavy smokers.

There is a relatively sparse representation of *ras* activations among bladder, breast and stomach cancers so far examined (2/38 transitional cell carcinoma of bladder; 1/21 mammary carcinoma; 1/27 gastric carcinoma). Furthermore, all types of base substitution have been found, and even those of the GC→AT type are not of course necessarily diagnostic for the mutagenic action of *N*-nitroso compounds. On the other hand, these compounds could induce other types of base substitution, or chromosomal rearrangements that activate oncogenes, as previously mentioned.

Table 7. Human tumour *ras* oncogenes activated by GC→AT transition mutations

Onco-gene	Activating mutation	Origin of tumour	Number	References
Ki-*ras*	12 (2), GGT→GAT, gly→asp	Bladder, lung, pancreas carcinomas	Single examples	NISHIMURA and SEKIYA[a] (1987)
		Bronchioloal-veolar carcinoma	1/10	RODENHUIS et al. (1987)
		Colon carcinoma[b]	5/27	BOS et al. (1987)
		Colon carcinoma	3/16	YANEZ et al. (1987)
Ha-*ras*	12 (2), GGC→GAC, gly→asp	Mammary carcinoma	Single example	NISHIMURA and SEKIYA[a] (1987)
Ki-*ras*	12 (1), GGT→AGT, gly→ser	Lung adeno-carcinoma	Single example	
		Gastric cancer	Single example	
N-*ras*	12 (2), GGT→GAT, gly→asp	Acute myeloid leukaemia	3/22	TOKSOZ et al. (1987)
	13 (2), GGT→GAT, gly→asp		2/22	

[a] Review article.
[b] In four cases, the mutation was found in adenomatous polyp precursors to carcinoma, i.e. deduced to be associated with initiation; in one case, possibly associated with progression from polyp to carcinoma.

In summary, evidence from studies of the activation of oncogenes in human cancer, at present limited, is consistent mainly with the participation of *N*-nitroso compounds as mutagens for colon cancer and acute myeloid leukaemia; evidence from other types of study has not particularly implicated these types of tumour. The concept that mutation could be involved at the stages of either initiation or progression is consistent with data for colon cancer.

D. Epidemiology

Although exposure to *N*-nitroso compounds appears to be ubiquitous, there are groups of people for which this exposure is believed to be much higher than average; this may occur in certain geographically limited areas. Also, there are analogous areas in which the initiating action of these compounds could be potentiated by modulatory dietary factors, of themselves not carcinogenic (co-carcinogens or promoters).

The high inherent sensitivity of the brain of the perinatal rat suggested a possible role for prenatal exposure to *N*-nitroso compounds in the aetiology of neurogenic tumours in children. An epidemiological study found evidence implicating certain sources of these compounds, such as burning incense, tobacco smoke and cured meats (PRESTON-MARTIN 1987) with relative risk factors of up to around 3. As with the proposed role of *N*-nitroso compounds in tobacco-induced cancer, a specific association is not yet proved.

As already noted, the finding of an elevated extent of methylation at O-6 of guanine in DNA of oesophageal and stomach mucosa from cancer patients in Linxian, North China, is perhaps the best evidence yet found aetiologically implicating higher than average dietary and endogenous levels of *N*-nitroso compounds (LU et al. 1986). Endogenous nitrosation was inhibited by vitamin C, and this enabled estimates of exposure to nitrosamine acids as 14 μg per day (endogenous) and 7 μg per day (in food). Other evidence showed that dietary vitamin C intake in general correlates negatively with cancer of the oesophagus, due to inhibition of endogenous nitrosation (MIRVISH 1986).

The oesophagus has been specifically favoured as a suggested target for *N*-nitroso compounds because only chemicals of this type have proved to be notably effective inducers of experimental oesophageal cancer. Relatively small geographical regions of specifically high oesophageal cancer are scattered throughout the world; apart from that in North China already mentioned, these include the Caspian littoral of Iran, the Transkei area of South Africa and, in Europe, parts of France and Scotland. It has been suggested that dietary factors specific to these areas modulate the initiating action of environmental or endogenous *N*-nitroso compounds in the direction of enhanced tumour promotion. It should be noted that dietary factors associated with cancer of the oesophagus (and stomach) show inverse relationships with cancer of the large bowel and breast (HILL 1986), e.g. fat consumption is associated positively with bowel cancer, negatively with oesophageal.

A suggested mechanism for oesophageal (as for gastric) carcinogenesis is that induction of hyperplasia would predispose the tissue to tumorigenesis; this is recognised as the earliest phase in the developmental sequence of oesophageal carcinoma (MING 1984). Although it also occurs as the first stage of complete experimental carcinogenesis by *N*-nitroso compounds, it can also be induced by dietary means, such as deficiencies in riboflavin (FOY and KONDI 1984) or zinc (SCHRAGER et al. 1986) or by the ingestion of *Fusaria* mycotoxins (MARASAS et al. 1984); these occur in certain areas of high incidence of oesophageal cancer (see e.g. CRADDOCK 1987). So far enhancement of experimental carcinogenesis in the oesophagus of the rat (by *N*-nitrosomethylbenzylamine) has been shown to result from zinc deficiency (SCHRAGER et al. 1986; BARCH and FOX 1987) but not from feeding a *Fusarium* mycotoxin (CRADDOCK et al. 1986).

Another dietary factor implicated in the aetiology of oesophageal cancer is consumption of alcoholic beverages, as noted e.g. for specific regions of France and Africa; the general conclusion is that the particular type of drink, rather than the overall consumption of alcohol, is implicated (see HOWE 1986). Nevertheless, ethanol itself markedly affects metabolism of *N*-nitrosodimethylamine: pretreatment induces oxidative demethylation, while simultaneous ingestion inhibits (SCHWARZ et al. 1980).

The inhibitory effect of ethanol, given to rats together with *N*-nitrosodiethylamine, on oxidative dealkylation in liver and kidney was thought to have caused the observed enhanced alkylation of oesophageal DNA (SWANN et al. 1984), but, as expected, chronic ethanol pretreatment, presumably inducing dealkylation, did not have this effect. There is evidence that the inhibitory effect occurs in humans (SPIEGELHALDER and PREUSSMANN 1985).

Suggested alternative modes of action of ethanol are to facilitate penetration of carcinogens into the oesophageal epithelium (SHIRAZI and PLATZ 1978) or to act as a surfactant promoting agent (BOYLAND 1983).

The urinary bladder also emerged from experimental studies as a site of complete carcinogenesis by *N*-nitroso compounds, and there has been some interest in the possibility that urinary nitrosamines could be carcinogenic in humans. Urinary tract infection by nitrate-reducing bacteria would be expected particularly to cause significant extents of formation of carcinogens, because of the relatively high content of nitrate and nitrosatable compounds in urine. This was first shown by BROOKS et al. (1972) who detected *N*-nitrosodimethylamine in the urine of women infected with *Proteus mirabilis*; HICKS et al. (1977) found nitrosamines in the urine of Egyptian bladder cancer patients, from areas where infection with *Schistosoma* (formerly *Bilharzia*) *haematobium* provides foci in the bladder for chronic bacterial superinfection; KAZIKOE et al. (1979) found *N*-nitrosodibutylnitrosamine, which is organotropic for the bladder in rats, in two cancer patients' urines at levels of 0.35 and 0.66 μg/l. Despite these promising results, it appears that no further extensive surveys of urinary nitrosamines in relation to human bladder carcinogenesis have been carried out. HILL (1986) points out that urinary tract infection is not likely to be generally important in Britain (as opposed to Africa) because women are more subject to infection but much less so to bladder cancer.

Specific groups at high risk are known, including paraplegics, who have chronic urinary tract infection with normally faecal organisms. Patients with benign bladder disease treated by ureterosigmoidostomy (transplant of the ureters into the sigmoid colon) show a highly significant elevated incidence (relative risk of several hundred) of carcinoma at the site of anastomosis. Rectal urine contains *N*-nitroso compounds and other mutagens (STEWART et al. 1981). The mean latent period to development of precancerous adenomatous polyps was 20 years and to carcinoma, 26 years. This evidence is therefore the strongest so far associating bacterially catalysed *N*-nitroso compound formation with initiation of colon cancer. More recently, the diversion of choice has been the colonic loop, which gives rise to urinary bacterial flora containing fewer types and apparently leads to less cancer (HILL 1986).

Some ability of *N*-nitroso compounds to induce experimental nasopharyngeal carcinoma has been found, especially in rats; an indication of the particular sensitivity of the nasal cavity of this species may be drawn from the finding that the carcinogenicity of formaldehyde (a metabolic product of α-oxidation of *N*-nitrosamines), evidently difficult to demonstrate, is sufficient to induce cancer at this site (SWENBERG et al. 1980). Geographical areas with a incidence of nasopharyngeal cancer associated with diet have been identified in Tunisia and South China. In Tunisia, stewing bases used daily and containing preserved mutton and peppers were found (POIRIER et al. 1987) to contain comparatively high levels of *N*-nitrosodimethylamine (up to 23 μg/kg) and cyclic nitrosamines, the latter including *N*-nitrosopiperidine (43 μg/kg), known to induce nasopharyngeal cancer in rats on oral administration (GARCIA and LIJINSKY 1972).

In China, steamed Cantonese-style fish preserved in salts containing nitrate was an early suspected source of carcinogens on epidemiological grounds (HO 1972), but since this was also common in the diet of lower-risk groups, other factors were implicated, notably infection with Epstein-Barr or a closely related virus (DETHÉ and ITO 1978). Some more recent epidemiological studies have supported the association with a diet rich in this type of fish, especially in childhood, but although the fish induced nasopharyngeal cancer when fed to rats, specific involvement of *N*-nitrosamines has not been established (summarized by YU and HENDERSON 1987).

It may be noted, in this regard, that the action of an accepted human carcinogen, the chloroethylnitrosourea methyl-CCNU, has also been associated with possible activation of latent Epstein-Barr virus. This follows from the specific types of tumour induced, mainly non-Hodgkin's lymphoma, and the short latent periods observed. As with other cytotoxic alkylating carcinogens, this could be ascribed to their immunosuppressive action (PENN 1986). In addition to acting as mutagens, it is possible therefore that *N*-nitroso compounds could act, in some as yet unspecified ways, together with tumorigenic virus in the induction of human cancers.

E. Conclusions

N-Nitroso compounds can act as complete carcinogens in experimental carcinogenesis and are considered to be initiators of cancer, mainly through their ability to alkylate DNA of target stem cells. This implies the ability to induce mutations, and extensive studies have established mechanisms, as predicted from knowledge of the chemistry of DNA alkylation. Specifically, *ras* proto-oncogenes are activated through base substitutions, principally GC→AT transitions, following O^6-alkylation of guanine in DNA. Repair mechanisms for removal of promutagenic alkylated bases, which prevent mutagenesis (and also cytotoxic action in certain instances), have been investigated in some detail.

As potential human carcinogens, *N*-nitroso compounds are ubiquitous in the environment and include the principal specific carcinogens found in tobacco. They are also formed endogenously, since nitrate, as a source of nitrite, is also an inescapable dietary constituent. But the search for alkylation of human DNA has so far yielded few positive results (principally in oesophageal DNA from cancer patients in an area associated with specifically high exposure to dietary *N*-nitrosamines), possibly reflecting lack of appropriate sensitivity of methods for its detection.

Epidemiological evidence is broadly consistent with the participation of *N*-nitroso compounds as human carcinogens, and specific oncogene-activating mutations identical with those induced by one of these compounds in rats or mice have been found in a few cases of human cancer. But it has proved difficult to implicate *N*-nitroso compounds as specific causative agents. An exception is non-Hodgkin's lymphoma induced in patients treated with a chloroethylnitrosurea (semustine), a cancer-chemotherapeutic alkylating agent.

Epidemiological studies point to particular instances in which *N*-nitroso compounds are rather strongly, but not conclusively, implicated as dietary carcinogens. These include oesophageal and nasopharyngeal cancer in specific geographical areas. The endogenous formation of *N*-nitroso compounds through bacterially catalysed nitrosation has been suggested but not yet very convincingly supported to be aetiologically involved in cancer of the gastrointestinal tract. The best evidence for this came from the observed very high risk factor for cancer and precancerous adenomatous polyps at the site of anastomosis of the ureter in ureterosigmoidostomy.

Activation of latent tumour virus (Epstein-Barr or related) by *N*-nitroso compounds has been indicated to explain features of lymphoma induction by semustine and of the epidemiology of nasopharyngeal cancer, but the mechanisms are as yet unknown.

References

Abbott PJ, Saffhill R (1979) DNA synthesis with methylated poly(dC-dG) templates. Evidence for a competitive nature to miscoding by O^6 methylguanine. Biochim Biophys Acta 562:51–61

Albert RE, Burns FJ (1977) Carcinogenic atmospheric pollutants and nature of low level risks. In: Hiatt HH, Watson JD, Winsten JA (eds) Origins of human cancer, book A. Cold Spring Harbor Laboratory, New York, pp 289–295

Ames BN, Durston WE, Yamasaki E, Lee FD (1973) Carcinogens are mutagens: simple test system combining liver bacteria for detection. Proc Natl Acad Sci USA 70:2281–2285

Ames BN, McCann J, Yamasaki E (1975) Methods for detecting carcinogens with the *Salmonella*/mammalian microsome mutagenicity test. Mutat Res 31:347–364

Appel KE, Graf H (1982) Metabolic nitrite formation from *N*-nitrosamines: evidence for a cytochrome P-450 dependent reaction. Carcinogenesis 3:293–296

Araki A, Muramatsu M, Matsushima T (1984) Comparison of mutagenicities of *N*-nitrosamines on *Salmonella typhimurium* TA 100 and *Escherichia coli* WP2 *UVRA*/PKM 101 using rat and hamster liver S9. Gann 75:8–16

Ashby J (1981) Triplet formed by three nitrosamines (NMorph, dPhNO and DNPT). In: De Serres FJ, Ashby J (eds) Evaluation of short-term tests for carcinogens. Elsevier/North-Holland, New York, pp 138–142

Auerbach C, Robson JM (1946) Chemical production of mutations. Nature 157:302

Bailey E, Connors TA, Farmer PB, Gorf SM, Rickard J (1981) Methylation of cysteine in hemoglobin following exposure to methylating agents. Cancer Res 41:2514–2517

Barbacid M (1987) *ras* Genes. Ann Rev Biochem 56:779–827

Barch DH, Fox CC (1987) Dietary zinc deficiency increases methylbenzylnitrosamine-induced formation of O^6-methylguanine in oesophageal DNA of the rat. Carcinogenesis 8:1461–1464

Bargmann CI, Hung M-C, Weinberg RA (1986) Multiple independent activation of *neu* oncogene by point mutation altering the transmembrane domain. Cell 45:649–657

Barnes JM (1974) Nitrosamines. Essays in Toxicology 5:1–15

Barnes JM, Magee PN (1954) Some toxic properties of dimethylnitrosamine. Br J Ind Med 11:167–174

Bartsch H, Malveille C, Camus AM, Martel-Planche G, Brun G, Hautefeuille A, Sabadie N, Barbin A, Kuroki T, Drevon C, Piccoli C, Montesano R (1980) Bacterial and mammalian mutagenicity tests: validation and comparative studies on 180 chemicals. In: Montesano R, Bartsch H, Tomatis L (eds) Molecular and cellular aspects of carcinogen screening tests. IARC, Lyon, pp 179–241 (IARC scientific vol no 27)

Bartsch H, O'Neill IK, Schulte-Hermann R (1987) The relevance of *N*-nitroso compounds to human cancer: exposures and mechanisms. IARC, Lyon (IARC scientific publication no 84)

Belinsky SA, Walker VE, Maronpot RR, Swenberg JA, Anderson MW (1987) Molecular dosimetry of DNA adduct formation and cell toxicity in rat nasal mucosa following exposure to the tobacco-specific nitrosamine 4-(*N*-methyl-*N*-nitrosamine)-1-(3-pyridyl)-1-butanone and the relationship to induction of neoplasia. Cancer Res 47:6058–6065

Beranek DT, Weiss CC, Swenson DH (1980) A comprehensive quantitative analysis of methylated and ethylated DNA using high pressure liquid chromatography. Carcinogenesis 1:595–606

Berenblum I (1941) Co-carcinogenic action of croton resin. Cancer Res 1:44–48

Bohr VA, Okumoto DS, Hanawalt PC (1986) Survival of UV-irradiated mammalian cells correlates with efficient DNA repair in an essential gene. Proc Natl Acad Sci USA 83:3830–3833

Boice JD Jr, Greene MH, Killen JY Jr, Ellenberg SS, Keehn RJ, McFadden E, Chen TT, Fraumeni JF Jr (1983) Leukaemia and pre-leukaemia after adjuvant therapy of gastro-intestinal cancer with semustine (methyl-CCNU). N Engl J Med 309: 1079–1084

Bos JL, Verlaan-de Vries M, Jansen AM, Veeneman GH, van Boom JH, van der Eb AJ (1984) Three different mutations in codon 61 of the human N-*ras* gene detected by synthetic oligonucleotide hybridization. Nucleic Acids Res 12:9155–9163

Bos JL, Fearon ER, Hamilton SR, Verlaan-de Vries M, van Boom JH, van der Eb AJ, Vogelstein B (1987) Prevalence of *ras* gene mutations in human colorectal cancers. Nature 327:293–297

Boyland E (1983) Surface active agents as tumour promoters. Environ Health Perspect 50:347–350

Boyland E, Nice E, Williams K (1971) Catalysis of nitrosation by thiocyanate from saliva. Food Cosmetic Toxicol 9:639–643

Boyle JM, Margison GP, Saffhill R (1986) Evidence for excision repair of O^6-*n*-butyldeoxyguanosine in human cells. Carcinogenesis 7:1987–1990

Brambilla G, Cavanna M, Parodi S, Baldini L (1972) Induction of tumours in newborn and adult Swiss mice by *N*-diazoacetylglycine amide. Eur J Cancer 8:127–129

Brennan RG, Pyzalska D, Blonski WJP, Hruska FE, Sundaralinjam M (1986) Crystal structure of the promutagen O^4-methylthymidine: importance of the anti conformation of the *O*(4) methoxy group and possible mispairing of O^4-methylthymidine with guanine. Biochemistry 25:1181–1185

Briscoe WH, Cotter L-E (1985) DNA sequence has an effect on extent and kinds of alkylation of DNA by a potent carcinogen. Chem Biol Interact 56:321–331

Brookes P, Lawley PD (1960) Reaction of mustard gas with nucleic acids in vitro and in vivo. Biochem J 77:478–484

Brookes P, Lawley PD (1964) Evidence for binding of polycyclic aromatic hydrocarbons to nucleic acids of mouse skin: relation between carcinogenic power of hydrocarbons and their binding to DNA. Nature 202:781–784

Brooks JB, Cherry WB, Thacker L, Alley CC (1972) Analysis by gas chromatography of amines and nitrosamines produced in vivo and in vitro by *Proteus mirabilis*. J Infect Dis 126:143–153

Bulbrook RD, Moore JW, Wang DY, Clark GMG (1984) Oestrogens and etiology and clinical course of breast cancer. In: Borzonyi M, Day NE, Lapis K, Yamasaki H (eds) Models, mechanisms and etiology of tumour promotion. IARC, Lyon, pp 385–395 (IARC publication no 56)

Butterworth BE, Slaga TJ (eds) (1987) Nongenotoxic mechanisms in carcinogenesis. Cold Spring Harbor Laboratory, New York (Banbury Report no 25)

Casanova-Schmitz M, Starr TB, Heck H d'A (1984) Differentiation between metabolic incorporation and covalent binding in labelling of macromolecules in rat nasal mucosa and bone marrow by inhaled [^{14}C]- and [^{3}H]formaldehyde. Toxicol Appl Pharmacol 76:26–44

Challis BC, Kyrtopoulos SA (1979) Nitrosation of amines by two-phase interaction of amines in solution with gaseous oxides of nitrogen. J Chem Soc Perkin Trans I, pp 299–304

Challis BC, Fernandes MHR, Glover BR, Latif S (1987) Formation of diazopeptides by nitrogen oxides. In: Bartsch H, O'Neill IK, Schulte-Herman R (eds) The relevance of *N*-nitroso compounds to human cancer: exposures and mechanisms. IARC scientific publication no 84. IARC, Lyon, pp 308–314

Chan FPH, Ens B, Frei JV (1981) Cytogenesis of murine thymic lymphomas induced by *N*-methyl-*N*-nitrisourea: difference in incidence of trisomy 15 in thymic lymphomas induced in neonatal and adult mice. Cancer Genet Cytogenet 4:337–344

Chang MJW, Webb TE, Koestner A (1979) Distribution of O^6-methylguanine in rat DNA following pretreatment in vivo with methylnitrosourea. Cancer Lett 6:123–127

Chaw YFM, Crane LE, Lange P, Shapiro R (1980) Isolation and identification of cross-links from formaldehyde-treated nucleic acids. Biochemistry 19:5525–5531

Clanton DJ, Lu Y, Blair DG, Shih TY (1987) Structural significance of the GTP-binding domain of *ras* p21 studied by site-directed mutagenesis. Mol Cell Biol 7:3092–3097

Cohen SM, Ellwein LB, Johansson SL (1987) Bladder tumour promotion. In: Butterworth BE, Slaga TJ (eds) Nongenotoxic mechanisms in carcinogenesis. Cold Spring Harbor Laboratory, New York, pp 55–68 (Banbury Report no 25)

Connell JR, Medcalf ASC (1982) Induction of SCE and chromosomal aberrations with relation to specific base methylation of DNA in Chinese hamster cells by *N*-methyl-*N*-nitrosourea and dimethyl sulphate. Carcinogenesis 3:385–390

Correa P, Haenszel W, Cuello C, Tannenbaum S, Archer M (1975) A model for gastric cancer epidemiology. Lancet ii:58–60

Coulondre C, Miller JH (1977) Genetic studies of the *lac* repressor. IV. Mutagenic specificity in the *lac* I gene of *Escherichia coli*. J Mol Biol 117:577–606

Coulondre C, Miller JH, Farabaugh PJ, Gilbert W (1978) Molecular basis of base substitution hotspots in *Escherichia coli*. Nature 274:775–780

Craddock VM (1971) Liver carcinomas in rats by single administration of dimethylnitrosamine after partial hepatectomy. JNCI 47:899–905

Craddock VM (1975) Effect of a single treatment with the alkylating carcinogens dimethylnitrosamine, diethylnitrosamine and methyl methanesulphonate on liver regenerating after partial hepatectomy. I. Test for induction of liver carcinomas. Chem Biol Interact 10:313–321

Craddock VM (1983) Nitrosamines and human cancer: proof of an association? Nature 306:638

Craddock VM (1987) Nutritional approach to oesophageal cancer in Scotland. Lancet i:217

Craddock VM, Driver HE (1987) Sequential histological studies of rat oesophagus during rapid initiation of cancer by repeated injection of *N*-methyl-*N*-benzylnitrosamine. Carcinogenesis 8:1129–1132

Craddock VM, Henderson AR (1984) Repair and replication of DNA in rat and mouse tissues in relation to cancer induction by *N*-nitroso-*N*-alkylureas. Chem Biol Interact 52:223–231

Craddock VM, Henderson AR (1986) Effect of *N*-nitrosamines carcinogenic for oesophagus on O^6-alkylguanine-DNA-methyltransferase in rat oesophagus and liver. J Cancer Res Clin Oncol 111:229–236

Craddock VM, Henderson AR, Gash S (1982) Nature of constitutive and induced mammalian O^6-methylguanine DNA repair system. Biochem Biophys Res Commun 107:546–553

Craddock VM, Sparrow S, Henderson AR (1986) Effect of the trichothecene mycotoxin diacetoxyscirpenol on nitrosamine-induced oesophageal cancer in the rat. Cancer Lett 31:197–204

Crespi M, Ohshima H, Ramazzotti V, Munoz N, Grassi A, Casale V, Leclerc H, Calmels S, Cattoen C, Kaldor J, Bartsch H (1987) Intragastric nitrosation and precancerous lesions of the gastrointestinal tract: testing of an etiological hypothesis. In: Bartsch H, O'Neill IK, Schulte-Herman R (eds) The relevance of N-nitroso compounds to human cancer: exposures and mechanisms. IARC, Lyon, pp 511–517 (IARC scientific publication no 84)

De Munter HK, Den Engelse L, Emmelot P (1979) Studies on lung tumours. IV. Correlation between [^{3}H]thymidine labelling of lung and liver cells and tumor formation in GRS/A and C3Hf/A male mice following administration of dimethylnitrosamine. Chem Biol Interact 24:299–316

Den Engelse L, Oomen LC, Van der Valk MA, Hart AA, Dux A, Emmelot P (1981) Studies on lung tumours. V. Susceptibility of mice to dimethylnitrosamine-induced tumour formation in relation to H-2 haplotype. Int J Cancer 28:199–208

De Serres FJ, Ashby J (1981) Evaluation of short-term tests for carcinogens. Elsevier/North-Holland, New York, Chaps. 6, 11

DeThé G, Ito Y (1978) Nasopharyngeal carcinoma: etiology and control. IARC, Lyon (IARC scientific publication no 20)

Dipple A, Moschel RC, Bigger CAH (1984) Polynuclear aromatic hydrocarbons. In: Searle CE (ed) Chemical carcinogens, 2nd edn, part I. American Chemical Society, Washington DC, pp 41–163 (ACS Monograph no 182)

Dipple A, Michejda CJ, Weisburger EK (1987) Metabolism of chemical carcinogens. In: Grunberger D, Goff SP (eds) Mechanisms of cellular transformation by carcinogenic agents. Pergamon, Oxford, pp 1–32

Doniger J, Notario V, DiPaolo JA (1987) Carcinogens with diverse mutagenic activities initiate neoplastic guinea pig cells that acquire the same N-*ras* point mutation. J Biol Chem 262:3813–3819

Driver HE, White INH, Steren FS, Butler WH (1987) Possible mechanism for the dose-response relationship observed for renal mesenchymal tumours induced in the rat by a single dose of *N*-nitrosodimethylamine. In: Bartsch H, O'Neill IK, Schulte-Hermann R (eds) The relevance of N-nitroso compounds to human cancer: exposures and mechanisms. IARC, Lyon, pp 253–255 (IARC scientific publication no 84)

Druckrey H (1975) Chemical carcinogenesis on *N*-nitroso derivatives. Gann Monographs Cancer Res 17:107–132

Druckrey H, Preussmann R (1962) Zur Entstehung carcinogener Nitrosamine am Beispiel des Tabakrauchs. Naturwiss 49:498

Druckrey H, Preussmann R, Ivankovic S, Schmidt CH, Mennel HD, Stahl KW (1964) Selective Erzeugung von Blasenkrebs an Ratten durch Dibutyl- und *N*-Butyl-*N*-butanol(4)-nitrosamin. Z Krebsforschung 66:280–290

Druckrey H, Ivankovic S, Preussmann R (1966) Teratogenic and carcinogenic effects in offspring after single injection of ethylnitrosourea to pregnant rats. Nature 210:1378–1379

Druckrey H, Preussmann R, Ivankovic S, Schmähl D (1967) Organotrope carcinogene Wirkungen bei 65 verschiedenen *N*-nitroso-Verbindungen an BP-Ratten. Z Krebsforschung 69:103–201

Dunn BP, Stich HF (1986) ^{32}P-Postlabelling analysis of aromatic DNA adducts in human oral mucosa. Carcinogenesis 7:1115–1120

Eisenstadt E, Warren AJ, Porter J, Atkins D, Miller JH (1982) Carcinogenic epoxides of benzo[*a*]pyrene and cyclopenta[*c,d*]pyrene induce base substitutions. Proc Natl Acad Sci USA 79:1945–1949

Elder JB, Ganguli PC, Gillespie JE (1979) Cimetidine and gastric cancer. Lancet i:1005

Ender F, Havre G, Helgebostad A, Koppang M, Madsen R, Ceh L (1964) Isolation and identification of a hepatotoxic factor in herring meal produced from sodium nitrate-preserved herring. Naturwiss 51:637–638

Farber E (1984) Cellular biochemistry of the stepwise development of cancer with chemicals: GHA Clowes memorial lecture. Cancer Res 44:5463–5474

Farmer PB, Shuker DEG, Bird I (1986) DNA and protein adducts as indicators of in vivo methylation of nitrosatable drugs. Carcinogenesis 7:49–52

Feldman MY (1973) Reactions of nucleic acids and nucleoproteins with formaldehyde. Prog Nucl Acid Res Mol Biol 13:1–47

Fialkow PJ (1979) Clonal origin of human tumors. Ann Rev Med 30:135–143

Fleig WE, Fussgaenger RD, Ditschuneit H (1982) Pathological changes in a human subject chronically exposed to dimethylnitrosamine. In: Magee PN (ed) Nitrosamines and cancer. Cold Spring Harbor Laboratory, New York, pp 37–68 (Banbury Report no 12)

Fong LYY, Lin HJ, Lee CLH (1979) Methylation of DNA in target and non-target organs of the rat with methylbenzylnitrosamine and dimethylnitrosamine. Int J Cancer 23:679–682

Foy H, Kondi A (1984) The vulnerable oesophagus: riboflavin deficiency and squamous cell dysplasia of the skin and oesophagus. JNCI 72:941–948

Frei JV (1980) Methylnitrosourea induction of thymomas in AKR mice requires one or two "hits" only. Carcinogenesis 1:721–723

Frei JV, Lawley PD (1976) Tissue distribution and mode of DNA methylation in mice by methyl methanesulphonate and *N*-methyl-*N'*-nitro-*N*-nitrosoguanidine: lack of thymic lymphoma induction and low extent of methylation of target tissue DNA at O-6 of guanine. Chem Biol Interact 13:215–222

Frei JV, Lawley PD (1980) Thymomas induced by simple alkylating agents in C57BL/Cbi mice: kinetics of the dose-response. JNCI 64:845–856

Frei JV, Venitt S (1975) Chromosome damage in bone marrow of mice treated with the methylating agents methyl methanesulphonate and *N*-methyl-*N*-nitrosourea in presence or absence of caffeine and its relationship with thymoma induction. Mutat Res 29:89–96

Frei JV, Swenson DH, Warren W, Lawley PD (1978) Alkylation of DNA in vivo in various organs of C57BL mice by the carcinogens *N*-methyl-*N*-nitrosourea, *N*-ethyl-*N*-nitrosourea and ethyl methanesulphonate in relation to induction of thymic lymphoma: some applications of high-pressure liquid chromatography. Biochem J 174:1031–1034

Freund HA (1937) Clinical manifestations and studies in parenchymatous hepatitis. Ann Intern Med 10:144–155

Friedberg EC (1985) DNA repair. Freeman, New York

Funato T, Yokota J, Salamoto H, Kameya T, Fukushima S, Ito N, Terada M, Sugimura T (1987) Activation of N-*ras* gene in a rat hepatocellular carcinoma induced by dibutylnitrosamine and butylated hydroxytoluene. Gann 78:689–694

Fussgaenger RD, Ditschuneit H (1980) Lethal exitus of patient with *N*-nitrosodimethylamine poisoning 2.5 years following first ingestion and signs of intoxication. Oncology 37:273–277

Garcia H, Lijinsky W (1972) Tumorigenicity of five cyclic nitrosamines in MRC rats. Z Krebsforsch 77:257–261

Goth R, Rajewsky MF (1974a) Molecular and cellular mechanism associated with pulse-carcinogenesis in the rat nervous system by ethylnitrosourea: ethylation of nucleic acids and elimination rates of ethylated bases from DNA of different tissues. Z Krebsforschung 82:37–64

Goth R, Rajewsky MF (1974b) Persistence of O^6-ethylguanine in rat brain DNA; correlation with nervous system-specific carcinogenesis by ethylnitrosourea. Proc Natl Acad Sci USA 71:639–643

Guerrero I, Villasante A, Corces V, Pellicer A (1985) Loss of normal N-*ras* allele in mouse thymic lymphoma induced by a chemical carcinogen. Proc Natl Acad Sci USA 82:7810–7814

Guerrero I, Villasante A, Diamond L, Berman JW, Newcomb EW, Steinberg JJ, Lake R, Pellicer A (1986) Oncogen activation and surface markers in mouse lymphomas induced by radiation and nitrosomethylurea. Leuk Res 10:851–858

Gupta RC (1987) ^{32}P-Postlabelling assay to measure carcinogen-DNA adducts. Prog Exp Tumor Res 32:21–32

Habs M, Schmähl D, Wiessler M (1978) Carcinogenicity of acetoxymethylmethylnitrosamine after subcutaneous, intravenous and intrarectal application in rats. Z Krebsforsch 91:217–221

Haddow A (1938) Cellular inhibition and the origin of cancer. Acta Unio Int Contra Cancrum 3:342–352

Hall CN, Daikin D, Viney N, Cook A, Kirkham JS, Northfield TC (1987) Evaluation of the nitrosamine hypothesis of gastric carcinogenesis in man. In: Bartsch H, O'Neill IK, Schulte-Hermann R (eds) IARC scientific vol no 84. IARC, Lyon, pp 527–530

Hamilton A, Hardy HL (1949) Industrial toxicology, 2nd edn. Hoerber, New York, p 311

Hanski C, Lawley PD (1985) Urinary excretion of 3-methyladenine and 1-methylnicotinamide by rats following administration of [methyl-^{14}C]methyl methanesulphonate and comparison with administration of [^{14}C] methionine or formate. Chem Biol Interact 55:225–234

Harris G, Lawley PD, Asbery LJ, Chandler PM, Jones MG (1983) Autoimmune haemolytic disease in mice after exposure to a methylating carcinogen. Immunology 49:439–449

Heath DF (1962) Decomposition and toxicity of dialkylnitrosamines in rats. Biochem J 85:72–91

Hecht SS, Chen CB, Dong M, Ornaf RM, Hoffmann D (1977) Chemical studies on tobacco smoke: studies on nonvolatile nitrosamines in tobacco. Beitr Tabakforschung 9:1–6

Hecht SS, Castonguay A, Chung F-L, Hoffmann D, Stoner GD (1982) Recent studies on metabolic activation of cyclic nitrosamines. In: Magee PN (ed) Nitrosamines and human cancer. Cold Spring Harbor Laboratory, New York, pp 103–120 (Banbury Report no 12)

Hecht SS, Carmella SG, Trushin N, Foiles PG, Lin D, Rubin JM, Chung F-L (1987) Investigations on molecular disometry of tobacco-specific *N*-nitrosamines. In: Bartsch H, O'Neill IK, Schulte-Hermann R (eds) IARC scientific publications no 84. IARC, Lyon, pp 423–429

Hein GE (1963) Reaction of tertiary amines with nitrous acid. J Chem Educ 40:181–184

Heller EP, Goldthwait DA (1983) Release of 3-methyladenine from linker and core DNA of chromatin by a purified DNA glycolase. Cancer Res 43:5747–5753

Hennings H, Shores R, Wenk ML, Spangler EF, Tarone R, Yuspa SH (1983) Malignant conversion of mouse skin tumours is increased by tumour initiators and unaffected by tumour promoters. Nature 304:67–69

Herron DC, Shank RC (1980) Methylated purines in human liver DNA after probable dimethylnitrosamine poisoning. Cancer Res 40:3115–3117

Heston WE (1950) Carcinogenic action of the mustards. JNCI 11:415–423

Heyting C, Van der Laken CJ, Van Raamsdonk W, Pool CW (1983) Immunohistochemical detection of O^6-ethyldeoxyguanosine in rat brain after in vivo applications of *N*-ethyl-*N*-nitrosourea. Cancer Res 43:2935–2941

Hicks RM, Wakefield JStJ, Chowaniec J (1975) Evaluation of a new model to detect bladder carcinogens or cocarcinogens: results obtained with saccharin, cyclamate and cyclophosphamide. Chem-Biol Interact 11:225–233

Hicks RM, Walter CL, Elsebai I, El Aasser A-B, El Merzabani M, Gough TA (1977) Demonstrations of nitrosamines in human urine: preliminary observations on a possible aetiology for bladder cancer in association with chronic urinary tract infections. Proc R Soc Med 70:413–416

Hill MJ (1986) Microbes and human carcinogenesis. Arnold, London

Ho JHC (1972) Nasopharyngeal carcinoma. Adv Cancer Res 15:57–92

Howe GM (1986) Global geocancerology: a world geography of human cancers. Churchill Livingstone, Edinburgh

IARC (1976) Dinitrosopentamethylenetetramine. IARC, Lyon, pp 241–245 (IARC Monographs on evaluation of carcinogenic risk of chemicals to humans, vol 11)

IARC (1980) Some non-nutritive sweetening agents. IARC, Lyon (IARC Monographs on evaluation of carcinogenic risk of chemicals to humans, vol 22)

IARC (1982a) *N*-Nitrosodiphenylamine. IARC, Lyon, pp 213–225 (IARC Monographs on evaluation of carcinogenic risk of chemicals to humans, vol 27)

IARC (1982b) Formaldehyde. IARC, Lyon, pp 345–389 (IARC Monographs on evaluation of the carcinogenic risk of chemicals to humans, vol 29)

Jensen DE, Stelman GJ, Spiegel A (1987) Species differences in blood-mediated nitrosocimetidine denitrosation. Cancer Res 47:353–359

Karran P, Lindahl T, Øfsteng I, Evensen GB, Seeberg E (1980) *Escherichia coli* mutants deficient in 3-methyladenine-DNA glycosylase. J Mol Biol 140:101–127

Kazikoe T, Wang T-T, Eng VWS, Furrer R, Dion P, Bruce WR (1979) Volatile *N*-nitrosamines in urine of normal donors and of bladder cancer patients. Cancer Res 39:829–832

Keefer LF, Fodor CH (1970) Facile hydrogen isotope exchange as evidence for α-nitrosamine carbanion. J Am Chem Soc 92:5747–5748
Keefer L, Lijinsky W, Garcia H (1973) Deuterium isotope effect on the carcinogenicity of dimethylnitrosamine in rat liver. JNCI 51:299–302
Kelly MG, O'Hara RW, Adamson RH, Gadekar K, Botkin CC, Reese WH, Kerber HT (1966) Induction of hepatic cell carcinoma in monkeys with *N*-nitrosodiethylamine. JNCI 36:323–351
Kennaway EL, Hieger I (1930) Carcinogenic substances and their fluorescent spectra. Br Med J i:1044–1046
Kim S, Lotlikar PD, Chin W, Magee PN (1977) Protein-bound carboxymethyl ester as a precursor of methanol formation during oxidation of dimethylnitrosamine in vitro. Cancer Lett 2:279–284
Kimbrough RD (1982) Pathological changes in human beings acutely poisoned by dimethylnitrosamine. In: Magee PN (ed) Nitrosamine and human cancer. Cold Spring Harbor Laboratory, New York, pp 25–49 (Banbury Report no 12)
Kleihues P, Lantos PL, Magee PN (1976) Chemical carcinogenesis in the nervous system. Int Rev Exp Pathol 15:154–232
Knowles MA, Eydmann ME, Proctor A, Padua RA, Roberts J (1987) *N*-Methyl-*N*-nitrosourea-induced transformation of rat urothelial cells in vitro is not mediated by activation of *ras* oncogenes. Oncogene 1:143–148
Knudson AG Jr (1985) Hereditary cancer, oncogenes and antioncogenes. Cancer Res 45:1437–1443
Koepke SR, Kroeger-Koepke MB, Michejda CJ (1987) *N*-Nitrosomethylaniline: possible mode of DNA modification. In: Bartsch H, O'Neill IK, Schulte-Hermann R (eds) IARC scientific publication no 84. IARC, Lyon, pp 68–70
Krüger FW (1971) Metabolism of nitrosamines in vivo. I. Evidence for β-oxidation of aliphatic di-*n*-alkylnitrosamines: simultaneous formation of 7-methylguanine besides 7-propyl- and 7-butylguanine after application of di-*n*-propyl or di-*n*-butylnitrosamine. Z Krebsforsch 76:145–154
Krüger FW (1972) New aspects in metabolism of carcinogenic nitrosamines. In: Nakahara W, Takayama S, Sugimura T, Odashima S (eds) Topics in chemical carcinogenesis. University of Tokyo Press, Tokyo, pp 213–235
Ladd KF, Archer MC, Newmark HL (1984) Increased endogenous nitrosation in smokers. In: O'Neill IK, Von Borstel RC, Miller CT, Long J, Bartsch H (eds) N-nitroso compounds: occurrence, biological effects and relevance to human cancer. IARC scientific publication no 57. IARC, Lyon, pp 811–817
Lantos PL (1986) Development of nitrosourea-induced brain tumours with a special note on changes occurring during latency. Food Chem Toxicol 24:121–127
Lawley PD (1973) Reaction of *N*-methyl-*N*-nitrosourea with ^{32}P-labelled DNA: evidence for formation of phosphotriesters. Chem Biol Interact 7:127–130
Lawley PD (1976) Methylation of DNA by carcinogens: some applications of chemical analytical methods. In: Montesano R, Bartsch H, Tomatis L (eds) Screening tests in chemical carcinogenesis. IARC, Lyon, pp 181–208 (IARC scientific publication no 12)
Lawley PD (1984) Carcinogenesis by alkylating agents. In: Searle CE (ed) Chemical carcinogens, 2nd edn, vol 1. American Chemical Society, Washington DC, Chap. 7 (ACS Monograph no 182)
Lawley PD (1989) Mutagens as carcinogens: development of current concepts. Mutat Res 213:3–25
Lawley PD, Brookes P (1961) Acidic dissociation of 7:9-dialkylguanines and its possible relation to mutagenic properties of alkylating agents. Nature 192:1081–1082
Lawley PD, Brookes P (1968) Cytotoxicity of alkylating agents towards sensitive and resistant strains of *Escherichia coli* in relation to extent and mode of alkylation of cellular macromolecules and repair of alkylation lesions in DNA. Biochem J 109:433–447
Lawley PD, Orr DJ (1970) Specific excision of methylation products from DNA of *Escherichia coli* treated with *N*-methyl-*N'*-nitro-*N*-nitrosoguanidine. Chem Biol Interact 2:154–157

Lawley PD, Shah SA (1972) Reaction of alkylating mutagens and carcinogens with nucleic acids: detection and estimation of a small extent of methylation of *O*-6 of guanine by methyl methanesulphonate in vitro. Chem Biol Interact 5:286–288
Lawley PD, Shah SA (1973) Methylation of DNA ^{3}H-^{14}C-methyl-labelled *N*-methyl-*N*-nitrosourea: evidence for transfer of the intact methyl group. Chem Biol Interact 7:115–120
Lawley PD, Thatcher CJ (1970) Methylation of DNA in cultured mammalian cells by *N*-methyl-*N'*-nitro-*N*-nitrosoguanidine. Biochem J 116:693–707
Lawley PD, Wallick CA (1957) Action of alkylating agents on DNA and guanylic acid. Chem Ind p 633
Lawley PD, Warren W (1976) Removal of minor methylation products 7-methyladenine and 3-methyladenine from DNA of *Escherichia coli* treated with dimethyl sulphate. Chem Biol Interact 12:211–220
Lawley PD, Orr DJ, Shah SA, Farmer PB, Jarman M (1973) Reaction products from *N*-methyl-*N*-nitrosourea and DNA containing thymidine residues: synthesis and identification of a new methylation product O^4-methylthymidine. Biochem J 135:193–201
Lawley PD, Harris G, Phillips E, Irving W, Colaco CB, Lydyard PM, Roitt IM (1986) Repair of chemical carcinogen-induced damage in DNA of human lymphocytes and lymphoid cell lines – studies of the kinetics of removal of O^6-methylguanine and 3-methyladenine. Chem Biol Interact 57:107–121
Leung K-H, Archer MC (1984) Studies on the metabolic activation of *β*-ketonitrosamines: mechanisms of DNA methylation by *N*-(2-oxopropyl)-*N*-nitrosoureas and *N*-acetoxymethyl-*N*-(2-oxopropyl)amine. Chem Biol Interact 48:169–179
Leung K-H, Archer MC (1985) Mechanism of DNA methylation by *N*-nitroso(2-oxopropyl)-propylamine. Carcinogenesis 6:189–191
Lewis SE, Johnson FM, Skow LC, Popp D, Barnett LB, Popp RA (1985) A mutation in the *β*-globin gene detected in the progeny of a female mouse treated with ethylnitrosourea. Proc Natl Acad Sci USA 82:5829–5831
Li BFL, Reese CB, Swann PF (1987) Synthesis and characterization of oligodeoxynucleotides containing 4-*O*-methylthymine. Bichemistry 26:1086–1093
Likhachev AJ, Ivanov MN, Bresil H, Planche-Martel G, Montesano R, Margison GP (1983) Carcinogenicity of single doses of *N*-nitroso-*N*-methylurea and *N*-nitroso-*N*-ethylurea in Syrian golden hamster and persistence of alkylated purines in DNA of various tissues. Cancer Res 43:829–833
Lijinsky W (1986) Deuterium isotope effects in carcinogenesis by *N*-nitroso compounds and related carcinogens. J Cancer Res Clin Oncol 112:229–239
Lijinsky W (1987) Structure-activity relations in carcinogenesis by *N*-nitroso compounds. Cancer Metastasis Rev 6:301–356
Lijinsky W (1988) Nucleic acid alkylation by *N*-nitroso compounds related to organ-specific carcinogenesis. In: Politzer PA, Roberts L (eds) Chemical carcinogens: activation mechanisms, structural and electronic factors and reactivity. Elsevier, New York
Lijinsky W, Reuber MD (1984) Comparison of nitrosocimetidine with nitrosomethyl-nitroguanidine in chronic feeding tests in rats. Cancer Res 44:447–449
Lijinsky W, Ross AE (1969) Alkylation of rat liver nucleic acids not related to carcinogenesis by *N*-nitrosamines. JNCI 42:1095–1100
Lijinsky W, Loo J, Ross A (1968) Mechanism of alkylation of nucleic acids by nitrosodimethylamine. Nature 218:1174–1175
Lijinsky W, Saavedra JE, Reuber MD (1981) Induction of carcinogenesis in Fischer rats by methylalkylnitrosamines. Cancer Res 41:1288–1292
Lijinsky W, Saavedra JE, Reuber MD, Singer SS (1982) Esophageal carcinogenesis in Fischer 344 rats by nitrosomethylethylamines substituted in the ethyl group. JNCI 68:681–684
Loeppky RN, Tomasik W, Ovtram JR, Kovacs DA, Byington KH (1984) Alternative bioactivation routes for *β*-hydroxynitrosamines: biochemical and chemical model structures. IARC, Lyon, pp 429–436 (IARC scientific publication no 57)

Loveless A (1969) Possible relevance of O^6-alkylation of deoxyguanosine to mutagenicity and carcinogenicity of nitrosamines and nitrosamides. Nature 223:206–207
Loveless A, Hampton CL (1969) Inactivation and mutation of coliphage T2 by *N*-methyl- and *N*-ethyl-*N*-nitrosourea. Mutat Res 7:1–12
Lu S-H, Ohshima H, Fu H-M, Tian Y, Li F-M, Blettner M, Wahrendorf J, Bartsch H (1986) Urinary excretion of *N*-nitrosamino acids and nitrate by inhabitants of high- and low-risk areas for oesophageal cancer in Northern China: endogenous formation of nitrosoproline and its inhibition by vitamin C. Cancer Res 46:1485–1491
Ludlum DB (1986) Nature and biological significance of DNA modification by the haloethylnitrosoureas. In: Schmähl D, Kaldor JM (eds) Carcinogenicity of alkylating drugs. IARC, Lyon, pp 71–81 (IARC scientific publications no 78)
Ludlum DB, Kent S, Mehta JR (1986a) Formation of O^6-ethylthioethylguanine in DNA by reaction of sulfur mustard and its apparent lack of repair by O^6-alkylguanine alkyltransferase. Carcinogenesis 7:1203–1206
Ludlum DB, Mehta JR, Teng WP (1986b) Prevention of 1-(3-deoxycytidyl), 2-(1-deoxyguanosinyl)ethane cross-link formation in DNA by rat liver O^6-alkylguanine-DNA alkyltransferase. Cancer Res 46:3353–3357
Magee PN (1956) Toxic liver injury: the metabolism of dimethylnitrosamine. Biochem J 64:676–682
Magee PN (1982) Nitrosamines and human cancer. Cold Spring Harbor Laboratory, New York
Magee PN, Barnes JM (1956) Production of malignant primary hepatic tumours in the rat by feeding dimethylnitrosamine. Br J Cancer 10:114–122
Magee PN, Farber E (1962) Toxic liver injury and carcinogenesis. Methylation of rat liver nucleic acids by dimethylnitrosamine in vivo. Biochem J 83:114–124
Maher VM, Domoradzki J, Corner RC, McCormick JJ (1986) Correlation between O^6-alkylguanine-DNA alkyltransferase activity and resistance of human cells to the cytotoxic and mutagenic effects of methylating and ethylating agents. In: Harris CC (ed) Biochemical and molecular epidemiology of cancer. Liss, New York, pp 411–418
Malling HV (1966) Mutagenicity of two potent carcinogens, dimethylnitrosamine and diethylnitrosamine, in *Neurospora crassa*. Mutat Res 3:537–540
Malling HV (1971) Dimethylnitrosamine: formation of mutagenic compounds by interaction with mouse liver microsomes. Mutat Res 13:425–429
Marasas WFO, Kriek NPJ, Fincham JE, Van Rensburg SJ (1984) Primary liver cancer and oesophageal basal cell hyperplasia in rats caused by *Fusarium moniliforme*. Int J Cancer 34:383–387
Margison GP (1982) Chronic or acute administration of various dialkylnitrosamines enhances removal of O^6-methylguanine from rat liver DNA in vivo. Chem Biol Interact 38:189–201
Marushige K, Marushige Y (1983) Alkylation of isolated chromatin with *N*-methyl- and *N*-ethyl-nitrosourea. Chem Biol Interact 46:165–177
Masson HA, Ioannides C, Gibson GG (1983) Role of highly purified forms of rat liver cytochrome P-450 in demethylation of dimethylnitrosamine and its activation to mutagens. Toxicol Lett 17:131–135
McGarrity JF, Cox DP (1983) Protonation of diazomethane in superacid media. J Am Chem Soc 105:3961–3966
McGarrity JF, Smyth T (1980) Hydrolysis of diazomethane –kinetics and mechanisms. J Am Chem Soc 102:7303–7308
Michejda CJ, Andrews AW, Koepke SR (1979) Derivatives of side-chain hydroxylated nitrosamines: direct acting mutagens in *Salmonella typhimurium*. Mutat Res 67:301–308
Ming S-C (1984) Precancerous states of the oesophagus and stomach. In: Carter RL (ed) Precancerous states. Oxford University Press, Oxford, pp 185–229
Mirvish SS (1975) Formation of *N*-nitroso compounds: chemistry, kinetics and in vivo occurrence. Toxicol Appl Pharmacol 31:325–351
Mirvish SS (1986) Effects of vitamins C and E on *N*-nitroso compound formation, carcinogenesis and cancer. Cancer 58:1842–1850

Mohr U, Green U, Althoff J, Schneider P (1978) Syncarcinogenic action of saccharin and sodium cyclamate in the induction of bladder tumours in *N*-methyl-*N*-nitrosourea-pretreated rats. In: Guggenheim B (ed) Health and sugar substitutes. Karger, Basel, pp 64–69

Montesano R, Brésil H, Planche-Martel G, Margison GP, Pegg AE (1983) Stability and capacity of dimethylnitrosamine-induced O^6-methylguanine repair system in rat liver. Cancer Res 43:5808–5814

Moolgavkar SH, Knudson AG Jr (1981) Mutation and cancer: a model for human carcinogenesis. JNCI 66:1037–1052

Morimoto K, Tanaka A, Yamaha T (1983) Reaction of 1-*n*-propyl-1-nitrosourea with DNA in vitro. Carcinogenesis 4:1455–1458

Morimoto K, Dolan ME, Scicchitano D, Pegg AE (1985) Repair of O^6-propylguanine and O^6-butylguanine in DNA by O^6-alkylguanine-DNA alkyltransferase from rat liver and *E. coli*. Carcinogenesis 6:1027–1031

Moschel RC, Hudgins WR, Dipple A (1980) Aralkylation of guanosine by the carcinogen *N*-nitroso-*N*-benzylurea. Org Chem 45:533–535

Mottram JC (1945) Change from benign to malignant in chemically induced warts in mice. Br J Exp Pathol 26:1–4

Nicolson GL (1987) Tumor cell instability, diversification, and progression to the metastatic phenotype: from oncogene to oncofetal expression. Cancer Res 47:1473–1487

Nishimura S, Sekiya T (1987) Human cellular oncogenes. Biochem J 243:313–327

Nowell PC (1986) Mechanisms of tumor progression. Cancer Res 46:2203–2207

O'Connell JF, Klein-Szanto AJP, DiGiovanni DM, Fries JW, Slaga TJ (1986) Malignant progression of mouse skin papillomas treated with ethylnitrosourea, *N*-methyl-*N'*-nitro-*N*-nitrosoguanidine, or 12-*O*-tetradecanoylphorbol-13-acetate. Cancer Lett 30:269–274

O'Connor PJ, Capps MJ, Craig AW (1973) Comparative studies of the hepatocarcinogen *N,N*-dimethylnitrosamine in vivo: reaction sites in rat liver DNA and significance of their relative stabilities. Br J Cancer 27:153–166

Ohshima H, Bartsch H (1981) Quantitative estimation of endogenous nitrosation in humans by monitoring *N*-nitrosoproline excreted in urine. Cancer Res 41:3658–3662

Okada M, Suzuki E, Mochizuki M (1976) Possible important role of urinary *N*-methyl-*N*-(3-carboxypropyl)nitrosamine in induction of bladder tumors in rats by *N*-methyl-*N*-dodecylnitrosamine. Gann 67:771–772

Olsson M, Lindahl T (1980) Repair of alkylated DNA in *Escherichia coli*: methyl group transfer from O^6-methylguanine to a protein cysteine residue. J Biol Chem 255:10569–10571

O'Neill IK, Von Borstel RC, Miller CT, Long J, Bartsch H (eds) (1984) *N*-Nitroso compounds: occurrence, biological effects and relevance to human cancer. IARC, Lyon (IARC scientific publication no 57)

Osborne MR (1984) DNA interactions of reactive intermediates derived from carcinogens. In: Searle CE (ed) Chemical carcinogens, 2nd edn. American Chemical Society, Washington DC, pp 503–507 (ACS Monograph no 182)

Osterman-Golkar S, Ehrenberg L, Segerback D, Hallstrom I (1976) Evaluation of genetic risks of alkylating agents. Hemoglobin as a dose monitor. Mutat Res 34:1–10

Oyasu R, Hirao Y, Izumi K (1981) Enhancement by urine of urinary bladder carcinogenesis. Cancer Res 41:478–481

Park KK, Archer MC, Wishnok JS (1980) Alkylation of nucleic acids by *N*-nitroso-di-*n*-propylguanine: evidence that carbonium ions are not significantly involved. Chem Biol Interact 29:139–144

Park M, Dean M, Cooper CS, Schmidt M, O'Brien SJ, Blair DG, Vande Woude GF (1986) Mechanism of *met* oncogene activation. Cell 45:895–904

Parsa I, Friedman S, Cleary CM (1987) Visualization of O^6-methylguanine in target cell nuclei of dimethylnitrosamine-treated human pancreas by a murine monoclonal antibody. Carcinogenesis 8:839–846

Pasternak L (1962) Mutagene Wirkung von Dimethylnitrosaminen bei *Drosophila melanogaster*. Naturwiss 49:381

Pegg AE, Lijinsky W (1984) Saturation of repair system for O^6-methylguanine in rat liver DNA by pretreatment with cyclic nitrosamines. Chem Biol Interact 51:365–370
Pegg AE, Perry W (1981) Alkylation of nucleic acids and metabolism of small doses of dimethylnitrosamine in the rat. Cancer Res 41:3128–3132
Penn I (1986) Malignancies induced by drug therapy: a review. In: Schmähl D, Daldo JM (eds) Carcinogenicity of alkylating drugs. IARC, Lyon, pp 13–27 (IARC scientific publication no 78)
Perantoni AO, Rice JM, Reed CD, Watatani M, Wenk ML (1987) Activated *neu* oncogene sequences in primary tumors of the peripheral nervous system induced in rats by transplacental exposure to ethylnitrosourea. Proc Natl Acad Sci USA 84:6317–6321
Peto R, Gray R, Brantom P, Grasso P (1984) Nitrosamine carcinogens in 5120 rodents. In: O'Neill IK, Von Borstel RC, Miller CT, Long J, Bartsch H (eds) IARC, Lyon, pp 627–665 (IARC scientific publication no 57)
Pitot HC, Beer DG, Hendrich S (1987) Multistage carcinogenesis in the rat hepatocyte. In: Butterworth BE, Slaga TJ (eds) Nongenotoxic mechanisms in carcinogenesis. Cold Spring Harbor Laboratory, New York, pp 41–53 (Banbury Report no 25)
Poirier S, Hubert A, deThe G, Ohshima H, Bourgade M-C, Bartsch H (1987) Occurrence of volatile nitrosamines in food samples collected in three high-risk areas for nasopharyngeal carcinoma. In: Bartsch H, O'Neill IK, Schulte-Hermann R (eds) IARC scientific vol no 84. IARC, Lyon, pp 415–419
Popper H, Sternberg SS, Oser BC, Oser M (1960) Carcinogenic effect of aramite in rats. A study of hepatic nodules. Cancer 13:1035–1045
Pour P, Wallcave L, Gingell R, Nagel D, Lawson T, Salmasi S, Tines S (1979) Carcinogenic effect of *N*-nitroso(2-hydroxypropyl)(2-oxopropyl)amine, a postulated proximate pancreatic carcinogen in Syrian hamsters. Cancer Res 39:3828–3833
Preston-Martin S (1987) *N*-Nitroso compounds as a cause of human cancer. In: Bartsch H, O'Neill IK, Schulte-Hermann R (eds) IARC scientific publication no 84. IARC, Lyon, pp 477–484
Preussmann R, Eisenbrand G (1984) *N*-Nitroso compounds in the environment. In: Searle CE (ed) Chemical carcinogens, 2nd edn, vol 2. American Chemical Society, Washington DC, Chap. 13 (ACS Monograph no 182)
Preussmann R, Stewart BW (1984) *N*-Nitroso carcinogens. In: Searle CE (ed) Chemical carcinogens, 2nd edn, vol 2. American Chemical Society, Washington DC, Chap. 12 (ACS Monograph no 182)
Preussmann R, Würtele G, Eisenbrand G, Spiegelhalder B (1978) Urinary excretion of *N*-nitrosodiethanolamine administered orally to rats. Cancer Lett 4:207–209
Quintanilla M, Brown K, Ramsden M, Balmain A (1986) Carcinogen-specific mutation and amplification of Ha-*ras* during mouse skin carcinogenesis. Nature 322:78–80
Ramanathan R, Rajalakshmi S, Sarma DSR, Farber E (1976) Nonrandom nature of in vivo methylation by dimethylnitrosamine and subsequent removal of methylated products from rat liver chromatin DNA. Cancer Res 36:2073–2079
Randerath K, Reddy MV, Disher RM (1986) Age- and tissue-related DNA modifications in untreated rats: detection by ^{32}P-postlabelling assay and possible tumor induction and ageing. Carcinogenesis 7:1615–1617
Reddy EP, Reynolds RK, Santos E, Barbacid M (1982) A point mutation is responsible for the acquisition of transforming properties by the T24 human bladder carcinoma oncogene. Nature 300:149–152
Reddy MV, Gupta RC, Randerath E, Randerath K (1984) ^{32}P-Postlabelling test for covalent DNA binding of chemicals in vivo: application to a variety of aromatic carcinogens and methylating agents. Carcinogenesis 5:231–243
Reiner B, Zamenhof S (1957) Studies on chemically reactive groups of DNA. J Biol Chem 228:475–486
Renard A, Verly WG (1983) Repair of O^6-ethylguanine DNA lesions in isolated cell nuclei: presence of activity in chromatin proteins. Eur J Biochem 136:453–460
Reynolds SH, Stowers SJ, Patterson RM, Maronpot RR, Aaronson SA, Anderson MW (1987) Activated oncogenes in B6C3F1 mouse: implications for risk assessment. Science 237:1309–1316

Rice JM, Ward JM (1982) Age dependence of susceptibility to carcinogenesis in the nervous system. Ann NY Acad Sci 381:274–289
Rice S, Cheng MY, Cramer RE, Mandel M, Mower HF, Seff K (1984) Structure of *N*-methyl-*N′*-nitro-*N*-nitrosoguanidine. J Am Chem Soc 106:239–243
Richardson KK, Richardson FC, Crosby RM, Swenberg JA, Skopek TR (1987) DNA base changes and alkylation following in vivo exposure of *Escherichia coli* to *N*-methyl-*N*-nitrosourea. Proc Natl Acad Sci USA 84:344–348
Roberts JJ (1978) Repair of DNA modified by cytotoxic mutagenic and carcinogenic chemicals. Adv Radiation Biol 7:212–446
Rodenhuis S, van de Wetering ML, Moot WJ, Evers SG, van Zandwijk N, Bos JL (1987) Mutational activation of the K-*ras* oncogene, a possible pathogenetic factor in adenocarcinoma of the lung. New England J Med 317:929–935
Roscoe JP, Claisse PJ (1976) Sequential in vivo-in vitro study of carcinogenesis induced in rat brain by ethylnitrosourea. Nature 262:314–316
Rose FL (1958) Discussion to Magee PN: Some experimental studies on toxic liver injury. In: Walpole AL, Spinks A (eds) Symposium on evaluation of drug toxicity. J and A Churchill, London, p 116
Ryan AJ, Billett MA, O'Connor PJ (1986) Selective repair of methylated purines in region of chromatin DNA. Carcinogenesis 7:1497–1503
Saffhill R (1985) In vitro miscoding of alkylthymines with DNA and RNA polymerases. Chem Biol Interact 53:121–130
Sandberg AA (1980) Chromosomes in human cancer and leukemia. Elsevier, New York
Sander J (1968) Nitrosaminsynthese durch Bakterien. Z Physiol Chem 349:429–432
Sander J, Bürkle G (1969) Induktion maligner Tumoren bei Ratten durch gleichzeitige Verfütterung von Nitrit und sekundären Aminen. Z Krebsforschung 73:54–66
Saul RL, Kabir SH, Cohen Z, Bruce WR, Archer MC (1981) Re-evaluation of nitrate and nitrite levels in human intestine. Cancer Res 41:2280–2283
Scherer E, Emmelot P (1975) Kinetics of evolution and growth of precancerous liver-cell foci and liver tumor formation by diethylnitrosamine in the rat. Eur J Cancer 11:689–696
Schoental R, Rive DJ (1965) Interaction of *N*-alkyl-*N*-nitrosourethanes with thiols. Biochem J 97:466–474
Schrager TF, Busby WF, Goldman ME, Newberne PM (1986) Enhancement of methylbenzyl-nitrosamine-induced esophageal carcinogenesis in zinc-deficient rats: effects on incorporation of [^{3}H]thymidine into DNA of esophageal epithelium and liver. Carcinogenesis 7:1121–1126
Schubert D, Heinemann S, Carlisle W, Tarikas H, Kimes B, Patrick J, Steinbach JH, Culp W, Brandt BL (1974) Clonal cell lines from the rat central nervous system. Nature 249:224–227
Schulz U, McCalla DR (1969) Reactions of cysteine with *N*-methyl-*N*-nitroso-*p*-toluenesulphonamide and *N*-methyl-*N′*-nitro-*N*-nitroso-guanidine. Can J Chem 47:2021–2027
Schwarz M, Appel KE, Schrenk D, Kunz W (1980) Effect of ethanol on microsomal metabolism of dimethylnitrosamine. J Cancer Res Clin Oncol 97:233–240
Scribner JD, Ford GP (1982) *n*-Propyldiazonium ion alkylates *O*-6 of guanine with rearrangement but alkylates N-7 without rearrangement. Cancer Letters 16:51–56
Shibata M-A, Nakanishi K, Shibata M, Masui T, Miyata Y, Ito N (1986) Promoting effect of sodium chloride in two-stage urinary bladder carcinogenesis in rats initiated by *N*-butyl-*N*-(4-hydroxybutyl)nitrosamine. Urol Res 14:201–206
Shih TY, Weeks MO (1984) Oncogenes and cancer. Cancer Invest 2:109–123
Shirazi SS, Platz CE (1978) Effect of alcohol on canine oesophageal mucosa. J Surg Res 25:373–379
Shuker DEG, Tannenbaum SR, Wishnok JS (1981) *N*-Nitroso bile acid conjugates. I. Synthesis, chemical reactivity and mutagenic activity. J Org Chem 46:2092–2096
Shuker DEG, Bailey E, Parry A, Lamb J, Farmer PB (1987a) Determination of urinary 3-methyladenine in humans as potential monitor of exposure to methylating agents. Carcinogenesis 8:959–962

Shuker DEG, Howell JR, Street BW (1987b) Formation and fate of nucleic acid and protein adducts derived from *N*-nitroso-bile acid conjugates. In: Bartsch H, O'Neill IK, Schulte-Hermann R (eds) IARC scientific publication no 84. IARC, Lyon, pp 187–193

Sieber SM, Adamson RH (1979) Chemical carcinogenesis in non-human primates and attempts at prevention. In: Chandra P (ed) Antiviral mechanisms in the control of neoplasia. Plenum, New York, pp 455–479

Singer B, Kusmierek J (1982) Chemical mutagenesis. Ann Rev Biochem 52:655–693

Singer B, Spengler SJ, Fraenkel-Corat H, Kusmierek JT (1986) O^4-Methyl, -ethyl or -isopropyl substituents on thymidine in poly(dA-dT) all lead to transitions upon replication. Proc Natl Acad Sci USA 83:28–32

Singer GM, Lijinsky W, Buettner L, McClusky GA (1981) Relationship of rat urinary metabolites of *N*-nitrosomethyl-*N*-alkylamines to bladder carcinogenesis. Cancer Res 41:4942–4946

Solt D, Farber E (1976) New principle for analysis of chemical carcinogenesis. Nature 263:701–703

Spiegelhalder B, Preussmann R (1985) In vivo nitrosation of amidopyrine in humans: use of "ethanol effect" for biological monitoring of *N*-nitrosodimethylamine in urine. Carcinogenesis 6:545–548

Stewart BW, Swann PF, Holsman JW, Magee PN (1974) Cellular injury and carcinogenesis. Evidence for alkylation of rat liver nucleic acids in vivo by *N*-nitrosomorpholine. Z Krebsforsch 82:1–12

Stewart M, Hill MJ, Pugh CRB, Williams JP (1981) Role of *N*-nitrosamines in carcinogenesis at the ureterocolic anastomosis. Br J Urol 53:115–118

Stowers SJ, Wiseman RW, Ward JM, Miller EC, Miller JA, Anderson MW, Eva A (1988) Detection of activated proto-oncogenes in *N*-nitrosodiethylamine-induced liver tumours: a comparison between $B6C3F_1$ mice and Fischer 344 rats. Carcinogenesis 9:277–276

Strauss BS, Rabkin S, Sagher D, Moore P (1982) Role of DNA polymerase in base substitution mutagenesis on non-instructional templates. Biochimie 64:829–838

Strauss BS, Larson K, Rabkin S, Sahm J, Shenkar R (1986) In vitro models for mutagenesis: a role for lesion, polymerase and sequence. Prog Clin Biol Res 209A:149–159

Suzuki E, Okada M (1980) Metabolic fate of *N,N*-dibutylnitrosamine in the rat. Gann 71:863–870

Swann PF, Magee PN (1968) Nitrosamine-induced carcinogenesis. Alkylation of nucleic acids of the rat by *N*-methyl-*N*-nitrosourea, dimethylnitrosamine, dimethyl sulphate and methyl methanesulphonate. Biochem J 110:39–47

Swann PF, McLean AEM (1968) Effect of diet on the toxic and carcinogenic action of dimethylnitrosamine. Biochem J 107:14–15

Swann PF, Coe AM, Mace R (1984) Ethanol and dimethylnitrosamine and diethylnitrosamine metabolism and disposition in the rat. Possible relevance to the influence of ethanol on human cancer incidence. Carcinogenesis 5:1337–1343

Swenberg JA, Kerns WD, Mitchell RI, Caralla EJ, Pavkov KL (1980) Induction of squamous cell carcinomas of the rat nasal cavity by inhalation exposure of formaldehyde vapor. Cancer Res 40:3398–3402

Swenberg JA, Dyroff MC, Bedell MA, Popp JA, Huh N, Kirsten U, Rajewsky MF (1984) O^4-Ethyldeoxythymidine but not O^6-ethyldeoxyguanosine accumulates in hepatocyte DNA of rats continuously exposed to diethylnitrosamine. Proc Natl Acad Sci USA 81:1692–1695

Swenberg JA, Richardson FC, Tyeryar L, Deal F, Boucheron J (1987) Molecular dosimetry of DNA adducts formed by continuous exposure of rats to alkylating hepatocarcinogens. Prog Exp Tumor Res 31:42–51

Swenson DH, Farmer PB, Lawley PD (1976) Identification of the methyl phosphotriester of thymidylyl(3′-5′)thymidine as a product from reaction of DNA with the carcinogen *N*-methyl-*N*-nitrosourea. Chem Biol Interact 15:91–100

Swenson DH, Frei JV, Lawley PD (1979) Synthesis of 1-(2-hydroxy)-1-nitrosourea and comparison of its carcinogenicity with that of 1-ethyl-1-nitrosourea. JNCI 63:1469–1473

Swenson DH, Harbach PR, Trzos RT (1980) Relationship between alkylation of specific DNA bases and induction of sister chromatid exchange. Carcinogenesis 1:931–936

Swenson DH, Petzold GL, Harbach PR (1986) Binding of 1-(2-hydroxyethyl)-1-nitrosourea to DNA in vitro and to DNA of thymus and bone marrow in C57BL mice in vivo. Cancer Lett 33:75–81

Tempest PR, Reeves BR, Spurr NK, Rance AJ, Chan A M-L, Brookes P (1986) Activation of the *met* oncogene in the human MNNG-HOS cell line involves a chromosomal rearrangement. Carcinogenesis 7:2051–2057

Terracini B, Testa MC, Cabral JR, Rossi L (1976) Roles of age at treatment and dose in carcinogenesis in C3Hf/Dp mice with a single administration of *N*-nitroso-*N*-methylurea. Br J Cancer 33:427–439

Thompson HJ, Meeker LD (1983) Induction of mammary gland carcinomas by subcutaneous injection of 1-methyl-1-nitrosourea. Cancer Res 43:1628–1629

Thompson MH, Hill MJ (1987) Etiology and mechanism of carcinogenesis: diet, luminal factors and colorectal cancer. In: Faivre J, Hill MJ (eds) Causation and prevention of colorectal cancer. Elsevier, Amsterdam, pp 99–120

Toksoz D, Farr CJ, Marshall CJ (1987) *ras* Gene activation in a minor proportion of the blast population in acute leukaemia. Oncogene 1:409–413

Topal MD, Eadie JS, Conrad M (1986) O^6-Methylguanine mutation and repair is non-uniform. J Biol Chem 261:9879–9885

Umbenhauer D, Wild CP, Montesano R, Saffhill R, Boyle JM, Huh N, Kirstein V, Thomale J, Rajewsky MF, Lu SH (1985) O^6-methyldeoxyguanonine in oesophageal DNA among persons at high risk from oesophageal cancer. Int J Cancer 36:661–665

Von Hofe E, Kleihues P, Keefer LK (1986) Extent of DNA 2-hydroxylation by *N*-nitrosomethylethylamine and *N*-nitrosodiethylamine in vivo. Carcinogenesis 7: 1335–1337

Von Hofe E, Schmerold I, Lijinsky W, Jeltsch P, Kleihues P (1987) DNA methylation in rat tissues by a series of homologous aliphatic nitrosamines ranging from *N*-nitrosodimethylamine to *N*-nitrosomethyldodecylamine. Carcinogenesis 8:1337–1341

Wada S, Miyanishi M, Nishimato Y, Kambe S, Miller RW (1968) Mustard gas as a cause of respiratory neoplasia in man. Lancet i:1161–1163

Warren W (1984) Analysis of alkylated DNA by high pressure liquid chromatography. In: Venitt S, Parry JM (eds) Mutagenicity testing, a practical approach. IRL Press, Oxford, pp 25–44

Warren W, Lawley PD (1980) Removal of alkylation products from DNA of *Escherichia coli* cells treated with carcinogens *N*-ethyl- and *N*-methyl-*N*-nitrosourea: influence of growth condition and DNA repair defects. Carcinogenesis 1:67–78

Warren W, Lawley PD, Gardner E, Harris G, Ball JK, Cooper CS (1987) Induction of thymomas by *N*-methyl-*N*-nitrosourea in ARK mice: interaction between chemical carcinogen and endogenous murine leukaemia viruses. Carcinogenesis 8:163–172

Watson JD, Crick FHC (1953) Genetical implications of the structure of deoxyribose nucleic acids. Nature 171:964–967

Waynforth HB, Magee PN (1975) Effects of various doses and schedules of administration of *N*-methyl-*N*-nitrosourea with and without croton oil promotion on skin papilloma production in BALB/c mice. Gann Monographs Cancer Res 17:439–448

White INH, Smith AG, Farmer PB (1983) Formation of *N*-alkylated protoporphyrin IX in livers of mice after diethylnitrosamine treatment. Biochem J 212:599–608

Wild CP, Umbenhauer D, Chapot B, Montesano R (1986) Monitoring of individual human exposure to aflatoxins and *N*-nitrosamines by immunoassays. J Cell Biochem 30:171–179

Wild CP, Lu SH, Montesano R (1987) Radioimmunoassay used to detect DNA alkylation adducts in tissues from populations at high risk from oesophageal and stomach cancer. In: Bartsch H, O'Neill IK, Schulte-Hermann R (eds) IARC scientific vol no 84. IARC, Lyon, pp 534–537

Williams GM (1987) Definition of a human cancer hazard. In: Butterworth BE, Slaga TJ (eds) Nongenotoxic mechanisms in carcinogenesis. Cold Spring Harbor Laboratory, New York, pp 367–380 (Banbury Report no 25)

Williams JC, Gusterson B, Humphreys J, Monaghan P, Coombes R, Rudland P, Neville AM (1981) *N*-Methyl-*N*-nitrosourea-induced rat mammary tumors. Hormone responsiveness but lack of spontaneous metastasis. JNCI 66:147–151

Williams LD, Shaw BR (1987) Protonated base pairs explain the ambiguous pairing properties of O^6-methylguanine. Proc Natl Acad Sci USA 84:1779–1783

Woodruff RC, Mason JM, Valencia R, Zimmering S (1984) Chemical mutagenesis testing in *Drosophila*. I. Comparison of positive and negative control data for sex-linked recessive lethal mutations and reciprocal translocations in three laboratories. Environ Mutagen 6:189–202

Yanez L, Groffen J, Valenzuela DM (1987) c-K-*ras* mutations in human carcinomas occur preferentially in codon 12. Oncogene 1:315–318

Yano K, Sonoda M, Sakagishi Y (1987) Reactions of *N*-nitrosoureas with cell membranes. In: Bartsch H, O'Neill IK, Schulte-Hermann R (eds) IARC scientific vol no 84. IARC, Lyon, pp 202–205

Yarosh DB (1985) Role of O^6-methylguanine-DNA methyltransferase in cell survival, mutagenesis and carcinogenesis. Mut Res 145:1–16

Yarosh DB, Fornace AJ, Day RS III (1985) Human O^6-alkylguanine-DNA alkyltransferase fails to repair O^4-methylthymine and methyl phosphotriesters in DNA as efficiently as does alkyltransferase from *Escherichia coli*. Carcinogenesis 7:949–953

Yoo J-S H, Ning SM, Patten CJ, Yang CS (1987) Metabolism and activation of *N*-nitrosodimethylamine by hamster and rat microsomes: comparative study with weanling and adult animals. Cancer Res 47:992–998

Yu MC, Henderson BE (1987) Intake of Cantonese-style salted fish as a cause of nasopharyngeal carcinoma. In: Bartsch H, O'Neill IK, Schulte-Hermann R (eds) IARC scientific vol no 84. IARC, Lyon, pp 547–549

Yunis JJ, Soreng AL, Bowe AE (1987) Fragile sites are targets of diverse mutagens and carcinogens. Oncogene 1:59–69

Zarbl H, Sukumar S, Arthur AV, Martin-Zanca D, Barbacid M (1985) Direct mutagenesis of Ha-*ras*-1 oncogenes by *N*-methyl-*N*-nitrosourea during initiation of mammary carcinogenesis in rats. Nature 315:382–385

CHAPTER 12

Heterocyclic-Amine Mutagens/Carcinogens in Foods

J. S. FELTON and M. G. KNIZE

A. Introduction

Concern about the role of diet in human cancer has prompted the search for compounds in common foods that may act as tumor initiators by producing somatic cell mutations. In the past 10 years, analyses of pyrolized amino acids and proteins and of cooked, protein-containing foods has led to the discovery of several classes of highly mutagenic heterocyclic aromatic amines. This review will concentrate on the chemical identity, mechanism of formation, and spectrum of genotoxicity and DNA interaction of these compounds. Clearly, these heterocyclic amines are not the only mutagens/carcinogens in food; aflatoxin B_1, nitrosamines, PAHs, and hydrazines are biologically active compounds found in food, but they are being treated in other chapters of this handbook and/or have been thoroughly discussed in numerous recent reviews.

Analysis, using *Salmonella*, mutation assays of mutagens in major sources of cooked protein in the American diet (based on U.S.D.A. and U.S.D.H.E.W. surveys) showed significant mutagen content in beef, pork, ham, bacon, chicken that had been fried or boiled, and a lesser amount in fish (BJELDANES et al. 1982b). Tofu, beans, cheese, and some fish, when cooked under similar conditions, produced low or negligible mutagenic activity (BJELDANES et al. 1982c). Other researchers in Japan have shown that when fish is broiled to a well-done state, considerable mutagenic activity can be detected (SUGIMURA et al. 1977a). In addition, beef extract, whether bacterial grade (COMMONER et al. 1978) or food grade (HARGRAVES and PARIZA 1982) is also mutagenic in the Ames /*Salmonella* assay.

The isolation, identification, and assessment of the biological activity of these mutagens, which are present in cooked meat at part per billion concentrations and which require metabolic activation, has been a difficult problem requiring extensive efforts from researchers in Japan, USA, UK, FRG, the Netherlands, Denmark, Sweden, Norway, Switzerland, and Australia. From these laboratories over the past 10 years, 17 heterocyclic amine mutagens have been isolated and identified (some only partially) from cooked meat. The most common class of mutagens in foods in the Western diet appears to be aminoimidazoazaarenes (AIA) (FELTON et al. 1986a), characterized by having 1 or 2 heterocyclic rings fused to an aminoimidazo ring. This class of AIA compounds can be differentiated from the pyrolysate mutagens, formed from high-temperature heating of amino acids and some food, by the resistance of the amino group to deamination by nitric acid (YOSHIDA and MATSUMOTO 1978; TSUDA et al. 1980, 1985; FELTON et al. 1984b).

The biological activities of these heterocyclic amines include bacterial mutations (reverse and forward mutations), formation of chromosomal lesions, transformation, and mutations in mammalian cells in culture, chromosomal lesions in rodents in vivo, mutations in mice and *Drosophila*, and tumors in rats and mice following feeding (see HATCH et al. 1984; HATCH 1986, for reviews).

It is the aim of this chapter to describe the methods that have been used for the isolation, identification, and quantitation of these mutagens, the structure/mutagenicity relationships, and the mutagen-forming reactions. In addition, the current status of the genetic toxicity, carcinogenicity, and human risk estimates of these heterocyclic amines found in food will be discussed.

B. Chemical Analysis of Heterocyclic-Amine Mutagens in Cooked Foods

I. Food Mutagen Sources and Exposure

In a collaborative study between Lawrence Livermore National Laboratory and the Nutritional Sciences Department at University of California Berkeley, it was found that potent bacterial mutagens are produced at normal cooking temperatures in beef, chicken, and pork (BJELDANES et al. 1982b). These findings have also been confirmed more recently in several laboratories. Other cooked foods such as cheese, tofu, and meats derived from organs other than muscle showed much lower activities (BJELDANES et al. 1982c). These early studies made it clear that muscle meats are the major source of bacterial mutagens in the western diet.

The pan residues remaining after frying pork were shown to have mutagenic activity equal to that in the meat itself (ÖVERVIK et al. 1987). These findings were extended to ground (minced) and fried beef and to chicken in our laboratory (KNIZE et al. 1988b), and the pan residues were found to have from 21% to 39% of the total mutagenicity of the meat patties themselves. These findings suggest that meat gravies are also an important source of exposure.

It was shown that volatile mutagens are produced from the cooking of beef (RAPPAPORT et al. 1979) and, in the frying of pork, 3%–11% of the total mutagenic activity produced was measured in the smoke (BERG et al. 1988), with 0.2%–0.5% of the total activity being true volatiles. It was also shown, by sampling air in different rooms in the house, that kitchen air is the most mutagenic (VAN HOUDT et al. 1984). It is generally assumed that the mutagen exposure from ingesting the foods is greater than that for the volatiles, and the emphasis has therefore been placed on the foods themselves. But mutagens in smoke may be important in an occupational setting, and a recent study suggests that there is increased lung cancer mortality among cooks (LUND and BORGEN 1987). The importance of mutagens in the air from cooking should be studied further.

II. Chemical Extraction and Purification

1. Extraction

The speed and sensitivity of the Ames/*Salmonella* test allows the use of the biological activity to monitor the extraction of the mutagenic compounds from complex food materials. The original extraction of mutagens from solid, cooked food was carried out using acid (COMMONER et al. 1978) or mixed organic solvents (FELTON et al. 1981). With the acid extraction method, the resulting aqueous supernatant was adjusted after centrifugation to a pH of 10–12, extracted with an organic solvent, usually dichloromethane, and evaporated. In the organic extraction scheme, after the evaporation of the organic extractant, the sample was dissolved in sodium hydroxide (pH 12) and extracted into dichloromethane. The basic extract contained essentially all of the mutagenic activity, suggesting that the mutagens were organic bases.

A method using trisulfocopper phthalocyanine bound to cotton ("blue cotton") has been devised for the purification of the multi-ring aromatic mutagens in cooked ground beef (HAYATSU et al. 1983) and, when compared with the XAD-2 method described below, gives greater mutagen purity but slightly lower total mutagen recovery.

A widely used method, shown in Fig. 1 (center), uses hydrochloric acid (pH 2) extraction followed by concentration and further purification on an XAD-2 resin column (BJELDANES et al. 1982a; FELTON et al. 1984b). Recently, it was shown that significant additional mutagenic activity could be extracted using acetone (Fig. 1, left) after the acid extraction (BECHER et al. 1988; KNIZE et al. 1988b). Also shown is a scheme for the extraction of the pan residues following cooking (Fig. 1, right).

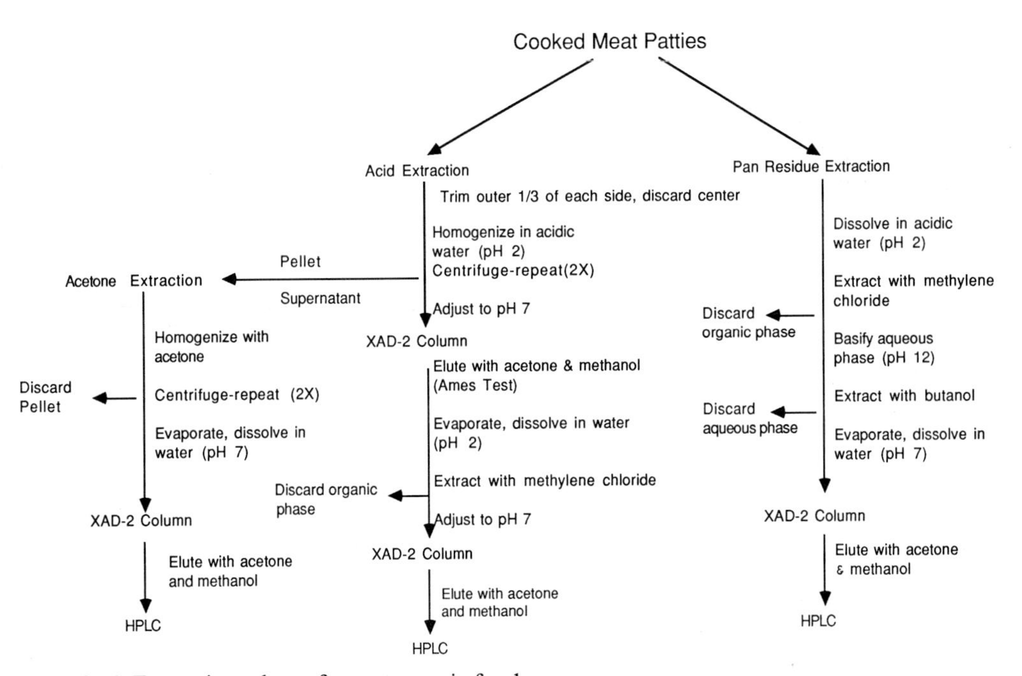

Fig. 1. Extraction scheme for mutagens in foods

2. Bioassay-Directed Purification

The purification of mutagens from a complex mixture has been best performed using a mutagenicity assay to guide the purification. An extract derived from cooked meat is chemically complex, containing thousands of components as seen by GC analysis (FELTON, unpublished data), each the result of the many heating and browning reactions known to occur in meat. Because of the complexity of reactions that can occur in meat at elevated cooking temperatures, our approach was to assume that any structural type of mutagen could be present.

The Ames/*Salmonella* mutation test using, specifically, the frameshift sensitive strains TA1538 and TA98 is the detection method of choice because the food extracts demonstrate tremendous mutagenic potency, so that only a small amount of sample needs to be used for mutagen detection. This leaves most of a sample for further purification and analysis. The Ames/*Salmonella* test requires only 48 h for each analysis, and numerous samples can be analysed in parallel each day, making it suitable for mutagen detection in the eluate from a high resolution chromatographic system (e.g. HPLC) that is being used to separate the many components in the complex mixture.

Although many foods have been examined for specific mutagens, only a few foods have been examined using the Ames/*Salmonella* test in conjunction with HPLC to detect all of the chromatographically separable mutagens. This type of analysis clearly shows that many mutagenic compounds are present (FELTON et al. 1984b).

III. Identification of Unknown Mutagens

1. General Approach

The identification and quantitation of mutagens in foods is required in order to determine accurately the dose of specific compounds in our diet that ultimately will lead to determination of the human risk of ingesting these compounds. After a mutagen is identified and characterized, its physical, chemical, and biological properties can then be used to reveal its presence and abundance in subsequent food extracts with much less effort.

The identification of unknown compounds requires knowledge of the atomic composition and the position of the atoms in the molecule. This, in turn, requires high resolution mass and NMR spectra coupled with synthesis of all possible isomers.

2. Characteristics of Food Mutagens

Table 1 lists the molecular weights, elemental compositions, and UV maxima (greater than 250 nm) of the mutagens found in heated food and food extracts. Also listed are the abbreviated name, the chemical name, the chemical abstract service number, and, in most cases, the reference for the synthesis of the compound.

Table 1. Summary of mutagens identified in heated food

Abbreviation	Chemical name	MW	Composition	UV max.	References	Chemical Abstract Service Ref. No
Phe-P-1	2-Amino-5-phenylpyridine	170	$C_{11}H_{10}N_2$	268	TSUJI et al. (1978)	33421-40-8
AαC	2-Amino-9H-pyrido[2,3-*b*]-indole	183	$C_{11}H_9N_3$	336	MATSUMOTO et al. (1979)	261148-68-5
Glu-P-2	2-Aminodipyrido-[1,2-*a*:3',2'-*d*]-imidazole	184	$C_{10}H_8N_4$	265	TAKEDA et al. (1978)	67730-10-3
Trp-P-2	3-Amino-1-methyl-5H-pyrido[4,3-*b*]-indole	197	$C_{12}H_{11}N_3$	264	TAKEDA et al. (1977)	62450-10-3
MeAαC	2-Amino-3-methyl-9H-pyrido[2,3-*b*]-indole	197	$C_{12}H_{11}N_3$	343	MATSUMOTO et al. (1981b)	68006-83-7
IQ	2-Amino-3-methyl-imidazo[4,5-*f*]-quinoline	198	$C_{11}H_{10}N_4$	264	KASAI et al. (1980a); LEE et al. (1982); WATERHOUSE and RAPOPORT (1985); ADOLFSSON and OLSSON (1983)	76180-96-6
IQx	2-Amino-3-methyl-imidazo[4,5-*f*]-quinoxaline	199	$C_{10}H_{13}N_3$	274	BECHER et al. (1988)	
Trp-P-1	3-Amino-1,4-dimethyl-5H-pyrido[4,3-*b*]-indole	211	$C_{13}H_{13}N_3$	264	AKIMOTO et al. (1977)	62450-06-0
4-MeIQ	2-Amino-3,4-dimethyl-imidazo[4,5-*f*]-quinoline	212	$C_{12}H_{12}N_4$	265	KASAI et al. (1980b); WATERHOUSE and RAPOPORT (1985); ADOLFSSON and OLSSON (1983)	77094-11-2
8-MeIQx	2-Amino-3,8-dimethyl-imidazo[4,5-*f*]-quinoxaline	213	$C_{11}H_{11}N_5$	274	KASAI et al. (1981a); GRIVAS and OLSSON (1985); GRIVAS (1986)	77500-04-0
4-MeIQx	2-Amino-3,4-dimethyl-imidazo[4,5-*f*]-quinoxaline	213	$C_{11}H_{11}N_5$	274	M. VAHL personal communication, (1987)	
PhIP	2-Amino-1-methyl-6-phenylimidazo-[4,5-*b*]-pyridine	224	$C_{13}H_{12}N_4$	315	KNIZE and FELTON (1986)	105650-23-5
4,8-DiMeIQx	2-Amino-3,4,8-trimethyl-imidazo-[4,5-*f*]-quinoxaline	227	$C_{12}H_{13}N_5$	274	Grivas (1985)	95896-78-9

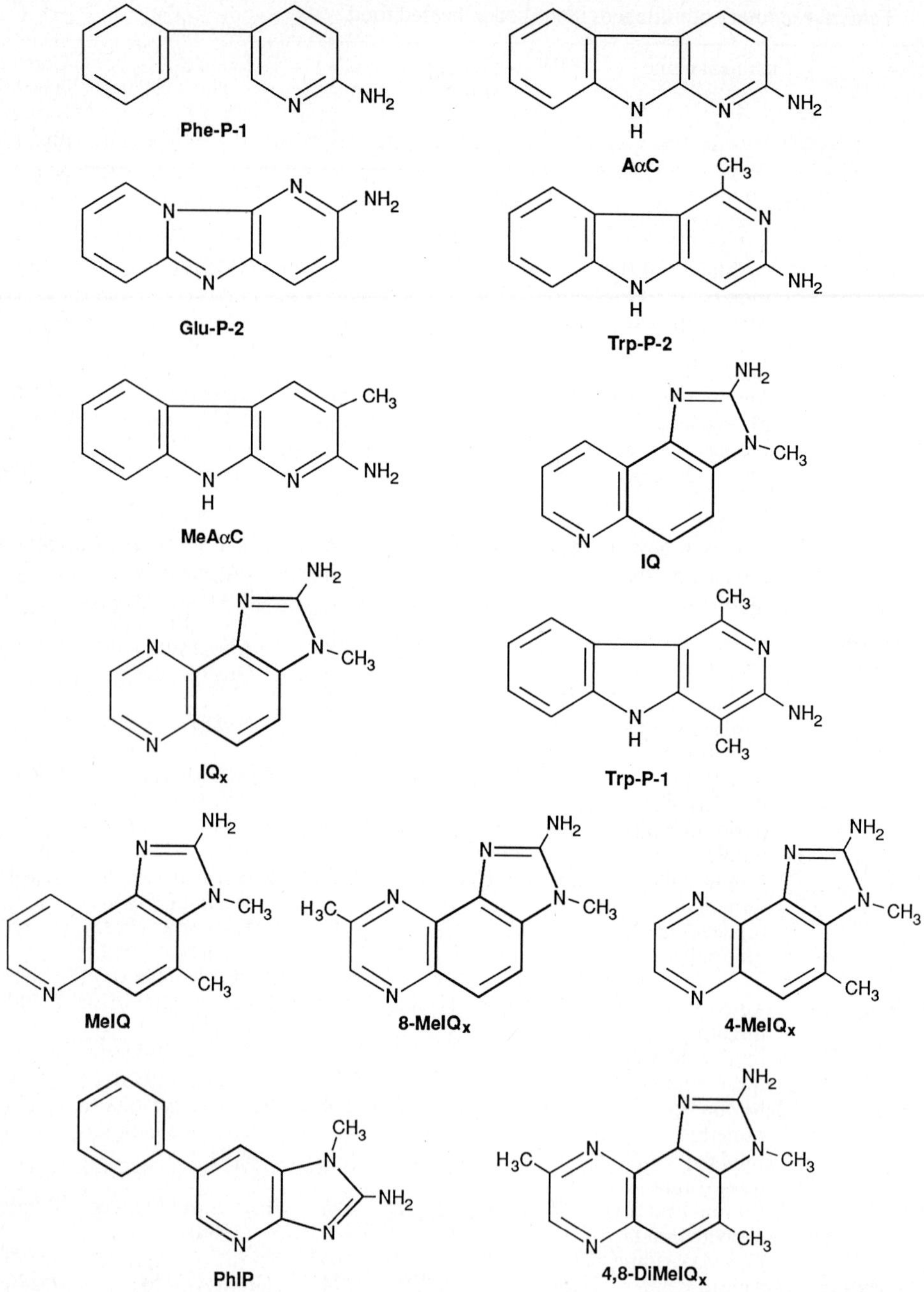

Phe-P-1

AαC

Glu-P-2

Trp-P-2

MeAαC

IQ

IQx

Trp-P-1

MeIQ

8-MeIQx

4-MeIQx

PhIP

4,8-DiMeIQx

Fig. 2. Chemical structures and common names of the 13 mutagens found in cooked foods

The structures of these mutagenic heterocyclic amines are shown in Fig. 2. Six of these compounds, Phe-P-1 (SUGIMURA et al. 1977b), AαC (YOSHIDA et al. 1978), Glu-P-2 (YAMAMOTO et al. 1978), Trp-P-2 (KOSAUGE et al. 1978; SUGIMURA et al. 1977b), MeAαC (YOSHIDA et al. 1978), and Trp-P-1 (SUGIMURA et al. 1977b), were first isolated and identified from pyrolysed amino acids and later found in at least one cooked food. Six others, IQ (KASAI et al. 1980a), IQx (BECHER et al. 1988), 4-MeIQ (KASAI et al. 1980b), 8-MeIQx (KASAI et al. 1981c), 4-MeIQx (VAHL, personal communication, 1987), and PhIP (FELTON et al. 1986b), were first isolated from a cooked food, and five of these were later made by heating simple mixtures in model systems. 4,8-DiMeIQx was independently isolated from a mutagen modeling system by boiling alanine, fructose, and creatinine (GRIVAS et al. 1985), and from fried beef (FELTON et al. 1984b). The synthetic mutagen isomers that had originally been made specifically to prove the structure of the 4,8-DiMeIQx isolated from the model boiling system were generously supplied by Drs. Grivas and Olsson to help prove the exact structure of that mutagen as purified from fried beef (KNIZE et al. 1987).

All of the compounds listed in Table I are *Salmonella* mutagens when activated by metabolism, some have identical molecular weights, some have identical UV absorbance spectra, and some mutagens are difficult to separate chromatographically. It is clear that no single property will uniquely identify all of the mutagenic compounds and that multiple criteria for identification should therefore be used.

a) Analytical Methods

Table 2 lists all of the mutagens that have been identified in at least one food, the method of analysis, the type of food, and the amount present in nanograms per gram.

With regard to the methods of analysis, nuclear magnetic resonance spectrometry (NMR) requires the highest quantity and purity and usually denotes the original identification.

High resolution mass spectrometry (HRMS) and mass spectrometry (MS) are analytical methods providing fairly specific detection endpoints for aromatic molecules, such as the molecular weight (as the base peak), a pattern of fragment ions, and quantitative information. Since isomer pairs have been identified for these compounds (8-MeIQx and 4-MeIQx; Trp-P-2 and MeAαC, for example) and since additional mutagenic synthetic isomers are known for most of the food mutagens, mass spectra need to be used in conjunction with other identification criteria.

Separation by gas chromatography (GC) or by liquid chromatography (LC) coupled with detection by mass spectrometry (MS), UV absorbance spectrometry (UV) or electrochemical detection (EC) are other identification methods that have been used. Chromatographic coelution and comparison of rentention times (RT) with known standards or, better still, the use of an internal standard (IS) of a radioactive or an isotopically labeled sample of the mutagen are the best ways to show identity (RT/IS). An internal standard is important because the sample matrix of the food can cause retention time shifts not seen with reference compounds alone. In addition, known isomers cannot always be resolved using a single chromatographic method, so coelution using two different methods of separation needs to be used in order to rule out closely eluting isomers and establish absolute identity (KNIZE et al. 1987).

Table 2. Amount and source of mutagens identified in heated foods and food extracts

Compound	Method	Food	Amount (ng/g)	References
Phe-P-1	GC/MS	Broiled sardine	8.6	YAMAIZUMI et al. (1980)
AαC	MS, UV	Grilled beef	651	MATSUMOTO et al. (1981a)
	MS, UV	Grilled chicken	180	
	MS, UV	Grilled mushroom	47	
	MS, UV	Grilled onion	1.5	
Glu-P-2	RT, UV	Broiled cuttlefish	280[a]	YAMAGUCHI et al. (1980a)
Trp-P-2	GC/MS	Broiled sardine	13.1	YAMAIZUMI et al. (1980)
	MS, RT-IS	30-h boiled beef extract	3.2	TAYLOR et al. (1985)
MeAαC	MS, UV	Grilled beef	63	MATSUMOTO et al. (1981a)
	MS, UV	Grilled chicken	15	
	MS, UV	Grilled mushroom	5.4	
IQ	NMR, HRMS	Broiled sardine	20	KASAI et al. (1981b)
	MS	Fried ground beef	0.6	BARNES et al. (1983)
	HRMS, UV	Fried ground beef	0.02	FELTON et al. (1984b)
	RT-IS, UV	Fried ground pork	nd	GRY et al. (1986)
	LC/MS	Broiled beef	0.5	YAMAIZUMI et al. (1986)
	LC/MS	Broiled sardine	4.9	
	LC/MS	Broiled salmon (skin)	1.1–1.7	
	LC/MS	Broiled salmon (flesh)	0.3–1.8	
	MS	Egg	0.1	GROSE et al. (1986)
	MS	Food grade beef extract	nd	HARGRAVES and PARIZA (1983)
	MS	Bact. grade beef extract	nd	
	RT, UV	Bact. grade beef extract	41 –142	TURESKY et al. (1983)
	RT, UV	Bact. grade beef extract	20 – 40	HAYATSU et al. (1983)
	MS	Creatine added meat product	17	BECHER et al. (1988)
	MS, RT-IS, UV	30-h boiled beef extract	0.5	TAYLOR et al. (1985)
IQx	MS, UV	Creatine added meat product	nd	BECHER et al. (1988)
Trp-P-1	MS, RT	Broiled beef	53[a]	YAMAGUCHI et al. (1980b)
	GC/MS	Broiled fish	13.3	YAMAIZUMI et al. (1980)
4-MeIQ	NMR, MS, UV	Broiled sardine	nd	KASAI et al. (1980b)
	MS	Fried ground beef	< 0.1	FELTON et al. (1986a)
	RT-IS, UV	Fried ground pork	nd	GRY et al. (1986)
	LC/MS	Broiled sardine	16.6	YAMAIZUMI et al. (1980)
	LC/MS	Broiled salmon (skin)	1.5–3.1	
	LC/MS	Broiled salmon (flesh)	0.6–2.8	
	MS	Bact. grade beef extract	nd	HARGRAVES and PARIZA (1983)

nd, not determined.
[a] Calculated from reference.

Table 2 (continued)

Compound	Method	Food	Amount (ng/g)	References
8-MeIQx	NMR, HRMS	Fried beef	nd	KASAI et al. (1981c)
	NMR, RT-IS	Fried ground beef	nd	KNIZE et al. (1987)
	HRMS, UV	Fried ground beef	0.1	FELTON et al. (1984b)
	RT	Fried ground beef	nd	HAYATSU et al. (1983)
	MS	Fried ground beef	0.45	HARGRAVES and PARIZA (1983)
	GC/MS	Fried ground beef	1.3 –2.4	MURRAY et al. (1988)
	RT-IS UV	Fried ground pork	nd	GRY et al. (1986)
	RT	Smoked dried mackerel	0.8	KATO et al. (1986)
	MS	Food grade beef extract	28	HARGRAVES and PARIZA (1983)
	RT, EC	Food grade beef extract	3.1	TAKAHASHI et al. (1985b)
	MS	Bact. grade beef extract	nd	HARGRAVES and PARIZA (1983)
	RT, UV	Bact. grade beef extract	142 –527	TURESKY et al. (1983)
	RT, UV	Bact. grade beef extract	222 –273	HAYATSU et al. (1983)
	RT, EC	Bact. grade beef extract	58.7	TAKAHASHI et al. (1985b)
	MS, UV	Creatine addet meat product	83	BECHER et al. (1988)
4-MeIQx	RT-IS, UV	Fried ground pork	nd	VAHL (personal communication, 1987)
PhIP	NMR, HRMS, UV	Fried ground beef	15	FELTON et al. (1986a)
	MS, UV	Fried ground pork	nd	GRY et al. (1986)
	MS, UV	Creatine added meat product	62	BECHER et al. (1988)
4,8-DiMeIQx	HRMS, UV	Fried ground beef	0.06	FELTON et al. (1984b)
	NMR, RT-IS	Fried ground beef	nd	KNIZE et al. (1987)
	MS	Fried ground beef	0.5	FELTON et al. (1986a)
	GC/MS	Fried ground beef	0.5–1.2	MURRAY et al. (1988)
	RT-IS	Fried ground pork	nd	GRY et al. (1988)
	RT	Smoked dried mackerel	0.08	KATO et al. (1986)
	RT, EC, UV	Bact. grade beef extract	10	TAKAHASHI et al. (1985a)
	MS, RT-IS	Creatine added meat product	15	BECHER et al. (1988)

b) Quantitation

The foods and cooking methods listed in Table 2 are as described in the original publications. It should be noted that large differences in the total mutagenicity and presumably in the amount of specific mutagens present are seen with changes in the cooking method, time, and temperature.

Two food-derived items, the creatinine-added meat product and the 30-h boiled beef supernatant, are included because although they are modifications of the standard household preparation and cooking of those foods, they contain only natural meat components.

The amount detected is included with the caveat noted above, that cooking practices can cause large variations in the total mutagenicity. In addition, starting weights are sometimes calculated before, and sometimes after, cooking. Also, the beef extracts are very concentrated compared with the beef muscle from which they were derived and are therefore hard to compare. In some cases the amount detected was not determined (nd). The large number of isolation steps required in order to separate the mutagens frequently makes accurate quantitation difficult. Now that there is some consensus in the mutagen types in foods, as illustrated in Table 2, specific assays for these mutagens need to be utilized to determine the amounts in our diet.

c) Identification

The mutagens found in cooked foods are of two types. One type has the amino group attached to the 2-position of an imidazole ring (imidazole type) and the other has the amino group attached to a pyridine ring (nonimidazole type) (see Fig. 2). A simple test was developed by TSUDA et al. (1980, 1985) to differentiate chemically the two mutagen types. All of the imidazole type mutagens are resistant to the effect of acid nitrite treatment. Assessing the acid nitrite resistance of some cooked food extracts, they showed that an extract from sardine is 88% resistant (88% imidazole type), beef is 75% resistant, and horse mackerel is 48% resistant. This suggests that the imidazole type of mutagens are the predominant structural type in foods.

Analysis of the data in Table 2 shows that the imidazole type has been found by many laboratories in a variety of foods. In contrast, the presence of some non-AIA compounds, such as AαC, has only been found in foods by one laboratory. IQ and 4-MeIQ are present in low amounts in fried beef and beef extracts and in two kinds of fish. 8-MeIQx and 4,8-DiMeIQX have been found in beef, beef extracts, pork, and mackerel. Other mutagens such as PhIP, IQx, and 4-MeIQx have been identified more recently, but it is quite likely that they will be found in other foods as synthetic reference compounds become available. All of these potent mutagens are found at low levels (0.1–15 ppb), but accurate determination of the amounts present in the many and varied foods in our diet still needs to be made.

Interestingly, all of the fried-muscle-food extracts that have been separated, using HPLC on a similar reversed-phase system, have a similar, but not identical, pattern of mutagenic peaks with respect to retention time and mutagenic response. Figure 3 shows the profile of mutagenic activity from a sample of fried ground beef with the retention times of synthetic-food-mutagen standards indicated by the arrows. Similar patterns are seen for beef fried at different temperatures (KNIZE et al. 1985), fried ground pork (GRY et al. 1986), fried ground fish (KNIZE et al. 1988 a), fried ground chicken (KNIZE et al. 1988 b), and a fried meat mixture (BECHER et al. 1988). This similarity in the mutagenic activity profiles of these foods suggests that all of the ground-fried-meat types contain a similar and limited set of mutagenic compounds.

d) Partially Characterized Mutagens from Cooked Foods

Because of the small amounts of the mutagens present in cooked foods and the difficulties with identification and proof of structure, some food mutagens have only been partially characterized. Table 3 lists four such mutagens and the avail-

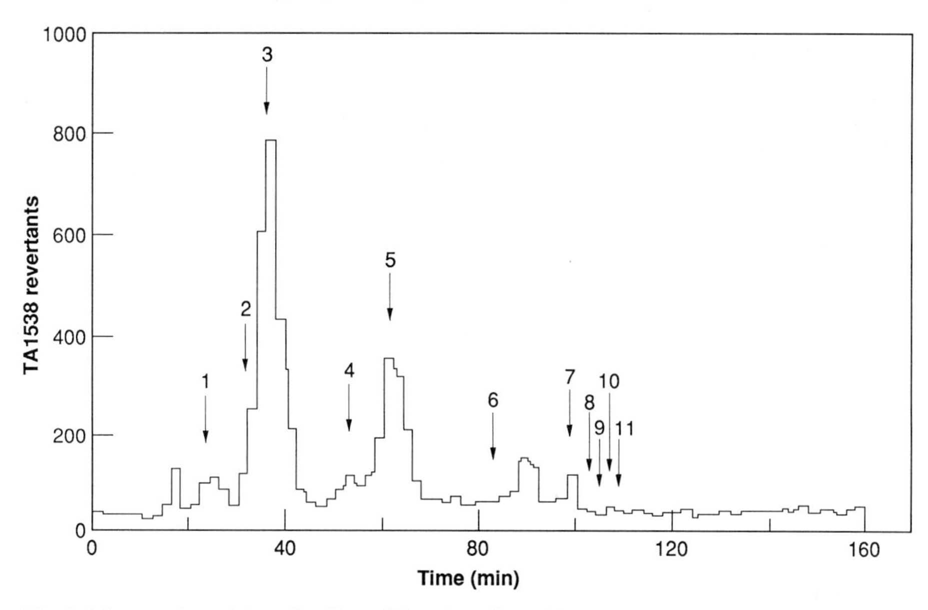

Fig. 3. Mutagenic activity of collected fractions from high pressure liquid chromatography of an extract of fried beef patties. Shown with *numbers* are the retention position of synthetic standard mutagens. *1*=IQx, *2*=IQ, *3*=8-MeIQx, *4*=4-MeIQ, *5*=4,8-DiMeIQx, *6*=4-MeIQx, *7*=PhIP, *8*=Trp-P-2, *9*=Trp-P-1, *10*=AAC, *11*=Phe-P-1

Table 3. Partially characterized mutagens in cooked foods

MW	Composition	Proposed structure	Useful UV max.	Food	References
162	nd	2-Amino-dimethyl imidazopyridine	nd	Creatine-added meat product	BECHER et al. (1988)
176	$C_9H_{12}N_4$	2-Amino-trimethyl imidazopyridine	299 nm	Fried beef	FELTON et al. (1984b)
176	nd	2-Amino-trimethyl imidazopyridine	nd	Creatine-added meat product	BECHER et al. (1988)
202	$C_{10}H_{10}N_4O$	2-Amino-methyl imidazo-benzoxazole	nd	Creatine-added meat product	BECHER et al. (1988)
202	nd	nd	328 nm	Fried pork	GRY et al. (1987)
216	$C_{11}H_{12}N_4O$	2-Amino-dimethyl imidazo-benzoxazole	326 nm	Fried beef	FELTON et al. (1986a)
216	$C_{11}H_{12}N_4O$	nd	nd	Creatine-added meat product	BECHER et al. (1988)
216	nd	nd	323 nm	Fried pork	GRY et al. (1986)

nd, not determined.

able data for each. The molecular weights have been determined by MS and the composition, by HRMS. It should be noted that two of the mutagens appear to contain an oxygen atom. Structures have been proposed based on both imidazole structures that predominate for these types of compounds and the high mutagenic response of the purified compounds. These molecules also have the electron-impact mass-spectral fragments typical of imidazole mutagens, which show losses of 15 (CH_3), 28 (CH_2N), and 42 (CH_2N_2) mass units. The UV absorbance maxima (greater than 250 nm) are also listed, and these differ from the maxima observed for the known imidazole mutagens in Table 1, suggesting that they are new types of mutagen. Clearly greater quantities of these mutagens will need to be purified from a cooked food or modeling system source. Only then can they be completely identified and their structures proven by comparison with a synthetic standard.

C. Mutagen Formation from Modeling Reactions

Understanding the precursors and reaction conditions for mutagen formation during cooking is a major concern in several laboratories. This information can be used to devise strategies to reduce or prevent mutagen formation and, if successful, to help to identify mutagens that are found in cooked foods but that are only present in small amounts.

Table 4 lists mutagens found in cooked foods that have been modeled from creatine or creatinine [collectively abbreviated creatin(in)e], an amino acid, and in some cases, a sugar. Details of the amino acid and sugar that were heated with creatin(in)e are given together with the heating conditions. Early reports suggested that water was essential for mutagen production (BJELDANES et al. 1983), although food mutagens have been made both with and without water being present. In comparisons of aqueous heating and dry heating of an uncooked beef extract at the same temperature, dry heating gave a greater percentage of the acid-nitrite resistant imidazole type of mutagen, similar to those seen in cooked-muscle foods (TAYLOR et al. 1986).

The heating temperature is important because an increase in temperature causes a large increase in total mutagen production in foods (COMMONER et al. 1978; BJELDANES et al. 1983; DOLARA et al. 1979; SPINGARN and WEISBURGER 1979; PARIZA et al. 1979; KNIZE et al. 1985).

The yield, in nanograms of mutagen per gram of total reactants, was calculated (if not given in the original reference) by adding the mass of amino acid, creatin(in)e, and sugar, if used, and dividing this into the amount of mutagen recovered. Most of the reaction mixtures were not originally designed to optimize the yield but were carried out in order to identify possible mutagen precursors.

In quantitative terms, dry heating the amounts of phenylalanine (5 mg) and creatine (440 mg) that are found in 100 g of raw beef yields 18 ppb of PhIP (TAYLOR et al. 1987). This is similar to the 15 ppb of PhIP found to be produced from 100 g of fried beef (FELTON et al. 1986a) and shows that simple dry heating produces yields comparable to those of the cooking process.

It is clear from Table 4 that IQ can be formed from any of four amino acids and that glucose or water are not required. There may be many routes of forma-

Table 4. Food mutagens modeled from creatin(in)e and amino acids

Mutagen	Method	Creatin(in)e	Amino acid	Sugar	Heating	Temperature, °C	Yield (ng/g)	References
IQ	NMR, MS	Creatinine	Glycine	Glucose	Water/gly	128	720	Grivas et al. (1986)
	MS, RT-IS	Creatinine	Phenylalanine	–	Dry	200	2 100	Taylor et al. (1987)
	MS, RT-IS	Creatinine	Phenylalanine	Glucose	Dry	200	7 100	Taylor et al. (1987)
	MS, RT, UV	Creatine	Proline	–	Dry	180	196	Yoshida et al. (1984)
	MS, RT, UV	Creatinine	Serine	–	Dry	200	3 200	Knize et al. (1988a)
IQx	MS, RT, UV	Creatinine	Serine	–	Dry	200	2 400	Knize et al. (1988a)
4-MeIQ	RT-IS, UV, EC	Creatinine	Alanine	Fructose	Water/gly	128	49	Grivas et al. (1985)
8-MeIQx	NMR, MS	Creatinine	Glycine	Glucose	Water/gly	128	4 830	Grivas et al. (1985)
	NMR/MS	Creatine	Glycine	Glucose	Water/gly	128	3 400	Jägerstad et al. (1984)
PhIP	MS, UV	Creatinine	Phenylalanine	Glucose	Water/gly	128	2 000[a]	Shioya et al. (1987)
	MS, RT-IS, UV	Creatinine	Phenylalanine	–	Dry	200	580 000	Taylor et al. (1987)
	MS, RT-IS, UV	Creatinine	Phenylalanine	Glucose	Dry	200	330 000	Taylor et al. (1987)
4,8-DiMeIQx	NMR, MS, UV	Creatinine	Alanine	Fructose	Water/gly	128	1 950	Grivas et al. (1986)

[a] Calculated from reference.

tion for IQ as well as for the other mutagens. All of these model reactions are low yielding and complex, i.e., heating either serine or phenylalanine with creatin(in)e gives at least two mutagenic products in each case.

The reactions producing mutagens are not merely the random coalescence of small fragments. Amongst the possible mutagenic quinoline and quinoxaline structures known, only mutagens with methyl groups at the 3-, 4-, or 8-position have been found. Isomers with methyl groups at the 1-, 5-, and 7-position are potent mutagens (NAGAO et al. 1981; KAISER et al. 1986; KNIZE et al. 1987) but have yet to be detected as products of natural reactions. There appears to be some specificity of the precursors that directs the reactions towards the formation of a limited set of mutagenic products.

Another mutagen, a 7,8-DiMeIQx, has been isolated and identified from the heating of creatinine, glucose, and glycine. Thus far, it has not been found in a heated food, but its production from meat precursors makes it a potential component of cooked food.

D. Heavy Isotope Labeling of Mutagens in Model Reactions

The relatively efficient formation of PhIP in dry heating reactions and the availability of heavy-isotope-labeled phenylalanine and creatine made it possible for TAYLOR et al. (1988) to show incorporation of specific atoms into the PhIP molecule.

Figure 4 shows the structures of creatine and phenylalanine and of the PhIP produced. Separate batches of PhIP were generated by heating creatine (for 2 h at 200 °C) with L-[ring U-^{13}C]phenylalanine; DL-[-3-^{13}C]phenylalanine and L-[$^{15}NH_2$]phenylalanine; or DL-[1-^{13}C]phenylalanine and, after the purification of PhIP, mass spectra were obtained. The first reaction gives a molecule with a mass 5–6 units higher than natural PhIP, showing that the phenyl ring from phenylalanine was incorporated intact. The other two reactions each give a

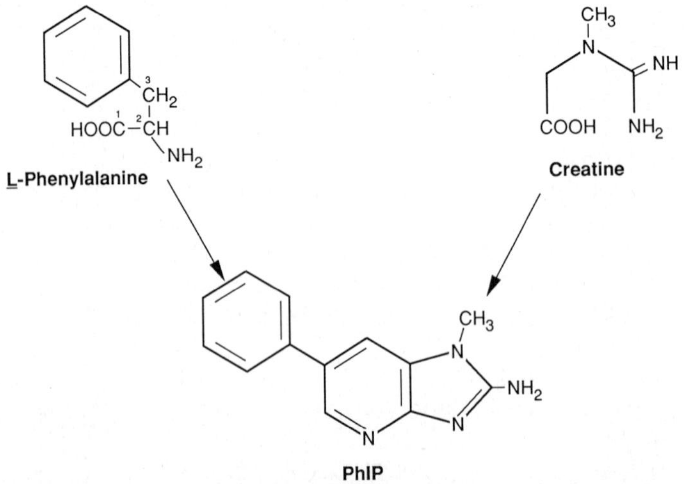

L-Phenylalanine

Creatine

PhIP

Fig. 4. Structures of L-phenylalanine, creatine, and PhIP

product one mass unit higher than natural PhIP, thus showing that the 3-carbon atom and the amino nitrogen from phenylalanine are incorporated into PhIP.

In a similar manner, [1-^{15}N]creatine, [methyl-^{13}C]creatine, or [$^{15}NH_2$]-creatine were heated with phenylalanine, purified, and analyzed by mass spectrometry. Each of the products from the isotopically labeled creatine reactions were one mass unit greater than the natural PhIP, showing that the 1-nitrogen, the methyl-carbon, and the amino-nitrogen from creatine are each incorporated into PhIP. Proton NMR spectrometry of the isotopically labeled PhIP molecules is underway in order to show the exact position of each isotope. This is possible because the heavy isotope will cause a splitting of neighboring proton peaks as compared with natural PhIP. Although it has been assumed that creatin(in)e and amino acids are the precursors for amino-imidazo mutagens in food, these are the first experiments reported that prove unequivocally the source of atoms that become incorporated into the mutagenic product.

E. Structure and Mutagenicity of the Aminoimidazoazaarene Compounds

There are several structural features that have a strong influence on the mutagenic activity of aromatic amines. Because of the metabolic interconversion of the amino group with the hydroxyamino, nitro, and nitroso groups, these latter three groups are called amine-generating groups, and the structural changes influencing carcinogenicity (ARCOS and ARGUS 1974) and mutagenicity (VANCE and LEVIN 1984) are consistent amongst them.

The numbers and positions of double bonds and aromatic rings have a large effect. Potent carcinogens and mutagens have at least one long uninterrupted conjugation system (ARCOS and ARGUS 1974). The isomeric position of the amine-generating group is also important, being optimally placed at the terminal end of the long conjugation system. Related to the above is the ability to resonance-stabilize the ultimate electrophile by distribution of the positive charge of a nitrinium ion (VANCE and LEVIN 1984). The conformation of the amine-generating group with respect to the plane of the aromatic rings (the "thickness" of the molecule) is also important.

Table 5 lists the mutagenic activity of 2-amino-*N*-methyl-imidazo molecules. It can be seen from this table that small structural changes can have large effects on bacterial mutagenicity. In a few cases a range of mutagenic activity per microgram is given. These interlaboratory differences may be the result of difficulties in accurately weighing very small amounts of mutagens or of differences in the amounts and activities of the S9 activating enzymes used in the Ames test.

In comparing the effects of the position and number of methyl groups, the data in Table 5 show that the mutagenic activity of IQ increases with the addition of a methyl group at the 4-position (4-MeIQ). A similar increase in mutagenicity can be seen by comparing 8-MeIQx with 4,8-DiMeIQx; 7-MeIQx with 4,7-DiMeIQx; and IQx with 4-MeIQx. A methyl group at the 5-position can decrease mutagenic activity (compare the data in Table 5 for IQ with 3,5-MeIQ; 8-MeIQx with 5,8-DiMeIQx; and iso-IQ with 1,5-MeIQ). For methyl groups in the 7- or 8-

Table 5. Mutagenicity of 2-amino-*N*-methylimidazoles in *Salmonella typhimurium*

Abbrev.	Chemical name	Strain TA98 (revertants per microgram)	Strain TA1538 (revertants per microgram)	MW	References
1,6-DMIP	2-Amino-1,6-dimethyl-imidazo[4,5-*b*]pyridine	–	8.0	162	Felton (unpublished data, 1988)
1,5,6-TMIB	2-Amino-1,5,6-trimethyl-imidazo[4,5-*b*]benzene	–	430	175	Felton (unpublished data, 1988)
NI	2-Amino-3-methyl-naphtho[1,2-*d*]imidazole	20	–	197	Kaiser et al. (1986)
linear-NI	2-Amino-1-methyl-naphtho[2,3-*d*]imidazole	1810	–	197	Kaiser et al. (1986)
Iso-NI	2-Amino-1-methyl-naphtho[1,2-*d*]imidazole	1060	–	197	Kaiser et al. (1986)
IQ	2-Amino-3-methyl-imidazo[4,5-*f*]quinoline	118000–433000	–	198	Kasai et al. (1980a); Jägerstad and Grivas (1985)
Iso-IQ	2-Amino-1-methyl-imidazo[4,5-*f*]quinoline	3116000	–	198	Kaiser et al. (1986)
IQx	2-Amino-3-methyl-imidazo[4,5-*f*]quinoxaline	–	100000	199	Becher et al. (1988)
4-MeIQ	2-Amino-3,4-dimethyl-imidazo[4,5-*f*]quinoline	253000–660000	–	212	Nagao et al. (1981) Jägerstad and Grivas (1985)
5-MeIQ	2-Amino-3,5-dimethyl-imidazo[4,5-*f*]quinoline	142000	–	212	Nagao et al. (1981)
1,4-MeIQ	2-Amino-1,4-dimethyl-imidazo[4,5-*f*]quinoline	750000	–	212	Nagao et al. (1981)

1,5-MeIQ	2-Amino-1,5-dimethyl-imidazo[4,5-*f*]quinoline	462000	–	212	NAGAO et al. (1981)
3-ethyl-IQ	2-Amino-3-ethyl-imidazo-[4,5-*f*]quinoline	38000	–	212	JÄGERSTAD and GRIVAS (1985)
8-MeIQx	2-Amino-3,8-dimethyl-imidazo[4,5-*f*]quinoxaline	37500–110000	99300	213	KATO et al. (1986); KNIZE et al. (1987)
7-MeIQx	2-Amino-3,7-dimethyl-imidazo[4,5-*f*]quinoxaline	233000	528000	213	KNIZE et al. (1987)
4-MeIQx	2-Amino-3,4-dimethyl-imidazo[4,5-*f*]quinoxaline	–	875000	213	FELTON (unpublished data, 1988)
PhIP	2-Amino-1-methyl-6-phenyl-imidazo[4,5-*b*]pyridine	2000	–	224	KNIZE and FELTON (1986)
3-MePhIP	2-Amino-3-methyl-6-phenyl imidazo[4,5-*b*]pyridine	22.3	–	224	KNIZE and FELTON (1986)
3-ethyl-MeIQ	2-Amino-3-ethyl-4-methyl-imidazo[4,5-*f*]quinoline	25500	–	226	JÄGERSTAD and GRIVAS (1985)
4,8-DiMeIQx	2-Amino-3,4,8-trimethyl-imidazo[4,5-*f*]quinoxaline	206000	320000	227	KNIZE et al. (1987); TAKAHASHI et al. (1985a)
4,7-DiMeIQx	2-Amino-3,4,7-trimethyl-imidazo[4,5-*f*]quinoxaline	351000	38700	227	JÄGERSTAD and GRIVAS (1985)
5,8-DiMeIQx	2-Amino-3,5,8-trimethyl-imidazo[4,5-*f*]quinoxaline	74000	3100	227	KNIZE et al. (1987)
5,7-DiMeIQx	2-Amino-3,5,7-trimethyl-imidazo[4,5-*f*]quinoxaline	243000	–	227	FELTON (unpublished data, 1988)
7,8-DiMeIQx	2-Amino-3,7,8-trimethyl-imidazo[4,5-*f*]quinoxaline	189000	–	227	KNIZE et al. (1987)
4,7,8-TriMeIQx	2-Amino-3,4,7,8-tetramethyl-imidazo[4,5-*f*]quinoxaline	8000	–	241	FELTON (unpublished data, 1988)

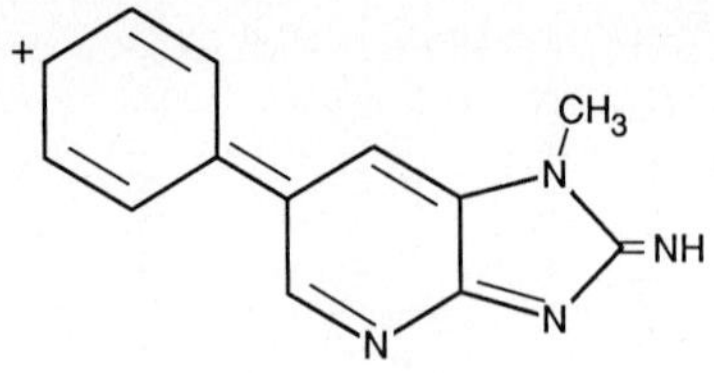

Fig. 5 a, b. Resonance stabilization of the PhIP (**a**) and 3-methyl PhIP (**b**) nitrinium ions. The position of the *N*-methyl group may determine the extent of conjugation with the phenyl ring

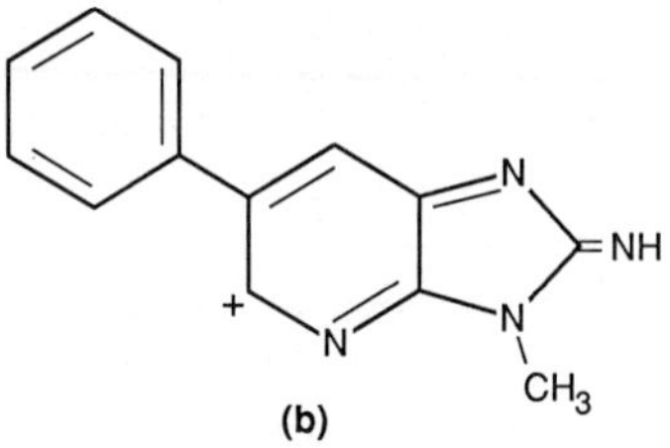

position in the quinoxaline-based mutagens, a 7-methyl quinoxaline is more mutagenic than the corresponding 8-methyl derivative (compare data for 8-MeIQx with 7-MeIQx or 4,8-DiMeIQx with 4,7-DiMeIQx, Table 5). Inconsistent with these results, however, is the data for 4,7,8-TriMeIQx which has surprisingly low mutagenic activity despite methyl groups at the optimal 4- and 7-positions. The effects of methyl groups on mutagenic activity may be based on a combination of steric interactions with macromolecules (e.g., cytochrome P-450) and resonance stabilization of the reactive intermediates.

Amongst the pyridine imidazoles, there is a 100-fold difference in mutagenic activity between PhIP and its 3-methyl isomer. This dramatic difference may be a result of resonance stabilization in PhIP since the 1-methyl group allows conjugation of the aminoimidazo group with the phenyl ring, whereas the 3-methyl isomer does not. Figure 5 shows the nitrinium ion of PhIP (a) and its 3-methyl isomer (b), represented using the scheme of VANCE and LEVIN (1984). A related mutagen, 1,6-DMIP, which has a methyl group in place of the phenyl group in PhIP, has mutagenic activity comparable to that of the 3-methyl isomer of PhIP. Thus, both the presence of the phenyl ring and its position with regard to alternating double bonds influence mutagenic activity in the pyridine imidazoles.

The presence of heterocyclic nitrogen atoms can also cause dramatic changes in mutagenic activity. The naphthoimidazole compound NI is 10000-fold less mutagenic when compared with IQ which has a nitrogen atom instead of one of the carbons present in NI. The addition of a second nitrogen atom to the quinoline ring (e.g., IQ) to make the quinoxaline (e.g., IQx) causes little further change in mutagenic activity, however.

The presence of the *N*-methyl group has been shown to be very important for the mutagenicity of IQ (NAGAO et al. 1981). The substitution of an *N*-ethyl for the *N*-methyl group lowers the mutagenic activity of IQ and, to a greater degree,

MeIQ. This may result from the bulky ethyl group causing steric interference with either the activating enzymes or the target macromolecules. The position of the *N*-methyl group has a consistent effect on the mutagens with three-fused rings. Compounds substituted in the 1-position are consistently more mutagenic than those methylated in the 3-position (compare iso-IQ to IQ and iso-NI to NI).

The orientation of the imidazole ring also has an important effect, and linear NI is about 100-fold more mutagenic than NI (Table 5). It is interesting to note that all of the three-ring aminoimidazole mutagens identified in foods to date have the angular orientation and the methyl group in the 3-position, although 1-position isomers and linear orientation are more mutagenic and would be detected using the Ames/*Salmonella* test as a guide for the purification at well below the ppb level. This may reflect some specificity in the reactivity of creatin(in)e with the meat components that combine to form these mutagens.

An understanding of the parameters that affect mutagenicity and carcinogenicity and that are based on molecular structure would be very useful in predicting the risk from new or inadequately tested chemicals. Much work has begun in this area (KLOPMAN et al. 1985; MAYNARD et al. 1986), but the effects of heterocyclic atoms in aromatic structures, which dramatically affect mutagenicity, have not yet been incorporated into these studies.

The large variations in mutagenicity caused by changes in structure, as mentioned above, are not always reflected in other test systems, as is discussed in the next section.

F. Genetic Toxicology

I. Microbial

Since the Ames/*Salmonella* microsome-mediated mutagenicity test is used to screen initially for these mutagens, all of the compounds are positive in this test. With the exception of PhIP, as was noted above, they are all extremely potent mutagens, active below 11 ng/plate. The greater response is clearly in the frameshift-sensitive strains (TA1538, TA98, and TA97) and with an S9 fraction derived from Aroclor-treated rodents (FELTON et al. 1981, 1984a).

Table 6. Comparison of the AIA mutagens isolated from cooked meat using 7 Ames/*Salmonella* strains [revertants per microgram (in thousands)][a]

Compound	TA96	TA97	TA98	TA100	TA102	TA104	TA1538
IQ	2	104	194	4	1	12	272
4-MeIQ	9	533	950	63	NS	67	1017
8-MeIQx	NS	56	83	1.5	<1.0	NS	99
4-MeIQx	20	600	1162	51	NS	< 1.0	1208
4,8-DiMeIQx	4	86	136	11	NS	7	208
PhIP	< 0.01	0.13	1.7	0.14	NS	0.05	2.0

NS, No statistically significant positive slope.
[a] Calculated from the linear portion of the dose-response curve (4 pts. minimum) by the method of MOORE and FELTON (1983). S9 (Aroclor-treated-rat liver) concentration was 2 mg/plate.

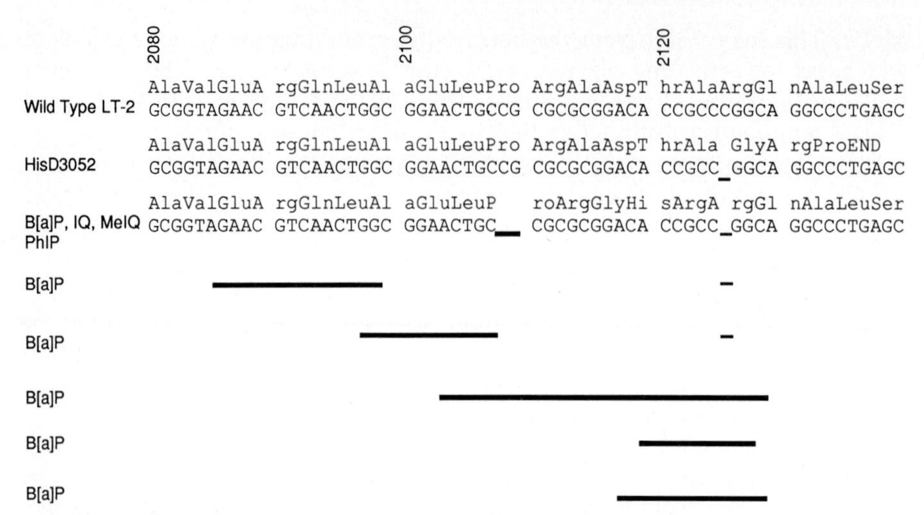

Fig. 6. DNA and amino acid sequences of the *Salmonella his*D gene immediately surrounding the *hisD3052* mutation (*thin bar* at base 2125) in strain TA1538. *Thick bars* indicate the extent of the deletions in the revertants. All deletions when coupled with the single C deletion at the *his*D3052 mutation give a corrected reading frame. The 2-base (CG) deletion shown in the third sequence is representative of all the CG or GC deletions in the 8 base-run of alternating CGs

Table 6 shows the results of a *Salmonella* strain comparison for six of the AIAs found in cooked meat. The most remarkable finding is that, although the compounds range over 3 orders of magnitude of potency, the ranking for each strain for any compound remains the same throughout the range. The strain specificity suggests that these compounds have a high affinity for alternating CG-base-sequence hotspots in DNA (strains TA1538 and TA98). Strain TA97, which has a hotspot containing a run of Cs, is also very responsive to this type of mutagen. In other bacterial strains, either specific for base substitutions or AT frameshift changes, they are weakly active or inactive.

In order to understand this relationship better, researchers in our laboratory developed a cloning method to analyze the DNA sequence in each of the revertants (Fuscoe et al. 1988). Every revertant colony induced by an AIA mutagen shows a 2-base deletion in the run of 4 CG pairs upstream from the original C deletion (*hisD3052*) in the *Salmonella* strain TA1538 (Fig. 6). In contrast, the spontaneous and benzo[*a*]pyrene-induced revertants show numerous other insertions and deletions that also result in viable colonies (Wu et al. 1987). Interestingly, when the repair system is functioning (strain TA1978), the AIA compounds produce large deletions and insertions in addition to the 2-base deletions in the alternating CG region (data not shown).

Although there seems to be specificity towards hotspots containing the bases G or C in *Salmonella*, both Trp-P-2 and IQ cause forward mutations in *Salmonella* strains TM677 and SV50, which detect 8-azaguanine and arabinose resistance, respectively (Felton et al. 1984a). The target sizes (number of DNA bases at risk to induce a mutant phenotype) are hard to compare between the different mutation test systems, but they all seem highly sensitive, with the CG-specific Ames strains providing the most potent target per nucleotide.

II. Nonmicrobial Genotoxicity

1. Mutation

a) Chinese Hamster Cells

Heterocyclic amines have been studied in CHO (Chinese hamster ovary) cells, both wild type and repair deficient, at the *hprt* and *aprt* loci, and in CHL (Chinese hamster lung) cells at the diphtheria-toxin resistance locus. In CHO cells the response was much greater in the repair-deficient line than in the wild type. The doses of the different compounds needed to give a significant mutation frequency are given in Table 7. The potent bacterial mutagens 4-MeIQ and 8-MeIQx are quite weak (THOMPSON et al. 1987). The relatively weaker bacterial mutagens Trp-P-2 and PhIP are the most potent (Trp-P-2 > PhIP > IQ > MeIQ > MeIQx). BROOKMAN et al. (1985) showed that the large difference in mutagenic potency between Trp-P-2 and IQ (THOMPSON et al. 1983) could be attributed to reduced binding of IQ to DNA in the CHO cell. In addition, WILD et al. (1988) showed a similarity of the adducts obtained from IQ in SALMONELLA and rat hepatocytes, thus supporting the argument that a difference in biological response is related to the total amount of DNA binding rather than to differential binding.

In contrast, Chinese hamster lung (CHL) cells show a good response for mutation (diphtheria-toxin resistance) after treatment with IQ, 4-MeIQ, or 8-MeIQx (NAKAYASU et al. 1983). In these experiments it was surprising to find that positive mutagenic responses were accompanied by very little cell killing. IQ and 4-MeIQ were found to give no induction of ouabain-resistant mutants in V79 cells in co-cultivation with Syrian hamster embryo cells (TAKAYAMA and TANAKA 1983). Possibly, differences in metabolism can explain both the differential hamster cell results and the inverse relationship between the bacterial results and most of the hamster data.

b) Drosophila

WILD et al. (1985) have shown that X-chromosomal recessive-lethal mutations in male germ cells can be induced by IQ in a dose-dependent manner. Surprisingly, FUJIKAWA et al. (1983) did not see significant sex-linked recessive-lethal mutations with Trp-P-2 but did see somatic mutations with Trp-P-1 and Trp-P-2. When IQ, 4-MeIQ, and 8-MeIQx were fed to the larvae of *Drosophila*, they all induced a mutagenic response in the wing spot test (YOO et al. 1985).

Table 7. Comparative mutagenic potency in *Salmonella* and Chinese hamster ovary cells (UV5)

Test system	IQ	4-MeIQ	8-MeIQx	PhIP	Trp-P-2
TA1538[a]	0.15	0.04	0.4	15	0.3
CHO UV5[b]	10	75	300	2	0.3

Data derived from THOMPSON et al. (1987).
[a] Dose of compound (μg) to induce 3×10^4 revertants.
[b] Lowest dose (μg/ml) to give a significant increase in mutation frequency.

c) Mouse

Doses of IQ (60 mg/kg over 3 days) and 4-MeIQ (51 mg/kg over 3 days) were not significant inducers of mutations (mouse spot test) in C57BL/6J mice (LARSEN and ANDERSEN 1984, personal communication). WILD et al. (1985) gave 20 mg/kg i.p. to NMRI mice and saw no significant increase over controls with the examination of 502 progeny. Since IQ clearly needs metabolism by specific cytochrome P-450s (MCMANUS et al. 1988) to become an active electrophile, it would be interesting to see whether injections given simultaneously with treatment with the correct enzyme inducer might give positive results.

2. DNA Repair and Damage

Early experiments showed positive effects with unscheduled DNA synthesis (UDS) for IQ with rat hepatocytes (WEISBURGER et al. 1983). Rat and hamster hepatocytes were subsequently used to assess UDS after exposure to IQ, 4-MeIQ, and 8-MeIQx (HOWES et al. 1986). All three compounds were positive in both species with the greater response seen in the hamster cells. The use of Aroclor 1254-pretreated rats as a source of hepatocytes increased the UDS response with IQ and 4-MeIQ over that seen in untreated rat hepatocytes (HOLME et al. 1987). A more recent and comprehensive study with IQ, 4-MeIQ, 8-MeIQx, 4,8-diMeIQx, and 7,8-diMeIQx in rat, mouse, and hamster hepatocytes showed a uniform positive result with the exception of male mice, which are negative for 8-MeIQx and 4,8-diMeIQx (YOSHIMI et al. 1988).

IQ and MeIQ have been reported to cause DNA damage as detected by alkaline elution from filters of DNA isolated from rat hepatocytes (HOLME et al. 1987) and mouse leukemia cells (DOLARA et al. 1985).

3. Clastogenesis

a) In Vitro Chromosome Effects

Excision-repair deficient CHO cells show a positive sister chromosome exchange (SCE) response for IQ with Aroclor-treated hamster liver S9 in the media (THOMPSON et al. 1983). MeIQx and MeIQ give weak SCE responses in the solubility range of 100–800 μg/ml and show no dose-dependent increase using similar methodology (THOMPSON et al. 1987). None of these three compounds give an increase in chromosomal aberrations following similar exposures. In contrast, PhIP, a much weaker bacterial mutagen, shows significant dose-response effects for SCE and aberrations, as was also shown (Sect. 1a) for mutation. In addition, the deficient nucleotide excision repair in the UV5 CHO cell line clearly increases the sensitivity of the cells, as measured by survival over the wild-type cells, and is probably the result of increased, covalently bound, bulky-base damage caused by the PhIP. It is important to remember that PhIP is a major mutagen (by mass) formed in cooked ground beef and thus should be of interest for further study on the basis of its potent clastogenic effects.

When V79 cells were co-cultured with Aroclor-pretreated hepatocytes and then exposed to IQ or 4-MeIQ, an increased incidence of SCEs was found (HOLME et al. 1987).

b) In Vivo Chromosome Effects

In the bone marrow of C57BL/6J mice preinduced with Aroclor 1254, IQ induces SCEs but not chromosomal aberrations (MINKLER and CARRANO 1984). In contrast, WILD et al. (1985) did not see micronuclei in the bone marrow of NMRI mice treated with IQ, but that were not preinduced with Aroclor. Most interestingly, PhIP caused increased SCEs in the bone marrow of the Aroclor-preinduced C57BL/6J mouse after 25 or 50 mg/kg i.p., but even at 100 mg/kg no aberrations were seen in the bone marrow or in the peripheral blood (TUCKER, personal communication, 1988).

G. DNA Binding

I. Guanine Adducts

HASHIMOTO et al. (1979, 1982) have shown that both Trp-P-2 and Glu-P-1 preferentially bind to the C-8 position of guanine to form bulky adducts. WILD et al. (1988) have suggested that the IQ reaction may also be specific for the same carbon on guanine. They used an innovative method for the generation of reactive and short-lived electrophilic arylamine derivatives. The reactive intermediates are derived from azido-IQ and form adducts in vitro with calf-thymus DNA which are identical (chromatographically) to those obtained from *Salmonella* DNA isolated after metabolic activation of IQ. SNYDERWINE et al. (1988) reacted *N*-hydroxy-IQ at neutral pH with single-stranded polynucleotides and found that the most extensive binding occurred with polyguanylic acid. The degree of binding could be enhanced up to sixfold by converting *N*-hydroxy-IQ to *N*-acetoxy-IQ. Finally, the *N*-hydroxy-IQ and *N*-acetoxy-IQ adducts were compared chromatographically with the synthetic *N*-(deoxyguanosin-8-yl)-IQ adduct and were found to have the same retention time.

II. Adduct Analysis by ^{32}P-Postlabeling

A number of laboratories have analyzed synthetic, in vitro, and in vivo generated DNA adducts by postlabeling with ^{32}P and subsequent multidirectional thin-layer chromatography (RANDERATH et al. 1981; GUPTA et al. 1982; REDDY et al. 1984). WILD and his co-workers (1988) used IQ-modified *Salmonella* DNA for comparison with the products formed from different IQ forms and azido-IQ-reacted calf-thymus DNA. They found similar patterns of adduct spots after reaction with IQ and nitro-IQ, suggesting that the same ultimate electrophile is formed. SCHUT et al. (1988) also looked at postlabeling patterns following IQ metabolism and interaction with DNA and found patterns of adduct spots very similar to those of WILD et al. (1988). They found four spots in IQ-exposed rat liver and large intestine and five spots in the small intestine. YAMASHITA et al. (1988) isolated rat liver DNA after intragastric exposure to 12 heterocyclic amines present in cooked foods and amino acid pyrolysates. The postlabeling patterns were different for each class of compound, but the patterns were very similar for the corresponding methyl analogues of Glu-P-2, IQ, and 8-MeIQx.

The IQ patterns were clearly different from those of WILD et al. (1988), which were generated synthetically or in vitro with *Salmonella*. A better understanding will be reached when particular spots can be identified as specific DNA adducts. This will then allow a much better insight into which electrophilic intermediates are important for DNA adduct formation and the DNA damage that leads to mutation and cancer.

H. Carcinogenesis

The majority of the research on the carcinogenicity of these heterocyclic amines has been carried out by researchers in Japan using mice and rats, although CORTESI and DOLARA (1983) have shown that IQ can cause in vitro transformation in a mouse embryo fibroblast cell line.

Table 8 shows a summary of the mouse tumour data after feeding IQ, 4-MeIQ, or 8-MeIQx in the diet. All three compounds induced hepatocellular carcinomas whilst lung adenomas and adenocarcinomas were increased in the mice treated with IQ and in the females treated with 8-MeIQx. IQ and 4-MeIQ also induced squamous cell carcinomas in the forestomach. 8-MeIQx induced intestinal tumors (35% for the males and 17% for the females), but the incidence was not statistically different from that in the controls. 8-MeIQx did induce a significantly higher incidence of lymphomas and leukemias in male mice (OHGAKI et al. 1987). 8-MeIQx induced numerous other tumors (16 in the treated group as against 7 in the control group); they included a number of adenomas of the harderian gland and two hemangioendotheliomas.

Rats (Fischer 344) when fed IQ (0.03% of diet) showed increased incidence of tumors of liver (hepatocellular carcinomas), zymbal gland (squamous cell), large and small intestine (adenocarcinomas), skin, and clitoral gland (squamous cell) (TAKAYAMA et al. 1984; SUGIMURA et al. 1988). The carcinogenicity of IQ was also assessed in the Sprague-Dawley rat and an increase in liver, mammary gland, and ear duct tumors was seen (TANAKA et al. 1985).

J. Significance of Aminoimidazoazaarenes in the Diet

Although the AIA mutagens identified in cooked beef and other meats now number more than a dozen compounds, the amounts and types of compounds generated in the many protein-containing foods included in the human diet under different cooking conditions, temperatures, and fat and moisture contents still need to be ascertained. Standard analytical chemical methods have proved successful but are labor intensive and have a slow throughput. Monoclonal antibodies to specific AIAs that can function in a complex mixture may be sensitive and specific enough to help with the quantitation of these compounds. Recently at M.I.T. (SKIPPER et al. 1987) and at LLNL (WATKINS et al. 1987; VANDERLAAN et al. 1988; FELTON et al. 1988) monoclonal antibodies were developed that can detect nanogram quantities of these heterocyclic amines in food extracts. These monoclonal antibodies should not only be valuable for cost-effective food analysis but also for obtaining structural information on new mutagens when the isolated quantities are not great enough for conventional physicochemical analysis (10 μg minimum for NMR spectroscopy).

Table 8. Induction of tumors in CDF_1 mice by IQ, 4-MeIQ, and 8-MeIQx

Chemical	Sex	Effective number of mice	Number of mice with tumors							
			Liver		Lung		Forestomach		Intestinal	
			Hepato-cellular adenoma	Hepato-cellular carcinoma	Adenoma	Adeno-carcinoma	Papil-loma	Squamous cell carcinoma	Adenoma	Adeno-carcin-oma
IQ (0.03%)[a]	M	39	8	8	13	14	11	5		
	F	36	5	22	7	8	8	3		
Control	M	33	2	0	4	3	1	0		
	F	38	0	0	3	4	0	0		
4-MeIQ (0.04%)	M	38	5	1			5	30		
	F	38	11	16			9	24		
Control	M	29	2	1			0	0		
	F	40	2	0			0	0		
8-MeIQx (0.06%)	M	37	5	10	5	11			11	2
	F	35	7	25	9	6			6	0
Control	M	36	5	0	3	7			8	1
	F	39	0	0	2	2			6	0

Derived from SUGIMURA et al. (1988) (IQ and 4-MeIQ); OHGAKI et al. (1987) (8-MeIQx).
[a] Percent by weight chemical in diet.

A real benefit/risk evaluation for these compounds may also evolve from the development of monoclonal antibody assays for protein and DNA adducts and for metabolites. Since these carcinogens/mutagens are in present in the human diet, the levels of the parent compounds and their metabolites in the urine, blood, feces, and bile can be analyzed (using ELISA assays) and pharmacokinetic models developed. In addition, estimation of their metabolism to electrophiles and the repair of DNA and protein damage may be possible in human autopsy material and in blood and surgical tissue.

There has been considerable work on the dietary factors that affect the metabolism and the cellular and molecular damage that results from ingested carcinogens/mutagens (PARIZA et al. 1986). These factors can include the protein and fat contents and the presence of trace substances, e.g., carotenoids and vitamins. The role of the entire diet on the activities of these potent mutagens, which are present in part per billion quantities, still needs to be determined.

The amounts and types of these AIA mutagens in specific ethnic diets also need to be determined in order to begin to make risk estimates. Clearly, diets that are rich in well-done meat cooked at temperatures over 200 °C will have significant levels of these carcinogenic, heterocyclic amines. Better animal cancer studies yielding more dose-response relationships will help with interspecies risk extrapolation. It is possible that, when the reactions responsible for the formation of these mutagens are better understood, specific inhibitors can be developed to prevent their formation. As more is known about the health risks of these dietary mutagens/carcinogens, it might be hoped that both the food processing industry and individuals preparing food will begin to follow methods designed to lower, and possibly to eliminate, these compounds from the diet.

Acknowledgements. We would like to thank Sandra Eyre for her excellent help in the preparation of the manuscript. Work was peformed under the auspices of the U.S. Department of Energy by the Lawrence Livermore National Laboratory under contract number W-7405-ENG-48 and was supported by the National Institute of Environmental Health Sciences/National Toxicology Program under IAG NIEHS 222Y01-ES-10063.

References

Adolfsson L, Olsson K (1983) A convenient synthesis of mutagenic ^{3}H-imidazo[4,5-*f*]quinoline-2-amines and their 2-^{14}C-labelled analogues. Acta Chem Scand [B]37:157–159

Akimoto H, Kawai A, Nomura H, Nagao M, Kawachi T, Sugimura T (1977) Synthesis of potent mutagens in tryptophan pyrolysates. Chem Lett 1061–1064

Arcos JC, Argus MF (1974) Chemical induction of cancer, structural bases and biological mechanisms, vol 11B. Academic, New York

Barnes WS, Maher JC, Weisburger JH (1983) High pressure liquid chromatographic method for the analysis of 2-amino-3-methylimidazo[4,5-*f*]quinoline, a mutagen formed from the cooking of food. J Agric Food Chem 31:883–886

Becher G, Knize MG, Nes IF, Felton JS (1988) Isolation and identification of mutagens from a fried Norwegian meat product. Carcinogenesis 9:247–253

Berg I, Övervik E, Nord C-E, Gustafsson J-A (1988) Mutagenic activity in smoke formed from broiling of lean pork. Mutat Res 207:199–204

Bjeldanes LF, Grose KR, Davis PH, Stuermer DH, Healy SK, Felton JS (1982a) An XAD-2 resin method for efficient extraction of mutagens from fried ground beef. Mutat Res 105:43–49

Bjeldanes LF, Morris MM, Felton JS, Healy SK, Stuermer DH, Berry P, Timourian H, Hatch FT (1982b) Mutagens from the cooking of food. II. Survey by Ames/*Salmonella* test of mutagen formation in the major protein-rich foods of the American diet. Food Chem Toxicol 20:357–363

Bjeldanes LF, Morris MM, Felton JS, Healy SK, Stuermer DH, Berry P, Timourian H, Hatch FT (1982c) Mutagens from the cooking of food. III. Secondary sources of cooked dietary protein. Food Chem Toxicol 20:365–369

Bjeldanes LF, Morris MM, Timourian H, Hatch FT (1983) Effects of meat composition and cooking conditions on mutagenicity of fried ground beef. J Agric Food Chem 31:18–21

Brookman KW, Salazar EP, Thompson LH (1985) Comparative mutagenic efficiencies of the DNA adducts from the cooked-food-related mutagens Trp-P-2 and IQ in CHO cells. Mutat Res 149:249–255

Commoner B, Vithayathil AJ, Dolara P, Nair S, Madyastha P, Cuca GC (1978) Formation of mutagens in beef and beef extract during cooking. Science 201:913–916

Cortesi E, Dolara P (1983) Neoplastic transformation of BALB 3T3 mouse embryo fibroblasts by the beef extract mutagen 2-amino-3-methylimidazo[4,5-*f*]quinoline. Cancer Lett 20:43–47

Dolara P, Commoner B, Vithayathil AJ, Cuca GC, Tuley E, Madyastha P, Nair S, Kriebel D (1979) The effect of temperature on the formation of mutagens in heated beef stock and cooked ground beef. Mutat Res 60:231–237

Dolara P, Salvadori M, Santoni G, Caderni G (1985) Mammalian cell DNA damage by some heterocyclic food mutagens is correlated with their potency in the Ames test. Mutat Res 144:57–58

Felton JS, Healy SK, Stuermer DH, Berry C, Timourian H, Hatch FT, Morris M, Bjeldanes LF (1981) Mutagens from the cooking of food (1). Improved isolation and characterization of mutagenic fractions from cooked ground beef. Mutat Res 88:33–44

Felton JS, Bjeldanes LF, Hatch FT (1984a) Mutagens in cooked foods: metabolism and genetic toxicity. In: Friedman M (ed) Nutritional and toxicological aspects of food safety, advances in experimental medicine and biology. Plenum, New York, 177:555–566

Felton JS, Knize MG, Wood C, Wuebbles BJ, Healy SK, Stuermer DH, Bjeldanes LF, Kimble BJ, Hatch FT (1984b) Isolation and characterization of new mutagens from fried ground beef. Carcinogenesis 5:95–102

Felton JS, Knize MG, Shen NH, Andresen BD, Bjeldanes LF, Hatch FT (1986a) Identification of the mutagens in cooked beef. Environ Health Perspect 67:17–24

Felton JS, Knize MG, Shen NH, Lewis PR, Andresen BD, Happe J, Hatch FT (1986b) The isolation and identification of a new mutagen from fried ground beef: 2-amino-1-methyl-6-phenylimidazo[4,5-*b*]pyridine (PhIP). Carcinogenesis 7:1081–1086

Felton JS, Watkins BE, Hwang M, Knize MG, Vanderlaan M (1988) Immunoassay of mutagenic/carcinogenic heterocyclic amines produced by cooking food. Proc Am Assoc Cancer Res 29:90

Fujikawa K, Inagaki E, Uchibori M, Kodo S (1983) Comparative induction of somatic eye-color mutations and sex-linked recessive lethals in *Drosophila melanogaster* by tryptophan pyrolysates. Mutat Res 122:315–320

Fuscoe JC, Wu R, Shen NH, Healy SK, Felton JS (1988) Base-change analysis of *Salmonella his*D gene revertant alleles. Mutat Res 201:241–251

Grivas S (1985) A convenient synthesis of the potent mutagen 3,4,8-trimethyl-3H-imidazo[4,5-*f*]quinoxaline-2-amine. Acta Chem Scand [B]39:213–217

Grivas S, Olsson K (1985) An improved synthesis of 3,8-dimethyl-3H-imidazo[4,5-*f*]quinoxaline-2-amine (MeIQx) and its 2-^{14}C-labelled analogue. Acta Chem Scand B39:31–34

Grivas S, Nyhammar T, Olsson K, Jagerstad M (1985) Formation of a new mutagenic DiMeIQx compound in a model system by heating creatinine, alanine and fructose. Mutat Res 151:177–183

Grivas S (1986) Efficient synthesis of mutagenic imidazo[4,5-*f*]quinoxalin-2-amines via readily accessible 2,1,3-benzoselenadiazoles. Acta Chem Scand [B]40:404–406

Grivas S, Nyhammar T, Olsson K, Jagerstad M (1986) Isolation and identification of the food mutagens IQ and MeIQx from a heated model system of creatinine, glycine and fructose. Food Chem 20:127–136

Grose KR, Grant JL, Bjeldanes LF, Andresen BD, Healy SK, Lewis PR, Felton JS, Hatch FT (1986) Isolation of the carcinogen IQ from fried egg patties. J Agric Food Chem 3:201–202

Gry J, Vahl M, Nielsen PA (1986) Mutagens in fried meat (in Danish). Ministry of the Environment, National Food Agency, Copenhagen publ no 139, and M Vahl (personal communication)

Gupta RC, Reddy MV, Randerath K (1982) ^{32}P-postlabeling analysis of non-radioactive aromatic carcinogen-DNA adducts. Carcinogenesis 3:1081–1092

Hargraves WA, Pariza MW (1982) Purification and characterization of bacterial mutagens from commercial beef extracts and fried ground beef. Proc Am Assoc Cancer Res 23:92

Hargraves WA, Pariza MW (1983) Purification and mass spectral characterization of bacterial mutagens from commercial beef extract. Cancer Res 43:1467–1472

Hashimoto Y, Shudo K, Okamoto T (1979) Structural identification of a modified base in DNA covalently bound with mutagenic 3-amino-1-methyl-5H-pyrido [A,3-b]. Indole Chem Pharm Bull 27:1058–1060

Hashimoto Y, Shudo K, Okamoto T (1982) Modification of DNA with potent mutacarcinogenic 2-amino-6-methyldipyrido [1,2-a:3′,2′-d] imidazole isolated from a glutamic acid pyrolysate: structure of the modified nucleic acid base and initial chemical event caused by the mutagen. J Am Chem Soc 104:7636–7640

Hatch FT, Felton JS, Stuermer DH, Bjeldanes LF (1984) Identification of mutagens from the cooking of food. In: de Serres FJ (ed) Chemical mutagens: principles and methods for their detection. 9:111–164

Hatch FT (1986) A current genotoxicity database for heterocyclic thermic food mutagens I. Genetically relevant endpoints. Environ Health Perspect 67:93–103

Hayatsu H, Matsui Y, Ohara Y, Oka T, Hayatsu T (1983) Characterization of mutagenic fractions in beef extract and in cooked ground beef. Use of blue cotton for efficient extraction. Gann 74:472–481

Holme JA, Hingslo JK, Soderlund E, Brunborg G, Christensen T, Alexander J, Dybing E (1987) Comparative genotoxic effects of IQ and MeIQ in *Salmonella typhimurium* and cultured mammalian cells. Mutat Res 187:181–190

Howes AJ, Beamand JA, Rowland IR (1986) Induction of unscheduled DNA synthesis in rat and hamster hepatocytes by cooked food mutagens. Food Chem Toxicol 24:383–387

Jagerstad M, Olsson K, Grivas S, Negishi C, Wakabayashi K, Tsuda M, Sato S, Sugimura T (1984) Formation of 2-amino-3,8-dimethylimidazo[4,5-*f*]quinoxaline in a model system by heating creatinine, glycine and glucose. Mutation Res 126:239–244

Jagerstad M, Grivas S (1985) The synthesis and mutagenicity of the 3-ethyl analogues of the potent mutagens IQ, MeIQ, MeIQx and its 3,7-dimethyl isomer. Mutat Res 144:131–136

Kaiser G, Harnasch D, King M-T, Wild D (1986) Chemical structure and mutagenic activity of aminoimidazoquinolines and aminonaphthimidazoles related to 2-amino-3-methylimidazo[4,5-*f*]quinoline. Chem Biol Interact 57:97–106

Kasai H, Nishimura S, Wakabayashi K, Nagao M, Sugimura T (1980a) Chemical synthesis of 1-amino-3-methylimidazo[4,5-*f*]quinoline (IQ), a potent mutagen isolated from broiled fish. Proc Jpn Acad 58:382–384

Kasai H, Yamaizumi Z, Wakabayashi K, Nagao M, Sugimura T, Yokoyama Miyazawa T, Nishimura S (1980b) Structure and chemical synthesis of ME-IQ, a potent mutagen isolated from broiled fish. Chem Lett 11:1391–1394

Kasai H, Shiomi T, Sugimura T, Nishimura S (1981a) Synthesis of 2-amino-3,8-dimethylimidazo[4,5-*f*]quinoxaline (Me-IQx), a potent mutagen isolated from fried beef. Chem Lett 675–678

Kasai H, Yamaizumi Z, Nishimura S, Wakabayashi K, Nagao M, Sugimura T, Spingarn NE, Weisburger JH, Yokoyama S, Miyazawa T (1981b) A potent mutagen in broiled fish. Part 1. 2-amino-3-methyl-^{3}H-imidazo[4,5-*f*]quinolinc. J Chem Soc Perkin I 2290–2293

Kasai H, Yamaizumi Z, Shiomi T, Yokoyama S, Miyazawa T, Wakabayashi K, Nagao M, Sugimura T, Nishimura S (1981c) Structure of a potent mutagen isolated from fried beef. Chem Lett 485–488

Kato T, Kikugawa K, Hayatsu H (1986) Occurrence of the mutagens 2-amino-3,8-dimethylimidazo[4,5-*f*]quinoxaline (MeIQx) and 2-amino-3,4,8-trimethylimidazo[4,5-*f*]quinoxaline (4,8-Me_2IQx) in some Japanese smoked, dried fish products. J Agric Food Chem 34:810–814

Klopman G, Frierson MR, Rosenkranz HS (1985) Computer analysis of toxicological data bases: mutagenicity of aromatic amines in *Salmonella* tester strains. Environ Mutagenesis 7:625–644

Knize MG, Andresen BD, Healy SK, Shen NH, Lewis PR, Bjeldanes LF, Hatch FT, Felton JS (1985) Effect of temperature, patty thickness, and fat content on the production of mutagens in fried ground beef. Food Chem Toxicol 23:1035–1040

Knize MG, Felton JS (1986) The synthesis of the cooked-beef mutagen 2-amino-1-methyl-6-phenylimidazo[4,5-*b*]pyridine and its 3-methyl isomer. Heterocycles 24: 1815–1819

Knize MG, Happe J, Healy SK, Felton JS (1987) Identification of the mutagenic quinoxaline isomers from fried ground beef. Mutat Res 178:25–32

Knize MG, Shen NH, Felton JS (1988a) The production of mutagens in foods. Proc Air Pollution Control Assoc 88-130.3:1–8

Knize MG, Shen NH, Felton JS (1988b) A comparison of mutagen production in fried-ground chicken and beef: effect of supplemental creatine. Mutagenesis 3:503–509

Kosuge T, Kawashi T, Nagao M, Yahagi T, Seino Y (1978) Isolation and structure studies of mutagenic principles in amino acid pyrolysates. Chem Pharm Bull (Tokyo) 26:611–619

Lee C-S, Hashimoto Y, Shudo K, Okamoto T (1982) Synthesis of mutagenic heteroaramatics: 2-aminoimidazo[4,5-*f*]quinolines. Chem Pharm Bull (Tokyo) 30:1857–1859

Lund E, Borgan JK (1987) Increased lung cancer mortality among Norwegian cooks. Scand J Work Environ Health 13:156

Matsumoto T, Yoshida D, Tomita H, Matsushita H (1979) Synthesis of 2-amino-9H-pyrido[2,3-*b*]indole isolated as a mutagenic principle from pyrolytic products of protein. Agric Biol Chem 43:675–677

Matsumoto T, Yoshida D, Tomita H (1981a) Determination of mutagens, amino-alpha-carbolines in grilled foods and cigarette smoke condensate. Cancer Lett 12:105–110

Matsumoto T, Yoshida D, Tomita H (1981b) Synthesis and mutagenic activity of alkyl derivatives of 2-amino-9H-pyrido[2,3-*b*]indole. Agric Biol Chem 45:2031–2035

Maynard AT, Pedersen LG, Posner HS, McKinney JD (1986) An ab initio study of the relationship between nitroarene mutagenicity and electron affinity. Mol Pharmacol 29:629–636

McManus ME, Burgess W, Snyderwine E, Stupans I (1988) Specificity of rabbit cytochrome P-450 isozymes involved in the metabolic activation of the food derived mutagen 2-amino-3-methylimidazo[4,5-*f*]guanoline. Cancer Res 48:513–519

Minkler JL, Carrano AV (1984) In vivo cytogenetic effects of the cooked-food-related mutagens Trp-P-2 and IQ in mouse bone marrow. Mutat Res Lett 140:49–53

Moore D, Felton JS (1983) A microcomputer program for analyzing Ames test data. Mutat Res 119:95–102

Murray S, Gooderham NJ, Boobis AR, Davies DS (1988) Measurement of MeIQx and DiMeIQx in fried beef by capillary column gas chromatography electron capture negative ion chemical ionisation mass spectrometry. Carcinogenesis 9:321–325

Nagao M, Wakabayashi K, Kasai H, Nishimura S, Sugimura T (1981) Effect of methyl substitution on mutagenicities of 2-amino-3-methylimidazo[4,5-*f*]quinoline, isolated from broiled sardine. Carcinogenesis 2:1147–1149

Nakayasu M, Makasato F, Sakamoto H, Terada M, Sugimura T (1983) Mutagenic activity of heterocyclic amines in Chinese hamster lung cells with diphtheria toxin resistance as a marker. Mutat Res 118:91–102

Ohgaki H, Hasegawa H, Suanaga M, Sato S, Takayama S, Sugimura T (1987) Carcinogenicity in mice of a mutagenic compound, 2-amino-3,8-dimethylimidazo[4,5-*f*]quinoxaline (MeIQx) from cooked foods. Carcinogenesis 8:665–668

Övervik E, Nilsson L, Fredholm L, Levin O, Nord C-E, Gustafsson J-A (1987) Mutagenicity of pan residues and gravy from fried meat. Mutat Res 187:47–53

Pariza MW, Ashoor SH, Chu FS, Lund DB (1979) Effects of temperature and time on mutagen formation in pan-fried hamburger. Cancer Lett 7:63–69

Pariza MW, Hargraves WA, Benjamin H, Christou M, Jefcoate CR, Storkson JM, Albright K, Draus D, Sharp P, Boissonneault GA, Elson CE (1986) Modulation of carcinogenesis by dietary factors. Environ Health Perspect 67:25–29

Randerath K, Reddy MV, Gupta RC (1981) ^{32}P-Labeling test for DNA damage. Proc Natl Acad Sci USA 78:6126–6129

Rappaport SM, McCartney MC, Wei ET (1979) Volatilization of mutagens from beef during cooking. Cancer Lett 8:139–145

Reddy MV, Gupta RC, Randerath E, Randerath K (1984) ^{32}P-Postlabeling test for covalent DNA binding of chemicals in vivo: application to a variety of aromatic carcinogens and methylating agents. Carcinogenesis 5:231–243

Schut HAJ, Putman KL, Randerath K (1988) ^{32}P-postlabeling analysis of DNA adducts in liver, small and large intestine of male Fischer-344 rats after intraperitoneal administration of 2-amino-3-methylimidazo[4,5-*f*]quinoline (IQ). In: King CM, Romano LJ, Schniltzle D (eds) Carcinogenic and mutagenic responses to aromatic amines and nitroarenes. Elsevier, New York, pp 265–269

Shioya M, Wakabayashi K, Sato S, Nagao M, Sugimura T (1987) Formation of a mutagen 2-amino-1-methyl-6-phenylimidazo[4,5-*b*]-pyridine (PhIP) in cooked beef, by heating a mixture containing creatinine, phenylalanine and glucose. Mutat Res 191:133–138

Skipper PL, Tannenbaum SR, Wogan GN (1987) Monoclonal antibodies recognizing 2-amino-3-methylimidazo[4,5-*f*]quinoline (IQ). Proc Am Assoc Cancer Res 28:128

Snyderwine EG, Roller PP, Adamson RH, Sato S, Thorgeirsson SS (1988) Reaction of *N*-hydroxylamine and *N*-acetoxy derivatives of 2-amino-3-methylimidazo[4,5-*f*]quinoline with DNA. Synthesis and identification of *N*-(deoxyguanosin-8-yl)-IQ. Carcinogenesis 9:1061–1065

Spingarn NE, Weisburger JH (1979) Formation of mutagens in cooked food. 1. Beef. Cancer Lett 7:259–264

Sugimura T, Nagao M, Kawachi T, Honda M, Yahagi T, Seino Y, Sato S, Matsukura N, Matsushima T, Shirai A, Sawamura M, Matsumoto H (1977a) Mutagen-carcinogens in foods with special reference to highly mutagenic pyrolytic products in broiled foods. In: Hiatt HH, Watson JD, Winsten JA (eds) Origins of human cancer. Cold Spring Harbor, New York, pp 1561–1577

Sugimura T, Kawachi T, Nagao M, Yahagi T, Okamoto T, Shudo K, Kosuge T, Tsuki K, Wakabayashi K, Litaka Y, Itai A (1977b) Mutagenic principles in tryptophan and phenylalanine pyrolysis products. Proc Jpn Acad 53:58–61

Sugimura T, Sato S, Wakabayashi K (1988) Mutagens/carcinogens in pyrolysates of amino acids and proteins and in cooked foods: heterocyclic aromatic amines. In: Woo YT, Lai DY, Arcos JC, Argue MF (eds) Chemical induction of cancer, structural bases, and biological mechanisms. Academic, New York, pp 681–710

Takahashi M, Wakabayashi K, Nagao M, Yamaizumi Z, Sato S, Kinae N, Tomita I, Sugimura T (1985a) Identification and quantification of 2-amino-3,4,8-trimethylimidazo[4,5-*f*]quinoxaline (4,8-DiMeIQx) in beef extract. Carcinogenesis 6:1537–1539

Takahashi M, Wakabayashi K, Nagao M, Yamamoto M, Masui T, Goto T, Kinae N, Tomita I, Sugimura T (1985b) Quantification of 2-amino-3-methylimidazo[4,5-*f*]-quinoline (IQ) and 2-amino-3,8-dimethylimidazo[4,5-*f*]quinoxaline (MeIQx) in beef extracts by liquid chromatography with electrochemical detection (LCEC). Carcinogenesis 6:1195–1199

Takayama S, Tanaka M (1983) Mutagenesis of amino acid pyrolysis products in Chinese hamster V79 cells. Toxicol Lett 17:23–28

Takayama S, Nakatsuru Y, Masuda M, Ohgaki H, Sato S, Sugimura T (1984) Demonstration of carcinogenicity in F344 rats of 2-amino-3-methylimidazo[4,5-*f*]quinoline from broiled sardine, fried beef and beef extract. Gann 75:467–470

Takeda K, Ohta T, Shudo K, Okamoto T, Tsuki K, Kosuga T (1977) Synthesis of a mutagenic principle isolated from tryptophan pyrolysate. Yakugaku Zasshi 97: 2145–2146

Takeda K, Shudo K, Okamoto T, Kosuge T (1978) Synthesis of mutagenic principles isolated from L-glutamic acid pyrolysate. Chem Pharm Bull (Tokyo) 26: 2924–2925

Tanaka T, Barnes WS, Williams GM, Weisburger JH (1985) Multipotential carcinogenicity of the fried food mutagen 2-amino-3-methylimidazo[4,5-*f*]quinoline in rats. Gann 76:570–576

Taylor RT, Fultz E, Knize MG (1985) Mutagen formation on a model beef boiling system. III. Purification and identification of three heterocyclic amine mutagens-carcinogens. J Environ Sci Health A20:135–148

Taylor RT, Fultz E, Knize MG (1986) Mutagen formation in a model beef supernatant fraction. IV. Properties of the system. Environ Health Perspect 67:59–74

Taylor RT, Fultz E, Knize MG, Felton JS (1987) Formation of the fried ground beef mutagens 2-amino-3-methylimidazo[4,5-*f*]quinoline (IQ) and 2-amino-1-methyl-6-phenylimidazo[4,5-*b*]pyridine (PhIP) from L-phenylalanine (Phe) + creatinine (Cre) (or creatine). Environ Mutagen 9 [Suppl 8]:106

Taylor RT, Fultz E, Morris C, Knize MG, Felton JS (1988) Model system phenylalanine (Phe) and creatine (Cr) heavy-isotope-labeling of the fried ground beef mutagen 2-amino-1-methyl-6-phenylimidazo[4,5-*b*]pyridine (PhIP). Environ Mutagen 11 [Suppl 11]:104

Thompson LH, Carrano AV, Salazar EP, Felton JS, Hatch FT (1983) Comparative genotoxic effects of the cooked food-related mutagens Trp-P-2 and IQ in bacteria and cultured mammalian cells. Mutat Res 117:243–257

Thompson LH, Tucker JD, Stewart SA, Christensen ML, Salazar EP, Carrano AV, Felton JS (1987) Genotoxicity of compounds from cooked beef in repair-deficient CHO cells versus *Salmonella* mutagenicity. Mutagenesis 2:483–487

Tsuda M, Takahashi Y, Nagao M, Hirayama T, Sugimura T (1980) Inactivation of mutagens from pyrolysates of tryptophan and glutamic acid by nitrite in acidic solution. Mutat Res 78:331–339

Tsuda M, Negishi C, Makino R, Sato S, Yamaizumi Z, Hirayama T, Sugimura T (1985) Uses of nitrite and hypochlorite treatments in determination of the contributions of IQ-type and non-IQ type heterocyclic amines to the mutagenicities in crude pyrolized materials. Mutat Res 147:335–341

Tsuji K, Yamamoto T, Zenda H, Kosuge T (1978) Studies on active principles of tar. VII. Production of biological active substances in pyrolysis of amino acids. Antifungal constituents in pyrolysis products of phenylaline. Yakugaku Zasshi 98:910–913

Turesky RJ, Wishnok JS, Tannenbaum SR, Pfund RA, Buchi GH (1983) Qualitative and quantitative characterization of mutagens in commercial beef extract. Carcinogenesis 4:863–866

Vance WA, Levin DE (1984) Structural features of nitroaromatics that determine mutagenic activity in salmonella typhimurium. Environ Mutagenesis 6:797–811

Vanderlaan M, Watkins BE, Hwang M, Knize MG, Felton JS (1988) Monoclonal antibodies for the immunoassay of mutagenic compounds produced by cooking beef. Carcinogenesis 9:153–160

van Houdt JJ, Jongen WMF, Alink GM, Boleil JSM (1984) Mutagenic activity of airborne particles inside and outside homes. Mutagenesis 6:861–869

Waterhouse AL, Rapoport H (1985) Synthesis and tritium labeling of the food mutagens IQ and methyl IQ. J Labeled Cpds and Radiopharm 22:201–216

Watkins BE, Knize MG, Morris CJ, Andresen BD, Happe J, Vandrelaan M, Felton JS (1987) The synthesis of derivatives of the cooked-food mutagens IQ, MeIQx and PhIP as haptemic compounds. Heterocycles 26:2069–2072

Weisburger JH, Horn CL, Barnes WS (1983) Possible genetoxic carcinogens in foods in relation to cancer causation. Seminars in Oncology 10:330–341

Wild D, Gocke E, Harnasch D, Kaiser G, King M-T (1985) Differential mutagenic activity of IQ (2-amino-3-methylimidazo[4,5-*f*]quinoline) in *Salmonella typhimurium* strains in vitro and in vivo, in *Drosophila* and in mice. Mutat Res 156:93–102

Wild D, Asan E, Dirr A, Fasshauer I, Henschler D (1988) DNA-adducts of amino-imidazoarenes and structurally analogous nitro and azidoimidazoarenes. In: King CCM, Romano LJ, Schnitzle D (eds) Carcinogen and mutagenic responses to aromatic amines and nitroarenes. Elsevier, New York, pp 73–85

Wu R, Shen NH, Healy SK, Fuscoe JC, Felton JS (1987) Analysis of DNA base changes in *Salmonella* revertants induced by mutagens derived from foods. Environ Mutagen 9:115

Yamaguchi K, Shudo K, Okamoto T, Sugimura T, Kosuga T (1980a) Presence of 2-aminodipyrido[1,2-*a*:3′,2′-*d*]imidazole in broiled cuttlefish. Gann 71:743–744

Yamaguchi K, Shudo K, Okamoto T, Sugimura T, Kosuge T (1980b) Presence of 3-amino-1,4-dimethyl-5H-pyrido[4,3-*b*]indole in broiled beef. Gann 71:745–746

Yamaizumi Z, Shiomi T, Kasai H, Nishimura S, Takahashi Y, Nagao M, Sugimura T (1980) Detection of potent mutagens, Trp-P-1 and Trp-P-2, in broiled fish. Cancer Lett 9:75–83

Yamaizumi Z, Kasai H, Nishimura S, Edmonds CG, McCloskey JA (1986) Stable isotope dilution quantification of mutagens in cooked foods by combined liquid chromatography-thermospray mass spectrometry. Mutat Res 173:1–7

Yamamoto T, Tsuji K, Kosuge T, Okamoto T, Shudo K, Takeda K, Litaka Y, Yamaguchi K, Seino Y, Yahagi T, Nagao M, Sugimura T (1978) Isolation and structure determination of mutagenic substances in L-glutamic acid pyrolysate. Proc Jpn Acad 54:248–250

Yamashita K, Umimoto A, Grivas S, Kato S, Sato S, Sugimura T (1988) Heterocyclic amines-DNA adducts analyzed by ^{32}P-postlabeling method. Nucleic Acids Symp Ser 19:111–114

Yoo MA, Ryo H, Todo T, Kondo S (1985) Mutagenic potency of heterocyclic amines in the *Drosophila* wing spot test and its correlation to carcinogenic potency. Gann 76:468–473

Yoshida D, Matsumoto T, Yoshimura R, Matsuzaki T (1978) Mutagenicity of amino-alpha-carbolines in pyrolysis products of soybean globulin. Biochem Biophys Res Comm 83:915–920

Yoshida D, Saito Y, Mizusaki S (1984) Isolation of 2-amino-3-methyl-imidazo-[4,5-*f*]quinoline as mutagen from the heated product of a mixture of creatine and proline. Agric Biol Chem 48:241–243

Yoshimi N, Sugie S, Iwata H, Mori H, Williams GM (1988) Species and sex differences in genotoxicity of heterocyclic amino pyrolysis and cooking products in the hepatocyte primary culture/DNA repair test using rat, mouse, and hamster hepatocytes. Environ Mol Mutagen 12:53–64

CHAPTER 13

Modern Methods of DNA Adduct Determination

D. H. PHILLIPS

A. Introduction

There is growing interest in the development of sensitive methods for detecting the covalent binding of chemical carcinogens to DNA in animals and humans. The formation of DNA adducts is widely believed to be a necessary early step in the process by which many carcinogens exert their biological effects, and an understanding of the parameters of carcinogen adduct formation, structure, persistence and repair has contributed fundamentally to the elucidation of the mechanisms of carcinogenesis and can be expected to continue to do so. There is also an urgent need for more accurate means of assessing the risk of human exposure to environmental carcinogens, and the presence of DNA adducts in tissues or cells from individuals suspected of being exposed to genotoxic agents is one of several ways by which the exposure can be monitored and, potentially at least, the risk assessed (GARNER 1985).

The first suggestion that mutagenic carcinogens interact with DNA arose from observations in the 1940s that nitrogen mustards altered the absorption spectra of DNA (CHANUTIN and GJESSING 1946). Subsequent studies by LAWLEY and others demonstrated the chemical reactivity of the alkylating agents towards nucleophilic sites in DNA bases (reviewed by LAWLEY 1989). The idea that the mode of action of other classes of chemical carcinogens, such as polycyclic aromatic hydrocarbons and aromatic amines, might also involve covalent interaction with cellular macromolecules arose from the demonstration of the persistent binding of carcinogens to proteins in vivo in susceptible tissues (MILLER and MILLER 1947; MILLER 1951). Subsequent demonstration of the covalent binding of dibenz[*a,h*]anthracene to DNA in mouse skin (HEIDELBERGER and DAVENPORT 1961) and of a correlation between the extent of DNA binding and the carcinogenic potency, in mouse skin, of polycyclic hydrocarbons (BROOKES and LAWLEY 1964) provided important evidence that DNA is the critical target for chemical carcinogenesis. The concept of different classes of chemical carcinogens, regardless of structure or reactivity, sharing a common mode of action was first formally proposed by the MILLERS (MILLER and MILLER 1969), who put forward the hypothesis that chemical carcinogens are or are converted by metabolism into electrophilic compounds (ultimate carcinogens) that exert their biological effects through covalent binding to cellular macromolecules. Alterations to DNA, if not recognised and removed by DNA repair processes, may then lead to alterations in the genome through point mutations, deletions, gene amplification or rearrangement. There is a current consensus that the demonstration that a chemical can form adducts with DNA is sufficient for it to be considered a potential mutagenic and carcinogenic hazard (DE SERRES 1988). It is, however, recognised that some classes of carcinogens may induce DNA damage indirectly, such as by altering DNA methylation patterns (BARROWS and SHANK 1981) or by compromising the fidelity of DNA polymerases (SIROVER and LOEB 1976), while other chemicals, such as the peroxisome proliferators, appear to be carcinogenic by mechanisms that involve oxidative stress (REDDY and LALWANI 1984).

Early studies of DNA adduct formation by carcinogens used, for the most part, radiolabelled compounds. Because only a very small fraction of an administered dose of a compound becomes bound to DNA in vivo, DNA binding could generally not be detected by standard physicochemical methods, thus necessitating synthesis of radioisotopically labelled compounds, generally containing tritium or carbon-14. Until recently, our knowledge of the covalent interactions or carcinogens with DNA has been based largely on studies of the separation of carcinogen-nucleoside or carcinogen-nucleotide adducts in digests of DNA from animal tissues exposed to radiolabelled compounds and the comparison of their chromatographic and other properties with those of nucleosides or nucleotides reacted in vitro with model reactive intermediates (BAIRD 1979).

Such studies have suffered from several limitations. Firstly, there is the requirement to undertake costly and time-consuming syntheses of radiolabelled test compounds. The half-lives of the isotopes place a natural limit on the specific activities of the compounds and thus on the sensitivity of detection that is achievable. In practice this has often meant that only the adducts of fairly potent carcinogens are formed with sufficient frequency in vivo to be readily detectable. Also, the amount of radioactive compound available usually precludes the administration of more than a few discrete doses to experimental animals, and it is seldom possible to monitor the formation of adducts during chronic administration of test compounds (i.e. under conditions frequently used in carcinogenicity testing studies) by this means. More importantly, it is not feasible to monitor DNA adducts in humans using radiolabelled compounds or to undertake any sort of retrospective analysis of DNA for prior environmental exposure to carcinogens. There has thus been a great deal of interest in the development of methods for DNA adduct detection that do not require the use of radiolabelled test compounds. Methods currently of interest include immunochemical, biochemical and physicochemical techniques, and it is the purpose of this article to review a number of these. The principles and applications of each method in laboratory studies will be considered individually, after which the application of the techniques to human studies will be discussed. The intention here will be to indicate the scope of the various methods, their relative advantages and limitations for different classes of carcinogens and the uses to which they have been, and can be, put.

B. Postlabelling Methods

I. Principles of ^{32}P-Postlabelling

The principle of postlabelling approaches to adduct detection, as the name implies, is that a radioisotope or other label is introduced into the adduct after it has been formed. The label can be incorporated either by chemical reaction or enzymically. The most sensitive method currently available is ^{32}P-postlabelling analysis, developed by RANDERATH and co-workers (RANDERATH K. et al. 1981; GUPTA et al. 1982) and recently reviewed by WATSON (1987).

^{32}P-Postlabelling analysis involves the following steps: a sample of DNA that contains adducts is digested enzymically to deoxyribonucleoside 3′-monophosphates by micrococcal nuclease and spleen phosphodiesterase. The DNA digest is then incubated with [γ-^{32}P]ATP in the presence of T4 polynucleotide kinase to yield [5′-^{32}P]-deoxyribonucleoside 3′,5′-bisphosphates. The ^{32}P-labelled adducts are then separated from the normal nucleotides and resolved chromatographically and detected by monitoring their radioactive decay.

In the standard procedure, approximately 200 ng of digested DNA is incubated with 50–150 μCi [γ-^{32}P]ATP in the presence of carrier ATP to give a slight molar excess of ATP. The reaction is terminated by the addition of apyrase, and the whole reaction mixture is applied to the origin of a polyethyleneimine-cellulose TLC sheet. The labelled adducts are then separated from the normal nucleotides and resolved in 2 dimensions using multidirectional chromatography. Location of the adducts is achieved by autoradiography using intensifying screens, the adducts are thus seen as dark spots on X-ray film (see Fig. 1). The adducts can be quantitated by scintillation or Cerenkov counting of

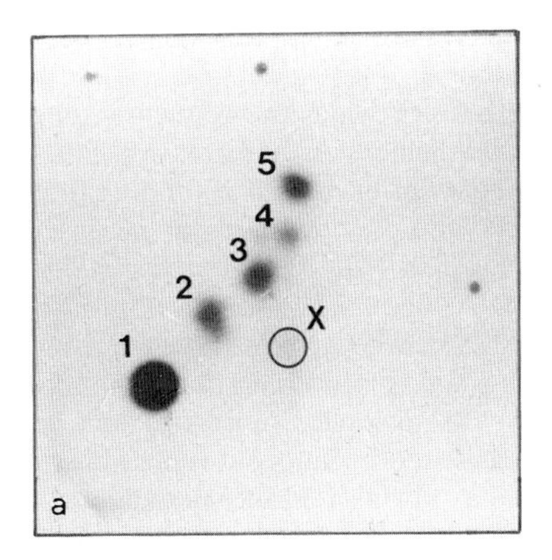

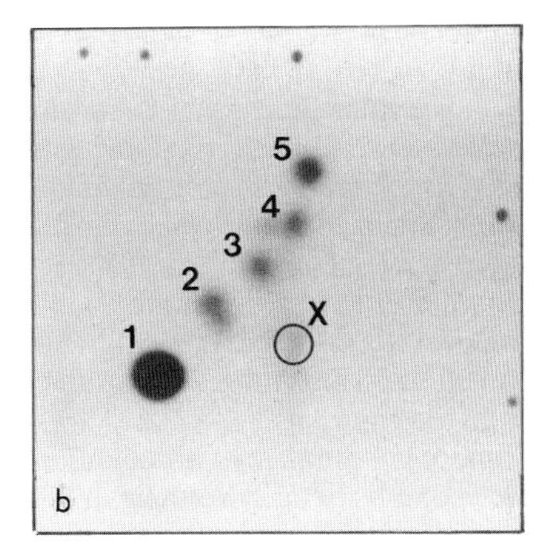

Fig. 1 a, b. ^{32}P-Postlabelling detection of DNA adducts formed by a pure carcinogen. Autoradiograms of 2-dimensional chromatograms of ^{32}P-labelled digests of DNA isolated from **a** the bone marrow and **b** the peripheral white blood cells of rats that had been treated orally with 7,12-dimethylbenz[*a*]anthracene are shown. The adducts (*1*–*5*) are detected as dark spots on the autoradiograms. The faint spot *X* was also detected in DNA samples from untreated rats, where the chromatograms were otherwise devoid of spots. (Reproduced from PHILLIPS et al. 1986b with permission)

the areas of the chromatograms corresponding to the spots on the autoradiogram and by comparing the values with the amount of radioactivity incorporated into an aliquot of the whole DNA digest (GUPTA et al. 1982).

The chromatographic conditions used to resolve the adducts depend on the chemical structure of the DNA-binding chemical whose adducts are being studied. Because it is necessary to remove a large excess of normal nucleotides, the procedure is most sensitive if the chromatographic properties of the adducts differ greatly from those of normal nucleotides. In practice this means that the greatest sensitivity is achievable with adducts containing a bound aromatic or other hydrophobic moiety such that it is possible to remove unmodified nucleotides completely from the chromatograms by running them on to a wick, while retaining the adducts at the origin for subsequent resolution using different solvent systems. Using the above procedures aromatic DNA adducts present at levels as low as 1 adduct in 10^7 nucleotides can be detected (GUPTA et al. 1982) (but see below). However, if the adduct contains only a small alkyl moiety, then its chromatographic properties are too similar to those of normal nucleotides to allow complete removal of the latter from the chromatograms without loss of the former, and chromatograms must therefore be obtained that contain both normal nucleotides and adducts (RANDERATH K. et al. 1981; REDDY et al. 1984). The level of sensitivity achievable in these instances is in practice about 2 orders of magnitude less (but see below).

A number of strategies are available for increasing the sensitivity of the technique. If a limiting amount of high specific activity [γ-^{32}P]ATP (it is commercially available at >6000 Ci/mmol and can also be readily synthesized in the laboratory from [^{32}P]-orthophosphate and ADP) is used instead of a molar excess of lower specific activity material, then certain adducts are labelled preferentially to normal nucleotides, which can result in a 5–50 fold increase in sensitivity for the same amount of radioactivity used (PHILLIPS et al. 1984; RANDERATH E. et al. 1985).

Of greater potential use, however, are methods by which the adducts can be separated from the normal nucleotides prior to labelling, or the latter prevented from being substrates for polynucleotide kinase by prior modification. The first approach has been described by GUPTA (1985), who used phase-transfer extraction, and by DUNN et al. (1987) and DUNN and SAN (1988), who used reverse-phase HPLC to concentrate adducts from DNA digests. The second approach was developed by REDDY and RANDERATH (1986), who demonstrated that incubation of DNA digests with nuclease P_1 prior to ^{32}P-labelling dephosphorylates the normal nucleotides (and thus they are no longer substrates for the kinase) but not many types of adduct. Using these modifications it is possible to detect adduct concentrations in 5–10 μg DNA as low as 1 adduct per 10^9–10^{10} nucleotides. However, these methods are, again, best suited to hydrophobic adducts and cannot be applied satisfactorily to the detection of adducts formed by simple alkylating agents.

For analysis of adducts formed by small aromatic compounds, recovery of labelled adducts can be improved by using a combination of reverse-phase and anion-exchange TLC (PHILLIPS et al. 1984; RANDERATH K et al. 1984a; REDDY et al. 1984). An alternative or additional method of adduct separation with potentially greater resolving power is HPLC, and its use with ^{32}P-postlabelled adducts (WEYAND et al. 1987; DIETRICH et al. 1987) and double-labelled (^{3}H and ^{32}P) adducts (SCHMEISER et al. 1988a) has been reported.

Where ^{32}P-postlabelling has been applied to the detection of adducts formed by small alkylating agents, improved resolution has been obtained by 3′-dephosphorylation of the nucleotides and nucleotide-adducts after 5′-^{32}P-labelling (HASELTINE et al. 1983; REDDY et al. 1984; HOLLSTEIN et al. 1986; WILSON et al. 1986). The adducted monophosphates can then be more readily separated from unmodified nucleotides by TLC or HPLC. Alternatively, HPLC separation of O^6-alkylated deoxyribonucleoside-3′-monophosphates from normal deoxyribonucleoside-3′-monophosphates prior to ^{32}P-postlabelling, combined with subsequent 3′-dephosphorylation by nuclease P_1 digestion and resolution of the labelled adducts by TLC, has achieved a level of sensitivity such that one O^6-alkyl-deoxyguanosine in 10^7 deoxyguanosine residues in a sample size of 100 μg DNA can be detected (WILSON et al. 1988). Procedures involving 3′-dephosphorylation of the ^{32}P-labelled DNA digests have also been used to determine bromodeoxyuridine incorporation into the DNA of rat brain tumour cells (BODELL and RASMUSSEN 1984) and could, in principle, be applied to the determination of incorporation of other base analogues into DNA.

The apparent inability of acrolein, a mutagenic constituent of cigarette smoke and a metabolite of cyclophosphamide, to react with poly-dC has been shown, using ^{32}P-postlabelling, to be a consequence of the rapid loss by cleavage of the *N*-glycosyl bond of the adduct from the polynucleotide (SMITH et al. 1988). A product that was released from the homopolymer was found to have identical chromatographic properties with the product of the reaction of acrolein with dCMP. In contrast, acrolein forms stable adducts with homopolymers of the other DNA nucleotides.

II. Applications in Animal and Tissue Culture Studies

The covalent binding of a large number of carcinogens of diverse chemical structure to DNA in vivo has been demonstrated by ^{32}P-postlabelling (Table 1). Many polycyclic aromatic hydrocarbons have been analysed, and in agreement with earlier, more limited studies using tritium-labelled compounds (BROOKES and LAWLEY 1964; PHILLIPS et al. 1979), a good correlation has been found to exist between carcinogenic potency on mouse skin and extent of covalent binding in that tissue (RANDERATH K. et al. 1985b). Binding of a series of naturally occurring alkenylbenzenes to mouse liver DNA in vivo has also been shown to correlate with the hepatocarcinogenic activity of the compounds (PHILLIPS et al. 1984; RANDERATH K. et al. 1984a).

Monitoring the persistence of adducts in vivo for long periods after treatment can be readily carried out by ^{32}P-postlabelling. The persistence of adducts in mouse skin has been demonstrated as long as 42 weeks after a single treatment with 7,12-dimethylbenz[*a*]anthracene (DMBA) (RANDERATH E. et al. 1985), and safrole-DNA adducts have been found to persist in mouse liver for at least 20 weeks after treatment (RANDERATH K. et al. 1984a). Other studies on adduct persistence have demonstrated differential rates of removal of adducts formed by

Table 1. Compounds tested by ^{32}P-postlabelling for DNA binding

Compound	Tissue or cell type	References
Polycyclic aromatic hydrocarbons		
Benzo[*a*]pyrene	Mouse skin	RANDERATH, E. et al. (1983); REDDY et al. (1984); WEYAND et al. (1987)
	Mouse intestine	REDDY and RANDERATH (1986)
	Mouse fetal tissues (8)	LU et al. (1986)
	Mouse C3H10T1/2 cells	GUPTA et al. (1987)
	Rat liver	GUPTA et al. (1982); REDDY et al. (1984)
	Rat mammary gland	PHILLIPS et al. (1985); SEIDMAN et al. (1988)
Benzo[*e*]pyrene	Mouse skin	REDDY et al. (1984)
Benz[*a*]anthracene	Mouse skin	REDDY et al. (1984); MCKAY et al. (1988)
7-Methylbenz[*a*]anthracene	Mouse skin	MCKAY et al. (1988)
7-Ethylbenz[*a*]anthracene	Mouse skin	MCKAY et al. (1988)
7,12-Dimethylbenz[*a*]anthracene	Mouse skin	RANDERATH, E. et al. (1983, 1985); RANDERATH, K. et al. (1985a); REDDY and RANDERATH (1986); REDDY et al. (1984); SCHOEPE et al. (1986)
	Mouse C3H10T1/2 cells	HEISIG et al. (1986)
	Mouse fetal cells	SCHMEISER et al. (1988a)
	Rat blood, bone marrow	PHILLIPS et al. (1986b)
7-Ethyl-12-methylbenz[*a*]antracene	Mouse C3H10T1/2 cells	HEISIG et al. (1986)
Dibenz[*a,c*]anthracene	Mouse skin	REDDY et al. (1984)
Dibenz[*a,h*]anthracene	Mouse skin	REDDY et al. (1984)
Chrysene	Mouse skin	REDDY et al. (1984); PHILLIPS et al. (1987)
3-Methylcholanthrene	Mouse skin	RANDERATH, E. et al. (1983); REDDY et al. (1984)
Benzo[*b*]fluoranthene	Mouse skin	WEYAND et al. (1987)
Benzo[*j*]fluoranthene	Mouse skin	WEYAND et al. (1987)
Benzo[*k*]fluoranthene	Mouse skin	WEYAND et al. (1987)
Indeno[1,2,3-*cd*]pyrene	Mouse skin	WEYAND et al. (1987)
Pyrene [a]	Mouse skin	RANDERATH, K. et al. (1985b)
Anthracene [a]	Mouse skin	RANDERATH, K. et al. (1985b)
Perylene [a]	Mouse skin	RANDERATH, K. et al. (1985b)
Benzo[*g,h,i*]perylene	Mouse skin	RANDERATH, K. et al. (1985b)
Aromatic amines		
2-Acetylaminofluorene (AAF)	Rat liver	GUPTA et al. (1985a, b);
	Rat liver, blood, spleen	WILLEMS et al. (1987)
	Mouse skin	REDDY et al. (1984)
	Chinese hamster ovary cells	ARCE et al. (1987)
N-Hydroxy-AAF	Rat liver	GUPTA (1984); GUPTA and DIGHE (1984); GUPTA et al. (1982); REDDY et al. (1984)
N-Hydroxy-2-aminofluorene	Mouse liver	LAI et al. (1987)
2-Acetylaminophenanthrene (AAP)	Rat liver	GUPTA (1985); GUPTA et al. (1985a, b, 1987)

[a] Adducts were not detected.

clofibrate, ethyl-α-p-chlorophenoxyisobutyrate; ciprofibrate, 2-[4-(2,2-dichlorocyclopropyl)phenoxy]2-methylpropionic acid; Wy-14643, [4-chloro-6(2,3-xylidino)2-pyrimidinylthio]acetic acid; DEHP, di(2-ethylhexyl)phthalate.

Table 1 (continued)

Compound	Tissue or cell type	References
N-Hydroxy-AAP	Rat liver	GUPTA and DIGHE (1984); REDDY et al. (1984)
4-Aminobiphenyl	Mouse liver	REDDY and RANDERATH (1986)
	Mouse skin	REDDY et al. (1984)
	Mouse fetal tissues (8)	LU et al. (1986)
4-Acetylaminobiphenyl (AAB)	Rat liver	GUPTA et al. (1987)
N-Hydroxy-AAB	Rat liver	GUPTA and DIGHE (1984); REDDY et al. (1984)
Benzidine	Mouse skin	REDDY et al. (1984)
N-Hydroxy-4-acetylamino-*trans*-stilbene	Rat liver	REDDY et al. (1984)
2-Amino-3-methyl-9*H*-pyrido-[2,3-*b*]indole (MeAαC)	Rat tissues (5)	YAMASHITA et al. (1986)
2-Amino-3-methylimidazo [4,5-*f*]quinoline (IQ)	Hamster embryo cells	ASAN et al. (1987)
Azo compounds		
4-Dimethylaminoazobenzene	Mouse skin	REDDY et al. (1984)
Congo red	Mouse skin	REDDY et al. (1984)
Evan's blue	Mouse skin	REDDY et al. (1984)
Nitro compounds		
4-Nitroquinoline-1-oxide	Mouse skin	REDDY et al. (1984)
2,6-Dinitrotoluene	Mouse skin	REDDY et al. (1984)
1-Nitro-9-aminoacridine	HeLa cells	BARTOSZEK and KONOPA (1987)
1-Nitrosopyrene	Human fibroblasts	BELAND et al. (1986)
Mycotoxins		
Aflatoxin B_1	Rat liver	RANDERATH, K. et al. (1984b, 1985b)
Sterigmatocystin	Rat liver	REDDY et al. (1985)
Alkenylbenzenes		
Safrole	Mouse liver	PHILLIPS et al. (1984); RANDERATH, K. et al. (1984a); REDDY and RANDERATH (1986)
	Mouse fetal tissues (8)	LU et al. (1986)
Estragole, methyleugenol, myristicin, elemicin, anethole, parsley apiol, dill apiol, eugenol[a]	Mouse liver	PHILLIPS et al. (1984); RANDERATH, K. et al. (1984a)
Isosafrole, allylbenzene	Mouse liver	RANDERATH, K. et al. (1984a)
1′-Hydroxy-2′,3′-dehydroestragole	Mouse liver	FENNELL et al. (1986)
Methylating agents		
N,*N*-Dimethylnitrosoamine	Mouse liver	REDDY et al. (1984)
1,2-Dimethylhydrazine	Mouse liver	REDDY et al. (1984)
N-Methyl-*N*-nitrosourea	Mouse liver	REDDY et al. (1984)
Streptozotocin	Mouse liver	REDDY et al. (1984)
Other compounds		
7*H*-Dibenz[*c*,*g*]carbazole	Mouse tissues (6)	SCHURDAK and RANDERATH (1985)
	Mouse liver, skin	SCHURDAK et al. (1987a, b)
	Mouse liver	REDDY and RANDERATH (1986); SCHURDAK and RANDERATH (1985)
	Human fibroblasts	PARKS et al. (1986)
Mitomycin C	Rat liver	REDDY and RANDERATH (1986)
Aristolochic acid I and II	Rat tissues (8)	SCHMEISER et al. (1988b)
Clofibrate[a], ciprofibrate[a], Wy-1463[a], DEHP[a]	Rat liver	GUPTA et al. (1985b)

N-hydroxy-2-acetylaminofluorene (*N*-OH-AAF) and *N*-hydroxy-4-acetylamino-bi-phenyl (*N*-OH-AAB) in rat liver DNA (GUPTA and DIGHE 1984). It is theoretically possible to monitor experimental animals for the presence of adducts at any stage in a carcinogenicity assay designed to run for the lifetime of the animals. The method has been used to demonstrate that the high incidence of hepatocellular carcinomas in rats chronically fed a diet devoid of choline is not associated with the formation of DNA adducts from chemicals of endogenous or dietary origin (GUPTA et al. 1987).

Due to the high sensitivity of the technique it can be used to provide data on whether or not a compound of known or unknown biological activity binds covalently to DNA. In a study of four peroxisome proliferators, no DNA adducts could be detected in the livers of rats fed the compounds (GUPTA et al. 1985b). As it was estimated that the limit of sensitivity was 1 adduct in 10^{10} nucleotides, the failure to detect any DNA binding supports the hypothesis that peroxisome proliferators exert their carcinogenic action by a mechanism that does not involve direct interaction with DNA (REDDY and LALWANI 1984).

Attempts to demonstrate the covalent binding of carcinogenic hormones to DNA by conventional means using radiolabelled compounds have produced, at best, equivocal results (LUTZ et al. 1982). However, using ^{32}P-postlabelling the formation of DNA adducts was demonstrated in hamster kidney DNA after several months of exposure of the animals to implanted pellets of diethylstilboestrol (LIEHR et al. 1985). Furthermore, when a series of structurally diverse oestrogenic carcinogens was investigated in the same system, the compounds were all found to give the same adduct profile (LIEHR et al. 1986), leading the investigators to conclude that the adducts were derived not from the carcinogens themselves but from an endogenous substance whose activation and DNA binding was induced by the hormone treatment. Although the nature of the DNA-binding species has not been identified, its formation appears to be cytochrome P-450-dependent, as adduct levels are highest in the cortex of hamster kidneys in which cytochrome P-450 levels are also highest and in which oestrogen-induced renal carcinomas develop (LIEHR et al. 1987).

Analysis of DNA from untreated rats has revealed the existence of age-related adducts in the liver, lung, kidney and heart (RANDERATH K. et al. 1986). The patterns of adducts were tissue-specific, and although they are, as yet, uncharacterised, they could conceivably be due to environmental agents or to reactive metabolites of endogenous origin. It can be speculated that such DNA damage may play a role in so-called "spontaneous" tumours (i.e. those for which no causative agent is known) or even in the aging process (RANDERATH K. et al. 1986).

^{32}P-Postlabelling has been effectively used to study pathways of metabolic activation of chemical carcinogens. Such studies are greatly facilitated by the lack of requirement for radiolabelled metabolites. The comparative DNA binding of DMBA and some of its metabolites in mouse skin revealed a high level of binding by the 3,4-dihydrodiol, the postulated proximate carcinogen, while other metabolites exhibited weaker binding than the parent hydrocarbon (SCHOEPE et al. 1986). A study of chrysene activation in mouse skin showed that DNA adducts were formed from the hydrocarbon through the formation of the 1,2-

dihydrodiol and to a lesser extent via a phenolic dihydrodiol, possibly the 9-hydroxy-1,2-dihydrodiol (PHILLIPS et al. 1987). The formation of adducts in mouse liver DNA by 7*H*-dibenzo[*c,g*]carbazole (DBC) was shown to involve intermediate formation of the 3-hydroxy derivative, but this metabolite was not involved in DNA binding in mouse skin (SCHURDAK et al. 1987a). Rat mammary epithelial cells exposed to benzo[*a*]pyrene(BP) in vitro and rat mammary glands exposed by site injection of the compounds in vivo revealed a different pattern of DNA adducts to that seen in other animal species and tissues (PHILLIPS et al. 1985; SEIDMAN et al. 1988), indicating a different pathway of activation from that involving formation of the *anti*-7,8-dihydrodiol-9,10-oxide(BPDE). LAI et al. (1987) have used ^{32}P-postlabelling to monitor the effect of enzyme inhibitors and genetic deficiency in metabolism on the formation of adducts in infant mice treated with *N*-hydroxy-2-aminofluorene (*N*-OH-AF) and to demonstrate the importance of *N*-sulphation in the activation of this compound. A study by FENNEL et al. (1986) compared the detection of the formation of adducts in mouse liver by 1′-hydroxy-2′,3′-dehydroestragole by the ^{32}P-postlabelling method with use of the tritium-labelled compound and showed that DNA binding resulted solely via formation of the 1′-sulphuric acid ester.

A number of studies have taken advantage of the requirement for only microgram quantities of DNA to monitor the levels of adducts formed in in vitro assays to determine some biological endpoint such as mutagenicity or transforming activity of genotoxic compounds. Thus, adduct level determination has been incorporated into studies on the cytotoxicity and mutagenicity of DBC in diploid human fibroblasts (PARKS et al. 1986), the cytotoxicity and transforming activity of 1-nitrosopyrene also in diploid human fibroblasts (BELAND et al. 1986), the mutagenicity of 2-acetylaminofluorene (AAF) in CHO cells (ARCE et al. 1987) and the mutagenic activity of several epoxide derivatives of chrysene in V79 Chinese hamster cells (PHILLIPS et al. 1986a). ^{32}P-Postlabelling was also used to measure the extent of DNA modification in experiments in which the c-Ha-*ras*-1 proto-oncogene was activated to a mutated, transforming oncogene by in vitro modification with ultimate carcinogens (MARSHALL et al. 1984; VOUSDEN et al. 1986).

III. Exposure to Environmental Carcinogens: Animal and Human Studies

The presence of aromatic adducts has been detected in DNA from the placentas of mothers who smoked cigarettes during pregnancy (EVERSON et al. 1986). ^{32}P-Postlabelling analysis revealed the presence of several adducts, the major one of which was chromatographically distinct from BP-DNA adducts, even though the DNA samples were positive when tested with antibodies raised against the latter (see Sect. C.III). This study demonstrates directly an association between cigarette smoking and DNA damage in human subjects, and a further study from the same group (EVERSON et al. 1988) has demonstrated an inverse relationship between the levels of smoking-related placental DNA adducts and the birthweight of the offspring. Which of the many chemicals present in tobacco smoke are responsible for the DNA binding is, as yet, unknown, for the adduct spots do not co-chromatograph with those chemical carcinogens so far studied (EVERSON et al. 1986). Application of cigarette smoke condensate to mouse skin, where it is known to be a tumour initiator, led to the formation of a similar pattern of DNA adducts as that seen in tissues isolated from smokers (RANDERATH E et al. 1986). A subsequent study has examined the tissue distribution of adducts in mice following topical application of cigarette smoke condensate and revealed, significantly, higher levels of DNA damage in the heart and lung than in skin (RANDERATH E et al. 1988). The ability to reproduce in experimental animals DNA adducts detected in humans suggests a possible means of identifying the individual components in the complex mixtures responsible for the DNA damage. In the case of cigarette smoke condensate it is possible to test subfractions by ^{32}P-postlabelling for their ability to form adducts with the eventual aim of isolating the pure substance or substances that form the same adduct pattern as observed with the unfractionated material (EVERSON et al. 1987). In fact, this is a similar approach, in principle, to that used by KENNAWAY and his colleagues in the first identification of pure carcinogens that were isolated from coal tar (see KENNAWAY 1955).

Exposure to polycyclic hydrocarbons in the workplace is thought to be a contributing factor to the increased incidence of lung cancer among iron foundry workers. ^{32}P-Postlabelling analysis of DNA from white blood cells of workers in a Finnish iron foundry revealed the presence of aromatic adducts at levels significantly higher than in control subjects (PHILLIPS et al. 1988). The more highly exposed individuals tended to have higher levels of adducts than those workers in low-exposure occupations, but some inter-individual variations were also evident. A study of mononuclear and non-mononuclear cell DNA from the bone marrow of healthy donors revealed the presence of aromatic adducts that were not detected in samples of human fetal bone marrow (PHILLIPS et al. 1986b). As the adult samples were from individuals not known to be occupationally exposed to carcinogens, the detection of adducts in their bone marrow DNA may be evidence of widespread exposure to as-yet-unidentified genotoxic agents.

Interestingly, in several of these studies on the detection of adducts in human DNA, no effect of smoking was seen on the levels of adducts in white blood cells (PHILLIPS et al. 1986b, 1988; EVERSON et al. 1987), although aromatic, smoking-related adducts are formed in other tissues (RANDERATH E et al. 1986; EVERSON et al. 1987). The reasons for this are at present unclear as white blood cells would be capable of metabolically activating aromatic carcinogenic components of cigarette smoke condensate if exposed to them. It may be that these components do not reach these cells in concentrations sufficient for higher levels of adducts to be formed in smokers than in non-smokers, although it is entirely possible that the same is not true for other, non-aromatic, tobacco carcinogens. In any case, present evidence suggests that the smoking status of individuals whose white blood cell DNA is analysed for occupational or other exposure to carcinogens by ^{32}P-postlabelling is unlikely to interfere with the interpretation of results. Another readily obtainable source of human DNA, that of exfoliated mucosal cells of the oral cavity, also did not show any differences between non-smokers, tobacco chewers and inverted smokers when analysed for aromatic adducts by ^{32}P-postlabelling (DUNN and STICH 1986).

The effect of exposure of animal or human tissue to other complex carcinogenic mixtures has also been determined using ^{32}P-postlabelling. Exposure of rats to diesel engine exhaust fumes has been demonstrated to result in the formation of a large number of different DNA adducts in the lungs of the animals (WONG et al. 1986). A study of liver DNA from fish dwelling in waters heavily polluted with polycyclic aromatic hydrocarbons showed a similarly complex pattern of adducts that was absent from the DNA of aquarium-raised fish (DUNN et al. 1987). Exposure of mouse skin to solutions of coal tar, creosote or bitumen resulted in the formation of aromatic adducts in the DNA of the skin and, to a lesser extent, lungs (SCHOKET et al. 1988a) (Fig. 2). The ability of human skin, maintained in organ culture, to activate to DNA-binding species components of these widely used materials that are suspected of being carcinogenic in humans has also been demonstrated (SCHOKET et al. 1988b).

Occupational exposure to styrene occurs as a result of the extensive use of this chemical in the manufacture of reinforced plastics. Styrene is metabolised in vivo to styrene-7,8-oxide, which is both mutagenic and carcinogenic. Detection of styrene oxide-DNA adducts by ^{32}P-postlabelling (LIU et al. 1988) should provide the means of monitoring human exposure in the workplace to styrene.

IV. Advantages and Limitations of ^{32}P-Postlabelling

At the present time, ^{32}P-postlabelling analysis is the most sensitive assay available for the detection of aromatic and lipophilic adducts, although it is not as sensitive for simple alkyl adducts, for reasons already discussed. Only small (mi-

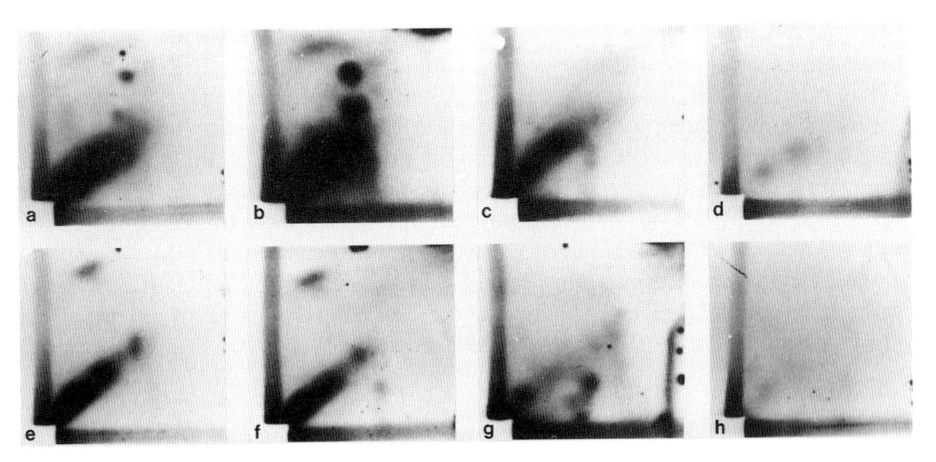

Fig. 2a–h. Detection by ^{32}P-postlabelling of the formation of DNA adducts in vivo in mouse skin (*upper line*, **a–d**) and lung (*lower line*, **e–h**) after the topical application of complex carcinogenic mixtures. The chromatograms are of ^{32}P-labelled digests of DNA from mice treated with: **a** and **e**, coal-tar; **b** and **f**, creosote; **c** and **g**, bitumen; **d** and **h**, solvent only. A band of radioactive material is evident in samples from treated animals indicative of the formation of a large number of different DNA adducts. The origins of the chromatograms, which are located at the *bottom left-hand corners*, were excised before autoradiography. (Reproduced from SCHOKET et al. 1988a with permission)

crogram) quantities of DNA are required to perform the analysis, there are no expensive items of equipment needed, and it can readily be applied to the analysis of adducts formed by complex mixtures. A major advantage of the technique is that it is not necessary to know the identity of an adduct in order to detect it, and the method can be used to search for the existence of previously unknown carcinogens or mutagens in the environment and to test whether a novel chemical forms DNA adducts in vivo. Most other techniques (see below) require the production of highly modified DNA as standards and/or extensive physico-chemical characterisation of an adduct before its detection and quantitation in biological samples can be attempted. ^{32}P-Postlabelling allows a broad approach to carcinogen-DNA adduct detection that is not limited to those relatively few compounds that have received extensive laboratory study. If, as epidemiological studies suggest (Doll and Peto 1981), the incidence of the majority of human cancers is influenced by environmental agents, then ^{32}P-postlabelling should be a useful tool in determining what role DNA-damaging chemicals may play in the aetiology of the common human cancers.

Because of the relatively large quantities of ^{32}P required, care must be exercised when performing the assay to avoid unacceptable exposure of researchers to ionising radiation or contamination of laboratories and equipment. Because hydrolysis of DNA samples is required, problems may be encountered if digestion is incomplete, leading to errors in adduct quantitation. Difficulties may also be encountered if an adduct is unstable as a mononucleotide or is a poor substrate for polynucleotide kinase. Although chemically synthesised standards are not obligatory, assessment of these effects cannot be reliably made without them. Nevertheless, in most cases in which standards were available, it has been found that near-quantitative labelling of adducts can be achieved. An exception appears to be the case of aflatoxin B_1-DNA adducts, for which problems of incomplete DNA digestion have been encountered (Randerath K. et al. 1984b, 1985b). There will be uncertainty in the quantitation of very low levels of adducts for which enhancement techniques such as nuclease P_1 digestion are used unless standardisation is possible with more highly modified DNA samples. Also, there is no straightforward way in which the chemical structures of adducts detected by ^{32}P-postlabelling can be characterised, but, in some respects, this can be regarded as a shortcoming of other analytical techniques rather than of ^{32}P-postlabelling alone. Certainly a multimethod approach to the detection and identification of carcinogen-DNA adducts, in particular those of environmental origin, is the one most likely to lead to full characterisation (see Sect. E).

V. ^{14}C-Postlabelling

An alternative form of postlabelling has recently been described by Watson et al. (1987). Detection of *N*-7-(2-oxoethyl)guanine, the principal adduct formed by vinyl chloride, in DNA was achieved by reducing and heating the DNA to release the adduct as *N*-7-(2-hydroxyethyl)guanine, which was then purified and reacted first with a large excess of [^{14}C]acetic anhydride and then with *n*-propylamine. The incorporation of radioactivity into *N*-acetylpropylamine gave a measure of the *N*-7-(2-oxoethyl)guanine in the DNA sample. The percentage recovery was determined by "spiking" the sample with [^{14}C]labelled adduct, and the procedure is estimated to have a sensitivity that allows detection of 1 adduct in 10^7 nucleotides.

VI. Fluorescent Postlabelling

A development of fluorescent labelling of mononucleotides and oligonucleotides that could be of use in the detection of carcinogen-DNA adducts has recently been reported by KELMAN et al. (1988). The procedure involves 5′-phosphoramidation with the nucleophile ethylenediamine followed by conjugation of the free amino group of the phosphoramidate with the fluorophore 5-methylamino-naphthalene 1-sulfonyl chloride (dansyl chloride). Using absorption detection a limit of sensitivity of 2 nmol was demonstrated, but with fluorescence detection this was lowered to 200 fmol. Reaction of the 5′-nucleotide with a polyamine was also demonstrated, leading to the possibility of enhanced sensitivity through the introduction of multiple fluorescent labels.

C. Immunochemical Methods

I. Principles

Antibodies specific for nucleic acid bases were first obtained in the early 1960s (BUTLER et al. 1962; TANENBAUM and BEISER 1963). Like other haptens, nucleosides must first be covalently bound to protein in order to elicit an immune response in animals. A number of methods, all involving reaction of the free amino groups of proteins, have been successfully employed and have been comprehensively reviewed by MÜLLER and RAJEWSKY (1981). A widely used procedure is that of ERLANGER and BEISER (1964) which involves periodate oxidation of the *cis*-diol groups of ribose and their reaction with the amino groups of proteins such as bovine serum albumin, chicken γ-globulin, keyhole limpet haemocyanin and horseshoe crab haemocyanin. Although the method is only applicable to carcinogen-modified ribonucleotides and ribonucleosides, antibodies raised in this way also recognise deoxyribonucleosides. Other coupling methods which are applicable to DNA fragments or where the carcinogen moiety contains *cis*-diol groups (e.g. BPDE-DNA adducts) include oxidation of the nucleoside to yield a 5′-carboxylic acid that is then complexed with protein as a mixed anhydride with ethylchlorocarbonate or by means of a carbodiimide (MÜLLER and RAJEWSKY 1981). Carbodiimides can also be used for the coupling of nucleoside-5′-monophosphates. Where antibodies to adduct-containing DNA are required (rather than to modified individual nucleosides), these can be readily obtained by coupling the nucleic acid with methylated bovine serum albumin.

Production of antibodies against carcinogen-modified DNA or nucleosides has been achieved by a variety of methods, commonly involving multiple injections administered to rabbits or, for the production of monoclonal antibodies, to mice of the emulsified protein conjugate over a period of several weeks or months (MÜLLER and RAJEWSKY 1981; POIRIER 1981; STRICKLAND and BOYLE 1984). Examples of antibodies raised against various carcinogen-modified nucleic acid components are shown in Table 2.

Table 2. Examples of carcinogen-DNA adducts against which antibodies have been raised

Adduct or modified DNA[a]	Antibody type	References
O^6-Methyldeoxyguanosine	Polyclonal	BRISCOE et al. (1978); KYRTOPOULOS and SWANN (1980); WILD et al. (1983)
	Monoclonal	WILD et al. (1983)
O^6-Ethyldeoxyguanosine	Polyclonal	MÜLLER and RAJEWSKY (1980); VAN DER LAKEN et al. (1982); WANI et al. (1984)
	Monoclonal	ADAMKIEWICZ et al. (1982); WANI et al. (1984)
O^6-*n*-Butyldeoxyguanosine	Polyclonal	RAJEWSKY et al. (1980)
	Monoclonal	ADAMKIEWICZ et al. (1982); SAFFHILL et al. (1982)
O^6-Isopropyldeoxyguanosine	Monoclonal	ADAMKIEWICZ et al. (1986)
O^6-(2-hydroxyethyl)-deoxyguanosine	Polyclonal	LUDEKE and KLEIHUES (1988)
O^4-Methylthymidine	Monoclonal	ADAMKIEWICZ et al. (1986)
O^4-Ethylthymidine	Polyclonal	RAJEWSKY et al. (1980); WANI and D'AMBROSIO (1987)
	Monoclonal	ADAMKIEWICZ et al. (1986)
O^4-*n*-Butylthymidine	Monoclonal	SAFFHILL et al. (1982)
O^2-*n*-Butylthymidine	Monoclonal	SAFFHILL et al. (1982)
1,N^2-Propanodeoxyguanosine	Monoclonal	FOILES et al. (1987)
N-(deoxyguanosin-8-yl)-AAF	Polyclonal	POIRIER et al. (1977, 1979); LENG et al. (1978), VAN DER LAKEN (1982)
N-(deoxyguanosin-8-yl)-AAF (ring-opened)	Polyclonal	RIO et al. (1982)
N-(deoxyguanosin-8-yl)-AF	Polyclonal	SPODHEIM-MAURIZOT and LENG (1980)
N-(deoxyguanosin-8-yl)-ABP	Polyclonal	ROBERTS et al. (1986)
Deoxyguanosin-N^2-yl-BP	Polyclonal	POIRIER et al. (1980); SLOR et al. (1981); TIERNEY et al. (1986); VAN SCHOOTEN et al. (1987)
	Monoclonal	SANTELLA et al. (1984, 1985); WALLIN et al. (1984); VAN SCHOOTEN et al. (1987)
Deoxyguanosin-7-yl-AFB_1	Monoclonal	HAUGEN et al. (1981)
Deoxyguanosin-7-yl-AFB_1 (ring-opened)	Monoclonal	HERTZOG et al. (1982)
1-Aminopyrene-DNA	Monoclonal	HSIEH et al. (1985)
Glu-P-3-DNA[b]	Polyclonal	HEBERT et al. (1985)
cis-Diamminedichloroplatinum(II)-DNA	Polyclonal	MALFOY et al. (1981); POIRIER et al. (1982a); FICHTINGER-SCHEPMAN et al. (1985a)
8-Methoxypsoralen-DNA	Monoclonal	YANG et al. (1987)
Phenylalanine mustard-DNA	Monoclonal	TILBY et al. (1987)
UV-irradiated DNA	Polyclonal	LEVINE et al. (1966)
Thymidine dimer	Monoclonal	STRICKLAND and BOYLE (1981)
Thymidine glycol	Polyclonal	WEST et al. (1982)
	Monoclonal	LEADON and HANAWALT (1983)

[a] Where known, the major carcinogen-DNA adduct recognised is indicated. In some cases antibodies were elicited against modified DNA, in others against the listed modified nucleoside itself.

[b] Glu-P-3, 3-amino-4,6-dimethyldipyrido(1,2-a:3′,2′-d)imidazole.

AAF, 2-acetylaminofluorene; ABP, aminobiphenyl; AFB, aflatoxin B_1.

Determination of the affinity and specificity of an antibody can be carried out by means of a competitive radioimmunoassay (RIA), in which radioactively labelled hapten (tracer) competes with unlabelled hapten (inhibitor) for the antibody binding sites under equilibrium conditions (FARR 1958). The affinity constant (K) of the antibody can then be determined from the formula:

$$K\,(\text{litres/mole}) = 1/([\text{I}]-[\text{T}])(1-1.5b+0.5b^2),$$

where [I] is the concentration of inhibitor required for 50% inhibition of tracer-antibody binding, [T] is the tracer concentration, and b is the fraction of tracer bound in the absence of inhibitor.

Antisera may be used unpurified in assays for DNA adducts or, alternatively, may be purified by ammonium sulphate precipitation or DEAE-cellulose chromatography (STRICKLAND and BOYLE 1984). Affinity chromatography using antigen immobilised on Sepharose is also employed to purify antibodies (LENG et al. 1978).

Monoclonal as well as polyclonal antibodies have been raised against many carcinogen-DNA adducts (Table 2). The production of monoclonals offers the advantage of homogeneity and of unlimited supply. However, with the exception of some antibodies against alkylated bases (MÜLLER and RAJEWSKY 1981), they are not more sensitive than polyclonals, nor are they more specific.

In solid-phase enzyme immunoassays, the amount of antibody bound to unlabelled antigen, the latter coated on plastic microtitre plates, is measured by using a second antibody that is either labelled with ^{125}I (radioimmunosorbent technique, RIST) (HARRIS et al. 1982) or linked to an enzyme, such as alkaline phosphatase (horseradish peroxidase is also used), that cleaves a chromogenic substance, for example p-nitrophenylphosphate (enzyme-linked immunosorbent assay, ELISA) (HARRIS et al. 1982), a fluorogenic substrate such as methylumbelliferyl-phosphate (high-sensitive ELISA, HS-ELISA) (VAN DER LAKEN et al. 1982) or a radiolabelled substrate (ultrasensitive enzymatic radioimmunoassay, USERIA) (HARRIS et al. 1982). Application of the solid-phase immunoassays to the detection of aromatic carcinogen-DNA adducts has resulted in a 10–100-fold increase in sensitivity over conventional RIA (HSU et al. 1980, 1981), whereas when RIST and ELISA were applied to smaller adducts such as O^6-ethyldeoxyguanosine, they were less sensitive than RIA (MÜLLER and RAJEWSKY 1980), although a subsequent study showed HS-ELISA to be more sensitive (VAN DER LAKEN et al. 1982). Although RIA is accurate and simple to perform, the solid-phase methods have found widespread favour because of their greater sensitivity in most instances and because of the development of instrumentation to facilitate rapid screening of large numbers of samples (KRIEK et al. 1984).

Further improvements to immunoassays have been made using the avidin-biotin system. Avidin, an inexpensive glycoprotein readily isolated in abundance from egg white, has a very high affinity for the vitamin biotin. Biotin can be readily coupled to the antibodies and avidin conjugated to the marker enzyme, and since avidin has four active sites for biotin and many biotin molecules can be bound to each protein molecule, premixing the biotinylated antibody and the enzyme-conjugated avidin offers the potential for amplifying the antigen-antibody reaction, which both improves the sensitivity and reduces the time needed to perform the assays (BA-ELISA, BA-USERIA) (SHAMSUDDIN and HARRIS 1983).

If the DNA modification is heat- or alkali-stable, then the immunoassay can be performed by immobilising the single-stranded DNA on nitrocellulose filters and reacting first with an antibody directed against the modified DNA and then with a second antibody, either radiolabelled or enzyme-linked, directed against the first one. This is the immuno-slot-blot (ISB) procedure (NEHLS et al. 1984a).

Visualisation of carcinogen-DNA binding in situ has been achieved by using indirect immunofluorescent localisation of adducts (POIRIER et al. 1982b). The procedure involves binding the adduct-specific antibody to fixed cells and then applying a second, fluorescent antibody directed against the first one. Alternatively, direct immunofluorescence detection is possible by linking the fluorescent material to the adduct-directed antibody (ADAMKIEWICZ et al. 1986). Either procedure makes possible the monitoring of carcinogen-adduct formation and repair in individual cells. The application of microfluorometry (SMITH et al. 1983) allows semi-quantitative analysis of the immunofluorescence of individual cells. Electron microscopy has been used to visualise the binding of antibodies to carcinogen-modified DNA and thus locate the sites of adduct formation (SLOR et al. 1981; NEHLS et al. 1984b; PAULES et al. 1985).

II. Applications in Animal and Tissue Culture Studies

There is now a large body of literature describing the use of antibodies to detect DNA adducts in carcinogen-treated cells and animal tissues, some of it the subject of earlier reviews (POIRIER 1981, 1984; MÜLLER and RAJEWSKY 1981; HARRIS et al. 1982; KRIEK et al. 1984; STRICKLAND and BOYLE 1984). The discussion here will be confined to representative examples, together with descriptions of more recent developments.

Heterogeneity of DNA adduct distribution by *N*-ethyl-*N*-nitrosourea (ENU) in fetal rat brain DNA has been demonstrated using a monoclonal antibody raised against O^6-ethyldeoxyguanosine (NEHLS et al. 1984b). Digestion of the DNA with a restriction enzyme allowed separation of a fraction of the DNA that contained tightly bound polypeptides and which was found by immune electron microscopy to contain clusters of 2–10 adduct-antibody binding sites, evidence that the adduct distribution in brain DNA is highly non-random. An earlier study of O^6-methyldeoxyguanosine distribution in liver DNA from mice injected with dimethylnitroamine (KYRTOPOULOS and SWANN 1980) demonstrated reduced adduct formation in satellite DNA compared with main band DNA, but similar rates of adduct removal in both fractions.

The developing rat brain is highly susceptible to the carcinogenic effects of ENU and has a low repair activity for O^6-ethyldeoxyguanosine, but further studies using RIA on the repair of O^6-ethyldeoxyguanosine in normal and malignant rat brain cells (HUH and RAJEWSKY 1986, 1988) have not entirely explained why this is so. Malignant cell lines efficiently removed O^6-ethylguanine from DNA, although some variability in the repair capacity was observed among subclones, suggesting an instability in the phenotype. Also, the same cells grown in vivo showed a lower repair capacity than when grown in monolayer culture.

Immunohistochemical analysis of the brains of rats that had been treated with ENU indicated that all brain cell nuclei contained O^6-ethyldeoxyguanosine shortly after a single injection of the compound, but when repeated small injections were given 1–2 weeks apart and the animals killed 1–2 weeks after the final dose, only the oligodendrocytes, granular neurons and endothelial cells, and part of the pyramidal neurons and astrocytes had accumulated the adduct (HEYTING et al. 1983). There was, however, no obvious correlation between those cells most likely to be target cells for ENU-induced brain tumours and those cells that accumulated the adduct.

O^4-Ethylthymidine is a quantitatively minor, but biologically important adduct formed when ethylating agents such as ENU react with DNA (SINGER et al. 1986). WANI and D'AMBROSIO (1987) were able to monitor the formation and repair of this adduct in human skin fibroblasts and kidney epithelial cells that had been treated with ENU. The adduct, which is apparently not removed by either the *O*-alkyl acceptor or glycosylase repair pathways, was found to be repaired to similar extents in the two cell types, with 50% of the damage removed in 3 days. The sensitivity of the assays, which incorporated the use of non-competitive ELISA and ISB techniques, was such that femtomolar quantities of adduct were detectable in less than 1 µg DNA. O^4-Ethylthimidine has been shown to assume greater quantitative importance if the experimental animals are receiving chronic exposure to an ethylating agent. SWENBERG et al. (1984) have demonstrated, using RIA, that O^4-ethylthymidine, but not O^6-ethyldeoxyguanosine, accumulates in hepatocyte DNA of rats given drinking water containing diethylnitrosamine (DEN). A similar, but less pronounced effect was observed for the equivalent *O*-methyl nucleosides in DNA from the livers of rats administered 1,2-dimethylhydrazine (RICHARDSON et al. 1985).

Polyclonal antibodies raised against O^6-methyldeoxyguanosine have been used in a BA-ELISA assay to study the ability of the tobacco carcinogen 4-(*N*-methyl-*N*-nitrosamino)-1-(3-pyridyl)-1-butanone (NNK) to form DNA adducts in rats (FOILES et al. 1985). In addition to confirming that NNK is a methylating agent, they demonstrated the formation of adducts in target tissues (nasal mucosa, lung and liver) but not in non-target tissues (oesophagus, spleen, heart and kidney). O^4-Methylthymidine and O^6-methyldeoxyguanosine formation and persistence were also studied by RIA in rats administered multiple doses of NNK (BELINSKY et al. 1986). In lung DNA the former adduct reached a steady state level after 4 days of treatment but decreased rapidly when carcinogen treatment ceased, while the level of the latter adduct increased throughout the treatment period and persisted after NNK treatment ended.

The ISB method has been used to follow the formation of O^6-(2-hydroxyethyl)deoxyguanosine in the tissues of rats treated with *N*-nitroso-*N*-(2-hydroxyethyl)urea (HENU) (LUDEKE and KLEIHUES 1988). Rabbit antibodies to the adduct detected the highest levels in the kidney after a single i.v. dose of HENU to male rats, and lower levels in the lung and liver. Repair of the adduct was particularly rapid in the liver. In contrast to *N*-methyl-*N*-nitrosourea (MNU) and ENU, HENU is a much weaker inducer of neural tumours despite being a potent carcinogen for other organs, and the levels of the adduct were found to be much lower in the brain than in other tissues. This study also demonstrated that prior saturation of the hepatic O^6-alkylguanine acceptor protein activity prevented the removal of O^6-(2-hydroxyethyl)deoxyguanosine, indicating that it is repaired by this pathway.

Similar studies on the localisation of adducts in liver slices from rats treated with either ENU or DEN have been reported (MENKVELD et al. 1985). Five hours after injection of DEN, O^6-ethyldeoxyguanosine was detectable in nuclei of the centrilobular hepatocytes, but not in the hepatocytes of peripheral regions of the liver lobules. Seven days after treatment all hepatocytes except those around the central veins were unstained. In contrast, 2 h after treatment with ENU, all cell types showed the presence of O^6-ethyldeoxyguanosine, with a homogeneous staining of hepatocytes throughout the liver. However, 24 h after treatment only the non-parenchymal cells, i.e. the target cells for ENU-induced liver tumours, showed significant presence of the adduct.

The interaction of AAF with DNA in mammalian tissues has been widely studied using antibodies raised against its nucleoside adducts. Injection of rats with AAF results in the formation of adducts with the *C*-8 position of guanine in which the acetyl group is either retained or lost and an acetylated N^2-guanine adduct. Using an antibody that recognises both *N*-(guanosin-8-yl)-2-acetylaminofluorene and the related deacetylated adduct, POIRIER et al. (1982c, 1984) measured adduct formation and removal during chronic dietary administration of AAF to rats. Hydrolysis of the DNA and separation by HPLC enabled the two adducts to be quantitated separately by RIA. These experiments demonstrated that the deacetylated (AF) adduct accumulates in rat liver DNA during a month of continuous feeding but that the acetylated (AAF) adduct does not. A similar picture of adduct accumulation was obtained by microfluorimetric determination of immunofluorescent-stained frozen liver sections (HUITFELDT et al. 1987). Returning the rats to a control diet resulted in a biphasic removal of adducts with time, consisting of a rapid removal of the majority of the adducts followed by a slow removal phase (POIRIER et al. 1984), with about one-third of the adducts present at the start of the switch to control diet still present 28 days later. Attempts to determine the reasons for the existence of both removable and persistent adducts were made by fractionating chromatin from liver nuclei into transcriptionally active (nuclease-sensitive), high and low salt-soluble, and matrix-bound DNA; although the initial levels of adducts in these fractions differed, the proportional rate of removal was the same (POIRIER et al. 1988).

Immunohistochemical localisation of AAF-DNA adducts in rat liver showed in one study (MENKVELD et al. 1985) that the distribution over the liver lobules was rather homogeneous 6 days after a single i.p. injection. Other studies (HUITFELDT et al. 1986, 1987), however, have revealed a more non-uniform distribution after dietary administration of the carcinogen for a month (see Fig. 3). Immunofluorescence was most intense in the nuclei of the periportal hepatocytes, intermediate in midzonal hepatocytes and weakest in centrilobar hepatocytes and bile duct epithelial cells. Adduct accumulation during the feeding of AAF was demonstrated by microfluorometry of the stained tissues (HUITFELDT et al. 1987), and the results obtained by this method were in agreement with the adduct measurements of liver DNA made using RIA (POIRIER et al. 1984).

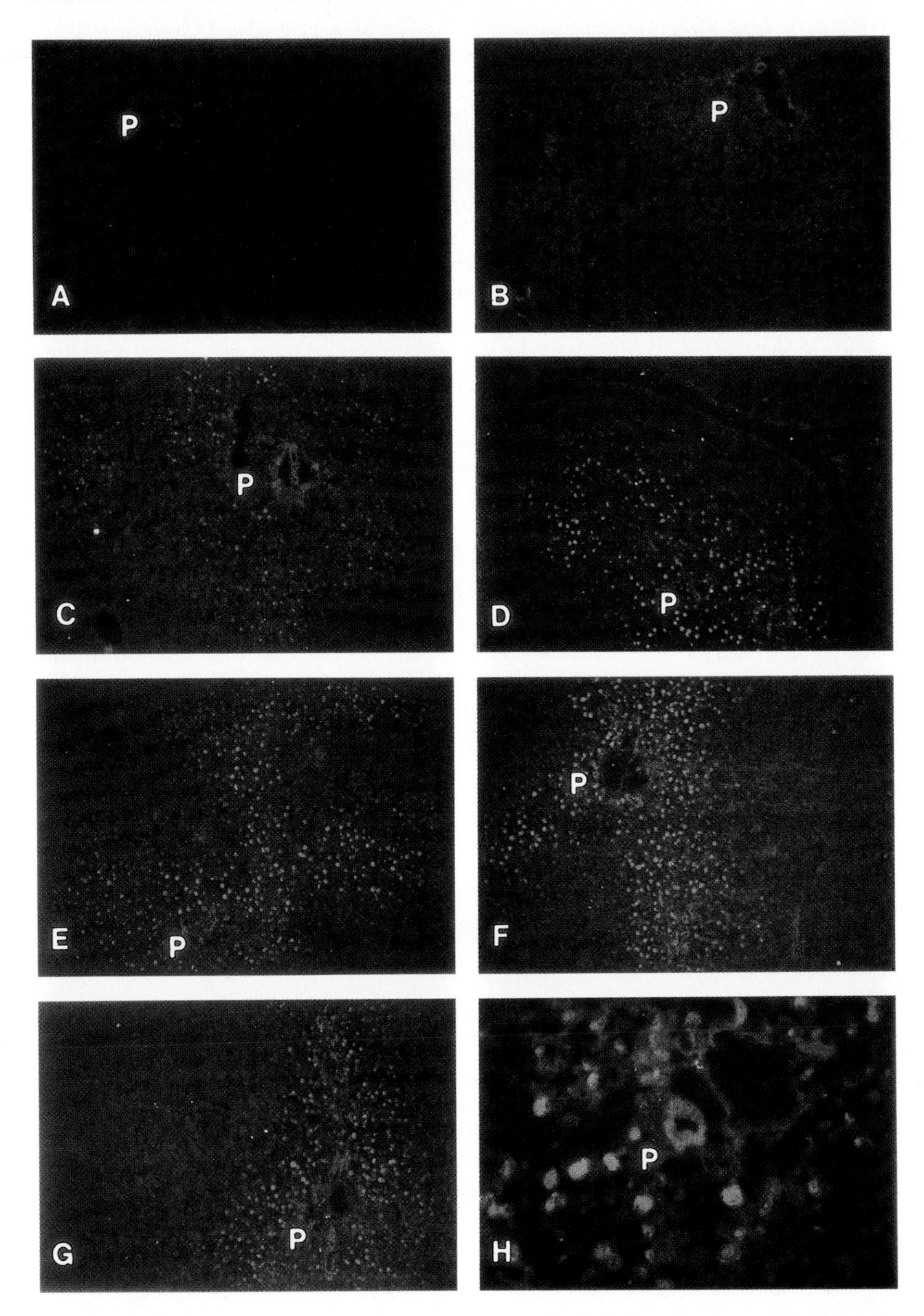

Fig. 3 A–H. Immunofluorescence of *N*-(deoxyguanosin-8-yl)-2-aminofluorene adducts in frozen liver sections from male Fischer rats fed either control diet (**A**) or 0.02% 2-acetylaminofluorene for 2, 4, 8, 12, 20 or 28 days (**B–G**, respectively). *P*, bile duct areas. **H**, Portal area with a bile duct, shown at 4 times the magnification of **A–G**. (Reproduced from HUITFELDT et al. 1987 with permission)

AAF-DNA antibodies have also been used to study the interaction of *N*-acetoxy-2-acetylaminofluorene (AAAF), a model electrophilic derivative of AAF, with DNA in cells. The dose-dependent binding of AAAF in HeLa cell nuclei has been detected by immunofluorscence (SAGE et al. 1981) and the binding sites of the compound in ColE1 DNA visualised by immunoelectron microscopy (DE MURCIA et al. 1979). Chromatin structure was found, by the latter technique, to influence the distribution of adducts in AAAF-treated chicken erythrocytes (LANG et al. 1982), DNA in linker regions being more highly modified than DNA associated with nucleosome core particles.

Visualisation by electron microscopy of AAF-DNA adducts in replicating simian virus 40 (SV40) indicated the presence of an adduct at halted replication forks (ARMIER et al. 1988) suggesting that the presence of adducts blocks the progression of replication. A similar study observing the distribution of BPDE-DNA adducts in synchronised C3H/10T$_{1/2}$ cells (PAULES et al. 1988) arrived at the same conclusion: that the fork junction is particularly sensitive to bulky adduct formation and that such adducts hinder the displacement of the forks during replication.

As an alternative to using radioactively labelled oligonucleotide probes in DNA hybridisation experiments for the detection of specific DNA sequences, immunochemical techniques can be used if the oligonucleotide is recognised by a specific antibody. TCHEN et al. (1984) have achieved this by modifying oligonucleotides with AAAF or its 7-iodo derivative (AAAIF). Once these probes are hybridised to their specific recognition DNA sequences, the hybridisation can be detected by binding the AAF-DNA antibody to the probes and then binding a peroxidase- or alkaline phosphatase-conjugated second antibody to the first and visualising its presence by histochemical staining (Fig. 4). This technique has been used, for example, to detect regions of DNA in the Z configuration in chromosomes of *Cebus* monkeys by using an AAAF-modified CA-rich probe that readily adopts the Z conformation in vitro and hybridises to similar regions in the chromosomal DNA (VIEGAS-PEQUIGNOT et al. 1986). The regions of the

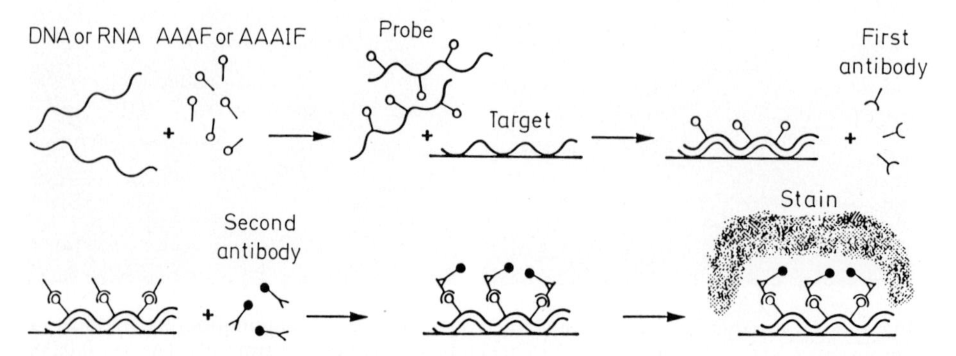

Fig. 4. Procedure for the preparation and detection of immunoreactive DNA or RNA probes. *AAAF*, *N*-acetoxy-2-acetylaminofluorene; *AAAIF*, *N*-acetoxy-2-acetylamino-7-iodofluorene. (Reproduced from TCHEN et al. 1984 with permission)

chromosomes recognised by this probe were the same as those recognised by Z-DNA-specific antibodies. The procedure is also sensitive enough to allow detection of single gene copies, as has been demonstrated by the location of the human thyroglobulin gene to the distal end of the long arm of chromosome 8 (LANDEGENT et al. 1985).

BP is the most widely studied of the carcinogenic polycyclic aromatic hydrocarbons and is the one upon which the most effort has been focused in raising antibodies against adducts formed by this class of compounds. Synthetically prepared adducts formed by BPDE, the major ultimate carcinogen of BP, have been used to prepare both polyclonal (POIRIER et al. 1980; TIERNEY et al. 1986) and monoclonal (SANTELLA et al. 1984, 1985) antibodies. Earlier publications reported that cross-reactivity with adducts of other classes of carcinogens, such as AAF, was minimal or undetectable, but more recently it has become apparent that the antibodies often recognise other polycyclic hydrocarbon DNA adducts (see Sect. C.IV). In addition to DNA adducts, antibodies to protein-bound BP have also been produced (WALLIN et al. 1984; SANTELLA et al. 1986). Some of the antibodies have a high affinity for BP tetraols and can therefore be exploited to measure adduct levels by releasing bound BP from macromolecules in the form of tetraols.

A study of BP-DNA adduct formation and removal in mouse skin found that similar levels of adducts were formed in the epidermis in vivo following exposure to initiating doses of BP as were formed in cultured keratinocytes treated with the compound at doses that caused resistance to calcium-induced terminal differentiation (NAKAYAMA et al. 1984). As with AAF-DNA adducts (see above), immunohistochemical methods have been developed to localise BP-DNA binding in cells. Early studies demonstrated, as expected, that immunofluorescence was confined to the nuclei of BPDE-treated human fibroblasts (SLOR et al. 1981) and mouse keratinocytes (POIRIER et al. 1982b). The distribution of BPDE-DNA adducts between regions of chromatin was determined using antibodies, and it was found that linker DNA contained approximately threefold higher levels than core DNA (SEIDMAN et al. 1983a, b). Immunoelectron microscopy has been used quantitatively for the determination of BPDE binding to DNA, and the results obtained were found to be in good agreement with those obtained by ELISA analysis of the same samples (PAULES et al. 1985) (Fig. 5).

An ELISA analysis of DNA isolated from mouse skin that had been treated in vivo with crude coal tar revealed the presence of adducts recognised by antibodies to BPDE-modified deoxyguanosine (MUKHTAR et al. 1986). Interestingly, the levels of adducts detected when BP and coal tar were both applied to mouse skin were lower than when BP was applied alone, suggesting that the complex mixture of chemicals in the coal tar fraction contains inhibitors of carcinogen-DNA adduct formation in addition to the DNA-binding compounds themselves.

Monoclonal antibodies have been used to detect aflatoxin B_1-DNA adducts in the livers of rats administered the compound (HAUGEN et al. 1981; GROOPMAN et al. 1982). The detection of 1 adduct in 1 355 000 nucleotides was achieved. The major DNA adduct formed by aflatoxin B_1 is at the *N*-7 position of guanine residues and is readily lost from the macromolecule by depurination. Methods

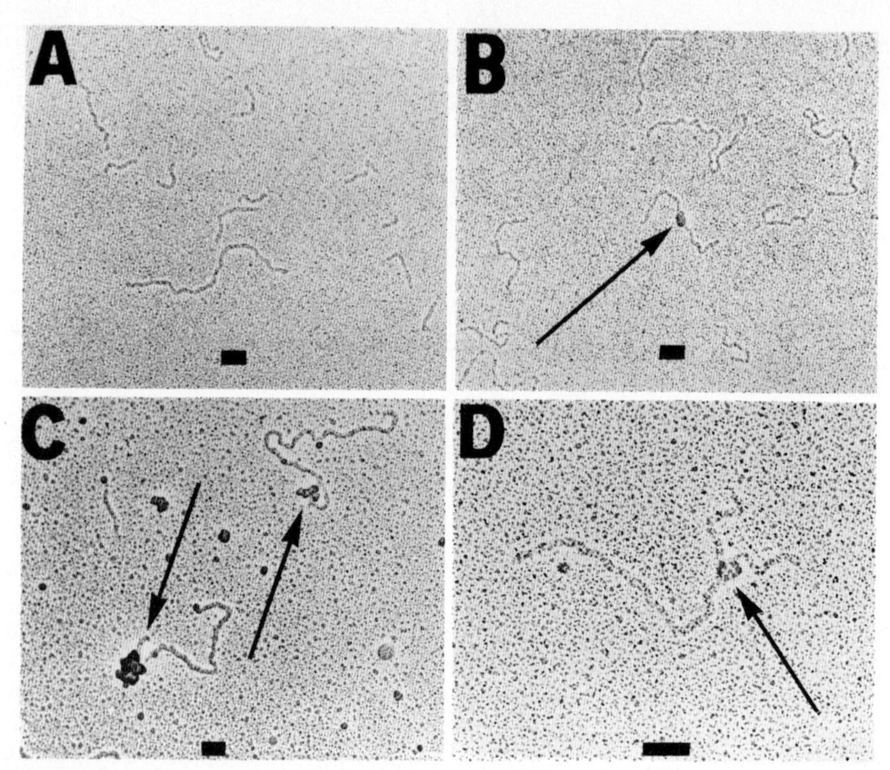

Fig. 5 A–D. Electron micrographs of DNA modified by benzo[*a*]pyrene diolepoxide (BPDE) and incubated with BPdG-DNA-specific rabbit serum followed by ferritin-labelled monovalent secondary antibody fragments. *Arrows* indicate ferritin molecules locating the sites of BPDE-DNA adducts. **A** Unmodified DNA; **B** 3.8 fmol BPDE bound/µg DNA; **C** 10.0 fmol BPDE bound/µg DNA; **D** 40.4 fmol BPDE bound/µg DNA. *Bar*, 0.1 µm. (Reproduced from PAULES et al. 1985 with permission)

have therefore been developed to detect the modified guanine in the urine of animals exposed to aflatoxin B_1. GROOPMAN et al. (1984, 1985) have used an antibody bound to a column support to concentrate the adduct and other aflatoxin derivatives from urine and to quantitate them by ELISA or HPLC. If DNA containing aflatoxin adducts is treated with alkali, opening of the adducted guanine imidazole ring occurs to yield a chemically stable adduct. Antibodies have been raised against this ring-opened form of the adduct and provide an alternative means of determining aflatoxin B_1 exposure (HERTZOG et al. 1982, 1983).

Prior to the production of antibodies to DNA adducts of platinum(II) complexes, detection of Pt in biological samples relied upon the use of either atomic absorption spectroscopy (FICHTINGER-SCHEPMAN et al. 1985b) or radiolabelled Pt compounds (EASTMAN 1983). Rabbit antisera elicited against DNA highly modified by reaction with the anti-tumour agent *cis*-diamminedichloroplatinum(II) (cisplatin) have now been made by several research groups (MALFOY et al. 1981; POIRIER et al. 1982a; FICHTINGER-SCHEPMAN et al. 1985a, b). The antibodies have been used to determine the levels of Pt-DNA adducts in cisplatin-treated mouse cells (POIRIER et al. 1982a), CHO cells (PLOOY et al. 1985) and cell lines derived from human testicular and bladder tumours (BEDFORD et al. 1988). REED et al. (1987a) used an ELISA assay to determine the levels of Pt-DNA adducts in rats administered cisplatin. A side effect of cisplatin is its renal toxicity and both i.p. and i.v. administration gave rise to high levels of adducts in the kidneys that, in contrast to adducts formed by most other carcinogenic chemicals, persisted for several weeks at the initially formed levels. Overnight fasting of animals increased the levels of adducts approximately twofold. TERHEGGEN et al. (1987) used immunocytochemical analysis to detect Pt-DNA adducts in cisplatin-treated rats and mice and were able to detect adducts in kidney tissue as long as 162 days after a single treatment. All tissues examined 6 h after treatment (kidney, liver, pancreas, heart, muscle, brain, testis, duodenum, spleen) contained detectable levels of DNA-bound Pt, with regional differences being evident within the kidneys (inner cortex > outer cortex > medulla). The antibodies used in these studies have also been applied to the detection of adducts in human subjects receiving cisplatin chemotherapy (see Sect. C. III).

Antibodies raised against thymine glycols in DNA (LEADON and HANAWALT 1983) have been used to demonstrate that oxidative damage to DNA can result from carcinogen treatment. Thus, treatment of human cells with *N*-hydroxy-2-naphthylamine resulted in the dose-dependent formation of thymine glycols in the cellular DNA (KANEKO and LEADON 1986). This chemical is therefore capable of causing both direct damage (by forming covalent adducts) and indirect damage through the generation of active oxygen species.

III. Human Studies

Inhabitants of Linxian County in the People's Republic of China have a high incidence of oesophageal cancer that is suspected of being due, in part, to a high dietary intake of nitrosamines (LU et al. 1985). UMBENHAUER et al. (1985) have used monoclonal antibodies to test for the presence of O^6-methyldeoxyguanosine in oesophageal and stomach DNA from individuals from Linxian County and from Europe. Higher levels were found in samples from the high-risk group than from the controls. Of 37 Chinese samples, 10 had levels of adduct below the detection limit of 12.5 fmol/mg DNA, 17 had levels from 15 to 50 fmol/mg DNA, and 10 had higher levels, up to 160 fmol/mg DNA. In contrast, 7 out of 12 European samples had levels below the limit of detection, and the other 5 contained levels of the adduct below 45 fmol/mg DNA. These results do not prove that the formation of the adducts was due to nitrosamine exposure, nor do they prove of themselves that nitrosamines are the agents responsible for the high incidence of oesophageal cancer in Linxian County. Nevertheless, studies such as these provide a powerful tool for epidemiological studies on the role of chemicals in the aetiology of human cancer.

Several studies have used antibodies raised against BPDE-DNA adducts to monitor human exposure to polycyclic aromatic hydrocarbons. All of the antibodies used have cross-reactivity to hydrocarbon-DNA adducts other than those formed by BP (see Sect. C. IV), which means that absolute values for adduct levels cannot be obtained. Nevertheless, comparison of exposed populations with unexposed or low exposure groups did show differences in some cases. Studies on white blood cell DNA from coke oven workers who had been occupationally exposed to polycyclic aromatic hydrocarbons revealed the presence of adducts that gave an immune response to antibodies raised against BPDE-DNA adducts in 18 out of 27 samples (HARRIS et al. 1985). In addition, serum from the workers was used in a USERIA assay, and 3 out of 11 samples were positive, indicating the presence of antibodies that recognise hydrocarbon-DNA adducts in the workers' blood. A second study of coke oven workers found that white blood cell DNA contained adducts and serum contained antibodies in about one-third of the samples tested (HAUGEN et al. 1986). ELISA and USERIA assays of white blood cell DNA from roofers and iron foundry workers revealed that 7 of 28 samples from the former group and 7 of 20 samples from the latter contained detectable levels of adducts recognised by BPDE-DNA antibodies. Another study of 35 Finnish iron foundry workers found that their white blood cell DNA had a significantly greater antigenicity towards the antibodies than did the DNA from control subjects (PERERA et al. 1988) (see also Sect. B. III). Comparisons using radioimmunoassay of white blood cell DNA from smokers and non-smokers have not revealed any significant differences between the two groups (PERERA et al. 1982, 1987), although hydrocarbon-adducts were detected in 5 of 27 lung samples, all of them from lung cancer patients. These findings are consistent with the analyses by ^{32}P-postlabelling of smokers' lymphocyte and lung DNA (Sect. B. III). Analysis of placental DNA from women who smoked during pregnancy revealed the presence of material recognised by BP-DNA antibodies (EVERSON et al. 1986). However, as discussed in Sect. B. III, the major aromatic adducts present were shown by ^{32}P-postlabelling to have chromatographic properties distinct from those of BP-DNA adducts. The positive response to the antibodies is thus probably a result of their cross-reactivity with other, structurally related adducts (see Sect. C. IV).

Immunocytochemical methods have been used to investigate the localisation of adducts recognised by BPDE-DNA antibodies in a number of human tissues (SHAMSUDDIN and GAN 1988). Positively staining areas in sections of human lung, ovary, uterine cervix and placenta were detected, although not in all specimens examined.

When the antibodies raised against aflatoxin adducts (see Sect. C. II) were used to analyse urine samples from humans exposed to aflatoxin B_1 through dietary contamination, the presence of the *N*-7 guanyl product was detected (GROOPMAN et al. 1985). The monoclonal antibody was covalently bound to Sepharose and used as a preparative column to isolate the adduct and other aflatoxin derivatives from the urine, and then the retained materials were eluted from the column and analysed by HPLC. This method, along with fluorescence detection of adducts (see Sect. D. I) and detection of non-adduct metabolites by either means, holds promise as a method of monitoring human exposure to mycotoxins.

One of the few instances in which human exposure to a genotoxic agent can be accurately assessed and where control individuals with zero exposure can be examined is in the treatment of cancer patients with DNA-damaging drugs.

There are several studies in which adduct formation has been monitored in patients receiving cisplatin or the structurally related compound diamminecyclobutane-dicarboxylato-platinum (CBDCA) using polyclonal antibodies to cisplatin-modified DNA. White blood cell DNA from seven patients who received a single dose of cisplatin was found to exhibit widely varying levels of Pt-DNA adducts (FICHTINGER-SCHEPMAN et al. 1987). Adduct levels diminished rapidly with time in those patients who were receiving cisplatin for the first time. In a study of testicular and ovarian cancer patients, the levels of adducts detected in white blood cell DNA samples correlated with the clinical response to cisplatin or CBDCA therapy (REED et al. 1986, 1987b). This raises the possibility of being able to predict, early in the course of treatment, which patients are likely to respond to platinum therapy. Examination of postmortem tissues has demonstrated persistent cisplatin-DNA adducts in several human organs, including kidney, spleen and liver, for as long as 22 months after the last treatment (POIRIER et al. 1987). Highly persistent Pt-DNA adducts were also noted in animal experiments (see Sect. **C.II**). Recently, monoclonal antibodies have been raised against adducts formed by another chemotherapeutic drug, phenylalanine mustard (melphalan) (TILBY et al. 1987), so patients treated with this drug may now be monitored for adduct formation.

IV. Advantages and Limitations of Immunochemical Methods

Most antibodies to carcinogen-DNA adducts are of high affinity and specificity and, when used in enzyme immunoassays, require only microgram quantities of DNA to detect adduct levels as low as 1 in 10^8 nucleotides. Large numbers of samples can be processed simultaneously, and their spectroscopic analysis can be automated. Antibodies are useful in studies on DNA structure and conformation and can visualise the location of specific carcinogen-DNA adducts within cells and tissues when combined with histochemical techniques. The methods can be equally sensitive and specific whether the adducts are derived from aliphatic or aromatic carcinogens, or from transition metal complexes.

However, the production of carcinogen-DNA adduct antisera requires the use of chemically synthesized nucleotide adducts or DNA highly modified (approximately 1% of bases) by reaction with an appropriate reactive intermediate of the carcinogen. A relatively large amount of effort is then needed to produce and determine the specificity of the antibodies produced. Thus, the number of different carcinogen-DNA adducts for which antibodies are available is still relatively small. Assays have frequently been calibrated using highly modified DNA diluted with different amounts of unmodified DNA, but two recent reports have concluded that the use of such DNA as a standard competitor in ELISA will lead to erroneous results when determining adduct levels in biological samples and other material with a low level of modification (VAN SCHOOTEN et al. 1987; SANTELLA et al. 1988). This casts uncertainty on some of the quantitative data on adduct levels so far derived using antibodies, and some examples of interlaboratory variations may be due, in part, to problems of this nature.

A further problem is cross-reactivity. All antibodies raised against a particular carcinogen-DNA adduct show a measure of response to other adducts

(SALIH and SWANN 1982; WILD et al. 1983; SANTELLA et al. 1985; EVERSON et al. 1986), which will create uncertainties in the analysis of DNA samples from tissues that may have been exposed to many genotoxic agents. While cross-reactivity with adducts formed by other classes of carcinogens is generally low, recognition by the antibodies of adducts formed by chemicals of the same class can be high (EVERSON et al. 1986; VAN SCHOOTEN et al. 1987; WESTON et al. 1987). Attempts in recent years to produce antibodies to BP-DNA adducts that do not cross-react appreciably with other hydrocarbon-DNA adducts have met with failure due, presumably, to the close structural similarity of these adducts. More recently, there has been a reversal of thinking in this area with the production of non-specific antibodies to hydrocarbon-DNA adducts now considered to be advantageous for environmental monitoring. Such antibodies can be used in immunoaffinity columns to extract hydrocarbon-DNA adducts from digested DNA samples for further analysis by other means (TIERNEY et al. 1986). However, the use of these antibodies to quantitate adducts in human DNA is still subject to a degree of uncertainty because without knowing all the components of a mixed adduct profile, the influence of cross-reactivity, and thus the level of total DNA damage, cannot be accurately assessed.

D. Physicochemical Methods

I. Fluorescence Spectroscopy

The earliest studies on polycyclic hydrocarbon activation made use of fluorescence spectroscopy to study covalent interactions with tissue components (see KRIEK et al. 1984). The method was frequently too insensitive to detect binding to DNA in vivo and lost favour with the advent of radiolabelled compounds. In the 1970s there was a revival of interest with the application of the use of photomultiplier detection of low levels of fluorescence (DUQUESNE et al. 1970; VIGNY and DUQUESNE 1974). With this photon-counting method it was possible to detect the fluorescence of polycyclic hydrocarbons bound to DNA after exposure in vivo and to determine, from the spectral characteristics, which ring of the compound had undergone metabolism and lost its aromaticity. Thus, it was demonstrated that BP bound to DNA in mouse skin retained an intact pyrene nucleus and was therefore metabolically activated in the 7,8,9,10-ring (DAUDEL et al. 1975) and that 7-methylbenz[*a*]anthracene underwent activation in the 1,2,3,4-ring, thereby retaining the anthracene chromophore (VIGNY et al. 1977a). Analysis can be performed either on hydrolysed (MOSCHEL et al. 1977) or unhydrolysed DNA (IVANOVIC et al. 1978; VIGNY et al. 1977b).

At normal temperatures, fluorescence spectra of DNA-bound carcinogens are generally broad and highly quenched. Low temperature fluorescence (77 K) considerably improves the sensitivity of the technique (IVANOVIC et al. 1976). At even lower temperatures (4.2 K), a phenomenon known as fluorescence-line-narrowing (FLN) is observed: a molecule embedded in an amorphous solid such as a glass or polymer and excited with a laser exhibits a dramatic narrowing of the fluorescence spectrum because only a narrow subset of vibrational states of the molecules becomes excited. HEISIG et al. (1984) have demonstrated the effect with BPDE-DNA adducts, and although some improvement in sensitivity will be needed for the purpose of analysing environmental exposures, the presence of multiple adducts in a complex mixture can be resolved because the spectrum of an individual adduct is highly characteristic (SANDERS et al. 1986).

Problems of fluorescence quenching can be overcome in another way. Acid hydrolysis of BPDE-containing DNA allows the free tetraol to be isolated by HPLC and analysed by fluorescence spectroscopy, resulting in a two- and tenfold enhancement of fluorescence yield over BPDE-modified nucleosides and adducts present in intact DNA, respectively (RAHN et al. 1982).

Spectra may also be simplified by synchronous fluorescence spectroscopy (SFS), in which the excitation and emission wavelengths are scanned with a fixed wavelength difference ($\Delta\lambda$) (VO-DINH 1978, 1982). For BPDE-DNA adducts, a $\Delta\lambda$ of 34 nm gives an emission maximum at 382 nm, while BP tetraols show a maximum at 379 nm (VAHAKANGAS et al. 1985). With standard DNA samples modified by BPDE, it was possible to detect adduct levels as low as 1 in 1.4×10^7 nucleotides, using 100 µg DNA. RAHN et al. (1980) used SFS with a $\Delta\lambda$ of 28 nm to detect BP-DNA adducts in mouse epidermis. Other polycyclic hydrocarbon-DNA adducts have different optimum $\Delta\lambda$ and emission maxima (VAHAKANGAS et al. 1985), so it is theoretically possible to analyse the components of complex adduct mixtures.

SFS analysis of white blood cell DNA from coke oven workers revealed that approximately 10% of samples had emission peaks at 379 nm (HAUGEN et al. 1986). This compares with one-third of the samples being positive when analysed by USERIA (see Sect. C. III). However, the four most positive samples were the same in both assays. In another study (HARRIS et al. 1985) the proportion of coke oven samples found to be positive by the two methods was similar. As the emission spectra were very broad in some samples, quantitation of adduct levels by SFS was not always possible.

SFS has also been applied to the detection of aflatoxin B_1 and its metabolites and adducts (HARRIS et al. 1986). Repeated scanning of samples at different $\Delta\lambda$ between 300 and 500 nm emission wavelengths produces contour maps that are specific for each aflatoxin derivative. Preliminary efforts have been made to identify individual components in mixtures by analysing fourth derivative spectra, and aflatoxin B_1 and aflatoxin M_1 were easily distinguishable in a 1:1 mixture by this means (HARRIS et al. 1986) (Fig. 6).

In addition to the immunochemical methods described for the detection of aflatoxin-guanine adducts excreted in urine (see Sects. C. II and III), fluorescence detection has also been used. The adduct was detected in samples from 6 of 81 individuals from an area of Kenya where food samples are known to be contaminated with aflatoxin B_1 (AUTRUP et al. 1983). The human urine samples were first purified and fractionated by reverse-phase HPLC. A subsequent study (AUTRUP et al. 1987) of a total of 983 urine samples from several districts of Kenya found that 12.6% contained detectable amounts of the aflatoxin-guanine adduct and that there was a moderate degree of correlation between aflatoxin exposure and liver cancer within certain ethnic groups. In a study of the urinary excretion of the adduct by rats administered doses as low as 0.125 mg/kg i.p., UV absorbance analysis after fractionation by HPLC was sufficiently sensitive to detect the adduct (BENNETT et al. 1981).

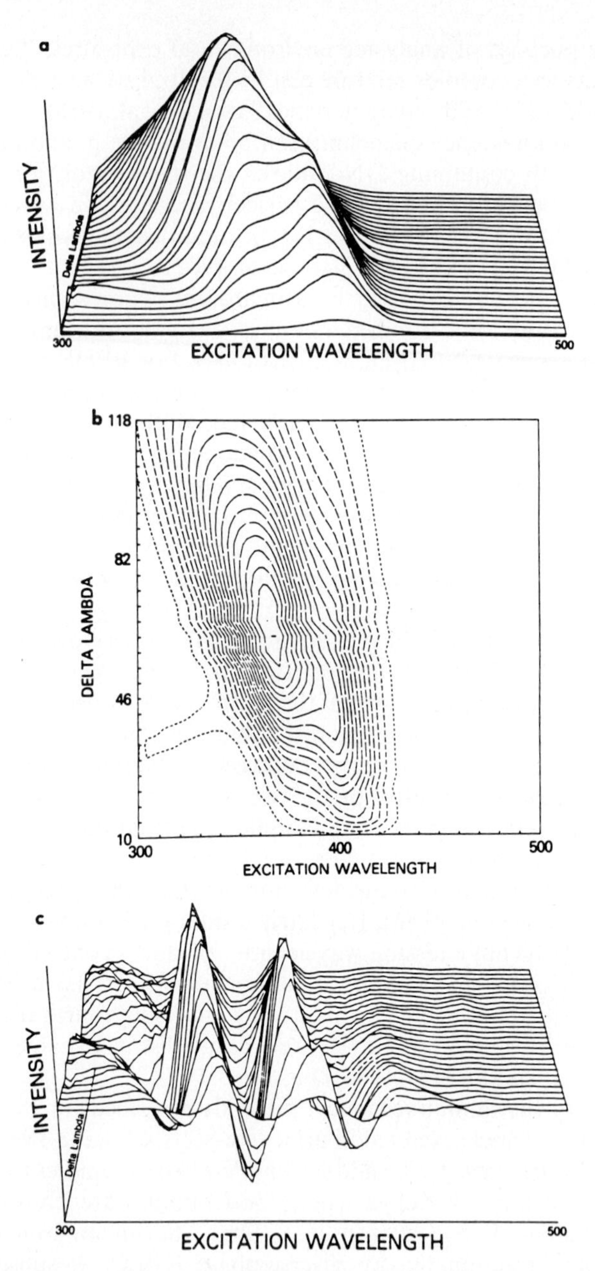

Fig. 6 a–c. Synchronous fluorescence spectroscopy of an equimolar mixture of 100 pmol of aflatoxin B_1 and aflatoxin M_1. The data is shown as a 3-dimensional graph in **a** and as a contour map in **b**. The fourth derivative of the original data **c** demonstrates the resolution of two peaks corresponding to the two components of the mixture. (Reproduced from HARRIS et al. 1986 with permission)

Combinations of fluorescent and UV-absorption detection have been used to determine levels of methylated bases in DNA from animals treated with alkylating agents. For example, the accumulation of O^6-methylguanine in non-parenchymal cells in preference to hepatocytes in the livers of rats chronically exposed to 1,2-dimethylhydrazine (BEDELL et al. 1982) was demonstrated by fluorescence detection of HPLC fractions of hydrolysed DNA. Similar levels of 7-methylguanine, determined by UV absorbance of the HPLC fractions, were present in both cell types. Since angiosarcomas occurred in over 90% of animals administered 1,2-dimethylhydrazine but hepatocellular carcinomas occurred in only 40%, there appears to be a correlation between cell specificity for carcinogenesis and O^6-methylguanine accumulation. Non-parenchymal cells are also more susceptible to tumour induction by dimethylnitrosamine, and the DNA of these liver cells was found by UV and fluorescence spectroscopy to accumulate O^6-methylguanine to a greater extent than hepatocyte DNA in mice fed the compound (LINDAMOOD et al. 1982).

Etheno adducts are also highly fluorescent and can be detected by fluorescence spectroscopy of DNA or RNA hydrolysates after HPLC separation. When DNA that had been reacted with chloroacetaldehyde, a reactive intermediate implicated in the metabolic activation of vinyl chloride, was analysed by this method, etheno adducts formed with different deoxyribonucleosides could be detected (OESCH et al. 1986; BEDELL et al. 1986). Detection of as little as 1 pmol ethenodeoxyadenosine has been claimed (BEDELL et al. 1986).

Fluorescence detection of adducts has been limited to those carcinogens whose DNA adducts are fluorescent, although the prospect of sensitive fluorescent postlabelling methods (see Sect. B. VI) may in future make this approach applicable to a larger number of adducts. In several applications to date the method is less sensitive than other techniques, such as immunoassay and postlabelling, but it does have the major advantage that it is non-destructive, and in determinations in which the intact DNA is analysed, the samples may be subsequently analysed by alternative methods.

II. Gas Chromatography and Mass Spectrometry

The release of methylated bases from DNA by depurination and their excretion in the urine has been studied by gas chromatography (GC) and mass spectrometry (MS). Because there is a normal daily excretion of 7-methylguanine (100–150 μg/day for rats), SHUKER et al. (1984) treated rats with deuterated alkylating agents and were thus able to distinguish between "natural" and carcinogen-derived 7-methylguanine in the urine. This method has been used to test three drugs, aminopyrine, cimetidine and pyrilamine, which are potential methylating agents if nitrosated in vivo (FARMER et al. 1986). Deuterated aminopyrine gave rise to deuterated 7-methylguanine when co-administered to rats with nitrite, but the other two compounds did not, nor did any of them if administered alone. The high background levels of 7-methylguanine in urine preclude its use as a marker of human environmental carcinogen exposure, but background levels of 3-methyladenine are reportedly much lower, and this may serve as a better indicator (FARMER et al. 1987).

The combination of GC with either electron capture detection (GC-ECD) or negative ion chemical ionisation mass spectrometry (GC-NCI-MS) offers the prospect of highly sensitive methods of adduct detection. Release of a base from a DNA nucleoside under mild conditions is desirable in these circumstances, and a general method for doing this, involving oxidation with DMSO-acetic anhydride has been proposed (MINNETIAN et al. 1987). Derivatisation of methylated or ethylated bases by cinnamoylation (ADAMS and GIESE 1985) or pentafluorobenzylation (NAZARETH et al. 1984; ADAMS et al. 1986) produces an electrophore that can be detected by GC-ECD in fmol amounts. In fact, as little as 0.04 fmol O^4-ethylthymidine has been detected by this method (KOCH et al. 1987). If the electrophore-labelled adducts are permethylated they may be analysed by GC-NCI-MS, as has been demonstrated for cytosine and 5-methylcytosine (MOHAMED et al. 1984). Alternatively, the GC phase of the procedure can be replaced with a liquid phase separation (moving belt liquid chromatography) of comparable sensitivity in selected cases (ANNAN et al. 1989). Electrophore labelling of nucleosides and nucleoside adducts is also reported to produce species readily suited to structural analysis by MS (TRAINOR et al. 1988), and this approach, not requiring base release, may be useful in future for identifying "unknown" adducts.

The impressive sensitivities achieved with these techniques have generally been made using standard reference compounds. The indications are that the detection and quantitation of adducts present in low concentrations in hydrolysates of DNA are not so straightforward or, as yet, so sensitive (KOCH et al. 1987; TURNER et al. 1987). Additional chromatographic procedures are likely to be necessary to prevent adduct sample contamination with unadducted bases interfering with the analyses. Despite this, the electrophore methods show considerable promise for the sensitive detection of carcinogen damage to DNA.

MS techniques have usually required that the chemical compound be derivatised to a volatile form, but the advent of fast atom bombardment (FAB) has removed the necessity for this and allows much larger molecules to be analysed than hitherto. Furthermore, the use of tandemly arranged mass spectrometers (MS-MS), whereby a fragment generated in the first machine can be selectively fragmented and monitored in the second may provide the eventual means for identifying adducts present in mixtures. Its use for analysing tetranucleotides modified with BPDE has been demonstrated (DINO et al. 1987).

E. Future Prospects

Developments in the field of carcinogen-DNA adduct detection have been rapid in recent years. Much of the impetus has come from the realisation of the usefulness of biochemical markers for monitoring human exposure to chemical carcinogens. Ambient monitoring of, for example, the workplace can provide information on external exposure levels but has its limitations because it does not take account of inter-individual variations in absorption, metabolism and excretion, or indeed of the "bioavailability" of the carcinogenic material. An illustration of this comes from studies with aluminium plant workers who are exposed to high levels of particulate polycyclic aromatic hydrocarbons. Urine samples from these workers show no evidence of increased levels of the compounds or their metabolites compared with non-exposed workers, nor are the frequencies of sister chromatid exchanges in blood lymphocytes raised (BECHER et al. 1984). Epidemiological studies indicate only limited evidence increased incidence of cancer among aluminium plant workers despite the high ambient levels of polycyclic hydrocarbons. This demonstrates the need for effective biomonitoring procedures in order to assess more reliably carcinogen exposure and to predict risk (rather than rely on retrospective epidemiology), and carcinogen-DNA adduct measurement is one of several complementary methods now available with the potential to provide useful information for this purpose (PERERA and WEINSTEIN 1982; SHERIDAN and DE SERRES 1985; GARNER 1985; GARNER et al. 1985; FARMER et al. 1987; DE SERRES 1988; DE SERRES and MATSUSHIMA 1988; PERERA 1988; SANTELLA 1988).

Some of the methods described in this chapter are sufficiently sensitive for monitoring DNA adduct formation in humans while others have still to achieve greater sensitivity before they can have useful applications in this area (Table 3). On current evidence a level of sensitivity that allows the detection of 1 adduct in 10^8 nucleotides would seem desirable, although application of techniques with lower levels of sensitivity will still be useful. The quantities of DNA required for the determination is also an important consideration, and those assays that require only microgram amounts have the greatest number of potential applications.

The rapid growth of research activity in this area has meant that some aspects of the techniques, especially with regard to quantitation and inter-laboratory reproducibility, remain to be properly validated. However, more cooperative efforts to standardise procedures are likely to be seen in the immediate future. With the techniques that require hydrolysis of DNA before assaying, there remains the uncertainty of whether or not digestion is complete. Also, the most readily avail-

Table 3. Sensitivity of DNA adduct detection methods

Method	Limit of detection (fmol)	µg DNA required	Adduct : nucleotides
^{32}P-Postlabelling	0.01	1–10	$1:3\times10^9$
Immunoassay			
RIA	40	Up to 10000	$1:10^8$
ELISA competitive	1	50	$1:6\times10^8$
ELISA non-competitive	3	0.1	$1:10^7$
USERIA	1	25	$1:10^7$
Slot-blot	1	1	$1:3\times10^6$
Fluorescence			
Low temperature	3	1000	$1:3\times10^8$
Synchronous scanning	20	100	$1:5\times10^6$
Line narrowing	1	1000	$1:10^8$
GC-MS	0.5	–	–

Data adapted from JEFFREY (1986) and FARMER et al. (1987).

able sources of human DNA for analysis (white blood cells, placenta, exfoliated oral mucosal cells, depurinated adducts in urine) are often not the target tissue for the carcinogenic activity of the compound or compounds under investigation, and the relationship between adducts in a non-target tissue and the probability of tumour initiation in the target tissue has yet to be established.

Methods of DNA adduct detection offer the means not only to measure human exposure to known carcinogenic sources but also to search for unidentified ones. In this area imaginative combinations of methods, such as the use of immunoaffinity columns combined with ^{32}P-postlabelling or fluorescence spectroscopy, are likely to become widely used and to lead to the identification of previously unrecognised sources of carcinogens. The rapid advances in mass spectrometry suggest that this technique will play an important role in the ultimate characterisation of the carcinogenic components of the sources thus identified.

Acknowledgements. Research in my laboratory is supported by grants to the Institute of Cancer Research from the Cancer Research Campaign, the UK Medical Research Council and the US National Cancer Institute. I would like to thank Drs. M. Poirier, C. Harris and R. Fuchs for permission to reproduce figures from their papers, and R. Giese for preprints of papers in press.

References

Adamkiewicz J, Drosdziok W, Eberhardt W, Langenberg U, Rajewsky MF (1982) High-affinity monoclonal antibodies specific for DNA components structurally modified by alkylating agents. In: Bridges BA, Butterworth BE, Weinstein IB (eds) Banbury report 13: indicators of genotoxic exposure. Cold Spring Harbor Laboratory, New York, pp 265–276

Adamkiewicz J, Ahrens O, Eberle G, Nehls P, Rajewsky MF (1986) Monoclonal antibody-based immunoanalytical methods for detection of carcinogen-modified DNA components. In: Singer B, Bartsch H (eds) The role of cyclic nucleic acid adducts in carcinogenesis and mutagenesis. IARC, Lyon, pp 403–411

Adams J, Giese RW (1985) Cinnamoylation of the sugar hydroxyls of 3-methylthymidine. Formation of a relatively stable ester derivative. J Chromatogr 347:99–107

Adams J, David M, Giese RW (1986) Pentafluorobenzylation of O^4-ethylthymidine and analogues by phase-transfer catalysis for determination by gas chromatography with electron capture detection. Anal Chem 58:345–348

Annan RS, Kresbach GM, Giese RW, Vouros R (1989) Trace analysis of derivatized 5-methylcytosine and 5-hydroxymethyluracil from DNA by moving belt liquid chromatography-negative chemical ionization mass spectrometry (in press)

Arce GT, Cline DT, Mead JE (1987) The ^{32}P-post-labeling method in quantitative DNA adduct dosimetry of 2-acetylaminofluorene-induced mutagenicity in Chinese hamster ovary cells and *Salmonella typhimurium* TA1538. Carcinogenesis 8:515–520

Armier J, Mezzina M, Leng M, Fuchs RPP, Sarasin A (1988) *N*-Acetoxy-*N*-2-acetyl-aminofluorene-induced damage on SV40 DNA: inhibition of DNA replication and visualization of DNA lesions. Carcinogenesis 9:789–795

Asan E, Fasshauer I, Wild D, Henschler D (1987) Heterocyclic aromatic amine-DNA-adducts in bacteria and mammalian cells detected by ^{32}P-postlabeling analysis. Carcinogenesis 8:1589–1593

Autrup H, Bradley KA, Shamsuddin AKM, Wakhisi J, Wasunna A (1983) Detection of putative adduct with fluorescence characteristics identical to 2,3-dihydro-2-(7′-guanyl)-3 hydroxyaflatoxin B_1 in human urine collected in Murang'a district, Kenya. Carcinogenesis 4:1193–1195

Autrup H, Seremet T, Wakhisi J, Wasunna A (1987) Aflatoxin exposure measured by urinary excretion of aflatoxin B_1-guanine adduct and hepatitis B virus infection in areas with different liver cancer incidence in Kenya. Cancer Res 47:3430–3433

Baird WM (1979) The use of radioactive carcinogens to detect DNA modifications. In: Grover PL (ed) Chemical carcinogens and DNA. CRC Press, Boca Raton, pp 59–83

Barrows LR, Shank RC (1981) Aberrant methylation of liver DNA in rats during hepatocarcinogenicity. Toxicol Appl Pharmacol 60:334–345

Bartoszek A, Konopa J (1987) ^{32}P-Post-labelling analysis of DNA adducts formed by antitumor 1-nitro-9-aminoacridines with DNA of HeLa S_3 cells. Biochem Pharmacol 36:4169–4171

Becher G, Haugen A, Bjorseth A (1984) Multimethod determination of occupational exposure to polycyclic aromatic hydrocarbons in an aluminum plant. Carcinogenesis 5:647–651

Bedell MA, Lewis JG, Billings KC, Swenberg JA (1982) Cell specificity in hepatocarcinogenesis: preferential accumulation of O^6-methylguanine in target cell DNA during continuous exposure of rats to 1,2-dimethylhydrazine. Cancer Res 42:3079–3083

Bedell MA, Dyroff MC, Doerjer G, Swenberg JA (1986) Quantitation of etheno adducts by fluorescence detection. In: Singer B, Bartsch H (eds) The role of cyclic nucleic acid adducts in carcinogenesis and mutagenesis. IARC, Lyon, pp 425–436

Bedford P, Fichtinger-Schepman AMJ, Shellard SA, Walker MC, Masters JRW, Hill BT (1988) Differential repair of platinum-DNA adducts in human bladder and testicular tumor continuous cell lines. Cancer Res 48:3019–3024

Beland FA, Ribovich M, Howard PC, Heflich RH, Kurian P, Milo GE (1986) Cytotoxicity, cellular transformation and DNA adducts in normal diploid fibroblasts exposed to 1-nitrosopyrene, a reduced derivative of the environmental contaminant, 1-nitropyrene. Carcinogenesis 7:1279–1283

Belinsky SA, White CM, Boucheron JA, Richardson FC, Swenberg JA, Anderson M (1986) Accumulation and persistence of DNA adducts in respiratory tissue of rats following multiple administrations of the tobacco specific carcinogen 4-(*N*-methyl-*N*-nitrosamino)-1-(3-pyridyl)-1-butanone. Cancer Res 46:1280–1284

Bennett RA, Essigmann JM, Wogan GN (1981) Excretion of an aflatoxin-guanine adduct in the urine of aflatoxin B_1-treated rats. Cancer Res 41:650–654

Bodell WJ, Rasmussen J (1984) A ^{32}P-postlabeling assay for determining the incorporation of bromodeoxyuridine into cellular DNA. Anal Biochem 142:525–528

Briscoe WT, Spizizen J, Tan EM (1978) Immunological detection of O^6-methylguanine in alkylated DNA. Biochemistry 17:1896–1901

Brookes P, Lawley PD (1964) Evidence for the binding of polynuclear aromatic hydrocarbons to the nucleic acids of mouse skin: relation between carcinogenic power of hydrocarbons and their binding to DNA. Nature 202:781–784

Butler VP, Beiser SM, Erlanger BF, Tanenbaum SW, Cohen S, Bendich A (1962) Purine-specific antibodies which react with deoxyribonucleic acid (DNA). Proc Natl Acad Sci USA 48:1597–1602

Chanutin A, Gjessing EC (1946) Effect of nitrogen mustards on the ultraviolet absorption spectrum of thymonucleate, uracil and purines. Cancer Res 6:599–601

Daudel P, Duquesne M, Vigny P, Grover PL, Sims P (1975) Fluorescence spectral evidence that benzo[*a*]pyrene-DNA products in mouse skin arise from diol-epoxides. FEBS Lett 57:250–253

de Murcia G, Lang ME, Freund A, Fuchs RPP, Daune MP, Sage E, Leng M (1979) Electron microscopic visualization of *N*-acetoxy-*N*-2-acetylaminofluorene binding sites in ColE1 DNA by means of specific antibodies. Proc Natl Acad Sci USA 76:6076–6080

de Serres FJ (1988) Meeting report: Banbury Center DNA adducts workshop. Mutation Res 203:55–68

de Serres FJ, Matsushima T (1988) Meeting report: approaches for human population monitoring. Mutation Res 203:185–199

Dietrich MW, Hopkins WE, Asbury KJ, Ridley WP (1987) Liquid chromatographic characterization of the deoxyribonucleoside-5′-phosphates and deoxyribonucleoside-3′-5′-biphosphates obtained by ^{32}P-postlabeling of DNA. Chromatographia 24:545–551

Dino JJ, Guenat CR, Tomer KB, Kaufman DG (1987) Analyses of carcinogen-modified oligonucleotides by fast atom bombardment/tandem mass spectrometry. Rapid Commun Mass Spec 1:69–71

Doll R, Peto R (1981) The causes of cancer. Oxford University Press, London

Dunn BP, San RHC (1988) HPLC enrichment of hydrophobic DNA adducts for enhanced sensitivity of ^{32}P-postlabeling analysis. Carcinogenesis 9:1055–1060

Dunn BP, Stich HF (1986) ^{32}P-Postlabelling analysis of aromatic DNA adducts in human oral mucosal cells. Carcinogenesis 7:1115–1120

Dunn BP, Black JJ, Maccubbin A (1987) ^{32}P-Postlabeling analysis of aromatic DNA adducts in fish from polluted areas. Cancer Res 47:6543–6548

Duquesne M, Vigny P, Gabillat N (1970) Detection de faibles luminescences en spectrophotometrie: comparaison entre deux techniques de mesure par photomultiplicateur "en courant" et "en impulsions". Photochem Photobiol 21:519–529

Eastman A (1983) Characterization of the adducts produced in DNA by *cis*-diamminedichloroplatinum(II) and *cis*-dichloro(ethylenediamine)-platinum(II). Biochemistry 22:3927–3933

Erlanger BF, Beiser SM (1964) Antibodies specific for ribonucleosides and ribonucleotides and their reaction with DNA. Proc Natl Acad Sci USA 52:68–74

Everson RB, Randerath E, Santella RM, Cefalo RC, Avitts TA, Randerath K (1986) Detection of smoking-related covalent DNA adducts in human placenta. Science 231:54–57

Everson RB, Randerath E, Avitts TA, Schut HAJ, Randerath K (1987) Preliminary investigations of tissue specificity, species specificity, and strategies for identifying chemicals causing DNA adducts in human placenta. Prog Exp Tumor Res 31:86–103

Everson RB, Randerath E, Santella RM, Avitts TA, Weinstein IB, Randerath K (1988) Quantitative associations between DNA damage in human placenta and maternal smoking and birth weight. JNCI 80:567–576

Farmer PB, Shuker DEG, Bird I (1986) DNA and protein adducts as indicators of in vivo methylation by nitrosatable drugs. Carcinogenesis 7:49–52

Farmer PB, Neumann H-G, Henschler D (1987) Estimation of exposure of man to substances reacting covalently with macromolecules. Arch Toxicol 60:251–260

Farr RS (1958) A quantitative immunological measure of the primary interaction between I BSA and antibody. J Infect Dis 103:239–262

Fennell TR, Juhl U, Miller EC, Miller JA (1986) Identification and quantitation of hepatic DNA adducts formed in B6C3F$_1$ mice from 1′-hydroxy-2′,3′-dehydroestragole: comparison of the adducts detected with the 1′-^{3}H-labelled carcinogen and by ^{32}P-postlabelling. Carcinogenesis 7:1881–1887

Fichtinger-Schepman AMJ, Baan RA, Luiten-Schuite A, Van Dijk M, Lohman PHM (1985a) Immunochemical quantitation of adducts induced in DNA by *cis*-diamminedichloroplatinum(II) and analysis of adduct-related DNA-unwinding. Chem Biol Interact 55:275–288

Fichtinger-Schepman AMJ, Lohman PHM, Reedijk J (1985b) Detection and quantification of adducts formed upon interaction of diamminedichloroplatinum(II) with DNA, by anion-exchange chromatography after enzymatic degradation. Nucleic Acids Res 10:5345–5356

Fichtinger-Schepman AMJ, Van Oosterom AT, Lohman PHM, Berends F (1987) *cis*-Diamminedichloroplatinum(II)-induced DNA adducts in peripheral leukocytes from seven cancer patients: quantitative immunochemical detection of the adduct induction and removal after a single dose of *cis*-diamminedichloroplatinum(II). Cancer Res 47:3000–3004

Foiles PG, Trushin N, Castonguay A (1985) Measurement of O^6-methyldeoxyguanosine in DNA methylated by the tobacco-specific carcinogen 4-(methylnitrosamine)-1-(3-pyridyl)-1-butanone using a biotin-avidin enzyme-linked immunosorbent assay. Carcinogenesis 6:989–993

Foiles PG, Chung F-L, Hecht SS (1987) Development of a monoclonal antibody-based immunoassay for cyclic DNA adducts resulting from exposure to crotonaldehyde. Cancer Res 47:360–363

Garner RC (1985) Assessment of carcinogen exposure in man. Carcinogenesis 6: 1071–1078
Garner RC, Ryder R, Montesano R (1985) Monitoring of aflatoxins in human body fluids and application to field studies. Cancer Res 45:922–928
Groopman JD, Haugen A, Goodrich GR, Wogan GN, Harris CC (1982) Quantitation of aflatoxin B_1-modified DNA using monoclonal antibodies. Cancer Res 42:3120–3124
Groopman JD, Trudel LJ, Donahue PR, Marshak-Rothstein A, Wogan GN (1984) High-affinity monoclonal antibodies for aflatoxins and their application to solid-phase immunoassays. Proc Natl Acad Sci USA 81:7728–7731
Groopman JD, Donahue PR, Zhu J, Chen J, Wogan GN (1985) Aflatoxin metabolism in humans: detection of metabolites and nucleic acid adducts in urine by affinity chromatography. Proc Natl Acad Sci USA 82:6492–6496
Gupta RC (1984) Nonrandom binding of the carcinogen *N*-hydroxy-2-acetylaminofluorene to repetitive sequences of rat liver DNA in vivo. Proc Natl Acad Sci USA 81: 6943–6947
Gupta RC (1985) Enhanced sensitivity of ^{32}P-postlabeling analysis of aromatic carcinogen: DNA adducts. Cancer Res 45:5656–5662
Gupta RC, Dighe NR (1984) Formation and removal of DNA adducts in rat liver treated with *N*-hydroxy derivatives of 2-acetylaminofluorene, 4-acetylaminobiphenyl, and 2-acetylaminophenanthrene. Carcinogenesis 5:343–349
Gupta RC, Reddy MV, Randerath K (1982) ^{32}P-postlabeling analysis of non-radioactive aromatic carcinogen-DNA adducts. Carcinogenesis 3:1081–1092
Gupta RC, Dighe NR, Randerath K, Smith HC (1985a) Distribution of initial and persistent 2-acetylaminofluorene-induced DNA adducts within DNA loops. Proc Natl Acad Sci USA 82:6605–6608
Gupta RC, Goel SK, Early K, Singh B, Reddy JK (1985b) ^{32}P-Postlabeling analysis of peroxisome proliferator-DNA adduct formation in rat liver in vivo and hepatocytes in vitro. Carcinogenesis 6:933–936
Gupta RC, Earley K, Locker J, Lombardi B (1987) ^{32}P-Postlabeling analysis of liver DNA adducts in rats chronically fed a choline-devoid diet. Carcinogenesis 8: 187–189
Harris CC, Yolken RH, Hsu I-C (1982) Enzyme immunoassays: applications in cancer research. Methods Cancer Res 20:213–243
Harris CC, Vahakangas K, Newman MJ, Trivers GE, Shamsuddin A, Sinopoli N, Mann DL, Wright WE (1985) Detection of benzo[*a*]pyrene diolepoxide-DNA adducts in peripheral blood lymphocytes and antibodies to the adducts in serum from coke oven workers. Proc Natl Acad Sci USA 82:6672–6676
Harris CC, LaVeck G, Groopman J, Wilson VL, Mann D (1986) Measurement of aflatoxin B_1, its metabolites, and DNA adducts by synchronous fluorescence spectrophotometry. Cancer Res 46:3249–3253
Haseltine WA, Franklin W, Lippke JA (1983) New methods for detection of low levels of DNA damage in human populations. Environ Health Perspect 48:29–41
Haugen A, Groopman JD, Hsu I-C, Goodrich GR, Wogan GN, Harris CC (1981) Monoclonal antibody to aflatoxin B_1-modified DNA detected by enzyme immunoassay. Proc Natl Acad Sci USA 78:4124–4127
Haugen A, Becher G, Benestad C, Vahakangas K, Trivers GE, Newman MJ, Harris CC (1986) Determination of polycyclic aromatic hydrocarbons in the urine, benzo[*a*]pyrene diol epoxide-DNA adducts in lymphocyte DNA, and antibodies to the adducts in sera from coke oven workers exposed to measured amounts of polycyclic aromatic hydrocarbons in the work atmosphere. Cancer Res 46:4178–4183
Hebert E, Saint-Ruf G, Leng M (1985) Immunological titration of 3-*N*-acetylhydroxyamino-4,6-dimethyldipyrido(1,2-*a*:3′,2′-*d*)imidazole-rat liver DNA adducts. Carcinogenesis 6:937–939
Heidelberger C, Davenport GR (1961) Local functional components of carcinogenesis. Acta Unio Internat Contra Cancrum 17:55–63
Heisig V, Jeffrey AM, McGlade MJ, Small GJ (1984) Fluorescence-line-narrowed spectra of polycyclic aromatic carcinogen-DNA adducts. Science 223:289–291

Heisig V, Harvey RG, Jeffrey AM (1986) Covalent binding of ethylated analogs of 7,12-dimethylbenz[*a*]anthracene to the DNA of mouse embryo fibroblast 10T1/2 cells. Cancer Lett 33:19–24

Hertzog PJ, Lindsay Smith JR, Garner RC (1982) Production of monoclonal antibodies to guanine imidazole ring-opened aflatoxin B_1DNA, the persistent DNA adduct in vivo. Carcinogenesis 3:825–828

Hertzog PJ, Shaw A, Lindsay Smith JR, Garner RC (1983) Improved conditions for the production of monoclonal antibodies to carcinogen-modified DNA, for use in enzyme-linked immunosorbent assays (ELISA). J Immunol Methods 62:49–58

Heyting C, Van der Laken CJ, Van Raamsdonk W, Pool CW (1983) Immunohistochemical detection of O^6-ethyldeoxyguanosine in the rat brain after in vivo applications of *N*-ethyl-*N*-nitrosourea. Cancer Res 43:2935–2941

Hollstein M, Nair J, Bartsch H, Bochner B, Ames BN (1986) Detection of DNA base damage by ^{32}P-postlabelling: tlc separation of 5′-deoxynucleoside monophosphates. In: Singer B, Bartsch H (eds) The role of cyclic nucleic acid adducts in carcinogenesis and mutagenesis. IARC, Lyon, pp 437–448

Hsieh LL, Jeffrey AM, Santella RM (1985) Monoclonal antibodies to 1-aminopyrene-DNA. Carcinogenesis 6:1289–1293

Hsu I-C, Poirier MC, Yuspa SH, Yolken RH, Harris CC (1980) Ultrasensitive enzymatic radioimmunoassay (USERIA) detects femtomoles of acetylaminofluorene-DNA adducts. Carcinogenesis 1:455–458

Hsu I-C, Poirier MC, Yuspa SH, Grunberger D, Weinstein IB, Yolken RH, Harris CC (1981) Measurement of benzo[*a*]pyrene-DNA adducts by enzyme immunoassays and radioimmunoassay. Cancer Res 41:1091–1095

Huh N, Rajewsky MF (1986) Enzymatic elimination of O^6-ethylguanine and stability of O^4-ethylthymine in the DNA of malignant neural cell lines exposed to *N*-ethyl-*N*-nitrosourea in culture. Carcinogenesis 7:435–439

Huh N, Rajewski MF (1988) Enzymatic elimination of O^6-ethylguanine from the DNA of ethylnitrosourea-exposed normal and malignant rat brain cells grown under cell culture versus in vivo conditions. Int J Cancer 41:762–766

Huitfeldt HS, Spangler EF, Hunt JM, Poirier MC (1986) Immunohistochemical localization of DNA adducts in rat liver tissue and phenotypically altered foci during oral administration of 2-acetylaminofluorene. Carcinogenesis 7:123–129

Huitfeldt HS, Spangler EF, Baron J, Poirier MC (1987) Microfluorometric determination of DNA adducts in immunofluorescent-stained liver tissue from rats fed 2-acetylaminofluorene. Cancer Res 47:2098–2102

Ivanovic V, Geacintov NE, Weinstein IB (1976) Cellular binding of benzo[*a*]pyrene to DNA characterised by low temperature fluorescence. Biochem Biophys Res Comm 70:1172–1179

Ivanovic V, Geacintov NE, Jeffrey AM, Fu PP, Harvey RG, Weinstein IB (1978) Cell and microsome mediated binding of 7,12-dimethylbenz[*a*]anthracene to DNA studied by fluorescence spectroscopy. Cancer Lett 4:131–140

Jeffrey AM (1986) Alternative methods for/to the analysis of carcinogen-DNA adducts. In: Friedberg T, Oesch F (eds) Primary changes and control factors in carcinogenesis. Deutscher Fachschriften-Verlag, Wiesbaden, pp 155–160

Kaneko M, Leadon SA (1986) Production of thymine glycols in DNA by *N*-hydroxy-2-naphthylamine as detected by a monoclonal antibody. Cancer Res 46:71–75

Kelman DJ, Lilga KT, Sharma M (1988) Synthesis and application of fluorescent labeled nucleotides to assay DNA damage. Chem Biol Interact 66:85–100

Kennaway EL (1955) The identification of a carcinogenic compound in coal-tar. Br Med J 2:749–752

Koch SAM, Turner MJ, Swenberg JA (1987) Quantitation of O^4-ethyldeoxythymidine using electrophore postlabeling. Proc Am Assoc Cancer Res 28:93

Kriek E, Den Engelse L, Scherer E, Westra JG (1984) Formation of DNA modifications by chemical carcinogens: identification, localization and quantification. Biochim Biophys Acta 738:181–201

Kyrtopoulos SA, Swann PF (1980) The use of radioimmunoassay to study the formation and disappearance of O^6-methylguanine in mouse liver satellite and main-band DNA following dimethylnitrosamine administration. J Cancer Res Clin Oncol 98:127–138

Lai C-C, Miller EC, Miller JA, Liem A (1987) Initiation of hepatocarcinogenesis in infant male B6C3F$_1$ mice by *N*-hydroxy-2-aminofluorene or *N*-hydroxy-2-acetylaminofluorene depends primarily on metabolism to *N*-sulfoxy-2-aminofluorene and formation of DNA-(deoxyguanosin-8-yl)-2-aminofluorene adducts. Carcinogenesis 8: 471–478

Landegent JE, Jansen in de Wal N, Baan RA, Hoeijmakers JHJ, Van der Ploeg M (1984) 2-Acetylaminofluorene-modified probes for the indirect hybridocytochemical detection of specific nucleic acid sequences. Exp Cell Res 153:61–72

Landegent JE, Jansen in de Wal N, Van Ommen G-JB, Baas F, de Vijlder JJM, Van Duijn P, Van der Ploeg M (1985) Chromosomal localization of a unique gene by non-autoradiographic in situ hybridization. Nature 317:175–177

Lang MC, de Murcia G, Mazen A, Fuchs RPP, Leng M, Daune M (1982) Nonrandom binding of *N*-acetoxy-*N*-2-acetylaminofluorene to chromatin subunits as visualized by immunoelectron microscopy. Chem Biol Interact 41:83–93

Lawley PD (1989) Mutagens as carcinogens: development of current concepts. Mutation Res 213:3–25

Leadon SA, Hanawalt PC (1983) Monoclonal antibody to DNA containing thymine glycol. Mutation Res 112:191–200

Leng M, Sage E, Fuchs RPP, Daune MP (1978) Antibodies to DNA modified by the carcinogen *N*-acetoxy-*N*-2-acetylaminofluorene. FEBS Lett 92:207–210

Levine L, Seaman E, Hammerschlag E, Van Vunakis H (1966) Antibodies to photoproducts of deoxyribonucleic acids irradiated with ultraviolet light. Science 153:1666–1667

Liehr JG, Randerath K, Randerath E (1985) Target organ-specific covalent DNA damage preceding diethylstilbestrol-induced carcinogenesis. Carcinogenesis 6:1067–1069

Liehr JG, Avitts TA, Randerath E, Randerath K (1986) Estrogen-induced endogenous DNA adduction: possible mechanism of hormonal cancer. Proc Natl Acad Sci USA 83: 5301–5305

Liehr JG, Hall ER, Avitts TA, Randerath E, Randerath K (1987) Localization of estrogen-induced DNA adducts and cytochrome P-450 activity at the site of renal carcinogenesis in the hamster kidney. Cancer Res 47:2156–2159

Lindamood C, Bedell MA, Billings KC, Swenberg JA (1982) Alkylation and de novo synthesis of liver cell DNA from C3H mice during continuous dimethylnitrosamine exposure. Cancer Res 42:4153–4157

Liu S-F, Rappaport SM, Rasmussen J, Bodell WJ (1988) Detection of styrene-oxide-DNA adducts by ^{32}P-postlabeling. Carcinogenesis 9:1401–1404

Lu L-JW, Disher RM, Reddy MV, Randerath K (1986) ^{32}P-Postlabeling assay in mice of transplacental DNA damage induced by the environmental carcinogens safrole, 4-aminobiphenyl, and benzo(*a*)pyrene. Cancer Res 46:3046–3054

Lu S-H, Ohshima H, Bartsch H (1985) Recent studies on *N*-nitroso compounds as possible etiological factors in oesophageal cancer. In: O'Neill IK, Von Borstel RC, Miller CT, Long J, Bartsch H (eds) *N*-Nitroso compounds: occurrence, biological effects and relevance to human cancer. IARC, Lyon, pp 947–953

Ludeke BI, Kleihues P (1988) Formation and persistence of O^6-(2-hydroxyethyl)-2′-deoxyguanosine in DNA of various rat tissues following a single dose of *N*-nitroso-*N*-(2-hydroethyl)urea. An immunoslot-blot study. Carcinogenesis 9:147–151

Lutz WK, Jaggi W, Schlatter C (1982) Covalent binding of diethylstilbestrol to DNA in rat and hamster liver and kidney. Chem Biol Interact 42:251–257

Malfoy B, Hartmann B, Macquet JP, Leng M (1981) Immunochemical studies of DNA modified by *cis*-diamminedichloroplatinum(II) in vivo and in vitro. Cancer Res 41:4127–4131

Marshall CJ, Vousden KH, Phillips DH (1984) Activation of c-Ha-*ras*-1 proto-oncogene by in vitro modification with a chemical carcinogen, benzo[*a*]pyrene diol-epoxide. Nature 310:586–589

McKay S, Phillips DH, Hewer AJ, Grover PL (1988) Metabolic activation of 7-ethyl- and 7-methylbenz[*a*]anthracene in mouse skin. Carcinogenesis 9:141–145

Menkveld GJ, Van der Laken CJ, Hermsen T, Kriek E, Scherer E, Den Engelse L (1985) Immunohistochemical localization of O^6-ethyldeoxyguanosine and deoxyguanosin-8-yl-(acetyl)aminofluorene in liver sections of rats treated with diethylnitrosamine, ethylnitrosourea or *N*-acetylaminofluorene. Carcinogenesis 6:263–270

Miller EC (1951) Studies on the formation of protein-bound derivatives of 3,4-benzpyrene in the epidermal fraction of mouse skin. Cancer Res 11:100–108

Miller EC, Miller JA (1947) The presence and significance of bound aminoazo dyes in the livers of rats fed *p*-dimethylaminoazobenzene. Cancer Res 7:468–480

Miller JA, Miller EC (1969) Metabolic activation of carcinogenic aromatic amines and amides via *N*-hydroxylation and *N*-hydroxyesterification and its relationship to ultimate carcinogens as electrophilic reactants. In: Bergmann ED, Pullman B (eds) The Jerusalem symposia on quantum chemistry and biochemistry, physicochemical mechanisms of carcinogenesis, vol 1. The Israel Academy of Sciences and Humanities, Jerusalem, pp 237–261

Minnetian O, Saha M, Giese RW (1987) Oxidation-elimination of a DNA base from its nucleoside to facilitate determination of alkyl chemical damage to DNA by gas chromatography with electrophore detection. J Chromatogr 410:453–457

Mohamed GB, Nazareth A, Hayes MJ, Giese RW, Vouros P (1984) Gas chromatography-mass spectrometry characteristics of methylated perfluoroacyl derivatives of cytosine and 5-methylcytosine. J Chromatogr 314:211–217

Moschel RC, Baird WM, Dipple A (1977) Metabolic activation of the carcinogen 7,12-dimethylbenz[*a*]anthracene for DNA binding. Biochem Biophys Res Commun 76: 1092–1098

Mukhtar H, Asokan P, Das M, Santella RM, Bickers DR (1986) Benzo[*a*]pyrene diol epoxide-I-DNA adduct formation in the epidermis and lung of Sencar mice following topical application of crude coal tar. Cancer Lett 33:287–294

Müller R, Rajewsky MF (1980) Immunological quantification by high-affinity antibodies of O^6-ethyldeoxyguanosine in DNA exposed to *N*-ethyl-*N*-nitrosourea. Cancer Res 40:887–896

Müller R, Rajewsky MF (1981) Antibodies specific for DNA components structurally modified by chemical carcinogens. J Cancer Res Clin Oncol 102:99–113

Nakayama J, Yuspa SH, Poirier MC (1984) Benzo[*a*]pyrene-DNA adduct formation and removal in mouse epidermis in vivo and in vitro: relationship of DNA binding to initiation of skin carcinogenesis. Cancer Res 44:4087–4095

Nazareth A, Joppich M, Abdel-Baky S, O'Connell K, Sentissi A, Giese RW (1984) Electrophore-labeling and alkylation of standards of nucleic acid pyrimidine bases for analysis by gas chromatography with electron-capture detection. J Chromatogr 314:201–210

Nehls P, Adamkiewicz J, Rajewsky MF (1984a) Immuno-slot-blot: a highly sensitive immunoassay for the quantitation of carcinogen-modified nucleosides in DNA. J Cancer Res Clin Oncol 108:23–29

Nehls P, Rajewsky MF, Spiess E, Werner D (1984b) Highly sensitive sites for guanine-O^6 ethylation in rat brain DNA exposed to *N*-ethyl-*N*-nitrosourea in vivo. EMBO J 3:327–332

Oesch F, Adler S, Rettelbach R, Doerjer G (1986) Repair of etheno DNA adducts by *N*-glycosylases. In: Singer B, Bartsch H (eds) The role of cyclic nucleic acid adducts in carcinogenesis and mutagenesis. IARC, Lyon, pp 373–379

Parks WC, Schurdak ME, Randerath K, Maher VM, McCormick JJ (1986) Human cell-mediated cytotoxicity, mutagenicity, and DNA adduct formation of 7*H*-dibenzo(*c,g*)-carbazole and its *N*-methyl derivative in diploid human fibroblasts. Cancer Res 46:4706–4711

Paules RS, Poirier MC, Mass MJ, Yuspa SH, Kaufman DG (1985) Quantitation by electron microscopy of the binding of highly specific antibodies to benzo[*a*]pyrene-DNA adducts. Carcinogenesis 6:193–198

Paules RS, Cordiero-Stone M, Mass MJ, Poirier MC, Yuspa SH, Kaufman DG (1988) Benzo[*a*]pyrene diol epoxide I binds to DNA at replication forks. Proc Natl Acad Sci USA 85:2176–2180

Perera FP (1988) The significance of DNA and protein adducts in human biomonitoring studies. Mutation Res 205:255–269

Perera FP, Weinstein IB (1982) Molecular epidemiology and carcinogen-DNA adduct detection: new approaches to studies of human cancer causation. J Chron Dis 35:581–600

Perera FP, Poirier MC, Yuspa SH, Nakayama J, Jaretzki A, Curnen MM, Knowles DM, Weinstein IB (1982) A pilot project in molecular epidemiology: determination of benzo[*a*]pyrene-DNA adducts in animals and human tissues by immunoassays. Carcinogenesis 3:1405–1410

Perera FP, Santella RM, Brenner D, Poirier MC, Munshi AA, Fischman HK, Van Ryzin J (1987) DNA adducts, protein adducts, and sister chromatid exchange in cigarette smokers and nonsmokers. JNCI 79:449–456

Perera FP, Hemminki K, Young TL, Brenner D, Kelly G, Santella RM (1988) Detection of polycyclic aromatic hydrocarbon-DNA adducts in white blood cells of foundry workers. Cancer Res 48:2288–2291

Phillips DH, Grover PL, Sims P (1979) A quantitative determination of the covalent binding of a series of polycyclic hydrocarbons to DNA in mouse skin. Int J Cancer 23: 201–208

Phillips DH, Reddy MV, Randerath LK (1984) ^{32}P-Post-labelling analysis of DNA adducts formed in the livers of animals treated with safrole, estragole and other naturally-occurring alkenylbenzenes. II. Newborn male $B6C3F_1$ mice. Carcinogenesis 5: 1623–1628

Phillips DH, Hewer A, Grover PL (1985) Aberrant activation of benzo[*a*]pyrene in cultured rat mammary cells in vitro and following direct application to rat mammary glands in vivo. Cancer Res 45:4167–4174

Phillips DH, Glatt HR, Seidel A, Bochnitschek W, Oesch F, Grover PL (1986a) Mutagenic potential of DNA adducts formed by diol-epoxides, triol-epoxides and the K-region epoxide of chrysene in mammalian cells. Carcinogenesis 7:1739–1743

Phillips DH, Hewer A, Grover PL (1986b) Aromatic DNA adducts in human bone marrow and peripheral blood leukocytes. Carcinogenesis 7:2071–2075

Phillips DH, Hewer A, Grover PL (1987) Formation of DNA adducts in mouse skin treated with metabolites of chrysene. Cancer Lett 35:207–214

Phillips DH, Hemminki K, Alhonen A, Hewer A, Grover PL (1988) Monitoring occupational exposure to carcinogens: detection by ^{32}P-postlabelling of aromatic DNA adducts in white blood cells from iron foundry workers. Mutation Res 204:531–541

Plooy ACM, Fichtinger-Schepman AMJ, Schutte HH, van Dijk M, Lohman PHM (1985) The quantitative detection of various Pt-DNA adducts in Chinese hamster ovary cells treated with cisplatin: application of immunochemical techniques. Carcinogenesis 6: 561–566

Poirier MC (1981) Antibodies to carcinogen-DNA adducts. JNCI 67:515–519

Poirier MC (1984) The use of carcinogen-DNA adduct antisera for quantitation and localization of genomic damage in animal models and the human population. Environ Mutagen 6:879–887

Poirier MC, Yuspa SH, Weinstein IB, Blobstein S (1977) Detection of carcinogen-DNA adducts by radioimmunoassay. Nature 270:186–188

Poirier MC, Dubin MA, Yuspa SH (1979) Formation and removal of specific acetylaminofluorene-DNA adducts in mouse and human cells measured by radioimmunoassay. Cancer Res 39:1377–1381

Poirier MC, Santella R, Weinstein IB, Grunberger D, Yuspa SH (1980) Quantitation of benzo[*a*]pyrene-deoxyguanosine adducts by radioimmunoassay. Cancer Res 40:412–416

Poirier MC, Lippard SJ, Zwelling LA, Ushay HM, Kerrigan D, Thill CC, Santella RM, Grunberger D, Yuspa SH (1982a) Antibodies elicited against *cis*-diamminedichloroplatinum(II)-modified DNA are specific for *cis*-diamminedichloroplatinum(II)-DNA adducts formed in vivo and in vitro. Proc Nat Acad Sci USA 79:6443–6447

Poirier MC, Stanley JR, Beckwith JB, Weinstein IB, Yuspa SH (1982b) Indirect immunofluorescent localization of benzo[*a*]pyrene adducted to nucleic acids in cultured mouse keratinocyte nuclei. Carcinogenesis 3:345–348

Poirier MC, True BA, Laishes BA (1982c) Formation and removal of (guan-8-yl)-DNA-2-acetylaminofluorene adducts in liver and kidney of male rats given dietary 2-acetylaminofluorene. Cancer Res 42:1317–1321

Poirier MC, Hunt JM, True BA, Laishes BA, Young JF, Beland FA (1984) DNA adduct formation, removal and persistence in rat liver during one month of feeding 2-acetylaminofluorene. Carcinogenesis 5:1591–1596

Poirier MC, Reed E, Ozols RF, Fasy T, Yuspa SH (1987) DNA adducts of cisplatin in nucleated peripheral blood cells and tissues of cancer patients. Prog Exp Tumor Res 31:104–113

Poirier MC, Fullerton NF, Beland FA (1988) DNA adduct formation and removal during chronic dietary administration of 2-acetylaminofluorene. In: King CM, Romano LJ, Schuetzle D (eds) Carcinogenic and mutagenic responses to aromatic amines and nitroarenes. Elsevier, New York, pp 321–328

Rahn RO, Chang SS, Holland JM, Stephens TJ, Smith LH (1980) Binding of benzo[*a*]pyrene to epidermal DNA and RNA as detected by synchronous luminescence spectrometry at 77 K. J Biochem Biophys Methods 3:285–291

Rahn RO, Chang SS, Holland JM, Shugart LR (1982) A fluorometric-HPLC assay for quantitating the binding of benzo[*a*]pyrene metabolites to DNA. Biochem Biophys Res Commun 109:262–268

Rajewsky MF, Müller R, Adamkiewicz J, Drosdziok W (1980) Immunological detection and quantification of DNA components structurally modified by alkylating carcinogens (ethylnitrosourea). In: Pullman B, Ts'o POP, Gelboin H (eds) Carcinogenesis: fundamental mechanisms and environmental effects. Reidel, Dordrecht, pp 207–218

Randerath E, Agrawal HP, Reddy MV, Randerath K (1983) Highly persistent polycyclic aromatic hydrocarbon-DNA adducts in mouse skin: detection by ^{32}P-postlabeling analysis. Cancer Lett 20:109–114

Randerath E, Agrawal HP, Weaver JA, Bordelon CB, Randerath K (1985) ^{32}P-Postlabeling analysis of DNA adducts persisting for up to 42 weeks in the skin, epidermis and dermis of mice treated topically with 7,12-dimethylbenz[*a*]anthracene. Carcinogenesis 6:1117–1126

Randerath E, Avitts TA, Reddy MV, Miller RH, Everson RB, Randerath K (1986) Comparative ^{32}P-analysis of cigarette smoke-induced DNA damage in human tissues and mouse skin. Cancer Res 46:5869–5877

Randerath E, Mittal D, Randerath K (1988) Tissue distribution of covalent DNA damage in mice treated dermally with cigarette "tar": preference for lung and heart DNA. Carcinogenesis 9:75–80

Randerath K, Reddy MV, Gupta RC (1981) ^{32}P-Labeling test for DNA damage. Proc Natl Acad Sci USA 78:6126–6129

Randerath K, Haglund RE, Phillips DH, Reddy MV (1984a) ^{32}P-Postlabelling analysis of DNA adducts formed in the livers of animals treated with safrole, estragole and other naturally-occurring alkenylbenzenes. I. Adult female CD-1 mice. Carcinogenesis 5: 1613–1622

Randerath K, Randerath E, Agrawal HP, Reddy MV (1984b) Biochemical (postlabelling) methods for analysis of carcinogen-DNA adducts. In: Berlin A, Hemminki K, Vainio H (eds) Monitoring human exposure to carcinogenic and mutagenic agents. IARC, Lyon, pp 217–231

Randerath K, Agrawal HP, Randerath E (1985a) 12-*O*-Tetradecanoylphorbol-13-acetate-induced rapid loss of persistent 7,12-dimethylbenz(*a*)anthracene-DNA adducts in mouse epidermis and dermis. Cancer Lett 27:35–43

Randerath K, Randerath E, Agrawal HP, Gupta RC, Schurdak ME, Reddy MV (1985b) Postlabeling methods for carcinogen-DNA adduct analysis. Environ Health Perspect 62:57–65

Randerath K, Reddy MV, Disher RM (1986) Age- and tissue-related DNA modifications in untreated rats: detection by ^{32}P-postlabeling assay and possible significance for spontaneous tumor induction and aging. Carcinogenesis 7:1615–1617

Reddy JK, Lalwani ND (1984) Carcinogenesis by hepatic peroxisome proliferators: evaluation of the risk of hypolipidemic drugs and industrial plasticizers to humans. CRC Crit Rev Toxicol 12:1–58

Reddy MV, Randerath K (1986) Nuclease P1-mediated enhancement of sensitivity of ^{32}P-postlabeling test for structurally diverse DNA adducts. Carcinogenesis 7:1543–1551
Reddy MV, Gupta RC, Randerath E, Randerath K (1984) ^{32}P-Postlabeling test for covalent DNA binding of chemicals in vivo: application to a variety of aromatic carcinogens and methylating agents. Carcinogenesis 5:231–243
Reddy MV, Irvin TR, Randerath K (1985) Formation and persistence of sterigmatocystin-DNA adducts in rat liver determined via ^{32}P-postlabeling analysis. Mutation Res 152: 85–96
Reed E, Yuspa SH, Zwelling LA, Ozols RF, Poirier MC (1986) Quantitation of *cis*-diamminedichloroplatinum II (cisplatin)-DNA-intrastrand adducts in testicular and ovarian cancer patients receiving cisplatin chemotherapy. J Clin Invest 77:545–550
Reed E, Litterst CL, Thill CC, Yuspa SH, Poirier MC (1987a) *cis*-Diamminedichloroplatinum(II)-DNA adduct formation in renal, gonadal, and tumor tissues of male and female rats. Cancer Res 47:718–722
Reed E, Ozols RF, Tarone R, Yuspa SH, Poirier MC (1987b) Platinum-DNA adducts in leukocyte DNA correlate with disease response in ovarian cancer patients receiving platinum-based chemotherapy. Proc Natl Acad Sci USA 84:5024–5028
Richardson FC, Dyroff MC, Boucheron JA, Swenberg JA (1985) Differential repair of O^4-alkylthymidine following exposure to methylating and ethylating hepatocarcinogens. Carcinogenesis 6:625–629
Rio P, Bazgar S, Leng M (1982) Detection of *N*-hydroxy-2-acetylaminofluorene-DNA adducts in rat liver measured by radioimmunoassay. Carcinogenesis 3:225–227
Roberts DR, Benson RW, Flammang TJ, Kadlubar FF (1986) Development of an avidin-biotin enzyme-linked immunoassay for detection of DNA adducts of the human bladder carcinogen 4-aminobiphenyl. Basic Life Sci 38:479–488
Saffhill R, Strickland PT, Boyle JM (1982) Sensitive radioimmunoassays for O^6-*n*-butylthymidine. Carcinogenesis 3:547–552
Sage E, Gabelman N, Mendez F, Bases R (1981) Immunocytological detection of AAF-DNA adducts in HeLa cell nuclei. Cancer Lett 14:193–204
Salih H, Swann PF (1982) Immunoassay of O^6-methyldeoxyguanosine in DNA: the use of polyethylene glycol to separate bound and free nucleoside. Chem Biol Interact 41: 169–180
Sanders MJ, Cooper RS, Jankowiak R, Small GJ, Jeffrey AM (1986) Identification of polycyclic aromatic hydrocarbon metabolites and DNA adducts in mixtures using fluorescence line narrowing spectrometry. Anal Chem 58:816–820
Santella RM (1988) Application of new techniques for the detection of carcinogen adducts to human population monitoring. Mutation Res 205:271–282
Santella RM, Lin CD, Cleveland WL, Weinstein IB (1984) Monoclonal antibodies to DNA modified by a benzo[*a*]pyrene diol epoxide. Carcinogenesis 5:373–377
Santella RM, Hsieh L-L, Lin C-D, Viet S, Weinstein IB (1985) Quantitation of exposure to benzo[*a*]pyrene with monoclonal antibodies. Environ Health Perspect 62:95–99
Santella RM, Lin CD, Dharmaraja N (1986) Monoclonal antibodies to a benzo[*a*]pyrene diolepoxide modified protein. Carcinogenesis 7:441–444
Santella RM, Weston A, Perera FP, Trivers GT, Harris CC, Young TL, Nguyen D, Lee BM, Poirier MC (1988) Interlaboratory comparison of antisera and immunoassays for benzo[*a*]pyrene-diol-epoxide-I-modified DNA. Carcinogenesis 9:1265–1269
Schmeiser H, Dipple A, Schurdak ME, Randerath E, Randerath K (1988a) Comparison of ^{32}P-postlabeling and high pressure liquid chromatographic analyses for 7,12-dimethylbenz[*a*]anthracene DNA adducts. Carcinogenesis 9:633–638
Schmeiser HH, Schoepe KB, Wiessler M (1988b) DNA adduct formation of aristolochic acid I and II in vitro and in vivo. Carcinogenesis 9:297–303
Schoepe K-B, Friesel H, Schurdak ME, Randerath K, Hecker E (1986) Comparative DNA binding of 7,12-dimethylbenz[*a*]anthracene and some of its metabolites in mouse epidermis in vivo as revealed by the ^{32}P-postlabeling technique. Carcinogenesis 7: 535–540
Schoket B, Hewer A, Grover PL, Phillips DH (1988a) Covalent binding of components of coal-tar, creosote and bitumen to the DNA of the skin and lungs of mice following topical application. Carcinogenesis 9:1253–1258

Schoket B, Hewer A, Grover PL, Phillips DH (1988b) Formation of DNA adducts in human skin maintained in short-term organ culture and treated with coal-tar, creosote or bitumen. Int J Cancer 42:622–626

Schurdak ME, Randerath K (1985) Tissue-specific DNA adduct formation in mice treated with the environmental carcinogen, 7*H*-dibenzo[*c,g*]carbazole. Carcinogenesis 6: 1271–1274

Schurdak ME, Stong DB, Warshawsky D, Randerath K (1987a) ^{32}P-Postlabeling analysis of DNA adduction in mice by synthetic metabolites of the environmental carcinogen, 7*H*-dibenzo[*c,g*]carbazole: chromatographic evidence for 3-hydroxy-7*H*-dibenzo[*c,g*]-carbazole being a proximate genotoxicant in liver but not skin. Carcinogenesis 8: 591–597

Schurdak ME, Stong DB, Warshawsky D, Randerath K (1987b) *N*-methylation reduces the DNA binding activity of 7*H*-dibenzo[*c,g*]carbazole ~300-fold in mouse liver but only ~2-fold in skin: possible correlation with carcinogenic activity. Carcinogenesis 8:1405–1410

Seidman LA, Moore CJ, Gould MN (1988) ^{32}P-Postlabeling analysis of DNA adducts in human and rat mammary epithelial cells. Carcinogenesis 9:1071–1077

Seidman M, Mizusawa H, Slor H, Bustin M (1983a) Immunological detection of carcinogen-modified DNA fragments after in vivo modification of cellular and viral chromatin. Cancer Res 43:743–748

Seidman M, Slor H, Bustin M (1983b) The binding of a carcinogen to the nucleosomal and non-nucleosomal regions of simian virus 40 chromosome in vivo. J Biol Chem 258: 5125–5220

Shamsuddin AKM, Gan R (1988) Immunocytochemical localization of benzo[*a*]pyrene-DNA adducts in human tissues. Hum Pathol 19:309–315

Shamsuddin AKM, Harris CC (1983) Improved enzyme linked immunoassays using biotin-avidin-enzyme complex. Arch Pathol Lab Med 107:514–517

Shamsuddin AKM, Sinopoli NT, Hemminki K, Boesch RR, Harris CC (1985) Detection of benzo[*a*]pyrene: DNA adducts in human white blood cells. Cancer Res 45:66–68

Sheridan W, de Serres FJ (1985) Report on the conference on DNA adducts: dosimeters to monitor human exposure to environmental mutagens and carcinogens. Mutation Res 147:59–63

Shuker DEG, Bailey E, Gorf SM, Lamb J, Farmer PB (1984) Determination of *N*-7-[$^2H_3$]methyl guanine in rat urine by gas chromatography-mass spectrometry following administration of trideuteromethylating agents or precursors. Anal Biochem 140: 270–275

Singer B, Spengler SJ, Fraenkel-Conrat H, Kusmierek JT (1986) O^4-methyl, -ethyl, or -isopropyl substituents on thymidine in poly(dA-dT) all lead to transitions upon replication. Proc Natl Acad Sci USA 83:28–32

Sirover MA, Loeb LA (1976) Infidelity of DNA synthesis in vitro: screening for potential metal mutagens or carcinogens. Science 194:1434–1436

Slor H, Mizusawa H, Niehart N, Kakefuda T, Day RS, Bustin M (1981) Immunochemical visualization of binding of the chemical carcinogen benzo[*a*]pyrene diol-epoxide 1 to the genome. Cancer Res 41:3111–3117

Smith MT, Redick JA, Baron J (1983) Quantitative immunohistochemistry: a comparison of microsensitometric analysis of unlabelled antibody peroxidase-antiperoxidase staining for nicotinamide adenosine dinucleotide phosphate (NADPH)-cytochrome c (P-450) reductase in rat liver. J Histochem Cytochem 31:1183–1189

Smith RA, Sysel IA, Tibbels TS, Cohen SM (1988) Implications for the formation of abasic sites following modification of polydeoxycytidylic acid by acrolein in vitro. Cancer Lett 40:103–109

Spodheim-Maurizot M, Leng M (1980) Antibodies to *N*-hydroxy-2-aminofluorene-modified DNA as probes in the study of DNA reacted with derivatives of acetylaminofluorene. Carcinogenesis 1:807–812

Strickland PT, Boyle JM (1981) Characterisation of two monoclonal antibodies specific for dimerised and non-dimerised adjacent thymidines in single stranded DNA. Photochem Photobiol 34:595–601

Strickland PT, Boyle JM (1984) Immunoassay of carcinogen-modified DNA. Prog Nucleic Acid Res Mol Biol 31:1–58

Swenberg JA, Dyroff MC, Bedell MA, Popp JA, Huh N, Kirstein U, Rajewsky MF (1984) O^4-Ethyldeoxythymidine, but not O^6-ethyldeoxyguanosine, accumulates in hepatocyte DNA of rats exposed continuously to diethylnitrosamine. Proc Natl Acad Sci USA 81: 1692–1695

Tanenbaum SW, Beiser SM (1963) Pyrimidine-specific antibodies which react with deoxyribonucleic acid (DNA). Proc Natl Acad Sci USA 49:662–668

Tchen P, Fuchs RPP, Sage E, Leng M (1984) Chemically modified nucleic acids as immunodetectable probes in hybridization experiments. Proc Natl Acad Sci USA 81: 3466–3470

Terheggen PMAB, Floot BGJ, Scherer E, Begg AC, Fichtinger-Schepman AMJ, den Engelse L (1987) Immunocytochemical detection of interaction products of *cis*-diamminedichloroplatinum(II) and *cis*-diammine-(1,1-cyclobutanedicarboxylato)-platinum(II) with DNA in rodent tissue sections. Cancer Res 47:6719–6725

Tierney B, Benson A, Garner RC (1986) Immunoaffinity chromatography of carcinogen DNA adducts with polyclonal antibodies directed against benzo[*a*]pyrene diol-epoxide-DNA. JNCI 77:261–267

Tilby MJ, Styles JM, Dean CJ (1987) Immunological detection of DNA damage caused by melphalan using monoclonal antibodies. Cancer Res 47:1542–1546

Trainor TM, Giese RW, Vouros P (1988) Mass spectroscopy of electrophore labeled nucleosides. Pentafluorobenzyl and cinnamoyl derivatives. J Chromatogr 452:369–376

Turner MJ, Koch SAM, Boucheron JA, Swenberg JA (1987) Methods for quantitative determination of the DNA adduct O^4-ethylthymidine by electron capture negative ion chemical ionization mass spectrometry. Presented at the 35th ASMS conference on mass spectrometry and allied topics, Denver, Colorado, May 24–29, 1987, pp 703–704

Umbenhauer D, Wild CP, Montesano R, Saffhill R, Boyle JM, Huh N, Kirstein U, Thomale J, Rajewsky MF, Lu SH (1985) O^6-Methyldeoxyguanosine in oesophageal DNA among individuals at high risk of oesophageal cancer. Int J Cancer 36:661–665

Vahakangas K, Haugen A, Harris CC (1985) An applied synchronous fluorescence spectrophotometric assay to study benzo[*a*]pyrene-diolepoxide-DNA adducts. Carcinogenesis 6:1109–1116

Van der Laken CJ, Hagenaars AM, Hermsen G, kriek E, Kuipers AJ, Nagel J, Scherer E, Welling M (1982) Measurement of O^6-ethyldeoxyguanosine and *N*-(deoxyguanosin-8-yl)-*N*-acetyl-2-aminofluorene in DNA by high-sensitive enzyme immunoassays. Carcinogenesis 3:569–572

Van Schooten FJ, Kriek E, Steenwinkel M-JST, Noteborn HPJM, Hildebrand MJX, Van Leeuwen FE (1987) The binding efficiency of polyclonal and monoclonal antibodies to DNA modified with benzo[*a*]pyrene diol epoxide is dependent on the level of modification. Implications for quantitation of benzo[*a*]pyrene-DNA adducts in vivo. Carcinogenesis 8:1263–1269

Viegas-Pequignot E, Malfoy B, Leng M, Dutrillaux B, Tchen P (1986) In situ hybridization of an acetylaminofluorene-modified probe recognized by Z-DNA antibodies in vitro. Cytogenet Cell Genet 42:105–109

Vigny P, Duquesne M (1974) A spectrophotofluorometer for measuring very weak fluorescences from biological molecules. Photochem Photobiol 20:15–25

Vigny P, Duquesne M, Coulomb H, Lacombe C, Tierney B, Grover PL, Sims P (1977a) Metabolic activation of polycyclic hydrocarbons. Fluorescence spectral evidence is consistent with metabolism at the 1,2- and 3,4-double bonds of 7-methylbenz[*a*]anthracene. FEBS Lett 75:9–12

Vigny P, Duquesne M, Coulomb H, Tierney B, Grover PL, Sims P (1977b) Fluorescence spectral studies on the metabolic activation of 3-methylcholanthrene and 7,12-dimethylbenz[*a*]anthracene in mouse skin. FEBS Lett 82:278–282

Vo-Dinh T (1978) Multicomponent analysis by synchronous luminescence spectrometry. Anal Chem 50:396–401

Vo-Dinh T (1982) Synchronous luminescence spectroscopy: methodology and applicability. Appl Spectrosc 36:576–581

Vousden KH, Bos JL, Marshall CJ, Phillips DH (1986) Mutations activating human c-Ha-*ras*1 protooncogene (HRAS) induced by chemical carcinogens and depurination. Proc Natl Acad Sci USA 83:1222–1226

Wallin H, Borrebaeck CAK, Gload C, Mattiasson B, Jergil B (1984) Enzyme immunoassay of benzo[*a*]pyrene conjugated to DNA, RNA and microsomal proteins using a monoclonal antibody. Cancer Lett 22:163–170

Wani AA, D'Ambrosio SD (1987) Immunological quantitation of O^4-ethylthymidine in alkylated DNA: repair of minor miscoding base in human cells. Carcinogenesis 8: 1137–1144

Wani AA, Gibson-D'Ambrosio RE, D'Ambrosio SM (1984) Quantitation of O^6-ethylguanosine in ENU alkylated DNA by polyclonal and monoclonal antibodies. Carcinogenesis 5:1145–1150

Watson WP (1987) Post-radiolabelling for detecting DNA damage. Mutagenesis 2: 319–331

Watson WP, Crane AE, Davis R, Smith RJ, Wright AS (1987) A postlabelling assay for N^7-(2-oxoethyl)guanine, the principal vinyl chloride-DNA adduct. Arch Toxicol [Suppl] 11:89–92

West GJ, West IW-L, Ward JF (1982) Radioimmunoassay of a thymine glycol. Radiation Res 90:595–608

Weston A, Trivers G, Vahakangas K, Newman M, Rowe M, Mann D, Harris CC (1987) Detection of carcinogen-DNA adducts in human cells and antibodies to these adducts in human sera. Prog Exp Tumor Res 31:76–85

Weyand EH, Rice JE, LaVoie EJ (1987) ^{32}P-Postlabeling analysis of DNA adducts from non-alternant PAH using thin-layer and high performance liquid chromatography. Cancer Lett 37:257–266

Wild CP, Smart G, Saffhill R, Boyle JM (1983) Radioimmunoassay of O^6-methyldeoxyguanosine in DNA in cells alkylated in vitro and in vivo. Carcinogenesis 4:1605–1609

Willems MI, de Raat WK, Baan RA, Wilmer JWGM, Lansbergen MJ, Lohman PHM (1987) Monitoring the exposure of rats to 2-acetylaminofluorene by the estimation of mutagenic activity in excreta, sister-chromatid exchanges in peripheral blood cells and DNA adducts in peripheral blood, liver and spleen. Mutation Res 176:211–223

Wilson VL, Smith RA, Autrup H, Krokan H, Musci DE, Le N-N-T, Longuria J, Ziska D, Harris CC (1986) Genomic 5-methylcytosine determination by ^{32}P-postlabeling analysis. Anal Biochem 152:275–284

Wilson VL, Basu AK, Essigmann JM, Smith RA, Harris CC (1988) O^6-Alkyldeoxyguanosine detection by ^{32}P-postlabeling and nucleotide chromatographic analysis. Cancer Res 48:2156–2161

Wong D, Mitchell CE, Wolff RK, Manderly JL, Jeffrey AM (1986) Identification of DNA damage as a result of exposure of rats to diesel engine exhaust. Carcinogenesis 7: 1595–1597

Yamashita K, Takayama S, Nagao M, Sato S, Sugimura T (1986) Amino-methyl-α-carboline-induced DNA modification in rat salivary glands and pancreas detected by ^{32}P-postlabeling method. Proc Japan Acad 62(B):45–48

Yang XY, DeLeo V, Santella RM (1987) Immunological detection and visualisation of 8-methoxypsoralen-DNA photoadducts. Cancer Res 47:2451–2455

CHAPTER 14

Biological Consequences of Reactions with DNA: Role of Specific Lesions

G. P. MARGISON and P. J. O'CONNOR

A. Introduction

This review will be restricted to a consideration of the alkylating agents since studies of the mechanisms in carcinogenesis induced by these compounds have tended to progress more rapidly than for those of other classes of chemical carcinogens. Progress in this area has been largely due to the relatively simple modifications produced and hence the early identification and characterisation of the products of the reactions of the alkylating agents with nucleic acids. This has enabled attempts to determine whether individual DNA lesions may give rise to specific biological effects and the possible molecular mechanisms involved (PEGG 1977; MARGISON and O'CONNOR 1979; SAFFHILL et al. 1985).

Amongst the alkylating agents the most widely studied groups are the *N*-nitroso compounds (NNC). These are well-established carcinogens in animals and provide many examples of organ-specific tumorigenesis (PREUSSMANN and STEWART 1984) that have been widely exploited in experimental systems. In addition, NNC and related alkylating agents can also be toxic, mutagenic, clastogenic and teratogenic, and a wide range of other biological effects have been reported (IARC 1978, 1985; TOMATIS and MOHR 1973; PARODI et al. 1983). However, the extent to which many of these effects can be attributed to DNA damage and/or reactions with other cellular targets has yet to be established.

Our understanding of the importance of environmental alkylating agents as potential human carcinogens has developed considerably over the past few years. The isolation and study of the tobacco-specific nitrosamines (IARC 1986) has recently provided stronger evidence of the carcinogenicity of this class of agents to humans, particularly where there is a close contact of tobacco with the exposed tissue, as with the habits associated with the chewing of tobacco (IARC 1985). The ubiquity of alkylating agents in the "normal" human environment has become widely appreciated, and this has been reviewed on several occasions (e.g. PREUSSMANN 1984; BARTSCH and MONTESANO 1984).

In addition to their presence in the environment, NNC can be formed endogenously, and although this has been demonstrated after ingestion of precursors such as nitrate/nitrite and secondary or tertiary amines (BARTSCH and MONTESANO 1984; MIRVISH 1982; PREUSSMANN 1984), the ability of macrophages to produce nitrosating species (STUEHR and MARLETTA 1985; MIWA et al. 1987) is currently a subject of considerable interest since local concentrations of nitrate/nitrite could be envisaged to occur at sites of infection, necrosis, wound healing, etc. The capacity for endogenous nitrosation can be assessed by a noninvasive test (i.e. the nitrosation of proline; OSHIMA and BARTSCH 1981), and this has permitted epidemiological studies (LU et al. 1986) which indicate that endogenous nitrosation, particularly of amino acids, represents a previously unrecognised and potentially major source of DNA-damaging agents. Moreover, it is one which may be independent of exogenous sources of nitrate or nitrite. Such studies have also revealed that the capacity for endogenous nitrosation varies greatly between individuals (WAGNER et al. 1985) so that genetic predisposition towards the production of potentially carcinogenic agents may, in some cases, prove to be a more important factor in human cancer incidence than exogenous exposure: the existence of high- and low-risk regions (see below) may also be indicative of unidentified factors that can affect endogenous nitrosation processes.

Apart from exposure arising from lifestyle, occupation and endogenous sources, other special situations may also be important for specific groups of individuals. Many drugs (and other chemicals such as herbicides) are prepared as tertiary amines for purposes of pharmacokinetics and/or solubility. Some may be nitrosated to yield conventional alkylating agents, e.g. aminopyrine which generates *N*-nitrosodimethylamine (NDMA) or disulfiram which produces *N*-nitrosodiethylamine (NDEA) (LIJINSKY and EPSTEIN 1970) and are carcinogenic in animals when administered in conjunction with nitrite (LIJINSKY 1982). Also, agents such as hydrazine and isoniazid which are not themselves alkylating agents may undergo putative condensation reactions with normal metabolites (e.g. formaldehyde) and subsequent metabolism to yield an alkylating intermediate (BOSAN et al. 1986; SAFFHILL et al. 1988 b).

The toxic effects of NNC and related alkylating agents have been exploited in the treatment of human cancer, e.g. 1,3-bis(2-chloroethyl)-1-nitrosourea (BCNU), 1-(2-chloroethyl)-3-cyclohexyl-1-nitrosourea (CCNU), 5-[3,3-dimethyl-1-triazino]-imidazole-4-carboxamide (DTIC), and although effective against a number of tumours, their application is limited by toxic side effects, especially on bone marrow (COLVIN 1981; REICH 1981). The high toxicity of NNC has also been illustrated by the dramatic effects in humans caused by NDMA poisoning (FLEIG et al. 1982; KIMBROUGH 1982).

Using sensitive radioimmunoassay techniques (WILD et al. 1983), the alkylation of human tissue DNA has so far been demonstrated in at least three separate studies. In patients from the Linxian district of north China, where there is a very high incidence of oesophageal cancer (UMBENHAUER et al. 1985), and from a group in Southeast Asia (SAFFHILL et al. 1988) where oesophageal cancer is also unusually common, the distribution of positive DNA samples indicates that sources of alkylating agents may be widespread within the environment. In a third group of patients, from Manchester, who presented with gastrointestinal problems, only 54% of the DNA samples were positive (SAFFHILL et al. 1988 a), suggesting some association with lifestyle or possibly medication. However, the probable variations in endogenous nitrosation reactions should be borne in mind when considering such data: clearly many more studies of this type are required to assess the relative contributions of exo- and endogenous factors in the alkylation of DNA in human tissues.

Collectively, the alkylating agents could be regarded as one of the most, if not the most, important group of carcinogens affecting the incidence of cancer in humans. There is thus a clear need not only to understand and to detect the sources of alkylating agents but also to gain sufficient knowledge of the very early stages of alkylation damage so that we may be able, eventually, to intervene and protect against its biological consequences. Such knowledge would not only be of value for the reduction of cancer incidence but also for preventing other deleterious effects of alkylating agents that would from our knowledge of their effects in animals also be suspected to arise in humans. This may be particularly important as many diseases have a multifactorial aetiology and may involve as yet undiscovered roles for the effects of alkylating agents. Several auto-immune disease conditions, for example, are known to be associated with partial DNA repair deficiencies for promutagenic alkyl lesions (HARRIS et al. 1982). On the other hand, a better understanding of the mechanisms of toxicity may allow improvements in chemotherapy in which more intensive treatment could be tolerated with fewer undue side effects (e.g. with BCNU and possibly also with isoniazid).

The following discussion is in two parts. Firstly, it provides a brief outline of the biological effects and end points which are amenable to investigation in the study of the actions of the alkylating agents. Secondly, it will summarise the results of recent attempts to attribute roles to specific DNA lesions and to correlate them with some of these biological effects.

B. Biological Effects

I. Effects on the Synthesis of Macromolecules

The inhibitory effects of alkylating agents on the synthesis of macromolecules have been recognised from the earliest studies of alkylation damage (MAGEE and BARNES 1967; see also PEGG 1977; MARGISON and O'CONNOR 1979). Although inhibition of protein synthesis may be the result of damage and possibly fragmentation of mRNA molecules, inhibition of both RNA synthesis and of semiconservative DNA synthesis has been attributed to DNA damage. Studies of single-stranded and double-stranded primed M13mp2 templates have shown that, whilst N7-methylguanine (7-MeG) and O^6-methylguanine (O^6-MeG) are the major products of DNA methylation and do not constitute blocks to DNA synthesis, N3-methyladenine does (LARSON et al. 1985). This is in keeping with the general remit that alkylations occurring within the narrow groove, in particular the 3-methylpurines, constitute lethal lesions via inhibition of DNA synthesis (EVENSEN and SEEBERG 1982; KARRAN et al. 1982; BOITEUX et al. 1984). The potential of an individual lesion for lethality may, however, vary with the cell type (see below for situations in which a lethal role for O^6-MeG is also implicated). Recently it has emerged that base sequence selectivity for sites of alkylation may vary with the nature of the alkylating agent itself (HARTLEY et al. 1988), and subtle differences of this kind will presumably be reflected in DNA chain termination-induced lethality.

Although one of the major toxic effects of alkylating agents is the inhibition of DNA synthesis, unscheduled ("repair") DNA synthesis is also readily induced, and this forms the basis of assays for DNA damaging agents (e.g. ROSSBERGER et al. 1987).

There are also examples of very low doses of agents such as UV light, *N*-methyl-*N'*-nitro-*N*-nitrosoguanidine (MNNG), or *N*-acetoxy-2-acetylaminofluorene recruiting quiescent human diploid fibroblast cell cultures into DNA synthesis of a replicative type (COHN et al. 1984). It is difficult to establish whether any of these changes in DNA synthesis patterns give rise to other biological effects such as mutations, clastogenesis or even teratogenesis: mechanisms can be envisaged, but the contribution of individual lesions remains undefined.

II. Promutagenicity and Alkylation-Induced Mutagenesis

Mutagenesis is one of the few biological effects of alkylating agents that can be directly attributed to specific DNA lesions (see SAFFHILL et al. 1985 for references). Originally, using alkylated RNA (GERCHMANN and LUDLUM 1973) and DNA (ABBOTT and SAFFHILL 1979) as templates in in vitro DNA replication assays and more recently with recombinant DNA procedures (ESSIGMAN et al. 1986), it has been found that among the 13 lesions produced in DNA the principal promutagenic bases are O^6-AG and O^4-AT (see SAFFHILL et al. 1985). During two rounds of DNA synthesis these primarily lead to GC→AT and AT→GC transitions, respectively, although some transversions at a much lower frequency are encountered (SAFFHILL et al. 1985). In the case of the methylating agents O^6-MeG is the major promutagenic base in DNA, being formed in relation to O^4-MeT at a ratio of about 100:1, but with the higher alkylating agents (e.g. $-C_2H_5$ to $-C_4H_9$), this ratio decreases progressively (SAFFHILL et al. 1985). Miscoding of O^6-AG is competitive with normal base pairing and is therefore precursor-concentration dependent (SAFFHILL et al. 1985) O^4-AT has a high tendency to code as C so that the mutagenic potential of higher alkylating agents would be expected to be considerably higher than for e.g. the methylating agents. However, the situation is complicated by the lower reactivity of the higher alkylating agents and their greater ability to inhibit DNA synthesis (MARGISON and O'CONNOR 1979).

The relative proportions of the promutagenic bases O^6-AG and O^4-AT in DNA can be profoundly affected by the efficiency of repair reactions for the individual products. In liver, for example, the ratio of O^6-EtG : O^4-EtT may change from 4:1 in the DNA of hepatocytes after an initial dose of NDEA to 1: >50 after continuous exposure to the nitrosamine for 28 days or more (SWENBERG et al. 1984), due principally to the induction of O^6-AG-alkyltransferase (ATase) activity in these cells. The numbers of promutagenic lesions induced in DNA may also vary at the cellular level in view of the differing capacities of individual cells to metabolise the carcinogen to the chemically active alkylating species (see Fig. 1) or their differing capacities for repair of specific lesions (see Fig. 1). Since it has been shown that alkylated DNA precursors can be incorporated into DNA during DNA synthesis, albeit relatively inefficiently (see SAFFHILL et al. 1985), the size of the nucleotide pools and the extents of normal and repair synthesis in different cell types may also influence the amounts of promutagenic bases introduced into DNA. At the level of chromatin, certain regions of DNA may be repaired more efficiently and others less so (RYAN et al. 1986), and in the case of UV damage for example, even regions of the same gene may be repaired to differing extents (BOHR et al. 1987). The DNA sequence itself may also influence mutagenesis. Sequence analysis of the first 540 base pairs of the *lac I* gene of *E. coli* revealed a strong influence of the 5′-flanking base on forward mutagenesis induced by MNNG. Guanine residues preceded by a guanine or adenine residue were 9 and 5 times, respectively, more likely to mutate than guanines preceded by a pyrimidine residue (BURNS et al. 1987). Sequence-dependent mutagenesis and other processes discussed above will clearly be influenced by the base sequence selectivity for sites of alkylation noted previously (HARTLEY et al. 1988). The potential to induce somatic mutations in tissues and cells, and probably also to elicit biological effects of the kind briefly discussed below, will thus depend on the extent, location and nature of the reaction as well as on the frequency and extent of DNA synthesis in specific cells in relation to the activity of those DNA repair processes which are capable of removing promutagenic lesions.

There is now a consensus of opinion that the balance between these activities will directly influence the extent to which an individual tissue is affected. The complexities of these interactions, however, and the subsequent dependence upon the action of promoting agents to expand the initiated population, have precluded any universal correlation between, for example, a single alkylation product and diffuse endpoints such as tumour induction.

Despite this, where extensive comparisons of the genetic effects of alkylating agents with their capacity to react with DNA have been examined (VAN ZEELAND et al. 1985), results with ethylating agents have indicated that O^6-EtG formation in DNA is closely related to the frequency of gene mutations in bacteria, cultured cells and mouse testis. It should also be considered together with such findings that the measured endpoint, mutation, is usually the loss of an enzyme activity, and, as pointed out below, this may occur by gene deletion or rearrangement and may not necessarily be a consequence of a point mutation event. This could explain the mutagenicity of methylmethanesulphonate (MMS), which gives rise to very small amounts of the classic promutagenic lesions (MARGISON and O'CONNOR 1979) and yet induces forward and reverse mutation in e.g. V79 cells (SUTER et al. 1980). Molecular analysis of $HPRT^-$ mutants induced by MMS indicates the presence of a high proportion of deletions (CHAUDRY and FOX 1988).

III. Clastogenesis

The alkylating agents are widely established as clastogenic agents in a variety of systems, but the mechanisms giving rise to this process appear to differ from those of mutagenesis.

In *Drosophila melanogaster*, spontaneous mutations are predominantly associated with transposable elements inserted into the affected gene, whereas only 5% of mutations induced at the Rp II 215 locus by ethylmethanesulphonate (EMS) and *N*-nitroso-*N*-ethylurea (NEU) were associated with transposable ele-

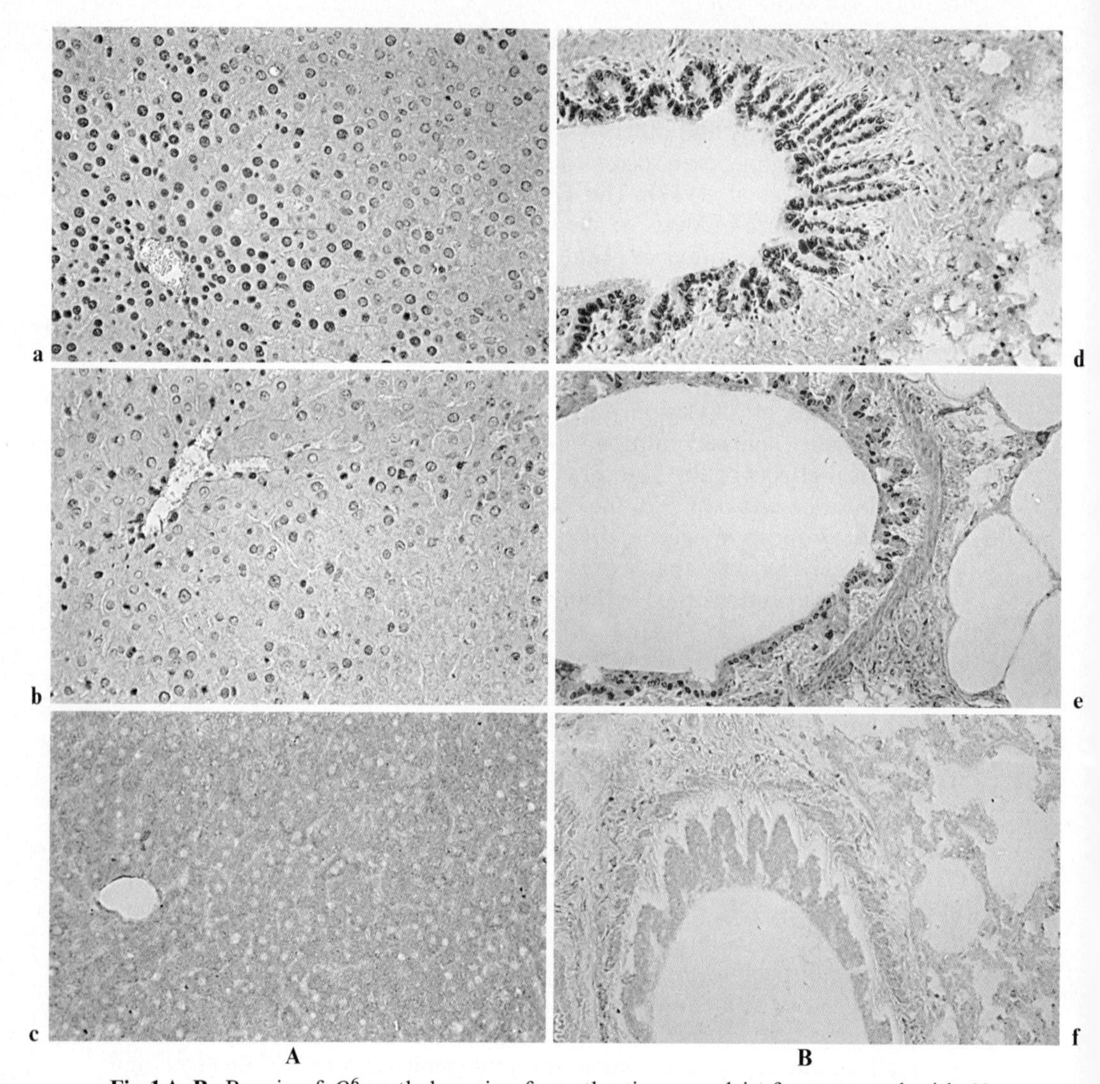

Fig. 1 A, B. Repair of O^6-methylguanine from the tissue nuclei of rats treated with *N*-nitrosodimethylamine. **A** The effects of phenobarbital: liver sections of rats treated with NDMA (2 mg/kg) 9 h before sampling. (*a*) Animal maintained on normal water and (*b*) animal exposed to phenobarbital in the drinking water (0.05%) for 3 weeks. The control section (*a*) shows differential staining for the presence of O^6-methylguanine in individual hepatocytes associated with the central vein. Although by 9 h some repair has already occurred, the differences in the intensity of staining still reflect differences in the capacity for the metabolism of the nitrosamine. Cells in the periportal region of the liver lobule are essentially negative, whereas both the centrilobular and periportal hepatocytes in a liver section (*c*) prepared from an animal not treated with the nitrosamine were completely negative. The fewer and less intensively staining O^6-methylguanine-positive nuclei in section (*b*) show the increased capacity for the repair of O^6-methylguanine in the hepatocytes of animals exposed to phenobarbital. Direct measurements of the O^6-methylguanine content of liver DNA (by radioimmunoassay or by radiochromatography) also show that exposure of animals to phenobarbital reduces the level of O^6-methylguanine in DNA. At 5 h after nitrosamine treatment this falls from ~60 μmoles/mole guanine to ~20 μmoles/mole

ments, implying a qualitative difference between these two types of mutagenesis (LACY et al. 1986). Point mutations induced in *Drosophila* by a range of alkylating agents showed a close correlation with the relative capacities of these compounds to react with oxygen atoms in DNA (VOGEL et al. 1985). However, when the postmeiotic cell stages of repair-proficient ring-X males were treated with MMS, EMS, NDEA or NEU and then mated to repair-defective mei-9^{LI} females, this resulted in a high sensitivity to chromosome loss induced by all the agents, irrespective of differences in the relative proportions of oxygen and nitrogen atom adducts in DNA (VOGEL 1986). Thus, whilst *N*-alkylation in *Drosophila* does not appear to contribute significantly to mutagenesis in the case of NEU-type mutagens, *O*-alkylation is clearly not exclusively, or even predominantly, related to the induction of clastogenic events. In the case of MMS-type mutagens, it is also evident that *N*-alkylation is potentially mutagenic, leading both to mutations and chromosome aberrations (CA) (VOGEL 1986).

Treatment of the human lymphoblast cell line TK6 with mitomycin C, *N*-nitroso-*N*-methylurea (NMU) or NEU at dose levels resulting in fewer than 2 lethal hits gave rise to linear dose-response curves for *hgprt* and *tk* gene mutants and also for CA, the number of mutants at either gene locus being similar for the three compounds. The relative amounts of mutagenic and clastogenic activity, however, were quite different for each of the three compounds investigated, supporting the generally held view that the inducers or mechanisms of mutagenesis and clastogenesis are non-identical (JENSEN and THILLY 1986), but giving no indication of the mechanism of the latter.

There is evidence also for divergence in the pathways leading to SCEs and CA. Treatment of the temperature-sensitive mouse FM3 A cell mutants with MNNG leads to the production of three groups of mutants conditional for the induction of SCEs and CA. At the non-permissive temperature, group 1 mutants manifested mainly SCEs, group 2 mutants showed both SCEs and CA, whereas mutants in group 3 showed only CA. This suggests the involvement of at least three pathways for the formation of these lesions and that only one may be common to both processes. The group 2 mutants could be further subdivided into 5

guanine in control and phenobarbital-treated animals, respectively. This does not, however, reflect differences in the capacity for metabolism as the amounts of 7-methylguanine in control and phenobarbital-treated animals are similar (O'CONNOR et al. 1988). **B** Normal rats given an LD_{50} dose of NDMA (40 mg/kg). Sections of lung showing the highly selective staining of the epithelial cells of some of the bronchioles 5 h after treatment with NDMA (*d*), and 12 days later (*e*) some of the epithelial cells still remain unrepaired. A control section (*f*) from an animal not treated with the nitrosamine does not show positive staining. (NB Lung tissue is alkylated at least 10 times less extensively than liver after the administration of NDMA. At this LD_{50} dose a marked centrilobular necrosis is observed in the liver, which is then followed by regenerative hyperplasia. After 12 days, the liver characteristically contains a few unrepaired cells containing O^6-methylguanine; FAN and O'CONNOR, unpublished observations). The sections were prepared as follows: paraffin wax sections (3 μ) from tissues fixed in 70% ethanol were stained for the presence of O^6-methylguanine using a rabbit polyclonal antibody. Primary antibody-positive cells were then identified using a rabbit peroxidase anti-peroxidase complex and 3,3-diaminobenzidine staining (O'CONNOR et al. 1988)

classes, each with different cytogenic properties, suggesting further that several gene products may be involved in the formation of SCEs or CA (TSUJI et al. 1986). A comparison of the induction of SCEs and CA in ionising radiation- and alkylating agent-sensitive cell lines led to the conclusion that, except for caffeine-induced CA in the radiation-sensitive cells, mutagen-induced lethal lesions were responsible for CA induction when a variety of agents were employed. In contrast SCE induction in these mutants is much more complex, indicating that the lesions involved in SCE production differ, at least in part, from those initiating CAs or lethality (TSUJI et al. 1987).

Clastogenic events in laboratory rodents have also been widely reported (see IARC 1978). Micronucleus formation occurs, for example, in the target tissues for carcinogenesis in rats treated with NNC: in the case of *N*-nitroso-methyl-benzylamine (NMBzA) and *N*-nitroso-methylamylamine (NMAA) these lesions were found in the oesophagus, after treatment with NDMA, in the liver and, with NDEA, which induces tumours in both liver and oesophagus, micronuclei were found in both tissues (MEHTA et al. 1987). In mice, trisomy 15 is frequently associated with nitrosourea-induced lymphomas and leukaemias (CARBONELL et al. 1982). In C57B1/6 × DBA_2 hybrids, trisomy 15 is present in virtually all the T-cell lymphomas induced by a single dose of NMU or by repeated doses of benzo[*a*]pyrene (BaP), and these aberrations could be detected in the cells of the thymus as early as 6 weeks into the latency period. The first transplantable cells are also found in the thymus (CARBONELL et al. 1987). Treatment of the same hybrid mouse strain with *N*-nitrosobutylurea (NBU) also induced trisomy 15, as well as trisomy 14, but at a much lower frequency (CARBONELL et al. 1982). Lymphomagenesis induced by radiation-induced leukemia virus (RadLv) (HAAS et al. 1984) on the other hand shows a variety of CA involving trisomies and chromosome loss. The differences between these two studies suggest that there is no obvious common pathway at the level of CA in the induction of leukaemogenesis. In general, however, there appears to be an involvement of trisomy 15 and trisomy 14 in the induction of murine T-cell leukaemogenesis, whereas in rats a different pattern is observed. NBU in rats, for example, induces myeloid/erythroid leukaemias which may exhibit a variety of CA including trisomy 2 (UENAKA et al. 1978).

Chromosomal changes have been studied extensively as a mechanism for altering the degree of gene expression, or of amplifying the effects of gene mutation (KLEIN 1981). Activation of the *met* oncogene involves chromosomal rearrangement (TEMPEST et al. 1986; DEAN et al. 1987), and trisomy 15 may alter the ratio of expression of the *myc* oncogene to those of other genes (UNO et al. 1987).

A number of attempts have been made to examine the possible relationships between specific DNA lesions and clastogenic effects, particularly SCE induction in mammalian cells, by measuring the amounts of various products in DNA under SCE-inducing conditions. It was proposed that O^6-AG may be an SCE-inducing lesion (WOLFF 1982), but this was not supported in later experiments using NMU, MMS and DMS (MORRIS et al. 1983; CONNELL and MEDCALF 1984). On the other hand, data described below suggest that O^6-AG or O^4-AT are SCE-inducing lesions although, with agents such as MMS, other lesions such as 3-MeG (WHITE et al. 1986) may be the principal cause of SCEs.

IV. Teratogenesis and Transplacental Effects

The teratogenic effects of alkylating agents in laboratory rodents have been well documented since the recognition of transplacental carcinogenesis in the early 1960s (TOMATIS and MOHR 1973). Depending on the stage of pregnancy at which the agent is administered, broadly speaking, three major biological effects are observed, i.e. embryotoxicity when the agent is administered early, and teratogenic or carcinogenic effects in the offspring when given mid-term or later during the gestational period.

During the early, predifferentiation, preimplantation period, the mammalian embryo has been traditionally regarded as refractory to teratogens. More recently, however, this concept has been modified. The in vitro treatment of preimplantation (4 day) embryos with NMU (BOSSERT and IANNOCCONE 1985) or treatment of pregnant females with NMU on gestational days 3.5–4.5 (TAKEUCHI 1984) leads to gross malformations of the fetus. In the latter experiments, although the incidence of fetal deaths was increased, the efficiency of implantation was unaffected, and at gestational day 1.5 or earlier no malformations were observed. Although the adults derived from preimplantation embryos exposed to NMU in vitro had a much higher overall mortality rate, up to 1 year after birth there were no gross malformations, histological abnormalities or chromosomal aberrations associated with NMU exposure (IANNOCCONE 1984).

In studies of chemically induced teratogenesis, special interest has focussed on unilateral defects in symmetrically developing organs. A left-sided preponderance of paw malformations is induced by *N*-nitroso-*N*-acetoxymethyl-*N*-methylamine when administered on day 11 or 12 of pregnancy (BOCHERT et al. 1985). This effect could not be reproduced by treatment of limb buds in organ culture but was found to be associated with a twofold higher level of DNA methylation when the limb bud DNA was isolated from embryos treated transplacentally and analysed.

When treatments are made from the middle and towards the later stages of pregnancy, carcinogenic effects in the offspring predominate (TOMATIS and MOHR 1973). Depending upon the day of treatment and the agent employed, the spectrum of tumours will vary. The rat embryonic nervous system is particularly sensitive to carcinogenesis induced by nitrosoureas (IVANKOVIC and DRUCKREY 1968). Treatment of pregnant female rats with NEU between 12 and 15 days of gestation is at least 50 times more effective in producing tumours in the offspring than are similar treatments in young adults, without taking into account the fact that the dose administered to the pregnant females was distributed between maternal and fetal tissue alike.

CA (BRAUN et al. 1986) and gene mutations (as detected by the mammalian spot test; BRAUN et al. 1984) have been observed following the transplacental administration of a series of monofunctional alkylating agents with varying proclivities for oxygen atoms in nucleic acids. As observed in other systems (see above), there was no correlation between the chemical reactivity of the compounds and their ability to induce chromosomal damage, although the methylating agents (DMS, MMS and NMU) were generally more efficient inducers of chromosomal damage than the ethylating agents (EMS and NEU; BRAUN et al.

1986). The mammalian spot test, on the other hand, indicated a clear mutagenic activity for compounds having high proclivity for O-atoms, while the agents reacting preferentially with N-atoms were without genetic effects.

Relatively few attempts have been made to examine the possible relationships between specific types of DNA damage and fetal malformation. However, with the development of in situ assays for specific lesions (see Fig. 1), experiments of this type may be more practical.

V. Transformation in Cultured Mammalian Cells

The tumour-inducing properties of alkylating agents and their function as initiating agents in the induction of preneoplastic changes in a variety of epithelial, e.g. liver, bladder, skin, colon (HICKS 1983; PITOT and SIRICA 1980; MASKEN 1981), and mesenchymal e.g. kidney, (HARD and BUTLER 1970) tissues have been well documented and will be considered below in relation to the alkylation of DNA. However, cell culture systems have also been widely used to investigate the ability of these agents to induce malignant transformation. NMU, for example, is a potent inducer of proliferating preneoplastic rat urothelial foci from which rapidly proliferating cell lines can be established. Some of these are tumorigenic after transplantation into either syngeneic rats or nude mice, but not in both (KNOWLES and JANI 1986). Treatment of cultures of human fetal lung diploid fibroblasts with NDEA or with a novel NNC, *N*-nitroso-*N*-1-methylacetonyl-*N*-3-methylbutylamine, have produced transformed cells with altered morphology, prolonged lifespan and an ability to form anchorage-independent colonies in soft agar (HUANG et al. 1986).

The transformation of $C_3H10T_{1/2}$ cells by NEU has been investigated extensively in relation to the number of post-treatment cell divisions, and it was found that the transformed focus forming ability of NEU was completely fixed within four generations. Although the phenotypic expression of NEU-induced transformants may be influenced by clone size at confluence, the dose dependency of transformation was qualitatively and quantitatively similar to that of mutation at the Na-K-ATPase locus, suggesting that transformation is the result of a single, low frequency event, possibly a gene mutation (DEKOK et al. 1981). When three methylating agents were compared for their capacity to induce a dose-dependent transformation of Syrian hamster embryo cells, the relative transformation efficiencies for MNNG, NMU and MMS were 500:5:1. At concentrations that induced equivalent transformation frequencies, the amounts of O^6-MeG, but not of 7-MeG, formed in DNA were the same for all three carcinogens. The data therefore give support to the role of O^6-MeG as a critical lesion for the initiation of carcinogenesis by these methylating agents, although other lesions such as O^4-MeT were not measured in these experiments. In these studies the transformation frequency relative to the amount of O^6-MeG was many-fold higher than the mutation frequency, but the degree of excess depended upon the gene locus under investigation. The minimal target size for transformation was $\sim 10^4$ nucleotides, indicating that several base mutations were required for transformation and that no one specific gene was critical for carcinogenesis (DONIGER et al. 1985).

Indications that a single mutational event can, under the appropriate circumstances, confer the potential for malignant transformation have come from experiments in which the mutated H-*ras*-1 gene has been employed. When a non-tumorigenic cell line established from primary hamster epidermal cells by exposure to MNNG was co-transfected with pEJ and pSV2-gpt, 60% of the transformants formed colonies in soft agar and carcinomas on transplantation into nude mice (STORER et al. 1986). Southern blot analysis of the transformants indicated that rapid malignant transformation was associated with the integration of the 6.6-kb *Bam*H1 fragment of pEJ which contains the activated H-*ras*-1 gene (i.e. point mutated at codon 12; TABIN et al. 1982). Such studies indicate that a mutant H-*ras*-1 gene, under the control of its normal cellular promotor, can rapidly transform a non-tumorigenic cell line and that activation of an endogenous H-*ras*-1 gene may function as the final completing event in the progress of cells to the malignant phenotype.

Given that carcinogenesis is a multi-step process, the number of critical events that are required to initiate transformation in these model systems will depend upon the number of changes which predispose towards transformation that have already taken place during the establishment (or construction) and maintenance of the particular cell line.

C. Correlation of Promutagenic Lesions with Carcinogenesis

Correlations have been made between the extent of DNA damage, the persistence of promutagenic bases (see above) and the tissue-specific sites of tumour formation (PEGG 1977; O'CONNOR et al. 1979; MARGISON and O'CONNOR 1979).

In extensive studies of the dose-dependent induction of thymic lymphomas in mice by NMU, NEU and EMS (FREI et al. 1978), the carcinogenic effectiveness of these agents has been positively correlated with the extent of alkylation at the O^6-atom, but not the N7-atom of guanine, in the DNA of the target tissues, thymus and bone marrow. While such studies point to a role in the initiation of carcinogenesis for O^6-AG, it should be noted that the DNA of other, non-target tissues in these same animals also received comparable amounts of O^6-alkylation. Further, analyses of tissue DNA from rats treated with NMBzA (KLEIHUES et al. 1983), which is specifically an oesophageal carcinogen, show that although a much higher level of O^6-MeG is formed in osesophagus than in liver doses of NDMA which produced similar levels of hepatic methylation lead to the dose-dependent formation of preneoplastic foci whilst methylation arising from NMBzA was without effect. (SILINSKAS et al. 1985 and references therein). Simple comparisons of the level of alkylation at specific sites in DNA based on mean average values for the whole tissue can be misleading as they take no account of differences in the levels of alkylation in the specific target cells. Depending on their capacity for metabolism, even adjacent cells may be highly alkylated or, negative (see Fig. 1). Similarly, the constitutive (or induced) repair capacity and the requirement for DNA synthesis may affect both the persistence and biological effectiveness of individual DNA lesions.

Whilst there seems to be general agreement that the replication of alkylated DNA is an essential event in initiation (see e.g. CRADDOCK et al. 1984; SCHUSTER et al. 1985), such is the complexity of the carcinogenic process that even when systematic evaluations have been made of the persistence of O^6-MeG after treatment with several specific methylating agents, no truly universal correlations emerge (KLEIHUES et al. 1983). When ethylating agents are employed, the amounts of O^4-EtT relative to O^6-EtG produced in DNA are much higher than with methylating agents (see above), and a very good correlation exists between the formation of preneoplastic lesions, liver cell carcinoma and the level of O^4-EtT (SWENBERG et al. 1985). This product accumulates during continuous treatment with an ethylating agent due to an ineffective repair process in the hepatocytes, whilst the repair of O^6-EtG is further induced, and the amounts of this product decrease correspondingly. These studies also implicated the role of DNA replication and were all the more impressive because the level of O^4-alkylation, foci formation and tumour production correlated with the heterogeneous response of the liver lobes with respect to these parameters (DRYOFF et al. 1986; RICHARDSON et al. 1986). When a methylating agent, 1,2-dimethylhydrazine (SDMH) was used in the same strain of animals, haemangiomas were produced rather than liver cell cancers, and O^6-MeG as well as O^4-MeT accumulated within the non-parenchymal liver cell fraction due to the lower capacity for the repair of O^6-MeG in these cells (SWENBERG et al. 1985). In attempting to draw such correlations, therefore, it is now more appropriate to consider the total number of promutagenic lesions in target cells (SAFFHILL et al. 1985) due to the differences in repair capacities referred to above which may radically alter the proportions of the individual promutagenic bases in some tissues or cells, yet not in others.

Whilst correlations of the kind outlined above are steadily building up a picture of cause and effect, the complexities of the carcinogenic process, coupled with the variety of lesions introduced into DNA by the initial reactions of alkylating agents, as well as the reactions with other cellular target molecules, virtually preclude any definitive conclusions. If we are ever to be able to investigate mechanisms of carcinogenesis in greater detail and thereby to derive practical benefits, methods must be adopted which will precisely define the relationships between the initial lesions in DNA and the plethora of biological effects that can ensue. For example, it has already been established by comparisons of the alkylating agents themselves that those agents which introduce a preponderance of modifications at O-atoms are better carcinogens than those which preferentially modify N-atoms (MARGISON and O'CONNOR 1979). Using in vitro systems we have identified the two principal promutagenic bases found in DNA (see SAFFHILL et al. 1985), and it has been shown that mutagenesis can be correlated in a dose-dependent manner with the formation of O^6-AG in the DNA of bacteria, cultured cells and mouse testis (VAN ZEELAND et al. 1985). Further, in bacteria, mutations induced by some methylating agents can be virtually prevented by the induction of high levels of the repair protein which eliminates the two promutagenic bases from DNA (SCHENDEL and ROBINS 1978). Recently, the isolation of the corresponding bacterial DNA repair genes has permitted the introduction of specific repair functions into repair-deficient mammalian cells. This now offers a way of exploring the role of specific DNA lesions in mammalian systems both in vitro and in vivo. These techniques, combined with the use of the appropriate immunohistochemical procedures (see above), should eventually permit studies of carcinogenesis at the level of individual cells within the target tissues in animal models.

D. Role of Specific Lesions: Effects of *E. coli* Alkyltransferase Gene Expression in Mammalian Cells

A number of groups have succeeded in obtaining expression of ATase activity encoded by the *E. coli ada* gene in various forms in mammalian cells, either transiently or in permanent cell lines.

Initial experiments involved a transient-expression system in which the feasibility of obtaining a bacterial ATase in an active form in mammalian cells was explored (BRENNAND and MARGISON 1986a). Functionally active bacterial ATase was demonstrated by the presence of dual alkylphosphotriester- (AP) and O^6-AG ATase activity and by fluorography. Higher levels of expression were obtained when the 5′ untranslated region of the *ada* gene was deleted from the construct. The section of the *E. coli* gene used in this case was the protein coding region together with six 5′ and five 3′ nucleotides and was denoted the "C" or coding fragment (Table 1 and Fig. 2; BRENNAND and MARGISON 1986a). The in-

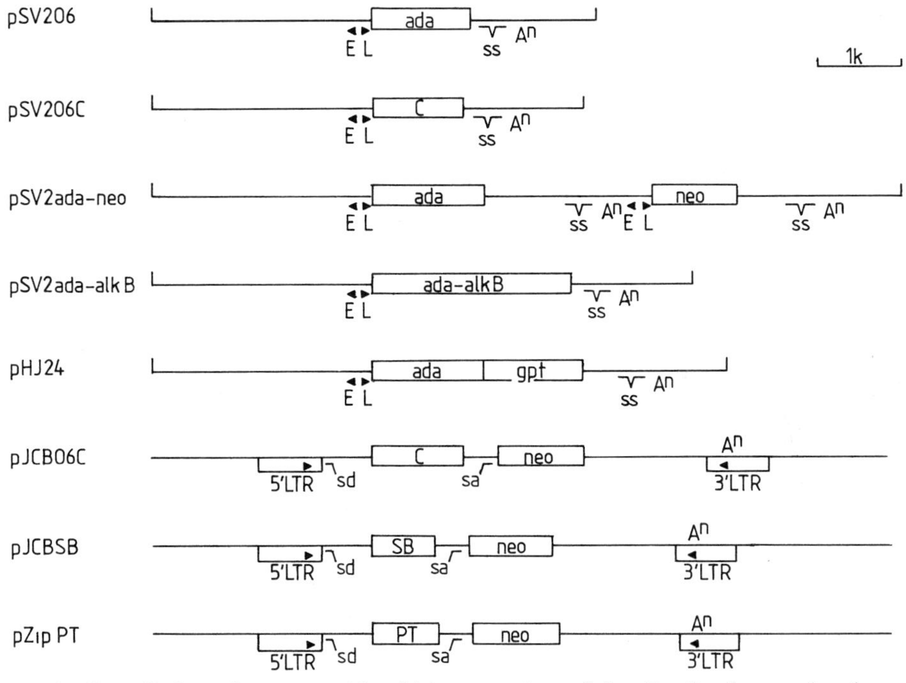

Fig. 2. Compilation of vectors with which expression of the *E. coli ada* gene has been achieved in mammalian cells. Constructs shown with *ada* all contain the 5′-untranslated region of the bacterial gene that includes the -35, -10 and Shine-Delgano consensus sequences and a 9-amino acid open reading frame (see text). *C* denotes a truncated version of the *ada* gene containing essentially the protein coding region; *SB* and *PT* denote truncated versions that give rise to the O^6-AG and AP ATase functions, respectively. For other details see text. ▶, promotor region; *E, L,* the early and late promotor regions of SV40; *ss,* splice region; A^n, polyadenylation signal; *sd, sa,* splice donor and acceptor sites; *LTR,* long terminal repeat. For ease of display the SV40-based plasmids are shown linearised at the *Eco*-RI site of the vector, and only a section of the 10–11-kb pZipneoSV(x)1-based plasmids (pJCBO6C, pJCBSB, pZipPT) are presented.

Table 1. Expression of *E. coli ada* gene in ATase-deficient mammalian cells

Cell type	Vector	Promotor	*ada* fragment	Selection	ATase activity (pm/mg)	References
COS7	pSV206	$SV40_E$	1.1kb HindIII-XhoII	none [a]	0.3	BRENNAND and MARGISON (1986a)
COS7	pSV206C	$SV40_E$	Coding [b]	none [a]	0.7	BRENNAND and MARGISON (1986a)
HeLaMR	pSV2ada-neo	$SV40_E$	1.3kb HindIII-Sma I	G418/ACNU [c]	2.2	ISHIZAKI et al. (1986)
HeLaS3	pSV2ada-alkB	$SV40_E$	3.0kb HindIII-BamHI	G418 [d]	+ [e]	SAMSON et al. (1986)
V79	pJCBO6C	MoMLV 5'LTR	Coding [b]	G418	1.6	BRENNAND and MARGISON (1986b); FOX et al.(1987)
V79	pJCBSB	MoMLV 5'LTR	0.8kb SalI-BamHI	G418	0.3	BRENNAND and MARGISON (1986c); FOX et al. (1987)
V79	pZipPT	MoMLV 5'LTR	0.7kb CR [f]	G418	0.3	KLEIBL, WHITE, OCKEY and MARGISON (unpublished results)
V79TG11	pJCBSB	MoMLV 5'LTR	0.8kb SalI-BamHI	G418	0.4	FOX and MARGISON (1988)
V79TG15	pJCBSB	MoMLV 5'LTR	0.8kb SalI-BamHI	G418	1.0	FOX and MARGISON (1988)
CHO	pHJ2	$SV40_E$	1.3kb HindIII	gpt/CNU [g]	0.5	KATAOKA et al. (1986)
CHO	pHJ24	$SV40_E$	1.3kb HindIII F [h]	gpt	0.5	KATAOKA et al. (1986)
FDCPmix	pJCBO6C	MoMLV 5'LTR	Coding [b]	G418/Mz	2.1	JELINEK et al. (1988)
FDCPmix	pJCBSB	MoMLV 5'LTR	0.8kb SalI-BamHI	G418/Mz	0.5	JELINEK et al. (unpublished results)
FDCPmix	pZipPT	MoMLV 5'LTR	0.7kb CR [f]	G418	0.3	JELINEK et al. (unpublished results)
DM	pJCBO6C	MoMLV 5'LTR	Coding [b]	G418	0.08	MUSK et al. (1989)

[a] Transient expression system.
[b] see text.
[c] 1-(4-amino-2-methyl-5-pyrimidinyl)methyl-3-(2-chloroethyl)-3-nitrosourea.
[d] Cotransfected with pSV2neo.
[e] Quantitative data not available.
[f] Truncated version of the coding region excluding the O^6-AG ATase region.
[g] 2-chloroethyl-3-nitrosourea.
[h] Insert contains frameshift and does not code for O^6-AG ATase activity.

creased expression was probably due to the elimination of a short open reading frame (ORF) coding for 10 amino acids, which is upstream of and out of the reading frame with the ATase translation initiation codon (DEMPLE et al. 1985; NAKABEPPU et al. 1985). However, high levels of expression of the *ada* gene have been achieved even when DNA sequences containing the intact 5′ untranslated region were used in vectorconstruction (ISHIZAKI et al. 1986; SAMSON et al. 1986; KATAOKA et al. 1986; Fig. 2). The short ORF or the presence of the bacterial gene transcription-controlling elements therefore do not preclude ATase expression in mammalian cells.

As with the transient expression system described above, many of the permanent cell lines have utilised the SV40 early region promotor to drive the expression of the *E. coli* gene(s). In such constructs the 3′ untranslated region (splice site and polyadenylation signal) were also SV40 derived (see e.g. Fig. 2). In the case of the retrovirus-based expression vector pZipneoSV(x)1 (Fig. 2) the transcription promotor is in the 5′ long terminal repeat (LTR) region and the polyadenylation signal is in the 3′ LTR. In this case the same message encodes the ATase gene and an *E. coli* aminoglycoside phosphotransferase gene *neo*, the expression of which in mammalian cells provides resistance to the antibiotic G418.

In this relatively complex system G418 resistance occurs if the full-length message is modified by splicing out the ATase sequence between 5′ donor and 3′ acceptor splice sites located upstream and downstream, respectively, of the "C" sequence (see Fig. 2). ATase activity on the other hand arises from mRNAs which remain intact (see MARGISON and BRENNAND 1987), although read through of intact message may also occur. In other experiments in which the ATase gene and an antibiotic selectable marker were also contained in the same vector, the SV40 early promotors were duplicated upstream of *neo* and *ada*, and the appropriate 3′ regions probably gave rise to individual polyadenylated mRNAs for both the *ada* and *neo* gene products (ISHIZAKI et al. 1986; Fig. 2). Alternatively, the *ada* gene was inserted between the SV40 promotor and the guanine phosphoribosyl transferase (gpt) gene which, in this particular case, provided the selection method (KATAOKA et al. 1986). The advantage of such constructs is that all cells that are antibiotic resistant will probably contain the ATase gene.

Another group has used cotransfection of different plasmids containing the *ada* gene and the selectable gene *neo* (SAMSON et al. 1986). In this case not all cells that express the selectable marker will necessarily contain the ATase gene. In addition to selection exploiting expression of *gpt* or *neo*, selection with chloroethylating agents has also been used to eliminate non-ATase expressing cells (ISHIZAKI et al. 1986; KATAOKA et al. 1986; JELINEK et al. 1988). Since selection by these methods has been shown in certain cases to select cells expressing the endogenous ATase genes (MORTEN and MARGISON 1988), other evidence of *E. coli* ATase gene expression (e.g. fluorography) ideally should be provided in order to ensure that a proportion of the total ATase activity is not derived from non-*ada* gene sources. Southern analysis has been used to demonstrate incorporation of the *ada* gene into the host genome, and in one case up to 30 copies of the gene were reported (ISHIZAKI et al. 1986). No studies on insertion sites or chromosome location have been published, and the size, processing, abundance and stability of the various mRNAs have not yet been delineated.

Separation of the O^6-AG-O^4-AT ATase function and the AP-ATase functions of the *ada* gene has been achieved by truncation to a 0.8-kb *Sal*1-*Bam*H1 fragment that codes only for the former function (BRENNAND and MARGISON 1986c). Introduction of a frameshift mutation results in the loss of O^6-AG-ATase activity (KATAOKA et al. 1986), and this has also been achieved by truncation of the coding region to an *Rsa*1 site upstream of the O^6-AG ATase alkyl acceptor cysteine residue coding region (KLEIBL, WHITE, OCKEY and MARGISON, manuscript in preparation).

The size of the *ada* protein synthesised in mammalian cells is ~39 kDa. It is indistinguishable from that produced in *E. coli*, and, as in *E. coli*, degradation can occur to produce 18- and 20-kDa fragments (SAMSON et al. 1986; MARGISON and BRENNAND 1987). Although the stability of the *E. coli* ATase protein in mammalian cells is unknown, synthesis of the protein occurs efficiently, i.e. within 8–12 h following depletion by feeding with O^6-MeG (KLEIBL, WHITE, OCKEY and MARGISON, unpublished results). This implies that some feedback control of the level of expression of the bacterial gene occurs in a mammalian cell environment, but it is unlikely that the mechanism resembles that present in *E. coli* (TEO et al. 1986).

Various methods have been used to demonstrate that cells produce functional ATase, but only in one case has the repair of alkylation damage in host cell DNA been shown (BRENNAND and MARGISON 1986b). Thus, penetration of the nuclear envelope by the repair protein occurs, and although the extent of ablation of some of the biological effects of alkylating agents (see below) seems to be related to the levels of ATase activity found in cell extracts (BRENNAND and MARGISON 1986c; FOX et al. 1987), this gives no indication of the rate, mechanism or extent of transport of the ATase across the nuclear membrane. Such parameters and also the access of the bacterial ATase to damage in different regions of the chromatin are currently being examined. In this context it is of interest to note that in CHO cells carrying the *Den V* gene of bacteriophage T4, the normal pattern of preferential repair of the active dihydrofolate reductase gene is lost and that UV damage is repaired equally effectively in a non-coding sequence downstream from the gene (BOHR and HANAWALT 1987).

I. Mutagenesis

Protection against the mutagenic (BRENNAND and MARGISON 1986b; KATAOKA et al. 1986; FOX et al. 1987) and, more specifically, the revertagenic effects (FOX and MARGISON 1988) of alkylating agents occurs in mammalian cells expressing *E. coli ada* ATases as determined by forward mutation to resistance to 6-thioguanine (TG) (BRENNAND and MARGISON 1986b; FOX et al. 1988) or 8-azadenine (AA) (KATAOKA et al. 1986) or reversion to TG sensitivity (FOX and MARGISON 1988). Such effects may have been predicted from previous studies on the mispairing characteristics of O^6-AG and O^4-AT (SAFFHILL et al. 1985). Although mutation to TG or AA resistance can occur by deletions or frameshifts, in addition to the transitions or transversions arising from single base changes, the decreased reversion frequency of spontaneous hypoxanthine phosphoribosyl transferase (HPRT)-mutants expressing *E. coli* ATase provides more conclusive evidence for the role of point mutations in this process (Fox and Margison 1988). The decrease in NMU-induced reversion frequency in *E. coli* ATase-expressing cells could be explained simply by the repair of O^6-MeG and/or O^4-MeT by the ATase. This was also the case after exposure to low doses of ENU. However, after higher doses of NEU, reversion frequency increased when selection was applied 48 h or later after exposure to the agent: by 72 h the reversion frequency was higher than in the control cells. One interpretation of this observation is that a complex forms between the O^6-EG or O^4-ET and the ATase, and this is recognised by a relatively slow endogenous repair system that may be error prone (Fox and Margison 1988). Alternatively, higher levels of O^6-EG and O^4-ET may be repaired predominantly by the nucleotide excision mechanism. Whether the repair of O^6-AG or O^4-AT is predominantly responsible for prevention or augmentation of these revertants will have to await sequencing of the *HPRT* gene in the mutants and revertants that have been obtained. An alternative approach to the resolution of this problem may come from the studies of the expression of the ATase encoded by the *E. coli ogt* gene, which appears to act $\sim$80 times faster on O^4-MeT in DNA than does the *ada* gene product (unpublished data).

As yet, relatively few compounds have been assayed using either mutation or reversion systems, but, with the possible exception of streptozotocin, the *E. coli* ATase provides a greater degree of protection against methylating and ethylating agents than against chloroethylating agents, all of which show a proclivity for reaction with the oxygen atoms in DNA.

II. Sister Chromatid Exchanges and Other Clastogenic Events

The frequency of SCE (Ishizaki et al. 1986; Samson et al. 1986; White et al. 1986), chromosome fragmentation and micronuclei induction (White et al. 1986) by methylating agents is much reduced in cells expressing the *E. coli* ATase compared with control cells. Thus, as previously suggested (Wolff 1982), O^6-MeG (or O^4-MeT) is apparently associated with the generation of these events. Protection against SCE induction was greater after exposure of cells to NMU than after MMS, but since MMS also induced large numbers of SCE, these data, by comparison with the methylation patterns for these two agents, provide indirect evidence that 3-MeG may also be an SCE-inducing lesion (White et al. 1986).

Although these results give no indication of the mechanism of SCE induction, it has been speculated that strand interruptions produced during the excision repair of O^6-MeG (White et al. 1986) may be responsible. Thus, providing the cells with the rapidly acting ATase might reduce the need for excision-mediated events. On the other hand, while ATase expression prevents the majority of SCE after low doses of NMU, it also prevents a large number of these events after MMS treatment despite the preponderance of damage that is not ATase-reparable. One interpretation of this is that strand interruptions produced by, or during, the repair of other lesions are relatively inefficient inducers of SCE. This would suggest that non-ATase-mediated repair of O^6-AG involves a system which differs from the glycosylase/apurinic endonuclease-initiated mechanism that deals with lesions such as 3-methyladenine. The existence of bypass or tolerance repair systems has also been deduced from comparisons of repair capacities in relation to DNA synthesis (ROBERTS 1978), and these may be intermediaries in SCE induction by O^6-AG.

Interestingly, no protection against the SCE-inducing effect of NBU was provided by *E. coli* ATase expression (WHITE et al. 1986). This may be because the *E. coli* ATase acts very much more slowly on higher O^6-AG in vitro than on O^6-MeG (see SAFFHILL et al. 1985; COOPER, POTTER and MARGISON, unpublished). The repair of O^6-BuG has been suggested to occur by an excision mechanism (BOYLE et al. 1986b), and it may be that this acts more rapidly than does the *E. coli* ATase. Again, expression of the *ogt* gene, the product of which acts more rapidly on higher O^6-AG than does the *ada* ATase (COOPER, POTTER and MARGISON, unpublished), might be expected to have a greater protective effect, and this is being pursued.

Expression of the O^6-AG ATase function of the *ada* gene results in a reduction in the frequency of micronuclei and other chromosome damage (deletions, rearrangements or fragmentation) after exposure to NMU (WHITE et al. 1986). The conclusion from this limited study is that O^6-MeG (and or O^4-MeT) is able to elicit the response, but again the mechanisms of the processes remain obscure.

III. Toxicity

Reduction in the toxic effects of alkylating agents by expression of *E. coli* ATase activity is highly dependent not only on the agent itself but also on the type of cells used. In V79 cells, ATase expression has little or no effect on the toxicity of MMS (BRENNAND and MARGISON 1986b; FOX et al. 1987), whilst in murine haemopoeitic (JELINEK et al. 1988) and CHO (KATAOKA et al. 1986) cells, protection is provided. With NMU, protection in murine cells is much greater than in V79 cells. It is reasonable to suggest that these differences are due to the endogenous excision, by-pass or tolerance repair capacities of the various host cells, although other factors (see above) may be involved. These and other effects, such as the inhibition of protein synthesis by the administered agent, may also influence the appearance of the survival curves in different cell lines or after exposure to different agents. The conclusion from these data is that O^6-MeG or O^4-MeT are potentially toxic lesions in mammalian cells. However, the extent to which this toxicity is manifested depends on a number of factors that are host-cell dependent. This may explain why different groups have arrived at different conclusions with respect to the toxicity of O^6-MeG in mammalian cells.

In most cases, protection against the toxic effects of chloroethylating agents is the most extensive (BRENNAND and MARGISON 1986b; SAMSON et al. 1986; MARGISON and BRENNAND 1987; FOX et al. 1987; JELINEK et al. 1988). It has been postulated that the mechanism of cell killing by chloroethylating agents is the formation of a G-C crosslink which is derived in a two-step reaction from the mono-adduct O^6-chloroethylguanine (O^6-CletG) (TONG et al. 1982). Results using transfected cells strongly support this mechanism but do not preclude the idea that other crosslinks involving O^6-CletG or O^4-chloroethylthymine (O^4-CleT) may be partly, or even exclusively, responsible for cell killing.

Of equal interest is that, with some alkylating agents, no protection is provided in the transfected cell lines. Thus, with nitrogen mustard (BRENNAND and MARGISON 1986b; JELINEK et al. 1988), it can be concluded that if any O^6-AG or O^4-AT damage is produced, this is either not a substrate for the ATase or it does not contribute significantly to the lethal effects of the agent. This is an agreement with the postulated mechanism of cell killing by nitrogen mustard, i.e. crosslinking between the N7-atoms of guanine (KOHN et al. 1966). In the case of NBU (FOX et al. 1987) the lack of protection, as with SCE (WHITE et al. 1986), is probably a consequence of the slow rate of action of the ATase on O^6-BuG (and probably O^4-BuT) in DNA.

E. Conclusions

The generation of defined cell lines that differ only in the expression of one or two characterised DNA repair functions provides the most conclusive evidence for the potential biological effects of the lesion(s) that are substrates for the repair process. In the case of O^6-AG ATase, the existence of an analogous endogenous mammalian repair system indicates that variation in the expression of the endogenous gene may have a profound effect on the biological response of different cell types to certain alkylating agents. Thus in the majority of cases in which the toxic mutagenic or chromosome-damaging effects of alkylating agents have been compared in cells differing in the levels of O^6-AG ATase, the crucial role of this repair function would appear to have been correctly deduced even though numerous other occult differences that could contribute to the overall response of the cell lines may have been present (DAY et al. 1987).

It is also clear from these transfection experiments that, as expected, several different alkylation products can produce similar biological effects and that other endogenous repair systems can act on damage that is also reparable by ATases: it seems likely that these functions can exist together in the same cell types, giving some cells an additional line of defence.

Work with these transfected cells is at an early stage, and it is expected that the approach will be extended to explore the universality of the effects in different cell types using a variety of agents and examining a range of effects. Thus, expression of *E. coli* ATase in plant cells (VELEMINSKY et al., submitted for publication) and yeast (BROSMANOVA et al., submitted for publication) has been achieved, and protection against the toxic effects of alkylating agents has been produced. In addition, transgenic mice containing the ATase gene have recently been generated (SEARLE et al. 1988), and this may allow an examination to be made of the role of O^6-AG and O^4-AT in the wider range of biological effects produced by alkylating agents in intact animals.

Acknowledgements. We are grateful to Miss Sarah J. Morrissey for careful preparation of the manuscript and to Dr. W.H. Butler for comments on the pathology. Work carried out in this laboratory was supported by the Cancer Research Campaign.

References

Abbott PJ, Saffhill R (1979) DNA synthesis with methylated poly(dC-dG) templates: evidence for a competitive nature for miscoding by O^6-methylguanine. Biochim Biophys Acta 562:51–61

Bartsch H, Montesano R (1984) Relevance of nitrosamines to human cancer. Carcinogenesis 5:1381–1393

Bochert G, Platzek T, Blankenburg G, Wiessler M, Neubert D (1985) Embryotoxicity induced by alkylating agents: left sided preponderence of paw malformations induced by acetoxymethyl-methylnitrosamine in mice. Arch Toxicol 56:139–150

Bohr VA, Hanawalt PC (1987) Enchanced repair of pyrimidine dimers in coding and noncoding genomic sequences in CHO cells expressing a prokaryotic DNA repair gene. Carcinogenesis 8:1333–1336

Bohr VA, Phillips DH, Hanawalt PC (1987) Heterogeneous DNA damage and repair in the mammalian genome. Cancer Res 47:6426–6436

Boiteux S, Huisman O, Laval J (1984) 3-Methyladenine residues in DNA induce the SOS function *sfiA* in *Escherichia coli*. EMBO J 3:2569–2573

Bosan WS, Lambert CE, Shank RC (1986) The role of formaldehyde in hydrazine induced methylation of liver DNA guanine. Carcinogenesis 7:413–415

Bossert NL, Iannaccone PM (1985) Midgestational abnormalities associated with *in vitro* preimplantation *N*-methyl-*N*-nitrosourea exposure with subsequent transfer to surrogate mothers. Proc Natl Acad Sci USA 82:8757–8761

Boyle JM, Margison GP, Saffhill R (1986a) Evidence for the excision repair of O^6-*n*-butyldeoxyguanosine in human cells. Carcinogenesis 7:1987–1990

Boyle JM, Saffhill R, Margison GP, Fox M (1986b) A comparison of cell survival, mutation and persistence of putative promutagenic lesions in Chinese hamster cells exposed to BNU or MNU. Carcinogenesis 7:1981–1985

Braun R, Huttner E, Schoneich J (1984) Transplacental genetic and cytogenetic effects of alkylating agents in the mouse. I. Induction of somatic coat colour mutations. Teratogenesis Carcinog Mutagen 4:449–457

Braun R, Huttner E, Schoneich J (1986) Transplacental genetic and cytogenetic effects of alkylating agents in the mouse. II. Induction of chromosomal aberrations. Teratogenesis Carcinog Mutagen 6:69–80

Brennand J, Margison GP (1986a) Expression of the *E. coli* O^6-methylguanine-methylphosphotriester methyltransferase gene in mammalian cells. Carcinogenesis 7:185–188

Brennand J, Margison GP (1986b) Reduction of the toxicity and mutagenicity of alkylating agents in mammalian cells harboring the *E. coli* alkyltransferase gene. Proc Natl Acad Sci USA 83:6292–6296

Brennand J, Margison GP (1986c) Expression in mammalian cells of a truncated *Escherichia coli* gene coding for O^6-alkylguanine alkyltransferase reduces the toxic effects of alkylating agents. Carcinogenesis 7:2081–2084

Burns PA, Gordon AJ, Glickman BW (1987) Influence of neighbouring base sequence on *N*-methyl-*N'*-nitro-*N*-nitrosoguanidine mutagenesis in the *lac I* gene of *Escherichia coli*. J Mol Biol 194:385–390

Carbonell F, Seidel HJ, Saks S, Kreja L (1982) Chromosome changes in butylnitrosourea (BNU)-induced mouse leukaemia. Int J Cancer 40:540–549

Carbonell F, Eul J, Anselstetter V, Hameister H, Seidel HJ, Kreja L (1987) Trisomy 15 as a regular finding in chemically induced murine T-cell leukaemogenesis. Int J Cancer 39:534–537

Chaudry MA, Fox M (1988) Molecular analysis of induced mutations at the HPRT locus in Chinese hamster cells expressing and not expressing the *E. coli* alkyltransferase gene. Environ Molec Mutagen 11 [Suppl 11]:22

Cohn SM, Krawisz BR, Dresler SL, Lieberman MW (1984) Induction of replication DNA synthesis in quiescent human fibroblasts by DNA damaging agents. Proc Natl Acad Sci USA 81:4828–4832

Colvin M (1981) Molecular pharmacology of alkylating agents. In: Crooke ST, Prestayko AW (eds) Cancer and chemotherapy, vol III. Academic, New York, pp 287–302

Connell JR, Medcalf ASC (1984) The induction of SCE with relation to specific base methylation in Chinese hamster ovary cells by *N*-methyl-*N*-nitrosourea and dimethylsulphate. In: Tice RE, Hollander A (eds) Basic life science, vol 29A. Sister chromatid exchanges. Plenum, New York, pp 343–352

Craddock VM, Henderson AR, Gash S (1984) Repair and replication of DNA in rat brain and liver during foetal and post-natal development in relation to nitroso-alkylurea induced carcinogenesis. J Cancer Res Clin Oncol 108:30–35

Day RS III, Babich MA, Yarosh DB, Scuderio DA (1987) The role of O^6-methylguanine in human cell killing, sister chromatid exchange induction and mutagenesis: a review. J Cell Sci [Suppl] 6:333–353

Dean M, Park M, Vande-Woude GF (1987) Characterisation of the rearranged *tps-met* oncogene breakpoint. Mol Cell Biol 7:921–924

DeKok AJ, Sip H, Den Englese L, Simons JW (1981) Transformation of $C_3H10t_{1/2}$ cells by *N*-ethyl-*N*-nitrosourea occurs as a single, low frequency, mutation like event. Carcinogenesis 7:1387–1392

Demple B, Sedgwick B, Robins P, Totty N, Waterfield MD, Lindahl T (1985) Active site and complete sequence of the suicidal methyltransferase that counters alkylation mutagenesis. Proc Natl Acad Sci USA 82:2688–2692

Doniger J, Day RS, DiPaolo JA (1985) Quantitative assessment of the role of O^6-methylguanine in the initiation of carcinogenesis by methylating agents. Proc Natl Acad Sci USA 82:421–425

Dryoff MC, Richardson FC, Popp JA, Bedell MA, Swenberg JA (1986) Correlation of O^4-ethyldeoxythymidine accumulation, hepatic initiation and carcinoma induction in rats continuously administered diethylnitrosamine. Carcinogenesis 7:241–246

Essigman JM, Loechler EL, Green CL (1986) Genetic toxicology of environmental chemicals. Part A: basic principles and mechanisms of action. Liss, New York, pp 433–440

Evensen G, Seeberg E (1982) Adaptation to alkylation resistance involves the induction of a DNA glycosylase. Nature 296:773–775

Fleig WG, Fussgaenger RD, Ditschuneit H (1982) Pathological changes in a human subject chromically exposed to dimethylnitrosamine. In: Magee PN (ed) Nitrosamines and human cancer. Banbury report no 12. Cold Spring Harbor Laboratory, New York, pp 37–44

Fox M, Margison GP (1988) Expression of an *E. coli* O^6-alkylguanine alkyltransferase gene in Chinese hamster cells protects against *N*-methyl and *N*-ethylnitrosourea-induced reverse mutation at the hypoxanthine phosphoribosyl transferase locus. Mutagenesis 3:409–413

Fox M, Brennand J, Margison GP (1987) Protection of Chinese hamster cells against the cytotoxic and mutagenic effects of alkylating agents by transfection of the *Escherichia coli* alkyltransferase gene and a truncated derivative. Mutagenesis 2:491–496

Frei JV, Swenson DH, Warren W, Lawley PD (1978) Alkylation of deoxyribonucleic acid in vivo in various organs of C57Bl mice by the carcinogens *N*-methyl-*N*-nitrosourea, *N*-ethyl-N-nitrosourea and ethylmethanesulphonate in relation to the induction of thymic lymphoma. Biochem J 174:1031–1044

Gerchmann LL, Ludlum DB (1973) The properties of O^6-methylguanine in templates for RNA polymerase. Biochem Biophys Acta 308:310–316

Haas M, Altman A, Rothenberg E, Bogart MH, Jones OW (1984) Mechanism of T-cell lymphomagenesis: transformation of growth factor dependent T-lymphoblastoma cells to growth factor-independent T-lymphoma cells. Proc Natl Acad Sci USA 81: 1742–1746

Hard GC, Butler WH (1970) Cellular analysis of renal neoplasia: light microscope study of the development of interstitial lesion induced in the rat kidney by a single dose of dimethylnitrosamine. Cancer Res 30:2806–2815

Harris G, Asberry L, Lawley PD, Denam AW, Hylton W (1982) Defective repair of O^6-methylguanine in autoimmune diseases. Lancet ii:952–956

Hartley JA, Mattes WB, Vaughan K, Gibson NW (1988) DNA sequence specificity of guanine N^7-alkylations for a series of structurally related triazenes. Carcinogenesis 9:669–674

Hicks RM (1983) Pathological and biochemical aspects of tumour promotion. Carcinogenesis 4:1209–1214

Huang M, Wang Z-H, Wang X-Q, Wu M (1986) Malignant transformation of human fetal lung fibroblasts induced by nitrosamine compounds in vitro. Sci Sin 29:1192–2000

Iannacconne PM (1984) Long-term effects of exposure to methylnitrosourea on blastocytes following transfer to surrogate female mice. Cancer Res 44:2785–2789

IARC (1978) Monographs on the evaluation of the carcinogenic risk of chemicals to humans: vol 17. Some *N*-nitrosocompounds. IARC, Lyon

IARC (1985) Monographs *ibid*: vol 37. Tobacco habits other than smoking; betelquid and areca nut chewing and some related nitrosamines. IARC, Lyon

IARC (1986) Monographs *ibid*: vol 38. Tobacco smoking. IARC, Lyon

Ishizaki K, Tsujimura T, Yawata M, Fujio C, Makabeppu Y, Sekiguchi M, Ikenaga M (1986) Transfer of the *E. coli* O^6-methylguanine methyltransferase gene into repair deficient human cells and restoration of cellular resistance to *N*-methyl-N^1-nitro-*N*-nitrosoguanidine. Mutation Res 166:135–141

Ivankovic S, Druckrey H (1968) Transplacentare Erzeugung maligner Tumoren des Nervensystems. I. Athyl-nitroso-Harnstoff (ANH) an BDIX-Ratten. Z Krebsforsch 71:320–360

Jelinek J, Kleibl K, Dexter TM, Margison GP (1988) Transfection of murine multipotent haemopoietic stem cells with an *E. coli* DNA alkyltransferase gene confers resistance to the toxic effects of alkylating agents. Carcinogenesis 9:81–87

Jensen JC, Thilly WG (1986) Spontaneous and induced chromosomal aberrations and gene mutations in human lymphoblasts: mitomycin C, methylnitrosourea and ethylnitrosourea. Mutation Res 160:95–102

Karran P, Hjelmgren T, Lindahl T (1982) Induction of a DNA glycosylase for *N*-methylated purines is part of the adaptive response to alkylating agents. Nature 296:770–773

Kataoka M, Mall J, Karran P (1986) Complementation of sensitivity to alkylating agents in *E. coli* and Chinese hamster ovary cells by expression of cloned bacterial DNA repair gene. EMBO J 5:3195–3200

Kimbrough RD (1982) Pathological changes in human beings acutely poisoned by dimethylnitrosamine. In: Magee PN (ed) Nitrosamines and human cancer. Banbury report no. 12. Cold Spring Harbor Laboratory, New York, pp 25–34

Kleihues P, Hodgson RM, Veit C, Schweinsberg F, Wiessler M (1983) DNA modification and repair in vivo: towards a biochemical basis of organ specific carcinogenesis by methylating agents. Basic Life Sci 24:509–524

Klein G (1981) The role of gene dosage and genetic transpositions in carcinogenesis. Nature 294:313–318

Knowles MA, Jani H (1986) Multistage transformation of cultured rat urothelium: the effects of *N*-methyl-*N*-nitrosourea, sodium saccharin, sodium cyclamate and 12-*O*-tetradecanoylphorbol-13-acetate. Carcinogenesis 7:2059–2065

Kohn KW, Spears CJ, Doty P (1966) Interstand crosslinks of DNA by nitrogen mustard. J Mol Biol 19:266–273

Lacy LR, Eisenberg MT, Osgood CJ (1986) Molecular analysis of chemically induced mutations with *RpII215* locus of *Drosophila melanogaster*. Mutation Res 162:47–54

Larson K, Sahm J, Shenkar R, Strauss B (1985) Methylation induced blocks to in vitro DNA replication. Mutation Research 150:77–84

Lijinsky W (1982) Carcinogenesis by exposure to nitrites and amines. In: Magee PN (ed) Nitrosamines and human cancer. Banbury report no. 12. Cold Spring Harbor Laboratory, New York, pp 257–263

Lijinsky W, Epstein SS (1970) Nitrosamines as environmental carcinogens. Nature 225:21–23

Lu SH, Oshima H, Fu H-M, Tian Y, Li F-M, Blettner M, Wahrendorf J, Bartsch H (1986) Urinary excretion of *N*-nitrosamino acids and nitrate by inhabitants of high and low-risk areas for esophageal cancer in N China: endogenous formation of nitrosoproline and its inhibition by Vitamin C. Cancer Res 46:1485–1491

Magee PN, Barnes JM (1967) Carcinogenic nitroso compounds. Adv Cancer Res 10:163–246

Margison GP, Brennand J (1987) Functional expression of the *E. coli* alkyltransferase gene in mammalian cells. J Cell Sci [Suppl] 6:83–96

Margison GP, O'Connor PJ (1979) Nucleic acid modification by *N*-nitroso compounds. In: Grover PL (ed) Chemical carcinogens and DNA, vol I. CRC Press, Boca Raton, pp 111–159

Masken AP (1981) Confirmation of the two-step nature of chemical carcinogenesis in the rat colon adenocarcinoma model. Cancer Res 41:1240–1245

Mehta R, Silinskas KC, Zucker PF, Ronen A, Heddle J, Archer MC (1987) Micronucleus formation induced in rat liver and esophagus by nitrosamines. Cancer Lett 35:313–320

Mivish S (1982) In vivo formation of *N*-nitrosocompounds: formation from nitrite and nitrogen dioxide and relation to gastric cancer. In: Magee PN (ed) Nitrosamines and human cancer. Banbury report no. 12. Cold Spring Harbor Laboratory, New York, pp 227–236

Miwa M, Stuehr DJ, Marletta MA, Wishnok JS, Tannenbaum SR (1987) Nitrosation of amines stimulated by macrophages. Carcinogenesis 8:955–958

Morris SM, Beranek DT, Heflich RH (1983) The relationship between sister chromatid exchange induction and the formation of specific methylated DNA adducts in Chinese hamster ovary cells. Mutation Res 121:261–266

Morten JEN, Margison GP (1988) Increased O^6-alkylguanine activity in Chinese hamster V79 cells following selection with chloroethylating agents. Carcinogenesis 9:45–49

Musk SRR, Hatton DM, Bouffler SD, Margison GP, Johnson RT (1989) Molecular mechanisms of alkylation sensitivity in indian muntjak cell lines. Carcinogenesis 10: (in press)

Nakabeppu Y, Kondo M, Kawabala S, Iwanaga S, Sekiguchi M (1985) Purification and structure of the intact Ada regulatory protein of *E. coli* K12, O^6-methylguanine-DNA methyltransferase. J Biol Chem 260:7281–7288

O'Connor PJ, Saffhill R, Margison GP (1979) *N*-nitroso compounds: biochemical mechanisms of action. In: Emelot P, Kriek E (eds) Environmental carcinogenesis: occurrence, risk evaluation and mechanisms. Elsevier/North-Holland, Amsterdam, pp 73–96

O'Connor PJ, Fida S, Fan CY, Bromley M, Saffhill R (1988) Phenobarbital: a non-genotoxic agent which induces the repair of O^6-methylguanine from hepatic DNA. Carcinogenesis 9:2033–2038

Oshima H, Bartsch H (1981) Quantitative estimation of endogenous nitrosation in humans by monitoring *N*-nitrosoproline excreted in the urine. Cancer Res 41:3658–3662

Parodi S, Zumino A, Ottagio L, DeFerrari M, Santi L (1983) Quantitative correlation between carcinogenicity and sister chromatid exchange induction in vivo for a group of 11 *N*-nitroso compounds. J Toxicol Environ Health 11:337–346

Pegg AE (1977) Metabolism of alkylated nucleosides: possible role in carcinogenesis by nitroso compounds and alkylating agents. Adv Cancer Res 25:195–270

Pitot HC, Sirica AE (1980) The stages of initiation in hepatocarcinogenesis. Biochim Biophys Acta 605:191–215

Preussmann R (1984) Occurrence and exposure to *N*-nitroso compounds and precursors. In: O'Neill IK, VonBorstel RC, Miller CT, Long J, Bartsch H (eds) *N*-nitroso compounds: occurrence, biological effects and relevance to human cancer. IARC, Lyon, pp 3–15 (IARC Sci Pub no 57)

Preussmann R, Stewart BW (1984) *N*-Nitroso carcinogens. In: Searle CE (ed) Chemical carcinogens, 2nd edn. American Chemical Society, Washington DC, pp 643–828 (ACS Monograph 182, vol 2)

Reich SD (1981) Chemical pharmacology of nitrosoureas. In: Crooke ST, Prestayko AW (eds) Cancer and chemotherapy. Academic, New York, pp 377–387

Richardson FC, Boucheron JA, Dryoff MC, Popp JA, Swenberg JA (1986) Biochemical and morphologic studies of heterogeneous lobe responses in hepatocarcinogenesis. Carcinogenesis 7:247–251

Roberts JJ (1978) The repair of DNA modified by cytotoxic, mutagenic and carcinogenic chemicals. Adv Radiat Biol 7:211–436

Rossberger S, Andrae H, Wiebel FJ (1987) Comparison of the continuous rat hepatoma cell line 2sFou with primary rat hepatocyte cultures for the induction of DNA repair synthesis by nitrosamines, benzo(*a*)pyrene and hydroxyurea. Mutat Res 182:41–51

Ryan AJ, Billett MA, O'Connor PJ (1986) Selective repair of methylated purines in regions of chromatin DNA. Carcinogenesis 7:1497–1503
Saffhill R, Margison GP, O'Connor PJ (1985) Mechanisms of carcinogenesis induced by alkylating agents. Biochim Biophys Acta 823:111–145
Saffhill R, Badawi AF, Hall CN (1988a) Detection of O^6-methylguanine in human DNA. In: Bartsch H, Hemminki K, O'Neill IK (eds) Methods for detecting DNA damaging agents in humans: applications in cancer epidemiology and prevention. IARC, Lyon, pp 301–305 (IARC Sci Pub no 89)
Saffhill R, Fida S, Bromley M, O'Connor PJ (1988b) Promutagenic alkyl lesions are induced in the tissue DNA of animals treated with isoniazid. Hum Toxicol 7:331–317
Samson L, Derfler B, Waldstein E (1986) Suppression of human DNA alkylation repair defects by *Escherichia coli* DNA repair genes. Proc Natl Acad Sci USA 83:5607–5610
Schendel PF, Robbins PE (1978) Repair of O^6-methylguanine in adapted *Escherichia coli*. Proc Natl Acad Sci USA 75:6017–6020
Schuster C, Rode G, Rabes HM (1985) O^6-methylguanine repair of methylated DNA in vitro: cell cycle-dependence of rat liver methyltransferase activity. J Cancer Res Clin Oncol 110:98–102
Searle P, Tinsley JM, O'Connor PJ, Margison GP (1988) Generation of transgenic mice containing a DNA repair gene from *Escherichia coli*. Proc Am Assoc Cancer Res 29:111
Silinskas KC, Zucker PF, Archer MC (1985) Formation of O^6-methylguanine in rat liver DNA by nitrosamines does not predict initiation of preneoplastic foci. Carcinogenesis 6:773–775
Storer RD, Stein RB, Sina JF, DeLuca JG, Allen HL, Bradley MO (1986) Malignant transformation of a preneoplastic hamster epidermal cell line by the EJ *c-Ha-ras*-oncogene. Cancer Res 46:1458–1464
Stuehr DJ, Marletta MA (1985) Mammalian nitrate biosynthesis: mouse macrophages produce nitrite and nitrate in response to *Escherichia coli* lipo-polysaccharide. Proc Natl Acad Sci USA 82:7738–7742
Suter W, Brennand J, McMillan S, Fox M (1980) Relative mutagenicity of antineoplastic drugs and other alkylating agents in V79 Chinese hamster cells: independence of cytotoxic and mutagenic responses. Mutat Res 73:171–181
Swenberg JA, Dryoff MC, Bedell MA, Popp JA, Huh N, Kirstein H, Rajewski MF (1984) O^4-ethyldeoxythymidine, but not O^6-ethyldeoxyguanosine, accumulates in hepatocyte DNA of rats exposed continuously to diethylnitrosamine. Proc Natl Acad Sci USA 81:1692–1695
Swenberg JA, Richardson FC, Boucheron JA, Dryoff MC (1985) Relationships between DNA adduct formation and carcinogenesis. Environ Health Perspect 62:177–183
Tabin CJ, Bradley SM, Bargman CI, Weinberg RA, Papageorge AG, Scolnick EM, Dhar H, Lowy DR, Chang EH (1982) Mechanism of activation of a human oncogene. Nature 300:143–149
Takeuchi IK (1984) Teratogenic effects of methylnitrosourea on pregnant mice before implantation. Experientia 40:879–881
Tempest PR, Reeves BR, Spurr NK, Rance AJ, Chan AM, Brookes P (1986) Activation of the *met*-oncogene in the human MNNG-HOS cell line involves chromosomal rearrangement. Carcinogenesis 7:2051–2057
Teo I, Sedgwick B, Kilpatrick MW, McCarthy TV, Lindahl T (1986) The intracellular signal for induction of resistance to alkylating agents in *E. coli*. Cell 45:315–324
Tomatis L, Mohr H (eds) (1973) Transplacental carcinogenesis. IARC, Lyon (IARC Sci Pub no 4)
Tong WP, Kirk MC, Ludlum DB (1982) Formation of the crosslink 1-[N^3-deoxycytidyl],2-[N^1-deoxyguanosinyl]-ethane in DNA treated with N,N^1-bis(2-chloroethyl)-N-nitrosourea. Cancer Res 42:3102–3105
Tsuji H, Hyodo M, Tsuji S, Hori T, Tobari I, Sato K (1986) Isolation of temperature-sensitive mouse FM3A cell mutants exhibiting conditional chromosome instability. Somatic Cell Mol Genet 12:595–610

Tsuji H, Takahashi E, Tsuji S, Tobari I, Shiomi T, Sato K (1987) Chromosomal instability in mutagen sensitive mutants isolated from mouse lymphoma L51798Y cells. II. Abnormal induction of sister chromatid exchanges and chromosomal aberrations by mutagens in an ionising sensitive mutant (MIO) and an alkylating agent-sensitive mutant (MSI). Mutation Res 178:107–116

Uenaka H, Ueda N, Maeda S, Sugiyama T (1978) Involvement of chromosome 2 in primary leukaemia induced in rats by *N*-nitroso-*N*-butylurea. JNCI 60:1399–1404

Umbenhauer D, Wild CP, Montesano R, Saffhill R, Boyle JM, Huh N, Kirstien U, Thomale J, Rajewski MF, Lu SH (1985) O^6-methyldeoxyguanosine in oesophageal DNA among individuals at high risk of oesophageal cancer. Int J Cancer 36:53–57

Uno M, Wirschubsky Z, Babonits M, Wiener F, Sumegi J, Klein G (1987) The role of chromosome 15 in murine leukamogenesis II. Relationship between tumourigenicity and the dosage of lymphoma vs normal-parent-derived chromosomes 15 in somatic cell hybrids. Int J Cancer 40:540–549

Van Zeeland AA, Mohn G, Neuhauser-Klaus A, Ehling UH (1985) Qauntitative comparison of genetic effects of alkylating agents on the basis of DNA adduct formation. Use of O^6-ethylguanine as molecular dosimeter for extrapolation from cells in culture to the mouse. Environ Health Perspect 62:163–169

Vogel EW (1986) *O*-alkylation in DNA does not correlate with the formation of chromosome breakage events in *D. melanogaster*. Mutation Res 162:201–213

Vogel EW, Dusenberry RL, Smith PD (1985) The relationship between kinetic reactions and mutagenic action of monofunctional alkylating agents in higher eukaryotic systems: the effects of excision defective *mei-9*^{LI} and *mus (2) 201*^{DI} mutants on alkylation induced damage in *Drosophila*. Mutation Res 149:193–207

Wagner DA, Shuker DEG, Bilmazes C, Obiedzinski M, Baker I, Young VR, Tannenbaum SR (1985) Effects of Vitamins C and E in endogenous synthesis of *N*-nitrosamino acids in humans: precursor-product studies with [^{15}N]-nitrate. Cancer Res 45:6519–6522

White GRM, Ockey CH, Brennand J, Margison GP (1986) Chinese hamster cells harbouring the *Escherichia coli* O^6-alkylguanine alkyltransferase gene are less susceptible to sister chromatid exchange induction and chromosome damage by methylating agents. Carcinogenesis 7:2077–2080

Wild CP, Smart G, Saffhill R, Boyle JM (1983) Radioimmunoassay of O^6-methyldeoxyguanosine in DNA of cells alkylated in vitro and in vivo. Carcinogenesis 4:1605–1609

Wolff S (1982) Chromosome aberrations, SCE and the lesions that produce them. In: Wolff S (ed) Sister chromatid exchange. Wiley, New York, pp 41–47

Subject Index

Springer

Springer